Left tail

α

Critical t score
(negative)

Two tails

α

Critical t score
(positive)

α/2

Critical t score
(negative)

α/2

Critical t score
(positive)

TABLE A-3	t Distribution					
				α		
Degrees of Freedom	.005 (one tail) .01 (two tails)	.01 (one tail) .02 (two tails)	.025 (one tail) .05 (two tails)	.05 (one tail) .10 (two tails)	.10 (one tail) .20 (two tails)	.25 (one tail) .50 (two tails)
1	63.657	31.821	12.706	6.314	3.078	1.000
2	9.925	6.965	4.303	2.920	1.886	.816
3	5.841	4.541	3.182	2.353	1.638	.765
4	4.604	3.747	2.776	2.132	1.533	.741
5	4.032	3.365	2.571	2.015	1.476	.727
6	3.707	3.143	2.447	1.943	1.440	.718
7	3.500	2.998	2.365	1.895	1.415	.711
8	3.355	2.896	2.306	1.860	1.397	.706
9	3.250	2.821	2.262	1.833	1.383	.703
10	3.169	2.764	2.228	1.812	1.372	.700
11	3.106	2.718	2.201	1.796	1.363	.697
12	3.054	2.681	2.179	1.782	1.356	.696
13	3.012	2.650	2.160	1.771	1.350	.694
14	2.977	2.625	2.145	1.761	1.345	.692
15	2.947	2.602	2.132	1.753	1.341	.691
16	2.921	2.584	2.120	1.746	1.337	.690
17	2.898	2.567	2.110	1.740	1.333	.689
18	2.878	2.552	2.101	1.734	1.330	.688
19	2.861	2.540	2.093	1.729	1.328	.688
20	2.845	2.528	2.086	1.725	1.325	.687
21	2.831	2.518	2.080	1.721	1.323	.686
22	2.819	2.508	2.074	1.717	1.321	.686
23	2.807	2.500	2.069	1.714	1.320	.685
24	2.797	2.492	2.064	1.711	1.318	.685
25	2.787	2.485	2.060	1.708	1.316	.684
26	2.779	2.479	2.056	1.706	1.315	.684
27	2.771	2.473	2.052	1.703	1.314	.684
28	2.763	2.467	2.048	1.701	1.313	.683
29	2.756	2.462	2.045	1.699	1.311	.683
Large (z)	2.575	2.326	1.960	1.645	1.282	.675

ELEMENTARY
STATISTICS *using* EXCEL®

ELEMENTARY
STATISTICS *using* EXCEL®

MARIO F. TRIOLA

Addison-Wesley

Boston San Francisco New York
London Toronto Sydney Tokyo Singapore Madrid
Mexico City Munich Paris Cape Town Hong Kong Montreal

Publisher: Greg Tobin
Sponsoring Editor: Deirdre Lynch
Editorial Assistant: Rebecca Martin
Development Editor: Elka Block
Executive Marketing Manager: Brenda Bravener
Senior Marketing Coordinator: Laura Potter Walton
Managing Editor: Karen Guardino
Senior Production Supervisor: Peggy McMahon
Text and Cover Designer: Barbara T. Atkinson
Cover photo: ©Kelly-Mooney Photography/CORBIS
Art Editor: Meredith Nightingale
Photo Research: Beth Anderson
Prepress Service Manager: Caroline Fell
Manufacturing Buyer: Evelyn Beaton
Technical Illustrator: Academy Artworks
Composition: WestWords, Inc.
Printer: VonHoffman Press

Library of Congress Cataloging-in-Publication Data

Triola, Mario F.
    Elementary statistics using Excel / Mario F. Triola
        p. cm.
    ISBN 0-201-69942-7
        1. Statistics. 2. Microsoft Excel (Computer file)   I.
Title.

    QA276.4.T75 2000
    519.5'0285'5369--dc21   00-058276

1 2 3 4 5 6 7 8 9 10—VHP—03020100

To Marc and Scott

About the Author

Mario F. Triola is a Professor of Mathematics at Dutchess Community College, where he has taught statistics for over 30 years. Marty is the author of *Elementary Statistics, Mathematics and the Modern World,* and *A Survey of Mathematics* and is coauthor of *Introduction to Technical Mathematics* and *Business Statistics*. Outside of the classroom, Marty's consulting work includes the mathematical design of casino slot machines and fishing rods, and he has worked with attorneys in determining probabilities in paternity lawsuits, identifying salary inequities based on gender, and analyzing disputed election results. Marty was recently a writing team member of the Project Coalition writing team with NASA and the American Mathematics Association of Two-Year Colleges.

When he's not working, Marty enjoys playing golf and tennis, running, hiking, and anything that flies. He has a commercial pilot's license with an instrument rating, and has flown airplanes, helicopters, sail planes, hang gliders, and hot air balloons. His passion for flying has included parachute jumps, flying in a Goodyear blimp, and parasailing.

The Text and Academic Authors Association has awarded Mario F. Triola a "Texty" for Excellence for his work on *Elementary Statistics*.

Preface

WHY EXCEL?

Statistics professors are now changing to Microsoft Excel® at a rate that exceeds the change to any other technology. A key factor fostering that trend is the recognition that corporate America has embraced the spreadsheet as an efficient and effective tool for the analysis of data. Excel has become the premier program for working with those spreadsheets. Motivated by a desire to better serve their students by better preparing them for their professional careers, many professors now include Excel as the medium of technology that is used throughout the statistics course. Students can now learn important professional skills along with concepts of statistics. This marriage of statistics concepts and spreadsheet applications is giving birth to a generation of students who can enter professional careers armed with knowledge and skills that were once desired, but are now demanded.

USING EXCEL

Elementary Statistics Using Excel® is designed to be a complete introduction to basic statistics. Instead of being a manual of computer instructions, this book places strong emphasis on understanding concepts of statistics, with Excel included throughout as the key supplement. Topics are presented with illustrative examples, identification of required assumptions, and underlying theory. Excel instructions are provided along with typical displays of results. Those instructions and displays are based on Excel 97 and Excel 2000. Because Excel was not designed to be a complete tool for statistical analysis, it is important to know its limitations as well as alternative approaches. For example, it is important to know how Excel's graphs of histograms are misleading, and it is important to know how to correct those graphs. Excel's ZTEST command for hypothesis testing provides results that are inconsistent, so it is important to know how to avoid the pitfalls of that command. *Elementary Statistics Using Excel®* guides the reader through procedures that result in correct and consistent results.

Excel lacks some important features, such as the ability to generate confidence intervals or to conduct hypothesis tests involving proportions. *Elementary Statistics Using Excel®* includes tools that enable these important features. This book has key features that are so critical to an effectice use of Excel as a statistical tool:

- Detailed **Excel instructions** are included throughout the book.
- Approximately 125 samples of Excel **displays** are included throughout the book.
- All **data sets** in Appendix B (except for Data Set 6) are available as Excel workbooks on the CD-ROM packaged with this book. Many exercises use

those data sets, so students can open Excel files instead of manually entering long lists of data.

- Exercises that use Excel are identified with a special icon.

- Appendix F answers with different values (from Excel and tables) include both values.

- Excel **workbook names** are identified for those exercises that refer to data sets in Appendix B, so that students can easily access and open the relevant Excel file.

- The CD-ROM packaged with this book includes **Data Desk/XL®**, a software add-in that enhances and expands the capability of Excel. Data Desk/XL can be installed to become part of the Excel program, and is easily accessed by clicking on DDXL, which appears on the main menu bar. Data Desk/XL includes many important procedures that are not already part of Excel.

- To make full use of Excel's statistical features, instructions are provided for installing the **Data Analysis** add-in, and instructions are provided for using important features of that Excel supplement.

- **Appendix C** lists basic functions along with references of key Excel commands.

- **Excel Projects** are included among the variety of the end-of-chapter features.

- For those who choose to further supplement Excel, the CD-ROM includes additional **software** that some students may find helpful. STATDISK is a separate program designed specifically for the content of this book. There are separate programs for the TI-83 Plus calculator.

- For those who choose to supplement Excel with other software, this book's CD-ROM includes the Appendix B **data sets** in a variety of formats. In addition to Excel, the data sets are available in formats for Minitab, SPSS, and the TI-83 Plus calculator, and they are also available as text files.

AUDIENCE/PREREQUISITES

Elementary Statistics Using Excel® is written for students majoring in any field. Although the use of algebra is minimal, students should have completed at least an elementary algebra course. In many cases, underlying theory is included, but this book does not stress the mathematical rigor more suitable for mathematics majors. Because the many examples and exercises cover a wide variety of different and interesting statistical applications, *Elementary Statistics Using Excel®* is appropriate for students pursuing careers in disciplines ranging from the social sciences of psychology and sociology to areas such as education, the allied health fields, business, economics, engineering, the humanities, the physical sciences, journalism, communications, and liberal arts.

FLEXIBLE SYLLABUS

The organization of this book reflects the preferences of most statistics instructors, but there are two common variations that can be used easily:

- **Early coverage of correlation/regression:** Some instructors prefer to cover the basics of correlation and regression early in the course, such as immediately following the topics of Chapter 2. *Sections 9-2 (Correlation) and 9-3 (Regression) can be covered early in the course.* Simply omit the subsection in Section 9-2 clearly identified as "Formal Hypothesis Test (Requires Coverage of Chapter 7)."

- **Minimum probability:** Some instructors feel strongly that coverage of probability should be extensive, while others feel just as strongly that coverage should be kept to a bare minimum. Instructors preferring minimum coverage can include Section 3-2 while skipping the remaining sections of Chapter 3. Sections 3-3 through 3-7 are not essential for the chapters that follow.

EXERCISES

There are **1500** exercises. Many of the exercises require interpretation of results. Because exercises are of such critical importance to any statistics book, great care has been taken to ensure their usefulness, relevance, and accuracy. Three statisticians have read carefully through the final stages of the book to verify accuracy of the text material and exercise answers. Exercises are arranged in order of increasing difficulty by dividing them into two groups: (1) Basic Skills and Concepts and (2) Beyond the Basics. The Beyond the Basics exercises address more difficult concepts or require a somewhat stronger mathematical background. In some cases, these exercises also introduce a new concept.

    Real data: *More than half of the exercises use real data.* Because the use of real data is such an important consideration for students, hundreds of hours have been devoted to finding real, meaningful, and interesting data. In addition to the real data included throughout the book, many exercises refer to the 20 data sets listed in Appendix B. Except for Data Set 6, those Appendix B data sets are available as Excel workbooks on the CD-ROM packaged with this book.

FEATURES

Beyond an interesting and accessible (and sometimes humorous) writing style, great care has been taken to ensure that each chapter of *Elementary Statistics Using Excel*® will help students understand the concepts presented. The following features are designed to help meet that objective:

- **Excel** instructions and examples of Excel displays are included throughout the book.

- **Chapter-opening features:** A list of chapter sections previews the chapter for the student; a chapter-opening problem, using real data, then motivates the chapter material; and the first section is a chapter overview that provides a statement of the chapter's objectives.

- **End-of-chapter features:**
  Statistics in the News encourages analysis of timely applications of statistics.
  A **Vocabulary List** of important terms (a full glossary is found in Appendix D);
  A **Chapter Review** summarizes the key concepts and topics of the chapter;
  Review Exercises provide practice on chapter concepts and procedures;
  Cumulative Review Exercises reinforce earlier material;
  From Data to Decision: Critical Thinking, a capstone problem, requires critical thinking and a writing component;
  Cooperative Group Activities encourage active learning in groups;
  Excel Projects are specifically designed to incorporate an application using technology;
  Internet Projects apply important chapter concepts.

- **Margin Essays:** The text includes 106 margin essays, which illustrate uses and abuses of statistics in real, practical, and interesting applications. Topics include

"Six Degrees of Separation," "Statistics and Land Mines," "Using Statistics to Identify Thieves," and "Choosing Personal Security Codes."

- **Flowcharts:** These appear throughout the text to simplify and clarify more complex concepts and procedures.

- **Real Data Sets:** Real data are used extensively throughout the entire book. Appendix B lists 20 data sets. These data sets are provided in printed form in Appendix B at the back of the book, and in electronic form on the CD packaged with the book. The data sets include such varied topics as eruptions of the Old Faithful geyser, nicotine contents of cigarettes, diamond prices and characteristics, weights of diet and regular Coke and Pepsi, movie financial and rating data, and rainfall amounts.

- **Excel Icons** mark exercises where it would be appropriate to use Excel. **Excel file names** are included with exercises that refer students to the data sets located on the CD-ROM and Web site.

- **Interviews:** Every chapter of the text includes interviews with professional men and women in a variety of fields who use statistics in their day-to-day work.

- **Appendices:** Appendix A contains tables; Appendix B lists 20 data sets; Appendix C is a summary of important Excel commands; Appendix D is a glossary of important terms; Appendix E is a bibliography of recommended text and reference books; Appendix F contains answers to all the odd-numbered section exercises, as well as *all* answers to Review Exercises and Cumulative Review Exercises.

- **Quick-Reference Endpapers:** Tables A-2 and A-3 (the normal and *t* distributions) are reproduced on the front inside cover pages. A symbol table is included on the last pages for quick and easy reference to key symbols.

- **Detachable Formula/Table Card:** This insert, organized by chapter, gives students a quick reference for studying, or for use when taking tests (if allowed by the instructor). The formula/table card is also available on the book Web site.

SUPPLEMENTS

Important supplements are available from Addison Wesley Longman Publishing Company. Instructors should contact the local AWL sales consultant, or e-mail the company directly at exam@awl.com for examination copies.

FOR THE INSTRUCTOR

- *Instructor's Guide and Solutions Manual*, contains solutions to all the text exercises, quizzes (with answers), and sample course syllabi. ISBN: 0-201-71052-8.

- **Testing System:** Not only is there a printed test bank, but there is also a computerized test generator, **TestGen-EQ 3.0 and Quizmaster 2.0**, that lets you view and edit testbank questions, transfer them to tests, and print in a variety of formats. The program offers many options for organizing and displaying test banks and tests. A built-in random number and test generator makes **TestGen-EQ** ideal for creating multiple versions of tests and provides more possible test items than print testbank questions. Powerful search and sort functions let the instructor easily locate questions and arrange them in the preferred order. Users can export tests as HTML text files so they can be viewed with a Web browser. Printed Testbank ISBN: 0-201-71053-6; TestGen-EQ for Mac and Windows ISBN: 0-201-71139-7.

- **PowerPoint® Lecture Presentation CD:** Free classroom lecture presentation software geared specifically to the sequence and philosophy of *Elementary Statistics Using Excel®*, has been prepared by Cheryl Slayden of Pellissippi State Technical Community College. Key graphics from the book are also included. Mac and Windows ISBN: 0-201-70462-5.

- **Triola Web Site:** The Triola Web site can be accessed at **http://www.awl.com/triolaexcel.** It provides dynamic resources for instructors and students. Some of the resources include internet projects keyed to every chapter of the text, data downloads, sample syllabi, and more. Many student resources are also available on the Web site, including practice quizzes, study tips, instructional video clips, and more.

FOR THE STUDENT

- **Videos,** designed to supplement the sections in the book, with many topics presented by Mario F. Triola. The videos feature professor and student interaction, and all technologies in the book are represented on the videos. This is an excellent resource for students who have missed class or wish to review a topic. It is also a valuable resource for instructors involved with distance learning, individual study, or self-paced learning programs. Videotapes ISBN: 0-201-70470-6. Available on CD-ROM. CD-ROM ISBN: 0-201-70966-X.

- **Data Software,** prepared by Mario F. Triola, includes the data sets (except for Data Set 6) from Appendix B in the textbook. These data sets are stored as text files, Minitab worksheets, TI Data application for the TI-83 Plus, STATDISK files, SPSS files, and Excel workbooks and are included on the CD-ROM located in the back of the book. The CD-ROM also includes programs for the TI-83 Plus® graphing calculator.

- **STATDISK Statistical Software,** specifically for the Triola text, is a statistical software package (for both Windows and Macintosh) licensed free to adopters of *Elementary Statistics Using Excel®*. This software is included on the CD-ROM in the back of the book.

STATDISK

- **Excel Add-In:** Free software, called Data Desk/XL®, designed to enhance the capabilities of Excel's statistics programs. Available on the CD-ROM bound in the back of the book.

- **CD-ROM:** A free CD-ROM, packaged with every new copy of the text, contains the text data sets in formats for various technologies, a TI-83 Plus data application, TI-83 Plus programs, STATDISK software, and the Excel add-in.

- *Student Solutions Manual,* provides detailed, worked-out solutions to odd-numbered exercises. ISBN: 0-201-71138-9.

- *STATDISK Student Laboratory Manual and Workbook,* Eighth edition, written by Mario F. Triola, includes instructions and experiments to be conducted by students using STATDISK software, either in the computer lab, or for out-of-class assignments. ISBN: 0-201-70466-8.

- *Excel® Student Laboratory Manual and Workbook,* written by Johanna Halsey and Ellena Reda (Dutchess Community College), includes instructions on and examples of Excel in use. The manual includes many examples and problems from the book as well as other appropriate exercises, and it encourages further exploration of statistical concepts. ISBN: 0-201-70459-5.

EXCEL

MINITAB

- *Minitab® Student Laboratory Manual and Workbook,* Eighth edition, written by Mario F. Triola, includes instructions on and examples of Minitab use. It also supplies many computer experiments and allows further exploration of statistical concepts. ISBN: 0-201-70461-7.

- *SPSS® Student Laboratory Manual and Workbook,* written by Roger Peck (California State University-Bakersfield), includes instructions on and examples of SPSS use. It also includes appropriate exercises from the text. ISBN: 0-201-70464-1.

TI-83 Plus

- *TI-83 Plus® Companion to Elementary Statistics,* Eighth edition, by Larry Morgan (Montgomery County Community College) is organized to follow the sequence of topics in the text, and it is an easy-to-follow, step-by-step guide on how to use the TI-83 Plus® graphing calculator. It provides worked-out examples to help students fully understand and use the graphing calculator. This supplement is also suitable for use with a TI-83 calculator. ISBN: 0-201-70469-2.

- **Triola Version of** *ActivStats®*, developed by Paul Velleman and Data **AS** Description, Inc., provides complete coverage of introductory statistics topics on CD-ROM. ActivStats integrates video, simulation, animation, narration, text, interactive experiments, World Wide Web access, and Data Desk®, a statistical software package. Homework problems and data sets from the text are included on the CD-ROM. Also available are *ActivStats for Excel* and *ActivStats for SPSS*. *ActivStats for Excel:* Windows and Macintosh ISBN: 0-201-70861-2. *ActivStats for SPSS:* Windows ISBN: 0-201-70860-4. Windows and Macintosh ISBN: 0-201-70859-0.

- *AWL Tutor Center:* For qualified adopters, free tutoring is available to students who purchase a new copy of *Elementary Statistics Using Excel®*. The Addison Wesley Longman Tutor Center (AWLTC) is staffed by qualified statistics and mathematics instructors who provide students with tutoring on text examples, problems, and odd-numbered exercises. Tutoring assistance is provided by telephone, fax, and e-mail and is available five days a week, seven hours a day. Each new book can be bundled with a registration number that provides each student with a free six-month subscription to the service. Request ISBN 0-201-72698-x (text bundled with AWLTC registration). Students who already have their text may purchase a subscription to the AWLTC by having their bookstore order ISBN 0-201-71049-8. For more information, please contact your Addison Wesley Longman sales representative.

Any of these products may be purchased separately or bundled with Addison Wesley Longman texts. Instructors can contact local AWL sales consultants for details or contact the company at exam@awl.com for examination copies of any of these items.

M.F.T.

LaGrange, New York

ACKNOWLEDGMENTS

It is an honor and a privilege to be part of a team so committed to the goal of publishing the best possible introductory statistics textbook. It is a pleasure working with the truly exceptional Addison Wesley Longman publishing team, and I thank Deirdre Lynch, Greg Tobin, Elka Block, Rebecca Martin, Peggy McMahon, Barbara Atkinson, Barbara Pendergast, Meredith Nightingale, Brenda Bravener, Laura Potter Walton, Joe Vetere, Trish Mescall, Holly Rioux, and the entire Addison Wesley Longman staff.

I thank the professionals who agreed to be interviewed for this book, and I thank Heather Carielli and Ashley McConnaughhay for their help in constructing new data sets. I thank Karen Estes for her work with icon placement, and I thank Paul Lorczak for his work on the Internet Projects. I am grateful for the many suggestions and comments provided by instructors and students, and I am grateful to the hundreds of researchers who did studies, conducted surveys, and compiled interesting and meaningful data used in examples and exercises throughout this book.

I also take great pride in thanking my wife, Ginny, and my sons, Marc and Scott, for their help and encouragement.

I would like to thank reviewers of *Elementary Statistics Using Excel*® for their helpful suggestions:

Judy Adams, North Idaho College
Abraham Biggs, Broward Community College
Michael Brands, Columbia Basin College
Raj Chhikara, University of Houston at Clear Lake
Cheryl Hawkins, Greenville Technical College
Cathey Jordan, Wake Tech Community College
Jeanne Osborne, Middlesex County College
Susan Schott, University of Central Florida
Jane West, Trident Technical College

I would also like to thank the following for their help with the Eighth edition of *Elementary Statistics*®, which is the basis for this book:

Text Accuracy Reviewers:
Tommy Leavelle, Mississippi College
Milton Loyer
David Lund, University of Wisconsin at Eau Claire
John Ritschdorff, Marist College

Reviewers:
William A. Ahroon, Plattsburgh State
Scott Albert, College of Du Page
John Bray, Broward Community College–Central
Patricia Buchanan, Pennsylvania State University
Paul Cox, Ricks College
Nirmal Devi, Embry Riddle Aeronautical University
Joseph DeMaio, Kennesaw State University
Dennis Doverspike, University of Akron
Lauren Johnson, Inver Hills Community College
John Klages, County College of Morris

Marlene Kovaly, Florida Community College at Jacksonville
Christopher Jay Lacke, Rowan University
Tommy Leavelle, Mississippi College
Sergio Loch, Grand View College
Rhonda Magel, North Dakota State University–Fargo
Kathleen Mittag, University of Texas–San Antonio
Lyn Noble, Florida Community College at Jacksonville–South
Keith Oberlander, Pasadena City College
Lindsay Packer, College of Charleston
Fabio Santos, LaGuardia Community College
Laura Snook, Blackhawk Community College
Tom Sutton, Mohawk College

For Submitting Student Reviews:
Laura Kincaid, Baldwin–Wallace College

For their work on the videos:
Randall Allbritton, Daytona Beach Community College
Elka Block, Twin Prime Editorial
Nick Gellar, Collin County Community College
Len Groeneveld, Springfield Technical Community College
Denise Heban
Sergio Loch, Grand View College
Frank Purcell, Twin Prime Editorial
John Ritschdorff, Marist College
Jacci White, St. Leo University

Contents

Introduction to Statistics

As a student, are you in the most dangerous profession?

Two employees were installing aluminum siding on a farmhouse when it became necessary to remove a 36-foot CB antenna fastened at the top of the house. One employee disconnected the antenna and handed it to the other employee who was standing on the ground. When the second employee attempted to lay the antenna on the ground, it made contact with a 7200-volt power transmission line. This second employee received a fatal shock (based on information from the Occupational Safety and Health Administration).

We all know that some professions are inherently more dangerous than others. Fatal injuries are sometimes suffered by police officers involved in car crashes or gunfights. Firefighters sometimes die as a result of smoke inhalation, collapsing buildings, or explosions. Taxicab drivers are sometimes killed by passengers determined to steal the money collected as fares. Coal miners have suffered shortened lives as a result of their exposure to dangerous levels of coal dust typically found in mines. Some seemingly safe jobs, such as those in the U.S. Postal Service, have gained notoriety from deaths that resulted from employee rage.

The Swiss physician H. C. Lombard once compiled longevity data for different professions. He used death certificates that included name, age at death, and profession. He then proceeded to compute the average (mean) length of life for the different professions, and he found that students were lowest with a mean of only 20.7 years! (See "A Selection of Selection Anomolies" by Wainer, Palmer, and Bradlow in *Chance,* Volume 11, No. 2.) Similar results would be obtained if the same data were collected today in the United States. Is being a student really more dangerous than working near power lines, or being a police officer, a taxicab driver, or a postal employee? We will address this issue later in the chapter.

1-1 Overview

We begin our study of *statistics* by noting that the word has two basic meanings. In one sense, the word refers to specific numbers, such as this published result: "Twenty-three percent of people polled believed that there are too many polls." A second meaning refers to statistics as a method of analysis. It is this second meaning that will concern us here.

> ### DEFINITION
>
> **Statistics** is a collection of methods for planning experiments, obtaining data, and then organizing, summarizing, presenting, analyzing, interpreting, and drawing conclusions based on the data.

In this book we introduce the methods of statistics and show their application in many fields of study. You will see that statistics is much more than the mere calculation of averages accompanied by colorful graphs. It is a powerful analytical tool that you can use to develop more generalized and meaningful conclusions about populations that go beyond the original sample data.

 In statistics we commonly use the terms *population* and *sample*. Because these terms are central to our study, we define them now.

> ### DEFINITIONS
>
> A **population** is the complete collection of all elements (scores, people, measurements, and so on) to be studied. The collection is complete in the sense that it includes all subjects to be studied.
>
> A **census** is the collection of data from *every* element in a population.
>
> A **sample** is a subcollection of elements drawn from a population.

For example, a typical Nielsen television survey uses a *sample* of 4000 households, and the results are used to form conclusions about the *population* of all 103,215,027 households in the United States. Or the *New York Times* might commission a survey of 1000 people to estimate the proportion of the population favoring legalization of marijuana for medical purposes. The 1000 people constitute the *sample,* and all of us constitute the *population.* Every 10 years our government tries to obtain a *census*, but fails because it is impossible to reach everyone.

In this chapter we present some basic concepts about the nature of data, and some examples of uses and abuses of statistics. We conclude with a discussion of the design of experiments. The following important points will be emphasized in this chapter:

- Sample data must be collected in an appropriate way, such as through a process of *random* selection.
- If sample data are not collected in an appropriate way, no amount of statistical torturing can salvage the data.

What we ask is that you begin your study of statistics with an open mind. We are convinced that by the time you complete this introductory course, you will be firm in your belief that statistics is an interesting and rich subject with applications that are extensive, real, and meaningful. We are also convinced that with regular class attendance and appropriate diligence, you will succeed in mastering the basic concepts presented in this course.

 # 1-2 The Nature of Data

Data are observations (such as measurements, genders, and survey responses) that have been collected. Data are sometimes used to find statistics. An isolated list of lifeless numbers might appear to be a data set awaiting some statistical manipulation, but the effective use of statistics requires that we know the context of the data, how the data were obtained, and the population from which the data were obtained. We will demonstrate later the importance of sampling methods and the principle of randomness. Now, however, we will introduce some key terms.

DEFINITIONS

A **parameter** is a numerical measurement describing some characteristic of a *population*.

A **statistic** is a numerical measurement describing some characteristic of a *sample*.

▷ EXAMPLES

1. **Parameter:** When Lincoln was first elected to the presidency, he received 39.82% of the 1,865,908 votes cast. If we consider the collection of all of those votes to be the population being considered, then the 39.82% is a *parameter*, not a statistic.
2. **Statistic:** Based on a sample of 877 surveyed executives, it was found that 45% of them would not hire anyone whose job application contained a typographical error. The figure of 45% is a *statistic* because it is based on a sample, not the entire population of all executives. ◁

Some data sets consist of numbers (such as heights), and others are non-numerical (such as eye colors). The terms *quantitative data* and *qualitative data* are often used to distinguish between these types.

The State of Statistics

The word *statistics* is derived from the Latin word *status* (meaning "state"). Early uses of statistics involved compilations of data and graphs describing various aspects of a state or country. In 1662, John Graunt published statistical information about births and deaths. Graunt's work was followed by studies of mortality and disease rates, population sizes, incomes, and unemployment rates. Households, governments, and businesses rely heavily on statistical data for guidance. For example, unemployment rates, inflation rates, consumer indexes, and birth and death rates are carefully compiled on a regular basis, and the resulting data are used by business leaders to make decisions affecting future hiring, production levels, and expansion into new markets.

Interesting Statistics

The following claims were circulated by e-mail. They can be tested using methods presented in this book.

- Women blink nearly twice as much as men.
- Right-handed people live, on average, nine years longer than left-handed people.
- You share your birthday with at least 9 million other people.
- China has more English-speaking people than the United States.
- American Airlines saves $40,000 in a year by eliminating one olive from each salad served in first-class.
- The three most valuable brand names on earth are Marlboro, Coca-Cola, and Budweiser, in that order.

Do you think that all of these claims are true?

DEFINITIONS

Quantitative data consist of numbers representing counts or measurements.

Qualitative (or **categorical** or **attribute**) **data** can be separated into different categories that are distinguished by some nonnumeric characteristic.

EXAMPLES

1. **Quantitative data:** The incomes of college graduates
2. **Qualitative data:** The genders (male/female) of college graduates

When working with quantitative data, it is important to use the appropriate units of measurement, such as dollars, feet, hours, and so on. In some cases, quantitative data are scaled to be larger or smaller so that we can work with more convenient numbers. We should be especially careful to observe such notations as "all amounts are in thousands of dollars," or "all times are in hundredths of a second," or "units are in kilograms." Ignoring such units of measurement can lead to very wrong conclusions.

 We can further describe quantitative data by distinguishing between *discrete* and *continuous* types.

DEFINITIONS

Discrete data result when the number of possible values is either a finite number or a "countable" number. (That is, the number of possible values is 0 or 1 or 2 and so on.)

Continuous (numerical) data result from infinitely many possible values that correspond to some continuous scale that covers a range of values without gaps, interruptions, or jumps.

EXAMPLES

1. **Discrete data:** The numbers of eggs that hens lay are *discrete* data because they represent *counts*.
2. **Continuous data:** The amounts of milk that cows produce are *continuous* data because they are *measurements* that can assume any value over a continuous span. During a given time interval, a cow might yield an amount of milk that can be any value between 0 gallons and

5 gallons. It would be possible to get 2.343115 gallons, because the cow is not restricted to the discrete amounts of 0, 1, 2, 3, 4, or 5 gallons.

Another common way of classifying data is to use four levels of measurement: nominal, ordinal, interval, and ratio. In applying statistics to real problems, the level of measurement of the data is an important factor in determining which procedure to use. (See Figure 14-1.) There will be some references to these levels of measurement in this book, but the important point here is based on common sense: Never do computations and never use statistical methods with data that are not appropriate. For example, it would not make sense to compute an average of social security numbers, because those numbers are data that are used for identification; they don't represent measurements or counts of anything.

DEFINITION

The **nominal level of measurement** is characterized by data that consist of names, labels, or categories only. The data cannot be arranged in an ordering scheme (such as low to high).

 EXAMPLES   The following are examples of sample data at the nominal level of measurement.

1. Survey responses of yes, no, and undecided
2. The names of the television shows watched at 10:00 P.M. tonight

Because nominal data lack any ordering or numerical significance, they cannot be used for calculations. Numbers are sometimes assigned to the different categories (especially when data are computerized), but these numbers have no real computational significance and any average calculated with them is meaningless.

DEFINITION

Data at the **ordinal level of measurement** can be arranged in some order, but differences between data values either cannot be determined or are meaningless.

Measuring Disobedience

How are data collected about something that doesn't seem to be measurable, such as people's level of disobedience? Psychologist Stanley Milgram devised the following experiment: A researcher instructed a volunteer subject to operate a control board that gave increasingly painful "electrical shocks" to a third person. Actually, no real shocks were given, and the third person was an actor. The volunteer began with 15 volts and was instructed to increase the shocks by increments of 15 volts. The disobedience level was the point at which the subject refused to increase the voltage. Surprisingly, two-thirds of the subjects obeyed orders even though the actor screamed and faked a heart attack.

▶ **EXAMPLES** The following are examples of sample data at the ordinal level of measurement.

1. **Course grades:** A college professor assigns grades of A, B, C, D, or F. These grades can be arranged in order, but we can't determine differences between the grades. Thus we know, for example, that A is higher than B (so there is an ordering), but we cannot subtract B from A (so the difference cannot be found).

2. **Rankings:** Based on several criteria, a magazine ranks cities according to their "livability." Those rankings (first, second, third, and so on) determine an ordering. However, the differences between rankings are meaningless. For example, a difference of "second minus first" might suggest $2 - 1 = 1$, but this difference of 1 is meaningless because it is not an exact quantity that can be compared to other such differences. The difference between the first city and the second city is not the same as the difference between the second city and the third city. Using the magazine rankings, the *difference* between New York City and Boston cannot be quantitatively compared to the *difference* between St. Louis and Philadelphia. ◀

Ordinal data provide information about relative comparisons, but not the magnitudes of the differences. They should not be used for calculations.

DEFINITION

The **interval level of measurement** is like the ordinal level, with the additional property that the difference between any two data values is meaningful. However, there is no natural zero starting point (where *none* of the quantity is present).

▶ **EXAMPLES** The following are examples of data at the interval level of measurement.

1. **Temperatures:** Body temperatures of 98.2°F and 98.6°F. Those values are ordered, and we can determine their difference (often called the *distance* between the two values). However, there is no natural starting point. The value of 0°F might seem like a starting point, but it is arbitrary and does not represent the total absence of heat. It is wrong to say that 50°F is twice as hot as 25°F. (Temperature readings on the Kelvin scale are at the ratio level of measurement; that scale has an absolute zero.)

2. **Years:** The years 1000, 2000, 1776, and 1492. (Time did not begin in the year 0, so the year 0 is arbitrary instead of being a natural zero starting point.) ◀

DEFINITION

The **ratio level of measurement** is the interval level modified to include the natural zero starting point (where zero indicates that *none* of the quantity is present). For values at this level, differences and ratios are both meaningful.

▷ **EXAMPLES**   The following are examples of data at the ratio level of measurement. Note the presence of the natural zero value, and note the use of meaningful ratios of "twice" and "three times."

1. **Weights:** Weights (in carats) of diamond engagement rings (0 does represent no weight, and 4 carats is twice as heavy as 2 carats)

2. **Prices:** Prices of college textbooks ($0 does represent no cost, and a $90 book is three times as costly as a $30 book)   ◀

This level of measurement is called the ratio level because the starting point makes ratios meaningful. Because a 200-lb weight is twice as heavy as a 100-lb weight, but 50°F is *not* twice as hot as 25°F, weights are at the ratio level while Fahrenheit temperatures are at the interval level. For a concise comparison and review, study Table 1-1 to see the differences among the four levels of measurement.

TABLE 1-1	Levels of Measurement of Data	
Level	Summary	Example
Nominal	Categories only. Data cannot be arranged in an ordering scheme.	Student states: 5 Californians ⎫ 20 Texans ⎬ Categories or names only 40 New Yorkers ⎭
Ordinal	Categories are ordered, but differences cannot be determined or they are meaningless.	Student cars: An order is 5 compact ⎫ determined by 20 mid-size ⎬ "compact, 40 full-size ⎭ mid-size, full-size."
Interval	Differences between values are meaningful, but there is no natural starting point. Ratios are meaningless.	Campus temperatures: 5°F ⎫ "0°F doesn't mean no heat." 20°F ⎬ 40°F is not twice as 40°F ⎭ hot as 20°F.
Ratio	Like interval level, but there is a natural zero starting point. Ratios are meaningful.	Student commuting distances: 5 mi ⎫ 40 mi is 20 mi ⎬ *twice* 40 mi ⎭ as far as 20 mi.

1-2 Basic Skills and Concepts

Understanding a Statistic Versus a Parameter.   *In Exercises 1–4, determine whether the given value is a statistic or a parameter.*

1. A sample of students is selected, and the average (mean) age is 20.7 years.

2. After checking computer records for every commercial movie made last year, the longest running time is found to be 187 minutes.

3. All of the state governors are surveyed, and 30 of them are found to be Democrats.

4. Several Domino sugar packs are randomly selected, and the average (mean) weight of the contents is 3.647 g.

Understanding Discrete Versus Continuous.   *In Exercises 5–8, determine whether the given values are from a discrete or continuous data set.*

5. A statistics professor counts 3 absent students.

6. A statistics professor finds that on the first test, the first paper is turned in 39.627 minutes after the test began.

7. In a survey of 1068 Americans, 673 state that they own answering machines.

8. A manufacturer of rechargeable calculator batteries finds that one batch consists of 850 good batteries and 7 that are defective.

Determining Appropriate Measurements.   *In Exercises 9–18, determine which of the four levels of measurement (nominal, ordinal, interval, ratio) is most appropriate.*

9. Heights of women basketball players in the WNBA

10. Ratings of superior, above average, average, below average, or poor for blind dates

11. Noon temperatures (in degrees Fahrenheit) in Death Valley this week

12. Social security numbers

13. The years in which new editions of this book were published

14. Zip codes

15. *Consumer Reports* magazine ratings of "best buy, recommended, not recommended"

16. Distances traveled by students who commute to college

17. The actual contents (in ounces) of cola in Coke cans labeled 12 oz

18. The number of errors made when a Circuit City store scans 8000 different purchases

1-2 Beyond the Basics

19. Temperature Increase In the "Born Loser" cartoon strip by Art Sansom, Brutus expresses joy over an increase in temperature from 1° to 2°. When asked what is so good about 2°, he answers that "It's twice as warm as this morning." Explain why Brutus is wrong yet again.

20. Interpreting Political Polling A pollster surveys 200 people and asks them their preference of political party. He codes the responses as 0 (for Democrat), 1 (for Republican), 2 (for Independent), or 3 (for any other responses). He then calculates the average (mean of the numbers) and gets 0.95. How can that value be interpreted?

 # 1-3 Uses and Abuses of Statistics

Uses of Statistics

Many students study statistics because they know that it impresses employers to see a statistics course on a job applicant's transcript, and because virtually any field of study benefits from the application of statistical methods. In this book we include an abundance of practical uses of statistics. Here are a few brief examples:

- Poll results are used to determine the television shows we watch and the products we buy.

- Manufacturers provide better products at lower costs by using statistical quality control tools, such as control charts.

- Diseases are controlled through analyses designed to anticipate epidemics.

- Endangered species of fish and other wildlife are protected through regulations and laws that react to statistical estimates of changing population sizes.

- Through statistical analysis of fatality rates, legislators can better justify laws, such as those governing air pollution, auto inspections, seat belt and air bag use, and drunk driving.

We cite only these few examples, because a complete compilation of the uses of statistics would easily fill the remainder of this book (a prospect not totally unpleasant to some readers). As you proceed with this course, you will encounter many different applications of statistics.

Abuses of Statistics

Abuses of statistics are abundant. About a century ago, statesman Benjamin Disraeli famously said, "There are three kinds of lies: lies, damned lies, and statistics." It has also been said that "figures don't lie; liars figure." Historian Andrew Lang said that some people use statistics "as a drunken man uses lampposts—for support rather than illumination." Political cartoonist Don Wright encourages us to "bring back the mystery of life: lie to a pollster." These statements refer to uses of statistics in which data are presented in ways that are designed to be misleading. Some abusers of statistics are simply ignorant or careless, but others have personal objectives and are willing to suppress unfavorable data while emphasizing supportive data. We will now present several examples of ways in which data can be distorted.

Detecting Phony Data

A class is given the homework assignment of recording the results when a coin is tossed 500 times. One dishonest student decides to save time by just making up the results instead of actually flipping a coin. Because people generally cannot make up results that are really random, we can often identify such phony data. With 500 tosses of an actual coin, it is extremely likely that you will get a run of six heads or six tails, but people almost never include such a run when they make up results.

Another way to detect fabricated data is to establish that the results violate Benford's Law: For many collections of data, the leading digits are not uniformly distributed. Instead, the leading digits of 1, 2, ..., 9 occur with rates of 30%, 18%, 12%, 10%, 8%, 7%, 6%, 5%, and 5% respectively. (See "The Difficulty of Faking Data" by Theodore Hill, *Chance,* Vol. 12, No. 3.)

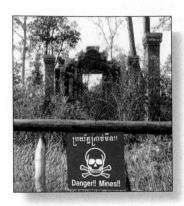

Statistics and Land Mines

The International Campaign to Ban Land Mines and the executive director of the Vietnam Veterans of America Foundation (VVAF) were recently awarded the Nobel Peace Prize. When VVAF asked for help in collecting data about land mines, a team of notable statisticians was assembled. Instead of working with intangible data, such as the value of a human life, they worked with tangible raw data, such as the area that a minefield makes unusable, and the cost of crops that cannot be grown. The data were included in *After the Guns Fall Silent: The Enduring Legacy of Landmines*, which became a key resource book in discussions of the land mine issue. The *AMSTAT News* quoted one of the book's editors: "... this data-gathering and analysis effort is what made it possible to put the issue before policy-makers. This work really made a difference."

Bad Samples   A major source of deceptive statistics is the use of inappropriate methods to collect data. One very common sampling method somehow allows the sample subjects to decide for themselves whether to be included.

> ### DEFINITION
>
> A **self-selected survey** (or **voluntary response sample**) is one in which the respondents themselves decide whether to be included.

In such surveys, people with strong opinions are more likely to participate, so the responses obtained are not necessarily representative of the whole population. *Self-selected samples* are obtained when a television news program asks viewers to call in with their preference for a political candidate or when a "researcher" mails a survey to people who can decide either to ignore it or to respond.

A self-selected sample is only one way in which the method of collecting data can be seriously flawed. As another example, consider our Chapter Problem, which described how Swiss physician H. C. Lombard once compiled longevity data for different professions by referring to death certificates that included name, age at death, and profession. He then computed the average (mean) length of life for the different professions and found that students ranked lowest with a mean of only 20.7 years. The problem is that the sample is not appropriate, because most people are students when they are relatively young; they are no longer students when they become a little older and are employed in some other profession. Also, all of the students included in the Lombard study *died*, so the 20.7-year longevity figure merely suggests that when students die, they are likely to be young. A similar error would be made if you computed the average age of teenagers who died. That figure could not possibly be more than 19 years, but this does not really mean that being a teenager is more dangerous than being a test pilot, a profession in which the members would surely die at ages with a mean considerably higher than 19 years.

Small Samples   It can be very misleading to make broad conclusions or inferences based on samples that are far too small. An example is the Children's Defense Fund's publication of *Children Out of School in America,* which reported that among secondary school students suspended in one region, 67% were suspended at least three times. That figure was based on a sample of only *three* students, and media reports failed to mention that the sample size was so small. Sometimes a sample might seem relatively large (as in a survey of "2000 randomly selected adult Americans"), but if conclusions are based on subgroups, such as the Catholic male Republicans from Maine, such conclusions might be based on samples that are too small. Although it is important to have a sample that is sufficiently large, size is not the only issue; it is just as important to have sample data that have been collected in an appropriate way, such as through random selection. Even large samples can be bad samples.

Loaded Questions   Survey questions can be worded to elicit a desired response. In a recent survey, different wordings of a question resulted in different responses:

- 97% yes: "Should the president have the line item veto to eliminate waste?"
- 57% yes: "Should the president have the line item veto, or not?"

Sometimes questions are unintentionally loaded by such factors as the order of the items being considered. Consider these questions from a poll conducted in Germany:

- Would you say that traffic contributes more or less to air pollution than industry?
- Would you say that industry contributes more or less to air pollution than traffic?

When traffic was presented first, 45% blamed traffic and 27% blamed industry; when industry was presented first, 24% blamed traffic and 57% blamed industry.

Misleading Graphs   Many visual devices—such as bar graphs and pie charts— **AS** can be used to exaggerate or diminish the true import of the data. (Such devices will be discussed in Chapter 2.) The two graphs in Figure 1-1 depict the *same data* from the Bureau of Labor Statistics, but part (a) is designed to exaggerate the difference between the salaries of people who have bachelor's degrees and the salaries of those who have only high school diplomas. By not starting the horizontal axis at zero, the graph in part (a) produces a misleading subjective impression. Figure 1-1 carries an important lesson: To interpret a graph correctly, we must analyze the *numerical* information provided with the graph, so that we won't be misled by its general shape.

Pictographs   Drawings of objects, called *pictographs,* may also be misleading. Some objects commonly used to depict data include three-dimensional

FIGURE 1-1
Salaries of People with Bachelor's Degrees and with High School Diplomas

FIGURE 1-2
Double the length, width, and height of a cube, and the volume increases by a factor of eight.

objects, such as moneybags, stacks of coins, army tanks (for military expenditures), barrels (for oil production), and houses (for home construction). When drawing such objects, artists can create false impressions that distort differences. If you double each side of a square, the area doesn't merely double; it increases by a factor of four. If you double each side of a cube, the volume doesn't merely double; it increases by a factor of eight. If taxes double over a decade, an artist may depict tax amounts with one moneybag for the first year and a second moneybag that is twice as deep, twice as tall, and twice as wide. Instead of appearing to double, taxes will appear to increase by a factor of eight, so the truth will be distorted by the drawing. See Figure 1-2.

Precise Numbers When a statement includes a very precise figure, such as "There are now 103,215,027 households in the United States," many people assume incorrectly that because it is precise it is also *accurate*. In this case, the number is an estimate, and it would be better to state that the number of households is about 103 million.

Distorted Percentages Misleading or unclear percentages are sometimes used. Continental Airlines ran ads boasting better service. In referring to lost baggage, the ads claimed that this was "an area where we've already improved 100% in the last six months." In an editorial criticizing this statistic, the *New York Times* interpreted the 100% improvement figure to mean that no baggage is now being lost—an accomplishment not yet achieved by Continental Airlines.

Partial Pictures "Ninety percent of all our cars sold in this country in the last 10 years are still on the road." Millions of consumers heard that impressive commercial message. What they were not told was that 90% of the cars the advertiser sold in this country were sold within the last three years. The claim was technically correct, but it was very misleading in that it did not present the complete story.

Deliberate Distortions In the book *Tainted Truth*, Cynthia Crossen cites an example of the magazine *Corporate Travel*, which published results showing that among car rental companies, Avis was the winner in a survey of people who rent cars. When Hertz requested detailed information about the survey, the actual survey responses disappeared and the magazine's survey coordinator resigned. Hertz sued Avis (for false advertising based on the survey) and the magazine; a settlement was reached.

Entire books have been devoted to abuses of statistics, including Darrell Huff's classic *How to Lie with Statistics*, Robert Reichard's *The Figure Finaglers*, and Cynthia Crossen's *Tainted Truth*. Understanding these practices will be extremely helpful in evaluating the statistical data found in everyday situations.

1-3 Basic Skills and Concepts

1. **"900" Numbers** In an ABC "Nightline" poll, 186,000 viewers each paid 50 cents to call a "900" telephone number with their opinion about keeping the United Nations headquarters in the United States. The results showed that 67% of those who called were in favor of moving the United Nations headquarters out of the United States. Interpret the results by identifying what we can conclude about the way the general population feels about keeping the United Nations headquarters in the United States.

2. **Telephone Surveys** The Hartford Insurance Company has hired you to poll a sample of adults about their car purchases. What is wrong with using telephone directories as the population from which the sample is drawn?

3. **CPI** The Consumer Price Index (CPI) is based on the cost of goods and services purchased by typical consumers. Assume that the cost is $500 this year.
   a. If there is inflation so that all costs rise 5% next year, what is next year's cost?
   b. Assume that in the year after next year, all costs drop by 5%. What is that year's cost? When the 5% increase is followed by the 5% decrease, do costs return to the original $500 level?

4. **Campus Crime** In a study of college campus crimes committed by students high on alcohol or drugs, a mail survey of 1875 students was conducted. A *USA Today* article noted, "Eight percent of the students responding anonymously say they've committed a campus crime. And 62% of that group say they did so under the influence of alcohol or drugs." Assuming that the number of students responding anonymously is 1875, how many actually committed a campus crime while under the influence of alcohol or drugs?

5. **Healthy Mothers** The *Newport Chronicle* claims that pregnant mothers can increase their chances of having healthy babies by eating lobsters. That claim is based on a study showing that babies born to lobster-eating mothers have fewer health problems than babies born to mothers who don't eat lobster. What is wrong with the claim?

6. **Job Applicants** "According to a nationwide survey of 250 hiring professionals, scuffed shoes was the most common reason for a male job seeker's failure to make a good first impression." Newspapers carried this statement based on a poll commissioned by Kiwi Brands, producers of shoe polish. Comment on why the results of this survey might be questionable.

7. **Motorcycle Helmets** The Hawaii State Senate held hearings when it was considering a law requiring that motorcyclists wear helmets. Some motorcyclists testified that they had been in crashes in which helmets would not have been helpful. Which important group was not able to testify? (See "A Selection of Selection Anomalies" by Wainer, Palmer, and Bradlow in *Chance*, Volume 11, No. 2.)

8. **Longevity** You need to conduct a study of longevity for people who were born after the end of World War II in 1945. If you were to visit graveyards and use the birth and death dates listed on tombstones, would you get good results? Why or why not?

9. Bad Question A survey includes this item: "Enter your height in inches." It is expected that actual heights of respondents can be obtained and analyzed, but there are two different major problems with this item. Identify them.

10. Mail Survey When author Shere Hite wrote *Women and Love: A Cultural Revolution in Progress*, she based her conclusions on 4500 replies that she received after mailing 100,000 questionnaires to various women's groups. Are her conclusions likely to be valid in the sense that they can be applied to the general population of all women?

11. Understanding Percentages Is a 10% price cut the same as two consecutive 5% price cuts? Why or why not?

12. Car Weights Refer to Data Set 18 in Appendix B and consider the weights of the randomly selected car models. If you were to calculate the average (mean) weight of those cars, is the result likely to be a good estimate of the weight of the cars being driven in the United States? Why or why not?

1-3 Beyond the Basics

13. Phony Data A researcher at the Sloan-Kettering Cancer Research Center was once criticized for falsifying data. Among his data were figures obtained from six groups of mice, with 20 individual mice in each group. These values were given for the percentage of successes in each group: 53%, 58%, 63%, 46%, 48%, 67%. What is the major flaw?

14. What's Wrong with This Picture? Try to identify each of the four major flaws in the following. A daily newspaper ran a survey by asking readers to call in their response to this question: "Do you support the development of atomic weapons that could kill millions of innocent people?" It was reported that 20 readers responded, and 87% said "no," while 13% said "yes."

15. Understanding Percentages A *New York Times* editorial criticized a chart caption that described a dental rinse as one that "reduces plaque on teeth by over 300%."
    a. If you remove 100% of some quantity, how much is left?
    b. What does it mean to reduce plaque by over 300%?

16. Interpreting Surveys In an angry letter to the Associated Press, a company president complained about a poll being unfair. He claimed that because the sample size of 1223 people was used to represent 120 million people, his letter represented 98,000 people who have the same views. Is his arithmetic correct? Is his claim valid?

 ## 1-4 Design of Experiments

Sometimes we have an interesting and important data set, but we don't have a particular objective in mind, so we want to *explore* the data to see what insights we might acquire. (In Chapter 2 we will introduce basic tools that can be used to explore or describe data sets.) More often, though, we do have a specific objective, and we want to collect the data and do the analysis that will help us to meet that objective. We typically get our data from two common sources: *observational studies* (such as polls) and *experiments* (such as using a treatment to improve hair growth).

DEFINITIONS

In an **observational study,** we observe and measure specific characteristics, but we don't attempt to *modify* the subjects being studied.

In an **experiment,** we apply some *treatment* and then proceed to observe its effects on the subjects.

An example of an observational study is a Nielsen poll. Each night, the Nielsen organization polls people to determine the percentage of the population that is watching NBC at 9:00 P.M. These people are merely counted; they do not receive any type of treatment. An example of an experiment was the testing of the Salk vaccine. In the famous 1954 experiment, some children were given the actual Salk vaccine and others were given a placebo that contained no medicine or drug. This was an experiment, as opposed to an observational study, because the subjects were given a treatment. In a well-designed experiment, like this one, we compare effects in the treatment group to effects in a group not given the treatment. The use of two groups (treatment group and placebo group) made it possible to evaluate the effectiveness of the Salk vaccine in preventing polio.

There are a few basic steps that should be followed in designing an experiment that is capable of yielding valid results.

1. *Identify your objective:* Identify the exact question to be answered, and clearly identify the relevant population. (For example, "Is the Salk vaccine effective in reducing the incidence of polio in the population of children?")

2. *Collect sample data:* The way in which sample data are collected is absolutely critical to the success of the experiment. The sample data must be representative of the population in question, the sample must be large enough so that the effects of the treatment can be known, and the question should be addressed without interference from extraneous factors.

3. *Use a random procedure that avoids bias.* (In the Salk vaccine experiment, for example, children were assigned to the two groups through a process of random selection.)

4. *Analyze the data and form conclusions.*

Controlling Effects of Variables

When conducting an experiment, it is all too easy to receive interference from variable factors that are not relevant to the issue being studied. These effects can be controlled through good experimental design. In the 1954 polio experiment, for example, a treatment group of children was given the actual vaccine and a control group was given a placebo that contained no

Sampling Rejected for the Census

It has been estimated that in the 1990 Census, roughly 10 million people were not counted, while 4 million others were counted twice. These errors can be largely corrected by applying the same sampling techniques used by pollsters. But there is a political issue. Population counts affect the number of seats in the House of Representatives, so Republicans oppose sampling because undercounted regions tend to be largely Democratic, while overcounted regions tend to be largely Republican. Democrats favor the use of sampling methods. Some people argue that the Constitution specifies that the census be an "actual enumeration," or a head count which does not allow for sampling methods, and the Supreme Court upheld that position. Will modern sampling methods be eventually used, or will the traditional census head count prevail?

Hawthorne and Experimenter Effects

The well-known placebo effect occurs when an untreated subject incorrectly believes that he or she is receiving a real treatment and reports an improvement in symptoms. The Hawthorne effect occurs when treated subjects somehow respond differently, simply because they are part of an experiment. (This phenomenon was called the "Hawthorne effect" because it was first observed in a study of factory workers at Western Electric's Hawthorne plant.) An experimenter effect (sometimes called a Rosenthall effect) occurs when the researcher or experimenter unintentionally influences subjects through such factors as facial expression, tone of voice, or attitude.

drug at all. In experiments of this type, a **placebo effect** occurs when an untreated subject incorrectly believes that he or she is receiving a treatment and reports an improvement in symptoms. The placebo effect can be countered by using **blinding,** a technique in which the subject doesn't know whether he or she is receiving a treatment or a placebo. Blinding is used so that we can determine whether the treatment effect is significantly greater than the placebo effect. The polio experiment was *double-blind*, meaning that blinding occurred at two levels: (1) The children being injected didn't know whether they were getting the Salk vaccine or a placebo, and (2) the doctors who gave the injections and evaluated the results didn't know either.

When designing an experiment to test the effectiveness of one or more treatments, it is important to put the subjects (often called *experimental units*) in different groups (or *blocks*) in such a way that those groups are very similar. A **block** is a group of subjects (or experimental units) that are similar. (They need to be similar only in the ways that might affect the outcome of the experiment.)

> **When testing one or more different treatments, form blocks so that each one consists of subjects that are similar.**

When deciding how to assign the subjects to different blocks, you can use random selection or you can try to carefully control the assignment so that the subjects within each block are similar. One approach is to use a **completely randomized experimental design,** in which subjects are put into different blocks through a process of *random selection*. An example of a completely randomized experimental design is, again, the polio experiment. Children were assigned to the treatment group or the placebo group through a process of random selection (equivalent to flipping a coin). Another approach is to use a **rigorously controlled design,** in which experimental units are very *carefully chosen* so that the subjects in each block are similar in the ways that are important. With a rigorously controlled design, you might start by giving the Salk vaccine to a healthy seven-year-old girl from Texas, while another healthy seven-year-old girl from Texas would be given the placebo. To make this work, you would have to go beyond health, age, and state of residence and identify other relevant factors to be considered.

When conducting experiments, the results are sometimes ruined because of *confounding*.

DEFINITION

Confounding occurs in an experiment when the effects from two or more variables cannot be distinguished from each other.

For example, suppose a professor in Vermont experiments with a new attendance policy ("your course average drops one point for each class cut"), but an exceptionally mild winter moderates the discomforts (snow and cold temperatures) that have reduced attendance in the past. If attendance does improve, we can't tell whether the improvement is attributable to the new attendance policy or to the mild winter—the effects of the attendance policy and the weather have been confounded.

Sample Size

Another important consideration in conducting experiments is the size of your sample. It must be large enough so that erratic behavior of very small samples will not produce misleading results. Repetition of an experiment is called *replication*.

A large sample is not necessarily a good sample. Although it is important to have a sample that is sufficiently large, it is more important to have a sample in which the elements have been chosen in an appropriate way, such as random selection.

> **Use a sample size large enough so that we can see the true nature of any effects, and obtain the sample using an appropriate method, such as one based on *randomness*.**

In the testing of the Salk vaccine, a sample size of only four people in the treatment group and four other people in the control group would not reveal whether the vaccine is effective. Fortunately, the actual experiment involved large samples: 200,000 children were given the actual Salk vaccine and 200,000 other children were given a placebo. Because the actual experiment used sufficiently large sample sizes, the effectiveness of the vaccine could be seen. Nevertheless, even though the treatment and placebo groups were very large, the experiment would have failed if subjects had not been assigned to the two groups in an appropriate way.

Randomization

One of the worst mistakes is to collect data in a way that is inappropriate. We cannot overstress this very important point:

> **Data carelessly collected may be so completely useless that no amount of statistical torturing can salvage them.**

We noted in Section 1-3 that a self-selected survey (or voluntary response sample) is one in which people decide themselves whether to respond. Self-selected samples are very common, but their results are generally useless for making valid inferences about larger populations.

We now define and describe some of the more common methods of sampling.

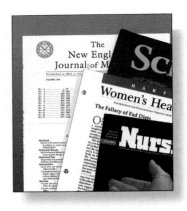

Meta-Analysis

The term *meta-analysis* refers to a technique of doing a study that essentially combines results of other studies. It has the advantage that separate smaller samples can be combined into one big sample, making the collective results more meaningful. It also has the advantage of using work that has already been done. Meta-analysis has the disadvantage of being only as good as the studies that are used. If the previous studies are flawed, the "garbage in, garbage out" phenomenon can occur. The use of meta-analysis is currently popular in medical research and psychological research. This study used data from 46 other studies: "Meta-Analysis of Migraine Headache Treatments: Combining Information from Heterogeneous Designs" by Dominici et al, *Journal of the American Statistical Association*, Vol. 94, No. 445.

DEFINITIONS

In a **random sample** members of the population are selected in such a way that each has an *equal chance* of being selected.

A **simple random sample** of size n subjects is selected in such a way that every possible sample of size n has the same chance of being chosen.

The Literary Digest Poll

In the 1936 presidential race, *Literary Digest* magazine ran a poll and predicted an Alf Landon victory, but Franklin D. Roosevelt won by a landslide. Maurice Bryson notes, "Ten million sample ballots were mailed to prospective voters, but only 2.3 million were returned. As everyone ought to know, such samples are practically always biased." He also states, "Voluntary response to mailed questionnaires is perhaps the most common method of social science data collection encountered by statisticians, and perhaps also the worst." (See Bryson's "The *Literary Digest* Poll: Making of a Statistical Myth," *The American Statistician*, Vol. 30, No. 4.)

With random sampling we expect all groups of the population to be (approximately) proportionately represented. Random samples are selected by many different methods, including using computers to generate random numbers. (Before computers, tables of random numbers were often used instead. For truly exciting reading, see this book consisting of 1 million digits that were randomly generated: *A Million Random Digits,* published by Free Press. The *Cliff Notes* summary of the plot is not yet available.) Careless or haphazard sampling can easily result in a biased sample with characteristics unlike the population from which the sample was drawn. Random sampling, in contrast, is very carefully planned and executed. There are other sampling techniques in use, and we describe the common ones here, but *only random sampling and simple random sampling will be used throughout the remainder of the book.*

DEFINITION

In **systematic sampling,** we select some starting point and then select every *k*th (such as every 50th) element in the population.

For example, if Coca Cola managers wanted to poll the 29,500 employees, they could begin with a complete employee roster, then select every 50th person to obtain a sample of size 590. This method is simple and is often used. (If we randomly pick the starting point, then randomly pick every 50th person on an alphabetized list, we have a *random sample* because everyone has the same chance of being chosen, but we do not have a *simple random sample* because not every possible sample of 590 employees has the same chance. For example, there is no chance of getting the first 590 people on the alphabetized list.)

DEFINITION

With **convenience sampling,** we simply use results that are readily available.

In some cases, results from convenience sampling may be quite good, but in many other cases they may be seriously biased. In investigating the proportion of left-handed people, it would be convenient for a student to survey his or her classmates, because they are readily available. Even though such a sample is not random, the results should be quite good. In contrast, it might be convenient (and perhaps also profitable) for ABC News to conduct a poll by asking audience members to call a "900" telephone number to register their opinions, but this would be a self-selected sample and the results would likely be biased.

> **DEFINITION**
>
> With **stratified sampling,** we subdivide the population into at least two different subgroups (or strata) that share the same characteristics (such as gender or age bracket), then we draw a sample from each stratum.

For example, using the 50 states as strata, we might select a random sample of voters in each state. If the different strata have sample sizes that are in the same proportion as in the population, we say that we have *proportionate* sampling. If it should happen that some strata are not represented in the proper proportion, then the results can be adjusted or weighted accordingly. For a fixed sample size, if you randomly select subjects from different strata, you are likely to get more consistent (and less variable) results than by simply selecting a random sample from the general population. For that reason, stratified sampling is often used to reduce the variation in the results.

> **DEFINITION**
>
> In **cluster sampling,** we first divide the population area into sections (or clusters), then randomly select some of those clusters, and then choose *all* the members from those selected clusters.

Note that stratified sampling and cluster sampling both involve the formation of subgroups, but cluster sampling uses *all* members from a sample of clusters, whereas stratified sampling uses a *sample* of members from all strata. An example of cluster sampling can be found in a pre-election poll, in which we randomly select 30 election precincts and then survey all the people from each of those precincts. This would be much faster and much less expensive than selecting one person from each of the many precincts in the population area. The results can be adjusted or weighted to correct for any disproportionate representations of groups. Cluster sampling is used extensively by government and private research organizations.

Figure 1-3 on page 22 illustrates the five common methods of sampling just described. In practice, professionals often collect data by using some combination of the five methods. Here is one typical example of what is called a *multistage sample design*: First randomly select a sample of counties from all 50 states, then randomly select cities and towns in those counties, then randomly select residential blocks in each city or town, then randomly select households in each block, then randomly select someone from each household. We will not use such a sample design in this book. We should again stress that the methods of this book typically require that we have a *simple random sample*.

Random Sampling:
Each member of the population has an equal chance of being selected. Computers are often used to generate random telephone numbers.

Systematic Sampling:
Select every kth member.

Convenience Sampling:
Use results that are readily available.

Hey! Do you believe in the death penalty?

Women   Men

Stratified Sampling:
Classify the population into at least two strata, then draw a sample from each.

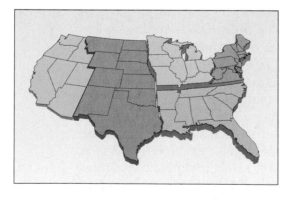

Cluster Sampling:
Divide the population area into sections, randomly select a few of those sections, and then choose all members in them.

FIGURE 1-3
Common Sampling Methods

No matter how well you plan and execute the sample collection process, there is likely to be some error in the results. For example, you might randomly select 1000 adults, ask them if they graduated from high school, and record the sample percentage of "yes" responses. If you randomly select another sample of 1000 adults, it is likely that you will obtain a *different* sample percentage.

> ### DEFINITIONS
>
> A **sampling error** is the difference between a sample result and the true population result; such an error results from chance sample fluctuations.
>
> A **nonsampling error** occurs when the sample data are incorrectly collected, recorded, or analyzed (such as by selecting a biased sample, using a defective measurement instrument, or copying the data incorrectly).

If we carefully collect a sample so that it is representative of the population, we can use the methods in this book to analyze the sampling error, but we must exercise extreme care so that nonsampling error is minimized.

1-4 Basic Skills and Concepts

Distinguishing Between Observational Study and Experiment.   In Exercises 1–4, determine whether the given description corresponds to an observational study or an experiment.

1. Cans of Coke are opened, and the volumes (in ounces) of the contents are measured.

2. The new drug Statisticzene is tested by recording its effects on the students who are given the drug.

3. The effectiveness of multimedia teaching is tested with a sample of students who complete a course of study using the multimedia approach.

4. Much controversy arose over a study of patients with syphilis who were *not* given a treatment that could have cured them.

Identifying Types of Sampling.   In Exercises 5–16, identify which of these types of sampling is used: random, systematic, convenience, stratified, or cluster.

5. Telephone Surveys The Gallup Organization plans to conduct a poll of New York City residents with the "212" area code. Computers are used to randomly generate telephone numbers that are automatically called.

6. MTV Survey A marketing expert for MTV is planning a survey in which 500 people will be randomly selected from each age group of 10–19, 20–29, and so on.

7. News Reporting An ABC news reporter polls people as they pass him on the street.

8. Medical Research A Johns Hopkins University researcher surveys all cardiac patients in each of 30 randomly selected hospitals.

9. Car Ownership A General Motors researcher has partitioned all registered cars into categories of subcompact, compact, mid-size, intermediate, and full-size. She is surveying 200 randomly selected car owners from each category.

10. Jury Selection The Dutchess County Commissioner of Jurors obtains a list of 42,763 car owners and constructs a pool of jurors by selecting every 100th name on that list.

11. Student Drinking The College of Newport conducts a study of student drinking by randomly selecting 10 different classes and interviewing all of the students in each of those classes.

12. Lobbying A lobbyist for the tobacco industry obtains a sample of members of Congress by writing the 535 names on individual index cards, putting them in a box, mixing them, then selecting 50 different names.

13. Education and Salary An economist is studying the effect of education on salary and conducts a survey of 150 randomly selected workers from each of these categories: less than a high school diploma, high school diploma, more than a high school diploma.

14. Prisoners A prison guard surveys all prisoners in his ward.

15. Sobriety Checkpoint The author was an observer at a police sobriety checkpoint at which every fifth driver was stopped and interviewed. (He witnessed the arrest of a former student.)

16. Exit Poll CNN is planning an exit poll in which 100 polling stations will be randomly selected and all voters will be interviewed as they leave the premises.

Obtaining a Simple Random Sample.   In Exercises 17–20, describe a procedure for obtaining a simple random sample.

17. Select a simple random sample of 200 students from the population of full-time students at your college.

18. Select a simple random sample of 50 Bufferin aspirin tablets to be tested for the precise amount of aspirin that they contain.

19. Select a simple random sample of 250 college textbooks from the population of all college textbooks currently being used in your state.

20. Select a simple random sample of 100 households in your county.

1-4 Beyond the Basics

21. Reducing Crime Two categories of survey questions are *open* and *closed*. An open question allows a free response, while a closed question allows only a fixed response. The following examples are based on Gallup surveys.

    Open question: What do you think can be done to reduce crime?

    Closed question: Which of the following approaches would be most effective in reducing crime?
    • Hire more police officers.
    • Get parents to discipline children more.
    • Correct social and economic conditions in poor neighborhoods.
    • Improve rehabilitation efforts in jails.
    • Give convicted criminals tougher sentences.
    • Reform courts.

    a.   What are the advantages and disadvantages of open questions?
    b.   What are the advantages and disadvantages of closed questions?
    c.   Which type is easier to analyze with formal statistical procedures, and why is that type easier?

22. Sampling Design The Addison Wesley Longman Publishing Company has commissioned you to survey 100 students who use this book. Describe procedures for obtaining a sample of each type: random, systematic, convenience, stratified, cluster.

23. Confounding Give an example (different than the one in the text) illustrating how confounding occurs.

24. Random Selection Among the 50 states, one state is randomly selected. Then, a statewide voter registration list is obtained and one name is randomly selected. Does this procedure result in a randomly selected voter?

25. Simple Random Samples We noted that systematic sampling could result in a random sample, but not a simple random sample.
    a. Does stratified sampling result in a random sample? A simple random sample? Explain.
    b. Does cluster sampling result in a random sample? A simple random sample? Explain.

 # 1-5 Introduction to Excel

The transition to the twenty-first century was accompanied by an explosion in the collection and analysis of data. A *spreadsheet* is a popular way to organize such data.

> **DEFINITION**
>
> A **spreadsheet** is a collection of data organized in an array of cells arranged in rows and columns, and it is used to summarize, analyze, and perform calculations with the data.

Because spreadsheets are so common in workplace settings, the study of Excel is very helpful in preparing you for your future job.

In Figure 1-4 on page 26 we show the basic format of an Excel spreadsheet. Notice how the body of the spreadsheet consists of cells in columns labeled A, B, C, . . . , and rows labeled 1, 2, 3, . . . This individual spreadsheet is part of a larger Excel structure, described as follows. (The number of worksheets is 3 by default, but you can insert additional worksheets or delete existing worksheets.)

Excel file = Workbook = 3 worksheets (or 3 arrays of cells as in Figure 1-4)

Although Excel is called a spreadsheet program, we typically refer to *worksheets* instead of spreadsheets.

> **DEFINITION**
>
> An Excel **worksheet** is one page or sheet (or spreadsheet) consisting of cells arranged in an array of rows and columns. (See Figure 1-4.) The cells can contain text, numbers, or formulas.

FIGURE 1-4
A Spreadsheet in Excel

In addition to a blank worksheet, Figure 1-4 also includes the menu bar, tool bar, formatting tool bar, formula bar, and sheet tabs. Here are broad descriptions of some of those items:

Menu bar:	Pull-down menus that allow you to perform different tasks, such as open an Excel file, save an Excel file, print the current worksheet, or sort data (arrange in order).
Tool bar:	Buttons for performing tasks (such as print, copy, paste) with one click of the mouse.
Formatting tool bar:	Format the cell contents by selecting the type font, size, and other attributes, such as **bold** or *italics*.

The best way to learn Excel is to actually use it, so let's walk through the creation of a worksheet (or spreadsheet). Table 1-2 is a small part of Data Set 15 in Appendix A, and we will proceed to manually construct a

TABLE 1-2	Movie Data					
Title	Year	Rating	Budget ($) (in millions)	Gross ($) (in millions)	Length (in minutes)	Viewer Rating
Aliens	1986	R	18.5	81.843	137	8.2
Armageddon	1998	PG-13	140	194.125	144	6.7
As Good As It Gets	1997	PG-13	50	147.54	138	8.1
Braveheart	1995	R	72	75.6	177	8.3
Chasing Amy	1997	R	0.25	12.006	105	7.9

worksheet for Table 1-2. (Data Set 15 can be opened from an Excel workbook stored on the CD-ROM, but it is important to develop an ability to manually create an Excel worksheet.)

Creating a Worksheet

1. Run Excel to get the blank worksheet shown in Figure 1-4. This often involves double-clicking on the Excel icon. Shown below are the Windows desktop icons for Excel 97 and Excel 2000.

Excel 97

Excel 2000

   (Excel should be installed on your computer. If it is not installed, follow the directions that are provided with your software.)

2. The cell in the upper left corner should be highlighted, so type "Title" and press the **Enter** key three times to move down to the fourth cell in the first column. (*Note:* You can move to different cells by pressing **Enter** to move down, or pressing the Tab key to move to the right, or pressing the arrow keys, or you can simply click on the desired cell.)

3. With cell A4 (column A, fourth cell down) highlighted, type "Aliens" and press the **Enter** key. Then type "Armageddon" and press the **Enter** key. Continue until the five movie titles have been entered.

   Some of the titles seem too long for the space provided, and the complete titles might not all be visible, but they are there. If you want to make the complete title visible, change the column width as follows: With a cell in column A highlighted, click on **Format,** then select **Column,** then select **Width,** enter a new width such as 20, then click **OK.** Alternatively, use the mouse to highlight the entire table, then click on **Format** and select **Autoformat.** A third approach is to simply double-click on the right column boundary in the gray shaded cell at the top of the list.

4. Click on the top cell in column B to highlight that cell, then proceed to enter "Year." Complete the other entries in column B and the remaining columns by using the same procedure for entering column A. You have now created your own spreadsheet!

After creating the Excel worksheet for Table 1-2, we could save the worksheet, print it, edit it, perform calculations, do statistical analyses, and construct graphs. Some of these important functions will be discussed in the chapters that follow. For now, we consider a few of the more important general functions of Excel.

Saving an Excel File

1. Click on **File,** then scroll down and click on the menu item **Save As.**

2. You should now see a dialog box labeled "Save As."

a. In the "Save in" box at the top, select the directory location of where you want your file saved. To change the drive that is shown, click on the drop-down button ▼ and then click on the desired location, such as (C:) or $3\frac{1}{2}$ Floppy (A:).

Many students use Excel in a college computer lab and save files on their own floppy disks, so they select $3\frac{1}{2}$ Floppy (A:) by clicking on that option. (If you select C, you will get a list of folders available in directory C, and you should click on the desired folder, such as "My Documents.")

b. Enter a name (such as "Movies") in the File name box.

c. In the "Save as type" box, select **Microsoft Excel Workbook.**

d. Click on **Save.**

Getting Help   To get help while you are running Excel:

1. Click on **Help** located to the right of the menu bar at the top.

2. Click on **Contents and Index.**

3. Click on the tab labeled **Index** and proceed to enter a keyword, such as "print" or "save."

Undoing Your Work   When using Excel, it often happens that you do something terribly wrong and you wish that you hadn't done it. Great news: You can *undo* your last action. Simply click on the counterclockwise arrow located on the menu bar. It looks like this:

If only life had an undo button.

Opening an Excel Workbook from the CD-ROM   Data sets from Appendix B are used throughout this book. Some of those data sets are large, so much time and patience would be required to manually create Excel worksheets. Good news: Except for Data Set 6, those data sets have already been stored as Excel worksheets on the CD-ROM packaged with this book. You can retrieve or "open" them instead of manually entering them. Let's assume that the CD-ROM is in drive D of your computer, and you want to open the file COLA.xls (for Data Set 1 in Appendix B).

1. Click on **File** located on the menu bar.
2. Click on the submenu item of **Open.**
3. You should now see a dialog box with the label "Open."
   a. Begin with the "Look in" box at the top. To change the directory location, click on the drop-down button ▼ and then click on the desired location. If the CD-ROM is in drive D, click on D to get a listing of the folders on the CD-ROM.
   b. Click on **DataSets** to get into the folder containing data sets.
   c. Click on **Excel** to get into the folder containing Excel workbooks.
   d. Click on **COLA.xls** to highlight the data for Data Set 1 in Appendix B.
   e. Click on **Open.**

You may also open Excel workbooks that you have previously saved. For example, if you saved a workbook named GRADES.xls on a floppy disk in drive A, click on **File, Open,** and proceed to select drive A and the desired workbook.

Doing Calculations with Excel   By entering an equal sign followed by an arithmetic expression in the formula bar, you can do calculations. Because the standard keyboard does not have special keys for operations such as division, exponents, and square roots, we must use the keys or commands designated for those purposes. Here are a few of the more common operations:

Operation	Excel Expression	Excel Example	Result
Addition	+	=2+3	5
Subtraction	−	=6−4	2
Multiplication	*	=3*4	12
Division	/	=12/3	4
Exponentiation	^	=2^3	8
Square root	SQRT( )	=SQRT(49)	7
Absolute value	ABS( )	=ABS(−5)	5
Factorial	FACT( )	=FACT(3)	6

Using operations such as those listed above, we can use Excel to evaluate arithmetic expressions as follows.

1. Click on an empty cell or the formula bar.
2. Enter an equal sign (**=**) followed by the desired arithmetic expression.
3. Press the **Enter** key.

Whether using calculators or computer software, it often happens that the result of an arithmetic expression is given in a form of scientific notation, and we should be able to express such results in a standard format. See the following two examples.

Excel Expression	Excel Result	Result in Standard Format
=0.2^10	1.024E-07	0.0000001024
=3^25	8.47289E+11	847,289,000,000

Printing an Excel Worksheet To print all of the cells in the current worksheet, simply click on the **Print** button located on the tool bar. The **Print** button looks like a printer, and is located as the fourth item from the left. (If you click on **File,** then select **Print,** you will get a dialog box that allows you to change printing options, such as the number of copies or the printer being used.)

Closing Thoughts: Closing Files and Closing Excel It is the end of the day, your work is done, you have saved your Excel file, and you now want to close your Excel file so that you can proceed to shut down the whole Excel program, shut down the computer, turn out the lights, and go home to feed your pet iguana.

To close an Excel *file,* click on **File,** then click on **Close.** Alternatively, you can simply click on the **X** box located at the extreme right of the bar just above the column labels of A, B, C, ...

To close the entire *Excel program,* click on **File,** then click on **Exit.** Alternatively, you can simply click on the **X** box located in the upper right corner of the screen.

Installing Add-Ins

In the following chapters, we will describe the use of Excel for a variety of different statistical functions. Some of those functions are readily available by selecting the menu items of **Insert,** then **Function** (or by clicking on *fx* found on the tool bar). Other functions come from two sources: (1) the Data Analysis add-in; (2) the Data Desk/XL add-in. The Data Analysis add-in comes with Excel, but it may or may not be installed and ready for use. The Data Desk/XL add-in is on the disk that comes with this book. It is provided by Addison Wesley Longman and must be installed. It would be very helpful to go through the following two subsections to ensure that your version of Excel includes the programs that will be used later in this book.

Tools/Data Analysis Click on the menu item of **Tools** and determine whether **Data Analysis** is listed as one of the submenu options. Clicking on **Tools,** then **Data Analysis** should result in the screen display shown on the next page.

If the **Tools** menu does not include **Data Analysis,** install it by following these steps:

1. Click on **Tools.**
2. Click on **Add-ins.**
3. Click on the box next to **Analysis Tool Pak** and click on the box next to **Analysis Tool Pak-VBA.** (If these items are not included among the list of possible add-ins, you must add them by running the Excel setup pro-

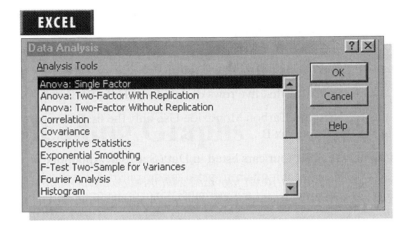

gram.) Click **OK.** After installing any add-ins, they remain available in Excel until they are removed.

Data Desk/XL (DDXL)   Data Desk/XL is a separate program provided by Addison Wesley Longman and must be installed as follows.

1. *Windows*: Copy the DDXL folder into Program Files (C:\Program Files\). *Macintosh*: Copy the DDXL folder into the folder containing the Microsoft Excel application.

2. Start Microsoft Excel.

3. Click on **Tools**, then select **Add-Ins...**

4. Select **Browse...**

5. Click on **DDXL,** then click **OK.**

1-5 Basic Skills and Concepts

EXCEL 1. a. What is a spreadsheet?
   b. In Excel, what term is commonly used in place of the term "spreadsheet"?
   c. What is an Excel workbook?
   d. How many worksheets are in an Excel workbook (before any worksheets are added or deleted)?

EXCEL 2. a. What is an Excel worksheet?
   b. An Excel worksheet can store data in as many as 16,777,216 cells. What is the total number of cells available in one Excel workbook (before adding any additional worksheets beyond those originally available)?

In Exercises 3–6, do the following in Excel.
   a. Create the described worksheet.
   b. Save the worksheet on a floppy disk.
   c. Print a copy of the worksheet.
   d. Close the worksheet.
   e. Open the worksheet.
   f. Using the mouse, click on the cell in the top left corner of the worksheet, then hold the mouse button and slide the mouse to the right and down so that the top half of

experiment	random sample
placebo effect	simple random sample
blinding	systematic sampling
block	convenience sampling
completely randomized experimental	stratified sampling
design	cluster sampling
rigorously controlled design	sampling error
confounding	nonsampling error
replication	

REVIEW

We began this chapter with a general description of the nature of statistics, then discussed different aspects of the nature of data. Uses and abuses of statistics were illustrated with examples. We then discussed the design of experiments, emphasizing the importance of good sampling methods. On completing this chapter, you should be able to do the following:

- Distinguish between a population and a sample.

- Distinguish between a parameter and a statistic.

- Identify the level of measurement (nominal, ordinal, interval, ratio) of a set of data.

- Recognize the importance of good sampling methods in general; recognize the importance of a *simple random sample* in particular; and recognize the serious deficiency of poor sampling methods, especially those that result in self-selected samples.

- Understand the importance of good experimental design, including the control of variable effects, sample size, and randomization.

- For Excel, you should know what a spreadsheet is, and you should be able to create a worksheet, save an Excel file, get help, undo work, open an Excel workbook, and print an Excel worksheet.

REVIEW EXERCISES

1. **Understanding Coke Data** The Coca Cola Company has 366,000 stockholders, and a poll is conducted by randomly selecting 30 stockholders from each of the 50 states. The number of shares held by each sampled stockholder is recorded.
   a. Are the values obtained discrete or continuous?
   b. Identify the level of measurement (nominal, ordinal, interval, ratio) for the sample data.
   c. Which type of sampling (random, systematic, convenience, stratified, cluster) is being used?
   d. If the average (mean) number of shares is calculated from the sample group, is the result a statistic or a parameter?
   e. If you are the chief executive officer of the Coca Cola Company, what characteristic of the data set would you consider to be extremely important?
   f. What is wrong with gauging stockholder views by mailing a questionnaire that stockholders could complete and mail back?

2. **More Coke** Identify the type of sampling (random, systematic, convenience, stratified, cluster) used when a sample of the 366,000 Coca Cola shareholders is obtained as described.
   a. A complete list of all stockholders is compiled, and every 500th name is selected.
   b. At the annual stockholders' meeting, a survey is conducted of all who attend.
   c. Fifty different stockbrokers are randomly selected, and a survey is made of all their clients who own shares of Coca Cola.
   d. A computer file of all stockholders is compiled so that they are all numbered consecutively, then random numbers generated by computer are used to select the sample of stockholders.
   e. All of the stockholders' zip codes are collected, and five stockholders are randomly selected from each zip code.

3. **Sampling Design** You plan to conduct a poll of the full-time students who attend your college. Describe a procedure for obtaining a sample of each type: random, systematic, convenience, stratified, cluster.

4. **Design of Experiment** You plan to conduct an experiment to test the effectiveness of Sleepeze, a new drug that allegedly cures insomnia. You will use a sample of subjects who are treated with the drug and another sample of subjects who are given a placebo.
   a. What is "blinding," and how might it be used in this experiment?
   b. Why is it important to use blinding in this experiment?
   c. What is a completely randomized block design?
   d. What is a rigorously controlled block design?
   e. What is replication, and why is it important?

5. **Dropout Rate** Your college dean has commissioned you to do a study of student dropout rates, and you plan to poll every 50th student who passes a central campus location. What is fundamentally wrong with this sampling plan?

6. Identify the level of measurement (nominal, ordinal, interval, ratio) used in each of the following.
   a. The amounts of tar (in milligrams) in a sample of cigarettes
   b. A movie critic's ratings of "must see, recommended, not recommended, don't even think about going"
   c. A movie critic's classification of "drama, comedy, adventure"
   d. IQ scores, where the score is considered to be a measure of intelligence (not the number of points scored on the IQ test)
   e. IQ scores, where the score is considered to be the number of points scored on the IQ test (not a measure of intelligence)

CUMULATIVE REVIEW EXERCISES

This book's Cumulative Review Exercises are designed to include topics from preceding chapters. For this chapter, we present exercises with expressions similar to those found throughout the book. Use Excel to obtain the indicated values.

1. $\dfrac{1.23 + 4.56 + 7.89}{3}$

EXCEL **2.** $\sqrt{\dfrac{(5-7)^2 + (12-7)^2 + (4-7)^2}{3-1}}$

EXCEL **3.** $\dfrac{1.96^2 \times (0.4)(0.6)}{0.025^2}$

EXCEL **4.** $\dfrac{98.20 - 98.60}{\dfrac{0.62}{\sqrt{106}}}$

EXCEL **5.** $\dfrac{25!}{16! \, 9!}$

EXCEL **6.** $\sqrt{\dfrac{10(513.27) - 71.5^2}{10(10-1)}}$

EXCEL **7.** $\dfrac{8(151,879) - (516.5)(2176)}{\sqrt{8(34,525.75) - 516.5^2} \, \sqrt{8(728,520) - 2176^2}}$

EXCEL **8.** $\dfrac{(183 - 137.09)^2}{137.09} + \dfrac{(30 - 41.68)^2}{41.68}$

In Exercises 9–12, some of the given expressions are designed to yield results expressed in a form of scientific notation. For example, the result of 1.23E + 5 can be expressed as 123,000, and the result of 4.56E − 4 can be expressed as 0.000456. Use Excel to perform the indicated operation and express the result as an ordinary number that is not in scientific notation.

EXCEL **9.** 0.95^{500} EXCEL **10.** 25^8 EXCEL **11.** 52^8 EXCEL **12.** 0.25^{10}

COOPERATIVE GROUP ACTIVITIES

1. *In-class activity:* From the cafeteria, obtain 18 straws. Cut 6 of them in half, cut 6 of them into quarters, and leave the other 6 as they are. There should now be 42 straws of different lengths. Put them in a bag, mix them up, then select one straw, find its length, then replace it. Repeat this until 20 straws have been selected. Important: Select the straws without looking into the bag, and select the first straw that is touched. Find the average (mean) length of the sample of 20 straws. Now remove all of the straws and find the mean length of the population. Did the sample provide an average that was close to the true population average? Why or why not?

2. *Out-of-class activity:* The following survey question raised concerns when responses suggested that about 22% of Americans thought that the Holocaust might not have occurred.

   "Does it seem possible or does it seem impossible to you that the Nazi extermination of the Jews never happened?"

   A subsequent poll showed that respondents were probably confused by the double negative in the wording of the question. Here is the wording used in a subsequent Roper poll:

"Does it seem possible to you that the Nazi extermination of the Jews never happened, or do you feel certain that it happened?"

Is this second version substantially less confusing? Write the question in a way that is clearer than both of these versions.

EXCEL PROJECT

If you have not done so already, install Excel's Data Analysis add-in and install Addison Wesley Longman's DDXL add-in as described in Section 1-5.

1. Click on **Tools**, then **Data Analysis** and record the first five items listed.
2. Click on **DDXL** and record the first five items listed.
3. Using the CD-ROM packaged with this textbook, open the Excel worksheet CIGARET.xls, then print the worksheet.

---

 from **DATA** *to* **DECISION**   *Critical Thinking*

Exercise 1 in Section 1-3 included a reference to an ABC "Nightline" poll in which 186,000 viewers each paid 50 cents to call a "900" telephone number with their opinion about keeping the United Nations in the United States. Of those who called, 67% were in favor of moving the United Nations out of the United States. The show's host, Ted Koppel, reported that a "scientific" poll of 500 people showed that 72% of us want the United Nations to *stay* in the United States.

Analyzing the results

Write a response to someone who claims that the first result is likely to be better because it is based on 186,000 people, whereas the second result is based on only 500 people.

Internet Project

In this section of each chapter, you will be directed to the Web page at

http://www.awlonline.com/triola

from which you can reach the page for all the Internet Projects given for *Elementary Statistics, Eighth Edition*. Go to this Web site now and familiarize yourself with all of the available features for the book.

Each Internet Project includes activities, such as exploring data sets, performing simulations, and researching true-to-life examples found at various Web sites. These activities will help you explore and understand the rich nature of statistics and its importance in our world. So visit this Web site often and enjoy the activities!

Statistics
at work

"We use statistics to determine the degree of isolation between putative stocks."

Sarah Mesnick

Behavioral and Molecular Ecologist

Sarah Mesnick is a National Research Council postdoctoral fellow. In her work as a marine mammal biologist, she conducts research at sea as well as in the Laboratory of Molecular Ecology. Her research focuses on the social organization and population structure of sperm whales. She received her doctorate in evolutionary biology at the University of Arizona.

What do you do?

My research focuses on the relationship between sociality and population structure in sperm whales. We use this information to build better management models for the conservation of this, and other, endangered marine mammal species.

What concepts of statistics do you use?

Currently, I use chi-square and *F*-statistics to examine population structure and regression measures to estimate the degree of relatedness among individuals within whale pods. We use the chi-square and *F*-statistics to determine how discrete populations of whales are in the Pacific. Discrete populations are managed as independent stocks. The regression analysis of relatedness is used to determine kinship within groups.

Could you cite a specific example illustrating the use of statistics?

I'm currently working with tissue samples obtained from three mass strandings of sperm whales. We use genetic markers to determine the degree of relatedness among individuals within the strandings. This is a striking behavior—entire pods swam up onto the beach following a young female calf, stranded, and subsequently all died. We thought that to do something as dramatic as this, the individuals involved must be very closely related. We're find-

ing, however, that they are not. The statistics enable us to determine the probability that two individuals are related given the number alleles that they share. Also, sperm whale—and many other marine mammal, bird, and turtle species—are injured or killed incidentally in fishing operations. We need to know the size of the population from which these animals are taken. If the population is small, and the incidental kill large, the marine mammal population may be threatened. We use statistics to determine the degree of isolation between putative stocks. If stocks are found to be isolated, we would use this information to prepare management plans specifically designed to conserve the marine mammals of the region. Human activities may need to protect the health of the marine environment and its inhabitants.

How do you approach your research?

We try not to have preconceived notions about how the animals are dispersed in their environment. In marine mammals in particular, because they are so difficult to study, there are generally accepted notions about what the animals are doing, yet these have not been critically investigated. In the case of relatedness among individuals within sperm whale groups, they were once thought to be matrilineal and accompanied by a "harem master." With the advent of genetic techniques, and dedicated field work, more open minds and more critical analyses—the statistics come in here—we're able to reassess these notions.

Describing, Exploring, and Comparing Data

Is there a keyboard configuration that is more efficient than the one that most of us now use?

The traditional typewriter keyboard configuration is called a *Qwerty* keyboard because of the position of the letters QWERTY in the top row of letters. Developed in 1872, the Qwerty configuration was supposed to force typists to slow down so that their machines would be less likely to jam. The Dvorak keyboard, developed in 1936, positioned the keys most frequently used in the middle (or "home") row, a move intended to improve efficiency. Both keyboard configurations are shown in the accompanying illustration.

QWERTY Keyboard

Dvorak Keyboard

   An article in the magazine *Discover* suggests that you can measure the ease of typing by using this point rating system: Count each letter on the home row as 0, each letter on the top row as 1, and each letter on the bottom row as 2. (See "Typecasting" by Scott Kim, *Discover*.) For example, the word *statistics* would have a rating of 7 on the Qwerty keyboard and a rating of 1 on the Dvorak keyboard:

```
         s t a t i s t i c s
Qwerty   0 1 0 1 1 0 1 1 2 0   (sum = 7)
keyboard
Dvorak   0 0 0 0 0 0 0 0 1 0   (sum = 1)
keyboard
```

   Using this rating system with each of the 52 words in the Preamble to the Constitution, we get the rating values shown in Tables 2-1 and 2-2.

   Interpreting Results: A visual comparison of Tables 2-1 and 2-2 might not reveal much at first, but in this chapter we will present methods for gaining meaningful insights, which will enable us to make intelligent and productive comparisons.

TABLE 2-1		Qwerty Keyboard Word Ratings							
2	2	5	1	2	6	3	3	4	2
4	0	5	7	7	5	6	6	8	10
7	2	2	10	5	8	2	5	4	2
6	2	6	1	7	2	7	2	3	8
1	5	2	5	2	14	2	2	6	3
1	7								

TABLE 2-2		Dvorak Keyboard Word Ratings							
2	0	3	1	0	0	0	0	2	0
4	0	3	4	0	3	3	1	3	5
4	2	0	5	1	4	0	3	5	0
2	0	4	1	5	0	4	0	1	3
0	1	0	3	0	1	2	0	0	0
1	4								

2-1 Overview

The title of this chapter—Describing, Exploring, and Comparing Data—is intended to suggest that we sometimes describe a data set, sometimes explore a data set, and sometimes compare data sets. The following are examples of each function.

- *Describing a data set:* A statistics professor has graded the tests in her class, and she wants to better understand how the class performed. She describes the test scores by computing values that measure the center and spread of the data and constructing a graph showing the distribution of the scores.

- *Exploring a data set:* A researcher has obtained the annual incomes of people who have purchased new cars. He wants to explore that data set in an attempt to identify any notable characteristics, such as the average income and the range of incomes.

- *Comparing data sets:* A sociologist has obtained annual incomes of female and male statistics professors. She wants to identify similarities and differences in an attempt to determine whether both genders are compensated equally.

To work with data in any of the ways just listed, we need a variety of tools to help us *understand* the data set. This chapter describes those tools.

When analyzing a data set, we must first determine whether we have a *sample* or a complete *population*. That determination will affect both the methods we use and the conclusions we form. We use methods of **descriptive statistics** to summarize or describe the important characteristics of a set of data, and we use methods of **inferential statistics** when we use sample data to make inferences (or generalizations) about a population. When your professor calculates the final exam average for your statistics class, that result is an example of a descriptive statistic. However, if we state that the result is an estimate of the final exam average for all statistics classes, we are making an inference that goes beyond the known data.

Descriptive statistics and inferential statistics are the two general divisions of the subject of statistics. This chapter deals with the basic concepts of descriptive statistics.

Important Characteristics of Data

When describing, exploring, and comparing data sets, the following characteristics of data are usually most important:

1. *Center:* A representative or average value that indicates where the middle of the data set is located

2. *Variation:* A measure of the amount that the values vary among themselves

3. *Distribution:* The nature or shape of the distribution of the data (such as bell-shaped, uniform, or skewed)

4. *Outliers:* Sample values that lie very far away from the vast majority of the other sample values

5. *Time:* Changing characteristics of the data over time

In this book we will show how the tools of statistics can be applied to these characteristics of data to provide instructive insights.

   Manual Calculations: Although this chapter includes detailed steps for important procedures, it is not necessary to master all those steps in all cases. Technology now makes it easy for us to obtain results by calculator or computer so that we can focus on meaning and interpretation. Nevertheless, we recommend that in each case you perform a few manual calculations before using your computer or calculator. Your understanding will be enhanced, and you will acquire a better appreciation for the technology.

 ## 2-2 Summarizing Data with Frequency Tables

When working with large data sets, it is generally helpful to organize and summarize the data by constructing a frequency table.

> ### DEFINITION
>
> A **frequency table** lists classes (or categories) of values, along with frequencies (or counts) of the number of values that fall into each class.

Table 2-3 is a frequency table summarizing the Qwerty word ratings from Table 2-1. The **frequency** for a particular class is the number of original scores that fall into that class. For example, the first class in Table 2-3 has a frequency of 20, indicating that there are 20 values between 0 and 2 inclusive.

   We will first present some standard terms used in discussing frequency tables, and then we will describe how to construct and interpret them.

> ### DEFINITIONS
>
> **Lower class limits** are the smallest numbers that can belong to the different classes. (Table 2-3 has lower class limits of 0, 3, 6, 9, and 12.)
>
> **Upper class limits** are the largest numbers that can belong to the different classes. (Table 2-3 has upper class limits of 2, 5, 8, 11, and 14.)
>
> **Class boundaries** are the numbers used to separate classes, but without the gaps created by class limits. They are obtained as follows: Find the size of the gap between the upper class limit of one class and the lower class limit of the next class. Add half of that amount to each upper class limit to find

TABLE 2-3	
Frequency Table of Qwerty Word Ratings	
Rating	Frequency
0–2	20
3–5	14
6–8	15
9–11	2
12–14	1

Handwritten annotations:
- (−.5) ↑ upper class (2.5)
- lower class → 0–2
- 20 (# of 0,1,2 in table 2-1)
- 5 classes
- →diff_ of all is 1
- − take half of 1 = (.5)
- − sub .5 fr. LCL
- − add .5 to UCL

Authors Identified

the upper class boundaries; subtract half of that amount from each lower class limit to find the lower class boundaries. (Table 2-3 has gaps between classes of exactly 1 unit, so 0.5 is added to the upper class limits and subtracted from the lower class limits. The class boundaries are $-0.5, 2.5, 5.5, 8.5, 11.5,$ and $14.5.$)

Class midpoints are the midpoints of the classes. (Table 2-3 has class midpoints of 1, 4, 7, 10, and 13.) Each class midpoint can be found by adding the lower class limit to the upper class limit and dividing the sum by 2.

Class width is the difference between two consecutive lower class limits or two consecutive lower class boundaries. (Table 2-3 uses a class width of 3.)

The definitions of class width and class boundaries are tricky. Be careful to avoid the easy mistake of making the class width the difference between the lower class limit and the upper class limit. See Table 2-3 and note that the class width is 3, not 2. You can simplify the process of finding class boundaries by understanding that they basically fill the gaps between classes by splitting the difference between the end of one class and the beginning of the next class. Carefully examine the definition of class boundaries, and spend some time until you understand it well.

Constructing Frequency Tables

The main reason for constructing a frequency table is to use it for constructing a graph that effectively shows the distribution of the data (for example, see *histograms*, introduced in the following section). Figure 2-1 describes in detail the procedure for constructing a frequency table. Some instructors include those detailed steps, while others skip them. It is important to observe the following guidelines when constructing a frequency table.

1. *Be sure that the classes are mutually exclusive.* In other words, each of the original values must belong to only one class.

2. *Include all classes, even if the frequency is zero.*

3. *Try to use the same width for all classes.* Sometimes open-ended intervals, such as "65 years or older," are impossible to avoid.

4. *Select convenient numbers for class limits.* Round up to use fewer decimal places or use numbers relevant to the situation.

5. *Use between 5 and 20 classes.*

6. *The sum of the class frequencies must equal the number of original data values.*

> **EXAMPLE** **Qwerty Frequency Table** Use the 52 values in Table 2-1 for the Qwerty keyboard and follow the procedure shown in Figure 2-1 to construct the frequency table shown in Table 2-4, which appears on page 46. Assume that you want 5 classes.

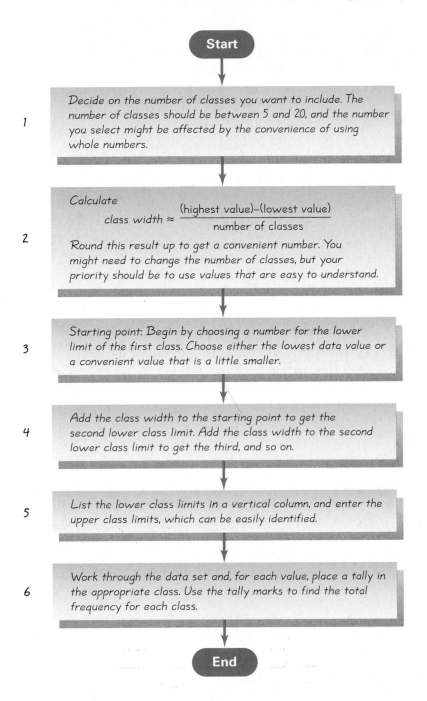

FIGURE 2-1
Procedure for Constructing
a Frequency Table

SOLUTION

Step 1:   Begin by selecting 5 as the number of desired classes.

Step 2:   Calculate the class width.

$$\text{class width} \approx \frac{(\text{highest value}) - (\text{lowest value})}{\text{number of classes}} = \frac{14 - 0}{5} = 2.8 \approx 3$$

Step 3:   Choose a starting point of 0, which is the lowest value in the list.

TABLE 2-4	Frequency Table of Qwerty Word Ratings	
x	Tally	*f*
0–2	‖ ‖ ‖ ‖ ‖ ‖ ‖ ‖ ‖ ‖	20
3–5	‖ ‖ ‖ ‖ ‖ ‖ ‖	14
6–8	‖ ‖ ‖ ‖ ‖ ‖ ‖	15
9–11	‖	2
12–14			1

0–

3–

6–

9–

12–

Step 4: Add the class width of 3 to the starting point of 0 to determine that the second lower class limit is 3. Continue to add the class width of 3 to get the remaining lower class limits of 6, 9, and 12.

Step 5: List the lower class limits vertically as shown in the margin. From this list, we can easily identify the corresponding upper class limits as 2, 5, 8, 11, and 14.

Step 6: After identifying the lower and upper limits of each class, proceed to work through the data set by entering a tally mark for each value, as shown in Table 2-4. When the tally marks are completed, add them to find the frequencies as shown. ◄

TABLE 2-5	
Relative Frequency Table of Qwerty Word Ratings	
Rating	Relative Frequency
0–2	38.5%
3–5	26.9%
6–8	28.8%
9–11	3.8%
12–14	1.9%

Relative Frequency Table

An important variation of the basic frequency table uses **relative frequencies,** which are easily found by dividing each class frequency by the total of all frequencies. A **relative frequency table** includes the same class limits as a frequency table, but relative frequencies are used instead of actual frequencies. The relative frequencies are often expressed as percents.

$$\text{relative frequency} = \frac{\text{class frequency}}{\text{sum of all frequencies}} \times 100$$

In Table 2-5 the actual frequencies shown in Table 2-4 are replaced by the corresponding relative frequencies expressed as percents. The first class has a relative frequency of 20/52 = 0.385, or 38.5%. The second class has a relative frequency of 14/52 = 0.269, or 26.9%, and so on. If constructed correctly, the sum of the relative frequencies should total 1 (or 100%), with some small discrepancies allowed for rounding errors.

Relative frequency tables make it easier for us to understand the distribution of the data and to compare different sets of data.

Cumulative Frequency Table

Another variation of the standard frequency table is used when cumulative totals are desired. The **cumulative frequency** for a class is the sum of the frequencies for that class and all previous classes. Table 2-6 is an example of a **cumulative frequency table,** with cumulative frequencies used instead of the

TABLE 2-6	Cumulative Frequency Table of Qwerty Word Ratings

Rating	Cumulative Frequency
no 0 because no # less than 0	
Less than 3	20
Less than 6	34 (20+14)
Less than 9	49 (34+15)
Less than 12	51 (20+14+15+2)
Less than 15	52 (20+14+15+2+1)

individual class frequencies. Using the original frequencies of 20, 14, 15, 2, and 1 from Table 2-3 or 2-4, we add 20 + 14 = 34 to get the second cumulative sum, then we add 20 + 14 + 15 = 49 to get the third, and so on.

Interpreting Frequency Tables

The following examples show how frequency tables can be used as tools for describing, exploring, and comparing data.

▷ **EXAMPLE   Describing Data** Mark McGwire hit 70 home runs in the 1998 season, and *USA Today* published the distances of those home runs. Table 2-7 summarizes the *last digits* of those distances. When such distances are actually measured, we usually find that the last digits occur with frequencies that are roughly the same. Because of its very uneven distribution, Table 2-7 strongly suggests that the home-run distances were not actually measured, but were estimated instead. ◀

▷ **EXAMPLE   Exploring Data** In studying the behavior of the Old Faithful geyser in Yellowstone National Park, geologists collect data for the times (in minutes) between eruptions. Table 2-8 summarizes actual data that were obtained. Examination of the frequency table reveals unexpected behavior: The distribution of times has two different peaks. This distribution led geologists to consider various possible explanations. ◀

 EXAMPLE   Comparing Data Sets The Chapter Problem at the beginning of this chapter includes two different data sets collected from the Qwerty and Dvorak keyboard arrangements. The Preamble to the Constitution was used as a sample, and each word was rated as described in the Chapter Problem. Table 2-9 combines the relative frequency tables for those two data sets. From Table 2-9 we see that there is a substantial

TABLE 2-7
Last Digits of Home-Run Distances

Last Digit	Frequency
0	55
1	2
2	1
3	1
4	0
5	3
6	0
7	2
8	4
9	2

TABLE 2-8
Times Between Old Faithful Eruptions

Time	Frequency
40–49	8
50–59	44
60–69	23
70–79	6
80–89	107
90–99	11
100–109	1

TABLE 2-9	Qwerty and Dvorak Keyboard Ratings	
Rating	Qwerty Relative Frequency	Dvorak Relative Frequency
0–2	38.5%	63.5%
3–5	26.9%	36.5%
6–8	28.8%	0%
9–11	3.8%	0%
12–14	1.9%	0%

difference between the two data sets. The Dvorak ratings are considerably lower, suggesting that the Dvorak keyboard is considerably more efficient than the Qwerty keyboard. This ability to create meaningful order from apparently chaotic data is truly wonderful. ◀

Frequency tables can be used to identify the general nature of the distribution of data, and they can also be used to construct graphs that visually display the distribution of data. Such graphs are discussed in the next section.

Using Excel to Construct a Frequency Table

You can use Excel to construct a frequency table from a list of sample data. In Excel, the process of constructing a frequency table is called **binning** of the data, because each category acts like a separate bin into which we can pour some of the individual data values. Correct interpretation of an Excel frequency table requires that you know this important principle:

Excel's bins (classes) are based on *upper class limits.*

The following Excel procedure for constructing a frequency table uses the menu item of "Histogram," which is a type of graph discussed in Section 2-3. However, the same dialog box is used to generate a frequency table or a histogram graph.

1. Enter your sample data in column A. Type the first value, then press the **Enter** key. Type the second value, then press the **Enter** key, and so on.

2. Click on **Tools,** found on the main menu bar.

3. Click on **Data Analysis,** found among the options under the Tools menu. (*Note:* If Data Analysis does not appear on the Tools menu, it must be installed as described in Section 1-5.)

4. Click on **Histogram,** then click **OK.**

5. You should now see a *Histogram* dialog box.

   a. First enter the *Input Range.* For example, if the data are listed in cells 1 through 52 of column A, enter the input range of A1:A52.

b. For the *Bin Range,* you have two options: (1) Leave the bin range blank and let Excel decide how to construct the frequency table. (2) If you want to have specific class limits, specify a range of cells that you have *previously* filled in with the values of the *upper class limits.* For Table 2-1, for example, you can see that the values range from 0 through 14, so you might enter these upper class limits in column B: 2, 5, 8, 11, 14. You would then enter a bin range of B1:B5 because that is where these bin values have been entered. Excel would determine the frequency for the values up to 2, then the frequency for the values above 2 and up to 5, and so on.

c. Select **Output Range,** then enter a cell location for the frequency table, such as E2. Finally, click OK. Shown below is the Excel display that results from the 52 Qwerty sample values that are summarized in Table 2-1. This Excel display results from specifying the class limits given in the second option in part (b) above. (That is, the upper class limits of 2, 5, 8, 11, 14 were entered in column B and the bin range of B1:B5 was used.) Compare Excel's display to Table 2-3 to see how Excel summarizes or "bins" the same data.

Excel Frequency Table of Qwerty Word Ratings

Bin	Frequency
2	20 ← There are 20 values between 0 and 2.
5	14
8	15
11	2 ← There are 2 values between 9 and 11.
14	1
More	0

TABLE 2-3

Frequency Table of Qwerty Word Ratings

Rating	Frequency
0–2	20
3–5	14
6–8	15
9–11	2
12–14	1

You can also use Excel to generate a frequency table with cumulative percentages. Follow the same five steps listed above, but in part (c) of Step 5, click on the box labeled "Cumulative Percentage" before clicking **OK.** See the following Excel result.

Bin	Frequency	Cumulative %
2	20	38.46%
5	14	65.38%
8	15	94.23%
11	2	98.08%
14	1	100.00%
More	0	100.00%

2-2 Basic Skills and Concepts

In Exercises 1 and 2, identify the class width, class midpoints, and class boundaries for the given frequency table, then reconstruct the frequency table to be in the bin format that Excel would display.

EXCEL **1.**

Height of Men (in.)	Frequency
55–59	1
60–64	3
65–69	49
70–74	46
75–79	1

EXCEL **2.**

Height of Women (in.)	Frequency
55–59	11
60–64	121
65–69	63
70–74	4
75–79	1

In Exercises 3 and 4, an Excel frequency table is given. Assuming that the frequency table has a uniform class width, reconstruct the Excel table to be in a frequency table format similar to Table 2–3, then identify the class width, class midpoints, and class boundaries..

EXCEL **3.** **GPA**

Bin	Frequency
0.49	72
0.99	23
1.49	47
1.99	135
2.49	288
2.99	276
3.49	202
3.99	97
More	0

EXCEL **4.** **Absences**

Bin	Frequency
9	21
19	10
29	3
39	2
49	14
More	0

Describing Data. *In Exercises 5–8, construct the relative frequency table that corresponds to the frequency table in the exercise indicated.*

5. Exercise 1   **6.** Exercise 2   **7.** Exercise 3   **8.** Exercise 4

Describing Data. *In Exercises 9–12, construct the cumulative frequency table that corresponds to the frequency table in the exercise indicated.*

EXCEL **9.** Exercise 1   **EXCEL** **10.** Exercise 2   **EXCEL** **11.** Exercise 3   **EXCEL** **12.** Exercise 4

Describing and Comparing Data. *In Exercises 13–16, use the data from Data Set 1 in Appendix B. The Excel workbook is COLA.xls.*

EXCEL **13.** Weights of Coke Construct a frequency table for the weights of regular Coke. Start the first class at 0.7900 lb and use a class width of 0.0050 lb. Does the result violate any of the guidelines for constructing frequency tables?

EXCEL **14.** Weights of Diet Coke Construct a frequency table for the weights of diet Coke. Start the first class at 0.7750 lb and use a class width of 0.0050 lb. Does the result violate any of the guidelines for constructing frequency tables?

EXCEL **15.** Comparing Weights of Pepsi and Coke Construct a frequency table for the weights of regular Pepsi. Start the first class at 0.8100 lb and use a class width of 0.0050 lb. Compare the frequency table to the result from Exercise 13. Is there a notable difference?

EXCEL 16. Comparing Weights of Diet Pepsi and Pepsi Construct a frequency table for the weights of diet Pepsi. Start the first class at 0.7700 lb and use a class width of 0.0050 lb. Compare the frequency table to the result from Exercise 15. Is there a notable difference?

EXCEL 17. Bears Refer to Data Set 7 in Appendix B and construct a frequency table of the weights of bears. The Excel workbook is BEARS.xls. Use 11 classes, beginning with a lower class limit of 0.

EXCEL 18. Exploring Body Temperature Data Refer to Data Set 6 in Appendix B and construct a frequency table of the body temperatures for midnight on the second day. Use eight classes, beginning with a lower class limit of 96.5. Describe two different notable features of the result.

EXCEL 19. Strengths of Aluminum Cans Refer to Data Set 12 in Appendix B. The axial loads are the weights that were applied before the cans collapsed. Construct a frequency table for the cans that are 0.0109 in. thick. Start the first class at 200 lb and use a class width of 20 lb. When the top lids of the cans are pressed into place, the applied pressure varies between 158 lb and 165 lb. Will these cans withstand that pressure? (The Excel workbook is CANS.xls.)

EXCEL 20. Exploring Strengths of Aluminum Cans Refer to Data Set 12 in Appendix B. The axial loads are the weights that were applied before the cans collapsed. Construct a frequency table for the cans that are 0.0111 in. thick, but do not include the value of 504 lb. (See Exercise 21.) Start the first class at 200 lb and use a class width of 20 lb. Compare the result to the frequency table for Exercise 19. Determine whether the thicker cans appear to be stronger. (The Excel workbook is CANS.xls.)

2-2   Beyond the Basics

EXCEL 21. Interpreting Outlier Effect Refer to Data Set 12 for the axial loads of the cans that are 0.0111 in. thick. The load of 504 pounds is called an *outlier* because it is far away from all of the other values. Repeat Exercise 20, but include the value of 504 pounds. Interpret the result by stating a generalization about how much of an effect an outlier might have on a frequency table.

22. Analyzing Effect of Classes Is it possible that the number of classes in a frequency table can have a dramatic effect on the apparent distribution of the data? If so, construct a data set for which a change from five classes to six results in a dramatic change in the distribution of the frequencies.

EXCEL 23. Phony Data Listed below are two sets of values that are supposed to be heights (in inches) of randomly selected adult males. One of the sets consists of heights actually obtained from randomly selected adult males, but the other set consists of numbers that were fabricated. Construct a frequency table for each set of heights. Compare the two frequency tables and interpret your results by identifying the set of data that you believe to be false and stating your reason.

a.	70	73	70	72	71	73	71	67	68	72	67	72	71	73
	72	70	72	68	71	71	71	73	69	73	71	66	77	67
b.	70	73	70	72	71	66	74	76	68	75	67	68	71	77
	66	69	72	67	77	75	66	76	76	77	73	74	69	67

Number of Values	Ideal Number of Classes
16–22	5
23–45	6
	7
	8
	9
	10
	11
	12

24. Analyzing the Number of Classes In constructing a frequency table, Sturges' guideline suggests that the ideal number of classes can be approximated by $1 + (\log n)/(\log 2)$, where n is the number of data values. Use this guideline to complete the table in the margin.

 ## 2-3 Pictures of Data

In Section 2-2 we showed how frequency tables can be used to describe, explore, or compare distributions of data sets. In this section we continue the study of distributions by introducing graphs that show the distributions of data in pictorial form. As you read through this section, keep in mind that the objective is not simply to construct a graph, but rather to learn something about a data set—that is, to understand the nature of the distribution.

Histograms

A common and important graphic device for presenting data is the histogram.

> **DEFINITION**
>
> A **histogram** is a bar graph in which the horizontal scale represents classes and the vertical scale represents frequencies. The heights of the bars correspond to the frequency values, and the bars are drawn adjacent to each other (without gaps).

We can construct a histogram after we have first completed a frequency table representing a data set. The histogram in Figure 2-2 corresponds directly to the frequency values in Table 2-3 (in the preceding section). Each bar of this histogram is marked with its lower class boundary at the left and

FIGURE 2-2
Histogram of Qwerty Word Ratings

FIGURE 2-3
Relative Frequency Histogram of Qwerty Word Ratings

its upper class boundary at the right. It is often more practical, however, to use class midpoint values instead of class boundaries. The use of class midpoint values is common in software packages that automatically generate histograms. (Excel does not use midpoint values.)

Before constructing a histogram from a completed frequency table, we must consider the scales used on the vertical and horizontal axes. The maximum frequency (or the next highest convenient number) should suggest a value for the top of the vertical scale; 0 should be at the bottom. In Figure 2-2 we designed the vertical scale to run from 0 to 20. The horizontal scale should be subdivided in a way that allows all the classes to fit well. Ideally, we should try to follow the rule of thumb that the vertical height of the histogram should be about three-fourths of the total width. Both axes should be clearly labeled.

Relative Frequency Histogram

A **relative frequency histogram** has the same shape and horizontal scale as a histogram, but the vertical scale is marked with *relative frequencies* instead of actual frequencies, as in Figure 2-3.

Using Excel for Histograms

Caution: Be vary careful when reading and interpreting an Excel-generated histogram, because the values shown on the horizontal scale might appear to be class midpoints, but they are actually upper class limits.

1. To generate a histogram, follow the same procedure described in Section 2-2 for constructing a frequency table, but make this addition: In part (c) of Step 5, click on the box labeled "Chart Output" before clicking **OK.**

2. Delete the spaces between bars by editing the histogram: Double-click on one of the bars, then click on the **Options** tab and proceed to change the "Gap width" to 0. (Decrease the gap width by holding the mouse button down while it is positioned on the arrow pointing downward.)

3. Change the size of the histogram by positioning the mouse pointer just inside the border, then clicking the mouse button. By clicking and dragging the black boxes on the edges, you can change the size of the histogram as you desire.

Excel's "Chart Wizard" is very flexible and allows you to generate a wide variety of different graphs. You can usually edit graphs by right-clicking or double-clicking on the element that you want to edit. In a histogram, for example, if you double-click on a number for the horizontal scale, you can select "Font" to make those values larger or smaller, or you can display them in a boldface font.

The accompanying Excel histograms result from the data in Table 2-1. The Excel graph on the top is the original histogram that results from the above procedure. The second histogram was modified so that the spaces between bars were deleted, and it was made larger. Compare this Excel display to Figure 2-2. Excel does not use class boundaries along the horizontal scale; *remember that the values along the horizontal scale are actually upper class limits, even though their placement suggests that they might be class midpoints.* The third histogram includes these additional improvements: The labels on the x-axis were edited so that the correct values of the class boundaries are used and they are located correctly, and the title of "Qwerty Word Rating" was inserted. The third histogram corresponds to Figure 2-2.

Excel's original histogram

Excel histogram enlarged with spaces deleted

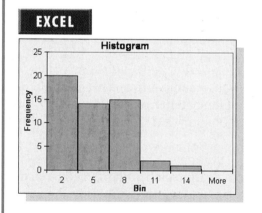

Excel histogram enlarged with spaces deleted and corrected class boundaries

Frequency Polygon

A **frequency polygon** uses line segments connected to points located directly above class midpoint values. Figure 2-4 shows the frequency polygon corresponding to Table 2-3. The heights of the points correspond to the class frequencies, and the line segments are extended to the right and left so that the graph begins and ends on the *x*-axis.

Using Excel for Frequency Polygons

You can use Excel to generate a frequency polygon, as follows.

1. In column A, enter 0 in the first cell, followed by the class frequencies, followed by a 0 at the end. For example, if you want to generate a frequency polygon based on Table 2-3, enter these frequencies under column A: 0, 20, 14, 15, 2, 1, 0.

FIGURE 2-4
Frequency Polygon of Qwerty
Word Ratings

2. Click on **Insert,** then click on **Chart** (or just click on the **Chart Wizard** icon on the tool bar, which looks like a bar graph).

3. Click **Line,** then click on the line graph in the upper left corner, then click **Next.**

4. For the data range, enter the range of values in column A, including the 0 at the beginning and the 0 at the end. For example, if column A contains 0, 20, 14, 15, 2, 1, 0, enter a data range of A1:A7. Also, click on the **Column** button.

5. Click on the **Series** tab and, in the box labeled "Category (X) axis labels," enter a space followed by a comma, then enter the first class midpoint value followed by a comma, the second class midpoint value followed by a comma, and so on. For the data in Table 2-3, enter the following: (space), 1, 4, 7, 10, 13, (space). Click **Next** when finished.

6. Click on the tab keys to delete the legend and grid lines. Click **Next.**

7. Click **Finish.** The result can be edited for improved appearance. Here is a frequency polygon that is similar to Figure 2-4.

Excel Frequency Polygon

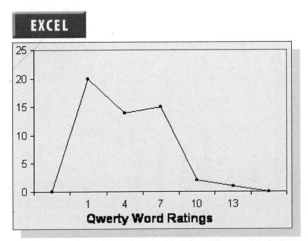

FIGURE 2-5
Ogive of Qwerty Word Ratings

[handwritten notes:]

Cumulative Frequency

Class limits	Less than 10 = 0
10-14	Less than 15 = 8
15-19	↓ wrk. w/ LCL
20-24	only
25-29	Less than = 80
30-34	
35-39	* add frequency

to get cum. freq.!

Ogive

An **ogive** (pronounced "oh-jive") is a line graph that depicts *cumulative* frequencies, just as the cumulative frequency table (see Table 2-6 in the preceding section) lists cumulative frequencies. Figure 2-5 is an ogive corresponding to Table 2-6. Note that the ogive uses class boundaries along the horizontal scale and that the graph begins with the lower boundary of the first class and ends with the upper boundary of the last class. Ogives are useful for determining the number of values less than some particular class boundary. For example, Figure 2-5 shows that 49 of the values are less than 8.5.

Dotplots

A **dotplot** consists of a graph in which each data value is plotted as a point (or dot) along a scale of values. Dots representing the same values are stacked. See Figure 2-6, which represents the 52 Qwerty keyboard word ratings listed in Table 2-1. The leftmost dot, for example, represents the value of 0. The four dots stacked above 1 show that there are four words with ratings of 1, and so on. The dotplot is similar to the histogram in that it shows the *distribution* of the data. (The DDXL Excel add-in can generate dotplots, but only distinct values are plotted as a single point. For example, DDXL would show Figure 2-6 with only 11 different points instead of 52 individual points.)

FIGURE 2-6
Dotplot of Qwerty Word Ratings

Stem-and-Leaf Plots

A **stem-and-leaf plot** represents data by separating each value into two parts: the stem (such as the leftmost digit) and the leaf (such as the rightmost digit). Although we could illustrate a stem-and-leaf plot with the same Qwerty keyboard data used in the other examples, the illustration below does a better job of showing just how a stem-and-leaf plot is constructed.

Raw Data (Test Grades)						Stem	Leaves	
67	72	85	75	89		6	7	← Grade of 67
89	88	90	99	100		7	25	← 72, 75
						8	5899	
						9	09	
						10	0	

From this illustration, you can see how 67 is separated into its stem (6) and its leaf (7). Each of the other values is broken up in a similar way. Note that the leaves are arranged in increasing order, *not* the order in which they occur.

By turning the page on its side, we can see a distribution of these data. The great advantage of the stem-and-leaf plot is that we can see the distribution of data and yet keep all the information in the original list. If necessary, we could reconstruct the original list of values.

The rows of digits in a stem-and-leaf plot are similar in nature to the bars in a histogram. One of the guidelines for constructing histograms is that the number of classes should be between 5 and 20, and that same guideline applies to stem-and leaf plots for the same reasons. Stem-and-leaf plots can be *expanded* to include more rows or *condensed* to include fewer rows. The stem-and-leaf plot in our example can be expanded by subdividing rows into those with the digits 0 through 4 and those with digits 5 through 9. This expanded stem-and-leaf plot is shown here.

Stem	Leaves	
6	7	
7	2	← Include leaves of 0, 1, 2, 3, 4
7	5	← Include leaves of 5, 6, 7, 8, 9
8		← No values between 80 and 84
8	5899	
9	0	
9	9	
10	0	

When it becomes necessary to *reduce* the number of rows, we can condense a stem-and-leaf plot by combining adjacent rows, as in the following illustration. Note that we use an asterisk to separate digits associated with the numbers in each stem. Every row in the condensed plot must include exactly one asterisk so that the shape of the plot is not distorted.

Stem	Leaves	
5-6	*7	← 67
7-8	25*5899	← 72, 75, 85, 88, 89, 89
9-10	09*0	← 90, 99, 100

AS   Another advantage of stem-and-leaf plots is that their construction provides a fast and easy procedure for *sorting* data (arranging data in order). Data must be sorted for a variety of statistical procedures, such as finding the median (discussed in Section 2-4) and finding percentiles or quartiles (discussed in Section 2-6).

Pareto Charts

Consider the following statement: Among 75,200 accidental deaths in the United States in a recent year, 43,500 were attributable to motor vehicles, 12,200 to falls, 6400 to poison, 4600 to drowning, 4200 to fire, 2900 to ingestion of food or an object, and 1400 to firearms (based on data from the National Safety Council). The information contained in this statement concerning the relationships among the data would be much more effectively conveyed in a Pareto chart.

A **Pareto chart** is a bar graph for qualitative data, with the bars arranged in order according to frequencies. As in histograms, vertical scales in Pareto charts can represent frequencies or relative frequencies. The tallest bar is at the left, and the smaller bars are farther to the right. By arranging the bars in order of frequency, the Pareto chart focuses attention on the more important categories. Figure 2-7 shows clearly that the number of motor vehicle accidental deaths is far greater than the numbers in the other categories. Although firearm accidental deaths attract considerable media attention, they are a relatively minor occurrence when compared to the other categories.

Handwritten note:

Frequency	Relative Freq
8	8/80 × 100
28	28/80 × 100
27	27/80 × 100
12	
4	
1	
80	

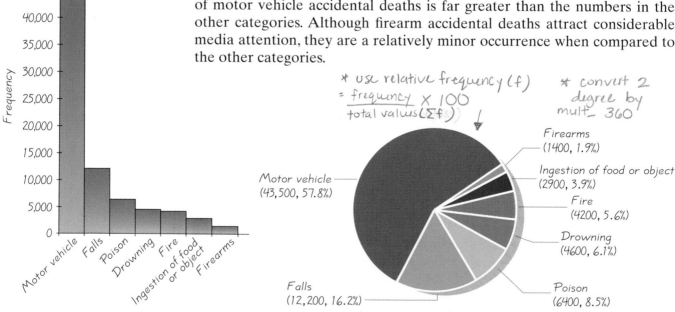

Handwritten note: * use relative frequency (f) = frequency/total values (Σf) × 100   * convert 2 degree by mult. 360

FIGURE 2-7
Pareto Chart of Accidental Deaths by Type

FIGURE 2-8
Pie Chart of Accidental Deaths by Type

Using Excel for Pareto Charts

1. Type the names of the different categories in column A, but arrange them in descending order according to the corresponding frequencies.
2. Enter the corresponding values in column B.
3. Click on **Insert,** then click on **Chart** (or just click on the **Chart Wizard** icon on the tool bar, which looks like a bar graph).
4. Click on the first chart type of **Column** (the default), then click on **Next.**
5. For the data range, enter the range starting with the first cell in column A to the last cell in column B. For example, enter a data range of A1:B7 if you have seven labels in column A and seven corresponding values in column B.
6. Click on the "columns" button, then click **Next.**
7. You have different options for displaying labels, values, and percents. When finished, click **Next** then click **Finish.** The accompanying Excel Pareto chart was obtained by using the accidental death data given in the subsection discussing Pareto charts.

Excel Pareto Chart

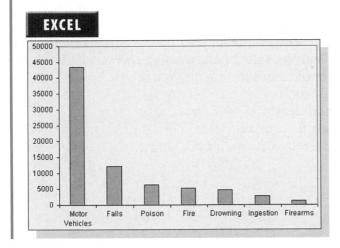

Pie Charts

Pie charts are yet another method used to represent data in pictorial form. Figure 2-8 is a **pie chart** that graphically represents as slices of pie the same accidental death data used for the Pareto chart. Construction of a pie chart involves slicing up the pie into the proper proportions. If the category of motor vehicles represents 57.8% of the total, then the wedge representing motor vehicles should be 57.8% of the total (with a central angle of $0.578 \times 360° = 208°$).

The Pareto chart (Figure 2-7) and the pie chart (Figure 2-8) depict the same data in different ways, but a comparison will probably show that the Pareto chart does a better job of showing the relative sizes of the different components.

Using Excel for Pie Charts

1. Type the names of the different categories in column A.
2. Enter the corresponding values in column B.
3. Click on **Insert,** then click on **Chart** (or just click on the **Chart Wizard** icon on the tool bar, which looks like a bar graph).
4. Click on the image of the **pie chart.**
5. Select the type of pie chart you want, then click **Next.**
6. For the data range, enter the range starting with the first cell in column A to the last cell in column B. For example, enter a data range of A1:B7 if you have seven labels in column A and seven corresponding values in column B.
7. Click on the "Columns" button, then click **Next.**
8. You have different options for displaying labels, values, and percents. The accompanying pie chart results from these choices: Click the tab for "Legend" and select no legend (with an empty box next to "Show Legend"), then click on the tab for "Data Labels" and select "Show label."
9. Click **Next,** then click **Finish** and the pie chart should be displayed. If it is too small, you can make it larger by clicking just inside the border. Click and drag the black boxes that appear on the border. The accompanying Excel pie chart was obtained by using the accidental death data given in the subsection discussing pie charts.

Excel Pie Chart

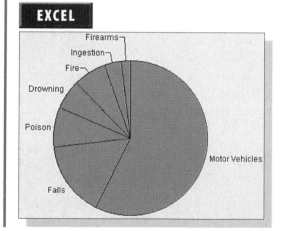

Scatter Diagrams

A **scatter diagram** is a plot of the paired (*x*, *y*) data with a horizontal *x*-axis and a vertical *y*-axis. The data are paired in a way that matches each value from one data set with a corresponding value from a second data set. To manually construct a scatter diagram, construct a horizontal axis for the values of the first variable, construct a vertical axis for the values of the second variable, then plot the points. The pattern of the plotted points is often helpful in determining whether there is some relationship between the two variables. (This issue is discussed at length when the topic of correlation is considered in Section 9-2.) Using the cigarette nicotine and tar data from Data Set 8 in Appendix B, we used Excel to generate the scatter diagram shown here. On the basis of that graph, there does appear to be a relationship between the nicotine and tar contents in cigarettes, as shown by the pattern of the points.

Excel Scatter Diagram

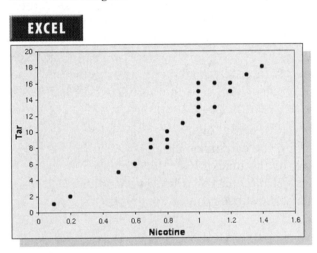

Using Excel for Scatter Diagrams

1. First enter the paired data in columns A and B. Enter the *x*-values in column A and enter the corresponding *y*-values in column B.

2. Click on **Insert,** then click on **Chart** (or just click on the **Chart Wizard** icon on the tool bar, which looks like a bar graph).

3. Click **XY (Scatter).**

4. Select the first type of scatter diagram (the default); click **Next.**

5. For the data range, enter the cell locations beginning with the first cell in column A to the last cell in column B. For example, if you have 29 pairs of data stored with the *x*-values in column A and the *y*-values in column B, enter A1:B29.

6. Click on the button for "Columns," then click **Next.**

7. At this point, you have options for deleting the legend, deleting the grid lines, and so on. When finished, click **Next,** then click **Finish.** The Excel scatter diagram shown above results from the paired nicotine/tar cigarette data from Data Set 8 in Appendix B.

Florence Nightingale

Florence Nightingale (1820–1910) is known to many as the founder of the nursing profession, but she also saved thousands of lives by using statistics. When she encountered an unsanitary and undersupplied hospital, she improved those conditions and then used statistics to convince others of the need for more widespread medical reform. She developed original graphs to illustrate that, during the Crimean War, more soldiers died as a result of unsanitary conditions than were killed in combat. Florence Nightingale pioneered the use of social statistics as well as graphics techniques.

Other Graphs

Numerous pictorial displays other than the ones just described can be used to represent data dramatically and effectively. In Section 2-7 we present box-plots, which are very useful for revealing the spread of data. Pictographs depict data by using pictures of objects, such as soldiers, tanks, airplanes, stacks of coins, or moneybags. Various graphs in Chapter 12 depict patterns of data over time.

The figure on page 63 has been described as possibly "the best statistical graphic ever drawn." It includes six different variables relevant to the march of Napoleon's army to Moscow and back during 1812–1813. The thick band at the left depicts the size of the army when it began its invasion of Russia from Poland. The lower band shows its size during the retreat, along with corresponding temperatures and dates. Although first developed in 1861 by Charles Joseph Minard, this graph is ingenious even by today's standards.

Another historically important graph is one developed by the world's most famous nurse, Florence Nightingale. This graph, shown in Figure 2-9, is particularly interesting because it actually saved lives when Nightingale used it to convince British officials that military hospitals needed to improve sanitary conditions, treatment, and supplies. It is drawn somewhat like a pie chart, except that the central angles are all the same and different radii are used to show changes in the numbers of deaths each month. The outermost regions represent deaths due to preventable diseases, the innermost regions represent deaths from wounds, and the middle regions represent deaths from other causes.

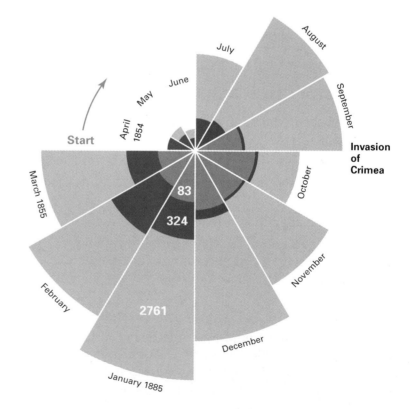

FIGURE 2-9
Deaths in British Military Hospitals During the Crimean War

Losses of Soldiers in Napoleon's Army During the Russian Campaign (1812–1813)
(Width of band shows size of army.)

FIGURE 2-10
Frequency Polygons for Qwerty and Dvorak Keyboards

Conclusion

The effectiveness of Florence Nightingale's graph illustrates well this important point: A graph is not in itself an end result; it is a tool for describing, exploring, and comparing data. When we create meaningful pictures of data, we are really doing the following:

Describing data: In a histogram, for example, we consider the shape of the distribution, whether there are extreme values, and any other notable characteristics. We note the overall shape of the distribution. Are the values evenly distributed? Is the distribution skewed (lopsided) to the right or left? Does the distribution peak in the middle? We observe any outliers (values located far away from most of the other values) that might hide the true nature of the distribution. (See Exercise 29.)

Exploring data: We look for any features of the graph that reveal useful and/or interesting characteristics of the data set. In Figure 2-9, for example, we see that more soldiers were dying from inadequate hospital care than were dying from battle wounds.

Comparing data: Figure 2-10, which combines the frequency polygons for the Qwerty keyboard word ratings and the Dvorak keyboard word ratings listed in Tables 2-1 and 2-2 in the Chapter Problem, makes the comparison easy. We can see clearly that the Dvorak word ratings have a distribution that is considerably different from the Qwerty word ratings. The Dvorak values are considerably lower, suggesting that the Dvorak keyboard is considerably more efficient, as it was designed to be.

2-3 Basic Skills and Concepts

In Exercises 1–4, refer to the Excel-generated histogram, which represents Boston rainfall amounts (in inches) on the Mondays in a recent year. Important: The values given on the horizontal axis have been edited to be the correct class boundaries in the correct positions.

1. How many Mondays had no rain at all?

2. How much rain fell on the Monday with the most rain?

3. How many Mondays had at least 3/4 in. of rain?

4. What is the class width?

5. **Planning to See Old Faithful** Visitors to Yellowstone National Park consider an eruption of the Old Faithful geyser to be a major attraction that should not be missed. The accompanying frequency table summarizes a sample of times (in minutes) between eruptions. Construct a histogram corresponding to the frequency table. If you're scheduling a bus tour of Yellowstone, what is the minimum time you should allocate to Old Faithful if you want to be reasonably sure that your tourists will see an eruption?

6. **Comparing Student and Faculty Cars** Samples of student cars and faculty/staff cars were obtained at the author's college, and their ages (in years) are summarized in the accompanying frequency table. Construct a relative frequency histogram for student cars and another relative frequency histogram for faculty cars. Compare the two frequency histograms. What are the noticeable differences between the two samples?

7. **Speeding Tickets** The accompanying frequency table describes the speeds of drivers ticketed by the Town of Poughkeepsie police. These drivers were traveling through a 30 mi/h speed zone on Creek Road, which passes the author's college. Construct a histogram corresponding to the frequency table. What does the distribution suggest about the enforced speed limit compared to the posted speed limit?

8. **Exploring Body Temperature Data** The frequency table on the next page summarizes a sample of human body temperatures. (See the temperatures for midnight on the second day, as listed in Data Set 6 in Appendix B.) Explore this data set by constructing a histogram corresponding to the frequency table. What does the distribution suggest about the common belief that the average body temperature is 98.6°F? If the subjects are randomly selected, the temperatures should have a distribution that is approximately bell-shaped. Is it? Why is the distribution of human temperatures so important?

Table for Exercise 5

Time	Frequency
40–49	8
50–59	44
60–69	23
70–79	6
80–89	107
90–99	11
100–109	1

Table for Exercise 6

Age	Students	Faculty/ Staff
0–2	23	30
3–5	33	47
6–8	63	36
9–11	68	30
12–14	19	8
15–17	10	0
18–20	1	0
21–23	0	1

Table for Exercise 7

Speed	Frequency
42–45	25
46–49	14
50–53	7
54–57	3
58–61	1

Temperature	Frequency
96.5–96.8	1
96.9–97.2	8
97.3–97.6	14
97.7–98.0	22
98.1–98.4	19
98.5–98.8	32
98.9–99.2	6
99.3–99.6	4

Comparing Data.   *In Exercises 9–12, make the comparisons by using the frequency tables developed for Data Set 1 in Appendix B. (See Exercises 13–16 in Section 2-2. The data are in the Excel workbook COLA.xls.*

9. Regular/Diet Coke Compare the weights of regular Coke and diet Coke by constructing two frequency polygons on the same set of axes, as in Figure 2-10. (See Exercises 13 and 14 in Section 2-2.)

10. Coke/Pepsi Compare the weights of regular Coke and regular Pepsi by constructing two frequency polygons on the same set of axes, as in Figure 2-10. (See Exercises 13 and 15 in Section 2-2.)

11. Diet Coke/Diet Pepsi Compare the weights of diet Coke and diet Pepsi by constructing two frequency polygons on the same set of axes, as in Figure 2-10. (See Exercises 14 and 16 in Section 2-2.)

12. Regular/Diet Pepsi Compare the weights of regular Pepsi and diet Pepsi by constructing two frequency polygons on the same set of axes, as in Figure 2-10. (See Exercises 15 and 16 in Section 2-2.)

Describing Data.   *In Exercises 13 and 14, list the original data represented by the given stem-and-leaf plots.*

13.

Stem	Leaves
60	0117
61	02889
62	13577
63	
64	0099

14.

Stem	Leaves
25	36 55 89
26	01 17 27 36
27	37 42 67
28	92

Describing Data.   *In Exercises 15 and 16, construct the dotplot for the data represented by the stem-and-leaf plot in the given exercise.*

15. Exercise 13

16. Exercise 14

Describing and Interpreting Data.   *In Exercises 17 and 18, construct the stem-and-leaf plots for the given data sets found in Appendix B.*

17. Bears The lengths (in inches) of the bears in Data Set 7. (*Hint:* First round the lengths to the nearest inch.)

18. Weights of Discarded Plastic Weights (in pounds) of plastic discarded by 62 households: Refer to Data Set 5, and start by rounding the listed weights to the nearest tenth of a pound (or one decimal place). (Use an expanded stem-and-leaf plot with about 11 rows.)

EXCEL 19. Job Hunting A study was conducted to determine how people get jobs. The table lists data from 400 randomly selected subjects. The data are based on results from the National Center for Career Strategies. Construct a Pareto chart that corresponds to the given data. If someone would like to get a job, what seems to be the most effective approach?

Job Sources of Survey Respondents	Frequency
Help-wanted ads	56
Executive search firms	44
Networking	280
Mass mailing	20

EXCEL 20. Job Sources Refer to the data given in Exercise 19, and construct a pie chart. Compare the pie chart to the Pareto chart. Can you determine which graph is more effective in showing the relative importance of job sources?

EXCEL 21. Causes of Train Derailments An analysis of train derailment incidents showed that 23 derailments were caused by bad track, 9 were due to faulty equipment, 12 were attributable to human error, and 6 had other causes (based on data from the Federal Railroad Administration). Construct a pie chart representing the given data.

EXCEL 22. Analyzing Causes of Train Derailments Refer to the data given in Exercise 21, and construct a Pareto chart. Compare the Pareto chart to the pie chart. Can you determine which graph is more effective in showing the relative importance of the causes of train derailments?

Describing and Interpreting Data. *In Exercises 23 and 24, use the given paired data from Appendix B to construct a scatter diagram.*

EXCEL 23. Cigarette Tar/CO In Data Set 8, use tar for the horizontal scale and use carbon monoxide for the vertical scale. Interpret the results by determining whether there appears to be a relationship between cigarette tar and carbon monoxide. If so, describe the relationship. (See the Excel workbook CIGARET.xls.)

EXCEL 24. Bear Neck/Weight In Data Set 7, use the distances around bear necks for the horizontal scale and use the bear weights for the vertical scale. Based on the result, what is the relationship between a bear's neck size and its weight? (See the Excel workbook BEARS.xls.)

In Exercises 25–28, refer to the figure on page 63, which describes Napoleon's 1812–1813 campaign to Moscow and back. The thick band at the left depicts the size of the army when it began its invasion of Russia from Poland, and the lower band describes its retreat.

25. Find the percentage of men who survived the entire campaign.

26. Find the number of men and the percentage of men who died crossing the Berezina River.

27. How many men died on the return from Moscow during the time when the temperature dropped from 16° F to −6° F?

28. Of the men who made it to Moscow, how many died on the return trip between Moscow and Botr? (Note that 33,000 men did not go to Moscow, but they joined the returning men who did.)

2-3 Beyond the Basics

EXCEL 29. a. Constructing Histograms and Comparing Data Refer to Data Set 12 in Appendix B and construct a histogram for the axial loads of cans that are 0.0111 in. thick. That data set includes an outlier of 504 lb. (An outlier is a value that is far away from the other values.) The Excel workbook is CANS.xls.
  b. Repeat part (a) after excluding the outlier of 504 lb.
  c. How much of an effect does an outlier have on the shape of the histogram?

EXCEL 30. Constructing and Analyzing Histograms Using a collection of sample data, we construct a frequency table with 10 classes and then construct the corresponding histogram. How is the histogram affected if the number of classes is doubled but the same vertical scale is used?

31. Comparing Ages of Oscar Winners In "Ages of Oscar-Winning Best Actors and Actresses" (*Mathematics Teacher* magazine) by Richard Brown and Gretchen Davis, stem-and-leaf plots are used to compare the ages of actors and actresses at the time they won Oscars. Here are the results for recent winners from each category.

Actors':	32 37 36 32 51 53 33 61 35 45 55 39
	76 37 42 40 32 60 38 56 48 48 40 43
	62 43 42 44 41 56 39 46 31 47 45 60

Actresses':	50 44 35 80 26 28 41 21 61 38 49 33
	74 30 33 41 31 35 41 42 37 26 34 34
	35 26 61 60 34 24 30 37 31 27 39 34

```
Actors'          Actresses'
Ages    Stem       Ages

         2
72       3
         4           4
         5           0
         6
         7
         8
```

  a. Construct a back-to-back stem-and-leaf plot for the above data. The first two ages from each group have been entered in the margin.
  b. Compare the stem-and-leaf plots for the two different sets of data and explain any differences.

32. Analyzing Graphs Figure 2-9, a *polar-area diagram*, has been called misleading because it uses varying areas to represent information that is one-dimensional. How might the data be graphed so that the information can be better understood?

 ## 2-4 Measures of Center

The main objectives of this section are to present the important measures of center and to show how to compute them.

> **DEFINITION**
>
> A **measure of center** is a value at the center or middle of a data set.

In Sections 2-2 and 2-3 we considered frequency tables and graphs that reveal the nature or shape of the *distribution* of a data set. In this section we focus on finding values that are at the *center* of a data set. There are several different ways to determine the center, so we have different definitions of measures of center, including the mean, median, mode, and midrange. We begin with the mean.

Mean

The (arithmetic) mean is generally the most important of all numerical descriptive measurements, and it is what most people call an average.

DEFINITION

The **arithmetic mean** of a set of values is the number obtained by adding the values and dividing the total by the number of values. This measure of center will be used often throughout the remainder of this text, and it will be referred to simply as the **mean.**

This definition can be expressed as Formula 2-1, in which the Greek letter Σ (uppercase Greek sigma) indicates that the data values should be added, Σx represents the sum of all data values, and n denotes the **sample size** (the number of values in the data set).

Formula 2-1
$$\text{mean} = \frac{\Sigma x \ (ungrouped\ data)}{n}$$

The mean is denoted by $\bar{x}$ (pronounced "x-bar") if the data set is a *sample* from a larger population; if all values of the population are used, then we denote the mean by μ (lowercase Greek mu). (Sample statistics are usually represented by English letters, such as $\bar{x}$, and population parameters are usually represented by Greek letters, such as μ.)

Notation	
Σ	denotes the *addition* of a set of values.
x	is the *variable* usually used to represent the individual data values.
n	represents the *number of values in a sample*.
N	represents the *number of values in a population*.
$\bar{x} = \dfrac{\Sigma x}{n}$	is the *mean of a set of sample values*.
$\mu = \dfrac{\Sigma x}{N}$	denotes the mean of all values in a *population*.

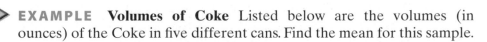

▷ **EXAMPLE   Volumes of Coke** Listed below are the volumes (in ounces) of the Coke in five different cans. Find the mean for this sample.

12.3   12.1   12.2   12.3   12.2

Marion: Average City USA

Because it closely parallels the makeup of the nation, Marion, Indiana, is a favorite choice of researchers who must conduct market testing. When researchers wanted to gauge reaction to potato chips made with the fat substitute olestra, they went to Marion in search of opinions that are likely to reflect the sense of the nation.

Marion has a median household income of $37,396, compared to the national median of $38,738. Marion is also ethnically diverse with more than 40 nationalities represented.

SOLUTION

The mean is computed by using Formula 2-1. First add the values, then divide by the number of values:

$$\bar{x} = \frac{\sum x}{n} = \frac{12.3 + 12.1 + 12.2 + 12.3 + 12.2}{5} = \frac{61.1}{5} = 12.22$$

The mean volume is 12.22 oz.

◄

At the end of this section we will describe the Excel procedure for getting the mean along with several other descriptive statistics. However, the Excel command **AVERAGE** can be used by itself. For example, the entry of **=AVERAGE(2, 3, 7)** will yield 4, which is the mean of 2, 3, and 7. The command of **=AVERAGE(A1:A500)** yields the mean of the 500 values in cells A1 through A500. The command of **=AVERAGE(A1, B1, C1)** yields the mean of the values in cells A1, B1, and C1. The advantage of expressions like **=AVERAGE(A1, B1, C1)** is that by clicking and dragging the mouse button, it becomes possible to repeat similar calculations without making a separate entry for each case.

One disadvantage of the mean is that it is sensitive to every value, so one exceptional value can affect the mean dramatically. The median largely overcomes that disadvantage.

Median

> **DEFINITION**
>
> The **median** of a data set is the middle value when the original data values are arranged in order of increasing (or decreasing) magnitude. The median is often denoted by $\tilde{x}$ (pronounced "x-tilde").

The end of this section includes the Excel procedure for getting the median along with several other descriptive statistics, but the Excel command **MEDIAN** can be used by itself. For example, the entry of **=MEDIAN(A1:A500)** yields the median of the values in cells A1 through A500.

To find the median, first sort the values (arrange them in order), then follow one of these two procedures:

1. If the number of values is odd, the median is the number located in the exact middle of the list.

2. If the number of values is even, the median is found by computing the mean of the two middle numbers.

Figure 2-11 summarizes this procedure for finding the median.

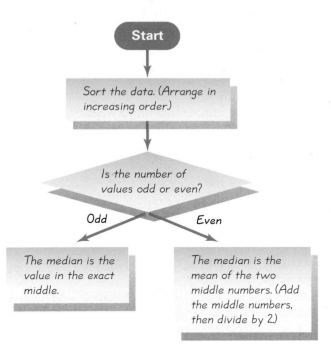

FIGURE 2-11
Procedure for Finding the
Median

▷ **EXAMPLE** **Executive Women** Find the median of the following salaries (in millions of dollars) paid to female executives (based on data from *Working Woman* magazine):

$$6.72 \qquad 3.46 \qquad 3.60 \qquad 6.44$$

SOLUTION First arrange the values in order:

$$3.46 \qquad 3.60 \qquad 6.44 \qquad 6.72$$

Because the number of values is an even number (4), the median is found by computing the mean of the two middle values, 3.60 and 6.44.

$$\text{median} = \frac{3.60 + 6.44}{2} = \frac{10.04}{2} = 5.02$$

The median is $5.02 million.   ◁

▷ **EXAMPLE** **Executive Women** Repeat the preceding example, this time including another salary of $26.70 million. That is, find the median of the following salaries (in millions of dollars):

$$6.72 \qquad 3.46 \qquad 3.60 \qquad 6.44 \qquad 26.70$$

SOLUTION First arrange the values in order:

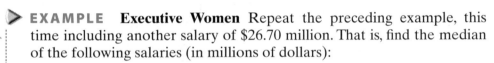

$$3.46 \qquad 3.60 \qquad 6.44 \qquad 6.72 \qquad 26.70$$

Class Size Paradox

There are at least two ways to obtain the mean class size, and they can have very different results. At one college, if we take the numbers of students in 737 classes, we get a mean of 40 students. But if we were to compile a list of the class sizes for each student and use this list, we would get a mean class size of 147. This large discrepancy is due to the fact that there are many students in large classes, while there are few students in small classes. Without changing the number of classes or faculty, we could reduce the mean class size experienced by students by making all classes about the same size. This would also improve attendance, which is better in smaller classes.

Because the number of values is an odd number (5), the median is in the exact middle of the list. The median is therefore $6.44 million. ◄

In Table 2-10, we show the means and medians for the data sets used in the two preceding examples. Examining that table, we see that when we include the extreme value of $26.70 million, the mean changes dramatically, but the median changes by only a small amount. Because the median is relatively insensitive to extreme values, it is often used for data sets with a relatively small number of extreme values. For example, the U.S. Census Bureau recently reported that the *median* household income is $36,078 annually. The median is used here because there is a small number of households with really high incomes.

TABLE 2-10	Means and Medians for Two Data Sets		
Data Set		Mean	Median
3.46 3.60 6.44 6.72		5.055	5.02
3.46 3.60 6.44 6.72 26.70		9.384	6.44

Mode

DEFINITION

The **mode** of a data set is the value that occurs most frequently. When two values occur with the same greatest frequency, each one is a mode and the data set is **bimodal.** When more than two values occur with the same greatest frequency, each is a mode and the data set is said to be **multimodal.** When no value is repeated, we say that there is no mode. The mode is often denoted by M.

Excel has a command **MODE** that can be used, but be very careful: *If there is more than one mode, Excel yields only the first mode.* For example, the command =**MODE(5, 5, 6, 6, 2, 2)** yields only 5. To find all modes in a large data set, you could find the first mode, delete it from the list, then find the next mode, and so on. If there is no mode, Excel displays #N/A.

▷ **EXAMPLE**   Find the modes of the following data sets.

a. 5   5   5   3   1   5   1   4   3   5
b. 1   2   2   2   3   4   5   6   6   6   7   9
c. 1   2   3   6   7   8   9   10

SOLUTION

a. The number 5 is the mode because it is the value that occurs most often.

b. The numbers 2 and 6 are both modes because they occur with the same greatest frequency. This data set is bimodal.

c. There is no mode because no value is repeated.

In fact, the mode is not often used with numerical data. But among the different measures of center we are considering, the mode is the only one that can be used with data at the nominal level of measurement. (Recall that the nominal level of measurement applies to data that consist of names, labels, or categories only.) The following example illustrates this point.

 EXAMPLE   College Majors A study of college students included 10 psychology majors, 20 English majors, and 5 math majors. We can't compute a mean or median of these majors, but we can report that the mode is English major, because that is the one occurring with the greatest frequency.

Midrange

> ### DEFINITION
>
> The **midrange** is the value midway between the highest and lowest values in the original data set. It is found by adding the highest data value to the lowest data value and then dividing the sum by 2, as in the following formula.
>
> $$\text{midrange} = \frac{\text{highest value} + \text{lowest value}}{2}$$

The midrange is seldom used. Because it is derived from only the highest and the lowest values, it is too sensitive to those extremes. Nevertheless, it does have two redeeming features: (1) It is easy to compute; and (2) it helps to reinforce the important point that there are several different ways to define the center of a data set. Figure 2-12 illustrates the differences among the mean, median, mode, and midrange.

FIGURE 2-12
Measures of Center

EXAMPLE   Ages of Thieves Find the midrange of the ages of people arrested on theft charges at the Dutchess County jail (based on data from a student of the author).

18	16	23	25	19	18	20	38

SOLUTION   The midrange is found as follows:

$$\frac{\text{highest value} + \text{lowest value}}{2} = \frac{38 + 16}{2} = 27 \text{ years}$$

Unfortunately, the term *average* is sometimes used for any measure of center, and it is sometimes used for the mean. Because of this ambiguity, we should avoid using the term *average* when referring to a particular measure of center and use instead the specific term, such as mean, median, mode, or midrange. When encountering a value reported as being an *average*, we should be aware that the value could have been computed by any one of several different approaches.

In the spirit of describing, exploring, and comparing data, we provide Table 2-11, which summarizes the different measures of center for the Qwerty and Dvorak keyboard word ratings listed in Tables 2-1 and 2-2 in the Chapter Problem. Recall that these two data sets consist of the ratings for each of the 52 words in the Preamble to the Constitution. Lower ratings represent words that are easier to type; higher ratings result from words that are more difficult to type. A comparison of the measures of center suggests that the Qwerty keyboard is more difficult to use. Its mean of 4.4 appears to be substantially greater than the mean of 1.7 obtained with the Dvorak keyboard. In later chapters of this text we will introduce methods for objectively determining whether such apparent differences are actually significant.

TABLE 2-11	Comparison of Qwerty and Dvorak Keyboards	
	Qwerty	Dvorak
Mean	4.4	1.7
Median	4.0	1.0
Mode	2	0
Midrange	7.0	2.5

Round-Off Rule

A simple rule for rounding answers is this:

Carry one more decimal place than is present in the original set of values.

When implementing this rule, we round only the final answer, not the intermediate values. Thus the mean of 2, 3, 5 is 3.333333. . . , which is rounded to 3.3. Because the original values were whole numbers, we rounded to the nearest tenth. As another example, the mean of 2.1, 3.4, 5.7 is 3.7333333. . . , which is rounded to 3.73 (one more decimal place than was used for the original values).

Mean from a Frequency Table

When data are summarized in a frequency table, we don't know the exact values falling in a particular class. To make calculations possible, we pretend that within each class, all sample values are equal to the class midpoint. Because each class midpoint is repeated a number of times equal to the class frequency, the sum of all sample values becomes $\Sigma(f \cdot x)$, where f denotes frequency and x represents the class midpoint. The total number of sample values is the sum of frequencies Σf. Formula 2-2 is used to compute the mean when the sample data are summarized in a frequency table. Formula 2-2 is not really a new concept; it is simply a variation of Formula 2-1.

First multiply each frequency and class midpoint, then add the products.

↓

Formula 2-2
$$\bar{x} = \frac{\Sigma(f \cdot x)}{\Sigma f}$$ (grouped data)   mean from frequency table

↑

sum of frequencies

In Table 2-12, the first two columns from Frequency Table 2-3 (on page 43) have been entered. Table 2-12 illustrates the procedure for using Formula 2-2 when calculating a mean from data summarized in a frequency table. (In reality, software or a calculator are generally used in place of manual calculations.

TABLE 2-12	Finding the Mean from a Frequency Table		
Rating	Frequency f	Class Midpoint x	$f \cdot x$
0–2	20	1	20
3–5	14	4	56
6–8	15	7	105
9–11	2	10	20
12–14	1	13	13
Totals:	$\Sigma f = 52$		$\Sigma(f \cdot x) = 214$

$$\bar{x} = \frac{\Sigma(f \cdot x)}{\Sigma f} = \frac{214}{52} = 4.1$$

YOU

Six Degrees of Separation

Social psychologists, historians, political scientists, and communications specialists are among those interested in "The Small World Problem": Given any two people in the world, how many intermediate links are required in order to connect the two original people? Social psychologist Stanley Milgram conducted an experiment using the U.S. mail system. Subjects were instructed to try and contact other target people by mailing an information folder to an acquaintance who they thought would be closer to the target. Among 160 such chains that were initiated, only 44 were completed. The number of intermediate acquaintances varied from 2 to 10, with a median of 5. A mathematical model was used to show that if those missing chains were completed, the median would be slightly greater than 5. (See "The Small World Problem," by Stanley Milgram, *Psychology Today,* May 1967.)

Not At Home

Pollsters cannot simply ignore those who were not at home when called the first time. One solution is to make repeated callback attempts until the person can be reached. Alfred Politz and Willard Simmons describe a way to compensate for those missing results without making repeated callbacks. They suggest weighting results based on how often people are not at home. For example, a person at home only two days out of six will have a 2/6 or 1/3 probability of being at home when called the first time. When such a person is reached the first time, his or her results are weighted to count three times as much as someone who is always home. This weighting is a compensation for the other similar people who are home two days out of six and were not at home when called the first time. This clever solution was first presented in 1949.

Using Table 2-12 we get $\bar{x} = 4.1$, but we get $\bar{x} = 4.4$ if we use the original list of 52 values. Remember, the frequency table yields an *approximation* of $\bar{x}$, because it is not based on the exact original list of sample values.

Weighted Mean

In some situations, the values vary in their degree of importance, so we may want to compute a **weighted mean,** which is a mean computed with the different scores assigned different weights. In such cases, we can calculate the weighted mean by assigning different weights to different values, as shown in Formula 2-3.

Formula 2-3   weighted mean: $\bar{x} = \dfrac{\Sigma(w \cdot x)}{\Sigma w}$

For example, suppose we need a mean of three test scores (85, 90, 75), but the first test counts for 20%, the second test counts for 30%, and the third test counts for 50% of the final grade. We can assign weights of 20, 30, and 50 to the test scores, then proceed to calculate the mean by using Formula 2-3 as follows:

$$\begin{aligned} \bar{x} &= \frac{\Sigma(w \cdot x)}{\Sigma w} \\ &= \frac{(20 \times 85) + (30 \times 90) + (50 \times 75)}{20 + 30 + 50} \\ &= \frac{8150}{100} = 81.5 \end{aligned}$$

As another example, college grade-point averages can be computed by assigning each letter grade the appropriate number of points (A = 4, B = 3, etc.), then assigning to each number a weight equal to the number of credit hours. Again, Formula 2-3 can be used to compute the grade-point average.

The Best Measure of Center

So far, we have considered the mean, median, mode, and midrange as measures of center. Which one of these is best? Unfortunately, there is no single best answer to that question, because there are no objective criteria for determining the most representative measure for all data sets. The different measures of center have different advantages and disadvantages, some of which are summarized in Table 2-13. An important advantage of the mean is that it takes every value into account, but an important disadvantage is that it is sometimes dramatically affected by a few extreme values. This disadvantage can be overcome by using a trimmed mean, as described in Exercise 23.

| TABLE 2-13 | Comparison of Mean, Median, Mode, and Midrange |

Average	Definition	How Common?	Existence	Takes Every Value into Account?	Affected by Extreme Values?	Advantages and Disadvantages
Mean	$\bar{x} = \dfrac{\Sigma x}{n}$	most familiar "average"	always exists	yes	yes	used throughout this book; works well with many statistical methods
Median	middle score	commonly used	always exists	no	no	often a good choice if there are some extreme values
Mode	most frequent score	sometimes used	might not exist; may be more than one mode	no	no	appropriate for data at the nominal level
Midrange	$\dfrac{\text{high} + \text{low}}{2}$	rarely used	always exists	no	yes	very sensitive to extreme values

General comments:

- For a data collection that is approximately symmetric with one mode, the mean, median, mode, and midrange tend to be about the same.
- For a data collection that is obviously asymmetric, it would be good to report both the mean and median.
- The mean is relatively *reliable*. That is, when samples are drawn from the same population, the sample means tend to be more consistent than the other averages (consistent in the sense that the means of samples drawn from the same population don't vary as much as the other averages).

Skewness

A comparison of the mean, median, and mode can reveal information about the characteristic of skewness, defined below and illustrated in Figure 2-13.

DEFINITION

A distribution of data is **skewed** if it is not symmetric and if it extends more to one side than the other. (A distribution of data is **symmetric** if the left half of its histogram is roughly a mirror image of its right half.)

Data skewed to the *left* are said to be **negatively skewed;** the mean and median are to the left of the mode. Although not always predictable, negatively skewed data generally have the mean to the left of the median. [See Figure 2-13(a).] Data skewed to the *right* are said to be **positively skewed;** the mean and median are to the right of the mode. Again, although not

FIGURE 2-13
Skewness

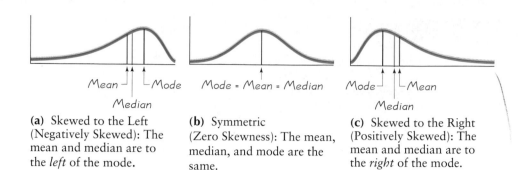

(a) Skewed to the Left (Negatively Skewed): The mean and median are to the *left* of the mode.

(b) Symmetric (Zero Skewness): The mean, median, and mode are the same.

(c) Skewed to the Right (Positively Skewed): The mean and median are to the *right* of the mode.

always predictable, positively skewed data generally have the mean to the right of the median. [See Figure 2-13(c).]

Lopsided to the right	**= Skewed to left**	**= Negatively skewed**
Lopsided to the left	**= Skewed to right**	**= Positively skewed**
Data not lopsided	**= Symmetric**	**= Zero skewness**

If we examine the histogram in Figure 2-2 for the word ratings using the Qwerty keyboard, we see a graph that appears to be skewed to the right. In practice, many distributions of data are symmetric and without skewness. Distributions skewed to the right are more common than those skewed to the left because it's often easier to get exceptionally large values than values that are exceptionally small. With annual incomes, for example, it's impossible to get values below the lower limit of zero, but there are a few people who earn millions of dollars in a year. Annual incomes therefore tend to be skewed to the right, as in Figure 2-13(c).

Using Excel for Descriptive Statistics

Important note regarding Excel's mode calculations: *If there is more than one mode value, Excel displays only the first mode.* If there is no mode, Excel displays #N/A.

You can easily obtain individual statistics, such as the mean, median, or mode. (Simply enter the data in column A, click on *fx,* select the function category of **Statistical,** then select the function name, such as "Average" for the mean, or "Max" for the maximum value, then complete the dialog box.) Because we usually want more than one individual statistic, we describe a procedure that will provide several important statistics at the same time.

1.   Enter the sample data in column A.
2.   Click on **Tools,** then click on **Data Analysis.** (If *Data Analysis* does not appear on the *Tools* menu, it must be installed as described in Section 1-5.)
3.   In the dialog box, click on **Descriptive Statistics,** then click **OK.**

4. You should now see a "Descriptive Statistics" dialog box. Begin by entering the range of data. For example, if you have 52 values listed in rows 1 through 52 of column A, enter A1:A52.

5. Click on the box next to "Summary Statistics." The dialog box should appear as shown above.

6. Click **OK** and you will get a display that includes several statistics. For example, if you use the Qwerty word ratings listed in Table 2-1, the Excel display will be as follows. (The original display had narrower columns, which were widened by using the mouse to click and drag the borders. Also, we highlighted the statistics that are relevant to this section.) Be aware that there may be other modes not included in the display. *If there is more than one mode, only the first of them will be displayed.*

	Column1
Mean	4.403846
Standard Error	0.394399
Median	4
Mode	2
Standard Deviation	2.844049
Sample Variance	8.088612
Kurtosis	1.091589
Skewness	0.923492
Range	14
Minimum	0
Maximum	14
Sum	229
Count	52

2-4 Basic Skills and Concepts

Finding Measures of Centers of Data. *In Exercises 1–4, find the (a) mean, (b) median, (c) mode, and (d) midrange for the given sample data. These data sets were provided by students of the author.*

EXCEL 1. Car Crashes The given values are the numbers of Dutchess County car crashes for each month of a recent year.

> 27   8   17   11   15   25   16   14   14   14   13   18

EXCEL 2. Eyeglass Fittings The given values are the measurements of distances (in millimeters) between the pupils of adult patients being fitted for eyeglasses.

> 67   66   59   62   63   66   66   55

EXCEL 3. McDonald's Service Times The given values are the service times (in seconds) of McDonald's drive-through customers. McDonald's wants to maintain a mean of 90 sec or less. Are these results acceptable?

> 88   107   35   93   65   55   119   83   99   74   46   108

EXCEL 4. Ordering Steaks by Weight The given values are the weights (in ounces) of steaks listed on a restaurant menu as "20-ounce Porterhouse" steaks. The weights are supposed to be 21 oz because the steaks supposedly lose an ounce when cooked. Do these steaks appear to weigh enough?

> 17   20   21   18   20   20   20   18   19   19
>
> 20   19   21   20   18   20   20   19   18   19

Finding and Comparing Measures of Centers of Data. *In Exercises 5–8, find the mean, median, mode, and midrange for each of the two samples, then compare the two sets of results.*

EXCEL 5. Customer Waiting Times Waiting times of customers (in minutes) at the Jefferson Valley Bank (where all customers enter a single waiting line) and the Bank of Providence (where customers wait in individual lines at three different teller windows):

| Jefferson Valley: | 6.5 | 6.6 | 6.7 | 6.8 | 7.1 | 7.3 | 7.4 | 7.7 | 7.7 | 7.7 |
| Providence: | 4.2 | 5.4 | 5.8 | 6.2 | 6.7 | 7.7 | 7.7 | 8.5 | 9.3 | 10.0 |

Interpret the results by determining whether there is a difference between the two data sets that is not apparent from a comparison of the measures of center. If so, how are the data sets different?

EXCEL 6. Regular/Diet Coke Weights (in pounds) of samples of the contents in cans of regular Coke and diet Coke:

| Regular: | 0.8192 | 0.8150 | 0.8163 | 0.8211 | 0.8181 | 0.8247 |
| Diet: | 0.7773 | 0.7758 | 0.7896 | 0.7868 | 0.7844 | 0.7861 |

Does there appear to be a significant difference between the two data sets? How might such a difference be explained?

EXCEL 7. Comparing Coke and Pepsi Weights (in pounds) of samples of the contents in cans of regular Coke and regular Pepsi:

| Coke: | 0.8192 | 0.8150 | 0.8163 | 0.8211 | 0.8181 | 0.8247 |
| Pepsi: | 0.8258 | 0.8156 | 0.8211 | 0.8170 | 0.8216 | 0.8302 |

Does there appear to be a significant difference in the weights of the two different brands?

EXCEL **8.** Male Skull Breadths Maximum breadth of samples of male Egyptian skulls from 4000 B.C. and 150 A.D.:

4000 B.C.:	131	119	138	125	129	126	131	132	126	128	128	131
150 A.D.:	136	130	126	126	139	141	137	138	133	131	134	129

(Based on data from *Ancient Races of the Thebaid* by Thomson and Randall-Maciver.)

Changes in head sizes over time suggest interbreeding with people from other regions. Do the head sizes appear to have changed from 4000 B.C. to 150 A.D.?

Finding the Mean of Data.   *In Exercises 9–12, find the mean of the data summarized in the accompanying frequency table.*

9. Old Faithful Eruptions Visitors to Yellowstone National Park consider an eruption of the Old Faithful geyser to be a major attraction that should not be missed. The accompanying frequency table summarizes a sample of times (in minutes) between eruptions.

10. Comparing Student and Faculty Cars Samples of student cars and faculty/staff cars were obtained at the author's college, and their ages (in years) are summarized in the accompanying frequency table. Compare the two means.

11. How Fast Were They Driving? The accompanying frequency table describes the speeds of drivers ticketed by the Town of Poughkeepsie police. These drivers were traveling through a 30 mi/h speed zone on Creek Road, which passes the author's college. How does the mean compare to the posted speed limit of 30 mi/h?

12. Interpreting Body Temperature Differences The accompanying frequency table summarizes a sample of human body temperatures. (See the temperatures for midnight on the second day, as listed in Data Set 6 in Appendix B.) How does the mean compare to the value of 98.6°F, which is the value that most people assume to be the mean?

Finding the Mean and Median of Data.   *In Exercises 13–16, refer to the data set in Appendix B and find the mean and median.*

EXCEL **13.** Comparing Textbook Prices Data Set 2 in Appendix B: Find the mean and median of the new textbook prices at the author's college and find the mean and median of new textbook prices at the University of Massachusetts. Compare and interpret the results. The Excel workbook is TEXTBOOK.xls.

EXCEL **14.** Cigarette Nicotine Content Data Set 8 in Appendix B: Find the mean and median of the nicotine contents of all cigarettes listed. The Excel workbook is CIGARET.xls.

EXCEL **15.** Interpreting Boston Rainfall Amounts Data Set 17 in Appendix B: Find the mean and median of the rainfall amounts in Boston on Thursday and find the mean and median of the rainfall amounts in Boston on Sunday. One report claimed that it rains more on weekends than during the week. Do these results appear to support that claim? The Excel workbook is BOSTRAIN.xls.

EXCEL **16.** Comparing Home-Run Distances Data Set 19 in Appendix B: Find the mean and median for the home-run distances of Mark McGwire and find the mean

Table for Exercise 9

Time	Frequency
40–49	8
50–59	44
60–69	23
70–79	6
80–89	107
90–99	11
100–109	1

Table for Exercise 10

Age	Students	Faculty/Staff
0–2	23	30
3–5	33	47
6–8	63	36
9–11	68	30
12–14	19	8
15–17	10	0
18–20	1	0
21–23	0	1

Table for Exercise 11

Speed	Frequency
42–45	25
46–49	14
50–53	7
54–57	3
58–61	1

Table for Exercise 12

Temperature	Frequency
96.5–96.8	1
96.9–97.2	8
97.3–97.6	14
97.7–98.0	22
98.1–98.4	19
98.5–98.8	32
98.9–99.2	6
99.3–99.6	4

and median for the home-run distances of Sammy Sosa. Compare and interpret the results. The Excel workbook is HOMERUNS.xls.

2-4 Beyond the Basics

17. The **harmonic mean** is often used as a measure of center for data sets consisting of rates of change, such as speeds. It is found by dividing the number of values (n) by the sum of the *reciprocals* of all values, expressed as

$$\frac{n}{\sum \frac{1}{x}}$$

(No value can be zero.) Four students drive from New York to Florida (1200 miles) at a speed of 40 mi/h (yeah, right!) and return at a speed of 60 mi/h. What is their average speed for the round trip? (The harmonic mean is used in averaging speeds.)

18. The **geometric mean** is often used in business and economics for finding average rates of change, average rates of growth, or average ratios. Given n values (all of which are positive), the geometric mean is the nth root of their product. The *average growth factor* for money compounded at annual interest rates of 10%, 8%, 9%, 12%, and 7% can be found by computing the geometric mean of 1.10, 1.08, 1.09, 1.12, and 1.07. Find that average growth factor.

19. The **quadratic mean** (or **root mean square,** or **R.M.S.**) is usually used in physical applications. In power distribution systems, for example, voltages and currents are usually referred to in terms of their R.M.S. values. The quadratic mean of a set of values is obtained by squaring each value, adding the results, dividing by the number of values (n), and then taking the square root of that result, expressed as

$$\text{quadratic mean} = \sqrt{\frac{\sum x^2}{n}}$$

Find the R.M.S. of these power supplies (in volts): $110, 0, -60, 12$.

20. Finding the Weighted Mean Kelly Bell gets quiz grades of 65, 83, 80, and 90. She gets a 92 on her final exam. Find the weighted mean if the quizzes each count for 15% and the final counts for 40% of the final grade.

21. Analyzing Transformed Data In each of the following, describe how the mean, median, mode, and midrange of a data set are affected.
   a.   The same constant k is added to each value of the data set.
   b.   Each value of the data set is multiplied by the same constant k.

22. Finding the Median When data are summarized in a frequency table, the median can be found by first identifying the *median class* (the class that contains the median). We then assume that the values in that class are evenly distributed and we can interpolate. This process can be described by

$$(\text{lower limit of median class}) + (\text{class width})\left(\frac{\left(\frac{n+1}{2}\right) - (m+1)}{\text{frequency of median class}} \right)$$

where n is the sum of all class frequencies and m is the sum of the class frequencies that *precede* the median class. Use this procedure to find the median of the data set summarized in Frequency Table 2-3.

23. Finding and Comparing the Trimmed Mean Because the mean is very sensitive to extreme values, we say that it is not a *resistant* measure of center. The **trimmed mean** is more resistant. To find the 10% trimmed mean for a data set, first arrange the data in order, then delete the bottom 10% of the values and the top 10% of the values, and calculate the mean of the remaining values. For the weights of the bears in Data Set 7 in Appendix B, find (a) the mean, (b) the 10% trimmed mean, (c) the 20% trimmed mean. How do the results compare?

24. Average Teacher's Salary Using an almanac, a researcher finds the average teacher's salary for each state. He adds those 50 values, then divides by 50 to obtain their mean. Is the result equal to the national average teacher's salary? Why or why not?

 # 2-5 Measures of Variation

In this section we introduce the characteristic of variation. Because variation is so important in statistics, this is one of the most important sections in the entire book. We will discuss in detail the following key concepts: (1) Variation refers to the amount that values vary among themselves, and it can be measured with specific numbers; (2) values that are relatively close together have lower measures of variation, and values that are spread farther apart have measures of variation that are larger; (3) the standard deviation, which is a particularly important measure of variation, can be computed; and (4) the values of standard deviation must be *interpreted* correctly.

Many banks once required that customers wait in separate lines at each teller's window, but most have now changed to a single main waiting line. Why did they make that change? The mean waiting time didn't change, because the waiting-line configuration doesn't affect the efficiency of the tellers. They changed to the single line because customers prefer waiting times that are more *consistent* with less variation. Thus thousands of banks made a change that resulted in lower variation (and happier customers), even though the mean was not affected. Let's now consider the same bank sample data used in Exercise 5 in the preceding section. The listed values are waiting times (in minutes) of customers.

Jefferson Valley Bank (Single waiting line)	6.5	6.6	6.7	6.8	7.1	7.3	7.4	7.7	7.7	7.7
Bank of Providence (Multiple waiting lines)	4.2	5.4	5.8	6.2	6.7	7.7	7.7	8.5	9.3	10.0

In Figure 2-14 we show dotplots of both data sets, and we include the measures of center. A comparison of the measures of center does not reveal any differences between the two data sets, but a comparison of the dotplots provides clear and strong visual evidence that the Jefferson Valley Bank has waiting times with substantially less variation than the times for the Bank of values. For example, if customer waiting times are in minutes, the standard deviation of those times will also be in minutes.

When applying Formula 2-4, it is best to take an approach that is often successful in working with mathematical expressions: Start on the inside and proceed outward. The procedure for calculating the standard deviation can be summarized as follows.

Procedure for Finding the Standard Deviation with Formula 2-4

Step 1: Find the mean of the values ($\bar{x}$).

Step 2: Subtract the mean from each individual value to get a list of deviations of the form $(x - \bar{x})$.

Step 3: Square each of the differences obtained from Step 2. [This produces numbers of the form $(x - \bar{x})^2$.]

Step 4: Add all of the squares obtained from Step 3 to get $\Sigma(x - \bar{x})^2$.

Step 5: Divide the total from Step 4 by the number $(n - 1)$; that is, 1 less than the total number of values present.

Step 6: Find the square root of the result of Step 5.

▶ **EXAMPLE** **Using Formula 2-4** Use Formula 2-4 to find the standard deviation of the Jefferson Valley Bank customer waiting times. Those times (in minutes) are reproduced below:

6.5   6.6   6.7   6.8   7.1   7.3   7.4   7.7   7.7   7.7

SOLUTION   We will follow the six steps in the procedure just given. Refer to those steps and to Table 2-14, which shows the detailed calculations.

Step 1: Obtain the mean of 7.15 by adding the values and then dividing by the number of values:

$$\bar{x} = \frac{\Sigma x}{n} = \frac{71.5}{10} = 7.15 \text{ min}$$

Step 2: Subtract the mean of 7.15 from each value to get these values of $(x - \bar{x})$: $-0.65, -0.55, \ldots, 0.55$.

Step 3: Square each value obtained in Step 2 to get these values of $(x - \bar{x})^2$: $0.4225, 0.3025, \ldots, 0.3025$.

Step 4: Sum all of the preceding values to get the value of

$$\Sigma(x - \bar{x})^2 = 2.0450$$

Step 5: With $n = 10$ values, divide by 1 less than 10:

$$\frac{2.0450}{9} = 0.2272$$

Step 6: Find the square root of 0.2272. The standard deviation is

$$\sqrt{0.2272} = 0.48 \text{ min}$$

◀

TABLE 2-14	Calculating Standard Deviation for Jefferson Valley Bank Customers	
x	$x - \bar{x}$	$(x - \bar{x})^2$
6.5 ~~-42.5~~	(6.5 - 7.15) −0.65	0.4225
6.6 43.56	(6.6 - 7.15) −0.55	0.3025
6.7 44.89	(6.7 - 7.15) −0.45	0.2025
6.8 46.24	−0.35	0.1225
7.1 50.41	−0.05	0.0025
7.3 53.29	0.15	0.0225
7.4 54.76	0.25	0.0625
7.7 59.29	0.55	0.3025
7.7 59.29	0.55	0.3025
7.7 59.29	0.55	0.3025
Totals: 71.5		2.0450

$$\bar{x} = \frac{71.5}{10} = 7.15 \text{ min}$$

$$s = \sqrt{\frac{2.0450}{10-1}} = \sqrt{0.2272} = \overset{0.476...}{0.48} \text{ min}$$

Where Are the 0.400 Hitters?

Ideally, we would now interpret the meaning of the result, but such interpretations will be discussed later in this section.

> **EXAMPLE** **Using Formula 2-5** The preceding example used Formula 2-4 for finding the standard deviation of the Jefferson Valley Bank customer waiting times. Using the same data set, find the standard deviation by applying Formula 2-5.

SOLUTION To use Formula 2-5 we first must find values for n, Σx, and Σx^2.

$n = 10$ (because there are 10 values in the sample)

$\Sigma x = 71.5$ (the sum of the 10 sample values)

$\Sigma x^2 = 513.27$ (the sum of the squares of the sample values: $6.5^2 + 6.6^2 + 6.7^2 + \cdots + 7.7^2$)

Formula 2-5 can now be used to find the standard deviation.

$$s = \sqrt{\frac{n(\Sigma x^2) - (\Sigma x)^2}{n(n-1)}} = \sqrt{\frac{10(513.27) - (71.5)^2}{10(10-1)}} = \sqrt{\frac{20.45}{90}} = 0.48 \text{ min}$$

⟵ short cut

A great self-test is to stop here and calculate the standard deviation of the waiting times for the Bank of Providence. Follow the same procedures used in the preceding two examples and verify that, for the Bank of

Providence, $s = 1.82$ min. Although the interpretations of these standard deviations will be discussed later, we can now compare them to see that the standard deviation of the times for the Jefferson Valley Bank (0.48 min) is much lower than the standard deviation for the Bank of Providence (1.82 min). This supports our subjective conclusion that the waiting times at the Jefferson Valley Bank have much less variation than those at the Bank of Providence.

Standard Deviation of a Population

In our definition of standard deviation, we referred to the standard deviation of *sample* data. A slightly different formula is used to calculate the standard deviation σ (lowercase Greek sigma) of a *population:* Instead of dividing by $n - 1$, divide by the population size N, as in the following expression.

$$\sigma = \sqrt{\frac{\Sigma(x - \mu)^2}{N}} \qquad \text{population standard deviation}$$

Because we generally deal with sample data, we will usually use Formula 2-4, in which we divide by $n - 1$. Many calculators do both the sample standard deviation and the population standard deviation, but they use a variety of different notations. Be sure to identify the notation used by your calculator, so that you get the correct result.

Variance of a Sample and Population

We are using the term *variation* as a general description of the amount that values vary among themselves. The term *variance* refers to a specific definition.

DEFINITION

The **variance** of a set of values is a measure of variation equal to the square of the standard deviation.

Sample variance: Square of the standard deviation s.

Population variance: Square of the population standard deviation σ.

The sample variance s^2 is an **unbiased estimator** of the population variance σ^2, which means that values of s^2 tend to target the value of σ^2 instead of systematically tending to overestimate or underestimate σ^2. (See Exercise 35.)

▶ **EXAMPLE**   In the preceding example, we used the Jefferson Valley Bank customer waiting times to find that the standard deviation is given by $s = 0.48$ min. Find the variance of that same sample.

SOLUTION   Instead of using the rounded value of $s = 0.48$ min, we will use the more precise value of $s = 0.4767$ min. Because the variance is the square of the standard deviation, we get this result:

$$\text{sample variance} = s^2 = (0.4767\ \text{min})^2 = 0.23\ \text{min}^2$$

The variance is an important statistic used in some important statistical calculations, such as analysis of variance (discussed in Chapter 11). For our present purposes, the variance has a serious disadvantage: The units of variance are different than the units of the original data set. For example, if the original customer waiting times are in minutes, the units of the variance are square minutes (min^2). What is a square minute? (Have some fun constructing a creative answer to that question.) Because the variance uses different units, it is extremely difficult to understand variance by relating it to the original data set. Because of this property, we will focus on the standard deviation as we try to develop an understanding of variation.

We now present the notation and the round-off rule we are using.

Notation

s = *sample* standard deviation

s^2 = *sample* variance

σ = *population* standard deviation

σ^2 = *population* variance

Note: Articles in professional journals and reports often use SD for standard deviation and VAR for variance.

Round-Off Rule

As discussed in Section 2-4, we use this rule for rounding final results:

Carry one more decimal place than is present in the original set of values.

Round only the final answer, never in the middle of a calculation. (If it becomes absolutely necessary to round in the middle, carry at least twice as many decimal places as will be used in the final answer.)

Finding Standard Deviation from a Frequency Table

We sometimes need to compute the standard deviation of a data set that is summarized in the form of a frequency table, such as Table 2-3 in Section 2-2. If the original list of sample values is available, use those values with

Data Mining

The term *data mining* is commonly used to describe the now popular practice of analyzing an existing large set of data for the purpose of finding relationships, patterns, or any interesting results that were not found in the original studies of the data set. Some statisticians express concern about ad hoc inference—a practice in which a researcher goes on a fishing expedition through old data, finds something significant, and then identifies an important question that has already been answered. Robert Gentleman, a column editor for *Chance* magazine, writes that "there are some interesting and fundamental statistical issues that data mining can address. We simply hope that its current success and hype don't do our discipline (statistics) too much damage before its limitations are discussed."

Formula 2-4 or 2-5 so that the result will be more exact. If the original data are not available, use one of these two methods:

1. If the total number of values is not too large, use your calculator or software program by entering each class midpoint a number of times equal to the class frequency.

2. Calculate the standard deviation using Formula 2-6.

(grouped data)

Formula 2-6
$$s = \sqrt{\frac{n[\Sigma(f \cdot x^2)] - [\Sigma(f \cdot x)]^2}{n(n-1)}}$$

standard deviation for frequency table

$n = \Sigma f$ (sum of frequency)/add

 EXAMPLE   **Qwerty Keyboard Ratings** Find the standard deviation of the 52 values summarized in Table 2-3, assuming that the original data set is not available.

SOLUTION

Method 1:   Table 2-3 has class midpoints of 1, 4, 7, 10, and 13. Using a calculator or software program, enter the value of 1 twenty times (because the frequency of the first class is 20), enter 4 fourteen times, and so on. Find the standard deviation of this set of 52 class midpoints. The result should be 3.0.

Method 2:   Use Formula 2-6. To do this, we first need to find the values of n, $\Sigma(f \cdot x)$, and $\Sigma(f \cdot x^2)$. After obtaining those values from Table 2-15, we apply Formula 2-6 as follows:

$$s = \sqrt{\frac{n[\Sigma(f \cdot x^2)] - [\Sigma(f \cdot x)]^2}{n(n-1)}} = \sqrt{\frac{52[1348] - [214]^2}{52(52-1)}}$$

$$= \sqrt{\frac{24{,}300}{2652}} = \sqrt{9.1628959} = 3.0$$

◀

$\overline{x} = \dfrac{\Sigma(f \cdot x)}{\Sigma f}$

$= \dfrac{(214)}{52} = 4.11$

TABLE 2-15	Calculating Standard Deviation from Frequency Table			
Word Rating	Frequency, f	Class Midpoint, x	$f \cdot x$	$f \cdot x^2$
0–2	20	1	20	20
3–5	14	4	56	224
6–8	15	7	105	735
9–11	2	10	20	200
12–14	1	13	13	169
Totals:	$\Sigma f = 52$		$\Sigma(f \cdot x) = 214$	$\Sigma(f \cdot x^2) = 1348$

Interpreting and Understanding Standard Deviation

We will now attempt to make some intuitive sense of standard deviation. First, we should clearly understand that the standard deviation measures the variation among values. Values close together will yield a small standard deviation, whereas values spread farther apart will yield a larger standard deviation. Refer again to Figure 2-14, which is a dotplot showing different amounts of variation.

Because variation is such an important concept and because the standard deviation is such an important tool in measuring variation, we will consider three different ways of developing a sense for values of standard deviations. The first one, the **range rule of thumb,** is based on the principle that for many data sets, the vast majority (such as 95%) of sample values lie within 2 standard deviations of the mean. (We could improve the accuracy of this rule by taking into account such factors as the size of the sample and the nature of the distribution, but we prefer to sacrifice accuracy for the sake of simplicity. We want a simple rule that will help us interpret values of standard deviations; later methods will produce more accurate results.)

Range Rule of Thumb

For Estimation: To obtain a rough estimate of the standard deviation s, use the equation

$$s \approx \frac{\text{range}}{4}$$

where range $=$ (highest value) $-$ (lowest value).

For Interpretation: If the standard deviation s is known, we can use it to find rough estimates of the minimum and maximum "usual" sample values as follows.

minimum "usual" value $\approx$ (mean) $- 2 \times$ (standard deviation)
maximum "usual" value $\approx$ (mean) $+ 2 \times$ (standard deviation)

When calculating a standard deviation using Formula 2-4 or 2-5, you can use the range rule of thumb as a check on your result, but you must realize that although the approximation will get you in the general vicinity of the answer, it can be off by a fairly large amount.

> **EXAMPLE   Customer Waiting Times** For the Jefferson Valley Bank customer waiting times (6.5, 6.6, 6.7, 6.8, 7.1, 7.3, 7.4, 7.7, 7.7, 7.7) we used Formulas 2-4 and 2-5 to compute the standard deviation as $s = 0.48$ min. Find the estimated standard deviation by using the range rule of thumb.

> **SOLUTION**   Scanning the list of values, we find a high of 7.7 and a low of 6.5, so the range is $7.7 - 6.5 = 1.2$. We use the range rule of thumb to get a rough estimate of s as follows:

$$s \approx \frac{\text{range}}{4} = \frac{1.2}{4} = 0.3 \text{ min}$$

Mail Consistency

A recent survey of 29,000 people who use the U.S. Postal Service revealed that they wanted better consistency in the time it takes to make a delivery. Now, a local letter could take one day or several days. *USA Today* reported a common complaint: "Just tell me how many days ahead I have to mail my mother's birthday card."

The level of consistency can be measured by the standard deviation of the delivery times. A lower standard deviation reflects more consistency. The standard deviation is often a critically important tool used to monitor and control the quality of goods and services.

INTERPRETATION   Previously, we found that the standard deviation is 0.48, so the range-rule-of-thumb estimate of 0.3 is a bit too low. Nevertheless, this estimate confirms that we are in the ballpark, and we would know that a value for *s* such as 7 is probably not correct.

EXAMPLE   Qwerty Keyboard Ratings Use the range rule of thumb to find a rough estimate of the standard deviation of the sample of 52 word ratings from the Qwerty keyboard, as listed in Table 2-1.

SOLUTION   In using the range rule of thumb to estimate the standard deviation of sample data, we find the range and divide by 4. By scanning the list of ratings, we can see that the lowest is 0 and the highest is 14, so the range is 14. The standard deviation *s* is estimated as follows:

$$s \approx \frac{\text{range}}{4} = \frac{14}{4} = 3.5$$

INTERPRETATION   This result is in the ballpark of the correct value of 2.8 that is obtained by calculating the exact value of the standard deviation with Formula 2-4 or 2-5.

The preceding examples illustrated how we can use known information about the range to estimate the standard deviation. The following example is particularly important as an illustration of one way to *interpret* the value of a standard deviation.

EXAMPLE   Heights of Men Previous results from the National Health Survey show that the heights of men have a mean of 69.0 in. and a standard deviation of 2.8 in. Use the range rule of thumb to find the minimum and maximum "usual" heights.

SOLUTION   With a mean of 69.0 in. and a standard deviation of 2.8 in., we use the range rule of thumb to find the minimum and maximum usual heights as follows:

$$\text{minimum} \approx (\text{mean}) - 2 \times (\text{standard deviation})$$
$$= 69.0 - 2(2.8) = 69.0 - 5.6 = 63.4 \text{ in.}$$
$$\text{maximum} \approx (\text{mean}) + 2 \times (\text{standard deviation})$$
$$= 69.0 + 2(2.8) = 69.0 + 5.6 = 74.6 \text{ in.}$$

INTERPRETATION   Based on these results, we expect that typical men will range in height between 63.4 in. and 74.6 in. The heights of some men do not fall between those limits, but those men are unusually short or unusually tall. We now have a much better understanding of how men's heights vary.

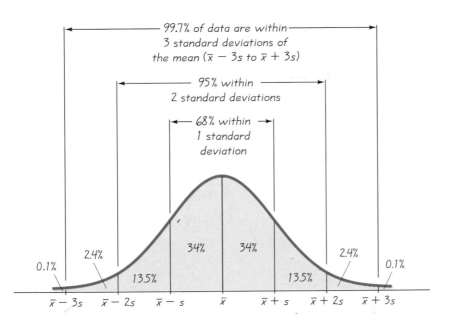

FIGURE 2-15
The Empirical Rule

Empirical (or 68–95–99.7) Rule for Data with a Bell-Shaped Distribution

Another rule that is helpful in interpreting values for a standard deviation is the **empirical rule.** This rule states that for data sets having a distribution that is approximately bell-shaped, the following properties apply. (See Figure 2-15.)

- About 68% of all values fall within 1 standard deviation of the mean.
- About 95% of all values fall within 2 standard deviations of the mean.
- About 99.7% of all values fall within 3 standard deviations of the mean.

> **EXAMPLE**   **Heights of Men** The heights of men have a bell-shaped distribution with a mean of 69.0 in. and a standard deviation of 2.8 in. (based on data from the National Health Survey). What percentage of men have heights between 60.6 in. and 77.4 in.?

SOLUTION   The key to solving this problem is to recognize that 60.6 in. and 77.4 in. are each exactly 3 standard deviations away from the mean of 69.0 in., as shown below.

$$3 \text{ standard deviations} = 3s = 3(2.8) = 8.4$$

Therefore, 3 standard deviations from the mean is

$$69.0 - 8.4 = 60.6$$

or

$$69.0 + 8.4 = 77.4$$

Buying Cars

For buying a new or used car, an excellent reference is the reliability data compiled and reported by *Consumer Reports* magazine. Frequency-of-repair data are based on 10 million pieces of data collected from thousands of readers. Statisticians analyze the data for patterns that lead to lists of both reliable cars and cars that should be avoided. Consumers Union President Rhoda Karpatkin writes, "Because numbers describe so much of our work, it should be no surprise that statisticians are key to that process."

The empirical rule tells us that about 99.7% of all values are within 3 standard deviations of the mean, so about 99.7% of all men's heights are between 60.6 in. and 77.4 in.   ◀

Hint: Difficulty in applying the empirical rule usually stems from confusion over the meaning of such phrases as "within 3 standard deviations of the mean." If you are at all uncertain, you should stop here and review the preceding example. The following table should help as well.

Phrase	Meaning
Within 1 standard deviation of the mean	Between $(\bar{x} - s)$ and $(\bar{x} + s)$
Within 2 standard deviations of the mean	Between $(\bar{x} - 2s)$ and $(\bar{x} + 2s)$
Within 3 standard deviations of the mean	Between $(\bar{x} - 3s)$ and $(\bar{x} + 3s)$

A third concept that is helpful in understanding or interpreting a value of a standard deviation is **Chebyshev's theorem.** The empirical rule applies only to data sets with a bell-shaped distribution. Chebyshev's theorem applies to *any* data set, but its results are very approximate.

Chebyshev's Theorem

The proportion (or fraction) of any set of data lying within K standard deviations of the mean is always *at least* $1 - 1/K^2$, where K is any positive number greater than 1. For $K = 2$ and $K = 3$, we get the following results:

- At least 3/4 (or 75%) of all values lie within 2 standard deviations of the mean.
- At least 8/9 (or 89%) of all values lie within 3 standard deviations of the mean.

▷ **EXAMPLE   Heights of Men** Heights of men have a mean of 69.0 in. and a standard deviation of 2.8 in. What can we conclude from Chebyshev's theorem?

SOLUTION   Applying Chebyshev's theorem with a mean of 69.0 and a standard deviation of 2.8, we can reach the following conclusions:

- At least 3/4 (or 75%) of all men have heights within 2 standard deviations of the mean (between 63.4 in. and 74.6 in.).
- At least 8/9 (or 89%) of all men have heights within 3 standard deviations of the mean (between 60.6 in. and 77.4 in.).   ◀

After studying this section, you should understand that the standard deviation is a measure of variation among scores. Given sample data, you should be able to compute the value of the standard deviation. You should be able to interpret the values of standard deviations that you compute. You should recognize that for typical data sets, it is unusual for a score to differ from the mean by more than 2 or 3 standard deviations.

Using Excel for Measures of Variation

Follow the same procedure described in Section 2-4. The displayed statistics will include the range, standard deviation, and variance. The standard deviation and variance are computed with the assumption that we have *sample* data, not *population* data, which are the results that we typically want throughout this book. (Use the Excel commands **STDEVP** and **VARP** for population parameters.) Shown here is the display that results from the data in Table 2-1. The statistics relevant to this section have been highlighted.

Column1	
Mean	4.403846
Standard Error	0.394399
Median	4
Mode	2
Standard Deviation	2.844049
Sample Variance	8.088612
Kurtosis	1.091589
Skewness	0.923492
Range	14
Minimum	0
Maximum	14
Sum	229
Count	52

2-5 Basic Skills and Concepts

Finding the Measures of Variation.   *In Exercises 1–4, find the range, variance, and standard deviation for the given data. (The same data were used in Section 2-4 where we found measures of center. Here we find measures of variation.)*

1. Car Crashes The given values are the numbers of Dutchess County car crashes for each month of a recent year.

27   8   17   11   15   25   16   14   14   14   13   18

2. Eyeglass Fittings The given values are the measurements of distances (in millimeters) between the pupils of adult patients being fitted for eyeglasses.

67   66   59   62   63   66   66   55

3. McDonald's Service Times The given values are the service times (in seconds) of McDonald's drive-through customers.

88   107   35   93   65   55   119   83   99   74   46   108

EXCEL

4. Actual Steak Weights The given values are the weights (in ounces) of steaks listed on a restaurant menu as "20-ounce Porterhouse" steaks.

17	20	21	18	20	20	20	18	19	19
20	19	21	20	18	20	20	19	18	19

Comparing Measures of Variation. *In Exercises 5–8, find the range, variance, and standard deviation for each of the two samples, then compare the two sets of results. (The same data were used in Section 2-4.)*

Table for Exercise 9

Time	Frequency
40–49	8
50–59	44
60–69	23
70–79	6
80–89	107
90–99	11
100–109	1

EXCEL

5. Customer Waiting Times Examples in this section included the following waiting times of customers at the Jefferson Valley Bank (where all customers enter a single waiting line) and the Bank of Providence (where customers wait in individual lines at three different teller windows):

Jefferson Valley:	6.5	6.6	6.7	6.8	7.1	7.3	7.4	7.7	7.7	7.7
Providence:	4.2	5.4	5.8	6.2	6.7	7.7	7.7	8.5	9.3	10.0

EXCEL

6. Comparing Regular and Diet Coke Data Weights (in pounds) of samples of the contents in cans of regular Coke and diet Coke:

Regular:	0.8192	0.8150	0.8163	0.8211	0.8181	0.8247
Diet:	0.7773	0.7758	0.7896	0.7868	0.7844	0.7861

Table for Exercise 10

Age	Students	Faculty/Staff
0–2	23	30
3–5	33	47
6–8	63	36
9–11	68	30
12–14	19	8
15–17	10	0
18–20	1	0
21–23	0	1

EXCEL

7. Comparing Coke and Pepsi Data Weights (in pounds) of samples of the contents in cans of regular Coke and regular Pepsi:

Coke:	0.8192	0.8150	0.8163	0.8211	0.8181	0.8247
Pepsi:	0.8258	0.8156	0.8211	0.8170	0.8216	0.8302

EXCEL

8. Ancient Skull Breadths Maximum breadth of samples of male Egyptian skulls from 4000 B.C. and 150 A.D.:

4000 B.C.:	131	119	138	125	129	126	131	132	126	128	128	131
150 A.D.:	136	130	126	126	139	141	137	138	133	131	134	129

(Based on data from *Ancient Races of the Thebaid* by Thomson and Randall-Maciver.)

Table for Exercise 11

Speed	Frequency
42–45	25
46–49	14
50–53	7
54–57	3
58–61	1

Finding the Standard Deviation. *In Exercises 9–12, find the standard deviation of the data summarized in the given frequency table. (The same frequency tables were used in Section 2-4.)*

9. Old Faithful Eruption The accompanying frequency table summarizes a sample of times (in minutes) between eruptions of the Old Faithful geyser in Yellowstone National Park.

10. Comparing Student and Faculty Cars Samples of students cars and faculty/staff cars were obtained at the author's college, and their ages (in years) are summarized in the accompanying frequency table.

Table for Exercise 12

Time	Frequency
96.5–96.8	1
96.9–97.2	8
97.3–97.6	14
97.7–98.0	22
98.1–98.4	19
98.5–98.8	32
98.9–99.2	6
99.3–99.6	4

11. Speeds of Ticketed Drivers The accompanying frequency table describes the speeds of drivers ticketed by the Town of Poughkeepsie police.

12. Body Temperature Differences The accompanying frequency table summarizes a sample of human body temperatures (from the temperatures for midnight on the second day, as listed in Data Set 6 in Appendix B).

Finding Standard Deviations Using Technology.   *In Exercises 13–16, refer to the data set in Appendix B. Use Excel to find the standard deviation. (The same data sets were used in Section 2-4.)*

EXCEL **13.** Comparing Textbook Prices Data Set 2 in Appendix B: Find the standard deviation of new textbook prices at the author's college and find the standard deviation of new textbook prices at the University of Massachusetts. The Excel workbook is TEXTBOOK.xls. Compare and interpret the standard deviations.

EXCEL **14.** Nicotine Content Data Set 8 in Appendix B: Find the standard deviation of the nicotine contents of all cigarettes listed. The Excel workbook is CIGARET.xls.

EXCEL **15.** Interpreting Boston Rainfall Amounts Data Set 17 in Appendix B: Find the standard deviation of the rainfall amounts in Boston on Thursday and find the standard deviation of the rainfall amounts in Boston on Sunday. The Excel workbook is BOSTRAIN.xls. Compare and interpret the standard deviations.

EXCEL **16.** Using the Range Rule of Thumb Data Set 19 in Appendix B: Find the standard deviation for the home-run distances of Mark McGwire and find the standard deviation for the home-run distances of Sammy Sosa. The Excel workbook is HOMERUNS.xls. Compare and interpret the standard deviations.

17. Using the Range Rule of Thumb Use the range rule of thumb to estimate the standard deviation of ages of your classmates at the time of your high school graduation.

18. Interpreting Test Scores Use the range rule of thumb to estimate the standard deviation of the scores on the first statistics test in your class.

19. Interpreting Test Scores The test scores in a statistics class have a mean of 75 and a standard deviation of 12. Use the range rule of thumb to estimate the minimum and maximum "usual" scores. Is a score of 50 considered unusual in this context?

20. Interpreting Heights of Women Heights of women have a mean of 63.6 in. and a standard deviation of 2.5 in. (based on data from the National Health Survey). Use the range rule of thumb to estimate the minimum and maximum "usual" heights of women. In this context, is it unusual for a woman to be 6 ft tall?

21. Estimating Heights of Women Heights of women have a bell-shaped distribution with a mean of 63.6 in. and a standard deviation of 2.5 in. Using the empirical rule, what is the approximate percentage of women with heights between

  a.   61.1 in. and 66.1 in.?

  b.   56.1 in. and 71.1 in.?

22. Estimating IQ Scores IQ scores have a bell-shaped distribution with a mean of 100 and a standard deviation of 15. Using the empirical rule, what is the approximate percentage of people with IQ scores between

  a.   70 and 130?

  b.   85 and 115?

23. Using Chebyshev's Theorem If heights of women have a mean of 63.6 in. and a standard deviation of 2.5 in., what can you conclude from Chebyshev's theorem about the percentage of women between 58.6 in. and 68.6 in. tall?

24. Using Chebyshev's Theorem If IQ scores have a mean of 100 and a standard deviation of 15, what can you conclude from Chebyshev's theorem about the percentage of IQ scores between 55 and 145?

25. Understanding Standard Deviation What do you know about the values in a data set having a standard deviation of $s = 0$?

26. Understanding Units of Measurement If a data set consists of the prices of textbooks expressed in dollars, what are the units used for standard deviation? What are the units used for variance?

27. Comparing Car Battery Lives The Everlast and Endurance brands of car battery are both labeled as lasting 48 months. In reality, they both have a mean life of 50 months, but the Everlast batteries have a standard deviation of 2 months, while the Endurance batteries have a standard deviation of 6 months. Which brand is the better choice? Why?

28. Interpreting Outliers A data set consists of 20 values that are fairly close together. Another value is included, but this new value is an outlier (very far away from the other values). How is the standard deviation affected by the outlier? No effect? A small effect? A large effect?

2-5 Beyond the Basics

EXCEL

29. Comparing Data Sets Two different sections of a statistics class take the same quiz and the scores are recorded below. Find the range and standard deviation for each section. What do the range values lead you to conclude about the variation in the two sections? Why is the range misleading in this case? What do the standard deviation values lead you to conclude about the variation in the two sections?

| Section 1: | 1 | 20 | 20 | 20 | 20 | 20 | 20 | 20 | 20 | 20 | 20 |
| Section 2: | 2 | 3 | 4 | 5 | 6 | 14 | 15 | 16 | 17 | 18 | 19 |

30. Transforming Data For each of the following, describe how the range and standard deviation of a data set are affected.
    a. The same constant k is added to each value of the data set.
    b. Each value of the data set is multiplied by the same constant k.
    c. For the body temperature data listed in Data Set 6 in Appendix B (12 A.M. on day 2), $\bar{x} = 98.20°$ F and $s = 0.62°$ F. Find the values of $\bar{x}$ and s after each temperature has been converted to the Celsius scale.
    [*Hint:* $C = 5(F - 32)/9$.]

31. a. The **coefficient of variation,** expressed as a percent, is used to describe the standard deviation relative to the mean. It allows us to compare variability of data sets with different measurement units (such as feet versus minutes), and it is calculated as follows:

$$\frac{s}{x} \cdot 100 \quad \text{or} \quad \frac{\sigma}{\mu} \cdot 100$$

Find the coefficient of variation for the following sample of car ages (in years):

$$0 \quad 1 \quad 3 \quad 3 \quad 5 \quad 6 \quad 6 \quad 6 \quad 6 \quad 8 \quad 12$$

    b. Genichi Taguchi developed a method of improving quality and reducing manufacturing costs through a combination of engineering and statistics. A key tool in the Taguchi method is the **signal-to-noise ratio.** The simplest way to calculate this ratio is to divide the mean by the standard deviation. Find the signal-to-noise ratio for the sample data given in part (a).

32. In Section 2-4 we introduced the general concept of skewness. Skewness can be measured by **Pearson's index of skewness:**

$$I = \frac{3(\bar{x} - \text{median})}{s}$$

If $I \geq 1.00$ or $I \leq -1.00$, the data can be considered to be *significantly skewed.* Find Pearson's index of skewness for the Qwerty keyboard word ratings listed in Table 2-1, and then determine whether there is significant skewness.

33. Understanding Standard Deviation A sample consists of six values that fall between 1 and 9 inclusive. What is the largest possible standard deviation?

34. Phony Data? For any data set of n values with standard deviation s, every value must be within $s\sqrt{n-1}$ of the mean. A statistics teacher reports that the test scores in her class of 17 students had a mean of 75.0 and a standard deviation of 5.0. Kelly, the class's self-proclaimed best student, claims that she received a grade of 97. Could Kelly be telling the truth?

35. Understanding Variance Let a population consist of the values 1, 2, and 3. Assume that samples of two different values are randomly selected *with replacement.*
   a.   Find the variance σ^2 of the population $\{1, 2, 3\}$.
   b.   List the nine different possible samples and find the sample variance s^2 for each of them. If you repeatedly select two different items, what is the mean value of the sample variances s^2?
   c.   For each of the nine samples, find the variance by treating each sample as if it is a population. (Be sure to use the formula for population variance.) If you repeatedly select two different items, what is the mean value of the population variances?
   d.   Which approach results in values that are better estimates of σ^2: part (b) or part (c)? Why? When computing variances of samples, should you use division by n or $n - 1$?
   e.   The preceding parts show that s^2 is an unbiased estimator of σ^2. Is s an unbiased estimator of σ?

2-6 Measures of Position

In this section we introduce measures that can be used to compare values from different data sets or to compare values within the same data set. The basic tools are z scores, quartiles, deciles, and percentiles. We begin with z scores.

z Scores

A z score (or standard score) is found by converting a value to a standardized scale, as given in the following definition.

If you use a calculator or computer software for exercises involving quartiles, you may get results that differ slightly from the answers given in the back of the book.

Just as there are three quartiles separating a data set into four parts, there are nine **deciles,** denoted by $D_1, D_2, D_3, \ldots, D_9$, which separate the data into 10 groups with about 10% of the values in each group. There are also 99 **percentiles,** denoted by $P_1, P_2, \ldots, P_{99}$, which partition the data into 100 groups with about 1% of the values in each group. (Quartiles, deciles, and percentiles are examples of *quantiles*—or *fractiles*—which partition data into parts that are approximately equal.)

The process of finding the percentile that corresponds to a particular value x is fairly simple, as indicated in the following expression.

$$\text{percentile of value } x = \frac{\text{number of values less than } x}{\text{total number of values}} \cdot 100$$

> **EXAMPLE**   **Weights of Coke** Table 2-16 lists 36 weights (in pounds) of the contents of 36 cans of regular Coke. Find the percentile corresponding to the weight of 0.8143 lb.
>
> **SOLUTION**   From Table 2-16 we see that there are 8 values less than 0.8143, so
>
> $$\text{percentile of } 0.8143 = \frac{8}{36} \cdot 100 = 22 \text{ (rounded)}$$
>
> **INTERPRETATION**   The weight of 0.8143 lb is the 22nd percentile.

TABLE 2-16	Sorted Weights (in pounds) of Regular Coke in 36 Cans				
0.7901	0.8044	0.8062	0.8073	0.8079	0.8110
0.8126	0.8128	0.8143	0.8150	0.8150	0.8152
0.8152	0.8161	0.8161	0.8163	0.8165	0.8170
0.8172	0.8176	0.8181	0.8189	0.8192	0.8192
0.8194	0.8194	0.8207	0.8211	0.8229	0.8244
0.8244	0.8247	0.8251	0.8264	0.8284	0.8295

The preceding example shows how to use a given sample value to find the corresponding percentile. There are several different methods for performing the reverse procedure, taking a given percentile and finding the corresponding sample value. The one we will use is summarized in Figure 2-17, which uses the following notation.

FIGURE 2-17
Finding the Value of the *k*th
Percentile

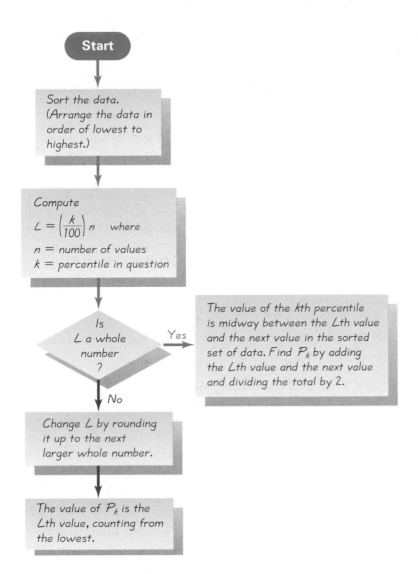

Notation

n = total number of values in the data set

k = percentile being used (Example: For the 25th percentile, $k = 25$.)

L = locator that gives the *position* of a value (Example: For the 25th value in the sorted list, $L = 25$.)

P_k = kth percentile (Example: P_{25} is the 25th percentile.)

▷ **EXAMPLE   Weights of Coke** Refer to the sample of Coke weights given in Table 2-16. Use Figure 2-17 to find the value of the 25th percentile, P_{25}.

SOLUTION   Referring to Figure 2-17, we see that the sample data are already sorted, so we can proceed to compute the value of the locator L. In this computation, we use $k = 25$ because we are attempting to find the value of the 25th percentile, and we use $n = 36$ because there are 36 sample values:

$$L = \frac{k}{100} \cdot n = \frac{25}{100} \cdot 36 = 9$$

Next, we are asked if L is a whole number. The answer is yes, so we proceed to the box located at the right. We now see that the value of the kth (25th) percentile is midway between the Lth (9th) value and the next value in the original set of data. That is, the value of the 25th percentile is midway between the 9th value and the 10th value. The 9th value is 0.8143 and the 10th value is 0.8150, so the value midway between them is 0.81465. We conclude that the 25th percentile is $P_{25} = 0.81465$ lb.

▶ **EXAMPLE**   **Decile** For the sample of Coke weights in Table 2-16, find the value of the 4th decile, D_4.

SOLUTION   First, we should note that the 4th decile, D_4, is the same as the 40th percentile, P_{40}, so we can proceed with the objective of finding the value of the 40th percentile.

Referring to Figure 2-17, the data are already sorted, so we compute the locator $L = (40/100)36 = 14.4$. Next, L is not a whole number, so we proceed straight down to where we change L by rounding it up to the next larger whole number: 15. The value of the 40th percentile (or the 4th decile) is the 15th value in the list, which is 0.8161 lb.

The preceding example showed that $D_4 = P_{40} = 0.8161$ lb, but you can verify that 0.8161 lb converts to the 36th percentile, not the 40th percentile. There is a discrepancy here, but such discrepancies become smaller as the sample size increases. We could eliminate the discrepancy by using a more exact and more complicated procedure that includes interpolation instead of rounding.

The preceding example showed that when finding a decile value (such as D_4), we can use the equivalent percentile value (such as P_{40}) instead. The table in the margin shows relationships relating quartiles and deciles to equivalent percentiles.

In earlier sections of this chapter we described several statistics, including the mean, median, mode, range, and standard deviation. Other statistics are sometimes defined using quartiles and percentiles, as in the following:

$$\text{interquartile range (or } IQR) = Q_3 - Q_1$$

$$\text{semi-interquartile range} = \frac{Q_3 - Q_1}{2}$$

$$\text{midquartile} = \frac{Q_1 + Q_3}{2}$$

$$\text{10--90 percentile range} = P_{90} - P_{10}$$

Quartiles	Deciles
$Q_1 = P_{25}$	$D_1 = P_{10}$
$Q_2 = P_{50}$	$D_2 = P_{20}$
$Q_3 = P_{75}$	⋮
	$D_9 = P_{90}$

After completing this section, you should be able to convert a value into its corresponding z score (or standard score) so that you can compare it to other values, which may be from different data sets. You should be able to convert a value into its corresponding percentile value so that you can compare it to other values in the same data set. Given a data set and some specified percentile score, you should be able to find the value corresponding to the given percentile. And, finally, you should understand the meanings of quartiles and deciles and be able to relate them to their corresponding percentile values.

Using Excel for z Scores

If you want to find the z score corresponding to a value of 130 from a population with a mean of 100 and a standard deviation of 15, enter the Excel command **=STANDARDIZE(130, 100, 15)** to get a result of 2. You could also use this procedure to convert a value to a corresponding z score:

1. Click on the toolbar item of *fx.*
2. Click on the function category of **Statistical.**
3. In the right column, scroll down and click on **STANDARDIZE,** then click **OK.**
4. You should now see a dialog box. For x, enter the value to be converted, then enter the values of the mean and standard deviation. Shown below is the dialog box for finding the standard score of Michael Jordan's height. (See the first example of this section, where his height is 78 in., the mean height of men is 69.0 in., and the standard deviation of men's heights is 2.8 in.) The display shows that Michael Jordan's height of 78 in. converts to a z score of 3.214285714, or 3.21 when rounded.

Using Excel to Find Percentile Values

Important note: When calculating percentiles and quartiles, Excel uses a slightly different approach than the simplified approach described in this book. There will often be a discrepancy between results obtained by using Excel and results obtained using the method described in the book. For example, the book includes an example in which the 25th percentile is found for the data in Table 2-16. The solution in the book shows an answer of 0.81465, but Excel produces an answer of 0.814825. Other software packages might result in other values.

Use this procedure to find the value of a given percentile:

1. Enter the data in column A.
2. Click on the tool bar item of **ƒx.**
3. Click on the function category of **Statistical.**
4. In the right column, scroll down and click on the function name of **PERCENTILE.**
5. You will get a dialog box requiring two items. For the "Array" box, enter the range of values, such as A1:A36 for 36 data values stored in column A.
6. For the value of k, enter the percentile in *decimal form.* For example, the 25th percentile would be entered as 0.25. Click **OK.**

Using Excel to Find Quartile Values

As with percentile values, there may be a discrepancy between quartile values found using the methods in this book and results found using Excel. Other software packages may give different results as well. Use the following procedure to find the value of a given quartile:

1. Enter the data in column A.
2. Click on the tool bar item of **ƒx.**
3. Click on the function category of **Statistical.**
4. In the right column, scroll down and click on the function name of **QUARTILE.**
5. You will get a dialog box requiring two items. For the "Array" box, enter the range of values, such as A1:A36 for 36 data values stored in column A.
6. Click on the empty box for Quart. Excel will display instructions indicating that an entry of 0 would result in the minimum value, an entry of 1 yields the first quartile Q_1, an entry of 2 yields Q_2, an entry of 3 yields Q_3, and an entry of 4 yields the maximum value. Enter 0, 1, 2, 3, or 4, then click **OK.**

2-6 Basic Skills and Concepts

Finding z Scores.   In Exercises 1–4, express all z scores with two decimal places.

1. IQ Scores IQ scores have a mean of 100 and a standard deviation of 15. Kelly Stewart has an IQ of 130.
   a. What is the difference between Kelly's IQ and the mean?
   b. How many standard deviations is that (the difference found in part a)?
   c. Convert Kelly's IQ score to a *z* score.
   d. How does the *z* score compare to the number of standard deviations by which a height differs from the mean?

EXCEL **2.** Heights of Women Women's heights have a mean of 63.6 in. and a standard deviation of 2.5 in. Jodi Taylor is 61.1 in. tall.
   a. What is the difference between Jodi's height and the mean?
   b. How many standard deviations is that (the difference found in part a)?
   c. Convert Jodi's height to a z score.
   d. How does the z score compare to the number of standard deviations by which a score differs from the mean?

EXCEL **3.** Heights of Men Adult males have heights with a mean of 69.0 in. and a standard deviation of 2.8 in. Find the z scores corresponding to the following.
   a. NBA basketball player Shaquille O'Neal, who is 7 ft 1 in. tall
   b. Bob Jenkins, who is 5 ft 4 in. tall
   c. The author, who is a 69.72-in.-tall golf and tennis "player"

EXCEL **4.** Body Temperatures Human body temperatures have a mean of 98.20° and a standard deviation of 0.62°. Convert the given temperatures to z scores.
   a. 100°   **b.** 96.96°   **c.** 98.20°

Interpreting z Scores.   In Exercises 5–8, express all z scores with two decimal places. Consider a value to be unusual if its z score is less than −2.00 or greater than 2.00.

EXCEL **5.** Heights of Women The Beanstalk Club is limited to women and men who are very tall. The minimum height requirement for women is 70 in. Women's heights have a mean of 63.6 in. and a standard deviation of 2.5 in. Find the z score corresponding to a woman with a height of 70 in. and determine whether that height is unusual.

EXCEL **6.** Length of Pregnancy A woman wrote to *Dear Abby* and claimed that she gave birth 308 days after a visit from her husband, who was in the Navy. Lengths of pregnancies have a mean of 268 days and a standard deviation of 15 days. Find the z score for 308 days. Is such a length unusual? What do you conclude?

EXCEL **7.** Body Temperatures Human body temperatures have a mean of 98.20° and a standard deviation of 0.62°. An emergency room patient is found to have a temperature of 101°. Convert 101° to a z score. Is that temperature unusually high? What does it suggest?

EXCEL **8.** Cholesterol Levels in Men For men aged between 18 and 24 years, serum cholesterol levels (in mg/100 mL) have a mean of 178.1 and a standard deviation of 40.7 (based on data from the National Health Survey). Find the z score corresponding to a male, aged 18–24 years, who has a serum cholesterol level of 259.0 mg/100 mL. Is this level unusually high?

EXCEL **9.** Comparing Test Scores Which is relatively better: a score of 75 on a history test or a score of 27 on a psychology test? Scores on the history test have a mean of 80 and a standard deviation of 12. Scores on the psychology test have a mean of 30 and a standard deviation of 8.

EXCEL **10.** Comparing Scores Three students take equivalent tests of a sense of humor and, after the laughter dies down, their scores are calculated. Which is the highest relative score?
   a. A score of 144 on a test with a mean of 128 and a standard deviation of 34
   b. A score of 90 on a test with a mean of 86 and a standard deviation of 18
   c. A score of 18 on a test with a mean of 15 and a standard deviation of 5

EXCEL **11.** Weights of Coke Refer to Data Set 1 in Appendix B for the sample of 36 weights of regular Coke. Convert the weight of 0.7901 to a z score. Is 0.7901 an unusual weight for regular Coke?

EXCEL **12.** Weights of Diet Coke Refer to Data Set 1 in Appendix B for the sample of 36 weights of diet Coke. Convert the weight of 0.7907 to a z score. Is 0.7907 an unusual weight for diet Coke?

Finding Percentiles. *In Exercises 13–16, use the 36 sorted weights of regular Coke listed in Table 2-16. Find the percentile corresponding to the given value.*

13. 0.8264   **14.** 0.8172   **15.** 0.8192   **16.** 0.8079

Finding Percentiles, Quartiles, and Deciles. *In Exercises 17–24, use the 36 sorted weights of regular Coke listed in Table 2-16. Find the indicated percentile, quartile, or decile.*

EXCEL **17.** P_{80} EXCEL **18.** D_8 EXCEL **19.** D_6 EXCEL **20.** D_3

EXCEL **21.** Q_3 EXCEL **22.** P_{10} EXCEL **23.** D_1 EXCEL **24.** P_{33}

Finding Percentiles. *In Exercises 25–28, use the weights (in pounds) of bears listed in Data Set 7 in Appendix B. Find the percentile corresponding to the given weight.*

25. 144   **26.** 212   **27.** 316   **28.** 90

Finding Percentiles, Quartiles, and Deciles. *In Exercises 29–36, use the weights (in pounds) of bears listed in Data Set 7 in Appendix B. The Excel workbook is BEARS.xls. Find the indicated percentile, quartile, or decile.*

EXCEL **29.** P_{85} EXCEL **30.** P_{35} EXCEL **31.** Q_1 EXCEL **32.** Q_3

EXCEL **33.** D_9 EXCEL **34.** D_3 EXCEL **35.** P_{50} EXCEL **36.** P_{95}

2-6 Beyond the Basics

37. Understanding Units of Measurement When finding a z score for the height of a basketball player in the NBA, how is the result affected if, instead of using inches, all values are expressed in centimeters? In general, how are z scores affected by the particular unit of measurement that is used?

38. Interpreting a z Score Heights of women have a mean of 63.6 in. and a standard deviation of 2.5 in. Erin Zito has a height described by $z = -1.25$. What is her height in inches?

39. Distribution of z Scores
   a. A data set has a distribution that is uniform. If all of the values are converted to z scores, what is the shape of the distribution of the z scores?
   b. A data set has a distribution that is bell-shaped. If all of the values are converted to z scores, what is the shape of the distribution of the z scores?
   c. In general, how is the shape of a distribution affected if all values are converted to z scores?

EXCEL **40.** Fibonacci Sequence The first several terms of the famous Fibonacci sequence are 1, 1, 2, 3, 5, 8, 13.
   a. Find the mean $\bar{x}$ and standard deviation s, then convert each value to a z score. Don't round the z scores; carry as many places as your calculator can handle.

TABLE A-2	Standard Normal (z) Distribution								

z	.00	.01	.02	.03	.04	.05	.06	.07	.08	.09
0.0	.0000	.0040	.0080	.0120	.0160	.0199	.0239	.0279	.0319	.0359
0.1	.0398	.0438	.0478	.0517	.0557	.0596	.0636	.0675	.0714	.0753
0.2	.0793	.0832	.0871	.0910	.0948	.0987	.1026	.1064	.1103	.1141
0.3	.1179	.1217	.1255	.1293	.1331	.1368	.1406	.1443	.1480	.1517
0.4	.1554	.1591	.1628	.1664	.1700	.1736	.1772	.1808	.1844	.1879
0.5	.1915	.1950	.1985	.2019	.2054	.2088	.2123	.2157	.2190	.2224
0.6	.2257	.2291	.2324	.2357	.2389	.2422	.2454	.2486	.2517	.2549
0.7	.2580	.2611	.2642	.2673	.2704	.2734	.2764	.2794	.2823	.2852
0.8	.2881	.2910	.2939	.2967	.2995	.3023	.3051	.3078	.3106	.3133
0.9	.3159	.3186	.3212	.3238	.3264	.3289	.3315	.3340	.3365	.3389
1.0	.3413	.3438	.3461	.3485	.3508	.3531	.3554	.3577	.3599	.3621
1.1	.3643	.3665	.3686	.3708	.3729	.3749	.3770	.3790	.3810	.3830
1.2	.3849	.3869	.3888	.3907	.3925	.3944	.3962	.3980	.3997	.4015
1.3	.4032	.4049	.4066	.4082	.4099	.4115	.4131	.4147	.4162	.4177
1.4	.4192	.4207	.4222	.4236	.4251	.4265	.4279	.4292	.4306	.4319
1.5	.4332	.4345	.4357	.4370	.4382	.4394	.4406	.4418	.4429	.4441
1.6	.4452	.4463	.4474	.4484	.4495 *	.4505	.4515	.4525	.4535	.4545
1.7	.4554	.4564	.4573	.4582	.4591	.4599	.4608	.4616	.4625	.4633
1.8	.4641	.4649	.4656	.4664	.4671	.4678	.4686	.4693	.4699	.4706
1.9	.4713	.4719	.4726	.4732	.4738	.4744	.4750	.4756	.4761	.4767
2.0	.4772	.4778	.4783	.4788	.4793	.4798	.4803	.4808	.4812	.4817
2.1	.4821	.4826	.4830	.4834	.4838	.4842	.4846	.4850	.4854	.4857
2.2	.4861	.4864	.4868	.4871	.4875	.4878	.4881	.4884	.4887	.4890
2.3	.4893	.4896	.4898	.4901	.4904	.4906	.4909	.4911	.4913	.4916
2.4	.4918	.4920	.4922	.4925	.4927	.4929	.4931	.4932	.4934	.4936
2.5	.4938	.4940	.4941	.4943	.4945	.4946	.4948	.4949 *	.4951	.4952
2.6	.4953	.4955	.4956	.4957	.4959	.4960	.4961	.4962	.4963	.4964
2.7	.4965	.4966	.4967	.4968	.4969	.4970	.4971	.4972	.4973	.4974
2.8	.4974	.4975	.4976	.4977	.4977	.4978	.4979	.4979	.4980	.4981
2.9	.4981	.4982	.4982	.4983	.4984	.4984	.4985	.4985	.4986	.4986
3.0	.4987	.4987	.4987	.4988	.4988	.4989	.4989	.4989	.4990	.4990
3.10 and higher	.4999									

NOTE: For values of z above 3.09, use 0.4999 for the area.
*Use these common values that result from interpolation:

z score	Area
1.645	0.4500 ←
2.575	0.4950 ←

From Frederick C. Mosteller and Robert E. K. Rourke, *Sturdy Statistics*, 1973, Addison-Wesley Publishing Co., Reading, MA. Reprinted with permission of Frederick Mosteller.

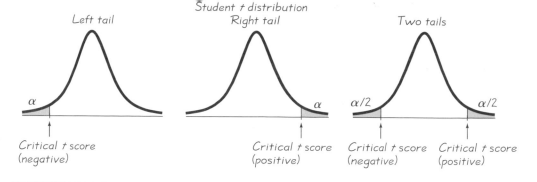

Student t distribution

Left tail

α

Critical t score
(negative)

Right tail

α

Critical t score
(positive)

Two tails

α/2   α/2

Critical t score   Critical t score
(negative)   (positive)

TABLE A-3	*t* Distribution					

	α					
Degrees of Freedom	.005 (one tail) 0.1 (two tails)	.01 (one tail) .02 (two tails)	.025 (one tail) .05 (two tails)	.05 (one tail) .10 (two tails)	.10 (one tail) .20 (two tails)	.25 (one tail) .50 (two tails)
1	63.657	31.821	12.706	6.314	3.078	1.000
2	9.925	6.965	4.303	2.920	1.886	.816
3	5.841	4.541	3.182	2.353	1.638	.765
4	4.604	3.747	2.776	2.132	1.533	.741
5	4.032	3.365	2.571	2.015	1.476	.727
6	3.707	3.143	2.447	1.943	1.440	.718
7	3.500	2.998	2.365	1.895	1.415	.711
8	3.355	2.896	2.306	1.860	1.397	.706
9	3.250	2.821	2.262	1.833	1.383	.703
10	3.169	2.764	2.228	1.812	1.372	.700
11	3.106	2.718	2.201	1.796	1.363	.697
12	3.054	2.681	2.179	1.782	1.356	.696
13	3.012	2.650	2.160	1.771	1.350	.694
14	2.977	2.625	2.145	1.761	1.345	.692
15	2.947	2.602	2.132	1.753	1.341	.691
16	2.921	2.584	2.120	1.746	1.337	.690
17	2.898	2.567	2.110	1.740	1.333	.689
18	2.878	2.552	2.101	1.734	1.330	.688
19	2.861	2.540	2.093	1.729	1.328	.688
20	2.845	2.528	2.086	1.725	1.325	.687
21	2.831	2.518	2.080	1.721	1.323	.686
22	2.819	2.508	2.074	1.717	1.321	.686
23	2.807	2.500	2.069	1.714	1.320	.685
24	2.797	2.492	2.064	1.711	1.318	.685
25	2.787	2.485	2.060	1.708	1.316	.684
26	2.779	2.479	2.056	1.706	1.315	.684
27	2.771	2.473	2.052	1.703	1.314	.684
28	2.763	2.467	2.048	1.701	1.313	.683
29	2.756	2.462	2.045	1.699	1.311	.683
Large (z)	2.575	2.326	1.960	1.645	1.282	.675

Formulas and Tables

for *Elementary Statistics Using Excel®* by Mario F. Triola
©2001 by Addison Wesley Longman Publishing Company, Inc.

TABLE A-4	Chi-Square (χ^2) Distribution

Degrees of Freedom	Area to the Right of the Critical Value									
	0.995	0.99	0.975	0.95	0.90	0.10	0.05	0.025	0.01	0.005
1	—	—	0.001	0.004	0.016	2.706	3.841	5.024	6.635	7.879
2	0.010	0.020	0.051	0.103	0.211	4.605	5.991	7.378	9.210	10.597
3	0.072	0.115	0.216	0.352	0.584	6.251	7.815	9.348	11.345	12.838
4	0.207	0.297	0.484	0.711	1.064	7.779	9.488	11.143	13.277	14.860
5	0.412	0.554	0.831	1.145	1.610	9.236	11.071	12.833	15.086	16.750
6	0.676	0.872	1.237	1.635	2.204	10.645	12.592	14.449	16.812	18.548
7	0.989	1.239	1.690	2.167	2.833	12.017	14.067	16.013	18.475	20.278
8	1.344	1.646	2.180	2.733	3.490	13.362	15.507	17.535	20.090	21.955
9	1.735	2.088	2.700	3.325	4.168	14.684	16.919	19.023	21.666	23.589
10	2.156	2.558	3.247	3.940	4.865	15.987	18.307	20.483	23.209	25.188
11	2.603	3.053	3.816	4.575	5.578	17.275	19.675	21.920	24.725	26.757
12	3.074	3.571	4.404	5.226	6.304	18.549	21.026	23.337	26.217	28.299
13	3.565	4.107	5.009	5.892	7.042	19.812	22.362	24.736	27.688	29.819
14	4.075	4.660	5.629	6.571	7.790	21.064	23.685	26.119	29.141	31.319
15	4.601	5.229	6.262	7.261	8.547	22.307	24.996	27.488	30.578	32.801
16	5.142	5.812	6.908	7.962	9.312	23.542	26.296	28.845	32.000	34.267
17	5.697	6.408	7.564	8.672	10.085	24.769	27.587	30.191	33.409	35.718
18	6.265	7.015	8.231	9.390	10.865	25.989	28.869	31.526	34.805	37.156
19	6.844	7.633	8.907	10.117	11.651	27.204	30.144	32.852	36.191	38.582
20	7.434	8.260	9.591	10.851	12.443	28.412	31.410	34.170	37.566	39.997
21	8.034	8.897	10.283	11.591	13.240	29.615	32.671	35.479	38.932	41.401
22	8.643	9.542	10.982	12.338	14.042	30.813	33.924	36.781	40.289	42.796
23	9.260	10.196	11.689	13.091	14.848	32.007	35.172	38.076	41.638	44.181
24	9.886	10.856	12.401	13.848	15.659	33.196	36.415	39.364	42.980	45.559
25	10.520	11.524	13.120	14.611	16.473	34.382	37.652	40.646	44.314	46.928
26	11.160	12.198	13.844	15.379	17.292	35.563	38.885	41.923	45.642	48.290
27	11.808	12.879	14.573	16.151	18.114	36.741	40.113	43.194	46.963	49.645
28	12.461	13.565	15.308	16.928	18.939	37.916	41.337	44.461	48.278	50.993
29	13.121	14.257	16.047	17.708	19.768	39.087	42.557	45.722	49.588	52.336
30	13.787	14.954	16.791	18.493	20.599	40.256	43.773	46.979	50.892	53.672
40	20.707	22.164	24.433	26.509	29.051	51.805	55.758	59.342	63.691	66.766
50	27.991	29.707	32.357	34.764	37.689	63.167	67.505	71.420	76.154	79.490
60	35.534	37.485	40.482	43.188	46.459	74.397	79.082	83.298	88.379	91.952
70	43.275	45.442	48.758	51.739	55.329	85.527	90.531	95.023	100.425	104.215
80	51.172	53.540	57.153	60.391	64.278	96.578	101.879	106.629	112.329	116.321
90	59.196	61.754	65.647	69.126	73.291	107.565	113.145	118.136	124.116	128.299
100	67.328	70.065	74.222	77.929	82.358	118.498	124.342	129.561	135.807	140.169

From Donald B. Owen, *Handbook of Statistical Tables*, ©1962 Addison-Wesley Publishing Co., Reading, MA. Reprinted with permission of the publisher.

HYPOTHESIS TEST: WORDING
OF FINAL CONCLUSION

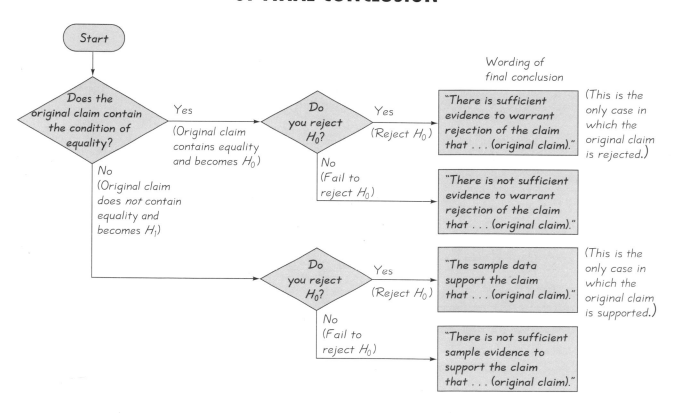

    b. Find the mean and standard deviation of the z scores found in part (a).

    c. If you use any other data set, will you get the same results obtained in part (b)?

41. Weights of Coke Use the so3rted weights of regular Coke listed in Table 2-16.

    a. Find the interquartile range.

    b. Find the midquartile.

    c. Find the 10–90 percentile range.

    d. Does $P_{50} = Q_2$? If so, does P_{50} *always* equal Q_2?

    e. Does $Q_2 = (Q_1 + Q_3)/2$? If so, does Q_2 *always* equal $(Q_1 + Q_3)/2$?

42. Interpolation When finding percentiles using Figure 2-17, if the locator L is not a whole number, we round it up to the next larger whole number. An alternative to this procedure is to interpolate so that a locator of 23.75 leads to a value that is 0.75 (or 3/4) of the way between the 23rd and 24th scores. Use this method of interpolation to find P_{35}, Q_1, and D_3 for the weights of bears listed in Data Set 7 in Appendix B.

2-7 Exploratory Data Analysis (EDA)

The general theme of this chapter is describing, exploring, and comparing data, and the focus of this section is exploration. We will first define exploratory data analysis, then introduce some new tools—outliers, 5-number summaries, and boxplots. We can then add these to the techniques presented earlier in this chapter.

> **DEFINITION**
>
> **Exploratory data analysis** is the process of using statistical tools (such as graphs, measures of center, and measures of variation) to investigate data sets in order to understand their important characteristics.

Recall that in Section 2-1 we listed five important characteristics of data beginning with these three: (1) *center,* a representative or average value; (2) *variation,* a measure of the amount of spread among the sample values; and (3) *distribution*, the nature or shape of the distribution of the data. When exploring a data set, we usually want to calculate the mean and the standard deviation and to generate a histogram, but we should not stop there. It is important to further examine the data set to identify any notable features, especially those that could have a strong effect on results and conclusions. One such feature is the presence of outliers.

Outliers

An **outlier** is a value that is located very far away from almost all of the other values. Relative to the other data, an outlier is an *extreme* value. When exploring a data set, outliers should be considered because they may reveal important information. The following example illustrates the effects of an outlier.

EXAMPLE   **Qwerty Keyboard Ratings** When using computer software, it is easy to make typing errors. Refer to the Qwerty word ratings listed in Table 2-1 with the Chapter Problem and assume that the first entry of 2 was incorrectly entered as 222 because the key was pressed too long when you were distracted by a meteorite landing on your porch. How does that error affect your exploration of the data set?

SOLUTION   After the first entry of 2 is replaced with the incorrect value of 222, the data in Table 2-1 will result in a mean of 8.6 (instead of the correct value of 4.4) and a standard deviation of 30.3 (instead of the correct value of 2.8). The accompanying **Excel** display shows the histogram for this modified data set.

Compare this histogram to the correct histogram shown in Figure 2-2 in Section 2-3 and you will see that this histogram is substantially different.

Our example demonstrates these important principles:

1. An outlier can have a dramatic effect on the mean.
2. An outlier can have a dramatic effect on the standard deviation.
3. An outlier can have a dramatic effect on the scale of the histogram so that the true nature of the distribution is totally obscured.

An easy way to find outliers is to examine a *sorted* list of the data. In particular, look at the minimum and maximum sample values and determine whether they are very far away from the other typical values. In the preceding example, the outlier of 222 is error. When such an error is identified, we should either correct it or delete it. Remember, however, that some data sets include outliers that are correct values, not errors. When exploring data, we might study the effects of outliers by constructing graphs and calculating statistics with and without the outliers included. (See Exercise 11 for a way to depict outliers on boxplots.)

Boxplots

Another frequently used graph is the boxplot. Boxplots are useful for revealing the center of the data, the spread of the data, the distribution of the data, and the presence of outliers. To construct a boxplot we first obtain the minimum value, the maximum value, and quartiles, as defined in the *5-number summary*.

> ### DEFINITIONS
>
> For a set of data, the **5-number summary** consists of the minimum value; the first quartile, Q_1; the median, or second quartile, Q_2; the third quartile, Q_3; and the maximum value.
>
> A **boxplot** (or **box-and-whisker diagram**) is a graph of a data set that consists of a line extending from the minimum value to the maximum value, and a box with lines drawn at the first quartile, Q_1; the median; and the third quartile, Q_3.

   Medians and quartiles are not very sensitive to extreme values. So boxplots, which use medians and quartiles, also have the advantage of not being as sensitive to extreme values as other devices based on the mean and standard deviation. Because boxplots don't show as much detailed information as histograms or stem-and-leaf plots, they might not be the best choice when dealing with a single data set. They are often terrific, though, for comparing two or more data sets. When using two or more boxplots for comparing different data sets, it is important to use the same scale so that correct comparisons can be made.

EXAMPLE   Qwerty Keyboard Ratings Refer to the 52 Qwerty word ratings in Table 2-1 (without the error of 222 used in the preceding example).

a.   Find the values constituting the 5-number summary.

b.   Construct a boxplot.

SOLUTION

a.   The 5-number summary consists of the minimum, Q_1, the median, Q_3, and the maximum. To find those values, first sort the data (by arranging them in order from lowest to highest). The minimum of 0 and the maximum of 14 are easy to identify from the sorted list. Now find the quartiles. Using the flowchart of Figure 2-17, we get $Q_1 = P_{25} = 2$, which is located by calculating the locator $L = (25/100)52 = 13$ and finding the value midway between the 13th and 14th values in the sorted list. The median is 4, which is the value midway between the 26th and 27th values. We also find that $Q_3 = 6$ by using Figure 2-17 for the 75th percentile. The 5-number summary is therefore 0, 2, 4, 6, and 14.

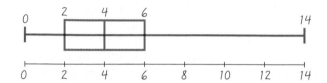

b. In Figure 2-18 we graph the boxplot for the data. We use the minimum (0) and the maximum (14) to determine a scale of values, then we plot the values from the 5-number summary as shown. ◀

In Figure 2-19 we show some generic boxplots along with common distribution shapes. It appears that the Qwerty word ratings have a skewed distribution.

In the accompanying **Excel** display, we show the Qwerty word ratings in the left boxplot and the Dvorak word ratings in the right boxplot. A comparison of the two boxplots reveals that the Dvorak ratings are grouped considerably lower, suggesting that the Dvorak keyboard is considerably easier to use. Also, the Qwerty ratings appear to have more spread because the Qwerty boxplot stretches farther.

Exploring

In this section we have added outliers, 5-number summaries, and boxplots to our list of tools useful for exploring data sets. We now have the following arsenal of exploration tools:

- *Measures of center*: mean, median, and mode
- *Measures of variation*: standard deviation and range
- *Measures of spread and relative location*: minimum value, maximum value, and quartiles
- *Unusual values*: outliers
- *Distribution*: histograms, stem-and-leaf plots, and boxplots

Now that you have all these tools in hand, you need to begin to *think creatively*! Rather than simply cranking out statistics and graphs, try to identify those that are particularly interesting or important. As a first step, investigate outliers and consider their effects by finding measures and graphs with and without the outliers included.

The following examples illustrate how these exploration tools work.

EXAMPLE   **Qwerty and Dvorak Keyboards: Did We Miss Anything?**
The Chapter Problem included 52 word ratings from the Qwerty keyboard
and 52 word ratings from the Dvorak keyboard. In preceding sections we
have compared those data sets using relative frequency tables (Table 2-9),
frequency polygons (Figure 2-10), and measures of center (Table 2-11). That
seems to be a reasonably thorough comparison, but apart from formal
methods to be considered later in this book, have we missed anything that is
really important? Yes! Our comparisons treated the two data sets as if they
were separate and independent, but the word ratings came from the *same*
52 words in the Preamble to the Constitution. We should therefore explore
the *differences* between the pairs of ratings corresponding to each of the 52
words. For example, the first word of the Preamble is "we," which has a
Qwerty rating of 2 and a Dvorak rating of 2, so the difference is $2 - 2 = 0$.
Explore the set of data consisting of the 52 differences listed here:

0 2   2 0 2 6 3 3 2 2 0   0 2 3 7 2 3 5 5   5 3 0 2 5 4 4

2 2 −1 2 4 2 2 0 2 2 3 2 2 5 1 4 2 2 2 13 0 2 6 3 0 3

SOLUTION   Keeping in mind our list of exploration tools, we find the
mean, standard deviation, 5-number summary, and outliers, and we generate a histogram. The results are shown below.

Mean: 2.7

Standard deviation: 2.3

5-number summary: minimum $= -1, Q_1 = 2$, median $= 2, Q_3 = 3.5$,
maximum $= 13$

Outlier: 13

INTERPRETATION   If the Qwerty and Dvorak keyboards had the same
level of typing difficulty, we would expect the differences between word
ratings to average around 0, but the mean difference is 2.7, and the histogram shows that the distribution is not centered about 0. Because the
differences between Qwerty and Dvorak appear to center around a value
well above 0, it appears that the Qwerty configuration results in greater

Good Advice for Journalists

typing difficulty. If we were to exclude the single outlier of 13, we would find that the mean does not change by much (it becomes 2.5). Excluding the outlier does not affect the histogram very much. The apparent difference is present with or without the outlier. This is convincing evidence that the Dvorak keyboard does in fact have a configuration that makes typing easier. (See also the end-of-chapter project "From Data to Decision: Critical Thinking.")

EXAMPLE   **Does It Rain More on Weekends?** Refer to Data Set 17 in Appendix B, which lists rainfall amounts (in inches) in Boston for every day of a recent year. The collection of this data set was inspired by media reports claiming that it rains more on weekends (Saturdays and Sundays) than on weekdays. Later in this book we will describe statistical methods that can be used to formally address that claim, but for now, let's explore the data set to see what can be learned. (Even if we already knew how to apply those formal statistical methods, we should first explore the data before proceeding with the formal analysis.)

SOLUTION   Referring again to our list of exploration tools, we find the mean and standard deviation for each day of the week, and we also find the 5-number summary for each day of the week. The results are summarized in the following table. Because the histograms for all seven days are pretty much the same, we show only the Excel histogram for the Monday rainfall amounts.

	Mean	Standard Deviation	Minimum	Q_1	Median	Q_3	Maximum
Monday	0.100	0.263	0.00	0.000	0.000	0.010	1.410
Tuesday	0.058	0.157	0.00	0.000	0.000	0.015	0.740
Wednesday	0.051	0.135	0.00	0.000	0.000	0.010	0.640
Thursday	0.069	0.167	0.00	0.000	0.000	0.040	0.850
Friday	0.095	0.228	0.00	0.000	0.000	0.040	0.960
Saturday	0.143	0.290	0.00	0.000	0.000	0.100	1.480
Sunday	0.068	0.200	0.00	0.000	0.000	0.010	1.280

Excel Histogram

INTERPRETATION   Examining and comparing the statistics and graphs, we make the following important observations.

- *Means:* The means vary from a low of 0.051 in. to a high of 0.143 in. The seven means vary by considerable amounts, and in later chapters of this book we will pursue whether these differences are *significant*. (Later methods will show that the means do not differ by significant amounts.) If we list the means in order from low to high, we get this sequence of days: Wednesday, Tuesday, Sunday, Thursday, Friday, Monday, Saturday. There does not appear to be a pattern of higher rainfall on weekends. Also, see the Excel graph of the seven means. In that graph, Monday is the first day, Tuesday is the second day, and so on. The Excel graph does not support the claim of more rainfall on Saturday and Sunday.

- *Variation:* The seven standard deviations vary from 0.135 in. to 0.290 in., but those values are not dramatically different. There does not appear to be anything highly unusual about the amounts of variation.

- The *minimums, first quartiles,* and *medians* are all 0.00 for each of the seven days. This is explained by the fact that for each day of the week, there are many days with no rain.

- *Outliers:* There are no outliers or unusual values. At the low end, there are many rainfall amounts of zero. At the high end, the sorted list of all 365 rainfall amounts ends with the high values of 0.92, 0.96, 1.28, 1.41, and 1.48.

- *Distributions:* The distributions of the rainfall amounts are skewed to the right. They are not bell-shaped as we might have expected. If the application of a particular method of statistics requires normally distributed (bell-shaped) populations, that requirement is not satisfied.

We have now acquired considerable insight into the nature of Boston rainfall amounts for different days of the week. Based on our exploration, we can conclude that Boston does not experience more rain on weekends than on the other days of the week.

Using Excel for Boxplots

Although Excel is not designed to generate boxplots, they can be generated using the Data Desk/XL add-in that is a supplement to this book. If DDXL does not appear on the menu bar, Data Desk/XL must be installed as described in Section 1-5. The quartile values from DDXL may differ from those found using Excel's **QUARTILE** command.

With DDXL installed, follow this procedure for generating a boxplot.

1. First enter the data in column A.
2. Click on **DDXL** and select **Charts and Plots.**
3. Under "Function Type," select the option of **Boxplot.**
4. In the dialog box, click on the pencil icon and enter the range of data, such as A1:A52 if you have 52 values listed in column A. Click **OK.** The result is a modified boxplot as described in Exercise 11. The values of the 5-number summary are also displayed. (Click on the "Summary Statistics" window, then scroll to the right to see the values in the 5-number summary.)

2-7 Basic Skills and Concepts

Exploring Data Using Boxplots.   *In Exercises 1–6, find the 5-number summary and construct a boxplot. Explore the data set and identify at least one characteristic that is particularly noteworthy.*

1. Employee Ages The following list shows the ages of most of the employees at the Vita Needle Company.

76	45	72	77	63	65	87	73	84	86	79	86	75
87	74	39	75	41	82	34	88	85	79	73	53	

(Based on data from "Where Retirement Became a Dirty Word" by Julie Flaherty, *New York Times.*)

2. Ages of Faculty Cars The following list shows the ages of cars in the faculty/staff parking lot at the author's college.

4	9	9	8	7	1	5	4	4	4	7	6	7	7	23

3. Pulse Rates Refer to Data Set 14 in Appendix B and explore the pulse rates. The Excel workbook is STATSURV.xls.

4. Old Faithful Eruptions Refer to Data Set 11 in Appendix B and explore the time intervals between eruptions of the Old Faithful geyser. The Excel workbook is OLDFAITH.xls.

5. Strengths of Aluminum Cans Refer to Data Set 12 in Appendix B and explore the axial loads of the cans that are 0.0111 in. thick. The Excel workbook is CANS.xls.

6. Body Temperatures Refer to Data Set 6 in Appendix B and explore the body temperatures for 12 A.M. on day 2.

Comparing Data Sets.   *In Exercises 7–10, find 5-number summaries, construct boxplots, and compare the data sets.*

7. Ages of Oscar Winners In "Ages of Oscar-Winning Best Actors and Actresses" (*Mathematics Teacher* magazine) by Richard Brown and Gretchen Davis, the authors compare the ages of actors and actresses at the time they won Oscars. The results for recent winners from each category are listed in the following table. Use boxplots to compare the two data sets.

Actors:	32 37 36 32 51 53 33 61 35 45 55 39
	76 37 42 40 32 60 38 56 48 48 40 43
	62 43 42 44 41 56 39 46 31 47 45 60

Actresses:	50 44 35 80 26 28 41 21 61 38 49 33
	74 30 33 41 31 35 41 42 37 26 34 34
	35 26 61 60 34 24 30 37 31 27 39 34

 8. Garbage Refer to Data Set 5 in Appendix B and use the weights of discarded paper and the weights of discarded plastic. The Excel workbook is GARBAGE.xls.

9. Textbook Prices Refer to Data Set 2 in Appendix B and use the new textbook prices at the author's college and the new textbook prices at the University of Massachusetts. The Excel workbook is TEXTBOOK.xls.

10. Regular/Diet Coke Refer to Data Set 1 in Appendix B and use the weights of regular Coke and the weights of diet Coke. The Excel workbook is COLA.xls.

2-7 Beyond the Basics

11. The boxplots discussed in this section are often called *skeletal* (or *regular*) boxplots. **Modified boxplots** are constructed as follows:

a. Find the *IQR*, which denotes the interquartile range defined by $IQR = Q_3 - Q_1$.

b. Draw the box with the median and quartiles as usual, but when drawing the lines to the right and left of the box, draw the lines only as far as the points corresponding to the largest and smallest values that are within 1.5 *IQR* of the box.

c. **Mild outliers** are sample values above Q_3 or below Q_1 by an amount that is greater than 1.5 *IQR* but not greater than 3 *IQR*. Plot mild outliers as *solid* dots.

d. **Extreme outliers** are values that are either above Q_3 by more than 3 *IQR* or below Q_1 by more than 3 *IQR*. Plot extreme outliers as small *hollow* circles.

The accompanying figure is an example of a modified boxplot. Refer to Data Set 12 in Appendix B for the axial loads of the cans that are 0.0111 in. thick. For that data set, the quartiles are 275, 285, and 295. Identify any mild outliers and extreme outliers.

STATISTICS IN THE NEWS

The Plague of Unwanted Calls

BOSTON All of us have had our dinners interrupted by telemarketers who seem to have honed the ability to call at the most inconvenient times. Legislators have already imposed some controls, so that calls in the middle of the night are now prohibited. Revised controls are constantly being considered. Instead of basing our analysis of the problem on anecdotal evidence, it is important to collect objective data that truly represent existing conditions. See the accompanying illustration from *USA Today*.

1. Refer to the *USA Today* illustration and assume that it is based on a survey of 1000 people. Construct the frequency table corresponding to the given data.

2. A major problem is caused by the two vague classes "don't know" and "6–up." Because those classes are not specific values, we cannot perform exact calculations. Delete the "don't know" class, and assume that the class of "6–up" is actually "6–10." Calculate the mean and standard deviation.

3. Repeat Exercise 2, but assume that the class of "6–up" is actually "6–12." Are the results dramatically affected?

Who can that be?
How many unwanted phone calls (telemarketers, donation solicitors, people they don't want to talk to), adults say they get each day at home:

1 call 22%
2 19%
3 14%
0 15%
4 6%
Don't know 4%
6-up 12%
5 8%

Source: Bruskin/Golding Research for Sony Electronics, Inc.

By Cindy Hall and Jerry Mosemak, USA TODAY

USA Today Snapshot. "A look at statistics that shape the nation" from 3/30/99

4. Can the *USA Today* illustration be used to obtain reasonably good estimates of the mean and standard deviation?

5. Do the results suggest that unwanted calls really are a problem?

VOCABULARY LIST

descriptive statistics
inferential statistics
frequency table
frequency
lower class limits
upper class limits
class boundaries
class midpoints
class width
relative frequencies
relative frequency table

cumulative frequency
cumulative frequency table
histogram
relative frequency histogram
frequency polygon
ogive
dotplot
stem-and-leaf plot
Pareto chart
pie chart
scatter diagram

measure of center	deviation
arithmetic mean	mean absolute deviation
mean	variance
sample size	unbiased estimator
median	range rule of thumb
mode	empirical (or 68–95–99.7) rule
bimodal	Chebyshev's theorem
multimodal	standard score
midrange	z score
weighted mean	quartiles
skewed	deciles
symmetric	percentiles
negatively skewed	exploratory data analysis (EDA)
positively skewed	outlier
range	5-number summary
standard deviation	boxplot (or box-and-whisker diagram)

REVIEW

In this chapter we considered methods for describing, exploring, and comparing data sets. When investigating a data set, the following characteristics are generally very important:

1. *Center:* A representative or average value

2. *Variation:* A measure of the amount that the values vary

3. *Distribution:* The nature or shape of the distribution of the data (such as bell-shaped, uniform, or skewed)

4. *Outliers:* Sample values that lie very far away from the vast majority of the other sample values

5. *Time:* Changing characteristics of the data over time

After completing this chapter, you should be able to do the following:

- Summarize the data by constructing a frequency table or relative frequency table (Section 2-2)

- Visually display the nature of the distribution by constructing a histogram, dotplot, stem-and-leaf plot, pie chart, or Pareto chart (Section 2-3)

- Calculate measures of center by finding the mean, median, mode, and midrange (Section 2-4)

- Calculate measures of variation by finding the standard deviation, variance, and range (Section 2-5)

- Compare individual values by using z scores, quartiles, deciles, or percentiles (Section 2-6)

- Investigate and explore the spread of data, the center of the data, and the range of values by constructing a boxplot (Section 2-7)

In addition to creating these tables, graphs, and measures, you should be able to understand and interpret the results. For example, you should clearly understand that the standard deviation is a measure of how much the data vary, and you should be able to use the standard deviation to distinguish between values that are usual and values that are unusual.

Statistics
at work

"The journalist must be able to cast a critical eye on research, divine the true context and significance of the work."

Mark Fenton

Editor at Large, Walking Magazine

Mark Fenton is also a walking advocate and is a champion walker. He was on the U.S. national racewalking team five times and has represented the U.S. in numerous international competitions. He has studied biomechanics and exercise physiology at the Olympic Training Center's Sports Science Laboratory in Colorado Springs, Colorado.

*W*hat is your job?

I'm editor-at-large for a health and fitness magazine, *Walking Magazine*. I spend a good deal of time as a public spokesman and public health advocate.

*W*hat concepts of statistics do you use?

I have to be familiar with all the common tools used in exercise science and public health research statistical analyses; from means and standard deviations, to statistical significance of differences, confidence intervals, and analyses of variance, as examples.

*H*ow do you use statistics in your job?

I regularly read medical research journals (*Medicine and Science in Sports and Exercise* and the *Journal of the American Medical Association* are two most notable) for exercise physiology, public health, and epidemiological research. Thus, I must be conversant in how studies are controlled and analyzed, and understand the relative power or value of a study's outcome. This information is used in my composition of articles and speeches on the findings of such work.

*P*lease describe one specific example illustrating how the use of statistics was successful in improving a product or service.

I regularly read research papers, and have to be cautious of holes in the evidence—dangerously small sample sizes; statistical significance but small absolute differences between groups; extremely large standard deviations, which can obscure the relevance of a finding for the average person.

A simple example is the problem of saying that the AVERAGE maximum heart rate for a 36-year-old woman is 190 beats per minute (226 minus age). That's helpful in calculating target heart rate, until you find that the deviation is quite high, and so the resulting figure can be off by as much as 10 beats per minute for roughly 1/3 of the population—an error large enough to be problematic when exercising. The study that pointed this out has altered how I offer exercise intensity recommendations.

*I*s your use of probability and statistics increasing, decreasing, or remaining stable?

With my growing work in public health, and the increasing complexity of the statistical tools being used in large population analyses, my conversance in the field MUST continue growing if I am to keep up.

Do you recommend statistics for today's college students? Why?

I absolutely recommend at least one course in statistics, as it is a useful tool in assessing the information with which we are bombarded—much of it without context—every day. At very least a comprehensive introductory course in statistics is a must for anyone interested in science journalism. The journalist must be able to cast a critical eye on research, divine the true context and significance of the work.

Do you feel job applicants are viewed more favorably if they have studied some statistics?

I suspect it is rarely asked of science journalists, but frankly should be one of the specific questions in an interview: Do you have the skills to assess the accuracy and power of the work you'll be reporting on? Keep in mind, one doesn't have to be a statistics expert. Rather, one should be able to ask and understand the answers to basic questions on analytical techniques, the true power, confidence, magnitude, and significance of findings.

Which other skills are important for today's college students?

Very high on my list of necessary attributes are excellent oral and written communications skills. These are a must in the workplace, and I am particularly concerned at the number of people unable to quickly and cogently construct and present an oral argument or presentation. It is the basis of a great deal of decision making in the business environment.

Please describe how you try to ensure objectivity.

I am usually assessing other's works, so I try to look for the simplest and clearest explanations of statistical work. It's my suspicion that the more convoluted an analysis, the further the researcher was from obtaining the desired result, and the more they felt obliged to "contrive" some positive outcome in the research. So, I think it's critical that I be able to knowledgeably question them about their methods. For example, in public health research it's common to analyze the effect of one condition on a population (obesity, for example) while controlling for others (smoking, diet, exercise habits, alcohol consumption), and, eventually, to try to assess the relative risks of disease and death for these various states. So, these are fairly involved processes, and it's important to at least be able to understand the premise and the balance of the researcher's approach.

3-1 Overview

In later chapters of this book, we will use sample data to make inferences (or conclusions) about populations. Many of those inferences will be based on *probabilities* of events. As a simple example, suppose that you were to win the top prize in a state lottery five consecutive times. There would be an uproar, and everyone would suspect that you were somehow cheating. They would know that even though there is a chance of someone winning five consecutive times, that chance is so incredibly low that they would reject it as a reasonable explanation. This is exactly how statisticians think: They reject explanations based on very low probabilities. Statisticians use the *rare event rule*.

> **Rare Event Rule for Inferential Statistics**
>
> If, under a given assumption (such as a lottery being fair), the probability of a particular observed event (such as five consecutive lottery wins) is extremely small, we conclude that the assumption is probably not correct.

Our primary objective in this chapter is to develop a sound understanding of probability values, which we will build upon in subsequent chapters. A secondary objective is to develop the basic skills necessary to solve probability problems.

 3-2 Fundamentals

In considering probability, we deal with procedures that produce outcomes.

> **DEFINITIONS**
>
> An **event** is any collection of results or outcomes of a procedure.
>
> A **simple event** is an outcome or an event that cannot be further broken down into simpler components.
>
> The **sample space** for a procedure consists of all possible *simple* events. That is, the sample space consists of all outcomes that cannot be broken down any further.

> **EXAMPLES**

Procedure	Example of Event	Sample Space
Roll one die	5 (simple event)	{1, 2, 3, 4, 5, 6}
Roll two dice	7 (not a simple event)	{1-1, 1-2, . . . , 6-6}

When rolling one die, 5 is a *simple event* because it cannot be broken down any further. When rolling two dice, 7 is *not a simple event* because it can be broken down into simpler events, such as 3-4 and 6-1. When rolling two dice, the *sample space* consists of 36 simple events: 1-1, 1-2, . . . , 6-6. When rolling two dice, the outcome of 3-4 is considered a simple event, because it is an outcome that cannot be broken down any further. We might incorrectly think that 3-4 can be further broken down into the individual results of 3 and 4, but 3 and 4 are not individual outcomes when two dice are rolled. When two dice are rolled, there are exactly 36 outcomes that are simple events: 1-1, 1-2, . . . , 6-6.   ◄

There are different ways to define the probability of an event, and we will present three approaches. First, however, we list some basic notation.

Notation for Probabilities

P denotes a probability.
A, B, and C denote specific events.
$P(A)$ denotes the probability of event A occurring.

Rule 1: Relative Frequency Approximation of Probability

Conduct (or observe) a procedure a large number of times, and count the number of times that event A actually occurs. Based on these actual results, $P(A)$ is *estimated* as follows:

$$P(A) = \frac{\text{number of times } A \text{ occurred}}{\text{number of times trial was repeated}}$$

Rule 2: Classical Approach to Probability (Requires Equally Likely Outcomes)

Assume that a given procedure has n different simple events and that each of those simple events has an *equal chance* of occurring. If event A can occur in s of these n ways, then

$$P(A) = \frac{\text{number of ways } A \text{ can occur}}{\text{number of different simple events}} = \frac{s}{n}$$

Rule 3: Subjective Probabilities

$P(A)$, the probability of event A, is found by simply guessing or estimating its value based on knowledge of the relevant circumstances.

Subjective Probabilities at the Racetrack

Researchers studied the ability of racetrack bettors to develop realistic subjective probabilities. (See "Racetrack Betting: Do Bettors Understand the Odds?", by Brown, D'Amato, and Gertner, *Chance* magazine, Vol. 7, No.3.) After analyzing results for 4400 races, they concluded that although bettors slightly overestimate the winning probabilities of "longshots" and slightly underestimate the winning probabilities of "favorites," their general performance is quite good. The subjective probabilities were calculated from the payoffs, which are based on the amounts bet, and the actual probabilities were calculated from the actual race results.

FIGURE 3-1
Three Approaches to Finding Probability

(a) Relative Frequency Approach (Rule 1): When trying to determine: P(tack lands point up), we must repeat the procedure of tossing the tack many times and then find the ratio of the number of times the tack lands with the point up to the number of tosses.

(b) Classical Approach (Rule 2): When trying to determine $P(2)$ with a balanced and fair die, each of the six faces has an equal chance of occurring.

$$P(2) = $$
$$\frac{\text{number of ways 2 can occur}}{\text{total number of simple events}}$$
$$= \frac{1}{6}$$

(c) Subjective Probability (Rule 3): When trying to estimate the probability of rain tomorrow, meteorologists use their expert knowledge of weather conditions to develop an estimate of the probability.

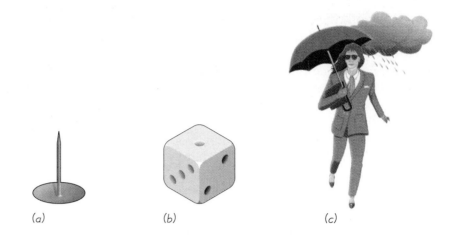

(a)   (b)   (c)

It is very important to note that *the classical approach* (*Rule 2*) *requires equally likely outcomes*. If the outcomes are not equally likely, either we must use the relative frequency estimate or we must rely on our knowledge of the circumstances to make an *educated guess*. Figure 3-1 illustrates the three approaches.

When finding probabilities with the relative frequency approach (Rule 1), we obtain an *approximation* instead of an exact value. As the total number of observations increases, the corresponding approximations tend to get closer to the actual probability. This property is stated as a theorem commonly referred to as the *law of large numbers*.

Law of Large Numbers

As a procedure is repeated again and again, the relative frequency probability (from Rule 1) of an event tends to approach the actual probability.

The law of large numbers tells us that the relative frequency approximations from Rule 1 tend to get better with more observations. This law reflects a simple notion supported by common sense: A probability estimate based on only a few trials can be off by substantial amounts, but with a very large number of trials, the estimate tends to be much more accurate. For example, an opinion poll of only a dozen people could easily be in error by large amounts, but a poll of thousands of *randomly selected* people will be much closer to the true population values.

Excel Demonstration of the Law of Large Numbers

The Excel display in Figure 3-2 illustrates the law of large numbers by showing results simulated on the computer. The simulation represents 100 consecutive births, where we plot the proportion of girls after each simulated birth. Note that as the number of births increases, the proportion of girls approaches the 0.5 value.

FIGURE 3-2
Illustration of the Law of
Large Numbers

Here is the Excel procedure for doing the simulation and generating a graph:

1. Enter the values of $1, 2, 3, \ldots 99, 100$ in column A. Instead of manually typing these entries, you could use this timesaving shortcut:

   a. Enter 1, then 2, then 3 in the first three cells of column A.

   b. Click on cell A1 and, while holding the mouse button down, drag the mouse downward so that cells A1, A2, and A3 are highlighted, then release the mouse button.

   c. Now use the mouse to click and drag the lower right corner of cell A3 downward so that all of the cells down to A100 are highlighted. Column A should now contain $1, 2, 3, \ldots, 99, 100$ in separate cells.

2. Click on cell B1.

3. Click **ƒx.** Select the function category of **Math & Trig,** then click on the function name of **RANDBETWEEN.**

4. You should now see a dialog box. In the "Bottom" box, enter 0. In the "Top" box, enter 1. Click **OK.** (You are telling Excel to select a random number between 0 and 1, so the result will be either a 0 or a 1.)

5. Cell B1 should be highlighted. Use the mouse to click and drag the lower right corner of cell B1 downward so that cells B1 through B100 are highlighted, then release the mouse button. The result should be a column of 0s and 1s, all randomly selected. This simulates the births of 100 babies, where the values of 0 and 1 represent genders of boy and girl.

6. Click on cell C1 and enter this expression: **=B1.** Then press the **Enter** key.

7. Click on cell C2 and enter this expression: **=B2+C1.** Then press the **Enter** key.

8. Click on cell C2 and use the mouse to click and drag its lower right corner downward, so cells C1 through C100 contain numbers. Examine column C and verify that it contains cumulative totals from column B. The second cell of column C is the sum of the first two cells from column B, the third cell of column C is the sum of the first three cells from column B, and so on.

How Probable?

How do we interpret such terms as *probable, improbable,* or *extremely probable*? The FAA interprets these terms as follows. *Probable:* A probability on the order of 0.00001 or greater for each hour of flight. Such events are expected to occur several times during the operational life of each airplane. *Improbable:* A probability on the order of 0.00001 or less. Such events are not expected to occur during the total operational life of a single airplane of a particular type, but may occur during the total operational life of all airplanes of a particular type. *Extremely improbable:* A probabilty on the order of 0.000000001 or less. Such events are so unlikely that they need not be considered to ever occur.

9. Click on cell D1 and enter this expression: **=C1/A1.** Press the **Enter** key.

10. Click on cell D1 and use the mouse to click and drag the lower right corner of the cell downward to cell D100. When you release the mouse button, you should see 100 values that are proportions. The entries of column D represent the proportions of girls. For example, cell D2 is the proportion of girls in two births, cell D3 is the proportion of girls in three births, and so on.

11. Construct a line graph of the column D entries. Click on **Insert,** then click on **Chart** (or just click on the **Chart Wizard** icon on the tool bar, which looks like a bar graph), then click **Line,** then click on the line graph in the upper left corner, then click **Next.** For the data range, enter D1:D100. Click on the **Column** button, then click **Next.** You can edit the graph and proceed to display it. Figure 3-2 was obtained by using this procedure. Your graph will not be exactly the same, because the simulation involves random selection of 0s (boys) and 1s (girls).

After examining the preceding three rules, it might seem that we should always use Rule 2 when a procedure has equally likely outcomes. Unfortunately, many procedures are so complicated that the classical approach (Rule 2) is impractical to use. In the game of solitaire, for example, the outcomes (hands dealt) are all equally likely, but it is extremely frustrating to try to use Rule 2 to find the probability of winning. In such cases we can more easily get good estimates by using the relative frequency approach (Rule 1). Simulations are often helpful when using this approach. (A **simulation** of a procedure is a process that behaves in the same ways as the procedure itself, so that similar results are produced.) For example, it's much easier to use Rule 1 for estimating the probability of winning at solitaire—that is, to play the game many times (or to run a computer simulation)—than to perform the extremely complex calculations required with Rule 2.

The following examples illustrate how the three rules work. In some of the examples we use the term *random.* Recall these definitions from Section 1-4: In a **random sample of one element** from a population, all elements available for selection have the same chance of being chosen; a sample of *n* items is a **random sample** (or a simple random sample) if it is selected in such a way that every possible sample of *n* items from the population has the same chance of being chosen. The general concept of randomness is of critical importance in statistics. When making inferences based on samples, we must have a sampling process that is representative, impartial, and unbiased. If a sample is not carefully selected, it may be totally worthless.

▷ **EXAMPLE** **Lightning Strikes** Find the probability that a randomly selected person will be struck by lightning this year.

SOLUTION The sample space consists of two simple events: The selected person is struck by lightning this year or is not. Because the sample space consists of events that are not equally likely, we can't use the classical approach (Rule 2). We must either use a relative frequency approximation (Rule 1) or subjectively estimate the probability (Rule 3). Before continuing, stop and try to guess the probability of a person being struck by lightning this year. Many people would make a reasonable guess that the probability is quite low, such as "one chance in a million." If we elect to use Rule 1, it isn't practical to conduct actual trials, but we can research past events. In a recent year, 377 people were struck by lightning in the United States. With a population of 274,037,295, the probability of being struck by lightning in a year is estimated to be

$$\frac{377}{274,037,295} \approx \frac{1}{727,000}$$

Probabilities that Challenge Intuition

In certain cases, our subjective estimates of probability values are dramatically different from the actual probabilities. Here is a classic example: If you take a deep breath, there is better than a 99% chance that you will inhale a molecule that was exhaled in dying Caesar's last breath. In that same morbid and unintuitive spirit, if Socrates' fatal cup of hemlock was mostly water, then the next glass of water you drink will likely contain one of those same molecules. Here's another less morbid example that can be verified: In classes of 25 students, there is better than a 50% chance that at least 2 students will share the same birthday.

▷ **EXAMPLE**   **Guessing Answers** On an SAT or ACT test, a typical multiple-choice question has 5 possible answers. If you make a random guess on one such question, what is the probability that your response is wrong?

SOLUTION   There are 5 possible outcomes or answers, and there are 4 ways to answer incorrectly. Random guessing implies that the outcomes in the sample space are equally likely, so we apply the classical approach (Rule 2) to get

$$P(\text{wrong answer}) = \frac{4}{5} = 0.8$$

In basic probability problems of the type we are now considering, it is very important to examine the available information carefully and to identify the total number of possible outcomes correctly. In some cases, the total number of possible outcomes is directly available, but in other cases we must process the available information to determine that total. The preceding example included the information that the total number of outcomes is 5, but the following example requires us to calculate the total number of possible outcomes.

▷ **EXAMPLE**   **Computer Ownership** In a statistics class, 18 students own their own computers and 7 do not. If one of the students is randomly selected, find the probability of getting one who does *not* own a computer.

SOLUTION   *Hint:* Instead of trying to formulate an answer directly from the written statement, summarize the given information in a format that allows you to better understand it. For example:

18 students own computers
<u> 7 students do not own computers</u>
25 total number of students

With random selection, the 25 students are equally likely to be selected, and Rule 2 applies as follows:

P(student does not own a computer)

$$= \frac{\text{number who do not own a computer}}{\text{total number of students}} = \frac{7}{25} = 0.28$$

There is a 0.28 probability that when one of the students is randomly selected, he or she does not own a computer. The way this solution was reached illustrates two important principles: First, you should list the given information in a clear and understandable format. Second, instead of finding probability values by merely dividing the smaller value by the larger value, you need to evaluate the available information carefully to identify the total number of possible outcomes and the number of ways in which the event can occur.

1st 2nd 3rd

boy-boy-boy

boy-boy-girl

boy-girl-boy

exactly → boy-girl-girl

2 boys

girl-boy-boy

girl-boy-girl

girl-girl-boy

girl-girl-girl

> **EXAMPLE**   **Gender of Children** Find the probability that a couple with 3 children will have exactly 2 boys. Assume that boys and girls are equally likely and that the gender of any child is not influenced by the gender of any other child.

SOLUTION   We first list the sample space that identifies the 8 outcomes. Those 8 outcomes are equally likely, so we use Rule 2. Of those 8 different possible outcomes, 3 correspond to exactly 2 boys, so

$$P(2 \text{ boys in 3 births}) = \frac{3}{8} = 0.375$$

INTERPRETATION   There is a 0.375 probability that if a couple has 3 children, exactly 2 will be boys.

> **EXAMPLE**   **Women College Graduates** If an American woman over age 25 is randomly selected, find the probability that she has a bachelor's degree.

SOLUTION   There are only two possible outcomes: Either the selected woman has a bachelor's degree or she does not. Because those two outcomes are not equally likely, we must either use the relative frequency approximation or settle for a subjective estimate of the probability. A study of 2500 randomly selected women over age 25 showed that 545 of them have bachelor's degrees (based on data from the Census Bureau). Based on that result, we *estimate* that the probability is 545/2500, or 0.218.

> **EXAMPLE**   **Thanksgiving Day** If a year is selected at random, find the probability that Thanksgiving Day will be on a (a) Wednesday, (b) Thursday.

SOLUTION

a. Thanksgiving Day always falls on the fourth Thursday in November. It is therefore impossible for Thanksgiving to be on a Wednesday. When an event is impossible, we say that its probability is 0.

b. It is certain that Thanksgiving will be on a Thursday. When an event is certain to occur, we say that its probability is 1.

Because any event imaginable is impossible, certain, or somewhere in between, it is reasonable to conclude that the mathematical probability of any event is 0, 1, or a number between 0 and 1 (see Figure 3-3).

- **The probability of an impossible event is 0.**
- **The probability of an event that is certain to occur is 1.**
- $0 \leq P(A) \leq 1$ **for any event** A.

In Figure 3-3, the scale of 0 through 1 is shown on the left, and the more familiar and common expressions of likelihood are shown on the right.

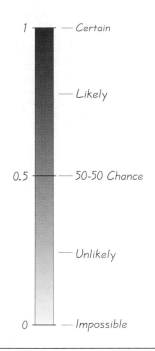

FIGURE 3-3
Possible Values for
Probabilities

Complementary Events

Sometimes we need to find the probability that an event A does *not* occur.

DEFINITION

The **complement** of event A, denoted by $\overline{A}$, consists of all outcomes in which event A does *not* occur.

> **EXAMPLE   Testing Corvettes** The General Motors Corporation wants to conduct tests of a new model of Corvette. A pool of 50 drivers has been recruited, 20 of whom are men. When the first person is selected from this pool, what is the probability of *not* getting a male driver?

SOLUTION   Because 20 of the 50 subjects are men, it follows that 30 of the 50 subjects are women, so

$$P(\text{not selecting a man}) = P(\overline{\text{man}})$$
$$= P(\text{woman})$$
$$= \frac{30}{50} = 0.6$$

Although it is difficult to develop a universal rule for rounding off probabilities, the following guide will apply to most problems in this text.

To Win, Bet Boldly

The *New York Times* published an article by Andrew Pollack in which he reported lower than expected earnings for the Mirage casino in Las Vegas. He wrote that "winnings for Mirage can be particularly volatile, because it caters to high rollers, gamblers who might bet $100,000 or more on a hand of cards. The law of averages does not work as consistently for a few large bets as it does for thousands of smaller ones…" This reflects the most fundamental principle of gambling: To win, place one big bet instead of many small bets! With the right game, such as craps, you have just under a 50% chance of doubling your money if you place one big bet. With many small bets, your chance of doubling your money drops substantially.

Rounding Off Probabilities

When expressing the value of a probability, either give the *exact* fraction or decimal or round off final decimal results to three significant digits. (*Suggestion:* When a probability is not a simple fraction such as 2/3 or 5/9, express it as a decimal so that the number can be better understood.)

All digits in a number are significant except for the zeros that are included for proper placement of the decimal point.

> **EXAMPLES**

- The probability of 0.000072098 has five significant digits (72098), and it can be rounded to three significant digits as 0.0000721.
- The probability of 1/3 can be left as a fraction, or rounded to 0.333. Do *not* round to 0.3.
- The probability of heads in a coin toss can be expressed as 1/2 or 0.5; because 0.5 is exact, there's no need to express it as 0.500.
- The fraction 7723/93725 is exact, but its value isn't obvious, so express it as the decimal 0.0824.

An important concept in this section is the mathematical expression of probability as a number between 0 and 1. This type of expression is fundamental and common in statistical procedures, and we will use it throughout the remainder of this text. A typical computer output, for example, may include a "*P*-value" expression such as "significance less than 0.001." We will discuss the meaning of *P*-values later, but they are essentially probabilities of the type discussed in this section. For now, you should recognize that a probability of 0.001 (equivalent to 1/1000) corresponds to an event so rare that it occurs an average of only once in a thousand trials.

Odds

Expressions of likelihood are often given as *odds,* such as 50:1 (or "50 to 1"). A serious disadvantage of odds is that they make many calculations extremely difficult. As a result, statisticians, mathematicians, and scientists prefer to use probabilities. The advantage of odds is that they make it easier to deal with money transfers associated with gambling, so they tend to be used in casinos, lotteries, and racetracks. There are three definitions that apply: the actual odds *against* event *A* occurring, the actual odds *in favor* of event *A* occurring, and the *payoff* odds. Note that in the three definitions that follow, the actual odds against and the actual odds in favor describe the actual likelihood of some event, but the payoff odds describe the relationship between the bet and the amount of the payoff. Racetracks and casinos are in business to make a profit, so the payoff odds will not be the same as the actual odds.

DEFINITION

The **actual odds against** event A occurring are the ratio $P(\overline{A})/P(A)$, usually expressed in the form of $a{:}b$ (or "a to b"), where a and b are integers having no common factors.

The **actual odds in favor** of event A are the reciprocal of the actual odds against that event. If the odds against A are $a{:}b$, then the odds in favor of A are $b{:}a$.

The **payoff odds** against event A represent the ratio of net profit (if you win) to the amount bet.

$$\text{payoff odds against event } A = (\text{net profit}){:}(\text{amount bet})$$

You Bet

In the typical state lottery, the "house" has a 65% to 70% advantage, since only 35% to 40% of the money bet is returned as prizes. The house advantage at racetracks is usually around 15%. In casinos, the house advantage is 5.26% for roulette, 5.9% for blackjack, 1.4% for craps, and 3% to 22% for slot machines. Some professional gamblers can systematically win at blackjack by using complicated card-counting techniques. They know when a deck has disproportionately more high cards, and this is when they place large bets. Many casinos react by ejecting card counters or by shuffling the decks more frequently.

> **EXAMPLE**   **Odds Against 5** Find the actual odds against rolling a 5 when a single die is rolled once.

SOLUTION $P(5) = 1/6$ because there is only one way to get a 5 and there are 6 simple events in the sample space. With $P(5) = 1/6$, we now proceed to find the actual odds against 5:

$$\text{actual odds against } 5 = \frac{P(\overline{5})}{P(5)} = \frac{5/6}{1/6} = \frac{5}{1}$$

We express this as $5{:}1$ or "5 to 1." Because the actual odds against getting a 5 are $5{:}1$, the actual odds in favor are $1{:}5$.

See Exercise 35 for conversions from odds into probabilities.

> **EXAMPLE**   If you bet \$5 on the number 13 in roulette, your probability of winning is $1/38$ and the payoff odds are given by the casino as $35{:}1$.

  a.   Find the actual odds against the outcome of 13.

  b.   How much net profit would you make if you win by betting on 13?

  c.   If the casino were operating just for the fun of it, and the payoff odds were changed to match the actual odds against 13, how much would you win if the outcome were 13?

SOLUTION

  a.   With $P(13) = 1/38$ and $P(\text{not } 13) = 37/38$, we get

$$\text{actual odds against } 13 = \frac{P(\text{not } 13)}{P(13)} = \frac{37/38}{1/38} = \frac{37}{1} \text{ or } 37{:}1$$

b. Because the payoff odds against 13 are 35 : 1, we have

$$35 : 1 = \text{(net profit)} : \text{(amount bet)}$$

so that there is a $35 profit for each $1 bet. For a $5 bet, the net profit is $175. The bettor would collect $175 plus the original $5 bet.

c. If the casino were operating for fun and not for profit, the payoff odds would be equal to the actual odds against the outcome of 13. If the payoff odds were changed from 35 : 1 to 37 : 1, you would obtain a net profit of $37 for each $1 bet. If you bet $5, your net profit would be $185. (The casino makes its profit by paying only $175 instead of the $185 that would be paid with a roulette game that is fair instead of favoring the casino.)   ◄

Remember, a major objective of this section is to develop the ability to interpret a probability value. Understand that values close to 0 reflect events that are very unlikely, but values close to 1 reflect events that are very likely.

3-2 Basic Skills and Concepts

1. Identifying Probability Values Which of the following values *cannot* be probabilities?

$$0, \quad 1, \quad -1, \quad 2, \quad 0.0123, \quad 3/5, \quad 5/3, \quad \sqrt{2}$$

2. Identifying Probability Values
   a. What is the probability of an event that is certain to occur?
   b. What is the probability of an impossible event?
   c. A sample space consists of 25 separate events that are equally likely. What is the probability of each?
   d. On a true/false test, what is the probability of answering a question correctly if you make a random guess?
   e. On a multiple-choice test with five possible answers for each question, what is the probability of answering a question correctly if you make a random guess?

3. Heads on a Coin When asked for the likelihood of tossing a coin and getting heads, many people respond with "50-50." What is the *probability* of getting heads?

4. Gender of Children In this section, we gave an example that included a list of the eight outcomes that are possible when a couple has three children. Refer to that list, and find the probability of getting exactly one girl among three children.

5. Finding Probability Values for M&Ms
   a. A package of M&M candies contains 144 that are orange and 1354 that are not. If one M&M candy is randomly selected, find the probability of getting an orange M&M.

b.   Refer to the sample of M&Ms in Data Set 10 in Appendix B. If one of those M&Ms is randomly selected, find the probability of getting one that is orange.

c.   The Mars candy company claims that 10% of its M&Ms are orange. Do the results from parts (a) and (b) agree with this claim?

6. Finding the Probability of Cheating In a survey of college students, 1162 stated that they cheated on an exam and 2468 stated that they did not (based on data from the Josephson Institute of Ethics). If one of these college students is randomly selected, find the probability that he or she cheated on an exam.

7. Finding the Probability of a Home Run Baseball player Mark McGwire broke a record when he hit 70 home runs in the 1998 season. During that season, he was at bat 509 times. If one of those "at bats" is randomly selected, find the probability that it is one of the times he hit a home run.

 8. Probability of Surviving the *Titanic* When the *Titanic* sank on April 15, 1912, 706 people survived and 1517 died. If someone aboard the *Titanic* was randomly selected, what is the probability that he or she survived the sinking?

Using Probability to Identify Unusual Events.   *In Exercises 9–16, consider an event to be "unusual" if its probability is less than or equal to 0.05.*

9. a.   Find the probability of getting a total of 2 when a pair of dice is rolled. (*Hint:* List the sample space of 36 outcomes, consisting of 1-1, 1-2, . . . , 6-6.)

b.   Is it "unusual" to get a total of 2 when a pair of dice is rolled?

10. a.   Find the probability of getting a 7 when a pair of dice is rolled. (See Exercise 9.)

b.   Is it "unusual" to get a 7 when a pair of dice is rolled?

11. a.   An instructor is obsessed with the metric system and insists that all multiple-choice questions must have 10 different possible answers, onc of which is correct. What is the probability of answering a question correctly if a random guess is made?

b.   Is it "unusual" to answer a question correctly by guessing?

12. a.   A roulette wheel has 38 slots: One slot is 0, another slot is 00, and the other slots are numbered 1 through 36. If you bet all of your textbook money on the number 13, what is the probability that you will win?

b.   Is it "unusual" to win when you bet on a single number in roulette?

13. a.   A study of 400 randomly selected American Airlines flights showed that 344 arrived on time (based on data from the Department of Transportation). What is the estimated probability of an American Airlines flight arriving late?

b.   Is it "unusual" for an American Airlines flight to arrive late?

14. Guessing Birthdays On their first date, Kelly asks Mike to guess the date of her birth, not including the year.

a.   What is the probability that Mike will guess correctly? (Ignore leap years.)

b.   Would it be "unusual" for him to guess correctly on his first try?

c.   If you were Kelly, and Mike did guess correctly on his first try, would you believe his claim that he made a lucky guess, or would you be convinced that he already knew when you were born?   (*Continued*)

d. If Kelly asks Mike to guess her age, and Mike's guess is too high by 15 years, what is the probability that Mike and Kelly will have a second date?

15. Lottery In the old New York State Lottery, you had to select six numbers between 1 and 54 inclusive. There were 25,827,165 different possible six-number combinations, and you had to select the correct combination of all six digits to win the grand prize. For a $1 bet, you selected two different six-number combinations. (You could not select a single six-number combination; you had to select two.)
    a. If you placed a $1 bet and selected two different six-number combinations, what was the probability of winning the grand prize?
    b. Was it unusual to win the grand prize?

16. Breaking Codes A code breaker for the National Security Agency is trying to decipher an intercepted message, and she is experimenting with different letter-substitution schemes.
    a. If she randomly selects a letter from the alphabet, what is the probability that she will select the correct letter, which is *e*?
    b. Would it be considered unusual to select *e* on her first try and find that it is the correct choice?

17. Probability of Alzheimer's Disease In a study of Americans over 65 years of age, it is found that 255 have Alzheimer's disease and 2302 do not (based on data from the Alzheimer's Association). If an American over 65 years of age is randomly selected, what is the estimated probability that he or she has Alzheimer's disease? Based on that probability, is Alzheimer's disease a major concern for those over 65 years of age?

18. Probability of Computer Failures A *PC World* survey of 4000 personal computer owners showed that 992 of them broke down during the first two years. (The computers broke down, not the owners.) In choosing among several computer suppliers, a purchasing agent wants to know the probability of a personal computer breaking down during the first two years. Use the survey results to estimate that probability.

19. Blood Groups In a study of blood donors, 225 were classified as group O and 275 had a classification other than group O (based on data from the Greater New York Blood Program). What is the approximate probability that a person will have group O blood?

20. TV Surveys A Nielsen survey of 3857 households shows that 463 have their televisions tuned to CBS on Monday night between 10:00 P.M. and 10:30 P.M. If a household is randomly selected, estimate the probability of getting one tuned to CBS in that time slot.

21. Probability of a Birthday
    a. If a person is randomly selected, find the probability that his or her birthday is October 18, which is National Statistics Day in Japan. Ignore leap years.
    b. If a person is randomly selected, find the probability that his or her birthday is in October. Ignore leap years.

22. Probability of Brand Recognition
    a. In a study of brand recognition, 831 consumers knew of Campbell's Soup, and 18 did not (based on data from Total Research Corporation). Use these results to estimate the probability that a randomly selected consumer will recognize Campbell's Soup.

b. *Estimate* the probability that a randomly selected adult American consumer will recognize the brand name of Bose. (Bose is a manufacturer of high-quality speakers and components.)

23. Fruitcake Survey In a Bruskin-Goldring Research poll, respondents were asked how a fruitcake should be used. One hundred thirty-two respondents indicated that it should be used for a doorstop, and 880 other respondents cited other uses, including birdfeed, landfill, and a gift. If one of these respondents is randomly selected, what is the probability of getting someone who would use the fruitcake as a doorstop?

24. Probability of IRS Correctness The U.S. General Accounting Office recently tested the IRS for correctness of answers to taxpayers' questions. For 1733 trials, the IRS was correct 1107 times. Use these results to estimate the probability that a random taxpayer's question will be answered correctly. Based on the result, would you say that the IRS does a good job answering taxpayers' questions correctly?

25. Probability of a Car Crash Among 400 randomly selected drivers in the 20–24 age bracket, 136 were in a car accident during the last year (based on data from the National Safety Council). If a driver in that age bracket is randomly selected, what is the approximate probability that he or she will be in a car accident during the next year? Is the resulting value high enough to be of concern to those in the 20–24 age bracket?

26. Probability of a Jail Sentence Data provided by the Bureau of Justice Statistics revealed that for a representative sample of convicted burglars, 76,000 were jailed, 25,000 were put on probation, and 2000 received other sentences. Use these results to estimate the probability that a convicted burglar will serve jail time. Does it seem that the result is high enough to deter burglars?

27. Probability of an Adverse Drug Reaction When the drug Viagra was clinically tested, 117 patients reported headaches and 617 did not (based on data from Pfizer, Inc.). Use this sample to estimate the probability that a Viagra user will experience a headache.

28. Probability of Credit-Card Fraud A study of credit-card fraud was conducted by Master Card International, and the accompanying table is based on the results. If one case of credit-card fraud is randomly selected from the cases summarized in the table, find the probability that the fraud resulted from a counterfeit card.

Method of Fraud	Number
Stolen card	243
Counterfeit card	85
Mail/phone order	52
Other	46

29. Gender of Children: Constructing Sample Space This section included a table summarizing the gender outcomes for a couple planning to have three children.
   a. Construct a similar table for a couple planning to have two children.
   b. Assuming that the outcomes listed in part (a) are equally likely, find the probability of getting two girls.
   c. Find the probability of getting exactly one child of each gender.

30. Genetics: Constructing Sample Space Both parents have the brown/blue pair of eye-color genes, and each parent contributes one gene to a child. Assume that if the child has at least one brown gene, that color will dominate and the eyes will be brown. (The actual determination of eye color is somewhat more complicated.)
   a. List the different possible outcomes. Assume that these outcomes are equally likely.
   b. What is the probability that a child of these parents will have the blue/blue pair of genes?
   c. What is the probability that the child will have brown eyes?

31. Finding Odds and Payoffs The American Statistical Association decides to invest some of its member revenue by buying a racehorse named Mean. Mean is entered in a race in which the actual probability of winning is 3/17.
   a. Find the actual odds against Mean winning.
   b. If the payoff odds are listed as 4:1, how much profit do you make if you bet $4 and Mean wins?

32. Finding Odds in Roulette A roulette wheel has 38 slots. One slot is 0, another is 00, and the others are numbered 1 through 36, respectively. You are placing a bet that the outcome is an odd number.
   a. What is your probability of winning?
   b. What are the actual odds against winning?
   c. When you bet that the outcome is an odd number, the payoff odds are 1:1. How much profit do you make if you bet $18 and win?
   d. How much profit would you make on the $18 bet if you could somehow convince the casino to change its payoff odds so that they are the same as the actual odds against winning? (*Recommendation:* Don't actually try to convince any casino of this; their sense of humor is remarkably absent when it comes to things of this sort.)

3-2 Beyond the Basics

33. Interpreting Effectiveness A double-blind experiment is designed to test the effectiveness of the drug Statisticzene as a treatment for number blindness. When treated with Statisticzene, subjects seem to show improvement. Researchers calculate that there is a 0.04 probability that the treatment group would show improvement if the drug has no effect. What should you conclude about the effectiveness of Statisticzene?

34. Determining Whether a Jury Is Random An attorney is defending a client accused of not meeting his alimony obligations. The pool of 20 potential jurors consists of all women, and the attorney calculates that there is a probability of 1/1,048,576 that 20 randomly selected people will be all women. Is there justification for arguing that the jury pool is unfair to his client?

35. Finding Probability from Odds If the actual odds against event A are $a:b$, then $P(A) = b/(a + b)$. Find the probability of Millenium winning his next race, given that the actual odds against his winning are $3:5$.

36. Finding Probability from Stem-and-Leaf Plot The stem-and-leaf plot summarizes the time (in hours) that managers spend on paperwork in one day (based on data from Adia Personnel Services). Use this sample to estimate the probability that a randomly selected manager spends more than 2.0 hours per day on paperwork.

0.	00
1.	0578
2.	00113449
3.	347
4.	445

37. Using Boxplot of IQ Scores After collecting IQ scores from hundreds of subjects, a boxplot is constructed with this five-number summary: 82, 91, 100, 109, 118. If one of the subjects is randomly selected, find the probability that his or her IQ score is greater than 109.

38. Leap Years and Guessing Birthdays In part (a) of Exercise 21, leap years were ignored in finding the probability that a randomly selected person will have a birthday on October 18.
   a. Recalculate this probability, assuming that a leap year occurs every four years. (Express your answer as an exact fraction.)
   b. Leap years occur in years evenly divisible by 4, except they are skipped in three of every four centesimal years (years ending in 00). The years

1700, 1800, and 1900 were not leap years, but 2000 is a leap year. Find the exact probability for this case, and express it as an exact fraction.

39. Flies on an Orange If two flies land on an orange, find the probability that they are on points that are within the same hemisphere.

40. Points on a Stick Two points along a straight stick are randomly selected. The stick is then broken at those two points. Find the probability that the three resulting pieces can be arranged to form a triangle. (This is a very difficult problem.)

 # 3-3 Addition Rule

The main objective of this section is to introduce the *addition rule* as a device for finding probabilities that can be expressed as $P(A \text{ or } B)$, the probability that either event A occurs or event B occurs (or they both occur) as the single outcome of a procedure. The key word to remember is *or*. Throughout this text we use the *inclusive or*, which means either one or the other or both. (Except for Exercise 27, we will not consider the *exclusive or*, which means either one or the other but not both.)

In the previous section we presented the fundamentals of probability and considered events categorized as *simple*. In this and the following section we consider *compound events*.

> ### DEFINITION
>
> A **compound event** is any event combining two or more simple events.

> **Notation for Addition Rule**
>
> $P(A \text{ or } B) = P(\text{event } A \text{ occurs or event } B \text{ occurs or they both occur})$

See Figure 3-4, which depicts a sample of 5 men and 10 women. Refer to that figure and answer this question: How many of the 15 subjects are men or red? (Remember, "men or red" really means "men, or red, or both.") Examination of Figure 3-4 should show that a total of 9 subjects are men or red. (*Important note:* It is *wrong* to add the 5 men to the 7 red people, because this total of 12 would have counted the 3 red men twice, but they are individual subjects that should be counted once each.) Because 9 of the 15 subjects are "men or red," the probability of randomly selecting someone who is a man or is red can be expressed as $P(\text{man or red}) = 9/15$.

This example suggests a general rule whereby we add the number of outcomes corresponding to each of the events in question:

When finding the probability that event *A* occurs or event *B* occurs, find the total of the number of ways *A* can occur and the number of ways *B* can occur, but *find that total in such a way that no outcome is counted more than once.*

Guess on SATs?

Students preparing for multiple-choice test questions are often told not to guess, but that's not necessarily good advice. Standardized tests with multiple-choice questions typically *compensate* for guessing, but they don't *penalize* it. For questions with five answer choices, one-fourth of a point is usually subtracted for each incorrect response. Principles of probability show that in the long run, pure random guessing will neither raise nor lower the exam score. Definitely guess if you can eliminate at least one choice or if you have a sense for the right answer, but avoid tricky questions with attractive wrong answers. Also, don't waste too much time on such questions.

FIGURE 3-4

One approach is to combine the number of ways event A can occur with the number of ways event B can occur and, if there is any overlap, compensate by subtracting the number of outcomes that are counted twice, as in the following rule.

Formal Addition Rule

$$P(A \text{ or } B) = P(A) + P(B) - P(A \text{ and } B)$$

where $P(A \text{ and } B)$ denotes the probability that A and B both occur at the same time as an outcome in a trial of a procedure.

Although the formal addition rule is presented as a formula, it is generally better to understand the spirit of the rule and apply it intuitively, as follows.

Intuitive Addition Rule

To find $P(A \text{ or } B)$, find the sum of the number of ways event A can occur and the number of ways event B can occur, *adding in such a way that every outcome is counted only once*. $P(A \text{ or } B)$ is equal to that sum, divided by the total number of outcomes.

Total Area = 1

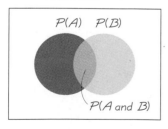

$P(A)$ $P(B)$

$P(A \text{ and } B)$

FIGURE 3-5
Venn Diagram Showing Overlapping Events

Figure 3-5 shows a Venn diagram that provides a visual illustration of the formal addition rule. In this figure we can see that the probability of A or B equals the probability of A (left circle) plus the probability of B (right circle) minus the probability of A and B (football-shaped middle region). This figure shows that the addition of the areas of the two circles will cause double-counting of the football-shaped middle region. This is the basic concept that underlies the addition rule. Because of the relationship between the addition rule and the Venn diagram shown in Figure 3-5, the notation $P(A \cup B)$ is sometimes used in place of $P(A \text{ or } B)$. Similarly, the notation $P(A \cap B)$ is sometimes used in place of $P(A \text{ and } B)$, so the formal addition rule can be expressed as

$$P(A \cup B) = P(A) + P(B) - P(A \cap B)$$

The addition rule is simplified whenever A and B cannot occur simultaneously, so $P(A \text{ and } B)$ becomes zero. Figure 3-6 illustrates that with no overlapping of A and B, we have $P(A \text{ or } B) = P(A) + P(B)$. The following definition formalizes the lack of overlapping shown in Figure 3-6.

Total Area = 1

FIGURE 3-6
Venn Diagram Showing Nonoverlapping Events

DEFINITION

Events A and B are **mutually exclusive** if they cannot occur simultaneously.

The flowchart of Figure 3-7 shows how mutually exclusive events affect the addition rule.

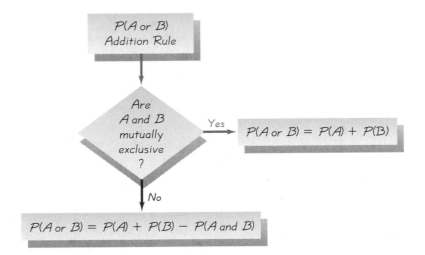

FIGURE 3-7
Applying the Addition Rule

① $P(A \text{ or } B) = P(A) + P(B)$
* mutually exculsive

② $P(A \overset{(U)}{or} B) = P(A) + P(B) - P(A \overset{(\cap)}{and} B)$
* not mutually exculsive

 EXAMPLE *Titanic* **Passengers** Refer to Table 3-1, reproduced on the next page for your convenience. Assuming that 1 person is randomly selected from the 2223 people aboard the *Titanic,* apply the addition rule to do the following.

a. Find P(selecting a man or a boy). mutually exculsive/union
b. Find P(selecting a man or someone who survived). not mutually exculsive/intersection

SOLUTION

a. From the table we see that among the 2223 people, the total number of men and boys is $1692 + 64 = 1756$. Because 1756 of the people

Shakespeare's Vocabulary

According to Bradley Efron and Ronald Thisted, Shakespeare's writings included 31,534 different words. They used probability theory to conclude that Shakespeare probably knew at least another 35,000 words that he didn't use in his writings. The problem of estimating the size of a population is an important problem often encountered in ecology studies, but the result given here is another interesting application. (See "Estimating the Number of Unseen Species: How Many Words Did Shakespeare Know?", in *Biometrika*, Vol. 63, No. 3.)

are men or boys and the total number of people is 2223, we get $P(\text{man or boy}) = 1756/2223 = 0.789924 = 0.790$ (rounded). Here, the event of selecting a man and the event of selecting a boy are mutually exclusive events, so we can add without concern for double-counting.

b. From the table we see that among the 2223 people, there are 1692 men and 706 people who survived, but we should not add 1692 and 706 because those events (men, survived) overlap and are not mutually exclusive. Using the intuitive addition rule, we must add in such a way that we do not double-count the 332 men who survived. The result should be 2066 individual people who were men or survived. We therefore get this result: $P(\text{man or survived}) = 2066/2223 = 0.929$. ◀

706 (total) − 332 (men) = 374
374 + 1692 = 2066
2066/2223 = .929

TABLE 3-1	*Titanic* Mortality				
	Men	Women	Boys	Girls	**Total**
Survived	332	318	29	27	**706**
Died	1360	104	35	18	**1517**
Total	**1692**	**422**	**64**	**45**	**2223**

There are several strategies you could use for counting people who are men or survivors. Any of the following would work.

- Color the cells representing people that are men or survivors, then add the numbers in those colored cells, being careful to add each number only once. This approach yields

$$1360 + 332 + 318 + 29 + 27 = 2066$$

- Add the 1692 men to the 706 survivors, but compensate for the double-counting by subtracting the 332 men survivors. This approach yields a result of

$$1692 + 706 - 332 = 2066$$

- Start with the total of 1692 men, then add the survivors who were not yet included in that total, to get a result of

$$1692 + 318 + 29 + 27 = 2066$$

Carefully study the preceding example, because it makes clear this essential feature of the addition rule: "Or" suggests addition, and the addition must be done without double-counting.

We can summarize the key points of this section as follows:

1. To find $P(A \text{ or } B)$, begin by associating "or" with addition.

2. Consider whether events A and B are mutually exclusive; in other words, can they happen at the same time? If they are not mutually exclusive (that is, if they can happen at the same time), be sure to avoid (or at least compensate for) double-counting when adding the relevant probabilities. If you understand the importance of not double-counting when you find $P(A \text{ or } B)$, you don't necessarily have to calculate the value of $P(A) + P(B) - P(A \text{ and } B)$.

Errors made when applying the addition rule often involve double-counting; that is, events that are not mutually exclusive are treated as if they were. One indication of such an error is a total probability that exceeds 1; however, errors involving the addition rule do not always cause the total probability to exceed 1.

Complementary Events

In Section 3-2 we defined the complement of event A and denoted it by $\overline{A}$. We said that $\overline{A}$ consists of all the outcomes in which event A does *not* occur. Events A and $\overline{A}$ must be mutually exclusive, because it is impossible for an event and its complement to occur at the same time. Also, we can be absolutely certain that A either does or does not occur, which implies that either A or $\overline{A}$ must occur. These observations enable us to apply the addition rule for mutually exclusive events as follows:

$$P(A \text{ or } \overline{A}) = P(A) + P(\overline{A}) = 1$$

We justify $P(A \text{ or } \overline{A}) = P(A) + P(\overline{A})$ by noting that A and $\overline{A}$ are mutually exclusive; we justify the total of 1 by our certainty that A either does or does not occur. This result of the addition rule leads to the following three equivalent forms.

Total Area = 1

FIGURE 3-8
Venn Diagram for the Complement of Event *A*

Rule of Complementary Events

$$P(A) + P(\overline{A}) = 1$$
$$P(\overline{A}) = 1 - P(A)$$
$$P(A) = 1 - P(\overline{A})$$

Figure 3-8 visually displays the relationship between $P(A)$ and $P(\overline{A})$.

▶ **EXAMPLE**   If $P(A) = 0.3$, find $P(\overline{A})$.

SOLUTION   Using the rule of complementary events, we get

$$P(\overline{A}) = 1 - P(A) = 1 - 0.3 = 0.7$$

◀

probability that when both sensors are triggered, water will get through the system so that cooling can occur.

Water   Reactor

17. Flat Tire Excuse A classic excuse for a missed test is offered by four students who claim that their car had a flat tire. On the makeup test, the instructor asks the students to identify the particular tire that went flat. If they really didn't have a flat tire and randomly select one that supposedly went flat, what is the probability that they will all select the same tire?

18. Voice Identification of Criminal In a Riverhead, New York, case, nine different crime victims listened to voice recordings of five different men. All nine victims identified the same voice as that of the criminal. If the voice identifications were made by random guesses, find the probability that all nine victims would select the same person. Does this constitute reasonable doubt?

19. Quality Control: Is the New Process Better? A quality-control manager claims that a new process for manufacturing cell phones is better because the rate of defects is lower than 4%, which had been the rate of defects in the past. When 25 cell phones are manufactured with the new process, there are no defects. Assuming that the new method has the same 4% defect rate as in the past, find the probability of getting no defects among the 25 cell phones. Based on the result, is there strong evidence to conclude that the new process is better?

20. Coincidences: Give Me a Brake The Rollins Trucking Company has a fleet of 16 trucks used to transport produce from Kansas. When 4 of the trucks are inspected at a police safety checkpoint, it is found that they all have defective brakes. The owner of the company, Ray Rollins, claims that the other 12 trucks have good brakes, and it is just a coincidence that police happened to select the faulty trucks. Assuming that Ray Rollins' claim is true, find the probability of randomly selecting 4 of his trucks and getting the 4 with defective brakes. Based on the result, does his claim seem plausible?

Accuracy of HIV Test. *Based on data from the New York State Health Department, there is a 0.3% HIV rate for the general population. Under certain conditions, a preliminary screening test for HIV is correct 95% of the time. (Subjects are not told that they are HIV infected until later tests verify the preliminary results.) If we randomly select 20,000 people from the general population, we expect results as summarized in the accompanying table. Use this table in Exercises 21–24.*

Sample from the General Population:
Preliminary Screening Test

	Positive	Negative
HIV virus infected	57	3
Not HIV infected	997	18,943

21. HIV Positive If two different people are randomly selected, find the probability that they both test positive for HIV.

22. HIV If someone is randomly selected, find the probability of getting someone who tests positive or someone who is HIV infected.

23. HIV If two different people are randomly selected, find the probability that they are both HIV infected.

24. HIV Negative If three different people are randomly selected, find the probability that they all test negative.

3-4 Beyond the Basics

25. Same Birthdays Find the probability that no two people have the same birthday when the number of randomly selected people is
   a. 3   **b.** 5   **c.** 25

26. Gender of Children If a couple plans to have eight children, find the probability that they are all of the same gender.

27. Drawing Cards Two cards are to be randomly selected without replacement from a shuffled deck. Find the probability of getting an ace on the first card and a spade on the second card.

28. Complements and the Addition Rule
   a. Develop a formula for the probability of not getting either A or B on a single trial. That is, find an expression for $P(\overline{A \text{ or } B})$.
   b. Develop a formula for the probability of not getting A or not getting B on a single trial. That is, find an expression for $P(\overline{A} \text{ or } \overline{B})$.
   c. Compare the results from parts (a) and (b). Does $P(\overline{A \text{ or } B}) = P(\overline{A} \text{ or } \overline{B})$?

3-5 Multiplication Rule: Complements and Conditional Probability

In this section we continue our discussion of the multiplication rule and show how the basic rule can be used in two other applications. We begin with the use of complements of events as a way to simplify certain calculations.

Complements: The Probability of "At Least One"

The multiplication rule and the rule of complements can be used together to greatly simplify certain types of calculations, such as those in which we want to find the probability that among several trials, *at least one* will result in some specified outcome. In such cases, the meaning of the language must be clearly understood:

• "At least one" is equivalent to "one or more."

• The complement of getting at least one item of a particular type is that you get no items of that type.

Composite Sampling

The U.S. Army once tested for syphilis by giving each inductee an individual blood test that was analyzed separately. One researcher suggested mixing pairs of blood samples. After the mixed pairs were tested, syphilitic inductees could be identified by retesting the few blood samples that were in the pairs that tested positive. The total number of analyses was reduced by pairing blood specimens, so why not put them in groups of three or four or more? Probability theory was used to find the most efficient group size, and a general theory was developed for detecting the defects in any population. This technique is known as *composite sampling*.

Suppose a couple plans to have three children and they want to know the probability of getting at least one girl. See the following interpretations.

At least 1 girl among 3 = 1 or more girls.

The complement of "at least 1 girl" = no girls = all 3 children are boys.

We could easily find the probability from a list of the entire sample space of eight outcomes, but we want to illustrate the use of complements, which can be used in many other problems that cannot be solved so easily.

▷ **EXAMPLE** **Gender of Children** Find the probability of a couple having at least 1 girl among 3 children. Assume that boys and girls are equally likely and that the gender of a child is independent of the gender of any brothers or sisters.

SOLUTION

Step 1: Use a symbol to represent the event desired. In this case, let A = at least 1 of the 3 children is a girl.

Step 2: Identify the event that is the complement of A.

$$\overline{A} = not \text{ getting at least 1 girl among 3 children}$$
$$= \text{all 3 children are boys}$$
$$= \text{boy and boy and boy}$$

Step 3: Find the probability of the complement.

$$P(\overline{A}) = P(\text{boy and boy and boy})$$
$$= \frac{1}{2} \cdot \frac{1}{2} \cdot \frac{1}{2} = \frac{1}{8}$$

Step 4: Find $P(A)$ by evaluating $1 - P(\overline{A})$.

$$P(A) = 1 - P(\overline{A}) = 1 - \frac{1}{8} = \frac{7}{8}$$

INTERPRETATION There is a 7/8 probability that if a couple has 3 children, at least 1 of them is a girl. ◁

The principle used in this example can be summarized as follows:

To find the probability of *at least one* of something, calculate the probability of *none,* then subtract that result from 1.

Conditional Probability

The probability of an event is often affected by knowledge of circumstances. For example, if you randomly select someone from the general population, the probability of getting a male is 0.5, but if you know that the selected person frequently changes TV channels with a remote control, the probability is 0.999 (OK, that might be a slight exaggeration). A *conditional probability* of an event occurs when the probability is affected by the knowledge of other circumstances. The conditional probability of event B occurring, given that event A has already occurred, can be found by using the multiplication rule $[P(A \text{ and } B) = P(A) \cdot P(B|A)]$ and solving for $P(B|A)$ by dividing both sides of the equation by $P(A)$.

DEFINITION

The **conditional probability** of event B occurring, given that event A has already occurred, can be found by dividing the probability of events A and B both occurring by the probability of event A:

$$P(B|A) = \frac{P(A \text{ and } B)}{P(A)}$$

This formula is a formal expression of conditional probability, but we can often use this intuitive approach:

Intuitive Approach to Conditional Probability

The conditional probability of B given A can be found by assuming that event A has occurred and, working under that assumption, calculating the probability that event B will occur.

 EXAMPLE   ***Titanic* Passengers** Refer to Table 3-1, reproduced here. Find the following:

a. If 1 of the 2223 people is randomly selected, what is the probability that this person survived, given that the selected person is a man?

b. If 1 of the 2223 people is randomly selected, what is the probability of getting a man, given that the selected person survived?

SOLUTION   **a.** We want $P(\text{survived}|\text{man})$, the probability of getting someone who survived, *given that the selected person is a man.* If we assume

Bayes' Theorem

Thomas Bayes (1702–1761) said that probabilities should be revised when we learn more about an event. Here's one form of Bayes' theorem:

$$P(A|B) = \frac{P(A) \cdot P(B|A)}{P(A) \cdot P(B|A) + P(\overline{A}) \cdot P(B|\overline{A})}$$

Suppose 60% of a company's computer chips are made in one factory (denoted by A) and 40% are made in its other factory (denoted by $\overline{A}$). For a randomly selected chip, the probability it came from factory A is 0.60. Suppose we learn that the chip is defective and the defect rates for the two factories are 35% (for A) and 25% (for $\overline{A}$). We can use the above formula to find that there is a 0.677 probability the defective chip came from factory A.

TABLE 3-1	*Titanic* Mortality				
	Men	Women	Boys	Girls	**Total**
Survived	332	318	29	27	**706**
Died	1360	104	35	18	**1517**
Total	**1692**	**422**	**64**	**45**	**2223**

that the selected person is a man, we are dealing with the 1692 people in the first column of Table 3-1. Among those 1692 people, 332 survived, so

$$P(\text{survived}|\text{man}) = \frac{332}{1692} = 0.196$$

The same result can be found by using the formula given with the definition of conditional probability:

$$P(\text{survived}|\text{man}) = \frac{P(\text{man and survived})}{P(\text{man})}$$

$$= \frac{332/2223}{1692/2223} = 0.196$$

b. Here we want $P(\text{man}|\text{survived})$. If we assume that the person selected is a survivor, we are dealing with the 706 people in the first row of Table 3-1. Among those 706 people, 332 are men, so

$$P(\text{man}|\text{survived}) = \frac{332}{706} = 0.470$$

Again, the same result can be found by applying the formula for conditional probability:

$$P(\text{man}|\text{survived}) = \frac{P(\text{survived and man})}{P(\text{survived})}$$

$$= \frac{332/2223}{706/2223} = 0.470$$

By comparing the results from parts (a) and (b), we see that $P(\text{survived}|\text{man})$ is very different from $P(\text{man}|\text{survived})$.

Does the probability of surviving the sinking of the *Titanic depend* on whether the person is a man, woman, or child? Let's list the relevant probabilities:

$$P(\text{survived}|\text{man}) = 0.196$$

$$P(\text{survived}|\text{woman or boy or girl}) = 0.704$$

INTERPRETATION   The rule of "women and children first" was not strictly followed on the *Titanic*, because some women and children died while some men survived. But the probabilities we just calculated show that if you were on the *Titanic*, you had a substantially better chance of surviving if you were a woman or child.   ◄

The calculation of conditional probabilities is not easy, but using the intuitive approach helps considerably. The next example would be quite difficult if a solution were attempted using the formal rules of probability, but if we carefully read the question and use the intuitive approach, the solution becomes much easier.

EXAMPLE   *Titanic* **Passengers** Refer to the *Titanic* mortality data in Table 3-1 and find the probability of getting a woman or child if a *Titanic* survivor is randomly selected.

SOLUTION   The question asks us to find $P(\text{woman or child}|\text{survivor})$. If we assume that a survivor is selected, we can restrict our sample space to the first row of 706 survivors. Among the 706 survivors, the number that are women or children is $318 + 29 + 27 = 374$. $P(\text{woman or child}|\text{survivor})$ is therefore $374/706 = 0.530$. Just try combining the formal addition rule and the formula for conditional probability to arrive at the same solution, and you will see the benefit of using this intuitive approach.

Sometimes the information at hand is confusing in its present format, but it can be simplified by constructing a table similar to Table 3-1. The next example shows the usefulness of such a table.

 EXAMPLE   **HIV** The New York State Health Department reports a 10% rate of the HIV virus for the "at-risk" population. Under certain conditions, a preliminary screening test for the HIV virus is correct 95% of the time. (Subjects are not told that they are HIV infected until additional tests verify the initial screening results.) If someone is randomly selected from the at-risk population, what is the probability that they have the HIV virus if it is known that they have tested positive in the initial screening?

SOLUTION   There are enough numbers floating around here to confuse the most able of us. Let's begin by imposing some order. Let's assume that the at-risk population consists of 5000 people, so we can now work with concrete numbers, and we can deduce the following.

1. Because there is a 10% rate of HIV, we expect to have 500 infected people (10% of 5000) and 4500 people (90% of 5000) not infected.

2. For the 500 infected people, the preliminary screening test will be correct 95% of the time, so 475 of them (95% of 500) will test positive and 25 of them (5% of 500) will test negative.

3. For the 4500 people not infected, the preliminary screening test will be correct 95% of the time, so 4275 of them (95% of 4500) will test negative and 225 will test positive.

These results can now be summarized in Table 3-2.

Coincidences?

John Adams and Thomas Jefferson (the second and third presidents) both died on July 4, 1826. President Lincoln was assassinated in Ford's Theater; President Kennedy was assassinated in a Lincoln car made by the Ford Motor Company. Lincoln and Kennedy were both succeeded by vice presidents named Johnson. Fourteen years *before* the sinking of the Titanic, a novel described the sinking of the Titan, a ship that hit an iceberg; see Martin Gardner's *The Wreck of the Titanic Foretold?* Gardner states, "In most cases of startling coincidences, it is impossible to make even a rough estimate of their probability."

TABLE 3-2	Results from the At-Risk Population: Preliminary Screening	
	Positive	Negative
HIV infected	475	25
Not HIV infected	225	4275

We can now address the key question: If someone from the at-risk population is randomly selected, what is the probability that this person has the HIV virus if it is known that he or she has tested positive in the preliminary screening? That is, find $P(\text{HIV}|\text{positive preliminary test})$. Table 3-2 shows that there are $475 + 225 = 700$ people who tested positive. Among those 700 people, 475 have the virus, so

$$P(\text{HIV}|\text{positive preliminary test}) = 475/700 = 0.679.$$

Testing for Independence

In the multiplication rule for dependent events, if $P(B|A) = P(B)$, then the occurrence of event A has no effect on the probability of event B. This is often used as a test for independence. If $P(B|A) = P(B)$, then A and B are independent events; however, if $P(B|A) \neq P(B)$, then A and B are dependent events. Another test for independence involves checking for the equality of $P(A \text{ and } B)$ and $P(A) \cdot P(B)$. If they are equal, events A and B are independent. If $P(A \text{ and } B) \neq P(A) \cdot P(B)$, then A and B are dependent events. These results are summarized as follows.

Two events A and B are *independent* if	Two events A and B are *dependent* if
$P(B|A) = P(B)$	$P(B|A) \neq P(B)$
or	or
$P(A \text{ and } B) = P(A) \cdot P(B)$	$P(A \text{ and } B) \neq P(A) \cdot P(B)$

3-5 Basic Skills and Concepts

Describing Complements.   *In Exercises 1–4, provide a written description of the complement of the given event.*

1. Gender of Children When a couple has three children, at least one of them is a girl.
2. Freshness When 12 apples are tested for freshness, all of them pass the test.
3. Statistics Grades In a statistics class, all students receive a grade of A.
4. Missile Hits When five missiles are fired, at least one of them strikes its target.
5. Probability of At Least One Girl If a couple plans to have three children, what is the probability that they will have at least one girl?
6. Probability of At Least One Girl If a couple plans to have 10 children (it could happen), what is the probability that there will be at least one girl?
7. At Least One 6 If you roll a die five times, what is the probability of getting at least one 6?
8. At Least One Correct Answer If you make random guesses for three multiple-choice test questions (each with five possible answers), what is the probability of getting at least one correct?
9. Tossing Coins Find the probability of getting heads on the third toss of a coin, given that the first two tosses resulted in heads.

10. Coin and Die A procedure consists of flipping a coin, then rolling a die. What is the probability of getting a 5, given that the coin flip resulted in heads?

11. Subjective Probability Use subjective probability to estimate the probability of randomly selecting an adult and getting a woman, given that the selected person is known to have given birth to a baby.

12. Subjective Probability Use subjective probability to estimate the probability of randomly selecting an adult and getting a male, given that the selected person owns a motorcycle.

13. Redundancy in Alarm Clocks A student experiences difficulties with malfunctioning alarm clocks. Instead of using one alarm clock, he decides to use three. What is the probability that at least one of his alarm clocks works correctly if each individual alarm clock has a 99% chance of working correctly?

14. Acceptance Sampling With one method of the procedure called *acceptance sampling,* a sample of items is randomly selected without replacement, and the entire batch is rejected if there is at least one defect. The Niko Electronics Company has just manufactured 5000 CDs, and 3% are defective. If 10 of the CDs are selected and tested, what is the probability that the entire batch will be rejected?

15. Using Composite Blood Samples When doing blood testing for HIV infections, the procedure can be made more efficient and less expensive by combining samples of blood specimens. If samples from three people are combined and the mixture tests negative, we know that all three individual samples are negative. Find the probability of a positive result for three samples combined into one mixture, assuming the probability of an individual blood sample testing positive is 0.1 (the probability for the "at-risk" population, based on data from the New York State Health Department).

16. Using Composite Water Samples The Home Testing Company tests water quality of homes that are in the process of being sold. To reduce laboratory costs, water samples from five homes are combined for one test, and further testing is done only if the combined sample fails. Based on past records, 4% of the homes fail. Find the probability that a combined sample from five homes will fail.

 Conditional Probabilities.   *In Exercises 17–20, refer to the Titanic mortality data in Table 3-1.*

17. If we randomly select someone who was aboard the *Titanic,* what is the probability of getting a man, given that the selected person died?

18. If we randomly select someone who died, what is the probability of getting a man?

19. What is the probability of getting a boy or girl, given that the randomly selected person is someone who survived?

20. What is the probability of getting a man or woman, given that the randomly selected person is someone who died?

3-5 Beyond the Basics

21. Determining Effect of Assumed Population Size Table 3-2 was constructed by assuming that the population size is 5000. (Recall that the table is based on a 10% rate of the HIV virus for the "at-risk" population, and that the preliminary

Monkey Typists

A classical claim is that a monkey randomly hitting a keyboard would eventually produce the complete works of Shakespeare, assuming that it continues to type century after century. The multiplication rule for probability has been used to find such estimates. One result of 1,000,000,000,000,000,000,000, 000,000,000,000 years is considered by some to be too short. In the same spirit, Sir Arthur Eddington wrote this poem: "There once was a brainy baboon, who always breathed down a bassoon. For he said, 'It appears that in billions of years, I shall certainly hit on a tune.'"

screening test for the HIV virus is currently correct 95% of the time.) Change the assumed population size from 5000 to 15,000.

    a. How are the entries in Table 3-2 affected?

    b. Using this new table based on a population size of 15,000, find $P(\text{HIV}|\text{positive})$, and compare the result to the one found in the example of this section.

    c. Does the assumed size of the population affect the probability values?

22. Using a Two-Way Table The New York State Health Department reports a 0.3% HIV rate for the general population, and under certain conditions, preliminary screening tests for the HIV virus are correct 95% of the time. Assume that the general population consists of 100,000 people.

    a. Construct a table for the general population that is similar to Table 3-2 for the "at-risk" population. (*Hint:* See this section for the procedure used to construct Table 3-2.)

    b. Using the table from part (a), find $P(\text{HIV}|\text{positive})$ for someone randomly selected from the general population. That is, find the probability of randomly selecting someone with HIV, given that this person tested positive.

23. Shared Birthdays Find the probability that of 25 randomly selected people,

    a. no 2 share the same birthday.

    b. at least 2 share the same birthday.

24. Unseen Coins A statistics professor tosses two coins that cannot be seen by any students. One student asks if one of the coins turned up heads. Given that the professor's response is "yes," find the probability that both coins turned up heads.

 ## 3-6 Probabilities Through Simulations

Finding probabilities of events sometimes seems very difficult. Sometimes a probability problem is important for us to solve, but we are not sure that our theoretical solution is correct. The following alternatives are often useful.

1. We can conduct actual trials or collect real data to find probability values or verify that our theoretical results are correct.

2. We can *simulate* the circumstances to learn how the results would occur in reality.

The first alternative is simply an application of the relative frequency approach to finding probabilities, as discussed in Section 3-2. For example, if we want the probability of rolling a pair of dice and getting a total of 7, we could actually roll a real pair of dice 100 times and record the number of 7s that occur. Very often, however, it is not practical to conduct actual trials, so we try instead to develop a *simulation*.

> **DEFINITION**
>
> A **simulation** of a procedure is a process that behaves the same way as the procedure, so that similar results are produced.

▷ **EXAMPLE** **Gender Selection** When testing techniques of gender selection, medical researchers need to know probability values of different outcomes, such as the probability of getting at least 60 girls among 100 children. Assuming that male and female births are equally likely, describe a simulation that results in the genders of 100 newborn babies.

SOLUTION   One approach is simply to flip a coin 100 times, with heads representing females and tails representing males. Another approach is to use a calculator or computer to randomly generate 0s and 1s, with 0 representing a male and 1 representing a female. The numbers must be generated in such a way that they are equally likely.   ◀

▷ **EXAMPLE** **Same Birthdays** Exercise 23 in Section 3-5 refers to the birthday problem, in which we find the probability that in a randomly selected group of 25 people, at least 2 share the same birthday. The theoretical solution is difficult. It isn't practical to survey many different groups of 25 people, so we develop a simulation instead.

SOLUTION   Begin by representing birthdays by integers from 1 through 365, where 1 = January 1, 2 = January 2, . . . , 365 = December 31. Then use a calculator or computer program to generate 25 random numbers, each between 1 and 365. Those numbers can then be sorted, so it becomes easy to survey the list to determine whether any 2 of the simulated birth dates are the same. We can repeat the process as many times as we like, until we are satisfied that we have a good basis for determining the probability. Our estimate of the probability is the number of times we did get at least 2 birth dates that are the same, divided by the total number of groups of 25 that were generated.   ◀

There are several ways of obtaining randomly generated numbers from 1 through 365, including the following.

- **A table of random digits:** Refer, for example, to the *CRC Standard Probability and Statistics Tables and Formulae,* which contains a table of 14,000 digits. (In such a table there are many ways to extract numbers from 1 through 365. One way is by referring to the digits in the first three columns and ignoring 000 as well as anything above 365.)

- **Excel:** Click on the cell in the upper left corner, then click on the function icon **fx.** Select **Math & Trig,** then select **RANDBETWEEN.** In the dialog box, enter 1 for bottom, and enter 365 for top. After getting the random number in the first cell, click and hold down the mouse button to drag the lower right corner of this first cell, and pull it down the column until 25 cells are highlighted. When you release the mouse button, all 25 random numbers should be present. The Excel display shows that the 1st and 3rd numbers are the same.

EXCEL

	A
1	15
2	3
3	15
4	362
5	164
6	184
7	158
8	59
9	143
10	85
11	134

Choosing Personal Security Codes

All of us use personal security codes for ATM machines, computer internet accounts, and home security systems. The safety of such codes depends on the large number of different possibilities, but hackers now have sophisticated tools that can largely overcome that obstacle. Researchers found that by using variations of the user's first and last names along with 1800 other first names, they could identify 10%-20% of the passwords on typical computer systems. When choosing a password, *do not* use a variation of any name, a word found in a dictionary, a password shorter than seven characters, telephone numbers, or social security numbers. Do include nonalphabetic characters, such as digits or punctuation marks.

When we use the term *permutations, arrangements,* or *sequences,* we imply that *order is taken into account* in the sense that different orderings of the same items are counted separately. The letters *ABC* can be arranged six different ways: *ABC, ACB, BAC, BCA, CAB, CBA.* (Later, we will refer to *combinations,* which do not count such arrangements separately.) In the following example, we are asked to find the total number of different sequences that are possible. That suggests use of the permutations rule.

> **EXAMPLE**   Singing legend Frank Sinatra recorded 381 songs. From a list of his top-10 songs, you must select 3 that will be sung in a medley as a tribute at the next MTV Music Awards ceremony. The order of the songs is important so that they fit together well. If you select 3 of Sinatra's top-10 songs, how many different sequences are possible?
>
> **SOLUTION**   We need to select $r = 3$ songs from $n = 10$ songs on Sinatra's top-10 list. The number of different arrangements is found as shown:
>
> $$_nP_r = \frac{n!}{(n-r)!} = \frac{10!}{(10-3)!} = 720$$
>
> There are 720 different possible arrangements of 3 songs selected from 10.   ◄

We sometimes need to find the number of permutations, but some of the items are identical to others. The following variation of the permutations rule applies to such cases.

Permutations Rule (When Some Items Are Identical to Others)

If there are n items with n_1 alike, n_2 alike, . . . , n_k alike, the number of permutations of all n items is

$$\frac{n!}{n_1! \, n_2! \cdots n_k!}$$

> **EXAMPLE**   **Letter Arrangements** The classic examples of the permutations rule are those showing that the letters of the word *Mississippi* can be arranged 34,650 different ways and that the letters of the word *statistics* can be arranged 50,400 ways. We will instead consider the letters *DDDDRRRRR*, which are included in a discussion of the runs test for randomness (Section 13-7). Those letters represent a sequence of diet (*D*) and regular (*R*) colas. How many ways can we arrange the letters *DDDDRRRRR*?

SOLUTION   In the sequence *DDDDRRRRR* we have $n = 9$ items, with $n_1 = 4$ alike and $n_2 = 5$ others that are alike. The number of permutations is computed as follows.

$$\frac{n!}{n_1!\, n_2!} = \frac{9!}{4!\, 5!} = \frac{362{,}880}{2880} = 126$$

In Section 13-7 we need the fact that there are 126 different possible sequences of *DDDDRRRRR,* and we can now see how that result is obtained.   ◄

The preceding example involved *n* items, each belonging to one of two categories. When there are only two categories, we can stipulate that *x* of the items are alike and the other $n - x$ items are alike, so the permutations formula simplifies to

$$\frac{n!}{(n - x)!\, x!}$$

This particular result will be used for the discussion of binomial probabilities, which are introduced in Section 4-3.

When we intend to select *r* items from *n* different items but *do not take order into account,* we are really concerned with possible combinations rather than permutations. That is, **when different orderings of the same items are counted separately, we have a permutation problem, but when different orderings of the same items are not counted separately, we have a combination problem** and may apply the following rule.

Combinations Rule

The number of **combinations** of *r* items selected from *n* different items is

$$_nC_r = \frac{n!}{(n - r)!\, r!}$$

— order is not important

It is very important to recognize that in applying the combinations rule, the following conditions apply:

- We must have a total of *n* different items available.
- We must select *r* of the *n* items (without replacement).
- We must consider rearrangements of the same items to be the same. (The combination *ABC* is the same as *CBA.*)

Because choosing between the permutations rule and the combinations rule can be confusing, we provide the following example, which is intended to emphasize the difference between them.

► **EXAMPLE   Elected Offices**   The Board of Trustees at the author's college has 9 members. Each year, they elect a 3-person committee to

Using Excel for Factorials, Permutations, and Combinations

Factorials: You can use the Excel command FACT. For example, evaluate 3! by entering the Excel command **=FACT(3).** The result will be 6. You can also use the following procedure.

1. Click *f*x, then click on the function category of **Math & Trig,** then click on the function name of **FACT,** then click **OK.**

2. You should now see a dialog box. In the "Number" box, enter the number *n*. For example, if you enter 10 in the number box, the result will be 3628800. That is, 10! = 3,628,800.

Permutations: You can use the Excel command PERMUT. For example, evaluate $_{10}P_3$ by entering the Excel command **=PERMUT(10, 3).** The result will be 720. You can also use the following procedure.

1. Click *f*x, then click on the function category of **Statistical,** then click on the function name of **PERMUT,** then click **OK.**

2. You should now see a dialog box. In the "Number" box, enter the total number *n* of available items. In the "Number_chosen" box, enter the number *r* of items selected. The result is the value of $_nP_r$. For example, if you enter 10 in the number box and 3 in the number chosen box, the result will be 720, indicating that $_{10}P_3 = 720$.

Combinations: You can use the Excel command COMBIN. For example, evaluate $_{10}C_3$ by entering the Excel command **=COMBIN(10, 3).** The result will be 120. You can also use the following procedure.

1. Click *f*x, then click on the function category of **Math & Trig,** then click on the function name of **COMBIN,** then click **OK.**

2. You should now see a dialog box. In the "Number" box, enter the total number *n* of available items. In the "Number_chosen" box, enter the number *r* of items selected. The result is the value of $_nC_r$. For example, if you enter 10 in the number box and 3 in the number chosen box, the result will be 120, indicating that $_{10}C_3 = 120$.

This section presented five different counting devices. When deciding which particular approach applies, we can reduce confusion by using a systematic approach that addresses key issues. The following summary may be helpful.

*   For a sequence of events in which the first can occur *m* ways, the second can occur *n* ways, and so on, use the fundamental counting rule: Multiply $m \cdot n$ and so on.

*   To find the number of arrangements of *n* *different* items, with *all* of them to be used in different arrangements, use the factorial rule: Evaluate *n*!.

*   For *n* *different* items with *some* of them to be used in different arrangements, use the permutations rule: Evaluate

$$_nP_r = \frac{n!}{(n-r)!}$$

- To find the total number of different arrangements when there are n items with n_1 of them *identical* to each other, n_2 of them identical to each other, and so on: Evaluate

$$\frac{n!}{n_1! \, n_2! \cdots n_k!}$$

- To find the number of different combinations when there are n different items and r of them are to be selected, use the combinations rule: Evaluate

$$_nC_r = \frac{n!}{(n-r)! \, r!}$$

3-7   Basic Skills and Concepts

Calculating Factorials, Combinations, Permutations.   *In Exercises 1–4, evaluate the given expressions and express all results using the usual format for writing numbers (instead of scientific notation).*

EXCEL **1.** 14!   EXCEL **2.** $_{20}C_5$ EXCEL **3.** $_{50}P_3$ EXCEL **4.** $_{50}P_4$

Probability of Winning the Lottery.   *This section included an example showing that the probability of winning the New York State lottery is 1/18,009,460. In Exercises 5–9, find the probability of winning the indicated lottery.*

EXCEL **5.** Florida: Select the winning six numbers from 1, 2, . . . , 49.

EXCEL **6.** New York regional lottery: Select the winning six numbers from 1, 2, . . . , 31.

EXCEL **7.** Connecticut: Select the winning six numbers from 1, 2, . . . , 44.

EXCEL **8.** Powerball (12 states): Select the winning five numbers from 1, 2, . . . , 49.

EXCEL **9.** New York Lottery The probability of winning the New York State lottery is 1/18,009,460. What is the probability of winning if the rules are changed so that in addition to selecting the correct six numbers from 1 to 51, you must now select them in the same order as they are drawn?

EXCEL **10.** Number of Security System Codes The author uses an ADT home security system that has a code consisting of four digits (0, 1, . . . , 9) that must be entered in the correct sequence. The digits can be repeated in the code.
   a.   How many different possibilities are there?
   b.   If it takes a burglar five seconds to try a code, how long would it take to try every possibility? Is this time long enough to be a deterrent?

EXCEL **11.** Elected Board of Directors There are 12 members on the board of directors for the Newport General Hospital.
   a.   If they must elect a chairperson, first vice chairperson, second vice chairperson, and secretary, how many different slates of candidates are possible?
   b.   If they must form an ethics subcommittee of four members, how many different subcommittees are possible?

variable names are A, BBB, and M3477K. How many different variable names are possible?

30. Handshakes and Round Tables
   a. Five managers gather for a meeting. If each manager shakes hands with each other manager exactly once, what is the total number of handshakes?
   b. If n managers shake hands with each other exactly once, what is the total number of handshakes?
   c. How many different ways can five managers be seated at a round table? (Assume that if everyone moves to the right, the seating arrangement is the same.)
   d. How many different ways can n managers be seated at a round table?

31. Evaluating Large Factorials Many calculators or computers cannot directly calculate 70! or higher. When n is large, $n!$ can be approximated by $n! = 10^K$, where $K = (n + 0.5) \log n + 0.39908993 - 0.43429448n$.
   a. Evaluate 50! using Excel's **FACT** command and also by using the approximation given here.
   b. The Bureau of Fisheries once asked Bell Laboratories for help finding the shortest route for getting samples from 300 locations in the Gulf of Mexico. If you compute the number of different possible routes, how many digits are used to write that number?

32. Computer Intelligence Can computers "think"? According to the *Turing test*, a computer can be considered to think if, when a person communicates with it, the person believes he or she is communicating with another person instead of a computer. In an experiment at Boston's Computer Museum, each of 10 judges communicated with four computers and four other people and was asked to distinguish between them.
   a. Assume that the first judge cannot distinguish between the four computers and the four people. If this judge makes random guesses, what is the probability of correctly identifying the four computers and the four people?
   b. Assume that all 10 judges cannot distinguish between computers and people, so they make random guesses. Based on the result from part (a), what is the probability that all 10 judges make all correct guesses? (That event would lead us to conclude that computers cannot "think" when, according to the Turing test, they can.)

Men, Women, and Other Planets

BOSTON The possibility of intelligent life on other planets intrigues us all. Interest is fueled by sporadic reports of unidentified flying objects—UFOs. Some people strongly believe that the U.S. Government is hiding direct knowledge of the existence of intelligent life from other planets, and that remains from crashed UFOs are being kept at a secret storage facility in New Mexico. If you're going to store UFO remains, New Mexico is probably as good a place as any. In one survey, registered voters were asked if they believe in the existence of intelligent life on other planets. See the *USA Today* illustration summarizing the results.

Refer to the *USA Today* illustration to answer the following questions.

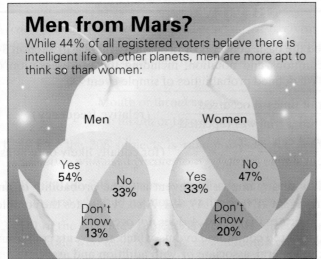

Men from Mars?

While 44% of all registered voters believe there is intelligent life on other planets, men are more apt to think so than women:

Men
Yes 54%
No 33%
Don't know 13%

Women
Yes 33%
No 47%
Don't know 20%

Source: Fox News/Opinion Dynamics poll   By Anne R. Carey and Jerry Mosemak, USA TODAY
USA Today Snapshot. "A look at statistics that shape the nation"

1. What is the probability that a male voter believes in intelligent life on other planets?

2. What is the probability that a female voter believes in intelligent life on other planets?

3. If two different male voters are randomly selected, what is the probability that they both believe in intelligent life on other planets?

4. Assuming that the population of voters includes an equal number of men and women, what is the probability that a randomly selected voter believes in intelligent life on other planets?

5. If a voter is randomly selected and believes in intelligent life on other planets, what is the probability that the selected person is a male?

This chapter deals exclusively with discrete random variables, but the following chapters will deal with continuous random variables.

(a) Discrete Random Variable: Count of the number of movie patrons.

(b) Continuous Random Variable: The measured voltage of a smoke detector battery.

FIGURE 4-2
Devices Used to Count and Measure Discrete and Continuous Random Variables

▷ **EXAMPLES** The following are examples of discrete and continuous random variables.

1. Let x = the number of eggs that a hen lays in a day. This is a *discrete* random variable because its only possible values are 0, or 1, or 2, and so on. No hen can lay 2.343115 eggs, which would have been possible if the data had come from a continuous scale.

2. The count of the number of patrons attending a rock concert is a whole number and is therefore a discrete random variable. The counting device shown in Figure 4-2(a) is capable of indicating only a finite number of values, so it can be used to obtain values for a *discrete* random variable.

3. Let x = the amount of milk a cow produces in one day. This is a *continuous* random variable because it can have any value over a continuous span. During a single day, a cow might yield an amount of milk that can be any value between 0 gallons and 5 gallons. (The author knows almost nothing about actual cow output per day, but 0 gallons to 5 gallons just seems like it's about right.) It would be possible to get 2.343115 gallons, because the cow is not restricted to the discrete amounts of 0, 1, 2, 3, 4, or 5 gallons.

4. The measure of voltage for a smoke-detector battery can be any value between 0 volts and 9 volts. It is therefore a continuous random variable. The voltmeter shown in Figure 4-2(b) is capable of indicating values on a continuous scale, so it can be used to obtain values for a *continuous* random variable. ◁

Graphs

There are various ways to graph a probability distribution, but we will consider only the **probability histogram.** Figure 4-3 is a probability histogram that is very similar to the relative frequency histogram discussed in Chapter 2, but the vertical scale shows *probabilities* instead of relative frequencies based on actual sample results.

In Figure 4-3, note that along the horizontal axis, the values of 0, 1, 2, . . . , 14 are located at the centers of the rectangles. This implies that the rectangles are each 1 unit wide, so the areas of the rectangles are 0.000, 0.001, 0.006, and so on. The areas of these rectangles are the same as the *probabilities* in Table 4-1. We will see in Chapter 5 and future chapters that this correspondence between area and probability is very useful in statistics.

Every probability distribution must satisfy each of the following two requirements.

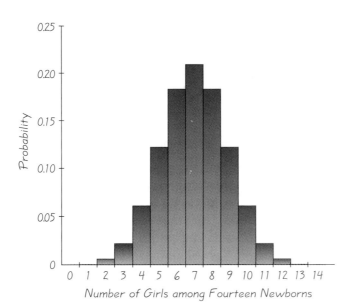

FIGURE 4-3
Probability Histogram for
Number of Girls Among
14 Newborn Babies

| Requirements for a Probability Distribution |

1. $\Sigma P(x) = 1$ where x assumes all possible values
2. $0 \le P(x) \le 1$ for every value of x

The first requirement states that the sum of the probabilities for all the possible values of the random variable must equal 1. This makes sense when we realize that the values of the random variable x represent all possible events in the entire sample space, so we are certain (with probability 1) that one of the events will occur. In Table 4-1, the sum of the probabilities is 0.999; it would be 1 if we eliminated the tiny rounding error by carrying more decimal places. Also the probability rule stating $0 \le P(A) \le 1$ for any event A (given in Section 3-2) implies that $P(x)$ must be between 0 and 1 for any value of x. Again, refer to Table 4-1 and note that each individual value of $P(x)$ does fall between 0 and 1 for any value of x. Because Table 4-1 does satisfy both of the requirements, it is an example of a probability distribution. A probability distribution may be described by a table, such as Table 4-1, or a graph, such as Figure 4-3, or a formula, as in the following two examples.

 EXAMPLE   Does $P(x) = x/5$ (where x can take on the values of 0, 1, 2, 3) describe a probability distribution?

SOLUTION   To be a probability distribution, $P(x)$ must satisfy the preceding two requirements. But

$$\Sigma P(x) = P(0) + P(1) + P(2) + P(3)$$
$$= \frac{0}{5} + \frac{1}{5} + \frac{2}{5} + \frac{3}{5}$$

could continue indefinitely. The uses of expected value (also called *expectation*, or *mathematical expectation*) are extensive and varied, and they play a very important role in an area of application called *decision theory*.

DEFINITION

The **expected value** of a discrete random variable is denoted by E, and it represents the average value of the outcomes. It is obtained by finding the value of $\Sigma[x \cdot P(x)]$.

$$E = \Sigma[x \cdot P(x)]$$

From Formula 4-1 we see that $E = \mu$. That is, the mean of a discrete random variable is the same as its expected value. Repeat the procedure of flipping a coin five times, and the *mean* number of heads is 2.5; when flipping a coin five times, the *expected value* of the number of heads is also 2.5.

> **EXAMPLE** **Lottery Numbers Game** Consider the numbers game started many years ago by organized crime groups and now run legally by many organized governments, as well as some governments that aren't too well organized. Often called a "Pick Three" game, you place a bet that the three-digit number of your choice will be the winning number selected. The typical winning payoff is 499 to 1, meaning that for each winning $1 bet, you would be given $500; your net return is therefore $499. Suppose that you bet $1 on the number 327. What is your expected value of gain or loss?

SOLUTION For this bet there are two simple outcomes: You win or you lose. Because you have selected the number 327 and there are 1000 possibilities (from 000 to 999), your probability of winning is 1/1000 (or 0.001) and your probability of losing is 999/1000 (or 0.999). Table 4-3 summarizes this situation.

From Table 4-3 we can see that when we bet $1 in the numbers game, our expected value is

$$E = \Sigma[x \cdot P(x)] = -50¢$$

This means that in the long run, for each $1 bet, we can expect to lose an average of 50¢. This is not a particularly sound investment scheme. ◄

Picking Lottery Numbers

In a typical state lottery, you select six different numbers. After a random drawing, any entries with the correct combination share in the prize. Since the winning numbers are randomly selected, any choice of six numbers will have the same chance as any other choice, but some combinations are better than others. The combination of 1, 2, 3, 4, 5, 6 is a poor choice because many people tend to select it. In a Florida lottery with a $105 million prize, 52,000 tickets had 1, 2, 3, 4, 5, 6; if that combination had won, the prize would have been only $1000. It's wise to pick combinations not selected by many others. Avoid combinations that form a pattern on the entry card.

TABLE 4-3	The Numbers Game		
Event	x	$P(x)$	$x \cdot P(x)$
Win	$499	0.001	$0.499
Lose	−$1	0.999	−$0.999
Total			−$0.50
			(or −50¢)

In the preceding example, a player will either lose $1 or win $499; there will never be a loss of 50¢, as the expected value of −50¢ might seem to suggest. The expected value of −50¢ is an average over a long run of bets placed. Even if we're thinking of placing only one bet, the expected value of −50¢ is a loss, so this is not a good bet. The potential gain is more than offset by the potential loss.

In this section we learned that a random variable has a numerical value associated with each outcome of some random procedure, and a probability distribution has a probability associated with each value of a random variable. We examined methods for finding the mean, variance, and standard deviation for a probability distribution. We saw that the expected value of a random variable is really the same as the mean. We also learned that lotteries are lousy investments.

4-2 Basic Skills and Concepts

Identifying Discrete and Continuous Random Variables.   *In Exercises 1 and 2, identify the given random variable as being discrete or continuous.*

1. **a.** The weight of the cola in a randomly selected can1.
   b. The cost of a randomly selected can of Coke
   c. The time it takes to fill a can of Pepsi
   d. The amount of cola (in ounces) in a can of Pepsi
   e. The number of cans of Coke on a beverage delivery truck

2. **a.** The cost of a randomly selected textbook
   b. The number of textbooks in a randomly selected bookstore
   c. The weight of a randomly selected textbook
   d. The number of pages in a randomly selected textbook
   e. The time it takes an author to write a textbook (which is a whole lot longer than you might think)

Identifying Probability Distributions.   *In Exercises 3–12, determine whether a probability distribution is given. In those cases where a probability distribution is not described, identify the requirements that are not satisfied. In those cases where a probability distribution is described, find its mean and standard deviation.*

3. Gender Selection In a study of the MicroSort gender-selection method, couples in a control group are not given a treatment, and they each have three children. The probability distribution for the number of girls is given in the accompanying table.

x	$P(x)$
0	0.125
1	0.375
2	0.375
3	0.125

4. TV Viewer Surveys When four different households are surveyed on a Monday night, the random variable x is the number of households with televisions tuned to *Monday Night Football* on ABC (based on data from Nielsen Media Research).

x	$P(x)$
0	0.522
1	0.368
2	0.098
3	0.011
4	0.001

5. Gender and Hiring If your college hires the next four employees without regard to gender, and the pool of applicants is large with an equal number of men and women, then the probability distribution for the number x of women hired is described in the accompanying table.

x	$P(x)$
0	0.0625
1	0.2500
2	0.3750
3	0.2500
4	0.0625

> **EXAMPLE** **Using Directory Assistance** Use the binomial probability formula to find the probability of getting exactly 3 correct responses among 5 different requests from AT&T directory assistance. Assume that in general, AT&T is correct 90% of the time. That is, find $P(3)$ given that $n = 5, x = 3, p = 0.9,$ and $q = 0.1.$

> **SOLUTION** Using the given values of $n, x, p,$ and q in the binomial probability formula (Formula 4-5), we get

$$P(3) = \frac{5!}{(5-3)!3!} \cdot 0.9^3 \cdot 0.1^{5-3}$$

$$= \frac{5!}{2!3!} \cdot 0.729 \cdot 0.01$$

$$= (10)(0.729)(0.01) = 0.0729$$

The probability of getting exactly 3 correct responses for 5 requests is 0.0729.

Calculation hint: When computing a probability with the binomial probability formula, it's helpful to get a single number for $n!/(n-x)!x!$, a single number for p^x, and a single number for q^{n-x}, then simply multiply the three factors together. Don't round too much when you find those three factors; round only at the end.

Method 2: Using Table A-1 in Appendix A In some cases, we can easily find binomial probabilities by simply referring to Table A-1 in Appendix A. First locate n and the corresponding value of x that is desired. At this stage, one row of numbers should be isolated. Now align that row with the proper probability of p by using the column across the top. The isolated number represents the desired probability. A very small probability, such as 0.000000345, is indicated by 0.0+.

Part of Table A-1 is shown in the margin. When $n = 5$ and $p = 0.90$ in a binomial distribution, the probabilities of 0, 1, 2, 3, 4, and 5 successes are $0.0+, 0.0+, 0.008, 0.073, 0.328,$ and $0.590,$ respectively.

> **EXAMPLE** Use the portion of Table A-1 (for $n = 5$ and $p = 0.90$) shown in the margin to find the following.

   a. The probability of exactly 3 successes

   b. The probability of *at least* 3 successes

SOLUTION

   a. The display from Table A-1 shows that when $n = 5$ and $p = 0.90$, the probability of $x = 3$ is given by $P(3) = 0.073$, which is the same value (except for rounding) computed with the binomial probability formula in the preceding example.

   b. "At least" 3 successes means that the number of successes is 3 or 4 or 5.

$$P(\text{at least } 3) = P(3 \text{ or } 4 \text{ or } 5)$$

$$= P(3) + P(4) + P(5)$$

From Table A-1:

n	x	p 0.90
5	0	0.0+
	1	0.0+
	2	0.008
	3	0.073
	4	0.328
	5	0.590

Binomial probability distribution for $n = 5$ and $p = 0.90$

x	$P(x)$
0	0.000
1	0.000
2	0.008
3	0.073
4	0.328
5	0.590

$$= 0.073 + 0.328 + 0.590$$
$$= 0.991$$

In part (b) of the preceding solution, if we wanted to find P(at least 3) by using the binomial probability formula, we would need to apply that formula three times to compute three different probabilities, which would then be added. Given this choice between the formula and the table, it makes sense to use the table. Unfortunately, Table A-1 includes only limited values of n as well as limited values of p, so the table doesn't always work, and we must then find the probabilities by using the binomial probability formula, software, or a calculator, as in the following method.

Method 3: Using Excel for Binomial Probabilities   Excel can be used to compute probabilities for a binomial distribution. You must know the values of n and p. To calculate binomial probabilities using Excel, you can use the command **BINOMDIST**. For example, to find the probability of $x = 3$ successes among $n = 5$ trials with probability of success $p = 0.9$, enter **=BINOMDIST(3, 5, .9, 0)**. The result is 0.0729. The last entry of 0 or "false" indicates that the probability of *exactly* 3 successes should be computed, so we have $P(3) = 0.0729$. A last entry of 1 or "true" would indicate that the *cumulative* probability of 3 or fewer successes should be computed. The command **=BINOMDIST(3, 5, .9, 1)** yields 0.08146 for the probability of 3 or fewer successes. You can also use this procedure for finding binomial probabilities:

1.   List the values of x in column A.
2.   Click on cell B1, then click on *f*x from the toolbar.
3.   Select the function category of **Statistical**.
4.   Select the function name of **BINOMDIST** and click **OK**.
5.   In the dialog box, enter A1 for the number of successes, enter the number of trials, enter the value of the probability p, and enter 0 or "false" for the binomial distribution (or enter 1 or "true" for the *cumulative* binomial distribution). The dialog box should appear as shown below.

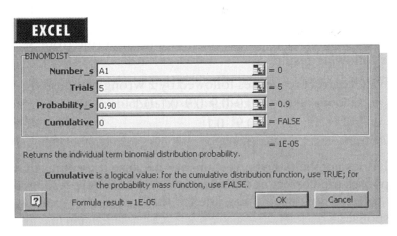

For Binomial Distributions

Formula 4-6	$\mu = n \cdot p$
Formula 4-7	$\sigma^2 = n \cdot p \cdot q$
Formula 4-8	$\sigma = \sqrt{n \cdot p \cdot q}$

EXAMPLE   Gender of Children   In Section 4-2 we included an example illustrating calculations for μ and σ. We used the example of the random variable x representing the number of girls in 14 births. (See Table 4-2 on page 187 for the calculations that illustrate Formulas 4-1 and 4-4.) Use Formulas 4-6 and 4-8 to find the mean and standard deviation for the numbers of girls in groups of 14 births.

SOLUTION   Using the values $n = 14$, $p = 0.5$, and $q = 0.5$, Formulas 4-6 and 4-8 can be applied as follows.

$$\mu = n \cdot p = (14)(0.5) = 7.0$$
$$\sigma = \sqrt{n \cdot p \cdot q}$$
$$= \sqrt{(14) \cdot (0.5) \cdot (0.5)} = 1.9 \quad \text{(rounded)}$$

If you compare these calculations to those required in Table 4-2, it should be obvious that Formulas 4-6 and 4-8 are substantially easier to use.   ◄

Formula 4-6 for the mean makes sense intuitively. If we were to ask any statistics student how many girls are expected in 100 births, the usual response would be 50, which can be easily generalized as $\mu = n \cdot p$. The variance and standard deviation are not so easily justified, and we will omit the complicated algebraic manipulations that lead to Formulas 4-7 and 4-8. Instead, refer again to the preceding example and Table 4-2 to verify that for a binomial distribution, Formulas 4-6, 4-7, and 4-8 will produce the same results as Formulas 4-1, 4-3, and 4-4.

EXAMPLE   Gender Selection   The Chapter Problem involved a preliminary trial with 14 couples who wanted to have baby girls. Although the result of 13 girls in 14 births makes it appear that the MicroSort method of gender selection is effective, we would have much more confidence in that conclusion if the sample size had been considerably larger than 14. Suppose the MicroSort method is used with 100 couples, each of whom will have 1 baby. Assume that the result is 68 girls among the 100 babies.

a.   Assuming that the MicroSort gender-selection method has no effect, find the mean and standard deviation for the numbers of girls in groups of 100 randomly selected babies.

b.   *Interpret* the values from part (a) to determine whether this result (68 girls among 100 babies) supports a claim that the MicroSort method of gender selection is effective.

SOLUTION

a. Assuming that the MicroSort method has no effect and that girls and boys are equally likely, we have $n = 100$, $p = 0.5$, and $q = 0.5$. We can find the mean and standard deviation by using Formulas 4-6 and 4-8 as follows:

$$\mu = n \cdot p = (100)(0.5) = 50$$
$$\sigma = \sqrt{n \cdot p \cdot q} = \sqrt{(100)(0.5)(0.5)} = 5$$

For groups of 100 couples who each have a baby, the mean number of girls is 50 and the standard deviation is 5.

b. We must now interpret the results to determine whether 68 girls among 100 babies is a result that could easily occur by chance, or whether that result is so unlikely that the MicroSort method of gender selection seems to be effective. We will use the range rule of thumb as follows:

$$\text{maximum usual value} = \mu + 2\sigma = 50 + 2(5) = 60$$
$$\text{minimum usual value} = \mu - 2\sigma = 50 - 2(5) = 40$$

INTERPRETATION   According to our range rule of thumb, values are considered to be usual if they are between 40 and 60, so 68 girls is an unusual result. It is very unlikely that we will get 68 girls in 100 births just by chance. If we did get 68 girls in 100 births, we should look for an explanation that is an alternative to chance. If the 100 couples used the MicroSort method of gender selection, it would appear to be effective in increasing the likelihood that a child will be a girl. ◄

You should develop the skills to calculate means and standard deviations using Formulas 4-6 and 4-8, but it is especially important to learn to *interpret* results by using those values. The range rule of thumb, as illustrated in part (b) of the preceding example, suggests that values are unusual if they lie outside of these limits:

$$\textbf{maximum usual value} = \boldsymbol{\mu + 2\sigma}$$

$$\textbf{minimum usual value} = \boldsymbol{\mu - 2\sigma}$$

4-4 Basic Skills and Concepts

Finding μ, σ, and Unusual Values.   *In Exercises 1–4, assume that a procedure yields a binomial distribution with n trials and the probability of success for one trial is p. Use the given values of n and p to find the mean μ and standard deviation σ. Also, use the range rule of thumb to find the minimum usual value $\mu - 2\sigma$ and the maximum usual value $\mu + 2\sigma$.*

1. $n = 100$, $p = 0.25$ **2.** $n = 1068$, $p = 0.88$

3. $n = 237$, $p = 2/3$ **4.** $n = 2500$, $p = 3/5$

5. Guessing Answers Several students are unprepared for a true/false test with 20 questions, and all of their answers are guesses.   (*continued*)

States Rig Lottery Selections

Many states run a lottery in which players select four digits, such as 1127 (the author's birthday). If a player pays $1 and selects the winning sequence in the correct order, a prize of $5000 is won. States monitor the number selections and, if one particular sequence is selected too often, players are prohibited from placing any more bets on it. The lottery machines are rigged so that once a popular sequence reaches a certain level of sales, the machine will no longer accept that particular sequence. This prevents states from paying out more than they take in. Critics say that this practice is unfair. According to William Thompson, a gambling expert at the University of Nevada in Las Vegas, "they're saying that they (the states) want to be in the gambling business, but they don't want to be gamblers. It just makes a sham out of the whole numbers game."

    a. Find the mean and standard deviation for the number of correct answers for such students.

    b. Would it be unusual for a student to pass by guessing and getting at least 12 correct answers? Why or why not?

6. **Guessing Answers** Several students are unprepared for a multiple-choice quiz with 20 questions, and all of their answers are guesses. Each question has five possible answers, and only one of them is correct.
    a. Find the mean and standard deviation for the number of correct answers for such students.
    b. Would it be unusual for a student to pass by guessing and getting at least 12 correct answers? Why or why not?

7. **Playing Roulette** If you bet on any single number in roulette, your probability of winning is 1/38. Assume that you bet on a single number in each of 100 consecutive spins.
    a. Find the mean and standard deviation for the number of wins.
    b. Would it be unusual to not win once in the 100 trials? Why or why not?

8. **Left-Handed People** Ten percent of American adults are left-handed. A statistics class has 25 students in attendance.
    a. Find the mean and standard deviation for the number of left-handed students in such classes of 25 students.
    b. Would it be unusual to survey a class of 25 students and find that 5 of them are left-handed? Why or why not?

9. **High School Graduates** The Census Bureau reports that 82% of Americans over the age of 25 are high school graduates. A survey of randomly selected Dutchess County residents included 1250 who were over the age of 25, and 1107 of them were high school graduates.
    a. Find the mean and standard deviation for the number of high school graduates in groups of 1250 Americans over the age of 25.
    b. Is the Dutchess County result of 1107 unusually high? Why or why not? How might such a result be explained?

10. **College Student Housing** In planning for enrollment, which is expected to grow by 200 students, the College of Newport has found that 9% of college students live in their own off-campus housing (based on data from the Independent Insurance Agents of America).
    a. For randomly selected groups of 200 college students, find the mean and standard deviation for the numbers who live in their own off-campus housing.
    b. Would it be unusual to find that 50 of the 200 new students live in their own off-campus housing? What is a likely cause of a number that is unusually high?

11. **Determining the Effectiveness of an HIV Training Program** The New York State Health Department reports a 10% rate of the HIV virus for the "at-risk" population. In one region, an intensive education program is used in an attempt to lower that 10% rate. After running the program, a follow-up study of 150 at-risk individuals is conducted.
    a. Assuming that the program has no effect, find the mean and standard deviation for the number of HIV cases in groups of 150 at-risk people.
    b. Among the 150 people in the follow-up study, 8% (or 12 people) tested positive for the HIV virus. If the program has no effect, is that rate unusually low? Does this result suggest that the program is effective?

12. Deciphering Messages The Central Intelligence Agency has specialists who analyze the frequencies of letters of the alphabet in an attempt to decipher intercepted messages. In standard English text, the letter r is used at a rate of 7.7%.
   a. Find the mean and standard deviation for the number of times the letter r will be found on a typical page of 2600 characters.
   b. In an intercepted message sent to Iraq, a page of 2600 characters is found to have the letter r occurring 175 times. Is this unusual?

13. Determining Whether Complaints Are Lower After a Training Program The Newtower Department Store has experienced a 3.2% rate of customer complaints and attempts to lower this rate with an employee training program. After the program, 850 customers are tracked and it is found that only 7 of them filed complaints.
   a. Assuming that the training program has no effect, find the mean and standard deviation for the number of complaints in such groups of 850 customers.
   b. Based on the results from part (a), is the result of 7 complaints unusual? Does it seem that the training program was effective in lowering the rate of complaints?

14. Are 10% of M&M Candies Blue? Mars, Inc., claims that 10% of its M&M plain candies are blue, and a sample of 100 such candies is randomly selected.
   a. Find the mean and standard deviation for the number of blue candies in such groups of 100.
   b. Data Set 10 in Appendix B consists of a random sample of 100 M&Ms in which only 5 are blue. Is this result unusual? Does it seem that the claimed rate of 10% is wrong?

15. Are TV Commercials Longer? Are television commercials getting longer, or does it just seem that way? In 1990, 6% of those commercials were one minute long. Recently, 50 commercials were randomly selected, and 16 of them were one minute long.
   a. If one-minute commercials continue to be 6% of the total, find the mean and standard deviation for the numbers of one-minute commercials that would be found in groups of 50.
   b. Is the recent result of 16 one-minute commercials unusually high, assuming that the same 6% rate applies? Does the recent result suggest that one-minute commercials are now more than 6% of the total?

16. Computer Warranty Repairs The Providence Computer Supply Company knows that 16% of its computers will require warranty repairs within one month of shipment. In a typical month, 279 computers are shipped.
   a. If x is the random variable representing the number of computers requiring warranty repairs among the 279 sold in one month, find its mean and standard deviation.
   b. For a typical month in which 279 computers are sold, what would be an unusually low figure for the number of computers requiring warranty repair within one month? What would be an unusually high figure? (These values are helpful in determining the number of service technicians the company will need to employ.)

4-4 Beyond the Basics

17. Using the Empirical Rule and Chebyshev's Theorem An experiment is designed to test the effectiveness of the MicroSort method of gender selection, and 100 couples try to have baby girls using the MicroSort method. In an example included in this section, the range rule of thumb was used to conclude that among 100 births, the number of girls should usually fall between 40 and 60.

a. The empirical rule (see Section 2-5) applies to distributions that are bell-shaped. Is the binomial probability distribution for this experiment (approximately) bell-shaped? How do you know?

b. Assuming that the distribution is bell-shaped, how likely is it that the number of girls will fall between 40 and 60 (according to the empirical rule)?

c. Assuming that the distribution is bell-shaped, how likely is it that the number of girls will fall between 35 and 65 (according to the empirical rule)?

d. Using Chebyshev's Theorem, what do we conclude about the likelihood that the number of girls will fall between 40 and 60?

18. Analyzing Results of Experiment in Gender Selection An experiment involving a gender-selection method includes a control group of 15 couples who are not given any treatment intended to influence the genders of the babies.

a. Construct a table listing the possible values of the random variable x (which represents the number of girls among the 15 births) and the corresponding probabilities.

b. Find the mean and standard deviation for the numbers of girls in such groups of 15.

c. Find the maximum and minimum usual values for the number of girls. (Use the range rule of thumb.)

d. Using the probabilities found in part (a), find the probability that the number of girls will be outside of the range of usual values found in part (c). That is, find the probability that the number of girls is less than the minimum usual value or greater than the maximum usual value. This result is the specific probability of an unusual outcome for this experiment.

19. Acceptable/Defective Products Mario's Pizza Parlor has just opened. Due to a lack of employee training, there is only a 0.8 probability that a pizza will be edible. An order for five pizzas has just been placed. What is the minimum number of pizzas that must be made in order to be at least 99% sure that there will be five that are edible?

 4-5 The Poisson Distribution

This section introduces the *Poisson distribution*. It is particularly important because it is often used as a mathematical model describing arrivals of people in a line, planes arriving at an airport, cars pulling into a gas station, diners arriving at a restaurant, students arriving at a bookstore line, and Internet users logging onto a Web site. We can use the Poisson distribution to find probabilities. For example, suppose your professor has an office hour scheduled every Monday at 11:00 and she finds that during that office hour, the mean number of students who come is 2.3. We can find the probability that for a randomly selected office hour on Monday at 11:00, exactly four students come. We use the Poisson distribution, defined as follows.

DEFINITION

The **Poisson distribution** is a discrete probability distribution that applies to occurrences of some event *over a specified interval*. The random variable x is the number of occurrences of the event in an interval. The interval can be time, distance, area, volume, or some similar unit. The probability of the event occurring x times over an interval is given by Formula 4-9.

Formula 4-9 $P(x) = \dfrac{\mu^x \cdot e^{-\mu}}{x!}$ where $e \approx 2.71828$

The Poisson distribution has the following requirements:

- The random variable x is the number of occurrences of an event *over some interval*.
- The occurrences must be *random*.
- The occurrences must be *independent* of each other.
- The occurrences must be *uniformly distributed* over the interval being used.

The Poisson distribution has these parameters:

- The mean is μ.
- The standard deviation is $\sigma = \sqrt{\mu}$.

A Poisson distribution differs from a binomial distribution in these fundamental ways:

1. The binomial distribution is affected by the sample size n and the probability p, whereas the Poisson distribution is affected only by the mean μ.

2. In a binomial distribution, the possible values of the random variable x are $0, 1, \ldots, n$, but a Poisson distribution has possible x values of $0, 1, 2, \ldots$, with no upper limit.

> **EXAMPLE   World War II Bombs** In analyzing hits by V-1 buzz bombs in World War II, South London was subdivided into 576 regions, each with an area of 0.25 km^2. A total of 535 bombs hit the combined area of 576 regions. If a region is randomly selected, find the probability that it was hit exactly twice.

SOLUTION   The Poisson distribution applies because we are dealing with the occurrences of an event (bomb hits) over some interval (a region with area of 0.25 km^2). The mean number of hits per region is

$$\mu = \frac{\text{number of bomb hits}}{\text{number of regions}} = \frac{535}{576} = 0.929$$

Because we want the probability of exactly two hits in a region, we let $x = 2$ and use Formula 4-9 as follows.

Probability of an Event That Has Never Occurred

Some events are possible, but are so unlikely that they have never occurred. Here is one such problem of great interest to political scientists: Estimate the probability that your single vote will determine the winner in a U.S. presidential election. Andrew Gelman, Gary King, and John Boscardin write in the *Journal of the American Statistical Association* (Vol. 93, No. 441) that "the exact value of this probability is of only minor interest, but the number has important implications for understanding the optimal allocation of campaign resources, whether states and voter groups receive their fair share of attention from prospective presidents, and how formal 'rational choice' models of voter behavior might be able to explain why people vote at all." The authors show how the probability value of 1 in 10 million is obtained for close elections.

f. Is it unusual to randomly select four homes and find that none of them have smoke detectors? Why or why not?

2. TV Ratings The television show *ER* has a 34 share, meaning that while it is being broadcast, 34% of the TV sets in use are tuned to *ER* (based on data from Nielsen Media Research). Assume that during a broadcast of *ER*, 20 TV sets are randomly selected from those that are in use.

   a. What is the expected number of sets tuned to *ER?*
   b. In such groups of 20, what is the mean number of sets tuned to *ER?*
   c. In such groups of 20, what is the standard deviation for the number of sets tuned to *ER?*
   d. For such a group of 20, find the probability that exactly 5 TV sets are tuned to *ER.*
   e. For such a group of 20, would it be unusual to find that 12 sets are tuned to *ER?* Why or why not?

3. Employee Drug Testing Among companies doing highway or bridge construction, 80% test employees for substance abuse (based on data from the Construction Financial Management Association). A study involves the random selection of 10 such companies.

   a. Find the probability that 5 of the 10 companies test for substance abuse.
   b. Find the probability that at least half of the companies test for substance abuse.
   c. For such groups of 10 companies, find the mean and standard deviation for the number (among 10) that test for substance abuse.
   d. Would it be unusual to find that 6 of 10 companies test for substance abuse? Why or why not?

4. Reasons for Being Fired Inability to get along with others is the reason cited in 17% of worker firings (based on data from Robert Half International, Inc.). Concerned about her company's working conditions, the personnel manager at the Boston Finance Company plans to investigate the five employee firings that occurred over the past year.

   a. Assuming that the 17% rate applies, find the probability that among those five employees, the number fired because of an inability to get along with others is at least four.
   b. If the personnel manager actually does find that at least four of the firings are due to an inability to get along with others, does this company appear to be very different from other typical companies? Why or why not?

5. Planning for a New Hospital A new hospital is being considered for the community of Newport, which does not yet have its own hospital. There is a need to know how many babies are born per day. Records show that Newport had 212 births last year.

   a. Find the average number of births per day.
   b. Find the probability of exactly two births on a given day.
   c. Find the probability of exactly three births on a given day. Is this value high enough to warrant the construction of a facility that can accommodate three births on the same day?

CUMULATIVE REVIEW EXERCISES

1. Home-Run Distances: Analysis of Last Digits The accompanying table lists the last digits of the 70 reported distances (in feet) of the 70 home runs hit by Mark McGwire in 1998 when he broke Roger Maris's home-run record, which had

x	f
0	55
1	2
2	1
3	1
4	0
5	3
6	0
7	2
8	4
9	2

lasted since 1961 (based on data from *USA Today*). The last digits of a data set can sometimes be used to determine whether the data have been measured or simply reported. The presence of disproportionately more 0s and 5s is often a sure indicator that the data have been reported instead of measured.

a. Find the mean and standard deviation of those last digits.

b. Construct the relative frequency table that corresponds to the given frequency table.

c. Construct a table for the probability distribution of randomly selected digits that are all equally likely. List the values of the random variable x $(0, 1, 2, \ldots,$ 9) along with their corresponding probabilities $(0.1, 0.1, 0.1, \ldots, 0.1)$, then find the mean and standard deviation of this probability distribution.

d. Recognizing that sample data naturally deviate from the results we theoretically expect, does it seem that the given last digits roughly agree with the distribution we expect with random selection? Or does it seem that there is something about the sample data (such as disproportionately more 0s and 5s) suggesting that the given last digits are not random? (In Chapter 10, we will present a method for answering such questions much more objectively.)

2. Proving That Dice Are Loaded A casino cheat is caught trying to use a pair of loaded dice. At his court trial, physical evidence reveals that some of the black dots were drilled, filled with lead, then repainted to appear normal. In addition to the physical evidence, the dice are rolled in court with these results:

```
12   8   9   12   12   9    8    7    12   10
12   3   2   12   10   9   12   11   11   12
```

A probability expert testifies that when fair dice are rolled, the mean should be 7.0 and the standard deviation should be 2.4.

a. Find the mean and standard deviation of the sample values obtained in court.

b. Based on the outcomes obtained in court, what is the probability of rolling a 12? How does this result compare to the probability of 1/36 (or 0.0278) for fair dice?

c. If the probability of rolling a 12 with fair dice is 1/36, find the probability of getting at least one 12 when fair dice are rolled 20 times.

d. If you are the defense attorney, how would you refute the results obtained in court?

COOPERATIVE GROUP ACTIVITIES

1. *In-class activity:* In Chapter 1 we gave several examples of misleading data sets. Suppose we want to identify the probability distribution for the number of children born to randomly selected couples. For each student in the class, find the number of brothers and sisters and record the total number of children (including the student) in each family. Construct the relative frequency table for the result obtained. (The values of the random variable x will be $1, 2, 3, \ldots$.) What is wrong with using this relative frequency table as an estimate of the probability distribution for the number of children born to randomly selected couples?

2. *In-class activity:* Divide into groups of three. Select one person who will be tested for extrasensory perception (ESP) by trying to correctly identify a digit randomly selected by another member of the group. Another group member should record the randomly selected digit, the digit guessed by the subject, and whether the guess was correct or wrong. Construct the table for the probability distribution of randomly generated digits, construct the relative frequency table for the random digits that were actually obtained, and construct a relative frequency table for the guesses. After comparing the three tables, what do you conclude? What proportion of guesses were correct? Does it seem that the subject has the ability to select the correct digit significantly more often than would be expected by chance?

EXCEL PROJECT

American Air Flight 2705 from New York to Boston has seats for 120 passengers. An average of 5% of people with reservations don't show up, so American Air overbooks by accepting 126 reservations for the 120 seats. We can analyze this system by using a binomial distribution with $n = 126$ and $p = 0.95$ (the probability that someone with a reservation does show up).

Find the probability that when 126 reservations are accepted for a particular flight, there are more passengers than seats. That is, find the probability of at least 121 people showing up with reservations, assuming that 126 reservations were accepted. Because of the value of n, Table A-1 cannot be used, and calculations with the binomial probability formula would be extremely time-consuming and painfully tedious. The best approach is to use Excel (Method 3 in Section 4-3).

Critical Thinking:
Is a transatlantic flight safe with two engines?

Statistics is at its best when it is used to benefit humanity in some way. Companies use statistics to become more efficient, increase shareholders' profits, and lower prices. Regulatory agencies use statistics to ensure the safety of workers and clients. This project involves a situation in which cost effectiveness and passenger safety are both critical factors. With new aircraft designs and improved engine reliability, airline companies wanted to fly transatlantic routes with twin-engine jets, but the Federal Aviation Administration required at least three engines for transatlantic flights. Lowering this requirement was of great interest to manufacturers of twin-engine jets (such as the Boeing 767). Also, the two-engine jets use about half the fuel of jets with three or four engines. Obviously, the key issue in approving the lowered requirement is the probability of a twin-engine jet making a safe transatlantic crossing. This probability should be compared to that of three- and four-engine jets. Such a study should involve a thorough understanding of the related probabilities.

Analyzing the results
A realistic estimate for the probability of an engine failing on a transatlantic flight is 1/14,000. Use this probability and the binomial probability formula to find the probabilities of 0, 1, 2, and 3 engine failures for a three-

engine jet and the probabilities of 0, 1, and 2 engine failures for a two-engine jet. Because of the numbers involved, carry all results to as many decimal places as your calculator will allow. Summarize your results by entering the probabilities in the two accompanying tables.

Use the results from the tables and assume that a flight will be completed if at least one engine works. Find the probability of a safe flight with a three-engine jet, and find the probability of a safe flight with a two-engine jet. Write a report for the Federal Aviation Administration that outlines the key issue, and include a recommendation. Support your recommendation with specific results.

Three Engines		Two Engines	
x	$P(x)$	x	$P(x)$
0	?	0	?
1	?	1	?
2	?	2	?
3	?		

Internet Project

Probability Distributions
and Simulation

Probability distributions are used to predict the outcome of the events they model. For example, if we toss a fair coin, the distribution for the outcome is a probability of 1/2 for heads and 1/2 for tails. If we toss the coin ten consecutive times, we expect five heads and five tails. We might not get this exact result, but in the long run, over hundreds or thousands of tosses, we expect the split between heads and tails to be very close to "50-50". Go to the site

http://www.awlonline.com/triola

and proceed to the Internet Project for Chapter 4 where you will find two explorations. In the first exploration you are asked to develop a probability distribution for a simple experiment, and use that distribution to predict the outcome of repeated trial runs of the experiment. In the second exploration, we will analyze a more complicated situation: the paths of rolling marbles as they move in pinball-like fashion through a set of obstacles. In each case, a dynamic visual simulation will allow you to compare the predicted results with a set of experimental outcomes.

Statistics _at work_

"Our program is really an education program, but it has wide recognition because the results are released publicly"

Barbara Carvalho
Director of the Marist College Poll

Lee Miringoff
Director of the Marist College Institute for Public Opinion

Barbara Carvalho and Lee Miringoff report on their poll results in many interviews for print and electronic media, including news programs for NBC, CBS, ABC, FOX, and public television. Lee Miringoff appears regularly on NBC's "Today" show.

What do you do?

We do public polling. We survey public issues, approval ratings of public officials in New York City, New York State, and nationwide. We don't do partisan polling for political parties, political candidates, or lobby groups. We are independently funded by Marist College and we have no outside funding that in any way might suggest that we are doing research for any particular group on any one issue. Reporters have come to depend on our results for its accuracy and professionalism, but also because they know that it is independent and is not commissioned by any one media source, as many polls are.

How do you select survey respondents?

For a statewide survey we select respondents in proportion to county voter registrations. Different counties have different refusal rates and if we were to select people at random throughout the state, we would get an uneven model of what the state looks like. We stratify by county and use random digit dialing so that we get listed and unlisted numbers.

You mentioned refusal rates. Are they a real problem?

One of the issues that we deal with extensively is the issue of people who don't respond to surveys. That has been increasing over time and there has been much attention from the survey research community. As a research center we do quite well when compared to others. But when you do face-to-face interviews and have refusal rates of 25% to 50%, there's a real concern to find out who is refusing and why they are not responding, and the impact that has on the representativeness of the studies that we're doing.

Is the political process actually influenced by poll results?

Although most polls that people see are public polls, the reality is that the political process is influenced by private polls that the public never sees. No one runs for high office today without using a private poll.

How do you use statistics in your work?

Although we use statistics in a variety of ways, there are two significant areas which are dependent on statistical concepts. First, and perhaps most central, is the concept of probability. The main idea behind doing polls is that by interviewing a few individuals one may generalize about a larger group. It is the method of sampling—determining who is selected to participate in a survey— that allows us to generalize from the opinions of a few to the larger group. Probability theory guides this task. Second, when analyzing survey results, it is essential to know if the differences we observe in our results or among groups are in fact real or simply owing to chance. Statistics provides the means by which we may address this issue.

What concepts of statistics do you use?

Statistics comes into play in our sampling, even before we get to data analysis. We use statistics to determine our sample size and to develop an estimate of what would be statistically significant. In the data analysis, most of our studies use basic descriptive statistics. Some of the academic studies get into regression analysis.

Would you recommend a statistics course for students?

Absolutely. All numbers are not created equally. Regardless of your field of study or career interests, an ability to critically evaluate research information that is presented to you, to use data to improve services, or to interpret results to develop strategies is a very valuable asset. Surveys, in particular, are everywhere. It is vital that as workers, managers, and citizens we are able to evaluate their accuracy and worth. Numbers are powerful. They are capable of providing support for your ideas or being used against you to undermine your position. Those who can use and understand them will have an advantage over those who can only take them at face value. The study of statistics is important for understanding one aspect of knowledge, and it's a key to opening up other avenues of pursuit. Statistics cuts across disciplines. Students will inevitably find it in their careers at some point.

Do you have any other recommendations for students?

It is important for students to take every opportunity to develop their communication and presentation skills. Sharpen not only your ability to speak and write, but also raise your comfort level with new technologies.

Normal Probability Distributions

Redesigning Ejection Seats

Ergonomics is the study of problems associated with people adjusting to their environments. The environments are varied and include manufacturing assembly lines, computer stations, cars, elevators, theater seats, and aircraft cockpits. Good ergonomic design results in an environment that is safe, functional, efficient, and comfortable.

Changing times often require ergonomic changes. When visiting homes that are hundreds of years old, you may have noticed that the door openings seem too low. The doorway heights that were comfortable for people a few hundred years ago are not so comfortable for the taller people of today. The United States Air Force recently encountered an ergonomic problem created by its recognition and acceptance of the fact that women make perfectly good pilots of fighter jets. Cockpits of fighter jets were originally designed for men only, so various cockpit changes were required to better accommodate the new women pilots. As this text was being written, the United States Air Force was redesigning its ACES-II ejection seat currently used in fighter jets. The ACES-II model was originally designed for men who weighed between 140 and 211 pounds. Based on data from the National Health and Examination Survey, weights of women have a distribution that is roughly bell-shaped with a mean of 143 pounds and a standard deviation of 29 pounds. Any women pilots weighing less than 140 pounds or more than 211 pounds would have a greater chance of injury if it became necessary to eject. What percent of women weigh between 140 and 211 pounds, so that they are within the design limits of the ACES-II ejection seats? Such questions are important for the ejection seat redesign, and we will answer such questions as we present the important concepts of this chapter.

5-1 Overview

Recall that in Chapter 4, we presented the following concepts:

- A *random variable* is a variable having a single numerical value, determined by chance, for each outcome of some procedure.
- A *probability distribution* describes the probability for each value of the random variable.
- A *discrete* random variable has either a finite number of values or a countable number of values. That is, the *number* of possible values that *x* can assume is 0, or 1, or 2, and so on.
- A *continuous* random variable has infinitely many values, and those values are often associated with measurements on a continuous scale with no gaps or interruptions.

In Chapter 4 we considered only *discrete* probability distributions, but in this chapter we present *continuous* probability distributions. Although we begin with a uniform distribution, most of the chapter will focus on *normal distributions*. Normal distributions are extremely important because they occur so often in real applications.

Curve is bell-shaped and symmetric

FIGURE 5-1
The Normal Distribution

> ### DEFINITION
>
> If a continuous random variable has a distribution with a graph that is symmetric and bell-shaped, as in Figure 5-1, we say that it has a **normal distribution.** The normal distribution can be described by the equation given as Formula 5-1.

Formula 5-1

$$y = \frac{e^{-\frac{1}{2}\left(\frac{x-\mu}{\sigma}\right)^2}}{\sigma\sqrt{2\pi}}$$

Don't be discouraged by the complexity of Formula 5-1, because it isn't really necessary for us to actually use it. What it shows is that any particular normal distribution is determined by two parameters: the mean, μ, and standard deviation, σ. Once specific values are selected for μ and σ, we can graph Formula 5-1 as we would graph any equation relating x and y; the result is a continuous probability distribution with a bell shape. This is arguably the single most important distribution in statistics, and we will use it often throughout the remainder of the book.

 ## 5-2 The Standard Normal Distribution

The focus of this chapter is the concept of a normal probability distribution, but we begin with a *uniform distribution*. The uniform distribution makes it easier for us to see some very important properties, which will be used with normal distributions.

DEFINITION

A continuous random variable has a **uniform distribution** if its values spread evenly over the range of possibilities. The graph of a uniform distribution results in a rectangular shape.

▷ **EXAMPLE   Fire Drills** A college dean must select a starting time for a required fire drill. She can select any time between 10:00 A.M. and 3:00 P.M., so we will represent that 5-hour period by the random variable *x*, which can be any value between 0 and 5. If she selects the time in such a way that every possible time is equally likely, then *x* is a continuous random variable with a uniform distribution. See Figure 5-2.   ◁

When we discussed *discrete* probability distributions in Section 4-2, we identified two requirements: (1) $\sum P(x) = 1$, and (2) $0 \le P(x) \le 1$ for all values of *x*. Also in Section 4-2, we stated that the graph of a discrete probability distribution is called a *probability histogram*. The graph of a continuous probability distribution, such as that of Figure 5-2, is called a *density curve,* and it must satisfy two properties similar to the requirements for discrete probability distributions, as listed in the following definition.

DEFINITION

A **density curve** (or **probability density function**) is a graph of a continuous probability distribution. It must satisfy the following properties:

1. The total area under the curve must equal 1.
2. Every point on the curve must have a vertical height that is 0 or greater.

By setting the height of the rectangle in Figure 5-2 to be 0.2, we force the enclosed area to be $5 \times 0.2 = 1$, as required. This property (area = 1) makes it very easy to solve probability problems, so the following statement is important:

Reliability and Validity

The reliability of data refers to the consistency with which results occur, whereas the validity of data refers to how well the data measure what they are supposed to measure. The reliability of an IQ test can be judged by comparing scores for the test given on one date to scores for the same test given at another time. To test the validity of an IQ test, we might compare the test scores to another indicator of intelligence, such as academic performance. Many critics charge that IQ tests are reliable, but not valid; they provide consistent results, but don't really measure intelligence.

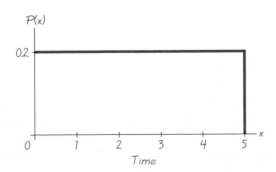

FIGURE 5-2
Uniform Distribution of Times

Because the total area under the density curve is equal to 1, there is a correspondence between area and probability.

> **EXAMPLE**   **Fire Drills** A college dean must select a starting time between 0 and 5 hours, and the selection is made in such a way that all possible times are equally likely. If she randomly selects a starting time, what is the probability that it is during the first half hour or the last half hour?
>
> **SOLUTION**   See Figure 5-3, where we shade the regions representing times that are in the first half hour or the last half hour. Because there is a correspondence between area and probability, we can find the probability by using areas as follows:
>
> $$P(\text{first or last half hour}) = P(\text{first half hour}) + P(\text{last half hour})$$
> $$= 0.1 + 0.1 = 0.2$$
>
> **INTERPRETATION**   The probability of randomly selecting a starting time during the first half hour or last half hour is 0.2.

FIGURE 5-3
Times in First or Last Half Hour

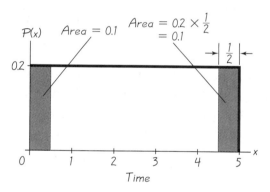

The density curve of a uniform distribution is a horizontal line, so it's easy to find the area of any underlying region by multiplying width and height. The density curve of a normal distribution has the more complicated bell shape shown in Figure 5-1, so it's more difficult to find areas, but the basic principle is the same: There is a correspondence between area and probability.

Just as there are many different uniform distributions (with different ranges of values), there are also many different normal distributions, with each one depending on two parameters: the population mean, μ, and the population standard deviation, σ. (Recall from Chapter 1 that a *parameter* is a numerical measurement describing some characteristic of a *population*.) Figure 5-4 shows density curves for heights of adult women and men. Because men have a larger mean height, the peak of the density curve for men is farther to the right. Because men's heights have a slightly larger standard deviation, the density curve for men is slightly wider. Figure 5-4 shows two different possible normal distributions. There are infinitely many other possibilities, but one is of special interest.

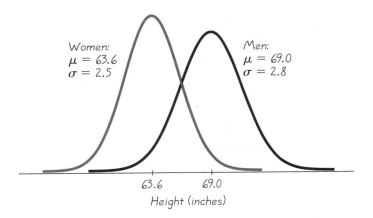

FIGURE 5-4
Heights of Adult Women and Men

> **DEFINITION**
>
> The **standard normal distribution** is a normal probability distribution that has a mean of 0 and a standard deviation of 1. (See Figure 5-5.)

Suppose that somehow we were forced to perform calculations using Formula 5-1. We would quickly see that the most workable values for μ and σ are $\mu = 0$ and $\sigma = 1$. By letting $\mu = 0$ and $\sigma = 1$, mathematicians have calculated areas under the curve. As shown in Figure 5-5, the area under the curve bounded by the mean of 0 and the value of 1 is 0.3413. Remember, the total area under the curve is always 1; this allows us to make the correspondence between area and probability, as we did in the preceding example with the uniform distribution.

Finding Probabilities When Given z Scores

Figure 5-5 shows that the area bounded by the curve, the horizontal axis, and the values of 0 and 1 is an area of 0.3413. Although the figure shows only one area, we can find areas (or probabilities) for many different regions. Such areas can be found by using Table A-2 (in Appendix A and the *Formulas and Tables* insert card), or software such as Excel. The key features of the different methods are summarized in Table 5-1.

FIGURE 5-5
Standard Normal Distribution, with Mean $\mu = 0$ and Standard Deviation $\sigma = 1$

TABLE 5-1	Normal Distribution Areas and Values

Table A-2

Gives area bounded on the left by a vertical line above the mean of 0 and bounded on the right by a vertical line above any specific positive score denoted by z.

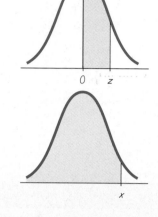

The procedures for using Table A-2 are described in the text.

Excel

Commands refer to the cumulative area from the left up to a vertical line above a specific value.

Standard normal distribution ($\mu = 0$ and $\sigma = 1$):

Given z score, find area to left   = **NORMSDIST (z)**

Given area p to left, find z score   = **NORMSINV (p)**

Nonstandard normal distribution (any mean, any standard deviation):

Given x value, find area to left   = **NORMDIST (x, μ, σ, 1)**

Given area p to left, find x value   = **NORMINV (p, μ, σ)**

If you are using Table A-2, it is essential to understand the following points.

1. Table A-2 is designed only for the *standard* normal distribution, which has a mean of 0 and a standard deviation of 1.

2. Each value in the body of the table is an area under the curve bounded on the left by a vertical line above the mean of 0 and bounded on the right by a vertical line above a specific positive score denoted by z, as illustrated in Figure 5-6.

3. When working with a graph, avoid confusing z scores and areas.

   z score: *Distance* along the horizontal scale of the standard normal distribution; refer to the leftmost column and top row of Table A-2.

FIGURE 5-6
The Standard Normal Distribution

The area of the shaded region bounded by the mean of 0 and the positive number z can be found in Table A-2.

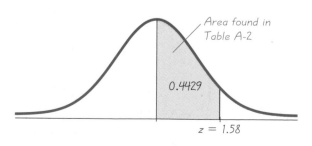

Area: *Region* **under the curve; refer to the values in the body of Table A-2.**

4. The part of the z score denoting hundredths is found across the top row of Table A-2.

The following example requires that we find the probability associated with a value between 0 and 1.58. Begin with the z score of 1.58 by locating 1.5 in the left column; next find the value in the adjoining row of probabilities that is directly below 0.08, as shown in this excerpt from Table A-2.

z	....	0.08
.		.
.		.
.		.
1.5	....	0.4429

The area (or probability) value of 0.4429 indicates that there is a probability of 0.4429 of randomly selecting a z score between 0 and 1.58. (The following sections will consider cases in which the mean is not 0 or the standard deviation is not 1.)

> **EXAMPLE** **Scientific Thermometers** The Precision Scientific Instrument Company manufactures thermometers that are supposed to give readings of 0°C at the freezing point of water. Tests on a large sample of these instruments reveal that at the freezing point of water, some thermometers give readings below 0° (denoted by negative numbers) and some give readings above 0° (denoted by positive numbers). Assume that the mean reading is 0°C and the standard deviation of the readings is 1.00°C. Also assume that the readings are normally distributed. If one thermometer is randomly selected, find the probability that, at the freezing point of water, the reading is between 0° and +1.58°.

SOLUTION The probability distribution of readings is a standard normal distribution, because the readings are normally distributed with $\mu = 0$ and $\sigma = 1$. We need to find the area between 0 and z (the shaded region) in Figure 5-6 with $z = 1.58$. From Table A-2 we find that this area is 0.4429.

If using Excel, the command **=NORMSDIST(1.58)** yields a value of 0.942947, but that is the cumulative area from the *left* up to $z = 1.58$, not the area between 0 and 1.58. Because the cumulative area from the left up to the center line is 0.5, subtract 0.5 from 0.942947 to get the result of 0.442947.

INTERPRETATION The probability of randomly selecting a thermometer with a reading between 0° and +1.58° is therefore 0.4429. Another way to interpret this result is to conclude that 44.29% of the thermometers will have readings between 0° and +1.58°. ◄

FIGURE 5-7
Using Symmetry to Find the Area to the Left of the Mean

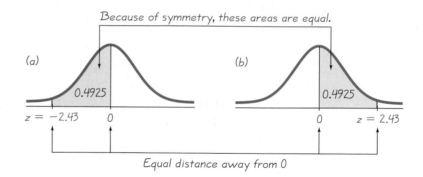

Because of symmetry, these areas are equal.

(a)   0.4925 $z = -2.43$ 0

(b)   0.4925   0 $z = 2.43$

Equal distance away from 0

AS

▷ **EXAMPLE**   **Scientific Thermometers** Using the thermometers from the preceding example, find the probability of randomly selecting one thermometer that reads (at the freezing point of water) between $-2.43°$ and $0°$.

SOLUTION   *Using Table A-2:* We are looking for the region that is shaded in Figure 5-7(a), but Table A-2 is designed to apply only to regions to the right of the mean (0), as in Figure 5-7(b). By comparing the shaded area in Figure 5-7(a) to the shaded area in Figure 5-7(b), we can see that those two areas are identical, because the density curve is symmetric. Referring to Table A-2, we can easily determine that the shaded area of Figure 5-7(b) is 0.4925, so the shaded area of Figure 5-7(a) must also be 0.4925.

If using Excel, the command **=NORMSDIST(−2.43)** yields a value of 0.007549. Because that is the cumulative area from the *left* up to $z = -2.43$, the desired area can be found by subtracting 0.007549 from 0.5. The result is 0.492451.

INTERPRETATION   The probability of randomly selecting a thermometer with a reading between $-2.43°$ and $0°$ is 0.4925. In other words, 49.25% of the thermometers have readings between $-2.43°$ and $0°$. ◁

The preceding solution illustrates an important principle:

Although a z score can be negative, the area under the curve (or the corresponding probability) can never be negative.

Now recall the empirical rule (presented in Section 2-5), which stated that for bell-shaped distributions,

- about 68% of all values fall within 1 standard deviation of the mean.
- about 95% of all values fall within 2 standard deviations of the mean.
- about 99.7% of all values fall within 3 standard deviations of the mean.

If we refer to Figure 5-5 with $z = 1$, Table A-2 or Excel can be used to find that the shaded area is 0.3413. It follows that the proportion of scores between $z = -1$ and $z = 1$ will be $0.3413 + 0.3413 = 0.6826$. That is, about 68% of all scores fall within 1 standard deviation of the mean. A similar calculation with $z = 2$ yields the value of $0.4772 + 0.4772 = 0.9544$ (or about 95%)

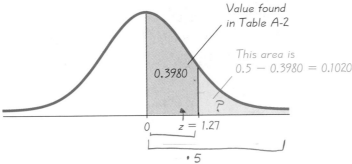

FIGURE 5-8
Finding the Area to the Right
of z = 1.27

as the proportion of scores between $z = -2$ and $z = 2$. Similarly, the proportion of scores between $z = -3$ and $z = 3$ is $0.4987 + 0.4987 = 0.9974$ (or about 99.7%). These exact values correspond very closely to those given by the empirical rule. In fact, the values of the empirical rule were found directly from the probabilities in Table A-2 and were slightly rounded for convenience. The empirical rule is sometimes called the *68-95-99.7 rule;* using exact values from Table A-2, it would be called the *68.26-95.44-99.74 rule,* but then it wouldn't sound as snappy.

Because we are dealing with a density curve for a probability distribution, the total area under the curve must be 1. Now refer to Figure 5-8, which shows that a vertical line directly above the mean of 0 divides the area under the curve into two equal parts, each containing an area of 0.5. The following example uses this characteristic.

▷ **EXAMPLE   Scientific Thermometers** Once again, make a random selection from the same sample of thermometers. Find the probability that the chosen thermometer reads (at the freezing point of water) more than $+1.27°$.

SOLUTION   We are again dealing with normally distributed values having a mean of $0°$ and a standard deviation of $1°$. The probability of selecting a thermometer that reads greater than $+1.27°$ corresponds to the green-shaded area of Figure 5-8.

Using Table A-2: Table A-2 cannot be used to find that green area directly, but we can use the table to find that $z = 1.27$ corresponds to the orange area of 0.3980, as shown in the figure. We now reason that because the area to the right of zero is one-half of the total area, it has an area of 0.5, and the green-shaded area is $0.5 - 0.3980$, or 0.1020.

Using Excel: The command **=NORMSDIST(1.27)** yields 0.897958. Because that is the cumulative area from the left up to $z = 1.27$, the desired area can be found by subtracting 0.897958 from 1. The result is 0.102042.

INTERPRETATION   We conclude that there is a probability of 0.1020 of randomly selecting one of the thermometers with a reading greater than $+1.27°$. Another way to interpret this result is to state that if many thermometers are selected and tested, then 0.1020 (or 10.20%) of them will read greater than $+1.27°$.   ◀

FIGURE 5-9
Finding the Area Between
$z = 1.20$ and $z = 2.30$

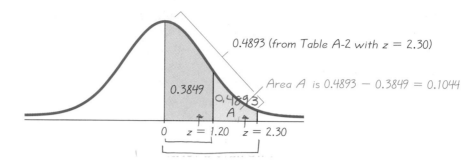

0.4893 (from Table A-2 with z = 2.30)

Area A is 0.4893 − 0.3849 = 0.1044

0.3849

0.4893

A

0   z = 1.20   z = 2.30

▷ **EXAMPLE**   **Scientific Thermometers** Assuming that one thermometer in our sample is randomly selected, find the probability that it reads (at the freezing point of water) between 1.20° and 2.30°.

SOLUTION   The probability of selecting a thermometer that reads between +1.20° and +2.30° corresponds to the green-shaded area of Figure 5-9.

Using Table A-2: Table A-2 is designed to provide only for regions bounded on the left by the vertical line above 0. We can use the table to find that $z = 1.20$ corresponds to an area of 0.3849 and that $z = 2.30$ corresponds to an area of 0.4893, as shown in the figure. If we denote the area of the green-shaded region by A, we can see from Figure 5-9 that

$$0.3849 + A = 0.4893$$

so

$$A = 0.4893 - 0.3849 = 0.1044$$

Using Excel: The commands **=NORMSDIST(1.20)** and **=NORMS-DIST(2.30)** yield areas of 0.884930 and 0.989276. Their difference is the desired area of 0.104346.

INTERPRETATION   If one thermometer is randomly selected, the probability that it reads (at the freezing point of water) between 1.20° and 2.30° is therefore 0.1044 (or 0.1043). ◁

The preceding example concluded with the statement that the probability of a reading between 1.20° and 2.30° is 0.1044. Such probabilities can also be expressed with the following notation.

Notation	
$P(a < z < b)$	denotes the probability that the z score is between a and b.
$P(z > a)$	denotes the probability that the z score is greater than a.
$P(z < a)$	denotes the probability that the z score is less than a.

Using this notation, we can express the result of the last example as $P(1.20 < z < 2.30) = 0.1044$, which states in symbols that the probability of a z score falling between 1.20 and 2.30 is 0.1044. With a continuous probability distribution such as the normal distribution, the probability of getting any single *exact* value is 0. That is, $P(z = a) = 0$. For example, there is a 0 probability of randomly selecting someone and getting a person whose height is exactly 68.12345678 in. In the normal distribution, any single point on the horizontal scale is represented not by a region under the curve, but by a vertical line above the point. For $P(z = 1.33)$, we have a vertical line above $z = 1.33$, but that vertical line by itself contains no area, so $P(z = 1.33) = 0$. With any continuous random variable, the probability of any one exact value is 0, and it follows that $P(a \leq z \leq b) = P(a < z < b)$. It also follows that the probability of getting a z score of *at most b* is equal to the probability of getting a z score *less than b*. It is important to correctly interpret key phrases such as *at most, at least, more than, no more than,* and so on. The illustrations in Figure 5-10 provide an aid to interpreting several of the most common phrases.

Finding z Scores When Given Probabilities

So far, the examples of this section involving the standard normal distribution have all followed the same format: Given z scores, we found areas under the curve. These areas represent probabilities. In many other cases, we want a reverse process because we already know the area (or probability), but we need to find the corresponding z score. In such cases, it is very important to avoid confusion between z scores and areas. Remember, z scores are *distances* along the horizontal scale, and they are represented by the numbers in

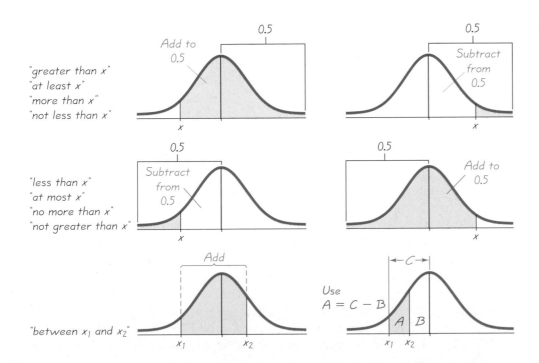

FIGURE 5-10
Interpreting Areas Correctly

Table A-2 that are in the extreme left column and across the top row. Areas (or probabilities) are regions under the curve, and they are represented by the values in the body of Table A-2. Also, z scores to the left of the centerline are always negative (as in Figure 5-7a). If we already know a probability and want to determine the corresponding z score, we find it as follows.

- **Using Table A-2:**

  1. Draw a bell-shaped curve, draw the centerline, and identify the region under the curve that corresponds to the given probability. If that region is not bounded by the centerline, work with a known region that is bounded by the centerline.

  2. Using the probability representing the area bounded by the centerline, locate the closest probability in the *body* of Table A-2 and identify the corresponding z score.

  3. If the z score is positioned to the left of the centerline, make it negative.

- **Using Excel:**

  1. Draw a bell-shaped curve, and identify the region under the curve that corresponds to the given probability. Remember that the Excel commands refer to cumulative areas from the *left* up to a particular value.

  2. Enter the command **=NORMSINV(p)**, where p is the cumulative area from the left.

▷ **EXAMPLE**   **Scientific Thermometers** Use the same thermometers as earlier, with temperature readings that are normally distributed with a mean of 0°C and a standard deviation of 1°C. Find the temperature corresponding to P_{95}, the 95th percentile. That is, find the temperature separating the bottom 95% from the top 5%. See Figure 5-11.

SOLUTION   Figure 5-11 shows the z score that is the 95th percentile, separating the top 5% from the bottom 95%.

Using Table A-2: We must refer to Table A-2 to find the unknown z score, and we must use a region bounded by the centerline (where $\mu = 0$) on one side, such as the shaded region of 0.45 in Figure 5-11. (Remember, Table A-2 is designed to directly provide only those areas that are bounded on the left by the centerline and bounded on the right by the z score.) We first search for the area of 0.45 *in the body of the table* and then find the corresponding z score. In Table A-2 the area of 0.45 is between the table values of 0.4495 and 0.4505, but there's an asterisk with a special note indicating that 0.4500 corresponds to a z score of 1.645. We can now conclude that the z score in Figure 5-11 is 1.645, so the 95th percentile is the temperature reading of 1.645°C.

Using Excel: Enter the command **=NORMSINV(0.95)** to get a result of 1.644853. Because that z score separates an area of 0.95 to the left, it is the 95th percentile.

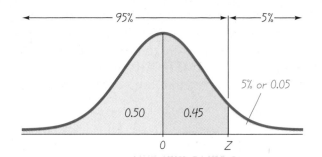

FIGURE 5-11
Finding the 95th Percentile

INTERPRETATION   When tested at freezing, 95% of the readings will be less than or equal to 1.645°C, and 5% of them will be greater than or equal to 1.645°C.

Note that in the preceding solution, Table A-2 led to a z score of 1.645, which is midway between 1.64 and 1.65. When using Table A-2, we can usually avoid interpolation by simply selecting the closest value. There are two special cases listed in the accompanying table that are important because they are used so often in a wide variety of applications. (The value of $z = 2.576$ is slightly closer to the area of 0.4950, but $z = 2.575$ has the advantage of being the value midway between $z = 2.57$ and $z = 2.58$.) Except in these two special cases, we can select the closest value in the table. (If a desired value is midway between two table values, select the larger value.) Also, for z scores above 3.09, we can use 0.4999 as an approximation of the corresponding area between the z score and the centerline.

z Score	Area
1.645	0.4500
2.575	0.4950

▷ **EXAMPLE   Scientific Thermometers** Using the same thermometers, find P_{10}, the 10th percentile. That is, find the temperature reading separating the bottom 10% of all temperatures from the top 90%.

SOLUTION   Refer to Figure 5-12, where the 10th percentile is shown as the z score separating the bottom 10% from the top 90%.
Using Table A-2: Table A-2 is designed for areas bounded by the centerline, so we refer to the shaded area of 0.40 (corresponding to 50% − 10%). In the *body of the table,* we select the closest value of 0.3997 and find that it corresponds to $z = 1.28$. However, because the

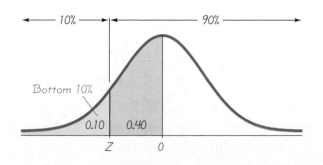

FIGURE 5-12
Finding the 10th Percentile

z score is below the mean of 0, it must be negative. The 10th percentile is therefore −1.28°C.

Using Excel: Because the area to the left of the desired value is 0.10, enter the command **=NORMSINV(0.10)** to get a result of −1.28155. Because that *z* score separates an area of 0.10 to the left, it is the 10th percentile.

INTERPRETATION   When tested at freezing, 10% of the thermometer readings will be equal to or less than −1.28°C, and 90% of the readings will be equal to or greater than −1.28°C.

The examples in this section were contrived so that the mean of 0 and the standard deviation of 1 coincided exactly with the parameters of the standard normal distribution. In reality, it is unusual to find such convenient parameters, because typical normal distributions involve means different from 0 and standard deviations different from 1. In the next section we introduce methods for working with such normal distributions.

5-2   Basic Skills and Concepts

Using a Continuous Uniform Distribution.   In Exercises 1–4, refer to the continuous uniform distribution depicted in Figure 5-2, assume that a time between 0 and 5 hours is randomly selected, and find the probability that the given time is selected.

1. Less than 4

2. Greater than 2

3. Between 3 and 4

4. Between 1.5 and 3.8

Using a Continuous Uniform Distribution.   In Exercises 5–8, assume that voltages in a circuit vary between 6 volts and 12 volts, and voltages are spread evenly over the range of possibilities, so that there is a uniform distribution. Find the probability of the given range of voltage levels.

5. Greater than 10 volts

6. Less than 11 volts

7. Between 7 volts and 10 volts

8. Between 6.5 volts and 8 volts

Using the Standard Normal Distribution.   In Exercises 9–28, assume that the readings on the thermometers are normally distributed with a mean of 0° and a standard deviation of 1.00°C. A thermometer is randomly selected and tested. In each case, draw a sketch, and find the probability of each reading in degrees.

9. Between 0 and 1.50

10. Between 0 and 1.28

11. Between −1.96 and 0

12. Between −0.48 and 0

13. Less than −1.79

14. Greater than 0.37

15. Greater than 2.05

16. Less than −0.92

17. Between 0.50 and 1.50

18. Between 1.50 and 2.50

19. Between −2.00 and −1.00

20. Between 2.00 and 2.34

EXCEL **21.** Less than 1.62   **EXCEL** **22.** Less than 2.44

EXCEL **23.** Greater than -0.27 **EXCEL** **24.** Greater than -2.09

EXCEL **25.** Between -1.08 and 0.33   **EXCEL** **26.** Between -0.90 and 1.95

EXCEL **27.** Greater than 0   **EXCEL** **28.** Less than 0

Finding Probability.   *In Exercises 29–32, assume that the readings on the thermometers are normally distributed with a mean of $0°$ and a standard deviation of $1.00°$. Find the indicated probability, where z is the reading in degrees.*

EXCEL **29.** $P(-1.96 < z < 1.96)$ **EXCEL** **30.** $P(z < 1.645)$

EXCEL **31.** $P(z > -2.575)$ **EXCEL** **32.** $P(1.96 < z < 2.33)$

Finding Temperature Values.   *In Exercises 33–40, assume that the readings on the thermometers are normally distributed with a mean of $0°$ and a standard deviation of $1.00°C$. A thermometer is randomly selected and tested. In each case, draw a sketch, and find the temperature reading corresponding to the given information.*

EXCEL **33.** Find P_{90}, the 90th percentile. This is the temperature reading separating the bottom 90% from the top 10%.

EXCEL **34.** Find P_{20}, the 20th percentile.

EXCEL **35.** Find Q_1, the temperature reading that is the first quartile.

EXCEL **36.** Find D_3, the temperature reading that is the third decile.

EXCEL **37.** If 5% of the thermometers are rejected because they have readings that are too low, but all other thermometers are acceptable, find the reading that separates the rejected thermometers from the others.

EXCEL **38.** If 12% of the thermometers are rejected because they have readings that are too high, but all other thermometers are acceptable, find the reading that separates the rejected thermometers from the others.

EXCEL **39.** A troubleshooter wants to examine thermometers that give readings in the bottom 3%. What reading separates the bottom 3% from the others?

EXCEL **40.** If 2.5% of the thermometers are rejected because they have readings that are too high and another 2.5% are rejected because they have readings that are too low, find the two readings that are cutoff values separating the rejected thermometers from the others.

5-2 Beyond the Basics

41. For a standard normal distribution, find the percentage of data that are
   a. within 1 standard deviation of the mean.
   b. within 1.96 standard deviations of the mean.
   c. between $\mu - 3\sigma$ and $\mu + 3\sigma$.
   d. between 1 standard deviation below the mean and 2 standard deviations above the mean.
   e. more than 2 standard deviations away from the mean.

42. If a continuous uniform distribution has parameters of $\mu = 0$ and $\sigma = 1$, then the minimum is $-\sqrt{3}$ and the maximum is $\sqrt{3}$.

(continued)

    a. For this distribution, find $P(-1 < x < 1)$.

    b. Find $P(-1 < x < 1)$ if you incorrectly assume that the distribution is normal instead of uniform.

    c. Compare the results from parts (a) and (b). Does the distribution affect the results very much?

43. Assume that z scores are normally distributed with a mean of 0 and a standard deviation of 1.

    a. If $P(0 < z < a) = 0.3907$, find a.

    b. If $P(-b < z < b) = 0.8664$, find b.

    c. If $P(z > c) = 0.0643$, find c.

    d. If $P(z > d) = 0.9922$, find d.

    e. If $P(z < e) = 0.4500$, find e.

44. In a continuous uniform distribution,

$$\mu = \frac{\text{minimum} + \text{maximum}}{2} \quad \text{and} \quad \sigma = \frac{\text{range}}{\sqrt{12}}$$

Find the mean and standard deviation for the uniform distribution represented in Figure 5-2.

5-3 Normal Distributions: Finding Probabilities

In Section 5-2 we presented the basic methods for working with normal distributions. Because all of the examples and exercises discussed there involved only the *standard* normal distribution (with a mean of 0 and a standard deviation of 1), they were necessarily unrealistic. In this section we include other normal distributions, so we will be able to work with cases that are more realistic and practical.

    The basic principle we will be explaining in this section is the following:

> **If we convert values to standard scores using Formula 5-2, then procedures for working with all normal distributions are the same as those for the standard normal distribution.**

Formula 5-2
$$z = \frac{x - \mu}{\sigma}$$

    If you used Table A-2 as your primary tool for finding probabilities or z scores in the Section 5-2, you will need to understand and apply the above principle. If you use certain calculators or software programs, the conversion to z scores is not necessary because probabilities can be found directly. Regardless of the method used, however, you need to clearly understand the basic principle, because it is an important foundation for concepts introduced in the following chapters.

    We illustrate the principle by showing that the area in Figure 5-13(a) is the same as the area in Figure 5-13(b). That is, the area bounded by values

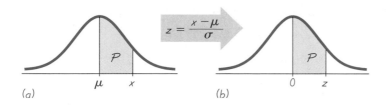

FIGURE 5-13
Converting to the Standard Normal Distribution

in any normal distribution (as in Figure 5-13a) is the same as the area bounded by the equivalent standardized scores in the standard normal distribution (as in Figure 5-13b). This means that when you are working with a normal distribution that is not the standard normal distribution, you can use Table A-2 the same way it was used in Section 5-2 as long as you first convert the values to z scores. We therefore recommend the following procedure for finding probabilities for values of a random variable with a normal probability distribution:

1. Sketch a normal curve, label the mean and the specific x values, then *shade* the region representing the desired probability.

2. For each relevant value x that is a boundary for the shaded region, use Formula 5-2 to convert that value to the equivalent z score.

3. Refer to Table A-2 to find the area of the shaded region. This area is the desired probability.

The following example applies these three steps, and it illustrates the relationship between a typical nonnormal distribution and the standard normal distribution.

Queues

EXAMPLE   **Jet Ejection Seats** In redesigning jet ejection seats to better accommodate women as pilots, it is found that women's weights are normally distributed with a mean of 143 lb and a standard deviation of 29 lb. If a woman is randomly selected, what is the probability that she weighs between 143 lb and 201 lb?

SOLUTION

Step 1:   See Figure 5-14, where we label the mean of 143 lb and the value of 201 lb, and we shade the area representing the probability we want.

Step 2:   To use Table A-2, we first must use Formula 5-2 to convert the distribution of weights to the standard normal distribution. The weight of 201 lb is converted to a z score as follows.

$$z = \frac{x - \mu}{\sigma} = \frac{201 - 143}{29} = \frac{58}{29} = 2.00$$

This result shows that the weight of 201 lb differs from the mean of 143 lb by 2.00 standard deviations.

FIGURE 5-14
Probability of Weight
Between 143 lb and 201 lb

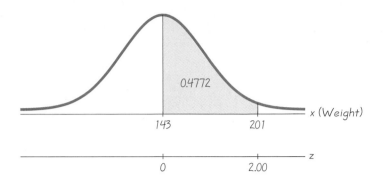

Step 3:   Referring to Table A-2, we find that $z = 2.00$ corresponds to an area of 0.4772.

INTERPRETATION   There is a probability of 0.4772 of randomly selecting a woman with a weight between 143 lb and 201 lb. This can be expressed in symbols as

$$P(143 < x < 201) = P(0 < z < 2.00) = 0.4772$$

Another way to interpret this result is to conclude that 47.72% of women have weights between 143 lb and 201 lb.

If using Excel, refer to Table 5-1 for the relevant command. For the preceding example, use the command **=NORMDIST(201, 143, 29, 1)** to get an area of 0.97725. Because that is the cumulative area to the left of 201 lb, subtract the area of 0.5 (the cumulative area from the left up to 143 lb). The result is 0.47725.

EXAMPLE   **Jet Ejection Seats** In the Chapter Problem, we reported that the Air Force designed the current ACES-II ejection seats for men weighing between 140 lb and 211 lb. What percentage of women have weights that are within those limits?

SOLUTION   As in the preceding example, women's weights are normally distributed with a mean of 143 lb and a standard deviation of 29 lb. See Figure 5-15, which shows the design limits on the distribution of women's weights. We need to find the area of regions A and B combined. Using Table A-2, we cannot find that combined region directly because Table A-2 isn't designed for such cases, but we can find it indirectly by using the basic procedures presented in Section 5-2. The way to proceed is first to find regions A and B separately, then add them.

For region A:

$$z = \frac{x - \mu}{\sigma} = \frac{140 - 143}{29} = -0.10$$

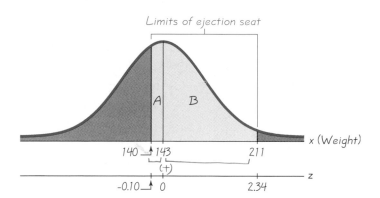

FIGURE 5-15
Weights of Women and
Ejection Seat Limits

Using Table A-2, we find that $z = -0.10$ corresponds to an area of 0.0398. Region *A* has an area of 0.0398.

For region B:

$$z = \frac{x - \mu}{\sigma} = \frac{211 - 143}{29} = 2.34$$

Using Table A-2, we find that $z = 2.34$ corresponds to an area of 0.4904. Region *B* has an area of 0.4904.

For regions A and B combined: fr, prior problem

area of *A* and *B* combined $= 0.0398 + 0.4904 = 0.5302$

INTERPRETATION   We found that 53.02% of women have weights between the ejection seat limits of 140 lb and 211 lb. This means that 46.98% of women do *not* have weights between the current limits, so far too many women pilots would risk serious injury if ejection became necessary.   ◀

If we solve the preceding problem using Excel, we should enter the commands **=NORMDIST(140, 143, 29, 1)** and **=NORMDIST(211, 143, 29, 1)** to get areas of 0.458804 and 0.990482. Because they are both cumulative areas from the left, the desired area is their difference: 0.531678.

Test your understanding of the methods of this section by working through the next example on your own. Write your own solution for each part before checking the solution given here.

▶ **EXAMPLE   Women's Weights** Again, assume that women's weights are normally distributed with a mean of 143 lb and a standard deviation of 29 lb. Also assume that we want to find the proportion of women weighing between 100 lb and 130 lb.

a. Draw the graph and shade the region representing the proportion of women weighing between 100 lb and 130 lb.

FIGURE 5-16
Weights of Women

b. Convert the limits of 100 lb and 130 lb to their corresponding *z* scores.

c. Use Table A-2 to find the two areas corresponding to the two *z* scores.

d. What is the proportion of women weighing between 100 lb and 130 lb?

SOLUTION

a. Start by sketching a normal distribution curve and entering the value of 143 lb in the middle. The desired limits of 100 lb and 130 lb are both below the mean of 143 lb, so those limits are to the left of the center, as shown in Figure 5-16.

b. We convert the values of 100 lb and 130 lb to *z* scores by using Formula 5-2.

$$z = \frac{x - \mu}{\sigma} = \frac{100 - 143}{29} = -1.48 \quad z = \cdot 4306$$

$$z = \frac{x - \mu}{\sigma} = \frac{130 - 143}{29} = -0.45 \quad z = \cdot 1736$$

c. Using Table A-2, the *z* score of 1.48 corresponds to an area of 0.4306. Using Table A-2 again, the *z* score of 0.45 corresponds to an area of 0.1736. Both of these areas are shown in Figure 5-16.

INTERPRETATION   The desired area is the green-shaded region, which is the difference between 0.4306 and 0.1736. Because 0.4306 − 0.1736 = 0.2570, we conclude that the proportion of women with weights between 100 lb and 130 lb is 0.2570.   ◄

In this section we have extended the concepts of Section 5-2 so that they apply to more realistic normal probability distributions. Still, all of the examples and exercises of this section are of the same type: We are given specific limit values and we must find an area (or probability or percentage). In many practical and real cases, the area (or probability or percentage) is known and we must find the relevant value(s). Problems of this type are discussed in the next section.

5-3 Basic Skills and Concepts

Using a Nonstandard Normal Distribution.   *In Exercises 1–4, assume that women's weights are normally distributed with a mean given by* μ *= 143 lb and a standard deviation given by* σ *= 29 lb (based on data from the National Health Survey). Also assume that a woman is randomly selected. Draw a graph, and find the indicated probability.*

 1. $P(143 \text{ lb} < x < 172 \text{ lb})$ **2.** $P(150 \text{ lb} < x < 180 \text{ lb})$

 3. $P(x > 150 \text{ lb})$ **4.** $P(x < 186.5 \text{ lb})$

Using a Nonstandard Normal Distribution.   *In Exercises 5–8, assume that heights of women are normally distributed with a mean given by* μ *= 63.6 in. and a standard deviation given by* σ *= 2.5 in. (based on data from the National Health Survey). Draw a graph, and find the indicated probability or percentage.*

 5. Rockette Height Requirement The heights of the Rockette dancers at New York City's Radio City Music Hall must be between 65.5 in. and 68.0 in. If a woman is randomly selected, find the probability that she meets the height requirement to be a Rockette.

 6. Beanstalk Club Height Requirement The Beanstalk Club, a social organization for tall people, has a requirement that women must be at least 70 in. (or 5 ft 10 in.) tall. What percentage of women meet that requirement?

 7. *Soyuz* Spacecraft Height Requirement In order to fit into a Russian *Soyuz* spacecraft, an astronaut must have a height between 64.5 in. and 72 in. What percentage of women meet that requirement?

 8. Height Requirement for Women Soldiers The U.S. Army requires women's heights to be between 58 in. and 80 in. Find the percentage of women meeting that height requirement. Are many women being denied the opportunity to join the Army because they are too short or too tall?

 9. SAT Test Required Scores The combined math and verbal scores for females taking the SAT-I test are normally distributed with a mean of 998 and a standard deviation of 202 (based on data from the College Board). If a college includes a minimum score of 900 among its requirements, what percentage of females do *not* satisfy that requirement?

 10. SAT Test Required Scores The scores for males on the math portion of the SAT-I test are normally distributed with a mean of 531 and a standard deviation of 114 (based on data from the College Board). If the College of Newport includes a minimum score of 600 among its requirements, what percentage of males do *not* satisfy that requirement?

11. Sugar Packets and Label Requirements Domino sugar packets are labeled as containing 3.5 g. Assume that those packets are actually filled with amounts that are normally distributed with a mean of 3.586 g and a standard deviation of 0.074 g (based on Data Set 4 in Appendix B). What percentage of packets have less than 3.5 g? Are many consumers being cheated?

12. Finding Percentage of Fever Temperatures Based on the sample results in Data Set 6 in Appendix B, assume that human body temperatures are normally distributed with a mean of 98.20°F and a standard deviation of 0.62°F. Bellevue Hospital in New York City uses 100.6°F as the lowest temperature considered to be a fever.

continued

What percentage of normal and healthy persons would be considered to have a fever? Does this percentage suggest that a cutoff of 100.6°F is appropriate?

13. *Oops* One classic use of the normal distribution is inspired by a letter to "Dear Abby" in which a wife claimed to have given birth 308 days after a brief visit from her husband, who was serving in the Navy. The lengths of pregnancies are normally distributed with a mean of 268 days and a standard deviation of 15 days. Given this information, find the probability of a pregnancy lasting 308 days or longer. What does the result suggest?

14. *Finding Percentage of Premature Births* The lengths of pregnancies are normally distributed with a mean of 268 days and a standard deviation of 15 days. If we stipulate that a baby is *premature* if born at least three weeks early, what percentage of babies are born prematurely? Such information is important so that a hospital can plan to have enough special equipment to handle premature babies.

15. *Finding Percentage of Coke Cans Under 12 oz* Cans of regular Coke are labeled as containing 12 oz. Assume that the actual contents are normally distributed with a mean of 12.19 oz and a standard deviation of 0.11 oz (based on Data Set 1 in Appendix B). What percentage of cans contain less than the 12 oz printed on the label? Are many consumers being cheated?

16. *Finding Percentage of Pepsi Cans Under 12 oz* Cans of regular Pepsi are labeled as containing 12 oz. Assume that the actual contents are normally distributed with a mean of 12.29 oz and a standard deviation of 0.09 oz (based on Data Set 1 in Appendix B). What percentage of cans contain less than the 12 oz printed on the label? Are many consumers being cheated?

17. *Marine Corps Height Requirements* The U.S. Marine Corps requires that men have heights between 64 in. and 78 in. Find the percentage of men meeting those height requirements. (The National Health Survey shows that heights of men are normally distributed with a mean of 69.0 in. and a standard deviation of 2.8 in.) Are too many men denied the opportunity to join the Marines because they are too short or too tall?

18. *Babies at Risk* Weights of newborn babies in the United States are normally distributed with a mean of 3420 g and a standard deviation of 495 g (based on data from "Birth Weight and Prenatal Mortality," by Wilcox et al., *Journal of the American Medical Association,* Vol. 273, No. 9). A newborn weighing less than 2200 g is considered to be at risk, because the mortality rate for this group is at least 1%. What percentage of newborn babies are in the "at-risk" category? If the Chicago General Hospital has 900 births in a year, how many of the babies are in the "at-risk" category?

19. *Babies at Risk* Do Exercise 18 for Norway, where weights of newborn babies are normally distributed with a mean of 3570 g and a standard deviation of 500 g (based on data from "Birth Weight and Prenatal Mortality," by Wilcox et al., *Journal of the American Medical Association,* Vol. 273, No. 9).

20. *Computer Chips Meeting Specs* An IBM subcontractor was hired to make ceramic substrates, which are used to distribute power and signals to and from computer silicon chips. Specifications require resistance between 1.500 ohms and 2.500 ohms, but the population has normally distributed resistances with a mean of 1.978 ohms and a standard deviation of 0.172 ohms. What percentage of the ceramic substrates will not meet the manufacturer's specifications? Does this manufacturing process appear to be working well?

EXCEL **21.** Cholesterol and Coronary Disease The serum cholesterol levels in men aged 18 to 24 are normally distributed with a mean of 178.1 and a standard deviation of 40.7. The units are mg/100 mL, and the data are based on the National Health Survey. One criterion for identifying risk of coronary disease is a cholesterol level above 300. If a man aged 18 to 24 is randomly selected, find the probability that his serum cholesterol level is above 300. Is this probability low enough to warrant serious concern?

EXCEL **22.** Dating Skulls Measurements of human skulls from different epochs are analyzed to determine whether they change over time. The maximum breadth is measured for skulls from Egyptian males who lived around 3300 B.C. Results show that those breadths are normally distributed with a mean of 132.6 mm and a standard deviation of 5.4 mm (based on data from *Ancient Races of the Thebaid* by Thomson and Randall-Maciver). An archeologist discovers a male Egyptian skull and a field measurement reveals a maximum breadth of 119 mm. Find the probability of getting a value of 119 mm or less if a skull is randomly selected from the period around 3300 B.C. Is the newly found skull likely to come from that era?

EXCEL **23.** Mensa IQ Requirement IQ scores are normally distributed with a mean of 100 and a standard deviation of 15. Mensa is an organization for people with high IQs, and eligibility requires an IQ above 131.5.
   a. If someone is randomly selected, find the probability that he or she meets the Mensa requirement.
   b. In a typical region of 75,000 people, how many are eligible for Mensa?

EXCEL **24.** Counterfeit Coins in Vending Machines Some vending machines are designed so that their owners can adjust the weights of the quarters that are accepted. If many counterfeit coins are found, adjustments are made to reject more coins, with the effect that most of the counterfeit coins are rejected along with many legal coins. Assume that quarters have weights that are normally distributed with a mean of 5.67 g and a standard deviation of 0.070 g. If a vending machine is adjusted to reject quarters weighing less than 5.50 g or more than 5.80 g, what is the percentage of legal quarters that are rejected? Is this result low enough so that not too many customers will be inconvenienced because their coins are rejected?

5-3 Beyond the Basics

25. Units of Measurement Weights of individual women are expressed in units of pounds. What are the units used for the z scores that correspond to individual weights?

26. z Scores Weights of women are normally distributed with a mean of 143 lb and a standard deviation of 29 lb. If weights of all woman are converted to z scores, what are the mean, standard deviation, and distribution of these z scores?

27. Transformations Weights of women are normally distributed with a mean of 143 lb and a standard deviation of 29 lb. What are the distribution, mean, and standard deviation after all of those weights have been converted to kilograms? (1 lb = 0.4536 kg)

28. *Using Continuity Correction* There are many situations in which a normal distribution can be used as a good approximation to a random variable that has only *discrete* values. In such cases, we can use this *continuity correction:* Represent each whole number by the interval extending from 0.5 below the number to 0.5 above it. Assume that IQ scores are all whole numbers having a distribution that is approximately normal with a mean of 100 and a standard deviation of 15.

    a. Without using any correction for continuity, find the probability of randomly selecting someone with an IQ score greater than 105.

    b. Using the correction for continuity, find the probability of randomly selecting someone with an IQ score greater than 105.

    c. Compare the results from parts (a) and (b).

5-4 Normal Distributions: Finding Values

In this section we consider problems such as this: If women's weights are normally distributed with a mean of 143 lb and a standard deviation of 29 lb, find the weight separating the top 10% from the others. In the examples and exercises in Section 5-3 we used a given value to find a probability or percentage. In this section we follow a reverse procedure—we use a given probability (or percentage) to find a specific value.

In working with problems of finding values when given probabilities, there are three important cautions to keep in mind.

1. *Don't confuse z scores and areas.* Remember, *z* scores are *distances* along the horizontal scale, but areas are *regions* under the normal curve. Table A-2 lists *z* scores in the left column and across the top row, but areas are found in the body of the table.

2. *Choose the correct (right/left) side of the graph.* A value separating the top 10% from the others will be located on the right side of the graph, but a value separating the bottom 10% will be located on the left side of the graph.

3. *A z score must be negative whenever it is located to the left of the centerline of 0.*

As in Section 5-3, graphs are extremely helpful, and we strongly urge you to use them. Also as in Section 5-3, we can use Table A-2 or Excel. We again recommend that you begin with Table A-2 and Formula 5-2 so that you can gain a clear understanding of what you are doing before using software.

Procedure for Finding Values Using Table A-2 and Formula 5-2

1. Sketch a normal distribution curve, enter the given probability or percentage in the appropriate region of the graph, and identify the *x* value(s) being sought.

2. Use Table A-2 to find the *z* score corresponding to the region bounded by *x* and the centerline of 0. Observe the following cautions:

- Refer to the *body* of Table A-2 to find the closest area, then identify the corresponding z score.
- Make the z score *negative* if it is located to the left of the centerline.

3. Using Formula 5-2, enter the values for μ, σ, and the z score found in Step 2, then solve for x. Based on the format of Formula 5-2, we can solve for x as follows.

$$x = \mu + (z \cdot \sigma) \quad \text{(another form of Formula 5-2)}$$

$\uparrow$

(Make z *negative* if it is to the left of the mean.)

4. Refer to the sketch of the curve to verify that the solution makes sense in the context of the graph and in the context of the problem.

The following example uses the procedure just outlined. Pay extra attention to Step 2, especially where we make the z score negative because it is to the left of the mean.

▶ **EXAMPLE** **Women's Weights** In Section 5-3 we used women's weights, which are normally distributed with a mean of 143 lb and a standard deviation of 29 lb (based on data from the National Health Survey). Find the value of P_{10}. That is, find the weight separating the bottom 10% from the top 90%.

SOLUTION

Step 1:   We begin with the graph shown in Figure 5-17. We have entered the mean of 143, shaded in green the area representing the bottom 10%, and identified the desired value as x. The orange area between 143 and x must be 40% of the total area (because the left half of the area must combine to be 50% of the total). Because the total area is 1, the orange area constituting 40% of the total must be 0.40.

Step 2:   We refer to Table A-2, but we look for an area of 0.4000 in the body of the table. (Remember, Table A-2 is designed to list areas only for those regions bounded on the left by the mean

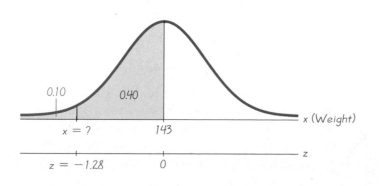

FIGURE 5-17
Finding P_{10} for Weights of Women

and on the right by some value.) The area closest to 0.4000 is 0.3997, and it corresponds to a z score of 1.28. Because the z score is to the left of the mean, we make it negative and use $z = -1.28$.

Step 3: With $z = -1.28$, $\mu = 143$, and $\sigma = 29$, we solve for x by using Formula 5-2 directly or by using the following version of Formula 5-2:

$$x = \mu + (z \cdot \sigma) = 143 + (-1.28 \cdot 29) = 105.88$$

Step 4: If we let $x = 105.88$ in Figure 5-17, we see that this solution is reasonable because the 10th percentile should be less than the mean of 143.

INTERPRETATION   The weight of 106 lb (rounded) separates the lowest 10% from the highest 90%. ◀

In this example, it would be very easy to forget the step of making the z score negative because it is below the mean. That error would have resulted in a weight of 180 lb, but in Step 4 we would see that 180 lb is not a reasonable answer. We could then proceed to correct that error.

Procedure for Finding Values Using Excel   To find values given a probability (or percentage) with Excel, use the command **=NORMINV(p, μ, σ)**, but remember that all of the Excel commands are based on cumulative areas to the *left*. For the preceding example, the total area to the left of the desired value is 0.10, the mean is 143, and the standard deviation is 29, so we enter **=NORMINV(0.10, 143, 29)** to get a result of 105.835.

 EXAMPLE   **Body Temperatures** Assume that body temperatures of healthy adults are normally distributed with a mean of 98.20°F and a standard deviation of 0.62°F (based on Data Set 6 in Appendix B). If a medical researcher wants to study people in the bottom 2.5% and people in the top 2.5%, find the temperatures separating those limits.

SOLUTION

Step 1: We begin with the graph shown in Figure 5-18. We have shaded in green the areas representing the bottom 2.5% and top 2.5% (or 0.025). The areas of 0.475 are found by using the fact that the centerline above the mean divides the total area of 1 into two parts, each with area 0.5. We get $0.5 - 0.025 = 0.475$.

Step 2: We refer to Table A-2, but we look for an area of 0.475 in the body of the table. (Remember, Table A-2 is designed to list areas only for those regions bounded on the left by the mean and on the right by some value.) The area of 0.4750 corresponds to $z = 1.96$. For the x value located on the right in Figure 5-18,

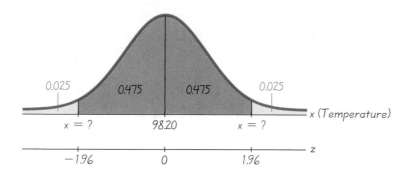

FIGURE 5-18
Body Temperatures

we use $z = 1.96$, but for the x value located on the left we use $z = -1.96$.

Step 3:   With $z = 1.96$, $\mu = 98.20$, and $\sigma = 0.62$, we solve for x using a variation of Formula 5-2:

$$x = \mu + (z \cdot \sigma) = 98.20 + (1.96 \cdot 0.62) = 99.42$$

With $z = -1.96$, $\mu = 98.20$, and $\sigma = 0.62$, we solve for x as follows:

$$x = \mu + (z \cdot \sigma) = 98.20 + (-1.96 \cdot 0.62) = 96.98$$

Step 4:   If we let $x = 99.42$ and 96.98 in Figure 5-18, we see that our solutions are reasonable.

INTERPRETATION   The researcher should select people with body temperatures below $96.98°$ F or above $99.42°$ F.   ◀

To solve the preceding problem using Excel, enter the commands **=NORMINV(0.025, 98.20, 0.62)** and **=NORMINV(0.975, 98.20, 0.62)** to get 96.98482 and 99.41518. When entering the probability value, remember to enter the cumulative area to the *left* of the given value.

5-4 Basic Skills and Concepts

Finding Values in a Normal Distribution.   *In Exercises 1–4, assume that a test is designed to measure a person's sense of humor and that scores on this test are normally distributed with a mean of 10 and a standard deviation of 2 (based on the author's fertile imagination). Draw a graph, find the relevant z score, then find the indicated value.*

  1. Find the score separating the top 10% from the bottom 90%.

  2. Find the score separating the top 25% from the bottom 75%.

3. Find the second decile, D_2, which is the score separating the bottom 20% from the top 80%.

4. Find the first quartile, Q_1, which is the score separating the bottom 25% from the top 75%.

Finding Weights in a Normal Distribution.   *In Exercises 5–8, assume that women's weights are normally distributed with a mean given by $\mu = 143$ lb and a standard deviation given by $\sigma = 29$ lb (based on data from the National Health Survey). Find the indicated value.*

EXCEL **5.** Find the third quartile, Q_3, which is the weight separating the bottom 75% from the top 25%.

EXCEL **6.** Find the sixth decile, D_6, which is the weight separating the bottom 60% from the top 40%.

EXCEL **7.** Find P_{15}, which is the weight separating the bottom 15% from the top 85%.

EXCEL **8.** Find P_{35}, which is the weight separating the bottom 35% from the top 65%.

In Exercises 9 and 10, assume that women's heights are normally distributed with a mean of 63.6 in. and a standard deviation of 2.5 in.

EXCEL **9.** Finding Beanstalk Club Height Requirement The Beanstalk Club, a social organization for tall people, has a requirement that women must be at least 70 in. (or 5 ft 10 in.) tall. Suppose that requirement is changed so that only the tallest 5% of women are eligible. What is the new minimum height requirement for women? Why isn't it practical to state the requirement for women as "a height that is in the top 5% of all women"?

EXCEL **10.** Finding Height Requirements for Women Soldiers To be eligible for the U.S. Army, a woman's height must be between 58 in. and 80 in. If that requirement is changed so that only the shortest 1% and tallest 1% are excluded, find the minimum and maximum acceptable heights.

EXCEL **11.** Finding SAT Test Requirement Scores for the combined math and verbal portions of the SAT-I test have a mean of 1017 and a standard deviation of 207, and the distribution is approximately normal (based on data from the College Board). If the College of Newport wants to consider only those applicants in the top 42%, what is the minimum required score? What is the problem with the college telling applicants that they must score in the top 42%?

EXCEL **12.** Study of Facial Behavior A study compared the facial behavior of nonparanoid schizophrenic persons with that of a control group of normal persons. The control group was timed for eye contact during a period of 5 minutes, or 300 seconds. The eye-contact times were normally distributed with a mean of 184 sec and a standard deviation of 55 sec (based on data from "Ethological Study of Facial Behavior in Nonparanoid and Paranoid Schizophrenic Patients," by Pitman, Kolb, Orr, and Singh, *Psychiatry,* 144:1). Because results showed that nonparanoid schizophrenic patients had much lower eye-contact times than did the control group, you have decided to further analyze people in the control group who are in the bottom 5%. For the control group, find P_5, the 5th percentile. That is, find the eye-contact time separating the bottom 5% from the rest.

EXCEL **13.** TV Warranty Replacement times for TV sets are normally distributed with a mean of 8.2 years and a standard deviation of 1.1 years (based on data from "Getting Things Fixed," *Consumer Reports*).
   a. Find the probability that a randomly selected TV will have a replacement time less than 5.0 years.
   b. If you want to provide a warranty so that only 1% of the TV sets will be replaced before the warranty expires, what is the time length of the warranty?

EXCEL **14.** CD Player Warranty Replacement times for CD players are normally distributed with a mean of 7.1 years and a standard deviation of 1.4 years (based on data from "Getting Things Fixed," *Consumer Reports*).

    a. Find the probability that a randomly selected CD player will have a replacement time less than 8.0 years.

    b. If you want to provide a warranty so that only 2% of the CD players will be replaced before the warranty expires, what is the time length of the warranty?

EXCEL **15.** Weights of Discarded Paper Weights of paper discarded by households each week are normally distributed with a mean of 9.4 lb and a standard deviation of 4.2 lb (based on data from the Garbage Project at the University of Arizona).

    a. For a randomly selected household, find the probability that the weight of the discarded paper is less than 15 lb.

    b. In planning for paper pickup and processing needs, the Orange County Resource Recovery Agency needs to know P_{99}, the weight that is exceeded by only 1% of households. Find that weight.

EXCEL **16.** Body Temperatures Healthy people have body temperatures that are normally distributed with a mean of 98.20°F and a standard deviation of 0.62°F (based on Data Set 6 in Appendix B).

    a. If a healthy person is randomly selected, what is the probability that he or she has a temperature below 100°F?

    b. The Newport Hospital wants to select a minimum temperature for requiring further medical tests. What should that temperature be, if we want only 1.5% of healthy people to exceed it? (Such a result is called a "false positive," meaning that the test result is positive, but the subject is not really sick.)

EXCEL **17.** Length of Pregnancies The lengths of pregnancies are normally distributed with a mean of 268 days and a standard deviation of 15 days. If we stipulate that a baby is *premature* if the length of pregnancy is in the lowest 4%, find the length that separates premature babies from those who are not premature. Premature babies often require special care, and this result could be helpful to hospital administrators in planning for that care.

EXCEL **18.** High IQ Scores IQ scores are normally distributed with a mean of 100 and a standard deviation of 15.

    a. People are considered to be "intellectually very superior" if their score is above 130. What percentage of people fall into that category?

    b. If we redefine the category of "intellectually very superior" to be scores in the top 2%, what does the minimum score become?

EXCEL **19.** High Cholesterol Levels The serum cholesterol levels in men aged 18 to 24 are normally distributed with a mean of 178.1 and a standard deviation of 40.7. Units are mg/100 mL, and the data are based on the National Health Survey.

    a. If a man aged 18 to 24 is randomly selected, find the probability that his serum cholesterol level is greater than 260, a value considered to be "moderately high."

    b. The Providence Health Maintenance Organization wants to establish a criterion for recommending dietary changes if cholesterol levels are in the top 3%. What is the cutoff for men aged 18 to 24?

EXCEL **20.** Filling Coke Cans Cans of regular Coke are labeled as containing 12 oz. Assume that the actual contents are normally distributed with a mean of 12.19 oz and a standard deviation of 0.11 oz (based on Data Set 1 in Appendix B).

    a. What percentage of cans contain less than 11.95 oz?

b. To save some of the Coke in cans that are filled too much, the quality-control manager is instituting a plan to remove some of the Coke from cans that are in the top 2%. What is the minimum amount of Coke in such cans that will be "reworked"?

5-4 Beyond the Basics

21. **Curving Test Scores** A teacher informs her psychology class that a test is very difficult, but the grades will be curved. Scores for the test are normally distributed with a mean of 25 and a standard deviation of 5.

 a. If she curves by adding 50 to each grade, what is the new mean? What is the new standard deviation?

 b. Is it fair to curve by adding 50 to each grade? Why or why not?

 c. If the grades are curved according to the following scheme (instead of adding 50), find the numerical limits for each letter grade.

  A: Top 10%

  B: Scores above the bottom 70% and below the top 10%

  C: Scores above the bottom 30% and below the top 30%

  D: Scores above the bottom 10% and below the top 70%

  F: Bottom 10%

 d. Which method of curving the grades is fairer: Adding 50 to each grade or using the scheme given in part (c)? Explain.

22. **SAT Scores** According to data from the College Entrance Examination Board, scores on the SAT-I test have a mean of 1017, and Q_1 is 880. The scores have a distribution that is approximately normal. Find the standard deviation, then use that result to find P_{99}.

23. **Women's Weights** Women's weights are normally distributed with a mean of 143 lb and a standard deviation of 29 lb.

 a. What weights are within 2 standard deviations of the mean?

 b. What percentage of weights are within 2 standard deviations of the mean?

 c. Find the two weights having these properties: The mean is midway between them, and 95% of all weights are between them.

24. **SAT and ACT Tests** Scores by women on the SAT-I test are normally distributed with a mean of 998 and a standard deviation of 202. Scores by women on the ACT test are normally distributed with a mean of 20.9 and a standard deviation of 4.6. Assume that the two tests use different scales to measure the same aptitude.

 a. If a woman gets an SAT score that is the 67th percentile, find her actual SAT score and her equivalent ACT score.

 b. If a woman gets an SAT score of 1220, find her equivalent ACT score.

 ## 5-5 The Central Limit Theorem

This section presents the central limit theorem, one of the most important and useful concepts in statistics. It forms a foundation for estimating population parameters and hypothesis testing—topics discussed at length in the following chapters. We will not present rigorous proofs in this section but will instead focus on the *concepts* and how to apply them. You will need to keep in mind the types of data sets we are considering. Now, instead of using sets of *individual* values, we will work with data sets in which each value is the *mean* of some other sample.

Recall that in Section 4-2 we defined a *random variable* to be a variable that has a single numerical value, determined by chance, for each outcome of a procedure. We also defined a *probability distribution* to be a graph, table, or formula that gives the probability for each value of the random variable. We now define a *sampling distribution* of the mean.

DEFINITION

The **sampling distribution** of the mean is the probability distribution of sample means, with all samples having the same sample size n. (In general, the sampling distribution of any particular statistic is the probability distribution of that statistic.)

Confused? Let's try to make sense of our abstract definition by relating it to **AS** something more concrete, as in the following example.

▷ **EXAMPLE**   **Random Digits** Consider the population of digits 0, 1, 2, 3, 4, 5, 6, 7, 8, 9, which are randomly selected with replacement.

a. *Random variable:* If we conduct trials that consist of randomly selecting a single digit, and if we represent the value of the selected digit by x, then x is a random variable (because its value depends on chance).

b. *Probability distribution:* Assuming that the digits are randomly selected, the probability of each digit is 1/10, which can be expressed as the formula $P(x) = 1/10$. This is a probability distribution (because it describes the probability for each value of the random variable x).

c. *Sampling distribution:* Now suppose that we randomly select many different samples, each of size $n = 4$. (Remember, we are sampling with replacement, so any particular sample might have the same digit occurring more than once.) In each sample, we calculate the sample mean $\bar{x}$ (which is itself a random variable because its value depends on chance). The probability distribution of the sample means $\bar{x}$ is a sampling distribution.

◀

Part (c) of the example illustrates a specific sampling distribution of sample means. Our main objective in this section is to present the important properties of such a sampling distribution. Those properties are included in the statement of the *central limit theorem*.

The Fuzzy Central Limit Theorem

In *The Cartoon Guide to Statistics* by Gonick and Smith, the authors describe the Fuzzy Central Limit Theorem as follows: "Data that are influenced by many small and unrelated random effects are approximately normally distributed. This explains why the normal is everywhere: stock market fluctuations, student weights, yearly temperature averages, SAT scores: All are the result of many different effects." People's heights, for example, are the results of hereditary factors, environmental factors, nutrition, health care, geographic region, and other influences which, when combined, produce normally distributed values.

Central Limit Theorem

Given:

1. The random variable x has a distribution (which may or may not be normal) with mean μ and standard deviation σ.

2. Samples all of the same size n are randomly selected from the population of x values. (The samples are selected so that all possible samples of size n have the same chance of being selected.)

Conclusions:

1. The distribution of sample means $\overline{x}$ will, as the sample size increases, approach a *normal* distribution.

2. The mean of the sample means will approach the population mean μ. (That is, the normal distribution from Conclusion 1 has mean μ.)

3. The standard deviation of the sample means will approach $\sigma/\sqrt{n}$. (That is, the normal distribution from Conclusion 1 has standard deviation $\sigma/\sqrt{n}$.)

Practical Rules Commonly Used:

1. For samples of size n larger than 30, the distribution of the sample means can be approximated reasonably well by a normal distribution. The approximation gets better as the sample size n becomes larger.

2. If the original population is itself normally distributed, then the sample means will be normally distributed for *any* sample size n (not just the values of n larger than 30).

The central limit theorem involves two different distributions: the distribution of the original population and the distribution of the sample means. As in previous chapters, we use the symbols μ and σ to denote the mean and standard deviation of the original population, but we now need new notation for the mean and standard deviation of the distribution of sample means.

AS

Notation for the Central Limit Theorem

If all possible random samples of size n are selected from a population with mean μ and standard deviation σ, the mean of the sample means is denoted by $\mu_{\overline{x}}$, so

$$\mu_{\overline{x}} = \mu$$

Also, the standard deviation of the sample means is denoted by $\sigma_{\overline{x}}$, so

$$\sigma_{\overline{x}} = \frac{\sigma}{\sqrt{n}}$$

$\sigma_{\overline{x}}$ is often called the **standard error of the mean.**

> **EXAMPLE**   **Random Digits** Again consider the population of digits 0, 1, 2, 3, 4, 5, 6, 7, 8, 9, which are randomly selected with replacement. Assume that we randomly select samples of size $n = 4$. In the original population of digits, all of the values are equally likely. Based on the "Practical Rules Commonly Used" (listed in the central limit theorem box), we *cannot* conclude that the sample means are normally distributed, because the original population does not have a normal distribution and the sample size of 4 is not larger than 30. However, we will explore the sampling distribution to see what can be learned.

Table 5-2 was constructed by recording the last four digits of social security numbers from each of 50 different students. The last four digits of social security numbers are random, unlike the beginning digits, which are used to code particular information. If we combine the four digits from each student into one big collection of 200 numbers, we get a mean of $\bar{x} = 4.5$, a standard deviation of $s = 2.8$, and a distribution with the graph shown in Figure 5-19. Now see what happens when we find the 50 sample means, as shown in Table 5-2. Even though the original collection of data does *not* have a normal distribution, the sample means have a distribution that is approximately *normal*. This can be a confusing concept, so you should stop right here and study this paragraph until its major point becomes clear: The original set of 200 individual numbers does *not* have a normal distribution (because the digits 0–9 occur with approximately equal frequencies), but the 50 sample means *do* have a normal distribution. It's a truly fascinating and intriguing phenomenon in statistics that by sampling from any distribution, we can create a distribution that is normal or at least approximately normal.   ◀

Figure 5-20 shows that the distribution of the sample means from the preceding example is approximately normal, even though the original population does not have a normal distribution and the sample size does not

FIGURE 5-19
Distribution of 200 Digits from Social Security Numbers (Last 4 Digits) of 50 Students

FIGURE 5-20
Distribution of 50 Sample Means for 50 Students

TABLE 5-2				
SSN digits				$\bar{x}$
1	8	6	4	4.75
5	3	3	6	4.25
9	8	8	8	8.25
5	1	2	5	3.25
9	3	3	5	5.00
4	2	6	2	3.50
7	7	1	6	5.25
9	1	5	4	4.75
5	3	3	9	5.00
7	8	4	1	5.00
0	5	6	1	3.00
9	8	2	2	5.25
6	1	5	7	4.75
8	1	3	0	3.00
5	9	6	9	7.25
6	2	3	4	3.75
7	4	0	7	4.50
5	7	5	6	5.75
4	1	5	7	4.25
1	2	0	6	2.25
4	0	2	8	3.50
3	1	2	5	2.75
0	3	4	0	1.75
1	5	1	0	1.75
9	7	4	0	5.00
7	3	1	1	3.00
9	1	1	3	3.50
8	6	5	9	7.00
5	6	4	1	4.00
9	3	9	5	6.50
6	0	7	3	4.00
8	2	9	6	6.25
0	2	8	6	4.00
2	0	9	7	4.50
5	8	9	0	5.50
6	5	4	9	6.00
4	8	7	6	6.25
7	1	2	0	2.50
2	9	5	0	4.00
8	3	2	2	3.75
2	7	1	6	4.00
6	7	7	1	5.25
2	3	3	9	4.25
2	4	7	5	4.50
5	4	3	7	4.75
0	4	3	8	3.75
2	5	8	6	5.25
7	1	3	4	3.75
8	3	7	0	4.50
5	6	6	7	6.00

exceed 30. If you closely examine Figure 5-20, you can see that it is not an exact normal distribution, but it would become closer to an exact normal distribution as the sample size increases far beyond 4.

 As the sample size increases, the sampling distribution of sample means approaches a *normal* distribution.

Applying the Central Limit Theorem

Many important and practical problems can be solved with the central limit theorem. When working on such problems, remember the following rules.

- *If you are working with a random sample of size n > 30, or if the original population is normally distributed, treat the distribution of sample means as a normal distribution.*
- *The distribution of the sample means has mean μ.*
- *The distribution of the sample means has a standard deviation found by computing $\sigma/\sqrt{n}$.*

In the following example, part (a) involves an *individual* value, so we use the same methods presented in Section 5-3; those methods apply to the normal distribution of the random variable *x*. Part (b), however, involves the mean for a *group* of 36 women, so we must use the central limit theorem in working with the random variable $\overline{x}$. Study this example carefully to understand the significant difference between the procedures used in parts (a) and (b).

 EXAMPLE **Jet Ejection Seats** In the Chapter Problem, we said that engineers were redesigning fighter jet ejection seats to better accommodate women. In human engineering and product design, it is often important to consider people's weights so that airplanes or elevators aren't overloaded, chairs don't break, and other such dangerous or embarrassing mishaps do not occur. Given that the population of women has normally distributed weights with a mean of 143 lb and a standard deviation of 29 lb (based on data from the National Health Survey), find the probability that

a. if 1 woman is randomly selected, her weight is greater than 150 lb.

b. if 36 different women are randomly selected, their mean weight is greater than 150 lb.

SOLUTION

a. *Approach: Use the methods presented in Section 5-3* (because we are dealing with an *individual* value from a normally distributed population). We seek the area of the green-shaded region in Figure 5-21(a). Before using Table A-2, we convert 150 to the corresponding *z* score:

$$z = \frac{x - \mu}{\sigma} = \frac{150 - 143}{29} = 0.24$$

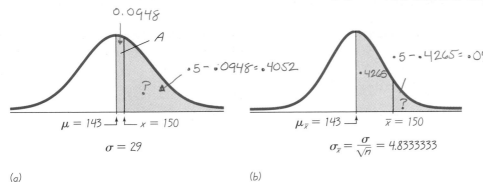

FIGURE 5-21
Distribution of (a) Individual Women's Weights and (b) Means of Samples of 36 Women's Weights

We now refer to Table A-2 and find that orange region A is 0.0948. The green-shaded region is therefore $0.5 - 0.0948 = \underline{0.4052}$. If using Excel, enter **=NORMDIST(150, 143, 29, 1)** to get 0.595369, which is the area to the *left* of 150 lb. It follows that the area to the *right* of 150 lb is $1 - 0.595369 = 0.404631$. The probability of the woman weighing more than 150 lb is 0.405.

b. *Approach: Use the central limit theorem* (because we are dealing with the *mean for a group* of 36 values, not an individual value). Because we are now dealing with a distribution of sample means, we must use the parameters $\mu_{\bar{x}}$ and $\sigma_{\bar{x}}$, which are evaluated as follows:

$$\mu_{\bar{x}} = \mu = 143$$

$$\sigma_{\bar{x}} = \frac{\sigma}{\sqrt{n}} = \frac{29}{\sqrt{36}} = 4.8333333$$

Here is a really important point: Whether we use Table A-2 or Excel, we must use the computed standard deviation of 4.8333333, not the original standard deviation of 29 (because we are working with the distribution of sample means for which the standard deviation is 4.8333333, not the distribution of individual weights for which the standard deviation is 29). We want to find the green-shaded area shown in Figure 5-21(b). If we are using Table A-2, we find the relevant z score, which is calculated as follows:

$$z = \frac{\bar{x} - \mu_{\bar{x}}}{\sigma_{\bar{x}}} = \frac{150 - 143}{\dfrac{29}{\sqrt{36}}} = \frac{7}{4.8333333} = 1.45$$

Referring to Table A-2, we find that $z = 1.45$ corresponds to an area of 0.4265, so the green-shaded region is $0.5 - 0.4265 = 0.0735$. If using Excel, enter **=NORMDIST(150, 143, 4.8333333, 1)** to get 0.92623, which is the area to the *left* of 150 lb. It follows that the area to the *right* of 150 lb is $1 - 0.92623 = 0.07377$. (The small discrepancy is due to rounding errors.) The probability that the 36 women have a mean weight greater than 150 lb is 0.0738, but the answer of 0.0735 found by using Table A-2 is quite close.

> **INTERPRETATION**   There is a 0.405 probability that a woman will weigh more than 150 lb, but there is only a 0.0738 probability that 36 women will have a mean weight of more than 150 lb. It's much easier for an individual to deviate from the mean than it is for a group of 36. A single extreme weight among the 36 weights will have reduced impact when it is averaged in with the other weights.   ◀

Interpreting Results

The next example illustrates another application of the central limit theorem, but carefully examine the conclusion that is reached. This example shows the type of thinking that is the basis for the important procedure of hypothesis testing (discussed in Chapter 7). This example illustrates the rare event rule for inferential statistics, first presented in Section 3-1.

> ### Rare Event Rule
>
> **If, under a given assumption, the probability of a particular observed event is exceptionally small, we conclude that the assumption is probably not correct.**

▷ **EXAMPLE   Body Temperatures** Assume that the population of human body temperatures has a mean of 98.6° F, as is commonly believed. Also assume that the population standard deviation is 0.62° F (based on data from University of Maryland researchers). If a sample of size $n = 106$ is randomly selected, find the probability of getting a mean of 98.2° F or lower. (The value of 98.2° F was actually obtained; see the midnight temperatures for Day 2 in Data Set 6 in Appendix B.)

SOLUTION   We weren't given the distribution of the population, but because the sample size $n = 106$ exceeds 30, we use the central limit theorem and conclude that the distribution of sample means is a normal distribution with these parameters:

$$\mu_{\bar{x}} = \mu = 98.6$$

$$\sigma_{\bar{x}} = \frac{\sigma}{\sqrt{n}} = \frac{0.62}{\sqrt{106}} = 0.06021972$$

Figure 5-22 shows the green-shaded area (see the tiny left tail of the graph) corresponding to the probability we seek. Having already found the parameters that apply to the distribution shown in Figure 5-22, we can now find the green-shaded area by using the same procedures developed in the preceding section. If using Table A-2, we first find the z score:

$$z = \frac{\bar{x} - \mu_{\bar{x}}}{\sigma_{\bar{x}}} = \frac{98.2 - 98.6}{0.06021972} = -6.64$$

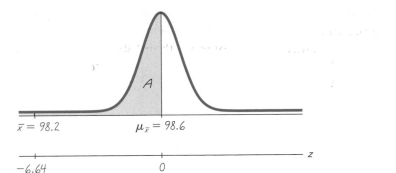

FIGURE 5-22
Distribution of Sample Mean
Body Temperatures ($n = 106$)

Referring to Table A-2, we find that $z = -6.64$ is off the chart, but for values of z above 3.09, we use an area of 0.4999. We therefore conclude that orange region A in Figure 5-22 is 0.4999 and the green-shaded region is $0.5 - 0.4999 = 0.0001$. (More precise results indicate that the area of the shaded region is closer to 0.00000000002, but even those results are only approximations. We can safely report that the probability is quite small, such as less than 0.001.)

INTERPRETATION   The result shows that if the mean of our body temperatures is really 98.6°F, then there is an extremely small probability of getting a sample mean of 98.2°F or lower when 106 subjects are randomly selected. University of Maryland researchers did obtain such a sample mean, and there are two possible explanations: Either the population mean really is 98.6°F and their sample represents a chance event that is extremely rare, or the population mean is actually lower than 98.6°F and so their sample is typical. Because the probability is so low, it seems more reasonable to conclude that the population mean is lower than 98.6°F. This is the type of reasoning used in *hypothesis testing,* to be introduced in Chapter 7. For now, we should focus on the use of the central limit theorem for finding the probability of 0.0001, but we should also observe that this theorem will be used later in developing some very important concepts in statistics.   ◀

Correction for a Finite Population

In applying the central limit theorem, our use of $\sigma_{\bar{x}} = \sigma/\sqrt{n}$ assumes that the population has infinitely many members. When we sample with replacement (that is, put back each selected item before making the next selection), the population is effectively infinite. Yet many realistic applications involve sampling without replacement, so successive samples depend on previous outcomes. In manufacturing, quality-control inspectors typically sample items from a finite production run without replacing them. For such a finite population, we may need to adjust $\sigma_{\bar{x}}$. Here is a common rule of thumb:

**When sampling without replacement and the sample size n is
greater than 5% of the finite population size N (that is, $n > 0.05N$),**

adjust the standard deviation of sample means $\sigma_{\bar{x}}$ by multiplying it by the *finite population correction factor:*

$$\sqrt{\frac{N-n}{N-1}}$$

Except for Exercises 21–23, the examples and exercises in this section assume that the finite population correction factor does *not* apply, because the population is infinite or the sample size doesn't exceed 5% of the population size.

It's very important to note that the central limit theorem works well when we are dealing with the distribution of sample means and either the sample size is greater than 30 or the original population has a normal distribution. Once we have established that the central limit theorem works well, we can determine values for $\mu_{\bar{x}} = \mu$ and $\sigma_{\bar{x}} = \sigma/\sqrt{n}$ and then proceed to use the methods presented in the preceding section.

The central limit theorem is so important because it allows us to use the basic normal distribution methods in a wide variety of different circumstances. In Chapter 6, for example, we will apply the theorem when we use sample data to estimate means of populations. In Chapter 7 we will apply it when we use sample data to test claims made about population means. Such applications of estimating population parameters and testing claims are extremely important uses of statistics, and the central limit theorem makes them possible.

5-5 Basic Skills and Concepts

Using the Central Limit Theorem. *In Exercises 1–6, assume that women's weights are normally distributed with a mean given by $\mu = 143$ lb and a standard deviation given by $\sigma = 29$ lb (based on data from the National Health Survey).*

1. **a.** If 1 woman is randomly selected, find the probability that her weight is between 143 lb and 155 lb.
   b. If 36 women are randomly selected, find the probability that they have a mean weight between 143 lb and 155 lb.

2. **a.** If 1 woman is randomly selected, find the probability that her weight is above 140 lb.
   b. If 100 women are randomly selected, find the probability that they have a mean weight greater than 140 lb.

3. **a.** If 1 woman is randomly selected, find the probability that her weight is above 160 lb.
   b. If 50 women are randomly selected, find the probability that they have a mean weight greater than 160 lb.

4. **a.** If 1 woman is randomly selected, find the probability that her weight is between 140 lb and 157 lb.
   b. If 75 women are randomly selected, find the probability that they have a mean weight between 140 lb and 157 lb.

5. **a.** If 16 women are randomly selected, find the probability that they have a mean weight between 150 lb and 155 lb.

    b. Why can the central limit theorem be used in part (a), even though the sample size does not exceed 30?

EXCEL **6. a.** If 4 women are randomly selected, find the probability that they have a mean weight between 135 lb and 140 lb.

    b. Why can the central limit theorem be used in part (a), even though the sample size does not exceed 30?

EXCEL **7.** Uniform Random-Number Generator The typical computer random-number generator yields numbers from a uniform distribution of values between 0 and 1, with a mean of 0.500 and a standard deviation of 0.289. If 50 random numbers are generated, find the probability that their mean is between 0.6 and 0.7. Would it be *unusual* to generate 50 such numbers and get a mean between 0.6 and 0.7?

EXCEL **8.** Mean IQ Score IQ scores are normally distributed with a mean of 100 and a standard deviation of 15. If 25 people are randomly selected for an IQ test, find the probability that their mean IQ score is between 95 and 105.

EXCEL **9.** Mean Replacement Times The manager of the Portland Electronics store is concerned that his suppliers have been giving him TV sets with lower than average quality. His research shows that replacement times for TV sets have a mean of 8.2 years and a standard deviation of 1.1 years (based on data from "Getting Things Fixed," *Consumer Reports*). He then randomly selects 50 TV sets sold in the past and finds that the mean replacement time is 7.8 years.

    a. Assuming that TV replacement times have a mean of 8.2 years and a standard deviation of 1.1 years, find the probability that 50 randomly selected TV sets will have a mean replacement time of 7.8 years or less.

    b. Based on the result from part (a), does it appear that the Portland Electronics store has been given TV sets with lower than average quality?

EXCEL **10.** Blood Pressure For women aged 18–24, systolic blood pressures (in mm Hg) are normally distributed with a mean of 114.8 and a standard deviation of 13.1 (based on data from the National Health Survey).

    a. If a woman between the ages of 18 and 24 is randomly selected, find the probability that her systolic blood pressure is above 120.

    b. If 12 women in that age bracket are randomly selected, find the probability that their mean systolic blood pressure is greater than 120.

    c. Given that part (b) involves a sample size that is not larger than 30, why can the central limit theorem be used?

EXCEL **11.** Iowa Rain The annual precipitation amounts for Iowa appear to be normally distributed with a mean of 32.473 in. and a standard deviation of 5.601 in. (based on data from the U.S. Department of Agriculture).

    a. If 1 year is randomly selected, find the probability that the annual precipitation is less than 29.000 in.

    b. If a decade of 10 years is randomly selected, find the probability that the annual precipitation amounts have a mean less than 29.000 in.

    c. Given that part (b) involves a sample size that is not larger than 30, why can the central limit theorem be used?

EXCEL **12.** Amounts of Coke Assume that cans of Coke are filled so that the actual amounts have a mean of 12.00 oz and a standard deviation of 0.11 oz.

    a. Find the probability that a sample of 36 cans will have a mean amount of at least 12.19 oz, as in Data Set 1 in Appendix B

(continued)

b. Based on the result from part (a), is it reasonable to believe that the cans are actually filled with a mean of 12.00 oz? If the mean is not 12.00 oz, are consumers being cheated?

EXCEL **13.** Amounts of Pepsi Assume that cans of Pepsi are filled so that the actual amounts have a mean of 12.00 oz and a standard deviation of 0.09 oz.

a. Find the probability that a sample of 36 cans will have a mean amount of at least 12.29 oz, as in Data Set 1 in Appendix B.

b. Based on the result from part (a), is it reasonable to believe that the cans are actually filled with a mean of 12.00 oz? If the mean is not 12.00 oz, are consumers being cheated?

EXCEL **14.** Coaching for the SAT Test Scores for women on the verbal portion of the SAT-I test are normally distributed with a mean of 502 and a standard deviation of 109 (based on data from the College Board). Randomly selected women are given the Columbia Review Course before taking the SAT test. Assume that the course has no effect.

a. If 1 of the women students is randomly selected, find the probability that her score is at least 535.

b. If 25 of the women are randomly selected, find the probability that their mean score is at least 535.

c. In finding the probability for part (b), why can the central limit theorem be used even though the sample size does not exceed 30?

d. If the random sample of 25 women does result in a mean score of 535, is there strong evidence to support the claim that the course is actually effective? Why or why not?

EXCEL **15.** Coaching for the SAT Test Scores for men on the verbal portion of the SAT-I test are normally distributed with a mean of 509 and a standard deviation of 112 (based on data from the College Board). Randomly selected men are given the Columbia Review Course before taking the SAT test. Assume that the course has no effect.

a. If 1 of the men is randomly selected, find the probability that his score is at least 590.

b. If 16 of the men are randomly selected, find the probability that their mean score is at least 590.

c. In finding the probability for part (b), why can the central limit theorem be used even though the sample size does not exceed 30?

d. If the random sample of 16 men does result in a mean score of 590, is there strong evidence to support the claim that the course is actually effective? Why or why not?

EXCEL **16.** Reduced Nicotine in Cigarettes The amounts of nicotine in Dytusoon cigarettes have a mean of 0.941 g and a standard deviation of 0.313 g (based on Data Set 8 in Appendix B). The Huntington Tobacco Company, which produces Dytusoon cigarettes, claims that it has now reduced the amount of nicotine. The supporting evidence consists of a sample of 40 cigarettes with a mean nicotine amount of 0.882 g.

a. Assuming that the given mean and standard deviation have not changed, find the probability of randomly selecting 40 cigarettes with a mean of 0.882 g or less.

b. Based on the result from part (a), is it valid to claim that the amount of nicotine is lower? Why or why not?

EXCEL **17.** Overloading of Waste Disposal Facility The town of Newport operates a rubbish waste disposal facility that is overloaded if its 4872 households discard waste with weights having a mean that exceeds 27.88 lb in a week. For many different weeks, it is found that the samples of 4872 households have weights that are normally distributed with a mean of 27.44 lb and a standard deviation of 12.46 lb (based on data from the Garbage Project at the University of Arizona). What is the proportion of weeks in which the waste disposal facility is overloaded? Is this an acceptable level, or should action be taken to correct a problem of an overloaded system?

EXCEL **18.** Effect of Diet on Length of Pregnancy The lengths of pregnancies are normally distributed with a mean of 268 days and a standard deviation of 15 days.
  a. If 1 pregnant woman is randomly selected, find the probability that her length of pregnancy is less than 260 days.
  b. If 25 randomly selected women are put on a special diet just before they become pregnant, find the probability that their lengths of pregnancy have a mean that is less than 260 days (assuming that the diet has no effect).
  c. If the 25 women do have a mean of less than 260 days, should the medical supervisors be concerned?

EXCEL **19.** Labeling of M&M Packages M&M plain candies have a mean weight of 0.9147 g and a standard deviation of 0.0369 g (based on Data Set 10 in Appendix B). The M&M candies used in Data Set 10 came from a package containing 1498 candies, and the package label stated that the net weight is 1361 g. (If every package has 1498 candies, the mean weight of the candies must exceed $1361/1498 = 0.9085$ g for the net contents to weigh at least 1361 g.)
  a. If 1 M&M plain candy is randomly selected, find the probability that it weighs more than 0.9085 g.
  b. If 1498 M&M plain candies are randomly selected, find the probability their mean weight is at least 0.9085 g.
  c. Given these results, does it seem that the Mars Company is providing M&M consumers with the amount claimed on the label?

EXCEL **20.** Elevator Overloading Women's weights are normally distributed with a mean of 143 lb and a standard deviation of 29 lb (based on data from the National Health Survey). An elevator in the Atlanta Women's Club is limited to 16 women, but it will be overloaded if the 16 women have a mean weight in excess of 157.5 lb (yielding a total weight in excess of $16 \cdot 157.5 = 2520$ lb). If 16 women occupy the elevator, find the probability that their mean weight will exceed 157.5 lb, causing the elevator to be overloaded. If the elevator is occupied by 16 women 12 times each day, how many times is it likely to be overloaded in a year? Is that number low enough to be of no concern?

5-5 Beyond the Basics

EXCEL **21.** Correcting for a Finite Population Repeat Exercise 20, assuming that the population size is $N = 300$ women. Also, because the women in the elevator must be different, assume the sampling is done without replacement. (*Hint:* See the discussion of the finite population correction factor.)

EXCEL **22.** Population Parameters A *population* consists of these values: 2, 3, 6, 8, 11, 18.
  a. Find μ and σ.
  b. List all samples of size $n = 2$ that are obtained without replacement.

(continued)

   c. Find the population of all values of $\bar{x}$ by finding the mean of each sample from part (b).

   d. Find the mean $\mu_{\bar{x}}$ and standard deviation $\sigma_{\bar{x}}$ for the population of sample means found in part (c).

   e. Verify that

$$\mu_{\bar{x}} = \mu \qquad \text{and} \qquad \sigma_{\bar{x}} = \frac{\sigma}{\sqrt{n}}\sqrt{\frac{N-n}{N-1}}$$

23. Using Finite Population Correction Factor The finite population correction factor can be ignored when sampling with replacement or when $n \leq 0.05N$. When collecting a sample (without replacement) that is 5% of the population N, what do the values of the finite population correction factor have in common for values of $N \geq 600$? (*Hint:* Find the value of the finite population correction factor for a few values of $N \geq 600$ with $n = 0.05N$. What do the results have in common?)

 24. Failure of Practical Rule According to a rule commonly used, the distribution of sample means can be considered to be normal provided that the sample size exceeds 30. There are rare cases where this rule does not work. Consider a binomial procedure with 36 trials and a probability of success equal to 0.002. (This is like randomly selecting 36 slips with replacement from a bowl of 500 slips with the number 1 on one slip and the number 0 on the others.) Show that even though the sample size exceeds 30, the means of the numbers of successes have a distribution that is *not* normal. For example, use Excel to generate 1000 results of a binomial procedure with $p = 0.002$ and $n = 36$. After loading Excel, click on **Tools,** then click on **Data Analysis,** then select **Random Number Generator** and click **OK.** In the dialog box, for the number of variables enter 1, for the number of random numbers enter 1000, for the distribution select "Binomial," enter a p value of 0.002, enter a number of trials of 36, then click **OK.** Excel will proceed to generate the 1000 values which will be listed in column A. Remember, each entry is the sum of 36 sample values. Convert the column A sums to means by dividing each entry by 36. (Click on cell B1 and enter $=$A1/36 and press the **Enter** key. Then click and drag the lower right corner of cell B1 down.) Construct a histogram of the 1000 means in column B (see Section 2-3) and verify that it is not a normal distribution.

5-6 Normal Distribution as Approximation to Binomial Distribution

In Section 4-3 we introduced the *binomial probability distribution,* which has these four requirements:

1. The procedure must have a *fixed number of trials.*

2. The trials must be *independent.*

3. Each trial must have all outcomes classified into *two categories.*

4. The probabilities must remain *constant* for each trial.

   In Section 4-3 we presented three methods for finding binomial probabilities: using the binomial probability formula, using Table A-1, and using Excel. In many cases, however, none of those methods are practical, because the calculations require too much time and effort. We now present a new method, which uses a normal distribution as an approximation of a binomial distribution. The following box summarizes the key point of this section.

Normal Distribution as Approximation to Binomial Distribution

If $np \geq 5$ and $nq \geq 5$, then the binomial random variable is approximately normally distributed with the mean and standard deviation given as

$$\mu = np$$
$$\sigma = \sqrt{npq}$$

See Figure 5-23, which shows the probability histogram for a binomial procedure in which there are $n = 100$ trials and the probability of success is $p = 0.5$. The binomial distribution is represented by the *bars* in Figure 5-23, but the approximating normal distribution has been superimposed so that you can clearly see how well it fits. The formal justification that allows us to use the normal distribution as an approximation to the binomial distribution results from more advanced mathematics, but Figure 5-23 is a convincing visual argument supporting that approximation.

When solving binomial probability problems, the normal approximation approach should generally be used after determining that other exact procedures either cannot be used or require too much time and effort. These procedures, taken from Section 4-3, should be tried first because they yield more exact results.

1. Use Excel if available.
2. If Excel is not available, use Table A-1, if possible.

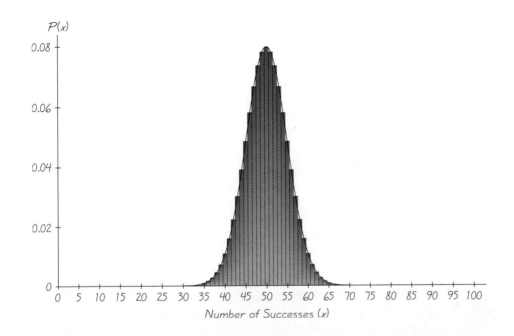

FIGURE 5-23
Probability Histogram for a Binomial Procedure with $n = 100$ and $p = 0.5$

3. If Excel is not available and the probabilities can't be found using Table A-1, use the binomial probability formula.

If the binomial probability cannot be found using these exact procedures, try the technique of using the normal distribution as an approximation to the binomial distribution. This approach involves the following procedure, which is also shown as a flowchart in Figure 5-24.

FIGURE 5-24
Solving Binomial Probability Problems Using a Normal Approximation

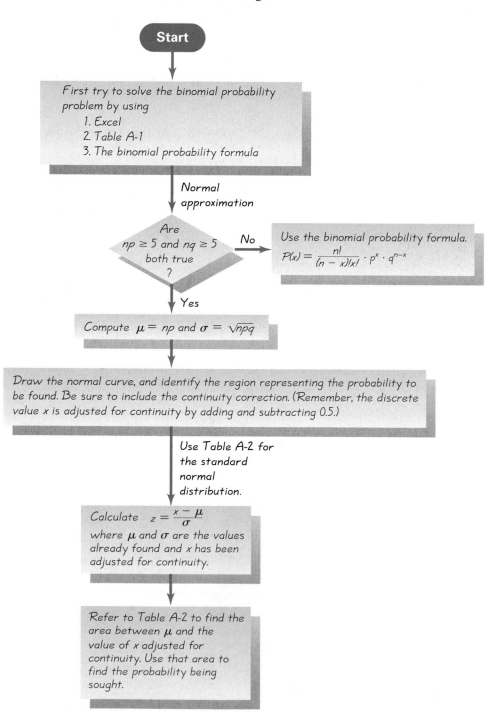

Procedure for Using a Normal Distribution to Approximate a Binomial Distribution

1. Establish that the normal distribution is a suitable approximation to the binomial distribution by verifying that $np \geq 5$ and $nq \geq 5$. (If these conditions are not both satisfied, then you will have to use software, or a calculator, or calculations with the binomial probability formula.)

2. Find the values of the parameters μ and σ by calculating $\mu = np$ and $\sigma = \sqrt{npq}$.

3. Identify the discrete value x (the number of successes). Change the *discrete* value x by replacing it with the *interval* from $x - 0.5$ to $x + 0.5$. (See the discussion under the subheading "Continuity Corrections" found later in this section.) Draw a normal curve and enter the values of μ, σ, and either $x - 0.5$ or $x + 0.5$, as appropriate.

4. Change x by replacing it with $x - 0.5$ or $x + 0.5$, as appropriate.

5. Find the area corresponding to the desired probability.

Using Table A-2 for the standard normal distribution: If you have been using Table A-2 in this chapter, find the z score: $z = (x - \mu)/\sigma$. Now use that z score to find the area between μ and either $x - 0.5$ or $x + 0.5$, as appropriate. That area can now be used to identify the area corresponding to the desired probability.

Using Excel: Use the command **=NORMDIST(x, μ, σ, 1),** where the values of x, μ, and σ are described in Steps 2, 3, and 4 above. Remember that this command will provide the cumulative area to the *left* of the x value, so adjust the result accordingly.

We will illustrate this normal approximation procedure with the following example.

> **EXAMPLE** **Gender Discrimination** The president of Portland College finds that applicants for admission are evenly divided between men and women, and both genders are equally qualified. She concludes that accepted students should be roughly 50% men and 50% women. She checks and finds that among the 1000 students accepted last year, 520 are men.
>
> Find the probability of randomly selecting *at least* 520 men, assuming that the very large pool of applicants consists of equal numbers of equally qualified men and women. (The probability of getting *exactly* 520 men doesn't really tell us anything, because with 1000 trials, the probability of any exact number of men is fairly small. Instead, we need the probability of getting a result *at least* as extreme as the one obtained.) Based on the probability of getting at least 520 men, does there appear to be discrimination based on gender?
>
> **SOLUTION** Refer to Figure 5-24 for the procedure followed in this solution. The given problem does involve a binomial distribution with a

She Won the Lottery Twice!

Evelyn Marie Adams won the New Jersey Lottery twice in four months. This happy event was reported as an incredible coincidence with a likelihood of only 1 chance in 17 trillion. But Harvard mathematicians Persi Diaconis and Frederick Mosteller note that there is 1 chance in 17 trillion that a particular person with one ticket in each of two New Jersey lotteries will win both times. However, there is about 1 chance in 30 that someone in the United States will win a lottery twice in a four-month period. Diaconis and Mosteller analyzed coincidences and conclude that "with a large enough sample, any outrageous thing is apt to happen."

fixed number of trials ($n = 1000$), which are presumably independent, two categories (man, woman) of outcome for each trial, and a probability of 0.5 that presumably remains constant from trial to trial.

We will assume that Exel is not available. Table A-1 does not apply, because it stops at $n = 15$. The binomial probability formula is not practical, because we would have to use it 481 times (once for each value of x from 520 to 1000 inclusive), and that would be crazy.

Let's proceed with the five-step approach of using a normal distribution to approximate the binomial distribution.

Step 1:   We must first verify that it is reasonable to approximate the binomial distribution by the normal distribution because $np \geq 5$ and $nq \geq 5$, as verified below.

$$np = 1000 \cdot 0.5 = 500 \qquad (\text{Therefore } np \geq 5.)$$
$$nq = 1000 \cdot 0.5 = 500 \qquad (\text{Therefore } nq \geq 5.)$$

Step 2:   We now proceed to find the values for μ and σ that are needed for the normal distribution. We get the following:

$$\mu = np = 1000 \cdot 0.5 = 500$$
$$\sigma = \sqrt{npq} = \sqrt{1000 \cdot 0.5 \cdot 0.5} = 15.8113883$$

Step 3:   The discrete value of 520 is represented by the vertical strip bounded by 519.5 and 520.5. (See the discussion of continuity corrections, which follows this example.)

Step 4:   Because we want the probability of *at least* 520 men, we want the area representing the discrete number of 520 (the region bounded by 519.5 and 520.5), as well as the area to the right, as shown in Figure 5-25.

Step 5:   We can now proceed to find the green shaded area of Figure 5-25 by using the same methods used in Section 5-3. If we plan to use Table A-2 for the standard normal distribution, we must first convert 519.5 to a z score, then use the table to find the

FIGURE 5-25
Finding the Probability of "At Least" 520 Men Among 1000 Accepted Applicants

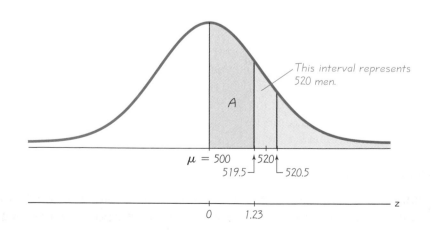

area of region A, which is then subtracted from 0.5. The z score is found as follows:

$$z = \frac{x - \mu}{\sigma} = \frac{519.5 - 500}{15.8113883} = 1.23$$

Using Table A-2, we find that $z = 1.23$ corresponds to an area of 0.3907, so the orange region A is 0.3907 and the green-shaded area is $0.5 - 0.3907 = 0.1093$. We can round the probability to three significant digits as 0.109.

INTERPRETATION   There is a 0.109 probability of getting at least 520 men among the 1000 applicants. That is, if applications from men and women are treated equally, then there is a 0.109 probability of getting at least 520 men. Because that probability of 0.109 is not very low, we conclude that the result of 520 men could easily happen by chance. There does not appear to be discrimination based on gender. (We would have concluded that there appears to be discrimination only if the probability had been very small, such as 0.01, indicating that it is very unlikely that we would get at least 520 men by chance.) The reasoning here will be considered in more detail in Chapter 7 when we discuss formal methods of testing hypotheses. For now, we should focus on the method of finding the probability by using the normal approximation technique.   ◀

Continuity Corrections

The procedure for using a normal distribution to approximate a binomial distribution includes a step in which we change a discrete number to an interval that is 0.5 below and 0.5 above the discrete number. See the preceding solution, where we changed 520 to the interval between 519.5 and 520.5. This particular step, called a *continuity correction,* is usually difficult to understand, so we will now consider it in more detail.

DEFINITION

When we use the normal distribution (which is continuous) as an approximation to the binomial distribution (which is discrete), a **continuity correction** is made to a discrete whole number x in the binomial distribution by representing the single value x by the *interval* from $x - 0.5$ to $x + 0.5$ (that is, adding and subtracting 0.5).

The following practical suggestions should help you use continuity corrections properly.

Procedure for Continuity Corrections

1. When using the normal distribution as an approximation to the binomial distribution, *always* use the continuity correction. (It is required because we are using the *continuous* normal distribution to approximate the *discrete* binomial distribution.)

2. In using the continuity correction, first identify the discrete whole number x that is relevant to the binomial probability problem. For example, if you're trying to find the probability of getting at least 520 men in 1000 randomly selected people, the discrete whole number of concern is $x = 520$. First focus on the x value itself, and temporarily ignore whether you want at least x, more than x, fewer than x, or whatever.

3. Draw a normal distribution centered about μ, then draw a *vertical strip area* centered over x. Mark the left side of the strip with the number equal to $x - 0.5$, and mark the right side with the number equal to $x + 0.5$. For $x = 520$, for example, draw a strip from 519.5 to 520.5. *Consider the entire area of the strip to represent the probability of the discrete number x itself.*

4. Now determine whether the value of x itself should be included in the probability you want. (For example, "at least x" does include x itself, but "more than x" does not include x itself.) Next, determine whether you want the probability of at least x, at most x, more than x, fewer than x, or exactly x. Shade the area to the right or left of the strip, as appropriate; also shade the interior of the strip itself *if and only if x itself* is to be included. This total shaded region corresponds to the probability being sought.

To see how this procedure results in continuity corrections, see the common cases illustrated in Figure 5-26. Those cases correspond to the statements in the following list.

Statement	Area
At least 520 (includes 520 and above)	To the *right* of 519.5
More than 520 (doesn't include 520)	To the *right* of 520.5
At most 520 (includes 520 and below)	To the *left* of 520.5
Fewer than 520 (doesn't include 520)	To the *left* of 519.5
Exactly 520	Between 519.5 and 520.5

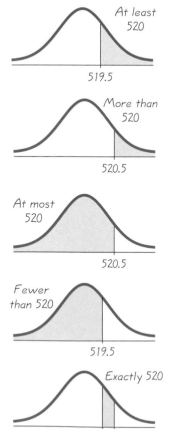

▷ **EXAMPLE   Defective Tires** About 4.4% of fatal motor vehicle crashes are caused by defective tires (based on data from the National Safety Council). If a highway safety study begins with the random selection of 750 cases of fatal motor vehicle crashes, estimate the probability that *exactly* 35 of them were caused by defective tires.

SOLUTION   The conditions described satisfy the criteria for the binomial distribution with $n = 750$, $p = 0.044$, $q = 0.956$, and $x = 35$. (The value of q is found from $q = 1 - p = 1 - 0.044 = 0.956$.) For the purposes of this example, we assume that Excel is not available. Table A-1 cannot be used, because $n = 750$ exceeds the largest table value of $n = 15$. The binomial probability formula applies, but

$$P(35) = \frac{750!}{(750 - 35)! \, 35!} \cdot 0.044^{35} \cdot 0.956^{750-35}$$

FIGURE 5-26
Identifying the Correct Area

is difficult to compute because many calculators cannot evaluate 70! or higher. We therefore proceed with the approach of using the normal distribution to approximate the binomial distribution.

Step 1: We begin by checking to determine whether the approximation is suitable:

$$np = 750 \cdot 0.044 = 33.0 \qquad \text{(Therefore } np \geq 5.\text{)}$$
$$nq = 750 \cdot 0.956 = 717.0 \qquad \text{(Therefore } nq \geq 5.\text{)}$$

Because np and nq are both at least 5, we conclude that the normal approximation to the binomial distribution is satisfactory.

Step 2: We now find the values of μ and σ needed for the normal distribution:

$$\mu = np = 750 \cdot 0.044 = 33.0$$
$$\sigma = \sqrt{npq} = \sqrt{(750)(0.044)(0.956)} = 5.6167606$$

Step 4: We draw the normal curve shown in Figure 5-27. The green-shaded region of the figure represents the probability we want. Use of the continuity correction results in the representation of 35 by the region between 34.5 and 35.5.

Step 5: If we use Table A-2 to find the green-shaded region, we must first find the probability corresponding to the region bounded on the left by the vertical line through the mean of 33.0 and on the right by the vertical line through 35.5. Therefore, one of the calculations required in this step is as follows:

$$z = \frac{35.5 - 33.0}{5.6167606} = 0.45$$

We also need the probability corresponding to the region bounded by 33.0 and 34.5, so we calculate

$$z = \frac{34.5 - 33.0}{5.6167606} = 0.27$$

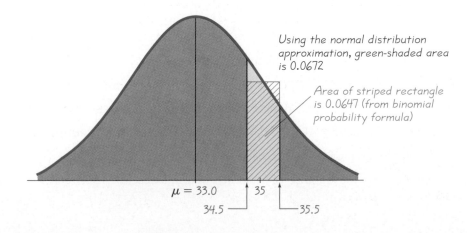

Using the normal distribution approximation, green-shaded area is 0.0672

Area of striped rectangle is 0.0647 (from binomial probability formula)

$\mu = 33.0$ 35

34.5 — 35.5

FIGURE 5-27
Distribution of the Number of Fatal Crashes Caused by Defective Tires

We use Table A-2 to find that $z = 0.45$ corresponds to a probability of 0.1736 and $z = 0.27$ corresponds to a probability of 0.1064. Consequently, the green-shaded region of Figure 5-27 depicts a probability of $0.1736 - 0.1064 = 0.0672$.

INTERPRETATION   The probability of getting exactly 35 cases (out of 750) of fatal crashes caused by defective tires is approximately 0.0672.   ◀

If we solve the preceding example using Excel, we get a result of 0.0647, but the normal approximation method resulted in a value of 0.0672. The discrepancy of 0.0025 occurs because the use of the normal distribution results in an *approximate* value that is the area of the green-shaded region in Figure 5-27, whereas the exact correct area is a rectangle centered above 35. (Figure 5-27 illustrates this discrepancy.) The area of the rectangle is 0.0647, but the area of the approximating green-shaded region is 0.0672.

Interpreting Results

In reality, when we use a normal distribution as an approximation to a binomial distribution, our ultimate goal is not simply to find a probability number. We often need to make some *judgment* based on the probability value, so we should understand that low probabilities correspond to events that are very unlikely, whereas large probabilities correspond to likely events. The probability value of 0.05 is often used as a cutoff to distinguish between unlikely events and likely events. Using 0.05 as a cutoff, the rare event rule can be restated as follows:

> **If, under a given assumption, the probability of a particular observed result is 0.05 or less, we conclude that the assumption is probably not correct.**

5-6   Basic Skills and Concepts

Applying Continuity Correction.   In Exercises 1–8, the given values are discrete. Use the continuity correction and describe the region of the normal curve that corresponds to the indicated probability. For example, the probability of "more than 47 defective items" corresponds to the area of the normal curve described with this answer: "the area to the right of 47.5."

1. Probability of more than 57 defective aircraft emergency locator transmitters

2. Probability of at least 57 defective aircraft emergency locator transmitters

3. Probability of fewer than 57 defective aircraft emergency locator transmitters

4. Probability of exactly 57 defective aircraft emergency locator transmitters

5. Probability of no more than 24 correct answers on an ACT test

6. Probability that the number of students who are absent is between 5 and 8 inclusive

7. Probability that the number of students with jobs is between 16 and 22 inclusive

8. Probability that the number of patients with group A blood is exactly 12

Using Normal Approximation.   In Exercises 9–12, do the following. (a) Find the indicated binomial probability by using Table A-1 in Appendix A. (b) If $np \geq 5$ and $nq \geq 5$, also estimate the indicated probability by using the normal distribution as an approximation to the binomial distribution; if $np < 5$ or $nq < 5$, then state that the normal approximation is not suitable.

9. With $n = 10$ and $p = 0.8$, find $P(7)$.

10. With $n = 14$ and $p = 0.40$, find $P(8)$.

11. With $n = 15$ and $p = 0.60$, find P(at least 9).

12. With $n = 12$ and $p = 0.90$, find P(fewer than 10).

EXCEL 13. Probability of At Least 52 Girls Estimate the probability of getting at least 52 girls in 100 births. Assume that boys and girls are equally likely.

EXCEL 14. Probability of Exactly 53 Girls Estimate the probability of getting exactly 53 girls in 100 births. Assume that boys and girls are equally likely.

EXCEL 15. Probability of At Least Passing Estimate the probability of passing a true/false test of 50 questions if 60% (or 30 correct answers) is the minimum passing grade and all responses are random guesses.

EXCEL 16. Multiple-Choice Test A multiple-choice test consists of 25 questions with possible answers of a, b, c, d, and e. Estimate the probability that with random guessing, the number of correct answers is between 3 and 10 inclusive.

EXCEL 17. Directory Assistance Errors An article in *USA Today* stated that "Internal surveys paid for by directory assistance providers show that even the most accurate companies give out wrong numbers 15% of the time." Assume that you are testing AT&T by making 50 requests and also assume that this provider gives the wrong number 15% of the time. Estimate the probability of getting fewer than three wrong numbers. If you did get fewer than three wrong numbers, what would you conclude about the 15% error rate?

EXCEL 18. Probability of At Least 50 Color-Blind Men Nine percent of men and 0.25% of women cannot distinguish between the colors red and green. This is the type of color blindness that causes problems with traffic signals. Researchers need at least 50 men with this type of color blindness, so they randomly select 600 men for a study of traffic-signal perceptions. Estimate the probability that at least 50 of the men cannot distinguish between red and green. Is the result high enough so that the researchers can be very confident of getting at least 50 men with red and green color blindness?

EXCEL 19. Medical School Admissions: Probability of Fewer Than 456 Graduates A study was conducted to determine whether there were significant differences between medical students admitted through special programs (such as affirmative action) and medical students admitted with the regular admissions criteria. It was found that the graduation rate was 94% for the medical students admitted through special programs (based on data from the *Journal of the American Medical Association*). Would it be unusual to randomly select 500 students from the special programs and get fewer than 456 who graduate? Why or why not?

EXCEL **20.** TV Share Value The CBS television show *60 Minutes* recently had a "share" of 20, meaning that among the TV sets in use, 20% were tuned to *60 Minutes* (based on data from Nielsen Media Research). An advertiser wants to verify that value by conducting its own survey of 200 households with TV sets in use at the time of a *60 Minutes* broadcast. Estimate the probability that fewer than 35 of the 200 households are tuned to *60 Minutes*. If fewer than 35 of the 200 households are tuned to *60 Minutes,* is there strong evidence to conclude that the share value of 20% is wrong?

EXCEL **21.** Overbooking Flights Air America is considering a new policy of booking as many as 400 persons on an airplane that can seat only 350. (Past studies have revealed that only 85% of the booked passengers actually arrive for the flight.) Estimate the probability that if Air America books 400 persons, not enough seats will be available. Is that probability low enough to be workable, or should the policy be changed?

EXCEL **22.** On-Time Flights Recently, Southwest Air had 80% of its flights arriving on time, which was the best rate among U.S. airlines (based on data from the U.S. Department of Transportation). In a check of 150 randomly selected Southwest Air flights, 113 arrived on time. Estimate the probability of getting 113 or fewer on-time flights among 150, assuming that the 80% rate is correct. Is it unusual to get 113 or fewer on-time flights among 150 randomly selected flights?

EXCEL **23.** Identifying Gender Discrimination After being rejected for employment, Kim Kelly learns that the Bellevue Advertising Company has hired only 21 women among its last 62 new employees. She also learns that the pool of applicants is very large, with an equal number of qualified men and women. Help her in her charge of gender discrimination by estimating the probability of getting 21 or fewer women when 62 people are hired, assuming no discrimination based on gender. Does the resulting probability really support such a charge?

EXCEL **24.** M&M Candies: Are 10% Blue? According to a consumer affairs representative from Mars (the candy company, not the planet), 10% of all M&M plain candies are blue. Data Set 10 in Appendix B shows that among 100 M&Ms chosen, 5 are blue. Estimate the probability of randomly selecting 100 M&Ms and getting 5 or fewer that are blue. Assume that the company's 10% blue rate is correct. Based on the result, is it very unusual to get 5 or fewer blue M&Ms when 100 are randomly selected?

EXCEL **25.** Employee Drug Testing Currently, about two-thirds of U.S. companies test newly hired employees for drugs, and 3.8% of those prospective employees test positive (based on data from the American Management Association). The Sigma Electronics Company tests 150 prospective employees and finds that 10 of them test positive for drugs. Estimate the probability of 10 or more positive results among 150 subjects. Based on that value, do the 10 positive test results seem to be unusually high?

EXCEL **26.** Blue Genes Some couples have genetic characteristics configured so that one-quarter of all their offspring have blue eyes. A study is conducted of 120 couples believed to have those characteristics, with the result that 20 of their 120 offspring have blue eyes. Estimate the probability that among 120 offspring, 20 or fewer have blue eyes. Based on that probability, does it seem that the one-quarter rate is wrong?

EXCEL 27. Blood Group Providence Memorial Hospital is conducting a blood drive because its supply of group O blood is low, and it needs 177 donors of group O blood. If 400 volunteers donate blood, estimate the probability that the number with group O blood is at least 177. Forty-five percent of us have group O blood, according to data provided by the Greater New York Blood Program.

EXCEL 28. Acceptance Sampling We stated in Section 3-4 that some companies monitor quality by using a method of acceptance sampling, whereby an entire batch of items is rejected if a random sample of a particular size includes more than some specified number of defects. The Dayton Machine Company buys machine bolts in batches of 5000 and rejects a batch if, when 50 of them are sampled, at least 2 defects are found. Estimate the probability of rejecting a batch if the supplier is manufacturing the bolts with a defect rate of 10%. Is this monitoring plan likely to identify the unacceptable rate of defects?

5-6 Beyond the Basics

EXCEL 29. Winning at Roulette Marc Taylor plans to place 200 bets of $1 each on the number 7 at roulette. A win pays off with odds of 35:1 and, on any one spin, there is a probability of 1/38 that 7 will be the winning number. Among the 200 bets, what is the minimum number of wins needed for Marc to make a profit? Estimate the probability that Marc will make a profit.

EXCEL 30. Interpreting Probability Value Twenty-six percent of college students live in campus housing (based on data from the Independent Insurance Agents of America). The Providence Insurance Company randomly selects 1000 college students. Estimate the probability that exactly 250 of them live in campus housing. If 250 of the selected students do live in campus housing, we should *not* conclude that the 26% rate is wrong, even though the probability of exactly 250 is low. Why not?

EXCEL 31. Replacement of TVs Replacement times for TV sets arc normally distributed with a mean of 8.2 years and a standard deviation of 1.1 years (based on data from "Getting Things Fixed," *Consumer Reports*). Estimate the probability that for 250 randomly selected TV sets, at least 15 of them have replacement times greater than 10.0 years.

EXCEL 32. Joltin' Joe Assume that a baseball player hits .350, so his probability of a hit is 0.350. (Ignore the complications caused by walks.) Also assume that his hitting attempts are independent of each other.
   a. Find the probability of at least 1 hit in 4 tries in 1 game.
   b. Assuming that this batter gets up to bat 4 times each game, estimate the probability of getting a total of at least 56 hits in 56 games.
   c. Assuming that this batter gets up to bat 4 times each game, find the probability of at least 1 hit in each of 56 consecutive games (Joe DiMaggio's 1941 record).
   d. What minimum batting average would be required for the probability in part (c) to be greater than 0.1?

33. Using Continuity Correction In each of the following, find the difference between the answers obtained with and without use of the continuity correction. What do you conclude from the results?

(continued)

   a. Estimate the probability of getting at least 11 girls in 20 births.

   b. Estimate the probability of getting at least 22 girls in 40 births.

   c. Estimate the probability of getting at least 220 girls in 400 births.

34. *Overbooking Flights* Vertigo Airlines works only with advance reservations and experiences a 7% rate of no-shows. How many reservations could be accepted for an airliner with a capacity of 250 if there is at least a 0.95 probability that all reservation holders who show will be accommodated?

 # 5-7 Determining Normality

Some important statistical methods require that the sample data were collected from a population that has a *normal* distribution. It is therefore very important to be able to determine whether we are working with data that are normally distributed. In this section we describe one of many different procedures that could be used for determining whether the population distribution is normal. That procedure follows the definition of the *normal quantile plot.*

> ### DEFINITIONS
>
> A **normal quantile plots** is a graph of points (x, y), where each point represents a pairing of these two values: (1) a value from the original set of sample data, and (2) a z score corresponding to a quantile value of the standard normal distribution. (See Step 3 in the following procedure for details on finding these z scores.) A **normal probability plot** is a similar graph that can be interpreted the same way.

Procedure for Determining Whether Data Have a Normal Distribution

1. *Histogram:* Construct a histogram. Reject normality if the histogram departs dramatically from a bell shape.

2. *Outliers:* Identify outliers. Reject normality if there is more than one outlier present. (Just one outlier could be an error or the result of chance variation, but be careful, because even a single outlier can have a dramatic effect on results.)

3. *Normal quantile plot:* If the histogram is basically symmetric and there is at most one outlier, construct a *normal quantile plot.* The following steps describe the construction of a normal quantile plot, but the procedure is messy enough so that we usually use software or a calculator to generate the graph.

   a. First sort the data by arranging the values in order from lowest to highest.

   b. With a sample of size n, each value represents a proportion of $1/n$ of the sample, so use the standard normal distribution to find the z score

20%	20%	20%	20%	20%

P_{10} P_{30} P_{50} P_{70} P_{90}

$z = -1.28$ $z = -0.52$ $z = 0$ $z = 0.52$ $z = 1.28$

FIGURE 5-28
Finding Five Quantiles

corresponding to these areas under the normal curve and extending from the extreme left: $1/2n, 3/2n, 5/2n, 7/2n$, and so on.

c. Plot the points (x, y), where each x is an original sample value and y is the corresponding z score found in Step (b). Some books or computer programs, such as the DDXL Excel add-in, reverse the roles of x and y.

Examine the normal quantile plot and reject normality if the points do not lie close to a straight line, or if the points exhibit some systematic pattern that is not a straight-line pattern.

Steps 1 and 2 are straightforward, but we will describe the normal quantile plot required in Step 3. Recall that in Section 2-6 we introduced percentiles, quartiles, and deciles, and we noted that they are all types of *quantiles,* which partition data into parts that are approximately equal. For example, suppose we have five sample values, so that each represents 20% of the sample. See Figure 5-28, where we show the five blocks of 20%. See the locations of the percentiles P_{10}, P_{30}, P_{50}, P_{70}, and P_{90}. The z score of -1.28 is found by using the standard normal distribution (from Table A-2 or Excel) to find the value separating an area of 10% in the left tail. The z score of -0.52 separates an area of 30% in the left tail, and so on. If the sample data come from a normal distribution, this plot of the original values against the z scores should result in points that lie close to a straight line.

Using Excel

The DDXL add-in can be used to construct a *normal probability plot.* (If DDXL does not appear on the menu bar, install the DDXL add-in by following the procedure described in Section 1-5.)

Note: The normal probability plot generated by the DDXL add-in has the sample data plotted on the y-axis; the x-axis is used to plot the corresponding values of the standard normal distribution that are expected for normally distributed data. Although the DDXL normal probability plot switches the role of the x- and y-axes as described in Step 3(c) of the above procedure, the interpretation is the same: Reject normality if the points do not lie close to a straight line, or if the points exhibit some systematic pattern that is not a straight-line pattern.

1. Click on **DDXL.**
2. Click on **Charts and Plots.**
3. Select the function type of **Normal Probability Plot.**
4. Click on the pencil icon for "Quantitative Variable" and Enter the range of values for the sample data. For example, if there are 36 values in rows 1 through 36 of column A, enter A1:A36.
5. Click **OK.**

A Professional Speaks About Sampling Error

▷ **EXAMPLE** **Diet Pepsi** A statistical procedure has a fairly strict requirement that a population must have a normal distribution. Use the sample of 36 weights of diet Pepsi listed in Data Set 1 in Appendix B and determine whether the sample appears to come from a population with a normal distribution.

SOLUTION

Step 1:   We begin by constructing a histogram of the 36 weights, with the result shown in the accompanying Excel display. An examination of the histogram shows that the data appear to be normally distributed.

Excel Histogram

Excel Normal Probability Plot

Step 2:   There do not appear to be any outliers.

Step 3:   The Excel screen display of the normal probability plot is also shown.

INTERPRETATION   Examination of the normal probability plot shows that the pattern of points is close to a straight line, and there is no systematic pattern that is not a straight-line pattern. We conclude that the sample data appear to come from a population with a normal distribution. ◀

EXAMPLE **Boston Rainfall** In Data Set 17 in Appendix B, use the 52 rainfall amounts listed for Sundays in Boston and test for normality.

▷ **SOLUTION**   The accompanying Excel screen displays show the histogram and normal probability plot. The histogram is extremely skewed, and the normal probability plot shows points that do not lie close to a straight line. Rainfall amounts in Boston on Sunday do not appear to be normally distributed. ◀

Excel Histogram

Excel Normal Probability Plot

We conclude this section with a brief summary:

- If the requirement of a normal distribution is not too strict, examination of a histogram and consideration of outliers may be all that you need to determine normality.

- Normal quantile plots or normal probability plots can be difficult to construct on your own, but they can be generated with Excel.

- In addition to the procedures discussed in this section, there are other more advanced procedures, such as the chi-square goodness-of-fit test, the Kolmogorov-Smirnov test, and the Lilliefors test. (See "Beyond Basic Statistics With the Graphing Calculator, Part I: Assessing Goodness-of-Fit," by Calzada and Scariano, *Mathematics and Computer Education*.)

5-7   Basic Skills and Concepts

Interpreting Normal Quantile Plots.   *In Exercises 1–4, examine the normal probability plot and determine whether it depicts data that have a normal distribution.*

Determining Normality.   *In Exercises 5–8, refer to the indicated data set and determine whether the requirement of a normal distribution is satisfied. Assume that this requirement is loose in the sense that the population distribution need not be exactly normal, but it must be a distribution that is basically symmetric with only one mode.*

5. Boston Rainfall The Wednesday rainfall amounts for Boston, as listed in Data Set 17 in Appendix B. The Excel workbook is BOSTRAIN.xls.

6. Weights of M&Ms The weights of the brown M&M candies, as listed in Data Set 10 in Appendix B. The Excel workbook is M&M.xls.

7. Old Faithful Geyser The durations of eruptions of the Old Faithful geyser, as listed in Data Set 11 in Appendix B. The Excel workbook is OLDFAITH.xls.

8. Weights of Sugar Packets The weights of Domino sugar packets, as listed in Data Set 4 in Appendix B. The Excel workbook is SUGAR.xls.

Generating Normal Quantile Plots.   *In Exercises 9–12, use the data from the indicated exercise in this section. Use Excel to generate a normal probability plot. Generate the graph, then determine whether the data come from a normally distributed population.*

9. Exercise 5   **10.** Exercise 6

11. Exercise 7   **12.** Exercise 8

5-7 Beyond the Basics

13. Constructing Normal Quantile Plot Use this sample of weights (in pounds) of randomly selected women: 154, 188, 103, 142, 137, 157, 102.
  a. Because there are seven values, the normal quantile z scores can be found by using the areas of $1/2n$, $3/2n$, $5/2n$, $7/2n$, $9/2n$, $11/2n$, and $13/2n$, where $n = 7$. Find those areas.
  b. Each area from part (a) is an area of the standard normal distribution extending from the left and bounded on the right by a vertical line above a standard z score. Find the z scores.
  c. After first sorting the original set of sample values, pair them with the z scores found in part (b) and construct the normal quantile plot.

14. Instead of using z scores for a normal quantile plot, suppose each value in a sample is converted to its corresponding standard score using $z = (x - \bar{x})/s$. If the (x, z) points are plotted in a graph, can this graph be used to determine whether the sample comes from a normally distributed population? Explain your answer.

IQs Are Rising

NEW YORK In the article "I.Q. Scores Are Up, and Psychologists Wonder Why," *New York Times* reporter Trish Hall noted that people are now scoring substantially higher on IQ tests than they did in the past. "Studies show that if children took that same (1932 IQ) test today, half would score above 120 on the 1932 scale," writes Hall.Currently, IQ scores have a mean of 100, a standard deviation of 15, and a normal distribution. According to Hall's statement, if today's American children took the same IQ test used in 1932, the mean score would be 120, instead of 100 as it was in 1932. The net effect is that today's IQ scores are 20 points higher than in 1932.

1. IQ scores below 70 are often categorized as "intellectually deficient" results. What percentage of people were considered to be intellectually deficient in 1932 (based on the 1932 test)?(Assume that in 1932 the standard deviation was 15.)

2. If the same people who took the 1932 IQ test were to take today's IQ test instead, what percentage of them would be considered to be intellectually deficient (because of scores below 70)?

3. By today's standards, half of the subjects who take IQ tests get scores that are above average (that is, above 100). Based on the 1932 standards, what percentage of us are above average?

4. Identify at least one factor that might partially explain the rise in IQ scores.

VOCABULARY LIST

normal distribution
uniform distribution
density curve
standard normal distribution
sampling distribution
central limit theorem

standard error of the mean
finite population correction factor
continuity correction
normal quantile plot
normal probability plot

REVIEW

We introduced the concept of probability distributions in Chapter 4, but included only *discrete* distributions. In this chapter we introduced *continuous* probability distributions and focused on the most important category: normal distributions. Normal distributions will be used extensively in the following chapters.

Normal distributions are bell-shaped when graphed. The total area under the density curve of a normal distribution is 1, so there is a correspondence between areas and probabilities. Specific areas can be found using Table A-2 or Excel. (We do not use Formula 5-1, the equation that is used to define the normal distribution.)

In this chapter we presented important methods for working with normal distributions, including those that use the standard score $z = (x - \mu)/\sigma$ for solving problems such as these:

- Given that IQ scores are normally distributed with $\mu = 100$ and $\sigma = 15$, find the probability of randomly selecting someone with an IQ above 90.

- Given that IQ scores are normally distributed with $\mu = 100$ and $\sigma = 15$, find the IQ score separating the bottom 85% from the top 15%.

In Section 5-5 we presented the following important points associated with the central limit theorem:

1. The distribution of sample means will, as the sample size n increases, approach a normal distribution.

2. The mean of the sample means is the population mean μ.

3. The standard deviation of the sample means will be $\sigma/\sqrt{n}$.

In Section 5-6 we noted that we can sometimes approximate a binomial probability distribution by a normal distribution. If both $np \geq 5$ and $nq \geq 5$, the binomial random variable x is approximately normally distributed with the mean and standard deviation given as $\mu = np$ and $\sigma = \sqrt{npq}$. Because the binomial probability distribution deals with discrete data and the normal distribution deals with continuous data, we apply the continuity correction, which should be used in normal approximations to binomial distributions.

Finally, in Section 5-7 we presented a procedure for determining whether sample data appear to come from a population that has a normal distribution. Some of the statistical methods covered later in this book have a loose requirement of a normally distributed population. In such cases, examination of a histogram and outliers might be all that is needed. In other cases, normal quantile plots might be necessary because there is a very strict requirement that the population must have a normal distribution.

REVIEW EXERCISES

1. Marine Height Requirement for Women To be eligible for the U.S. Marine Corps, a woman must have a height between 58 in. and 73 in. Women have normally distributed heights with a mean of 63.6 in. and a standard deviation of 2.5 in. (based on data from the National Health Survey).
   a. Find the percentage of women who satisfy the Marine Corps height requirement.
   b. If the requirement is changed to exclude the shortest 1% and the tallest 1%, find the heights that are acceptable.
   c. If nine women are randomly selected, find the probability that their mean height is between 63.0 in. and 65.0 in.

2. Men's Heights Men have normally distributed heights with a mean of 69.0 in. and a standard deviation of 2.8 in. (based on data from the National Health Survey).

a. The Beanstalk Club is a social organization for tall people. Men are eligible for membership if they are at least 74 in. tall. What percentage of men are eligible for membership in the Beanstalk Club?

b. If a man is randomly selected, what is the probability that he will satisfy the height limitation for the Newport Police Department, which requires men to have heights between 62.0 in. and 76.0 in.?

c. If the Newport Police Department wants to change its height limitation for men so that only the shortest 2% and the tallest 2% are excluded, what are the minimum and maximum acceptable heights?

d. If 25 men are randomly selected, what is the probability that their mean height is more than 68.0 in.?

 3. Skulls in Anthropology Measurements of human skulls from different epochs are analyzed to determine whether they change over time. The maximum breadth is measured for skulls from Egyptian males who lived around 3300 B.C. Results show that those breadths are normally distributed with a mean of 132.6 mm and a standard deviation of 5.4 mm (based on data from *Ancient Races of the Thebaid* by Thomson and Randall-Maciver).

a. Find the probability of getting a value greater than 140 mm if a skull is randomly selected from the period of around 3300 B.C.

b. Find the breadth value that is D_2, the second decile.

c. If 35 skulls are randomly selected from the 3300 B.C. population, what is the probability that their mean breadth will be between 133.0 mm and 134.0 mm?

 4. Magazine Sweepstakes: Probability of New Subscriptions *Entertainment Report* magazine runs a sweepstakes as part of a campaign to acquire new subscribers. In the past, 26% of those who received sweepstakes entry materials have ended up entering the contest and subscribing to the magazine (based on data reported in *USA Today*). Estimate the probability that when sweepstakes entry materials are sent to 500 randomly selected households, the resulting number of new subscriptions is between 125 and 150 inclusive.

 5. Measurement Errors The Orange County Bureau of Weights and Measures studies a scale used by the Shop Till U Drop supermarket. It is found that errors on the scale are normally distributed with a mean of 0 oz and a standard deviation of 1 oz. (The errors can be positive or negative.)

a. If an item is randomly selected and weighed, find the probability that the error is between 0 oz and 1.00 oz.

b. If an item is randomly selected and weighed, find the probability that the error is greater than 2.00 oz.

c. If an item is randomly selected and weighed, find the probability that the error is greater than −1.50 oz.

d. If 25 items are randomly selected and weighed, find the probability that the errors have a mean between 0.25 oz and 0.50 oz.

e. Find the fourth decile, D_4, which is the error value separating the bottom 40% of all errors from the top 60%.

6. Uniform Distribution The Royal Beverage Bottling Company has designed a machine that fills cola cans so that the contents are *uniformly* distributed with a minimum of 11.8 oz and a maximum of 12.2 oz. If one can is randomly selected, find the probability that the amount of cola is

a. at least 12.0 oz.

b. between 11.9 oz and 12.1 oz.

c. less than 11.9 oz.

7. MCAT Scores Scores on the biology portion of the Medical College Admissions Test are normally distributed with a mean of 9.1 and a standard deviation of 2.1.
  a. What percentage of scores are above 8.5, the minimum score required for admission to the Providence School of Medicine?
  b. What percentage of scores are between 7.5 and 9.0?
  c. What is the value of Q_3, the score separating the bottom 75% from the top 25%?
  d. What is the value of Q_1?
  e. If 35 scores are randomly selected, what is the probability that their mean is above 9.0?

8. Telemarketing When telephone solicitors call from the National Magazine Subscription Company, 18% of the people who answer stay on the line for more than one minute. If 800 people are called in one day, find the probability that at least 150 stay on the line for more than one minute.

9. Tire Wear The Chemco Company manufactures car tires that last distances having a normal distribution with a mean of 35,600 mi and a standard deviation of 4275 mi.
  a. If a tire is randomly selected, what is the probability it lasts more than 30,000 mi?
  b. If 40 tires are randomly selected, what is the probability they last distances that have a mean greater than 35,000 mi?
  c. If the manufacturer wants to guarantee the tires so that only 3% will be replaced because of failure before the guaranteed number of miles, for how many miles should the tires be guaranteed?

10. Testing Claimed Psychic Powers The Amazing Bob claims that he has psychic powers and can predict future events. In a test of his ability, he is asked to predict the outcome of a die before it is rolled, and this trial is repeated for a total of 36 rolls. If his predictions are actually random guesses, find the probability that he will be correct at least 13 times. If Bob is correct at least 13 times, is this strong evidence that he has an ability to correctly predict the outcomes? Why or why not?

CUMULATIVE REVIEW EXERCISES

1. Eye Measurement Statistics The listed sample distances (in millimeters) were obtained by using a pupilometer to measure the distances between the pupils of adults (based on data collected by a student of the author).

67 66 59 62 63 66 66 55 63 61 60 56 66 67 59 59 60 62 61 63 62 57

63 60 64 65 63 64 59 60 60 60 56 66 62 60 62 56 60 60 58 52 54

  a. Find the mean $\bar{x}$ of the distances in this sample.
  b. Find the median of the distances in this sample.
  c. Find the mode of the distances in this sample.
  d. Find the standard deviation s of this sample.
  e. Does it seem that the sample values come from a population with a distribution that is approximately normal? Why or why not?
  f. Convert the distance of 66 mm to a z score.
  g. Find the actual percentage of these sample values that exceed 59 mm.

h. Assuming a normal distribution, find the percentage of *population* distances that exceed 59 mm. Use the sample values of $\bar{x}$ and s as estimates of μ and σ.

i. What level of measurement (nominal, ordinal, interval, ratio) describes this data set?

j. The listed measurements appear to be rounded to the nearest millimeter, but are the exact unrounded distances discrete data or continuous data?

EXCEL **2.** Left-Handedness According to data from the American Medical Association, 10% of us are left-handed.

a. If three people are randomly selected, find the probability that they are all left-handed.

b. If three people are randomly selected, find the probability that at least one of them is left-handed.

c. Why can't we solve the problem in part (b) by using the normal approximation to the binomial distribution?

d. If groups of 50 people are randomly selected, what is the mean number of left-handed people in such groups?

e. If groups of 50 people are randomly selected, what is the standard deviation for the numbers of left-handed people in such groups?

f. Would it be unusual to get 8 left-handed people in a randomly selected group of 50 people? Why or why not?

COOPERATIVE GROUP ACTIVITIES

1. *Out-of-class activity:* Use the reaction timer shown on the right by following the instructions given below. Collect reaction times for a sample of at least 40 different subjects taken from a homogeneous group, such as right-handed college students. For each subject, measure the reaction time for each hand. Construct a histogram for the right-hand times and another histogram for the left-hand results. Based on the histogram shapes, do the two sets of times each appear to be normally distributed? Calculate the values of the mean and standard deviation for each of the two data sets and compare the two sets of results. Using the right-hand sample mean $\bar{x}$ as an estimate of the population mean μ, and using the right-hand sample standard deviation s as an estimate of the population standard deviation σ, find the quartiles Q_1, Q_2, and Q_3. Repeat that procedure to find Q_1, Q_2, and Q_3 for the left hand. If the right-hand reaction times are to be used to screen job applicants, what time is the cutoff separating the fastest 5% from the slowest 95%?

Instructions for Using the Reaction Timer

a. Cut the reaction timer along the dashed line.

b. Ask the subject to hold his or her thumb and forefinger horizontally; those fingers should be spread apart by a distance that is the same as the width of the reaction timer.

c. Hold the reaction timer so that the bottom edge is just above the subject's thumb and forefinger.

d. Ask the subject to catch the reaction timer as quickly as possible, then release it after a few seconds.

e. Record the reaction time (in seconds) corresponding to the point at which the subject catches it.

REACTION TIMER

−.204
−.202
−.199
−.196
−.194
−.191
−.188
−.185
−.183
−.180
−.177
−.174
−.171
−.168
−.165
−.161
−.158
−.155
−.151
−.148
−.144
−.140
−.137
−.133
−.129
−.125
−.121
−.116
−.112
−.107
−.102
−.097
−.091
−.085
−.079
−.072
−.065
−.056
−.046
−.032

    c. Press the **Enter** key, then click and drag the lower right corner of cell C1 down to highlight the same number of rows containing data in column A.

5. Column C should now contain the desired number of sample values having the desired mean and the desired standard deviation. Verify this by clicking on any empty cell, then using **Tools, Data Analysis, Descriptive Statistics.** Check that column C has the correct values for the sample size, mean, and standard deviation.

from DATA to DECISION

Critical Thinking:
Design a new fighter-jet ejection seat

In the Chapter Problem we stated that fighter jets currently have the ACES-II ejection seats, which were originally designed for men weighing between 140 lb and 211 lb. Now that women have been approved as pilots of fighter jets, those seats must be redesigned. Any redesign must take these population characteristics into account: Women's weights are normally distributed with a mean of 143 lb and a standard deviation of 29 lb, and men's weights are normally distributed with a mean of 172 lb and a standard deviation of 30 lb (based on data from the National Health Survey).

Analyzing the results

1. As of this writing, the goal is to redesign the ejection seats so that any weights between 103 lb and 245 lb can be safely accommodated. What percentage of

women's weights fall within those limits? What percentage of men's weights fall within those limits?

2. Plan your own design so that at least 95% of all women's weights will be accommodated and at least 95% of all men's weights will be accommodated. What is the minimum weight and what is the maximum weight that satisfies that 95% requirement simultaneously for both genders? Cost effectiveness requires that the difference between the minimum and maximum be as small as possible.

Internet Project

Exploring the Central Limit Theorem

The Central Limit Theorem is one of the most important results in statistics. It also may be one of the most surprising. Basically, the theorem says that the normal distribution is everywhere. No matter what probability disribution underlies an experiment, there is a corresponding distribution of means that will be aproximately normal in shape. The best way to both understand and appreciate the Central Limit Theorem, however, is to see it in action. The Internet Project for this chapter can be found at this site:

http://www.awlonline.com/triola

You will be asked to view, interpret, and discuss a demonstration of the Central Limit Theorem as part of a dice rolling experiment. In addition, you will be guided in a search through the internet for other such demonstrations.

Statistics *at work*

"It is possible to be a journalist and not be comfortable with statistics, but you're definitely limited in what you can do.

Joel B. Obermayer

Newspaper reporter for *The News & Observer*

Joel B. Obermayer writes about medical issues and health affairs for *The News & Observer,* a newspaper that covers the eastern half of North Carolina. He reports on managed care, public health, and research at academic medical centers, including Duke University and the University of North Carolina at Chapel Hill.

What concepts of statistics do you use?

I use ideas like statistical significance, error rates, and probability. I don't need to do anything incredibly sophisticated, but I need to be very comfortable with the math and with asking questions about it.

I use statistics to look at medical research to decide whether different studies are significant and to decide how to write about them. Mostly, I need to be able to read statistics and understand them, rather than develop them myself. I use statistics to develop good questions and to bolster the arguments I make in print. I also use statistics to decide if someone is trying to give me a positive spin on something that might be questionable. For example, someone at a local university once sent me a press release about miracle creams that supposedly help slim you down by dissolving fat cells. Well, I doubt that those creams work. The study wasn't too hot either. They were trying to make claims based on a study of only 11 people. The researcher argued that 11 people were enough to make good empirical health claims. Not too impressive. People try to manipulate the media all the time. Good verifiable studies with good verifiable statistical bases make it easier to avoid being manipulated.

Is your use of probability and statistics increasing, decreasing, or remaining stable?

Increasing. People's interest in new therapies that may only be in the clinical trial stage is increasing, partly because of the emphasis on AIDS research and on getting new drugs approved and delivered to patients sooner. It's more important than ever for a medical writer to use statistics to make sure that the studies really prove what the public relations people say they prove.

Should prospective employees have studied some statistics?

It is possible to be a journalist and not be comfortable with statistics, but you're definitely limited in what you can do. If you write about government-sponsored education programs and whether they're effective, or if you write about the dangers of particular contaminants in the environment, you're going to need to use statistics.

In my field, editors often don't think about statistics in the interview process. They worry more about writing skills. A knowledge of statistics is more important for what you can do once you are on the job.

*D*o you recommend statistics for today's college students? Why?

I think a basic mathematical understanding of the world is a great thing to have, and a knowledge of statistics is a part of that. At the very least, I think it's important to understand some of the basics of statistics. For certain types of jobs, it's critical.

*W*hich other skills are important for today's college students?

As a journalist, I think writing skills are pretty high up there.

6

Estimates and Sample Sizes

Is the mean body temperature really 98.6° F?

Table 6-1 lists 106 body temperatures (from Data Set 6 in Appendix B) obtained by University of Maryland researchers. Using the methods described in Chapter 2, we can obtain the following important characteristics of the sample data set:

- As revealed by a histogram, the distribution of the data is approximately bell-shaped.

- The mean is $\bar{x} = 98.20°$ F.

- The standard deviation is $s = 0.62°$ F.

- The sample size is $n = 106$.

- There are no outliers.

It is commonly believed that the mean body temperature is 98.6° F, but the data in Table 6-1 seem to suggest that it is actually 98.20° F. We know that samples tend to vary, so perhaps it is true that the mean body temperature is 98.6° F and the sample mean, $\bar{x} = 98.20°$ F, is the result of a chance sample fluctuation. On the other hand, perhaps the sample mean of 98.20° F is correct and the commonly believed value of 98.6° F is wrong. On the basis of an analysis of the sample data in Table 6-1, we will see whether the mean body temperature is or is not 98.6° F.

TABLE 6-1	Body Temperatures (°F) of 106 Healthy Adults								
98.6	98.6	98.0	98.0	99.0	98.4	98.4	98.4	98.4	98.6
98.6	98.8	98.6	97.0	97.0	98.8	97.6	97.7	98.8	98.0
98.0	98.3	98.5	97.3	98.7	97.4	98.9	98.6	99.5	97.5
97.3	97.6	98.2	99.6	98.7	99.4	98.2	98.0	98.6	98.6
97.2	98.4	98.6	98.2	98.0	97.8	98.0	98.4	98.6	98.6
97.8	99.0	96.5	97.6	98.0	96.9	97.6	97.1	97.9	98.4
97.3	98.0	97.5	97.6	98.2	98.5	98.8	98.7	97.8	98.0
97.1	97.4	99.4	98.4	98.6	98.4	98.5	98.6	98.3	98.7
98.8	99.1	98.6	97.9	98.8	98.0	98.7	98.5	98.9	98.4
98.6	97.1	97.9	98.8	98.7	97.6	98.2	99.2	97.8	98.0
98.4	97.8	98.4	97.4	98.0	97.0				

Sample temperatures were obtained by Dr. Philip Mackowiak, Dr. Steven Wasserman, and Dr. Myron Levine, University of Maryland researchers.

6-1 Overview

In Chapter 2 we noted that we use *descriptive statistics* to summarize or describe important characteristics of data, but with *inferential statistics* we use sample data to make inferences (or generalizations) about a population. The two major applications of inferential statistics involve the use of sample data to (1) estimate the value of a population parameter, and (2) test some claim (or hypothesis) about a population. In this chapter we introduce methods for estimating values of the following population parameters: population means, proportions, and variances. We also present methods for determining the sample sizes necessary to estimate those parameters. In Chapter 7 we will introduce the basic methods for testing claims (or hypotheses) that have been made about a population.

6-2 Estimating a Population Mean: Large Samples

A Study Strategy: It will become obvious that this section contains a great deal of information and introduces many concepts. The time you devote to this section will be well spent because we introduce the concept of a confidence interval, and that same general concept will be applied to the following sections of this chapter. We suggest that you proceed as follows: First, read this section with the limited objective of simply trying to understand what confidence intervals are, what they accomplish, and why they are needed. Second, try to develop the ability to construct confidence interval estimates of population means. Third, learn how to interpret a confidence interval correctly. Fourth, read the section once again and try to understand the underlying theory. You will always enjoy much greater success if you understand what you are doing, instead of blindly applying mechanical steps in order to obtain an answer.

Here is the main objective of this section: Given a collection of more than 30 sample values, develop an estimate of the value of the population mean μ. For example, the 106 body temperatures given in Table 6-1 in the Chapter Problem constitute a collection of sample values. By using statistics from that sample, we will proceed to estimate the mean μ of *all* body temperatures. This section will consider only samples with more than 30 values. (Section 6-3 will present methods for working with smaller samples.) This sample size condition is listed along with another assumption we make throughout this section:

Assumptions

1. $n > 30$ (The sample has more than 30 values.)
2. The sample is a simple random sample. (All samples of the same size have an equal chance of being selected.)

Recall from Section 1-4 that a simple random sample of n values is obtained if every possible sample of size n has the same chance of being chosen. This requirement of random selection means that the methods of this section cannot be used with some other types of sampling, such as stratified, cluster, and convenience sampling. We should be especially clear about this important point:

Data collected carelessly can be absolutely worthless, even if the sample is quite large.

We know that different samples naturally produce different results. The methods of this section assume that those sample differences are due to chance random fluctuations, not some unsound method of sampling. For example, if you were to collect a sample of IQ test results from 40 of your closest friends, you should not use the results to estimate the mean IQ of all Americans. When samples of friends are used, there are cultural and economic characteristics that cause differences other than the differences found with randomly selected groups of people. Instead of being randomly selected, your sample of friends would be a biased sample, and it would be totally worthless as a basis for estimating the mean IQ of all Americans. That's not what friends are for.

Assuming that we have a simple random sample with more than 30 values, we can now proceed with our major objective: using the sample as a basis for estimating the value of the population mean μ. We begin with some general definitions.

Estimating Wildlife Population Sizes

The National Forest Management Act protects endangered species, including the northern spotted owl, with the result that the forestry industry was not allowed to cut vast regions of trees in the Pacific Northwest. Biologists and statisticians were asked to analyze the problem, and they concluded that survival rates and population sizes were decreasing for the female owls, known to play an important role in species survival. Biologists and statisticians also studied salmon in the Snake and Columbia Rivers in Washington, and penguins in New Zealand. In the article "Sampling Wildlife Populations" (*Chance*, Vol. 9, No. 2), authors Bryan Manly and Lyman McDonald comment that in such studies, "biologists gain through the use of modeling skills that are the hallmark of good statistics. Statisticians gain by being introduced to the reality of problems by biologists who know what the crucial issues are."

DEFINITIONS

An **estimator** is a formula or process for using sample data to estimate a population parameter.

An **estimate** is a specific value or range of values used to approximate a population parameter.

A **point estimate** is a single value (or point) used to approximate a population parameter.

The sample mean $\overline{x}$ is the best point estimate of the population mean μ.

Although we could use another statistic such as the sample median, midrange, or mode as an estimate of the population mean μ, studies have shown that the sample mean $\overline{x}$ usually provides the best estimate, for two reasons.

1. For many populations, the distribution of sample means $\overline{x}$ tends to be more consistent (with *less variation*) than the distributions of other sample statistics. (That is, if you use sample means to estimate the population mean μ, those sample means will have a smaller standard deviation than would other sample statistics, such as the median or the mode. The differences between $\overline{x}$ and μ therefore tend to be smaller than the differences obtained with some other statistic, such as the median.)

2. For all populations, the sample mean $\overline{x}$ is an **unbiased estimator** of the population mean μ, meaning that the distribution of sample means tends to center about the value of the population mean μ. (That is, sample means do not systematically tend to overestimate the value of μ, nor do they systematically tend to underestimate μ. Instead, they tend to target the value of μ itself.)

Captured Tank Serial Numbers Reveal Population Size

During World War II, Allied intelligence specialists wanted to determine the number of tanks Germany was producing. Traditional spy techniques provided unreliable results, but statisticians obtained accurate estimates by analyzing serial numbers on captured tanks. As one example, records show that Germany actually produced 271 tanks in June 1941. The estimate based on serial numbers was 244, but traditional intelligence methods resulted in the extreme estimate of 1550. (See "An Empirical Approach to Economic Intelligence in World War II," by Ruggles and Brodie, *Journal of the American Statistical Association*, Vol. 42.)

 EXAMPLE **Body Temperatures** Use the sample body temperatures given in Table 6-1 to find the best point estimate of the population mean μ of all body temperatures.

SOLUTION For the sample data in Table 6-1, $\overline{x} = 98.20°$ F. Because the sample mean $\overline{x}$ is the best point estimate of the population mean μ, we conclude that the best point estimate of the population mean μ of all body temperatures is 98.20° F. ◄

Why Do We Need Confidence Intervals?

In the preceding example we saw that 98.20° F was our *best* point estimate of the population mean μ, but we have no indication of just how *good* our best estimate was. If we knew only the first four temperatures of 98.6, 98.6, 98.0, and 98.0, the best point estimate of μ would be their mean ($\overline{x} = 98.30°$ F), but we wouldn't expect this point estimate to be very good because it is based on such a small sample. Because the point estimate has the serious flaw of not revealing anything about how good it is, statisticians have cleverly developed another type of estimate that does reveal how good the estimate is. This estimate, called a *confidence interval* or *interval estimate*, consists of a range (or an interval) of values instead of just a single value.

> **DEFINITION**
>
> A **confidence interval** (or **interval estimate**) is a range (or an interval) of values used to estimate the true value of a population parameter.

A confidence interval is associated with a degree of confidence, such as 0.95 (or 95%). The degree of confidence tells us the percentage of times that the confidence interval actually does contain the population parameter, assuming that the estimation process is repeated a large number of times. The definition of degree of confidence uses α (lowercase Greek alpha) to represent a probability or area. The value of α is the complement of the *degree of confidence.* For a 0.95 (or 95%) degree of confidence, $\alpha = 0.05$. For a 0.99 (or 99%) degree of confidence, $\alpha = 0.01$.

DEFINITION

The **degree of confidence** is the probability $1 - \alpha$ (often expressed as the equivalent percentage value) that is the relative frequency of times that the confidence interval actually does contain the population parameter, assuming that the estimation process is repeated a large number of times. (The degree of confidence is also called the **level of confidence,** or the **confidence coefficient.**)

The most common choices for the degree of confidence are 90% (with $\alpha = 0.10$), 95% (with $\alpha = 0.05$), and 99% (with $\alpha = 0.01$). The choice of 95% is most common because it provides a good balance between precision (as reflected in the width of the confidence interval) and reliability (as expressed by the degree of confidence).

Here's an example of a confidence interval based on the sample data of 106 body temperatures given in Table 6-1:

The 0.95 (or 95%) degree of confidence interval estimate of the population mean μ is $98.08°$ F $< \mu < 98.32°$ F.

Interpreting a Confidence Interval

We must be careful to interpret confidence intervals correctly. There is a correct interpretation and a wrong interpretation of the confidence interval $98.08°$ F $< \mu < 98.32°$ F.

Correct: We are 95% confident that the interval from 98.08 to 98.32 actually does contain the true value of μ. This means that if we were to select many different samples of size 106 and construct the confidence intervals, 95% of them would actually contain the value of the population mean μ. (Note that in this correct interpretation, the level of 95% refers to the *process* being used to estimate the mean, and it does not refer to the population mean itself.)

Wrong: There is a 95% chance that the true value of μ will fall between 98.08 and 98.32.

At this point in time, there is a fixed and constant value of μ, the mean body temperature of the whole population. If we use sample data to find specific limits, such as 98.08 and 98.32, those limits either enclose the population mean μ or do not, and we cannot determine whether they do or do not without knowing the true value of μ. But it's wrong to say that μ has a 95% chance of falling within the specific limits of 98.08 and 98.32, because μ is a constant, not a random variable. Either μ will fall within these limits or it won't; there's no probability involved. This is a confusing concept, so consider the easier example in which we want to find the probability of a baby being born a girl. If the baby has already been born, but the doctor hasn't yet told us the gender, we can't say that there is a 0.5 probability that the baby is a girl, because the baby is already a girl or is not. There is no chance involved, because the gender has been determined. Similarly, a population mean μ is already determined, and the confidence interval limits either contain μ or do not, so it's wrong to say that there is a 95% chance that μ will fall between 98.08 and 98.32.

FIGURE 6-1
Confidence Intervals from
20 Different Samples

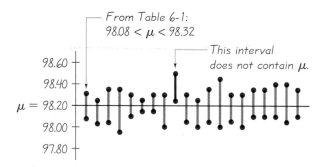

A confidence level of 95% tells us that the *process* we are using will, in the long run, result in confidence interval limits that contain μ 95% of the time. Suppose that body temperatures really came from a population with a true mean of 98.20° F. Then the confidence interval obtained from the given sample data would contain the population mean, because 98.20 is between 98.08 and 98.32. This is illustrated in Figure 6-1. Figure 6-1 shows the first confidence interval for the real data from Table 6-1, but the other 19 confidence intervals represent hypothetical samples. With 95% confidence, we expect that 19 out of 20 samples should result in confidence intervals that do contain the true value of μ, and Figure 6-1 illustrates this with 19 of the confidence intervals containing μ while one confidence interval does not contain μ.

Critical Values

Constructing a confidence interval requires that we find a standard z score that can be used to distinguish between sample statistics that are likely to occur and those that are unlikely. Such a z score is called a *critical value* (defined below), and it is based on the following observations.

1. We know from the central limit theorem that sample means tend to be normally distributed, as in Figure 6-2.

2. Sample means have a relatively small chance (with probability denoted by α) of falling in one of the orange extreme tails of Figure 6-2.

3. Denoting the area of each orange-shaded tail by $\alpha/2$, we see that there is a total probability of α that a sample mean will fall in either of the two tails.

FIGURE 6-2
The Standard Normal
Distribution: The Critical
Value $z_{\alpha/2}$

4. By the rule of complements (from Chapter 3), there is a probability of $1 - \alpha$ that a sample mean will fall within the green-shaded region of Figure 6-2.

5. The z score separating the right-tail region is commonly denoted by $z_{\alpha/2}$ and is referred to as a *critical value* because it is on the borderline separating sample means that are likely to occur from those that are unlikely to occur.

These observations can be formalized with the following notation and definition.

Notation for Critical Value

$z_{\alpha/2}$ is the positive z value that is at the vertical boundary separating an area of $\alpha/2$ in the right tail of the standard normal distribution. (The value of $-z_{\alpha/2}$ is at the vertical boundary for the area of $\alpha/2$ in the left tail.) The subscript $\alpha/2$ is simply a reminder that the z score separates an area of $\alpha/2$ in the right tail of the standard normal distribution.

DEFINITION

A **critical value** is the number on the borderline separating sample statistics that are likely to occur from those that are unlikely to occur. The number $z_{\alpha/2}$ is a critical value that is a z score with the property that it separates an area of $\alpha/2$ in the right tail of the standard normal distribution. (See Figure 6-2.)

▶ **EXAMPLE**   **Critical Value** Find the critical value $z_{\alpha/2}$ corresponding to a 95% degree of confidence.

SOLUTION   A 95% degree of confidence corresponds to $\alpha = 0.05$. See Figure 6-3, where we show that the area in each of the orange-shaded

FIGURE 6-3
Finding $z_{\alpha/2}$ for 95% Degree of Confidence

table at bottom

(.05/2 = .025)

work backward

$z \mid .06$

$1.9 \mid .4750$

tails is $\alpha/2 = 0.025$. We find $z_{\alpha/2} = 1.96$ by noting that the region to its left (and bounded by the mean of $z = 0$) must be $0.5 - 0.025$, or 0.475. We can refer to Table A-2 and find that the area of 0.4750 (found in the *body* of the table) corresponds exactly to a z score of 1.96. For a 95% degree of confidence, the critical value is therefore $z_{\alpha/2} = 1.96$.

The preceding example showed that a 95% degree of confidence results in a critical value of $z_{\alpha/2} = 1.96$. This is the most common critical value, and it is listed with two other common values in the table that follows.

Degree of Confidence	α	Critical Value, $z_{\alpha/2}$
90%	0.10	1.645
95%	0.05	1.96
99%	0.01	2.575

Margin of Error

When we collect a set of sample data, such as the set of 106 body temperatures listed in Table 6-1, we can calculate the sample mean $\overline{x}$, and that sample mean is typically different from the population mean μ. The difference between the sample mean and the population mean can be thought of as an error. In Section 5-5 we saw that $\sigma/\sqrt{n}$ is the standard deviation of sample means. Using $\sigma/\sqrt{n}$ and the $z_{\alpha/2}$ notation, we now define the *margin of error* E as follows.

DEFINITION

When data from a simple random sample are used to estimate a population mean μ, the **margin of error,** denoted by E, is the maximum likely (with probability $1 - \alpha$) difference between the observed sample mean $\overline{x}$ and the true value of the population mean μ. The margin of error E is also called the *maximum error of the estimate* and can be found by multiplying the critical value and the standard deviation of sample means, as shown in Formula 6-1.

Formula 6-1
$$E = z_{\alpha/2} \cdot \frac{\sigma}{\sqrt{n}} \quad \rightarrow \text{margin of error formula}$$

Given the way that the margin of error E is defined, there is a probability of $1 - \alpha$ that a sample mean will be in error (different from the population mean μ) by no more than E, and there is a probability of α that the sample mean will be in error by more than E. The calculation of the margin of error E as given in Formula 6-1 requires that you know the population standard deviation σ, but in reality we seldom know σ when the population mean μ is not known. The following method of calculation is common practice.

Calculating E When σ Is Unknown

If $n > 30$, we can replace σ in Formula 6-1 by the sample standard deviation s.

If $n \leq 30$, the population must have a normal distribution and we must know the value of σ to use Formula 6-1.

[Section 6-3 will present a method for calculating the margin of error E for small ($n \leq 30$) samples with σ unknown.]

On the basis of the definition of the margin of error E, we can now identify the confidence interval for the population mean μ. The three commonly used formats for expressing the confidence interval are shown in the following box.

Confidence Interval (or Interval Estimate) for the Population Mean μ (Based on Large Samples: $n > 30$)

$$\bar{x} - E < \mu < \bar{x} + E \qquad \text{where} \qquad E = z_{\alpha/2} \cdot \frac{\sigma}{\sqrt{n}}$$

confidence interval formula

or

$$\mu = \bar{x} \pm E$$

or

$$(\bar{x} - E, \bar{x} + E)$$

DEFINITION

The two values $\bar{x} - E$ and $\bar{x} + E$ are called **confidence interval limits.**

Procedure for Constructing a Confidence Interval for μ (Based on a Large Sample: $n > 30$)

1. Find the critical value $z_{\alpha/2}$ that corresponds to the desired degree of confidence. (For example, if the degree of confidence is 95%, the critical value is $z_{\alpha/2} = 1.96$.)
2. Evaluate the margin of error $E = z_{\alpha/2} \cdot \sigma / \sqrt{n}$. If the population standard deviation σ is unknown, use the value of the sample standard deviation s provided that $n > 30$.
3. Using the value of the calculated margin of error E and the value of the sample mean $\bar{x}$, find the values of $\bar{x} - E$ and $\bar{x} + E$. Substitute those values in the general format for the confidence interval:

$$\bar{x} - E < \mu < \bar{x} + E$$

Estimating Sugar in Oranges

In Florida, members of the citrus industry make extensive use of statistical methods. One particular application involves the way in which growers are paid for oranges used to make orange juice. An arriving truckload of oranges is first weighed at the receiving plant, then a sample of about a dozen oranges is randomly selected. The sample is weighed and then squeezed, and the amount of sugar in the juice is measured. Based on the sample results, an estimate is made of the total amount of sugar in the entire truckload. Payment for the load of oranges is based on the estimate of the amount of sugar because sweeter oranges are more valuable than those less sweet, even though the amounts of juice may be the same.

or

$$\mu = \bar{x} \pm E$$

or

$$(\bar{x} - E, \bar{x} + E)$$

4. Round the resulting values by using the following round-off rule.

Round-Off Rule for Confidence Intervals Used to Estimate μ

1. When using the *original set of data* to construct a confidence interval, round the confidence interval limits to one more decimal place than is used for the original set of data.

2. When the original set of data is unknown and only the *summary statistics* $(n, \bar{x}, s)$ are used, round the confidence interval limits to the same number of decimal places used for the sample mean.

The following example clearly illustrates the relatively simple(!) procedure for actually constructing a confidence interval. The original data from Table 6-1 are listed with one decimal place and the summary statistics use two decimal places, so the confidence interval limits will be rounded to two decimal places.

EXAMPLE   **Body Temperatures** For the body temperatures in Table 6-1, we have $n = 106$, $\bar{x} = 98.20°$ F, and $s = 0.62°$ F. Using a 0.95 degree of confidence, find both of the following:

a. The margin of error E

b. The confidence interval for μ

SOLUTION

a. The 0.95 degree of confidence implies that $\alpha = 0.05$, so $z_{\alpha/2} = 1.96$, as shown in the preceding example. The margin of error E is calculated by using Formula 6-1 as follows. (Note that σ is unknown, but we can use $s = 0.62$ for the value of σ because $n > 30$.)

$$E = z_{\alpha/2} \cdot \frac{\sigma}{\sqrt{n}} = 1.96 \cdot \frac{0.62}{\sqrt{106}} = 0.12 \quad \leftarrow \text{given}$$

b. With $\bar{x} = 98.20$ and $E = 0.12$, we construct the confidence interval as follows.

$$\bar{x} - E < \mu < \bar{x} + E$$
$$\text{given } (\bar{x}) \rightarrow 98.20 - 0.12 < \mu < 98.20 + 0.12$$
$$98.08 < \mu < 98.32$$

INTERPRETATION   This result could also be expressed as $\mu = 98.20 \pm 0.12$ or as $(98.08, 98.32)$. Based on the sample of 106 body temperatures listed in Table 6-1, the confidence interval for the population mean μ is 98.08° F $< \mu <$ 98.32° F, and this interval has a 0.95 degree of confidence. This means that if we were to select many different samples of size 106 and construct the confidence intervals as we did here, 95% of them would actually contain the value of the population mean μ.

Note that the confidence interval limits of 98.08° F and 98.32° F do not contain 98.6° F, the value generally believed to be the mean body temperature. Based on the sample data in Table 6-1, it seems very unlikely that 98.6° F is the correct value of μ.

Finding the Point Estimate and E from a Confidence Interval

AS

Later in this section we will describe how Excel can be used to find a confidence interval. A typical computer program requires that you enter a confidence level and sample data, and the display shows the confidence interval limits. The sample mean $\bar{x}$ is the value midway between those limits, and the margin of error E is one-half the difference between those limits (because the upper limit is $\bar{x} + E$ and the lower limit is $\bar{x} - E$, the distance separating them is $2E$).

Point estimate of μ:

$$\bar{x} = \frac{(\text{upper confidence interval limit}) + (\text{lower confidence interval limit})}{2}$$

Margin of error:

$$E = \frac{(\text{upper confidence interval limit}) - (\text{lower confidence interval limit})}{2}$$

 EXAMPLE   Body Temperatures A journal article includes this confidence interval estimate of the population mean: 98.08° F $< \mu <$ 98.32° F. Use the given confidence interval limits to find the point estimate $\bar{x}$ and the margin of error E.

SOLUTION

$$\bar{x} = \frac{(\text{upper confidence limit}) + (\text{lower confidence limit})}{2}$$

$$= \frac{98.32° \text{ F} + 98.08° \text{ F}}{2} = 98.20° \text{ F}$$

$$E = \frac{(\text{upper confidence interval limit}) - (\text{lower confidence interval limit})}{2}$$

$$= \frac{98.32° \text{ F} - 98.08° \text{ F}}{2} = 0.12° \text{ F}$$

Using a Confidence Interval to Describe, Explore, or Compare Data

In some cases, we might use a confidence interval to achieve an ultimate goal of estimating the value of a population parameter. For the body temperature data used in this section, an important goal might be to estimate the mean body temperature of healthy adults, and our results strongly suggest that the commonly used value of 98.6° F is incorrect (because we have 95% confidence that the limits of 98.08° F and 98.32° F contain the true value of the population mean). In other cases, a confidence interval might be one of several different tools used to describe, explore, or compare data sets. For example, consider two different data sets consisting of the heights of 100 randomly selected men and the heights of 100 randomly selected women. The accompanying graphs, statistics, and confidence intervals can be used to compare those two data sets.

Men

Women

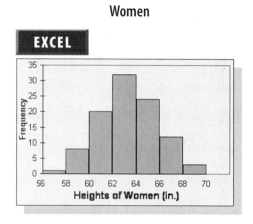

Descriptive Statistics

Men: $n = 100, \bar{x} = 68.76$ in., $s = 2.93$ in.

Women: $n = 100, \bar{x} = 63.39$ in., $s = 2.44$ in.

95% Confidence Intervals

Men: 68.19 in. $< \mu <$ 69.33 in.

Women: 62.91 in. $< \mu <$ 63.87 in.

The histograms suggest that heights of men and women have distributions that are approximately normal, and the heights of women appear to be generally lower. The descriptive statistics also suggest that the mean height of women is less than the mean height of men. Now carefully examine the confidence intervals and see that the upper limit for women is less than the lower limit for men. That is, the confidence intervals do not overlap at all, indicating that women have a mean height that is less than the mean height of men. This is not a big surprise, because most of us already know that women are typically shorter than men, but we now have solid statistical evidence supporting that belief, and we can see how confidence intervals can be used for such comparisons.

There is a $1 - \alpha$ probability that
a sample mean will be in error
by less than E or $z_{\alpha/2}\sigma/\sqrt{n}$

There is a probability of α that a
sample mean will be in error
by more than E (in one of the
red tails)

FIGURE 6-4
Distribution of Sample Means

What Is the Basis for the Procedure of Finding a Confidence Interval?

The basic idea underlying the construction of confidence intervals relates to the central limit theorem, which indicates that with large ($n > 30$) samples, the distribution of sample means is approximately normal with mean μ and standard deviation $\sigma/\sqrt{n}$. The confidence interval format is really a variation of the equation that was already used with the central limit theorem. Express $z = (\overline{x} - \mu_{\overline{x}})/\sigma_{\overline{x}}$ as shown below.

$$z = \frac{\overline{x} - \mu}{\dfrac{\sigma}{\sqrt{n}}}$$

If we solve this equation for μ, we get

$$\mu = \overline{x} - z\frac{\sigma}{\sqrt{n}}$$

Using the positive and negative values for z results in the confidence interval limits we are using.

Let's consider the specific case of a 95% degree of confidence, so $\alpha = 0.05$ and $z_{\alpha/2} = 1.96$. For this case, there is a probability of 0.05 that a sample mean will be more than 1.96 standard deviations (or $z_{\alpha/2}\sigma/\sqrt{n}$, which we denote by E) away from the population mean μ. Conversely, there is a 0.95 probability that a sample mean will be within 1.96 standard deviations (or $z_{\alpha/2}\sigma/\sqrt{n}$) of μ. (See Figure 6-4.) If the sample mean $\overline{x}$ is within $z_{\alpha/2}\sigma/\sqrt{n}$ of the population mean μ, then μ must be between $\overline{x} - z_{\alpha/2}\sigma/\sqrt{n}$ and $\overline{x} + z_{\alpha/2}\sigma/\sqrt{n}$; this is expressed in the general format of our confidence interval (with $z_{\alpha/2}\sigma/\sqrt{n}$ denoted as E): $\overline{x} - E < \mu < \overline{x} + E$.

Using Excel for Confidence Intervals

First find the value of the sample standard deviation s (which can be found using fx, **Statistical, STDEV**). Excel has a command **CONFIDENCE**, but there are two problems with it: (1) The command **CONFIDENCE** provides only the margin of error instead of the confidence interval limits, and

(2) it uses critical values based on the *normal* distribution only. The format is **=CONFIDENCE(α, s, n).** For example, the command **=CONFIDENCE(0.05, 0.623, 106)** yields a margin of error of 0.118599.

The Data Desk/XL add-in can be used to generate the complete confidence interval for the examples and exercises of this section as well as the following sections, but only the confidence levels of 90%, 95%, and 99% are available. Also, DDXL requires the original list of sample values. (If you know only the summary statistics of n, $\bar{x}$, and s, use the procedure given in the Excel Project at the end of Chapter 5.) Here is the procedure for using the Data Desk/XL add-in.

1. Click on **DDXL.** (If DDXL does not appear on the menu bar, it must be installed by following the procedure described in Section 1-5.)

2. Click on **Confidence Intervals.**

3. For the confidence intervals described in this section, select **1 Var z Interval,** which indicates that we have one variable (one sample) and we want a confidence interval constructed using the normal (z) distribution.

4. A dialog box will appear. Click on the pencil icon labeled "Quantitative Variable" and enter the range of values containing the sample data. For example, if there are 106 sample values listed in rows 1 through 106 of column A, enter A1:A106.

5. Click **OK** and you will get another dialog box.

6. Click on the box for the desired level of confidence. (Note that only the confidence levels of 90%, 95%, and 99% are available.)

7. Type in the value of the population standard deviation. If the sample size is greater than 30 and the population standard deviation is unknown, enter the value of the sample standard deviation.

8. Click **Compute Interval.**

On page 324 is the portion of the DDXL screen display that results from construction of a 95% confidence interval using the sample data in Table 6-1. The result at the bottom can be rounded to $98.08 < \mu < 98.32$, which is the same result we obtained in this section.

6-2 Basic Skills and Concepts

Interpreting Computer or Calculator Results.   *In Exercises 1–4, assume that a computer or calculator was used to generate the given confidence interval limits. Find the mean $\overline{x}$ and margin of error E.*

1. $72.6 < \mu < 82.6$

2. $(1007.3, 1009.4)$

3. $(207, 212)$

4. $2.17 < \mu < 2.57$

Finding Critical Values.   *In Exercises 5–8, find the critical value $z_{\alpha/2}$ that corresponds to the given degree of confidence.*

5. 99%

6. 96%

7. 97%

8. 92%

Finding Margin of Error and Confidence Interval.   *In Exercises 9–12, use the given degree of confidence and sample data to find (a) the margin of error E and (b) the confidence interval for the population mean μ.*

9. Salaries of statistics professors: 95% confidence; $n = 60$, $\overline{x} = \$85{,}678$, $s = \$12{,}345$ (we wish)

10. Distances driven by male students in one year: 90% confidence; $n = 100$, $\overline{x} = 11{,}878$ mi, $s = 3577$ mi

11. Times between uses of a TV remote control by males during commercials: 99% confidence; $n = 150$, $\overline{x} = 7.60$ sec, $s = 2.72$ sec

12. Starting salaries of college graduates who have taken a statistics course: 95% confidence; $n = 100$, $\overline{x} = \$43{,}704$, $s = \$9879$

13. Interpreting Confidence Interval A sample of 35 skulls is obtained for Egyptian males who lived around 1850 B.C. The maximum breadth of each skull is measured with the result that $\overline{x} = 134.5$ mm and $s = 3.48$ mm (based on data from *Ancient Races of the Thebaid* by Thomson and Randall-Maciver). Using these sample results and a confidence level of 95%, Excel provides this confidence interval: $133.35 < \mu < 135.65$. Write a statement that interprets the confidence interval.

14. Interpreting Confidence Interval The U.S. Department of Health, Education, and Welfare collected sample data for 1525 women, aged 18 to 24. That sample group has a mean serum cholesterol level (measured in mg/100 mL) of 191.7 with a standard deviation of 41.0 (see USDHEW publication 78-1652). Using these sample data and a 90% confidence level, Excel provides this confidence interval: $189.97 < \mu < 193.43$. Write a statement that interprets the displayed result. If a doctor claims that the mean serum cholesterol level for women in that age bracket is 200, does that claim seem to be consistent with the confidence interval?

15. Comparing Bowling Lanes Bowling league members bowled (what else?) at two different locations. At the location with oiled lanes, 40 bowling scores had a mean of 205.5 and a standard deviation of 35.4. At the location with dry lanes, 40 bowling scores had a mean of 185.0 and a standard deviation of 28.4 (based on data from a student of the author).

a. Construct a 95% confidence interval estimate of the mean for all scores at the location with oiled lanes.

b. Construct a 95% confidence interval estimate of the mean for all scores at the location with dry lanes.

c. Compare and interpret the results. Does it appear that the lane surface has an effect on the bowling scores?

16. Comparing Service Times

a. The drive-through service times were recorded for 52 randomly selected customers at a Burger King restaurant. Those times had a mean of 181.3 sec and a standard deviation of 82.2 sec. Construct a 95% confidence interval estimate of the population mean.

b. The drive-through service times were recorded for 95 randomly selected customers at a McDonald's restaurant. Those times had a mean of 127.8 sec and a standard deviation of 60.6 sec. Construct a 95% confidence interval estimate of the population mean.

c. Compare and interpret the results. Does it appear that either restaurant has better performance. Why or why not?

EXCEL 17. Shoplifting Listed below are the values (in dollars) of items stolen and recovered for a Sears store. Construct a 99% confidence interval estimate of the mean value of all such items that are stolen and recovered.

144.95	144.00	542.88	346.95	149.99	402.00	34.00	119.96	134.98
16.00	100.97	37.50	209.00	38.00	55.00	23.99	17.99	145.94
99.97	449.99	17.47	24.99	39.99	37.99	529.88	9.99	274.89
379.99	162.97	24.99	229.99	16.96	101.91	389.96	79.99	330.96

EXCEL 18. Lengths of CDs Listed below are the lengths (in minutes) of randomly selected CDs of country, rock, and movie soundtracks. Construct a 95% confidence interval for the mean length of all such CDs.

62.28	56.13	54.03	20.37	56.13	54.03	29.01	58.02	74.14	54.04
50.16	52.15	53.15	47.42	49.19	51.15	49.63	52.00	61.57	57.06
75.12	32.04	32.57	60.41	32.07	37.06	62.06	39.29	33.15	51.37
50.97	52.07	43.16	38.48	73.23	36.26	39.57	69.71	49.22	48.19

EXCEL 19. Bullet Production A Lyman Electronic Digital Caliper was used to accurately measure the lengths of 40 Triton Quik-Shok 380 ACP 90 grain cartridges (bullets). The results (in millimeters) are listed below. The cartridges are supposed to have a mean length of 0.950 mm and they must be between 0.945 mm long and 0.955 mm, otherwise they could be dangerous when fired. Construct a 90% confidence interval estimate of the mean length, then interpret the results.

0.951	0.954	0.954	0.953	0.953	0.953	0.952	0.955	0.951	0.949
0.950	0.950	0.954	0.950	0.948	0.948	0.945	0.947	0.950	0.954
0.951	0.951	0.952	0.943	0.952	0.952	0.949	0.946	0.949	0.949
0.953	0.953	0.953	0.950	0.949	0.950	0.950	0.949	0.950	0.951

EXCEL 20. Times in Voting Booth In studying the time that voters are in the polling booth, a student of the author recorded the given times (in seconds) at a polling station in Marlboro, New York. Construct a 95% confidence interval estimate of the population mean. Interpret the result.

85	60	65	83	45	30	22	43	30	46	115	86	52
100	50	50	35	51	110	15	45	63	40	110	18	34
37	25	55	48	120	45	80	63	44	93	51	57	48
29	83	48	50	110	114	38	60	37	53	70		

EXCEL 21. Comparing Regular and Diet Coke Refer to Data Set 1 in Appendix B. The Excel workbook is COLA.xls.
   a. Construct a 95% confidence interval estimate of the population mean weight of the cola in regular Coke cans.
   b. Construct a 95% confidence interval estimate of the population mean weight of the cola in diet Coke cans.
   c. Compare the confidence intervals from parts (a) and (b) and interpret the results.

EXCEL 22. Comparing Textbook Prices Refer to Data Set 2 in Appendix B. The Excel workbook is TEXTBOOK.xls. Construct a 99% confidence interval estimate of the mean cost of new textbooks at each of the two colleges. Compare and interpret the results.

EXCEL 23. Comparing Discarded Plastic and Paper Refer to Data Set 5 in Appendix B. The Excel workbook is GARBAGE.xls. Construct a 90% confidence interval estimate of the mean weight of discarded plastic, and construct a 90% confidence interval estimate of the mean weight of discarded paper. Compare and interpret the results. Which is more of an ecological problem: discarded paper or discarded plastic?

EXCEL 24. Second-Hand Smoke Refer to Data Set 9 in Appendix B. The Excel workbook is COTININE.xls. Construct a 95% confidence interval estimate of the mean amount of serum cotinine in smokers, then do the same for the ETS group, which consists of nonsmokers exposed to environmental tobacco smoke at home or work. Compare and interpret the results.

6-2   Beyond the Basics

EXCEL 25. Effect of Outlier Test the effect of an outlier as follows: Use the sample data in Table 6-1 to find a 95% confidence interval estimate of the population mean, but change the first entry from 98.6 to 986. Nobody can really have a body temperature of 986° F, but such an error can easily occur when a decimal point is inadvertently omitted when the sample values are entered.
   a. Find the 95% confidence interval.
   b. By comparing the result from part (a) to the confidence interval found in this section, describe the effect of an outlier on a confidence interval. Are the confidence interval limits sensitive to outliers?
   c. Based on part (b), how should you handle outliers when they are found in sample data sets that will be used for the construction of confidence intervals?

26. Using 5-Number Summary A simple random sample of 200 U.S. Air Force applicants is given standard IQ tests. The IQ scores are normally distributed with this 5-number summary: 80, 100, 108, 116, 136. Construct a 95% confidence interval estimate of the population mean of all such applicants.

27. A 95% confidence interval for the lives (in minutes) of Kodak AA batteries is $430 < \mu < 470$. (See Program 1 of *Against All Odds: Inside Statistics.*) Assume that this result is based on a sample of size 100.
   a. What is the value of the sample mean?
   b. What is the value of the sample standard deviation?
   c. Construct the 99% confidence interval.
   d. If the confidence interval $432 < \mu < 468$ is obtained from the same sample data, what is the degree of confidence?

28. Using Finite Population Correction Factor The standard error of the mean is $\sigma/\sqrt{n}$, provided that the population size is infinite. If the population size is finite and is denoted by N, then the correction factor

$$\sqrt{\frac{N - n}{N - 1}}$$

should be used whenever $n > 0.05N$. This correction factor multiplies the margin of error E given in Formula 6-1. Find the 95% confidence interval for the mean of 100 IQ scores if a sample of 31 of those scores produces a mean and standard deviation of 132 and 10, respectively.

 ## 6-3 Estimating a Population Mean: Small Samples

In Section 6-2 we introduced methods for estimating a population mean μ, but all of the examples and exercises were based on the following assumptions: (1) that the sample has more than 30 values ($n > 30$), and (2) that the sample is a simple random sample. Time, cost, and other constraints, however, often require us to work with a small sample. If the sample size n is 30 or fewer, the normal distribution might not be a suitable approximation of the distribution of means from small samples. In this section we describe a procedure for dealing with such cases. Here is the main objective of this section: Given a simple random sample of 30 or fewer values, develop an estimate of the value of the population mean μ.

Assumptions

1. $n \le 30$

2. The sample is a simple random sample.

3. The sample is from a normally distributed population. (This is a loose requirement, which can be met if the population has only one mode and is basically symmetric.)

If the first assumption is not satisfied, we have a large sample, and we can use the methods described in Section 6-2. If the second assumption is not satisfied, we cannot use the methods described in this book, but it might be possible to use more advanced methods. If the third assumption is not satisfied because the population has a distribution that is very nonnormal, we cannot use the methods of this section, but we might be able to use nonparametric methods (see Chapter 13) or bootstrap resampling methods (see the Excel Project at the end of this chapter).

As in Section 6-2, the sample mean $\bar{x}$ is the best point estimate (or single-valued estimate) of the population mean μ. As in Section 6-2, the distribution of sample means $\bar{x}$ tends to be more consistent (with *less variation*) than the distributions of other sample statistics, and the sample mean $\bar{x}$ is an *unbiased estimator* of the population mean μ, meaning that the distribution of sample means tends to center about the value of the population mean μ.

The sample mean $\bar{x}$ is the best point estimate of the population mean μ.

In Section 6-2 we noted that there is a serious limitation to the usefulness of a point estimate: The single value of a point estimate does not reveal how good that estimate is. Confidence intervals give us much more meaningful information by providing a range of values associated with a degree of likelihood that the range actually does contain the true value of μ. In developing confidence interval estimates of μ in this section, we have two cases: (1) The population standard deviation σ is known; (2) σ is not known.

Case 1 (σ Is Known):   The first case is largely unrealistic, because if we don't know the value of the population mean μ and we are trying to estimate that value, it's a pretty safe bet that we also do not know the value of the population standard deviation σ. If it should somehow happen that we satisfy the three assumptions listed near the beginning of this section, and we already know the value of σ, we can construct confidence interval estimates of μ by using the same methods described in Section 6-2. That is, the confidence interval limits are $\bar{x} - E$ and $\bar{x} + E$, where $E = z_{\alpha/2} \cdot \sigma/\sqrt{n}$.

Case 2 (σ Is Not Known):   The second case, when σ is not known, is much more realistic and practical, and it is the major focus of this section. Now, instead of using the normal distribution, we use the *Student t distribution* developed by William Gosset (1876–1937). Gosset was a Guinness Brewery employee who needed a distribution that could be used with small samples. The Irish brewery where he worked did not allow the publication of research results, so Gosset published under the pseudonym *Student*.

Student t Distribution

If the distribution of a population is essentially normal (approximately bell-shaped), then the distribution of

$$t = \frac{\bar{x} - \mu}{\dfrac{s}{\sqrt{n}}}$$

AS

is essentially a **Student t distribution** for all samples of size n. The Student t distribution, often referred to as the t **distribution,** is used to find critical values denoted by $t_{\alpha/2}$.

Excerpts from a Department of Transportation Circular

The following excerpts from a Department of Transportation circular concern some of the accuracy requirements for navigation equipment used in aircraft. Note the use of the confidence interval.

"The total of the error contributions of the airborne equipment, when combined with the appropriate flight technical errors listed, should not exceed the following with a 95% confidence (2-sigma) over a period of time equal to the update cycle."

"The system of airways and routes in the United States has widths of route protection used on a VOR system use accuracy of 64.5 degrees on a 95% probability basis."

We will soon discuss some of the important properties of the *t* distribution, but we will first present the components needed for the construction of confidence intervals. Let's start with the critical value denoted by $t_{\alpha/2}$. A value of $t_{\alpha/2}$ can be found in Table A-3 by locating the appropriate number of *degrees of freedom* in the left column and proceeding across the corresponding row until reaching the number directly below the applicable value of α for two tails.

DEFINITION

The number of **degrees of freedom** for a single data set is the number of sample values that can vary after certain restrictions have been imposed on all data values.

For example, if 10 students have quiz scores with a mean of 80, we can freely assign values to the first 9 scores, but the 10th score is then determined. The sum of the 10 scores must be 800, so the 10th score must equal 800 minus the sum of the first 9 scores. Because those first 9 scores can be freely selected to be any values, we say that there are 9 degrees of freedom available. For the applications of this section, the number of degrees of freedom is simply the sample size minus 1.

$$\text{degrees of freedom} = n - 1$$

 EXAMPLE **Finding a Critical Value** A sample of size $n = 15$ is a simple random sample selected from a normally distributed population. Find the critical value $t_{\alpha/2}$ corresponding to a 95% degree of confidence.

SOLUTION Because $n = 15$, the number of degrees of freedom is given by $n - 1 = 14$. Using Table A-3, we locate the 14th row by referring to the column at the extreme left. As in Section 6-2, a 95% degree of confidence corresponds to $\alpha = 0.05$, so we find the column with the heading "0.05 (two tails)." The value in the 14th row and the "0.05 (two tails)" column is 2.145, so $t_{\alpha/2} = 2.145$.

Now that we know how to find critical values denoted by $t_{\alpha/2}$, we can go on to describe the margin of error E and the confidence interval.

Margin of Error E for the Estimate of μ
[Based on an Unknown σ and a Small Simple Random Sample
($n \le 30$) from a Normally Distributed Population]

Formula 6-2 $$E = t_{\alpha/2}\frac{s}{\sqrt{n}}$$

where $t_{\alpha/2}$ has $n - 1$ degrees of freedom

> **Confidence Interval for the Estimate of μ**
> **[Based on an Unknown σ and a Small Simple Random Sample**
> **($n \leq 30$) from a Normally Distributed Population]**
>
> $$\overline{x} - E < \mu < \overline{x} + E$$
>
> where $\qquad E = t_{\alpha/2}\dfrac{s}{\sqrt{n}}$

▷ **EXAMPLE   Finding a Confidence Interval** Because cardiac deaths appear to increase after heavy snowfalls, a study was designed to compare cardiac demands of snow shoveling to those of using an electric snow thrower. Ten subjects cleared tracts of snow using both methods, and their maximum heart rates (beats per minute) were recorded during both activities. The following results were obtained for the snow shoveling portion of the experiment (based on data from "Cardiac Demands of Heavy Snow Shoveling," by Franklin et al., *Journal of the American Medical Association,* Vol. 273, No. 11):

Maximum Heart Rates During Manual Snow Shoveling:
$n = 10, \overline{x} = 175, s = 15$

Find the 95% confidence interval estimate of the population mean for those who shovel snow.

SOLUTION   We will proceed to construct a 95% confidence interval by using the t distribution. We select the t distribution because the following three conditions are met: (1) The sample is small ($n \leq 30$), (2) the population standard deviation σ is unknown, and (3) the population appears to have a normal distribution, because the sample data have a distribution that is approximately bell-shaped.

First we find the critical value $t_{\alpha/2}$, then we use that result to find the margin of error E, and then we use the value of E to find the confidence interval.

1. *Find $t_{\alpha/2}$:* The value of $t_{\alpha/2} = 2.262$ is found in Table A-3 as the critical value corresponding to $n - 1 = 9$ degrees of freedom (left column of Table A-3) and "0.05 (two tails)" (top column heading of Table A-3). Remember that a 95% confidence level corresponds to $\alpha = 0.05$.

2. *Find the margin of error E:* The margin of error $E = 10.7296$ is computed using Formula 6-2 as shown below.

$$E = t_{\alpha/2}\frac{s}{\sqrt{n}} = 2.262 \cdot \frac{15}{\sqrt{10}} = 10.7296$$

3. *Find the confidence interval:* The confidence interval can now be found by using $\bar{x} = 175$ and $E = 10.7296$ as shown below.

$$\bar{x} - E < \mu < \bar{x} + E$$
$$175 - 10.7296 < \mu < 175 + 10.7296$$
$$164 < \mu < 186$$

INTERPRETATION   This result could also be expressed in the format of $\mu = 175 \pm 11$ or $(164, 186)$. On the basis of the given sample results, we are 95% confident that the limits of 164 beats per minute and 186 beats per minute actually do contain the value of the population mean μ. As expected, these results suggest that snow shoveling dramatically increases the workload of the heart. (See Data Set 14 in Appendix B for the pulse rates of 100 statistics students. Those at-rest pulse rates have a mean of 70 beats per minute, and the highest is 100.) ◀

We now list the important properties of the t distribution that we are using in this section.

Important Properties of the Student t Distribution

1. The Student t distribution is different for different sample sizes. (See Figure 6-5 for the cases $n = 3$ and $n = 12$.)

2. The Student t distribution has the same general symmetric bell shape as the standard normal distribution, but it reflects the greater variability (with wider distributions) that is expected with small samples.

3. The Student t distribution has a mean of $t = 0$ (just as the standard normal distribution has a mean of $z = 0$).

4. The standard deviation of the Student t distribution varies with the sample size, but it is greater than 1 (unlike the standard normal distribution, which has $\sigma = 1$).

FIGURE 6-5
Student t Distributions for
$n = 3$ and $n = 12$

The Student t distribution has the same general shape and symmetry as the standard normal distribution, but it reflects the greater variability that is expected with small samples.

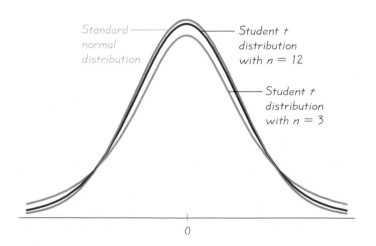

5. As the sample size n gets larger, the Student t distribution gets closer to the standard normal distribution. For values of $(n > 30)$, the differences are so small that we can use the critical z values instead of developing a much larger table of critical t values. (The values in the bottom row of Table A-3 are equal to the corresponding critical z values from the standard normal distribution.)

The following is a summary of the conditions indicating use of a t distribution instead of the standard normal distribution. (These same conditions will also apply in Chapter 7.)

Conditions for Using the Student t Distribution

1. The sample is small $(n \leq 30)$; and

2. σ is unknown; and

3. The parent population has a distribution that is essentially normal. (Because the distribution of the parent population is often unknown, we often estimate it by constructing a histogram of sample data.)

Choosing the Appropriate Distribution

It is sometimes difficult to decide whether to use the standard normal z distribution or the Student t distribution. The flowchart in Figure 6-6 and Table 6-2

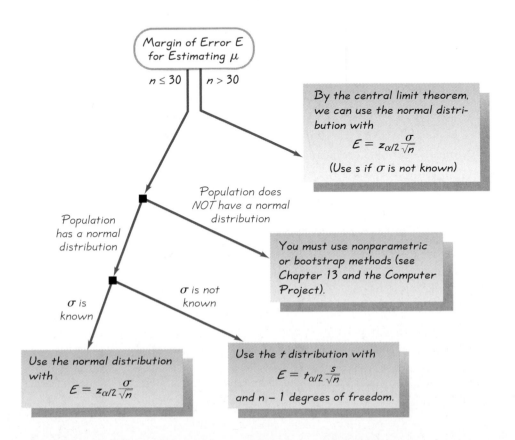

FIGURE 6-6
Using Normal and t Distributions

both summarize the key points to be considered when constructing confidence intervals for estimating μ, the population mean. In Figure 6-6 or Table 6-2, note that if we have a small ($n \leq 30$) sample drawn from a distribution that differs significantly from a normal distribution, we can't use the methods described in this chapter. One alternative is to use nonparametric methods (see Chapter 13), and another alternative is to use the computer bootstrap method, which makes no assumptions about the original population. This method is described in the Excel Project at the end of this chapter.

Another Approach   Rather than using *sample size* as the major criterion for choosing between the normal (z) and Student t distributions, some statisticians and software use *knowledge of the population standard deviation σ* as the major criterion, as follows:

- Use the *normal distribution* if the population standard deviation σ is known, and the sample is from a normally distributed population.

- Use the *Student t distribution* if the population standard deviation σ is *not* known, and the sample is from a normally distributed population.

In reality, we seldom know the value of σ, so this procedure suggests that we almost always use the Student t distribution, even if the sample size is large. But using the Student t distribution with a large sample size results in a large number of degrees of freedom, so the value of $t_{\alpha/2}$ is, for all practical purposes, the same as the value of $z_{\alpha/2}$. The confidence interval will be essentially the same with the two approaches. Therefore, we will base our choice on the sample size, as described in Figure 6-6 and Table 6-2.

TABLE 6-2	Margin of Error E for Confidence Interval Estimate of μ: $\overline{x} - E < \mu < \overline{x} + E$	
	Large Sample ($n > 30$)	Small Sample ($n \leq 30$)
σ known	$E = z_{\alpha/2}\dfrac{\sigma}{\sqrt{n}}$	Normally distributed population: Use $E = z_{\alpha/2}\dfrac{\sigma}{\sqrt{n}}$ Very nonnormal population distribution: Methods of this chapter don't apply. Use nonparametric methods or bootstrapping.
σ not known	Use s for σ in $E = z_{\alpha/2}\dfrac{\sigma}{\sqrt{n}}$	Normally distributed population: Use $E = t_{\alpha/2}\dfrac{s}{\sqrt{n}}$ Very nonnormal population distribution: Methods of this chapter don't apply. Use nonparametric methods or bootstrapping.

▷ **EXAMPLE** **Choosing Distributions** Assuming that you plan to construct a confidence interval for the population mean μ, use the given data to determine whether the margin of error E should be calculated using a critical value of $z_{\alpha/2}$ (from the normal distribution), a critical value of $t_{\alpha/2}$ (from a t distribution), or neither (so that the methods of this chapter cannot be used).

a. $n = 50, \bar{x} = 77.6, s = 14.2$, and the shape of the distribution is skewed.

b. $n = 25, \bar{x} = 77.6, s = 14.2$, and the distribution is bell-shaped.

c. $n = 25, \bar{x} = 77.6, \sigma = 14.2$, and the distribution is bell-shaped.

d. $n = 25, \bar{x} = 77.6, \sigma = 14.2$, and the distribution is extremely skewed.

SOLUTION   Refer to Figure 6-6 or Table 6-2 to determine the following.

a. Because the sample is large $(n > 30)$, the margin of error is calculated using $z_{\alpha/2}$ in Formula 6-1, with the sample standard deviation s used in place of σ.

b. Use $t_{\alpha/2}$ because (1) the sample is small, (2) the population appears to have a normal distribution, and (3) σ is unknown.

c. Use $z_{\alpha/2}$ because the population appears to have a normal distribution and the value of σ is known.

d. Because the sample is small and the population has a distribution that is very nonnormal, the methods of this chapter do not apply. Neither the normal nor the t distribution can be used.

◀

In Section 6-2 we presented the basic concepts of point estimate and confidence interval estimate. In this section we extended those concepts to small sample cases. In the following section we describe a method for determining the sample size needed to estimate a population mean.

Using Excel

Important: The methods of this section require that you use the DDXL add-in; do *not* use the Excel command CONFIDENCE, because it does not apply to the methods of this section.

To construct confidence interval estimates of a population mean based on a small $(n \leq 30)$ sample with an unknown population standard deviation, follow the same procedure given in Section 6-2, but in Step 3 select **1 Var t Interval.** (Step 7 will be skipped because there is no box for entering the population standard deviation.)

Here is a summary of the Excel procedure for the methods of this section: Select **DDXL, Confidence Intervals, 1 Var t Interval,** enter the range of sample data, click **OK,** click on the desired confidence level, then click on **Compute Interval.**

Note that selection of the options **1 Var z Interval** and **1 Var t Interval** corresponds to the use of the normal (z) distribution or the t distribution as shown below:

DDXL Option	Distribution	Section
1 Var z Interval	Normal	6-2
1 Var t Interval	t	6-3

6-3   Basic Skills and Concepts

Using Correct Distribution   *In Exercises 1–8, do one of the following, as appropriate: (a) Find the critical value $t_{\alpha/2}$; (b) find the critical value $z_{\alpha/2}$; (c) state that neither the normal nor the t distribution applies.*

1. 95%; $n = 20$; σ is unknown; population appears to be normally distributed.

2. 95%; $n = 40$; σ is unknown; population appears to be very skewed.

3. 99%; $n = 15$; σ is unknown; population appears to be normally distributed.

4. 90%; $n = 15$; σ is unknown; population appears to be very skewed.

5. 90%; $n = 12$; $\sigma = 24$; population appears to be very skewed.

6. 99%; $n = 100$; $\sigma = 4.2$; population appears to be very skewed.

7. 98%; $n = 9$; σ is unknown; population appears to be normally distributed.

8. 95%; $n = 30$; σ is unknown; population appears to be normally distributed.

Finding Confidence Intervals.   *In Exercises 9 and 10, use the given degree of confidence and sample data to find (a) the margin of error and (b) the confidence interval for the population mean μ. Assume that the population has a normal distribution.*

9. Math SAT scores for women: 95% confidence; $n = 15$, $\bar{x} = 496$, $s = 108$

10. Salaries of airline pilots: 99% confidence; $n = 12$, $\bar{x} = \$97,334$, $s = \$17,747$

TI-83 Plus   Exercise 11

Interpreting Calculator Display.   *In Exercises 11 and 12, use the given data and the TI-83 Plus calculator display to express the confidence interval in the Excel format of $\bar{x} - E < \mu < \bar{x} + E$. Also write a statement that interprets the confidence interval.*

11. IQ scores of professional athletes: 95% confidence; $n = 25$, $\bar{x} = 104$, $s = 12$

12. Final exam test scores in a statistics course: 99% confidence; $n = 27$, $\bar{x} = 78.8$, $s = 12.2$

Constructing Confidence Intervals.   *In Exercises 13–24, construct the confidence interval. Some of these exercises require use of the t distribution, but others require use of the normal distribution.*

TI-83 Plus   Exercise 12

13. Snow Shoveling and Heart Attacks Because cardiac deaths appear to increase after heavy snowfalls, a study was designed to compare cardiac demands of snow shoveling to those of using an electric snow thrower. Ten subjects cleared tracts of snow using both methods, and their maximum heart rates (beats per minute) were recorded during both activities. The following results were obtained from the portion of the experiment involving the use of an electric snow thrower (based on

data from "Cardiac Demands of Heavy Snow Shoveling," by Franklin et al., *Journal of the American Medical Association,* Vol. 273, No. 11):

Maximum Heart Rates During Automated Snow Removal:
$n = 10, \bar{x} = 124, s = 18$

Find the 95% confidence interval estimate of the population mean for those who use the electric snow thrower. Compare the result to the confidence interval found in the related example of this section and interpret your findings.

EXCEL **14.** Weights of Quarters If we refer to the weights (in grams) of quarters listed in Data Set 13 in Appendix B, we will find 50 weights with a mean of 5.622 g and a standard deviation of 0.068 g. The Excel workbook is QUARTERS.xls. Based on this simple random sample of quarters in circulation, construct a 95% confidence interval estimate of the population mean of all quarters in circulation. The U.S. Department of the Treasury claims that it mints quarters to yield a mean weight of 5.670 g. Is this claim consistent with the confidence interval? If not, what is a possible explanation for the discrepancy?

15. Destroying Dodge Vipers With *destructive testing,* sample items are destroyed in the process of testing them. Crash testing of cars is one very expensive example of destructive testing. If you were responsible for such crash tests, there is no way you would want to tell your supervisor that you must crash and destroy more than 30 cars so that you could use the normal distribution. Let's assume that you have crash tested 12 Dodge Viper sports cars (list price: $59,300) under a variety of conditions that simulate typical collisions. Analysis of the 12 damaged cars results in repair costs having a distribution that appears to be bell-shaped, with a mean of $\bar{x} = \$26,227$ and a standard deviation of $s = \$15,873$ (based on data from the Highway Loss Data Institute). Find the 95% interval estimate of μ, the mean repair cost for all Dodge Vipers involved in collisions, and then interpret the result.

16. Costs of Repairing Hondas In crash tests of 15 Honda Odyssey minivans, collision repair costs are found to have a distribution that is roughly bell-shaped, with a mean of $1786 and a standard deviation of $937 (based on data from the Highway Loss Data Institute). Construct the 99% confidence interval for the mean repair cost in all such vehicle collisions.

EXCEL **17.** Estimating Car Pollution In a sample of seven cars, each car was tested for nitrogen-oxide emissions (in grams per mile) and the following results were obtained: 0.06, 0.11, 0.16, 0.15, 0.14, 0.08, 0.15 (based on data from the Environmental Protection Agency). Assuming that this sample is representative of the cars in use, construct a 99% confidence interval estimate of the mean amount of nitrogen-oxide emissions for all cars.

18. Crash Hospital Costs A study was conducted to estimate hospital costs for accident victims who wore seat belts. Twenty randomly selected cases have a distribution that appears to be bell-shaped with a mean of $9004 and a standard deviation of $5629 (based on data from the U.S. Department of Transportation). Construct the 99% confidence interval for the mean of all such costs.

19. TV Replacement Times A sample consists of 75 TV sets purchased several years ago. The replacement times of those TV sets have a mean of 8.2 years and a standard deviation of 1.1 years (based on data from "Getting Things Fixed,"

Consumer Reports). Construct a 90% confidence interval for the mean replacement time of all TV sets from that era. Does the result apply to TV sets currently being sold?

20. **Interpreting Graphs** Statisticians at the College of Newport are conducting research into the ability of people to read and interpret graphs presented by the media. They have developed a comprehensive test of that ability, and they want to estimate the mean score for a simple random sample of people. Test subjects are paid $50 to participate, but the $1000 budget is enough to fund only 20 subjects. Those 20 test scores have a mean of 212 and a standard deviation of 67. Construct a 95% confidence interval for the population mean.

EXCEL 21. **Comparing Yellow and Brown M&M Candies** Refer to Data Set 10 in Appendix B. The Excel workbook is M&M.xls.
   a. Construct a 90% confidence interval for the mean weight of yellow M&M plain candies.
   b. Construct a 90% confidence interval for the mean weight of brown M&M plain candies.
   c. Compare the methods used to find the two confidence intervals, and interpret the results.

EXCEL 22. **Comparing Regular and Diet Pepsi** Refer to Data Set 1 in Appendix B. The Excel workbook is COLA.xls.
   a. Construct a 95% confidence interval estimate of the mean weight of cola in cans of regular Pepsi.
   b. Construct a 95% confidence interval estimate of the mean weight of cola in cans of diet Pepsi.
   c. Compare the results from parts (a) and (b) and interpret them.

EXCEL 23. **Comparing Textbook Prices** Refer to Data Set 2 in Appendix B. The Excel workbook is TEXTBOOK.xls.
   a. Construct a 99% confidence interval estimate of the mean cost of new hardcover textbooks at the University of Massachusetts.
   b. Construct a 99% confidence interval estimate of the mean cost of new paperback textbooks at the University of Massachusetts.
   c. Compare and interpret the results.

EXCEL 24. **Comparing Movies** Refer to Data Set 15 in Appendix B. The Excel workbook is MOVIES.xls.
   a. Construct a 99% confidence interval estimate of the mean length (in minutes) of movies with a rating of R.
   b. Construct a 99% confidence interval estimate of the mean length (in minutes) of movies with a rating different than R.
   c. Compare and interpret the results.

6-3 Beyond the Basics

25. **Using the Wrong Distribution** Assume that a small ($n \leq 30$) sample is a simple random sample selected from a normally distributed population for which σ is unknown. Construction of a confidence interval should use the t distribution, but how are the confidence interval limits affected if the normal distribution is incorrectly used instead?

26. **Effects of Units of Measurement** A confidence interval is constructed for a small simple random sample of temperatures (in degrees Fahrenheit) selected from a normally distributed population for which σ is unknown.

a.   How is the margin of error E affected if each temperature is converted to the Celsius scale? $\left[C = \dfrac{5}{9}(F - 32) \right]$

b.   If the confidence interval limits are denoted by a and b, find expressions for the confidence interval limits after the original temperatures have been converted to the Celsius scale.

c.   Based on the results from part (b), can confidence interval limits for the Celsius temperatures be found by simply converting the confidence interval limits from the Fahrenheit scale to the Celsius scale?

27. Finding Degree of Confidence A simple random sample of Boeing 747 aircraft is selected, and the times (in hours) required to test for structural stress fractures are as follows: 8.1, 9.9, 9.5, 6.9, 9.8. Based on these results, a confidence interval is found to be $\mu = 8.84 \pm 1.24$. Find the degree of confidence.

6-4 Determining Sample Size Required to Estimate μ

In this section we will address the following key question: When we plan to collect a simple random sample of data that will be used to estimate a population mean μ, *how many* sample values must be obtained? In other words, we will find the sample size n that is required to estimate the value of a population mean. For example, suppose we want to estimate the mean weight of airline passengers (an important value for reasons of safety). How many passengers must be randomly selected and weighed? Determining the size of a simple random sample is a very important issue, because samples that are needlessly large waste time and money, and samples that are too small may lead to poor results. In many cases we can find the minimum sample size needed to estimate some parameter, such as the population mean μ.

  If we begin with the expression for the margin of error E (Formula 6-1 in Section 6-2) and solve for the sample size n, we get the following.

Sample Size for Estimating Mean μ

Formula 6-3
$$ n = \left[\frac{z_{\alpha/2}\,\sigma}{E} \right]^2 $$

where $z_{\alpha/2}$ = critical z score based on the desired degree of confidence

    E = desired margin of error

    σ = population standard deviation

  Formula 6-3 is quite remarkable because it implies that the sample size does not depend on the size (N) of the population; the sample size depends on the desired degree of confidence, the desired margin of error, and the value of the standard deviation σ.

TV Sample Sizes

A *Newsweek* article described the use of people meters as a way of determining how many people are watching different television programs. The article stated, "Statisticians have long argued that the household samples used by the rating services are simply too small to accurately determine what America is watching. In that light, it may be illuminating to note that the 4000 homes reached by the people meters constitute exactly 0.0045% of the wired nation." This implied claim that the sample size of 4000 homes is too small is not valid. Methods of this chapter can be used to show that a random sample size of 4000 can provide results that are quite good, even though the sample might be only 0.0045% of the population.

to estimate a population mean μ. In this section we will take those same three concepts, but will now apply them to the population proportion p. Here are the main objectives of this section:

1. Given sample data with each item belonging to one of two categories, estimate the population proportion p for one of those categories.

2. Determine the sample size required to estimate a population proportion p.

Television ratings are important because they affect the billions of dollars spent for advertising. *Super Bowl* commercials are among the most expensive, and charges for commercials are based on a Nielsen Media Research estimate of the proportion of households that tune in. Nielsen must address two important questions: (1) How many households must be sampled, and (2) after the sample data have been collected, what is the estimate of the proportion of households tuned to the *Super Bowl?* This section provides the tools that you need to answer such questions.

Estimating the Proportion p

We begin by considering methods for using sample data to estimate the population proportion p. (Later in this section, we will consider the issue of determining sample size.) The following assumptions apply.

Assumptions

1. The sample is a simple random sample.

2. The conditions for the binomial distribution are satisfied. (See Section 4-3.) That is, there is a fixed number of trials, the trials are independent, there are two categories of outcomes, and the probabilities remain constant for each trial.

3. The normal distribution can be used to approximate the distribution of sample proportions because $np \geq 5$ and $nq \geq 5$ are both satisfied. (Because p and q are unknown, we use the sample proportion to estimate their values. Also, there are procedures for dealing with situations in which the normal distribution is not a suitable approximation. See Exercise 40.)

We now introduce the new notation $\hat{p}$ (called "p hat") for the sample proportion.

Notation for Proportions

$p = population$ proportion

$\hat{p} = \dfrac{x}{n} = sample$ proportion of x successes in a sample of size n

$\hat{q} = 1 - \hat{p} = sample$ proportion of x failures in a sample of size n

Probability and Percent   Although this chapter focuses on the population proportion p, the procedures discussed here can also be applied to probabilities or percentages, but percentages must be converted to proportions by dropping the percent sign and dividing by 100. For example, the percentage of people who don't buy books, 48.7%, can be expressed in decimal form as 0.487. The symbol p may therefore represent a proportion, a probability, or the decimal equivalent of a percent. For example, if you survey 200 statistics students and find that 80 of them have purchased TI-83 Plus calculators, then the sample proportion is $\hat{p} = x/n = 80/200 = 0.400$, and $\hat{q} = 0.600$ (calculated from $1 - 0.400$). Instead of computing the value of x/n, the value of $\hat{p}$ is sometimes already known because the sample proportion or percentage is given directly. For example, if it is reported that 1068 American television viewers are surveyed and 25% of them are college graduates, then $\hat{p} = 0.25$ and $\hat{q} = 0.75$.

If we want to estimate a population proportion with a single value, the best estimate is $\hat{p}$. As in Sections 6-2 and 6-3, when an estimate of a population parameter is a single value, we call it a *point estimate*.

> **The sample proportion $\hat{p}$ is the best point estimate of the population proportion p.**

We use $\hat{p}$ as the point estimate of p (just as $\bar{x}$ is used as the point estimate of μ) because it is unbiased and is the most consistent of the estimators that could be used. It is unbiased in the sense that the distribution of sample proportions tends to center about the value of p; that is, sample proportions $\hat{p}$ do not systematically tend to underestimate p, nor do they systematically tend to overestimate p. The sample proportion $\hat{p}$ is the most consistent estimator in the sense that the standard deviation of sample proportions tends to be smaller than the standard deviation of any other unbiased estimators.

The point estimate has a major disadvantage: Although the value of $\hat{p}$ is the *best* single value estimate of p, it gives no indication of just how good it is. What we therefore need is an estimate that *does* give us a sense of how accurate it is, and the *confidence interval* meets that objective. We first present the *margin of error* (defined in Section 6-2), which is used for finding a confidence interval, and then we present the format of the confidence interval itself.

Margin of Error of the Estimate of p

Formula 6-4
$$E = z_{\alpha/2}\sqrt{\frac{\hat{p}\hat{q}}{n}}$$

Confidence Interval (or Interval Estimate) for the Population Proportion p

$$\hat{p} - E < p < \hat{p} + E \qquad \text{where } E = z_{\alpha/2}\sqrt{\frac{\hat{p}\hat{q}}{n}}$$

Push Polling

"Push polling" is the practice of political campaigning under the guise of a poll. Its name is derived from its objective of pushing voters away from opposition candidates by asking loaded questions designed to discredit them. Here's an example of one such question that was used: "Please tell me if you would be more likely or less likely to vote for Roy Romer if you knew that Gov. Romer appoints a parole board which has granted early release to an average of four convicted felons per day every day since Romer took office." The National Council on Public Polls characterizes push polls as unethical, but some professional pollsters do not condemn the practice as long as the questions do not include outright lies.

The confidence interval is sometimes expressed in the following formats.

$$p = \hat{p} \pm E$$

or

$$(\hat{p} - E, \hat{p} + E)$$

In Chapter 3, when probabilities were given in decimal form, we rounded to three significant digits. We use that same rounding rule here.

Round-Off Rule for Confidence Interval Estimates of p

Round the confidence interval limits to three significant digits.

> **EXAMPLE** **Misleading Survey Responses** Do people lie about voting? In a survey of 1002 people, 701 people said that they voted in the recent presidential election (based on data from ICR Research Group). Voting records show that 61% of eligible voters actually did vote. Using these survey results,

a. find the point estimate of the proportion of people who say that they voted.

b. find the 95% confidence interval estimate of the proportion of people who say that they voted.

c. determine whether the survey results are consistent with the actual voter turnout of 61%.

SOLUTION

a. The point estimate of p is

$$\hat{p} = \frac{x}{n} = \frac{701}{1002} = 0.6996$$

b. Constructing the confidence interval requires that we first evaluate the margin of error E. The value of E can be found from Formula 6-4. We use $\hat{p} = 0.6996$ (found in part a), $\hat{q} = 0.3004$ (from $\hat{q} = 1 - \hat{p}$), and $z_{\alpha/2} = 1.96$ (from Table A-2, where 95% converts to $\alpha = 0.05$, which is divided equally between the two tails, so $z = 1.96$ is the boundary for the related area of 0.4750).

$$E = z_{\alpha/2}\sqrt{\frac{\hat{p}\hat{q}}{n}} = 1.96\sqrt{\frac{(0.6996)(0.3004)}{1002}} = 0.0283855$$

We can now find the confidence interval by using $\hat{p} = 0.6996$ and $E = 0.0283855$.

$$\hat{p} - E < p < \hat{p} + E$$
$$0.6996 - 0.0283855 < p < 0.6996 + 0.0283855$$
$$0.671 < p < 0.728 \text{ (rounded)}$$

This same result could be expressed in the format of $p = 0.700 \pm 0.028$ or $(0.671, 0.728)$. If we want the 95% confidence interval for the true population *percentage,* we could express the result as $67.1\% < p < 72.8\%$. This confidence interval is often reported with a statement such as this: "Among those eligible to vote, the percentage who say they voted is estimated to be 70%, with a margin of error of plus or minus 2.8 percentage points." That statement is a verbal expression of this format for the confidence interval: $p = 70\% \pm 2.8\%$. (The level of confidence should also be reported, but it rarely is in the media. The media typically use a 95% degree of confidence but omit any reference to it.)

c. Based on the survey results, we are 95% confident that the limits of 67.1% and 72.8% contain the true percentage of eligible voters who *say* that they voted. But we know that 61% of the eligible voters actually did vote. Because 61% does not fall within the 95% confidence interval, we can conclude that the percentage of people who say that they voted appears to be different than the percentage of people who actually did vote.

Interpreting a Confidence Interval   As in Sections 6-2 and 6-3, we must be careful to interpret the confidence interval correctly. Examples of correct and wrong interpretations follow.

Correct: We are 95% confident that the interval from 0.671 to 0.728 actually does contain the true value of p. (This means that if we were to select many different samples of size 1002 and construct the confidence intervals as we did in the preceding example, 95% of them would actually contain the value of the population proportion p.)

Wrong: There is a 95% chance that the true value of p will fall between 0.671 and 0.728.

The differences in wording between the correct and the wrong statements might seem subtle, but the real difference is substantial. The "wrong" statement is incorrect because the population proportion p is a fixed value, not a random variable. Either p falls between the confidence interval limits or it does not, and there is no probability or chance involved here.

Rationale for the Margin of Error   Because the sampling distribution of proportions is approximately normal (because the conditions $np \geq 5$ and $nq \geq 5$ are both satisfied), we can use results from Section 5-6 to conclude that μ and σ are given by $\mu = np$ and $\sigma = \sqrt{npq}$. Both of these parameters pertain to n trials, but we convert them to a per-trial basis by dividing by n as follows:

$$\text{Mean of sample proportions: } \mu = \frac{np}{n} = p$$

$$\text{Standard deviation of sample proportions: } \sigma = \frac{\sqrt{npq}}{n} = \sqrt{\frac{npq}{n^2}} = \sqrt{\frac{pq}{n}}$$

The first result may seem trivial, because we already stipulated that the true population proportion is p. The second result is nontrivial and is useful in describing the margin of error E, but we replace the product pq by $\hat{p}\hat{q}$ because we don't yet know the value of p (it is the value we are trying to estimate). Formula 6-4 for the margin of error reflects the fact that $\hat{p}$ has a probability of $1 - \alpha$ of being within $z_{\alpha/2}\sqrt{pq/n}$ of p. The confidence interval for p, as given previously, reflects the fact that there is a probability of $1 - \alpha$ that $\hat{p}$ differs from p by less than the margin of error $E = z_{\alpha/2}\sqrt{\hat{p}\hat{q}/n}$.

Determining Sample Size

Suppose we want to collect sample data with the objective of estimating some population proportion. How do we know *how many* sample items must be obtained? If we take the expression for the margin of error E (Formula 6-4), then solve for n, we get Formula 6-5. Formula 6-5 requires $\hat{p}$ as an estimate of the population proportion p, but if no such estimate is known (as is usually the case), we replace $\hat{p}$ by 0.5 and replace $\hat{q}$ by 0.5, with the result given in Formula 6-6.

Sample Size for Estimating Proportion p

When an estimate $\hat{p}$ is known: **Formula 6-5** $n = \dfrac{[z_{\alpha/2}]^2 \hat{p}\hat{q}}{E^2}$

When no estimate $\hat{p}$ is known: **Formula 6-6** $n = \dfrac{[z_{\alpha/2}]^2 \cdot 0.25}{E^2}$

Round-Off Rule for Determining Sample Size

If the computed sample size is not a whole number, round it up to the next *higher* whole number.

Use Formula 6-5 when reasonable estimates of $\hat{p}$ can be made by using previous samples, a pilot study, or someone's expert knowledge. When no such estimate can be made, we assign the value of 0.5 to each of $\hat{p}$ and $\hat{q}$, so the resulting sample size will be at least as large as it should be. The underlying reason for the assignment of 0.5 is this: The product $\hat{p} \cdot \hat{q}$ has 0.25 as its largest possible value, which occurs when $\hat{p} = 0.5$ and $\hat{q} = 0.5$. (Try experimenting with different values of $\hat{p}$ to verify that $\hat{p} \cdot \hat{q}$ has 0.25 as the largest possible value.) Note that Formulas 6-5 and 6-6 do not include the population size N, so the size of the population is irrelevant. (*Exception:* When sampling is without replacement from a relatively small finite population. See Exercise 35.)

▷ **EXAMPLE** **E-Mail** The use of answering machines, fax machines, voice mail, and e-mail is growing rapidly, and they are having a dramatic effect on the way we communicate. Suppose a sociologist wants to determine the current percentage of U.S. households using e-mail. How many households must be surveyed in order to be 90% confident that the sample percentage is in error by no more than four percentage points?

a. Use this result from an earlier study: In 1997, 16.9% of U.S. households used e-mail (based on data from *The World Almanac and Book of Facts*).

b. Assume that we have no prior information suggesting a possible value of $\hat{p}$.

SOLUTION

a. The prior study suggests that $\hat{p} = 0.169$, so $\hat{q} = 0.831$ (found from $\hat{q} = 1 - 0.169$). With a 90% level of confidence, we have $\alpha = 0.10$, so $z_{\alpha/2} = 1.645$. Also, the margin of error is $E = 0.04$ (the decimal equivalent of "four percentage points"). Because we have an estimated value of $\hat{p}$, we use Formula 6-5 as follows.

$$n = \frac{[z_{\alpha/2}]^2 \hat{p}\hat{q}}{E^2} = \frac{[1.645]^2(0.169)(0.831)}{0.04^2}$$

$$= 237.51965 = 238 \quad \text{(rounded up)}$$

We must survey at least 238 randomly selected households.

b. As in part (a), we again use $z_{\alpha/2} = 1.645$ and $E = 0.04$, but with no prior knowledge of $\hat{p}$ (or $\hat{q}$), we use Formula 6-6 as follows.

$$n = \frac{[z_{\alpha/2}]^2 \cdot 0.25}{E^2} = \frac{[1.645]^2 \cdot (0.25)}{0.04^2}$$

$$= 422.81641 = 423 \quad \text{(rounded up)}$$

INTERPRETATION   To be 90% confident that our sample percentage is within four percentage points of the true percentage for all households, we should randomly select and survey 423 households. By comparing this result to the sample size of 238 found in part (a), we can see that if we have no knowledge of a prior study, a larger sample is required to achieve the same results as when the value of $\hat{p}$ can be estimated. But now let's use common sense: We know that the use of e-mail is growing so rapidly that the 1997 estimate is too old to be of much use. Today, substantially more than 16.9% of households use e-mail. Realistically, we need a sample larger than 238 households. Assuming that we don't really know the current rate of e-mail usage, we should randomly select 423 households. With 423 households, we will be 90% confident that we are within four percentage points of the true percentage of households using e-mail.   ◁

Common Errors When calculating sample size using Formula 6-5 or 6-6, be sure to substitute the critical z score for $z_{\alpha/2}$. For example, if you are working with 95% confidence, be sure to replace $z_{\alpha/2}$ with 1.96. (Here is the logical sequence: $95\% \Rightarrow \alpha = 0.05 \Rightarrow z_{\alpha/2} = 1.96$ found from Table A-2.) Don't make the mistake of replacing $z_{\alpha/2}$ with 0.95 or 0.05. Also, don't make the mistake of using $E = 3$ as the margin of error corresponding to "three percentage points." When using Formula 6-5 or 6-6, the value of E never exceeds 1, and it's typically a number like 0.03. This error causes the sample size to be 1/10,000th of what it should be, so that you might end up with a sample size of only 1 when the answer is rounded up. You really can't estimate a population proportion by surveying only one person, regardless of how knowledgeable that person claims to be.

Population Size Part (b) of the preceding example involved application of Formula 6-6, the same formula frequently used by Nielsen, Gallup, and other professional pollsters. Many people incorrectly believe that the sample size should be some percentage of the population, but Formula 6-6 shows that the population size is irrelevant. (In reality, the population size is sometimes used, but only in cases in which we sample without replacement from a relatively small population. See Exercise 35.) Most of the polls featured in newspapers, magazines, and broadcast media involve sample sizes in the range of 1000 to 2000. Even though such polls may involve a very small percentage of the total population, they can provide results that are quite good. When Nielsen surveys 4000 TV households from a population of 104 million households, only 0.004% of the households are surveyed; still, we can be 95% confident that the sample percentage will be within one percentage point of the true population percentage.

Polls are now an integral part of American life. They affect the television shows we watch, the leaders we elect, the legislation that governs us, and the products we consume. An understanding of the concepts of this section is truly essential in this new millennium.

Finding the Point Estimate and E from a Confidence Interval As in the preceding sections, if we use Excel to find confidence interval limits, the sample proportion $\hat{p}$ and the margin of error E can be found as follows.

Point estimate of p:

$$\hat{p} = \frac{(\text{upper confidence interval limit}) + (\text{lower confidence interval limit})}{2}$$

Margin of error:

$$E = \frac{(\text{upper confidence interval limit}) - (\text{lower confidence interval limit})}{2}$$

> **EXAMPLE** A journal article includes this confidence interval estimate of the population proportion: $0.400 < p < 0.500$. Use the given confidence interval limits to find the point estimate $\hat{p}$ and the margin of error E.

SOLUTION

$$\hat{p} = \frac{(\text{upper confidence interval limit}) + (\text{lower confidence interval limit})}{2}$$

$$= \frac{0.500 + 0.400}{2} = 0.450$$

$$E = \frac{(\text{upper confidence interval limit}) - (\text{lower confidence interval limit})}{2}$$

$$= \frac{0.500 - 0.400}{2} = 0.050$$

Using Excel for Confidence Interval Estimates of p

To construct confidence interval estimates of a population proportion p, you must first identify the number of successes x and the number of trials n. Also, you must use the DDXL add-in. (The Excel command CONFI-DENCE does not apply to proportions.) Follow these steps.

1. Enter the number of successes x in cell A1 and enter the number of trials n in cell A2.

2. Click on **DDXL.** (If DDXL does not appear on the menu bar, install Data Desk/XL as described in Section 1-5.)

3. Click on **Confidence Intervals.**

4. Click on **Summ 1 Var Prop Interval,** which indicates that we have summary data (x and n) and we want to construct a confidence interval estimate of one population proportion.

5. Click on the pencil icon for the number of successes, then enter A1. Click on the pencil icon for the number of trials and enter A2.

6. Click **OK** and another dialog box appears.

7. Click on the desired confidence level (90% or 95% or 99%).

8. Click **Compute Interval.**

6-5 Basic Skills and Concepts

Interpreting Confidence Interval Limits.   *In Exercises 1–4, use the given confidence interval limits to find the point estimate $\hat{p}$ and the margin of error E.*

1. $0.800 < p < 0.840$

2. $(0.444, 0.474)$

3. $(0.432, 0.455)$

4. $0.560 < p < 0.598$

Finding Margin of Error.   *In Exercises 5–10, assume that a sample is used to estimate a population proportion p. Find the margin of error E that corresponds to the given statistics and the degree of confidence.*

5. $n = 500, x = 400, 95\%$

EXCEL **6.** $n = 1000, x = 250, 99\%$

EXCEL **7.** $n = 1068, x = 237, 98\%$

EXCEL **8.** $n = 777, x = 543, 90\%$

EXCEL **9.** 95% confidence; the sample size is 200, of which 35% are successes

EXCEL **10.** 99% confidence; the sample size is 550, of which 90% are successes

Constructing Confidence Intervals. In Exercises 11–14, use the given sample data and degree of confidence to construct the confidence interval estimate of the population proportion p.

EXCEL **11.** $n = 750, x = 150, 99\%$ confidence

EXCEL **12.** $n = 1200, x = 800, 90\%$ confidence

EXCEL **13.** $n = 1357, x = 222, 95\%$ confidence

EXCEL **14.** $n = 3622, x = 412, 95\%$ confidence

Determining Sample Size. In Exercises 15–18, use the given data to find the minimum sample size required to estimate a population proportion or percentage.

EXCEL **15.** Margin of error: 0.025; confidence level: 99%; $\hat{p}$ and $\hat{q}$ unknown

EXCEL **16.** Margin of error: 0.04; confidence level: 98%; $\hat{p}$ and $\hat{q}$ unknown

EXCEL **17.** Margin of error: three percentage points; confidence level: 95%; $\hat{p}$ is estimated to be 0.15 from a prior study

EXCEL **18.** Margin of error: four percentage points; confidence level: 97%; $\hat{p}$ is estimated to be 0.33 from a prior study

19. Interpreting Calculator Display In 1920 only 35% of U.S. households had telephones, but that rate is now much higher. A recent survey of 4276 randomly selected households showed that 4019 of them had telephones (based on data from the U.S. Census Bureau). Using those survey results and a 99% confidence level, the TI-83 Plus calculator display is as shown.
   a. Write a statement that interprets the confidence interval.
   b. Based on the preceding result, should pollsters be concerned about results from surveys conducted by telephone?

20. Interpreting Calculator Display The Hartford Insurance Company wants to estimate the percentage of drivers who change tapes or CDs while driving. A random sample of 850 drivers results in 544 who change tapes or CDs while driving (based on data from *Prevention* magazine). Using the sample data and a 90% confidence level, the TI-83 Plus calculator display is as shown.
   a. Write a statement that interprets the confidence interval.
   b. Based on the preceding result, does the practice of changing tapes or CDs while driving appear to be a threat to safety?

EXCEL **21.** Confidence Interval for Theme Park Attendance Each year, billions of dollars are spent at theme parks owned by Disney, Universal Studios, Sea World, Busch Gardens, and others. A survey of 1233 people who took trips revealed that 111 of them included a visit to a theme park (based on data from the Travel Industry Association of America).

TI-83 Plus

```
1-PropZInt
 (.93053,.94926)
 p=.9398971001
 n=4276
```

TI-83 Plus

```
1-PropZInt
 (.61292,.66708)
 p=.64
 n=850
```

a. Find the point estimate of the *percentage* of people who visit a theme park when they take a trip.

b. Find a 95% confidence interval estimate of the *percentage* of all people who visit a theme park when they take a trip.

EXCEL **22.** Confidence Interval for Voter Surveys In a recent presidential election, 611 voters were surveyed and 308 of them said that they voted for the candidate who won (based on data from the ICR Survey Research Group).

a. Find the point estimate of the *percentage* of voters who said that they voted for the candidate who won.

b. Find a 99% confidence interval estimate of the *percentage* of voters who said that they voted for the candidate who won.

c. Of those who voted, 43% actually voted for the candidate who won. Is this result consistent with the survey results? How might a discrepancy be explained?

EXCEL **23.** Pilot Fatalities Researchers studied crashes of general aviation (noncommercial and nonmilitary) airplanes and found that pilots died in 5.2% of 8411 crash landings (based on data from "Risk Factors for Pilot Fatalities in General Aviation Airplane Crash Landings," by Rostykus, Cummings, and Mueller, *Journal of the American Medical Association*, Vol. 280, No. 11).

a. Construct a 95% confidence interval estimate of the percentage of pilot deaths in all general aviation crashes.

b. Among crashes with an explosion or fire on the ground, the pilot fatality rate is estimated by the 95% confidence interval of (15.5%, 26.9%). Is this result substantially different from the result from part (a)? What can you conclude about an explosion or fire as a risk factor?

EXCEL **24.** Effectiveness of Gender Selection In a study of the Clark method of gender selection, 40 couples tried to have baby girls. Among the 40 babies, 62.5% were girls.

a. Construct a 95% confidence interval estimate for the proportion of girls from all couples who try to have baby girls with the Clark method of gender selection.

b. Based on the result from part (a), can we conclude that the Clark method is effective, with a rate of girls that is greater than 50%?

EXCEL **25.** Drug Testing The drug Ziac is used to treat hypertension. In a clinical test, 3.2% of 221 Ziac users experienced dizziness (based on data from Lederle Laboratories).

a. Construct a 99% confidence interval estimate of the percentage of all Ziac users who experience dizziness.

b. In the same clinical test, people in the placebo group didn't take Ziac, but 1.8% of them reported dizziness. Based on the result from part (a), what can we conclude about dizziness as an adverse reaction to Ziac?

EXCEL **26.** Smoking and College Education The tobacco industry closely monitors all surveys that involve smoking. One survey showed that among 785 randomly selected subjects who completed four years of college, 18.3% smoke (based on data from the American Medical Association).

a. Construct the 99% confidence interval for the true percentage of smokers among all people who completed four years of college.

b. Based on the result from part (a), does the smoking rate for those with four years of college appear to be substantially different than the 27% rate for the general population?

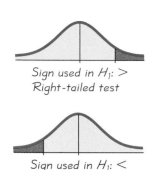

Sign used in H_1: >
Right-tailed test

Sign used in H_1: <
Left-tailed test

Sign used in H_1: ≠
Two-tailed test

By examining the null hypothesis, we should be able to deduce whether a test is right-tailed, left-tailed, or two-tailed. The tail will correspond to the critical region containing the values that would conflict significantly with the null hypothesis. A useful check is summarized in the margin figures, which show that the inequality sign in H_1 points in the direction of the critical region. The symbol ≠ is often expressed in programming languages as <>, and this reminds us that an alternative hypothesis such as $\mu \neq 98.6$ corresponds to a two-tailed test.

Conclusions in Hypothesis Testing

We have seen that the original claim sometimes becomes the null hypothesis and at other times becomes the alternative hypothesis. However, our standard procedure of hypothesis testing requires that we always test the null hypothesis, and we will see in the following section that our initial conclusion will always be one of the following:

1. Reject the null hypothesis.
2. Fail to reject the null hypothesis.

Wording the Final Conclusion: The conclusion of rejecting the null hypothesis or failing to reject it is fine for those of us with the wisdom to take a statistics course, but we should use simple, nontechnical terms in stating what the conclusion really means. Figure 7-4 shows how to formulate the correct wording of the final conclusion. Note that only one case leads to wording indicating that the sample data actually *support* the conclusion. If you want to support some claim, state it in such a way that it becomes the alternative hypothesis, and then hope that the null hypothesis gets rejected. For example, to support the claim that the mean body temperature is different from 98.6° F, make the claim that $\mu \neq 98.6$. This claim will be an alternative hypothesis that will be supported if you reject the null hypothesis, H_0: $\mu = 98.6$. If, on the other hand, you claim that $\mu = 98.6$, you will either reject or fail to reject the claim; in either case, you will not *support* the claim that $\mu = 98.6$.

Accept/Fail to Reject: Some texts say "accept the null hypothesis" instead of "fail to reject the null hypothesis." Whether we use the term *accept* or *fail to reject*, we should recognize that *we are not proving the null hypothesis;* we are merely saying that the sample evidence is not strong enough to warrant rejection of the null hypothesis. It's like a jury's saying that there is not enough evidence to convict a suspect. The term *accept* is somewhat misleading, because it seems to imply incorrectly that the null hypothesis has been proved. The phrase *fail to reject* says more correctly that the available evidence isn't strong enough to warrant rejection of the null hypothesis. In this text we will use the terminology *fail to reject the null hypothesis*, instead of *accept the null hypothesis*.

 EXAMPLE   **Stating the Final Conclusion** Suppose researchers claim that the mean body temperature is different from 98.6° F. This claim of

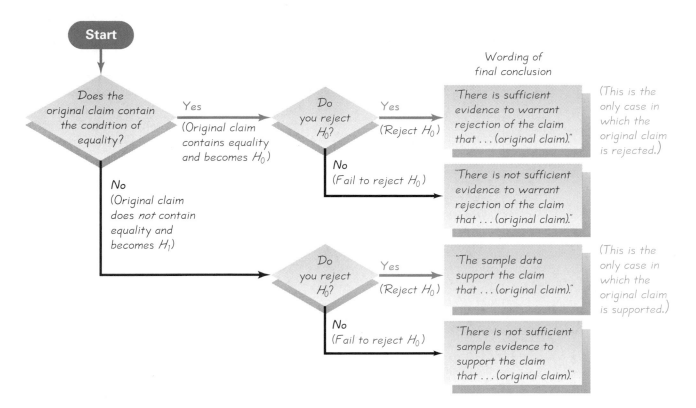

FIGURE 7-4
Wording of Final Conclusion

$\mu \neq 98.6$ becomes the alternative hypothesis (because it does not contain equality), while the null hypothesis becomes $\mu = 98.6$. Further suppose that the sample evidence causes us to reject the null hypothesis of $\mu = 98.6$. State the conclusion in simple, nontechnical terms.

SOLUTION   Refer to Figure 7-4. The original claim does not contain the condition of equality, and we do reject the null hypothesis. The wording of the final conclusion should therefore be as follows: "The sample data support the claim that the mean body temperature is different from 98.6° F."

Type I and Type II Errors   When testing a null hypothesis, we arrive at a conclusion of rejecting it or failing to reject it. Such conclusions are sometimes correct and sometimes wrong (even if we do everything correctly). Table 7-2 on the next page summarizes the two different types of errors that can be made, along with the two different types of correct decisions. We distinguish between the two types of errors by calling them *type I and type II errors*.

- **Type I error:** The mistake of rejecting the null hypothesis when it is actually true. The symbol α (alpha) is used to represent the probability of a type I error.

- **Type II error:** The mistake of failing to reject the null hypothesis when it is actually false. The symbol β (beta) is used to represent the probability of a type II error.

TABLE 7-2	Type I and Type II Errors		
		True State of Nature	
		The null hypothesis is true.	The null hypothesis is false.
Decision	We decide to reject the null hypothesis.	**Type I error** (rejecting a true null hypothesis) α	Correct decision
	We fail to reject the null hypothesis.	Correct decision	**Type II error** (failing to reject a false null hypothesis) β

Notation

α (alpha) = probability of a type I error (the probability of rejecting the null hypothesis when it is true)

β (beta) = probability of a type II error (failing to reject a false null hypothesis)

EXAMPLE   Identifying Type I and Type II Errors Assume that we are conducting a hypothesis test with the following null and alternative hypotheses:

$$H_0: \mu = 98.6$$
$$H_1: \mu \neq 98.6$$

Give statements identifying

a. a type I error.

b. a type II error.

SOLUTION

a. A type I error is the mistake of rejecting a true null hypothesis, so this is a type I error: Rejecting the claim that the mean is 98.6 when the mean really does equal 98.6.

b. A type II error is the mistake of failing to reject the null hypothesis when it is false, so this is a type II error: Failing to reject the claim that the mean equals 98.6 when the mean is really different from 98.6.

Controlling Type I and Type II Errors: One step in our standard procedure for testing hypotheses involves the selection of the significance level α, which is the probability of a type I error. However, we don't select β [P(type II error)]. It would be great if we could always have $\alpha = 0$ and $\beta = 0$, but in reality that is not possible, so we must attempt to manage the α and β error probabilities. Mathematically, it can be shown that α, β, and the sample size n are all related, so when you choose or determine any two of them, the third is automatically determined. The usual practice in research and industry is to select the values of α and n, so the value of β is determined. Depending on the seriousness of a type I error, try to use the largest α that you can tolerate. For type I errors with more serious consequences, select smaller values of α. Then choose a sample size n as large as is reasonable, based on considerations of time, cost, and other relevant factors. (Sample size determinations were discussed in Section 6-4.) The following practical considerations may be relevant:

1. For any fixed α, an increase in the sample size n will cause a decrease in β. That is, a larger sample will lessen the chance that you make the error of not rejecting the null hypothesis when it's actually false.

2. For any fixed sample size n, a decrease in α will cause an increase in β. Conversely, an increase in α will cause a decrease in β.

3. To decrease both α and β, increase the sample size.

To make sense of these abstract ideas, let's consider M&Ms (produced by Mars, Inc.) and Bufferin brand aspirin tablets (produced by Bristol-Myers Products).

- The mean weight of the M&M candies is supposed to be at least 0.9085 g (in order to conform to the weight printed on the package label).

- The Bufferin tablets are supposed to have a mean weight of 325 mg of aspirin.

Because M&Ms are candies used for enjoyment, whereas Bufferin tablets are drugs used for treatment of health problems, we are dealing with two very different levels of seriousness. If the M&Ms don't have a mean weight of 0.9085 g, the consequences are not very serious, but if the Bufferin tablets don't contain a mean of 325 mg of aspirin, the consequences could be very serious, possibly including consumer lawsuits and actions on the part of the Federal Drug Administration. Consequently, in testing the claim that $\mu = 0.9085$ g for M&Ms, we might choose $\alpha = 0.05$ and a sample size of $n = 100$; in testing the claim that $\mu = 325$ mg for Bufferin tablets, we might choose $\alpha = 0.01$ and a sample size of $n = 500$. The smaller significance level α and larger sample size n are chosen because of the more serious consequences associated with testing a commercial drug.

Power of a Test: We have presented notation indicating that β is the probability of failing to reject a false null hypothesis (type II error). It follows that $1 - \beta$ is the probability of rejecting a false null hypothesis. Statisticians refer to this probability as the *power* of a test, and they often use it to gauge the test's effectiveness in recognizing that a null hypothesis is false.

Large Sample Size Isn't Good Enough

Biased sample data should not be used for inferences, no matter how large the sample is. For example, in *Women and Love: A Cultural Revolution in Progress*, Shere Hite bases her conclusions on 4500 replies that she received after mailing 100,000 questionnaires to various women's groups. A *random* sample of 4500 subjects would usually provide good results, but Hite's sample is biased. It is criticized for overrepresenting women who join groups and women who feel strongly about the issues addressed. Because Hite's sample is biased, her inferences are not valid, even though the sample size of 4500 might seem to be sufficiently large.

Stating Conclusions.   *In Exercises 29–32, identify the null hypothesis, then state the final conclusion in simple, nontechnical terms. Be sure to address the original claim. (See Figure 7-4.)*

29. Original claim: Women have a mean height equal to 60.0 in.

Initial conclusion: Reject the null hypothesis.

30. Original claim: The mean IQ score of statistics professors is greater than 100.

Initial conclusion: Reject the null hypothesis.

31. Original claim: The mean amount of Coke in cans labeled 12 oz is actually less than 12 oz.

Initial conclusion: Fail to reject the null hypothesis.

32. Original claim: The mean body temperature of healthy adults is equal to 98.2° F.

Initial conclusion: Fail to reject the null hypothesis.

Identifying Type I and Type II Errors.   *In Exercises 33–36, identify the type I error and the type II error that correspond to the given hypothesis. (Hint: Begin by identifying the null hypothesis.)*

33. The mean amount of Coke in cans is equal to 12 oz.

34. The mean annual income of college graduates is greater than \$20,000.

35. The mean IQ of convicted felons is less than 100.

36. The mean weight loss of Slimnow diet pill users is at least 5 lb.

7-2 Beyond the Basics

37. Why Must Equality Be Assumed? When testing a claim that a mean is at least 100, the null hypothesis becomes $\mu \geq 100$. When actually conducting the hypothesis test, why is it necessary to assume that $\mu = 100$ instead of $\mu \geq 100$?

38. Proving Claims Under standard conditions, the Ford Taurus travels 21 mi while consuming exactly 1 gal of gas. You have developed a modification to the engine that you believe increases that distance. If you want to justify your claim of increased mileage by using a hypothesis test, is it possible to state your claim so that it becomes the null hypothesis? Why or why not?

39. Using $\alpha = 0$ Someone suggests that in testing hypotheses, you can eliminate a type I error by making $\alpha = 0$. In a two-tailed test, what critical values correspond to $\alpha = 0$? If $\alpha = 0$, will the null hypothesis ever be rejected?

40. Power of a Test Assume that you are using a significance level of $\alpha = 0.05$ to test the claim that $\mu < 2$ and that your sample is a simple random sample of 50 values.
   a. Find β, the probability of making a type II error, given that the population actually has a normal distribution with $\mu = 1.5$ and $\sigma = 1$. (*Hint:* With H_0: $\mu \geq 2$, begin by finding the values of the sample means that do not lead to rejection of H_0, then find the probability of getting a sample mean with one of those values.)

b. Find $1 - \beta$, which is the *power* of the test. If β is the probability of *failing* to reject a false null hypothesis, describe the probability of $1 - \beta$.

7-3 Testing a Claim about a Mean: Large Samples

In this section we combine the components introduced in Section 7-2 into a unified and systematic procedure for testing claims made about population means. We will present three methods of testing hypotheses that might appear to be different but are equivalent in the sense that they always lead to similar conclusions. The first procedure is the traditional method; the second procedure is the *P*-value method; and the third procedure is the confidence interval method. Because confidence intervals were thoroughly discussed in Chapter 6, our emphasis now will be placed on the traditional and *P*-value methods.

We begin by identifying the three assumptions that apply to the methods of this section.

Assumptions

1. The sample is a simple random sample. (Remember this very important point made in Chapter 1: *Data carelessly collected may be so completely useless that no amount of statistical torturing can salvage them.*)

2. The sample is large ($n > 30$). (This allows us to apply the central limit theorem so that we can use the normal distribution. *There is no requirement that the population data must have a normal distribution.*)

3. If the value of the population standard deviation σ is unknown, we can use the sample standard deviation s as an estimate of σ. (We can do this only if the sample size is large, with more than 30 sample values.)

Our first step is to *explore* the data set. Using the methods introduced in Chapter 2, we investigate center, variation, and distribution by drawing a graph; finding the mean, standard deviation, and 5-number summary; and identifying any outliers. In this section we will use the 106 body temperatures given in the Chapter Problem. Almost everyone assumes that the mean body temperature is 98.6° F, but we will test that belief. For now, we might note that the 106 body temperatures in Table 7-1 have a mean of 98.20° F. Those sample values also have the following 5-number summary: minimum = 96.5, Q_1 = 97.8, median = 98.4, Q_3 = 98.6, and maximum = 99.6. It's worth noting that 98.6 is the third quartile instead of being at the center. There do not appear to be any outliers that are far away from all of the other sample values. (Outliers can dramatically affect some results, such as those discussed in this chapter.) The accompanying Excel display of a histogram shows that the 106 body temperatures have a distribution that is roughly bell-shaped, and the histogram verifies that there are no outliers very far away from the other sample values. The histogram also shows that 98.6 is not a perfect balance point for the histogram; if a fulcrum is placed at 98.6, the histogram would tilt to the left, indicating that the mean appears to be lower than 98.6. (Remember that the original Excel histogram must be modified to show the corrected class boundaries. The graph shown here is a corrected version.)

The Traditional Method of Testing Hypotheses

Figure 7-5 summarizes the steps used in the **traditional (or classical) method** of testing hypotheses. This procedure uses the components described in Section 7-2 as part of a system to identify a sample result that is *significantly* different from the claimed value. The relevant sample statistic is converted to a test statistic, which we compare to a critical value. The following expression can be used for the examples and exercises of this section.

Test Statistic for Claims about μ When $n > 30$

$$z = \frac{\overline{x} - \mu_{\overline{x}}}{\dfrac{\sigma}{\sqrt{n}}}$$

Note that Step 7 in Figure 7-5 uses this decision criterion:

Reject the null hypothesis if the test statistic is in the critical region.

Fail to reject the null hypothesis if the test statistic is not in the critical region.

Section 7-2 presented an informal example of a hypothesis test; the following example formalizes that test.

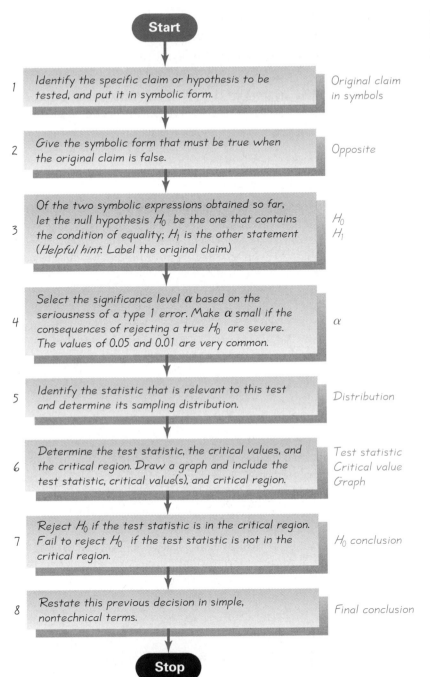

FIGURE 7-5
Traditional Method of
Hypothesis Testing

EXAMPLE   Body Temperatures Using the sample data given in the Chapter Problem ($n = 106$, $\bar{x} = 98.20$, $s = 0.62$) and a 0.05 significance level, test the claim that the mean body temperature of healthy adults is equal to $98.6°$ F. Use the traditional method by following the procedure outlined in Figure 7-5.

SOLUTION   Refer to Figure 7-5 and follow these steps.

Step 1: The claim that the mean is equal to 98.6 is expressed in symbolic form as $\mu = 98.6$.

Step 2: The alternative (in symbolic form) to the original claim is $\mu \neq 98.6$.

Step 3: The statement $\mu = 98.6$ contains the condition of equality, and so it becomes the null hypothesis. We have

$$H_0: \mu = 98.6 \text{ (original claim)} \qquad H_1: \mu \neq 98.6$$

Step 4: As specified in the statement of the problem, the significance level is $\alpha = 0.05$.

Step 5: Because the claim is made about the population mean μ, the sample statistic most relevant to this test is $\bar{x} = 98.20$. Because $n > 30$, the central limit theorem indicates that the distribution of sample means can be approximated by a *normal* distribution.

Step 6: In calculating the test statistic, we can use $s = 0.62$ as a reasonable estimate of σ (because $n > 30$), so the test statistic of $z = -6.64$ is found by converting the sample mean of $\bar{x} = 98.20$ to $z = -6.64$ through the following computation.

$$z = \frac{\bar{x} - \mu_{\bar{x}}}{\frac{\sigma}{\sqrt{n}}} = \frac{98.20 - 98.6}{\frac{0.62}{\sqrt{106}}} = -6.64$$

The critical z values are found by first noting that the test is two-tailed because a sample mean significantly less than or greater than 98.6 is strong evidence against the null hypothesis that $\mu = 98.6$. Because the test is two-tailed, we divide $\alpha = 0.05$ equally between the two tails to get 0.025 in each tail. See Figure 7-6. The critical values of $z = -1.96$ and $z = 1.96$ can be found by using Table A-2 or Excel. The test statistic, critical region, and critical values are shown in Figure 7-6.

FIGURE 7-6
Distribution of Means of Body Temperatures Assuming $\mu = 98.6$

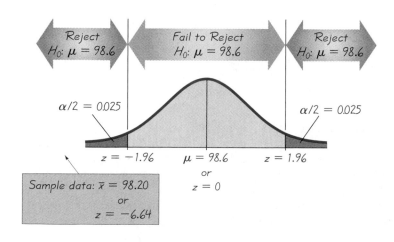

Step 7: Because the test statistic of $z = -6.64$ falls within the critical region, we reject the null hypothesis.

INTERPRETATION To restate the Step 7 conclusion in nontechnical terms, we might refer to Figure 7-4 in the preceding section. We are rejecting the null hypothesis, which is the original claim. We conclude that there is sufficient evidence to warrant rejection of the claim that the mean body temperature of healthy adults is 98.6° F. ◀

In the preceding example, the sample mean of 98.20 was converted to a test statistic of $z = -6.64$. That is, the sample mean is 6.64 standard deviations away from the claimed mean. In Chapter 2 we saw that a z score of 6.64 is so large that it represents an unusual result (if, in fact, $\mu = 98.6$). This evidence is strong enough to make us believe that the mean is actually different from 98.6. The critical values and critical region clearly identify the range of unusual values that cause us to reject the claimed value of μ.

The preceding example illustrated a two-tailed hypothesis test; the following example illustrates a right-tailed test. The following example also illustrates the difference between *statistical significance* and *practical significance.*

▷ **EXAMPLE** **Coke Volumes** Data Set 1 in Appendix B includes the volumes (in ounces) of the Coke in a sample of 36 different cans that are all labeled 12 oz. Here are the sample statistics: $n = 36$, $\bar{x} = 12.19$ oz, $s = 0.11$ oz. Upon seeing these statistics, a line manager claims that the mean amount of Coke is greater than 12 oz, causing lower company profits. Using a 0.01 significance level, test the manager's claim that the mean is greater than 12 oz.

SOLUTION We know that measured amounts are subject to random chance variation, so we need to determine whether the sample mean of 12.19 oz is *significantly* greater than 12 oz. Refer to Figure 7-5 and follow these steps.

Step 1: Express the manager's claim as $\mu > 12$.

Step 2: The opposite (in symbolic form) of the original claim is $\mu \leq 12$.

Step 3: The statement $\mu \leq 12$ contains the condition of equality, so it becomes the null hypothesis and we have

$$H_0: \mu \leq 12 \qquad H_1: \mu > 12 \text{ (original claim)}$$

Step 4: With a 0.01 significance level, we have $\alpha = 0.01$.

Step 5: We will use a *normal* distribution because the claim is directed at the population mean, so we are working with the distribution of sample means. Because $n > 30$, the central limit theorem indicates that the distribution of sample means can be approximated by a normal distribution.

FIGURE 7-7
Distribution of Means of
Amounts of Coke Assuming
$\mu = 12$ oz

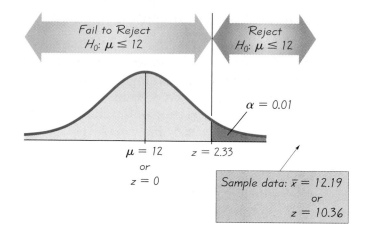

Step 6:   In calculating the test statistic, we can use $s = 0.11$ as a reasonable estimate of σ (because $n > 30$), so the test statistic of $z = 10.36$ is computed as follows:

$$.5 - .01 = .49$$
$$z = 2.33 \uparrow$$

$$z = \frac{\bar{x} - \mu_{\bar{x}}}{\frac{\sigma}{\sqrt{n}}} = \frac{12.19 - 12}{\frac{0.11}{\sqrt{36}}} = 10.36$$

The critical value of $z = 2.33$ is found in Table A-2 as the z score corresponding to an area of 0.49. (The test is right-tailed because the null hypothesis of $\mu \leq 12$ is rejected if the sample mean $\bar{x}$ is *greater than* 12 by a significant amount.) The test statistic, critical region, and critical value are shown in Figure 7-7.

Step 7:   From Figure 7-7 we see that the sample mean of 12.19 oz does fall within the critical region, so we reject the null hypothesis H_0.

INTERPRETATION   Our final conclusion is that the sample data support the claim that the mean amount of Coke is greater than 12 oz. (Refer to Figure 7-4 for help with wording this final conclusion.) The Coca Cola Company could increase profits by filling cans with slightly less Coke. As consumers, we know that we are not being cheated by the Coca Cola Company, but the extra Coke we are given is so small that there would not be a perceptible difference. From the perspective of consumers, even though the result is *statistically significant*, it does not appear to have *practical significance* because the mean appears to exceed 12 oz by a very small amount.

In presenting the results of a hypothesis test, it is not always necessary to show all of the steps included in Figure 7-5. However, the results should include the null hypothesis, the alternative hypothesis, the calculation of the test statistic, a graph such as Figure 7-7, the initial conclusion (reject H_0 or fail to reject H_0), and the final conclusion stated in nontechnical terms. The graph should show the test statistic, critical value(s), critical region, and significance level.

The *P*-Value Method of Testing Hypotheses

The *P*-value method of testing hypotheses is very similar to the traditional method, but the key difference is the way in which we decide whether to reject the null hypothesis. The approach used by the *P*-value method is to find the *probability* (*P*-value) of getting a result such as the one obtained, and reject the null hypothesis if that probability is very low. Let's start with a formal definition of the *P*-value.

> ### DEFINITION
>
> A ***P*-value** (or **probability value**) is the probability of getting a value of the sample test statistic that is *at least as extreme* as the one found from the sample data, assuming that the null hypothesis is true.

It is important to understand why the preceding definition uses the probability of a value *at least as extreme* as the one found. Why not simply find the probability of the particular result that occurred? Suppose you toss a coin 1000 times to determine whether it favors heads, and suppose you get 501 heads. Intuition should suggest that this is not evidence that the coin favors heads, because it is quite easy to get 501 heads in 1000 tosses. Yet, the probability of getting exactly 501 heads in 1000 tosses is actually quite small: 0.0252. With 1000 tosses, any specific number of heads will have a very low probability. However, the result of 501 heads among 1000 tosses is not *unusual* because the probability of getting *at least* 501 heads is high: 0.487. The point to keep in mind is that when you are deciding whether a sample result is unusual, you need to find the probability of getting a result *at least as extreme* as the result obtained, not the probability of getting the particular result by itself. This is the concept underlying *P*-values.

The *P*-value is the likelihood that a sample, such as the one obtained, will occur when the null hypothesis is actually true. A very small *P*-value (such as 0.05 or lower) suggests that the sample results are very unlikely under the assumption of the null hypothesis, so that such a small *P*-value is evidence against the null hypothesis. It would be helpful to remember these associations:

P-Value	Interpretation
Small *P*-values (such as 0.05 or lower)	*Unusual* sample results. *Significant* difference from the null hypothesis
Large *P*-values (such as above 0.05)	Sample results are *not unusual*. *Not a significant* difference from the null hypothesis

AS

If you are a medical researcher and you believe that you have developed a new drug to cure baldness, you want your hypothesis test to result in a *small P*-value, indicating that your results are highly significant.

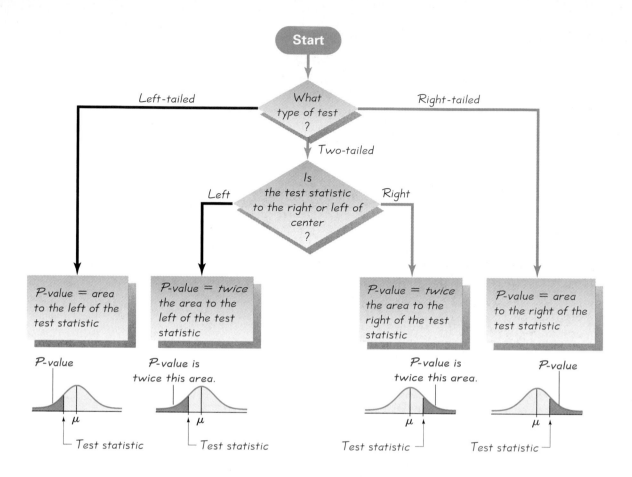

FIGURE 7-8
Finding *P*-Values

The procedure for finding *P*-values is summarized in Figure 7-8. The figure shows the following:

Right-tailed test: The *P*-value is the area to the right of the test statistic.

Left-tailed test: The *P*-value is the area to the left of the test statistic.

Two-tailed test: The *P*-value is *twice* the area of the extreme region bounded by the test statistic.

Figure 7-9 summarizes the *P*-value method for testing hypotheses. A comparison of the traditional method (summarized in Figure 7-5) and the *P*-value method (Figure 7-9) shows that they are essentially the same, but they differ in the decision criterion. The traditional method compares the test statistic to the critical values, whereas the *P*-value method compares the *P*-value to the significance level. However, the traditional and *P*-value methods are equivalent in the sense that they will always result in the same conclusion. They are simply different paths leading to the same final destination.

The *P*-value approach uses most of the same basic procedures as the traditional approach, but Steps 6 and 7 are different.

Step 6: Find the *P*-value (as shown in Figure 7-8).

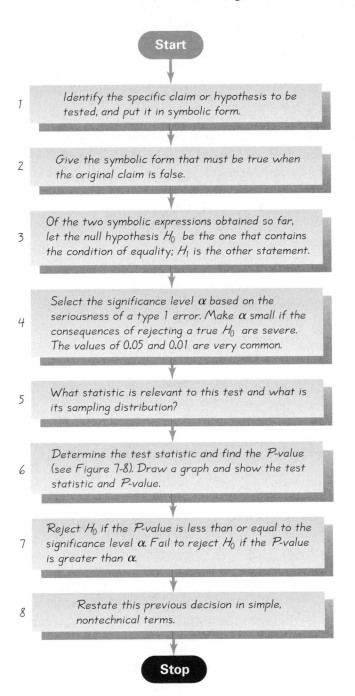

FIGURE 7-9
P-Value Method of Testing Hypotheses

Step 7:   Report the *P*-value. Some statisticians prefer to simply report the *P*-value and leave the conclusion to the reader. Others prefer to use the following decision criterion:

- *Reject the null hypothesis* if the *P*-value is less than or equal to the significance level α.

- *Fail to reject the null hypothesis* if the *P*-value is greater than the significance level α.

Many statisticians consider it good practice to always select a significance level *before* doing a hypothesis test. This is a particularly good procedure when using *P*-values because we may be tempted to adjust the significance level based on the results. For example, with a 0.05 significance level and a *P*-value of 0.06, we should fail to reject the null hypothesis, but it is sometimes tempting to say that a probability of 0.06 is small enough to warrant rejection of the null hypothesis. Other statisticians argue that prior selection of a significance level reduces the usefulness of *P*-values. They contend that no significance level should be specified and that the conclusion should be left to the reader. We will use the decision criterion that involves a comparison of a significance level and the *P*-value.

Ethics in Reporting

The American Association for Public Opinion Research developed a voluntary code of ethics to be used in news reports of survey results. This code requires that the following be included: (1) identification of the survey sponsor, (2) date the survey was conducted, (3) size of the sample, (4) nature of the population sampled, (5) type of survey used, (6) exact wording of survey questions. Surveys funded by the U.S. government are subject to a pre-screening that assesses the risk to those surveyed, the scientific merit of the survey, and the guarantee of the subject's consent to participate.

EXAMPLE   Body Temperatures Use the *P*-value method to test the claim that the mean body temperature of healthy adults is equal to $98.6°$ F. As before, use a 0.05 significance level and the sample data summarized in the Chapter Problem ($n = 106$, $\bar{x} = 98.20$, $s = 0.62$, a bell-shaped distribution).

SOLUTION Except for Steps 6 and 7, this solution is the same as the one developed earlier in this section using the traditional method. Steps 1, 2, and 3 led to the following hypotheses:

$$H_0: \mu = 98.6 \text{ (original claim)} \qquad H_1: \mu \neq 98.6$$

In Steps 4 and 5 we noted that the significance level is $\alpha = 0.05$ and that the central limit theorem indicates use of the normal distribution. We now proceed to Steps 6 and 7.

Step 6:   The test statistic of $z = -6.64$ was already found in a preceding example. We can now find the *P*-value by referring to Figure 7-8. Because the test is two-tailed, the *P*-value is twice the area to the left of the test statistic $z = -6.64$. Using Table A-2, we find that the area to the left of $z = -6.64$ is 0.0001, so the *P*-value is $2 \times 0.0001 = 0.0002$ (see Figure 7-10). (More precise results show that the area to the left of $z = -6.64$ is actually much less than 0.0001.)

Step 7:   Because the *P*-value of 0.0002 is less than the significance level of $\alpha = 0.05$, we reject the null hypothesis.

INTERPRETATION As with the traditional method, we conclude in Step 8 that there is sufficient evidence to warrant rejection of the claim that the mean body temperature is $98.6°$ F.

The following example applies the *P*-value approach to the Coke data discussed in an earlier example in this section.

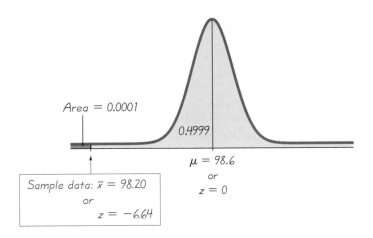

FIGURE 7-10
P-Value Method of Testing
H_0: $\mu = 98.6$

Because the test is two-tailed, the
P-value is *twice* the red-shaded
area.

▷ **EXAMPLE** **Coke Volumes** Use the *P*-value approach and a 0.01 significance level to test the claim that the amounts of Coke in cans have a
mean that is greater than 12 oz. Recall that the sample data consist of
36 cans with a mean of 12.19 oz and a standard deviation of 0.11 oz.

SOLUTION Again, refer to the steps used earlier to solve this problem
by the traditional method. Steps 1, 2, and 3 resulted in the following
hypotheses:

$$H_0: \mu \leq 12 \text{ oz} \qquad H_1: \mu > 12 \text{ oz} \text{ (original claim)}$$

Steps 4 and 5 led to a significance level of $\alpha = 0.01$ and the decision
that the normal distribution is relevant to this test of a claim about a
sample mean. We now proceed to Steps 6 and 7.

Step 6: The test statistic of $z = 10.36$ was computed in the earlier
solution to this same problem. To find the *P*-value, we refer to
Figure 7-8, which indicates that for this right-tailed test, the
P-value is the area to the right of the test statistic. Referring to
Table A-2, we find that the area to the right of $z = 10.36$ is
$0.5 - 0.4999$, or 0.0001. The *P*-value is therefore 0.0001. (Using
a reference more precise than Table A-2 shows that the *P*-value
is actually much smaller than 0.0001.)

Step 7: Because the *P*-value of 0.0001 is less than the significance level
of $\alpha = 0.01$, we reject the null hypothesis that $\mu \leq 12$.

INTERPRETATION As in the previous solution of this same problem, we
conclude that there is sufficient evidence to support the claim that the
mean is greater than 12 oz. The *P*-value of 0.0001 shows that it is very
unlikely that we would get a sample mean of 12.19 oz if the population
mean were really 12 oz. ◁

The next procedure for testing hypotheses is based on confidence intervals and therefore requires the concepts discussed in Section 6-2.

Testing Claims with Confidence Intervals

Confidence intervals can be used to identify results that are highly unlikely, so we can determine whether there is a *significant* difference between sample results and a claimed value of a parameter. For example, let's again consider the hypothesis-testing problem described at the beginning of the chapter. We want to test the claim that the mean body temperature of healthy adults is equal to 98.6° F. Sample data consist of $n = 106$ temperatures with mean $\overline{x} = 98.20$ and standard deviation $s = 0.62$. Chapter 6 described methods of constructing confidence intervals. In particular, we used the body temperature sample data to construct the following 95% confidence interval:

$$98.08 < \mu < 98.32$$

We are 95% confident that the limits of 98.08 and 98.32 contain the population mean μ. (This means that if we were to repeat the experiment of collecting a sample of 106 body temperatures, 95% of the samples would result in confidence interval limits that actually do contain the value of the population mean μ.) This confidence interval suggests that it is very unlikely that the population mean is equal to 98.6. That is, on the basis of the confidence interval given here, we reject the common belief that the mean body temperature of healthy adults is 98.6° F. We can generalize this procedure as follows: First use the sample data to construct a confidence interval, and then apply the following decision criterion.

> **A confidence interval estimate of a population parameter contains the likely values of that parameter. We should therefore reject a claim that the population parameter has a value that is not included in the confidence interval.**

Using this criterion, we see that the confidence interval of $98.08 < \mu < 98.32$ does not contain the claimed value of 98.6, and we therefore reject the claim that the population mean equals 98.6.

We can make a direct correspondence between a confidence interval and a hypothesis test only when the test is two-tailed. A one-tailed hypothesis test with significance level α corresponds to a confidence interval with degree of confidence $1 - 2\alpha$.

Hypothesis Test	**Confidence Interval**
Two-tailed; significance level α	Use degree of confidence $1 - \alpha$.
One-tailed; significance level α	Use degree of confidence $1 - 2\alpha$.

For example, if we want to conduct a two-tailed hypothesis test with a 0.05 significance level, we can use a 95% confidence interval. But if we want to conduct a right-tailed hypothesis test with a 0.05 significance level, we should use a 90% confidence interval.

In the remainder of the text, we will apply methods of hypothesis testing to other circumstances, such as those involving claims about proportions or standard deviations or those involving more than one population. It is easy to become entangled in a complex web of steps without ever understanding the underlying rationale of hypothesis testing. The key to that understanding lies in the rare event rule for inferential statistics: **If, under a given observed assumption, the probability of getting the observed sample is exceptionally small, we conclude that the assumption is probably not correct.** When testing a claim, we make an assumption (null hypothesis) that contains equality. We then compare the assumption and the sample results to form one of the following conclusions.

- If the sample results can easily occur when the assumption (null hypothesis) is true, we attribute the relatively small discrepancy between the assumption and the sample results to chance.

- If the sample results cannot easily occur when the assumption (null hypothesis) is true, we explain the relatively large discrepancy between the assumption and the sample results by concluding that the assumption is not true, so we reject the assumption.

Using Excel

Caution: Excel's built-in ZTEST function is extremely tricky to use, because the generated *P*-value is not always the same as the same standard *P*-value used in this book and in the statistical community at large. We strongly recommend against using the ZTEST function. Instead, use DDXL or use the command **NORMDIST** as described below.

Using DDXL: For the methods of this section, one approach is to use the Data Desk/XL add-in that is a supplement to this book. The procedure for using DDXL requires the original list of sample values. (If only the summary statistics of n, $\overline{x}$, and s are known, use the command **NORMDIST** as described below, or see the Excel Project at the end of Chapter 5 for a way to generate a list of values with the same sample statistics.) Here is the procedure for using DDXL.

1. Enter the sample data in column A.
2. For the large sample cases discussed in this section, first find the sample standard deviation using the command **STDEV.** For example, if you have 106 sample values listed in rows 1 through 106 of column A, enter the command **=STDEV(A1:A106)** to find the sample standard deviation s.
3. Click on **DDXL,** then select the option of **1 Var z Test.**
4. Click on the pencil icon and enter the range of cells containing the sample data. For example, if you have 106 values in the first 106 rows of column A, enter A1:A106. Click **OK.**
5. You will see a four-step dialog box. (See the display shown on the next page.) Start with Step 1 by clicking on the bar labeled **Set $\mu 0$ and sd.** You will get a smaller dialog box with a position for entering the value of the claimed population mean and the value of the population standard deviation.
   - Enter the claimed value of the population mean.

- Enter the value of the population standard deviation; if the population standard deviation is not known, enter the value of the sample standard deviation *s* (provided that the sample is large).
- Click **OK**.

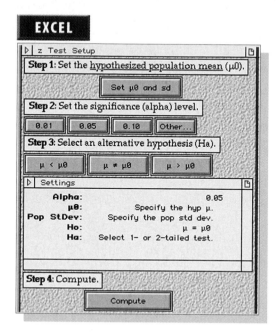

6. Proceed to Step 2 where you select the significance level.
7. In Step 3, select the form of the alternative hypothesis that is appropriate for the test.
8. In Step 4, click on the **Compute** bar. Shown below is the Excel display for the first example of this section. (The discrepancy between the test statistic value of -6.611 and the value of -6.64 in the example is due to a rounding error in the example. If the example had used a more precise standard deviation of 0.6229, the test statistic would become -6.61 and would agree with the test statistic in the Excel display.) Note that the display includes the *P*-value, but not the critical values.

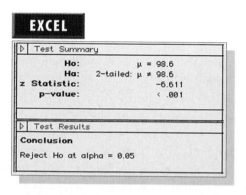

Using NORMDIST: For the methods of this section, we can use the *P*-value approach to hypothesis testing by using the Excel command **NORMDIST** that was described in Chapter 5. This approach can be used with sample data described with summary statistics of n, $\bar{x}$, and s. For purposes of typical hypothesis tests found in this section, the command **=NORMDIST($\bar{x}$, μ, s/sqrt(n), 1)** will yield the area to the *left* of the test statistic, so use the following for determining *P*-values:

Left-tailed test: *P*-value is the result of
    =NORMDIST($\bar{x}$, μ, s/sqrt(n), 1).

Right-tailed test: *P*-value is 1 minus the result of
    =NORMDIST($\bar{x}$, μ, s/sqrt(n), 1).

Two-tailed test: *P*-value is the smaller of these two values:

$$2 \times \textbf{NORMDIST}(\bar{x}, \mu, s/\textbf{sqrt}(n), \textbf{1})$$

$$2 \times [\textbf{1} - \textbf{NORMDIST}(\bar{x}, \mu, s/\textbf{sqrt}(n), \textbf{1})]$$

For example, given a sample of 106 body temperatures with a mean of 98.20 and a standard deviation of 0.62, we can test the claim that $\mu = 98.6$ by entering the command **=NORMDIST(98.20, 98.6, 0.62/sqrt(106), 1)**. The result of that command is 1.55195E-11, which is a tiny number. Because the test is two-tailed, the *P*-value is the smaller of 2(1.55195E-11) and 2(1 − 1.55195E-11), so the *P*-value is 2(1.55195E-11), or 0.0000000000310. Because the *P*-value is so small, we reject the null hypothesis of $\mu = 98.6$.

7-3 Basic Skills and Concepts

Finding P-Values.   *In Exercises 1–4, use the given claim and test statistic to find the P-value.*

1. Claim: The mean IQ score of convicted felons is equal to 100.

   Test statistic: $z = -1.03$

2. Claim: The mean height of men is equal to 70 in.

   Test statistic: $z = -2.00$

3. Claim: The mean score on the verbal portion of the Graduate Record Exam is at least 500.

   Test statistic: $z = -12.67$

4. Claim: The mean price of college textbooks is greater than $60.

   Test statistic: $z = 2.34$

In Exercises 5–8, identify the following components of the hypothesis test: (a)test statistic, (b)critical value(s), (c)P-Value, (d)final conclusion.

5. Exam Scores Test the claim that for the population of statistics final exams, the mean score is given by $\mu = 75$. Sample data are summarized as $n = 500$, $\bar{x} = 77$, and $s = 15$. Test at the $\alpha = 0.05$ significance level.

6. Heights of Males Test the claim that for the population of male college students, the mean height is given by $\mu = 69.0$ in. Sample statistics include $n = 50$, $\bar{x} = 69.6$ in., and $s = 3.0$ in. Use a significance level of $\alpha = 0.01$.

7. Grade-Point Average Test the claim that the population of sophomore college students has a mean grade point average greater than 2.25. Sample statistics include $n = 150$, $\bar{x} = 2.35$, and $s = 0.80$. Use a significance level of $\alpha = 0.02$.

8. Ages of Jets Test the claim that for the population of U.S. commercial jets, the mean age is less than 18 years. Sample statistics include $n = 85$, $\bar{x} = 15.8$ years, and $s = 8.3$ years. Use a significance level of $\alpha = 0.04$.

Traditional Method. *In Exercises 9–14, test the given claim by using the traditional method of testing hypotheses.*

 9. Body Temperatures In this section we presented an example about testing the claim that the mean body temperature equals 98.6° F. Use the same sample data ($n = 106$, $\bar{x} = 98.20$, $s = 0.62$ oz) to test the claim that the mean body temperature is *less than* 98.6° F. Use a significance level of $\alpha = 0.01$.

 10. Coke Volumes In this section we presented an example about testing the claim that the mean volume of Coke in cans is greater than 12 oz. Use the same sample data ($n = 36$, $\bar{x} = 12.19$ oz, $s = 0.11$ oz) to test the claim that the mean *equals* 12 oz. Use a significance level of $\alpha = 0.02$.

 11. Baseballs In previous tests, baseballs were dropped 24 ft onto a concrete surface, and they bounced an average of 92.84 in. In a test of a sample of 40 new balls, the bounce heights had a mean of 92.67 in. and a standard deviation of 1.79 in. (based on data from Brookhaven National Laboratory and *USA Today*). Use a 0.05 significance level to determine whether there is sufficient evidence to support the claim that the new balls have bounce heights with a mean different from 92.84 in. Does it appear that the new baseballs are different?

 12. Compulsive Buyers Researchers designed a questionnaire to identify compulsive buyers. For a sample of consumers who identified themselves as compulsive buyers, questionnaire scores have a mean of 0.83 and a standard deviation of 0.24 (based on data from "A Clinical Screener for Compulsive Buying," by Faber and Guinn, *Journal of Consumer Research,* Vol. 19). Assume that the subjects were randomly selected and that the sample size was 32. Using a 0.01 significance level, test the claim that the self-identified compulsive-buyer population has a mean greater than 0.21, the mean for the general population. Does the questionnaire seem to be effective in identifying compulsive buyers?

 13. Drug Production The nighttime cold medicine Dozenol bears a label indicating the presence of 600 mg of acetaminophen in each fluid ounce of the drug. The Food and Drug Administration randomly selected 40 one-ounce samples and found that the mean acetaminophen content is 593 mg, whereas the standard deviation is 21 mg. Use $\alpha = 0.01$ and test the claim of the Medassist Pharmaceutical Company that the population mean is equal to 600 mg. Would you buy this cold medicine?

14. Effectiveness of Seat Belts A study included 123 children who were wearing seat belts when injured in motor vehicle crashes. The amounts of time spent in an intensive care unit have a mean of 0.83 day and a standard deviation of 0.16 day (based on data from "Morbidity Among Pediatric Motor Vehicle Crash Victims: The Effectiveness of Seat Belts," by Osberg and Di Scala, *American Journal of Public Health,* Vol. 82, No. 3). Using a 0.01 significance level, test the claim that the seat belt sample comes from a population with a mean of less than 1.39 days, which is the mean for the population who were not wearing seat belts when injured in motor vehicle crashes. Do the seat belts seem to help?

P-Value Method.   *In Exercises 15–20, test the given claim by using the P-value method of testing hypotheses.*

EXCEL **15.** Testing Effectiveness of SAT Course The effectiveness of a test preparation course was studied for a simple random sample of 75 subjects who took the SAT before and after coaching. The differences between the scores resulted in a mean increase of 0.6 and a standard deviation of 3.8. (See "An Analysis of the Impact of Commercial Test Preparation Courses on SAT Scores," by Sesnowitz, Bernhardt, and Kwain, *American Education Research Journal,* Vol. 19, No. 3.) Using a 0.05 significance level, test the claim that the population mean increase is greater than 0, indicating that the course is effective in raising scores. Should people take this course?

EXCEL **16.** Prison Time for Convicted Embezzlers When 70 convicted embezzlers were randomly selected, the mean length of prison sentence was found to be 22.1 months and the standard deviation was found to be 8.6 months (based on data from the U.S. Department of Justice). Kim Patterson is running for political office on a platform of tougher treatment of convicted criminals. Test her claim that prison terms for convicted embezzlers have a mean of less than 2 years. Use a 0.01 significance level.

EXCEL **17.** Conductor Life Span A *New York Times* article noted that the mean life span for 35 male symphony conductors was 73.4 years, in contrast to the mean of 69.5 years for males in the general population. Assuming that the 35 males have life spans with a standard deviation of 8.7 years, use a 0.05 significance level to test the claim that male symphony conductors have a mean life span that is different from 69.5 years.

EXCEL **18.** Weights of Garbage Data Set 5 in Appendix B includes the *total* weights of garbage discarded by households in one week (based on data collected as part of the Garbage Project at the University of Arizona). The Excel workbook is GARBAGE.xls. For that data set, the mean is 27.44 lb and the standard deviation is 12.46 lb. Using a 0.01 significance level, test the claim of the city of Providence supervisor that the mean weight of all garbage discarded by households each week is less than 29 lb, the amount that can be handled by the town. Based on the result, is there any cause for concern that there might be too much garbage to handle?

EXCEL **19.** Weights of Quarters If we refer to the weights (in grams) of quarters listed in Data Set 13 in Appendix B, we find 50 weights with a mean of 5.622 g and a standard deviation of 0.068 g. The Excel workbook is QUARTERS.xls. The U.S. Department of the Treasury claims that the procedure it uses to mint quarters yields a mean weight of 5.670 g. Use a 0.01 significance level to test the claim that the mean weight of quarters in circulation is 5.670 g. If the claim is rejected, what is a possible explanation for the discrepancy?

EXCEL **20.** Testing Wristwatch Accuracy Forty people are selected and the accuracy of their wristwatches is checked, with positive errors representing watches that are ahead of the correct time and negative errors representing watches that are behind the correct time. The 40 values have a mean of 117.3 sec and a standard deviation of 185.0 sec. Use a 0.01 significance level to test the claim that the population of all watches has a mean equal to 0 sec. What do you conclude about the accuracy of people's wristwatches?

Testing Claims Using Sample Data.   *In Exercises 21–24, use the given data set to test the stated claim.*

EXCEL **21.** Weights of Sugar Packets Refer to Data Set 4 in Appendix B (Excel workbook SUGAR.xls) and test the claim that the weights of the sugar packets have a
(continued)

mean equal to 3.5 g, as indicated on the label. If the mean does not appear to equal 3.5 g, what do you conclude?

22. Second-Hand Smoke Refer to Data Set 9 in Appendix B (Excel workbook COTININE.xls) and use the measured levels of serum cotinine (in ng/ml) in people who are nonsmokers and have no exposure to environmental tobacco smoke at home or at work (NOETS). Test the claim of tobacco company spokesmen that these people have "positive" levels of cotinine, suggesting that this measure is not a good indicator of exposure to tobacco smoke. Use a significance level of 0.005.

23. Pepsi Volumes Refer to Data Set 1 in Appendix B (Excel workbook COLA.xls) and use the volumes (in ounces) of regular Pepsi. Test the claim that the mean is greater than 12 oz, as indicated on the can labels. Can Pepsi be supplied in a way that is more profitable to the company?

24. Textbook Prices Refer to Data Set 2 in Appendix B (Excel workbook TEXT-BOOK.xls) and use the prices of the 40 *new* textbooks at the University of Massachusetts. Test the claim that such textbooks have a mean price less than $70. Use a 0.05 significance level.

Interpreting Computer and Calculator Displays. In Exercises 25–28, use the computer or calculator display to form a conclusion.

25. Mean Weight of M&M Candies A package of M&M plain candies is labeled as containing 1361 g, and there are 1498 candies, so the mean weight of the individual candies should be 1361/1498, or 0.9085 g. In a test to determine whether consumers are being cheated, a sample of 33 brown M&Ms is randomly selected. (See Data Set 10 in Appendix B.) When the 33 weights are used with Excel, the display is as shown here. Interpret those results. Are consumers being cheated?

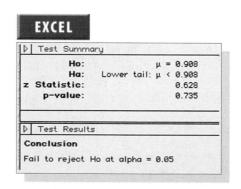

26. Analysis of Last Digits Analysis of the last digits of sample data values sometimes reveals whether the data have been accurately measured and reported. If the last digits are uniformly distributed from 0 through 9, their mean should be 4.5. Reported data (such as weights or heights) are often rounded so that the last digits include disproportionately more 0s and 5s. The last digits in the reported lengths (in feet) of the 70 home runs hit by Mark McGwire in 1998 are used to test the claim that they come from a population with a mean of 4.5 (based on data from *USA Today*). When Excel is used to test that claim, the display is as shown here. Using a 0.05 significance level, interpret the Excel results. Does it appear that the distances were accurately measured?

27. Testing Mean Body Temperature In an example in this section we tested the claim that the mean body temperature is equal to 98.6° F, but we used 106 readings taken at 12:00 A.M. on the second day of observations. If we use the 38 body temperatures taken at 8:00 A.M. on the first day, Excel will display the results as shown. Interpret those results. Are these results consistent with the results found in the example in this section? For these results, could there be a potential problem due to the large number of missing values?

28. Are Thinner Aluminum Cans Weaker? Data Set 12 in Appendix B includes the measured axial loads (in pounds) of 175 cola cans that use aluminum 0.0109 in. thick. Before obtaining these sample results, the standard cans had a thickness of 0.0111 in. and the mean axial load was 281.81 lb. When using the axial loads of the thinner cans in a test of the claim that the mean axial load is less than 281.81 lb, Excel provides the accompanying display. Assume that we are using a 0.01 significance level. Interpret the results. Do the thinner cans appear to have a mean axial load less than 281.81 lb?

EXCEL

Test Summary	
Ho:	$\mu = 281.81$
Ha:	Lower tail: $\mu < 281.81$
z Statistic:	-8.792
p-value:	$< .001$

Test Results
Conclusion
Reject Ho at alpha = 0.01

7-3 Beyond the Basics

29. Finding Standard Deviation A journal article reported that a null hypothesis of $\mu = 100$ was rejected because the P-value was less than 0.01. The sample size was given as 62, and the sample mean was given as 103.6. Find the largest possible standard deviation.

30. Axial Loads of Aluminum Cans In Exercise 28, find the *largest* sample mean below 281.81 lb that will support the claim that the mean axial load of the 0.0109-in. cans is less than 281.81 lb. (Use the same sample size and sample standard deviation.)

31. Finding Probability of Type II Error For a given hypothesis test, the probability α of a type I error is fixed, whereas the probability β of a type II error depends on the particular value of μ that is used as an alternative to the null hypothesis. For hypothesis tests of the type found in this section, we can find β as follows:

Step 1: Find the value(s) of $\overline{x}$ that correspond to the critical value(s). In

$$z = \frac{\overline{x} - \mu_{\overline{x}}}{\sigma_{\overline{x}}}$$

substitute the critical value(s) for z, enter the values for $\mu_{\overline{x}}$ and $\sigma_{\overline{x}}$, and solve for $\overline{x}$.

Step 2: Given a particular value of μ that is an alternative to the null hypothesis H_0, draw the normal curve with this new value of μ at the center. Also plot the value(s) of $\overline{x}$ found in Step 1.

Step 3: Refer to graph in Step 2, and find the area of the new critical region bounded by the value(s) of $\overline{x}$ found in Step 1. This is the probability of rejecting the null hypothesis, given that the new value of μ is correct.

Step 4: The value of β is 1 minus the area from Step 3. This is the probability of failing to reject the null hypothesis, given that the new value of μ is correct.

The preceding steps allow you to find the probability of failing to reject H_0 when it is false. You are determining the area under the curve that excludes the critical region in which you reject H_0; this area corresponds to a failure to reject a false H_0, because we use a particular value of μ that goes against H_0. Refer to the body-temperature example discussed in this section and find β (the probability of a type II error) corresponding to the following:

a. $\mu = 98.7$

b. $\mu = 98.4$

32. Power of Test The *power* of a test, expressed as $1 - \beta$, is the probability of rejecting a false null hypothesis. Refer to Exercise 28. If the test of that claim has a power of 0.2, find the mean μ that is being used as an alternative to H_0 (see Exercise 31).

 # 7-4 Testing a Claim about a Mean: Small Samples

One great advantage of learning the methods of hypothesis testing described in Sections 7-2 and 7-3 is that those same methods can be easily modified for use in many other circumstances, such as those discussed in this section. The examples and exercises in Sections 7-2 and 7-3 all use large samples in tests of claims about means, but in this section we will deal with small sample cases. The main objective of this section is to develop the ability to test claims made about population means when the number of sample values is 30 or fewer. The following assumptions apply to the methods described in this section.

Assumptions

1. The samples are simple random samples. (We stress again that if data are carelessly collected, they may be so completely useless that no amount of statistical torturing can salvage them.)
2. The sample is small ($n \leq 30$).
3. The value of the population standard deviation σ is unknown. (We seldom know the population standard deviation when we do not know the population mean.)
4. The sample values come from a population with a distribution that is approximately normal.

If we plan to test a hypothesis and the four conditions just listed are satisfied, we use the Student t (not the normal) distribution with the test statistic and critical values described as follows.

Better Results with Smaller Class Size

> **Test Statistic for Claims about μ When $n \leq 30$ and σ Is Unknown**
>
> If a population is essentially normal, then the distribution of
>
> $$t = \frac{\bar{x} - \mu_{\bar{x}}}{\dfrac{s}{\sqrt{n}}}$$
>
> is essentially a *Student t distribution* for all samples of size n. (The Student t distribution is often referred to as the t *distribution*.)

Critical Values

1. Critical values are found in Table A-3.
2. Degrees of freedom = $n - 1$.
3. After finding the number of degrees of freedom, refer to Table A-3 and locate that number in the column at the left. With a particular row of t values now identified, select the critical t value that corresponds to the appropriate column heading. If a critical t value is located at the left tail, be sure to make it negative.

We first used the Student t distribution in Section 6-3, where we developed confidence interval estimates of μ based on small samples. The following is a list of the important properties of the Student t distribution.

Important Properties of the Student t Distribution

1. The Student t distribution is different for different sample sizes (see Figure 6-5 in Section 6-3).
2. The Student t distribution has the same general bell shape as the standard normal distribution; its wider shape reflects the greater variability that is expected with small samples.

3. The Student t distribution has a mean of $t = 0$ (just as the standard normal distribution has a mean of $z = 0$).

4. The standard deviation of the Student t distribution varies with the sample size and is greater than 1 (unlike the standard normal distribution, which has $\sigma = 1$).

5. As the sample size n gets larger, the Student t distribution gets closer to the standard normal distribution. For values of $n > 30$, the differences are so small that we can use the critical z values instead of developing a much larger table of critical t values. (The values in the bottom row of Table A-3 are equal to the corresponding critical z values from the standard normal distribution.)

Choosing the Appropriate Distribution

When testing claims made about population means, sometimes the normal distribution applies, sometimes the Student t distribution applies, and sometimes neither applies, so we must use nonparametric methods. (Nonparametric methods, which do not require a particular distribution, are discussed in Chapter 13.)

Figure 7-11 summarizes the decisions to be made in choosing between the normal and Student t distrubutions. Studying the figure leads to the following observations.

FIGURE 7-11
Choosing Between the Normal and Student t Distributions When Testing a Claim about the Population Mean μ

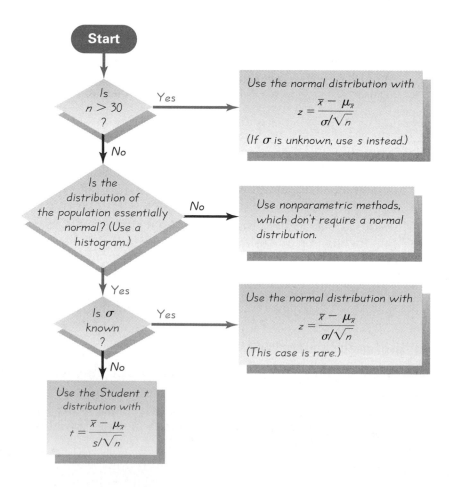

1. According to the central limit theorem, if we obtain a large ($n > 30$) sample (from any population with any distribution), the distribution of the sample means can be approximated by the normal distribution.

2. If we obtain a sample (of any size) from a population with a normal distribution, the distribution of the sample means will be approximately normal with mean μ and standard deviation $\sigma/\sqrt{n}$. In a hypothesis test, the value of μ corresponds to the null hypothesis, and the value of the population standard deviation σ must be known. If σ is unknown and the sample is large, we can use the sample standard deviation s as a substitute for σ because large simple random samples tend to be representative of the populations from which they come.

3. If the sample is small ($n \leq 30$), σ is unknown, and the parent population has a distribution that is essentially normal, we can use the Student t distribution.

4. If our simple random samples are small, σ is unknown, and the population distribution is grossly nonnormal, we cannot use the methods of this chapter. Instead, we can use nonparametric methods, some of which are discussed in Chapter 13.

The following example involves a small sample drawn from a normally distributed population for which the standard deviation σ is not known. These conditions require that we use the Student t distribution.

▷ **EXAMPLE** **Pulse Rates** The author, at the peak of an exercise program, claimed that his pulse rate was lower than the mean pulse rate of statistics students. The author's pulse rate was measured to be 60 beats per minute, and the 20 students in his class measured their pulse rates with the results listed below. Use a 0.01 significance level to test the author's claim that statistics students have a mean pulse rate greater than 60 beats per minute.

    61 74 72 77 79 90 69 79 90 72 68 64 83 82 83 62 88 74 55 65

SOLUTION   Before jumping to the hypothesis test, let's first explore the sample data. Using the methods of Chapter 2, we find these statistics: $n = 20$, $\bar{x} = 74.4$, $s = 10.0$. The 5-number summary is 55.0 (minimum), 66.5 (Q_1), 74.0 (median), 82.5 (Q_3), and 90.0 (maximum). There are no outliers very far away from all of the other sample values, and the histogram is roughly bell-shaped. It does appear that statistics students have a mean greater than 60, but we need to formally determine whether the sample evidence is sufficient to support that conclusion. Let's conduct a formal hypothesis test using the steps outlined in Figure 7-5.

As in the preceding section, Steps 1, 2, and 3 result in the following null and alternative hypotheses:

$$H_0: \mu \leq 60$$
$$H_1: \mu > 60 \quad \text{(original claim)}$$

Step 4:   The significance level is $\alpha = 0.01$.

Step 5: In this test of a claim about the population mean, the most relevant statistic is the sample mean. In selecting the correct distribution, we refer to Figure 7-11. The sample is small (because $n = 20$ and does not exceed 30), it's reasonable to conclude that the population distribution is normal (based on the shape of the histogram), and σ is unknown. Figure 7-11 shows that we should use the Student t distribution (not the normal distribution).

Step 6: The test statistic is

$$t = \frac{\bar{x} - \mu_{\bar{x}}}{\dfrac{s}{\sqrt{n}}} = \frac{74.4 - 60}{\dfrac{10.0}{\sqrt{20}}} = 6.440$$

The critical value of $t = 2.540$ is found by referring to Table A-3. First locate $n - 1 = 19$ degrees of freedom in the column at the left. Because this test is right-tailed with $\alpha = 0.01$, refer to the column with the heading of 0.01 (one tail). The critical value of $t = 2.540$ corresponds to 19 degrees of freedom and $\alpha = 0.01$ in one tail. In this right-tailed test, the test statistic of $t = 6.440$ exceeds the critical value of $t = 2.540$, so the test statistic falls in the critical region.

Step 7: Because the test statistic of $t = 6.440$ falls in the critical region, we reject H_0.

INTERPRETATION (Refer to Figure 7-4 for help in wording the final conclusion.) There is sufficient evidence to support the claim that the sample comes from a population with a mean pulse rate greater than 60 beats per minute.

The critical value in the preceding example was $t = 2.540$, but if the sample had been large ($n > 30$), the critical value would have been $z = 2.33$. The larger Student t critical value shows that with a small sample, the sample evidence must be *more extreme* before we consider the difference to be significant.

P-Values

The preceding example followed the traditional approach to hypothesis testing, but many statistics software packages and much of the literature will display P-values. Because the t distribution table (Table A-3) includes only selected values of the significance level α, we cannot usually find an exact P-value from Table A-3. Instead, we can use that table to identify limits that contain the P-value. In the last example we found the test statistic to be $t = 6.440$, and we know that the test is one-tailed with 19 degrees of freedom. By examining the row of Table A-3 corresponding to 19 degrees of freedom, we see that the test statistic of 6.440 exceeds the largest critical value in that row. Although we cannot pinpoint an exact *P*-value from Table A-3, we can conclude that it must be less than 0.005. That is, we conclude that *P*-value < 0.005. (Some calculators and computer programs allow us to find exact *P*-values. For the preceding example, using the DDXL add-in with Excel results in the statement that "*P*-value < .001"). With a significance level of 0.01 and a *P*-value less than 0.005, we reject the null hypothesis (because the *P*-value is less than the significance level), as we did using the traditional method in the preceding example.

> **EXAMPLE** **Finding a *P*-Value** Use Table A-3 to find the *P*-value corresponding to these results: The Student *t* distribution is used in a two-tailed test with a sample of *n* = 10 scores, and the test statistic is *t* = 2.567.

SOLUTION Refer to the row of Table A-3 with 9 degrees of freedom and note that the test statistic of 2.567 falls between the critical values of 2.821 and 2.262. Because the test is two-tailed, we consider the values of α at the top that are identified with "two tails." The critical values of 2.821 and 2.262 correspond to 0.02 (two tails) and 0.05 (two tails), so we express the *P*-value as follows:

$$0.02 < P\text{-value} < 0.05$$

With a true null hypothesis, the chance of getting a sample mean (from 10 sample values) that converts to a test statistic of *t* = 2.567 is somewhere between 0.02 and 0.05.

So far, we have discussed tests of hypotheses made about population means only. In the next section we will test hypotheses made about population proportions or percentages, and in the last section we will consider claims made about standard deviations or variances.

Using Excel

When using Excel for conducting hypothesis tests using the Student *t* distribution, use the DDXL add-in. Do not use the ZTEST function that is part of the standard Excel package. (Excel's ZTEST function often provides misleading *P*-values.) The procedure for using DDXL requires the original list of sample values. (If only the summary statistics of $n, \overline{x}$, and *s* are known, see the Excel Project at the end of Chapter 5 for a way to generate a list of values with the same sample statistics.) Here is the procedure for using DDXL:

1. Enter the sample data in column A.
2. Click on **DDXL,** then select the option of **1 Var t Test.**
3. Click on the pencil icon and enter the range of cells containing the sample data. For example, if you have 20 values in the first 20 rows of column A, enter A1:A20. Click **OK.**

Death Penalty as Deterrent

A common argument supporting the death penalty is that it discourages others from committing murder. Jeffrey Grogger of the University of California analyzed daily homicide data in California for a four-year period during which executions were frequent. Among his conclusions published in the *Journal of the American Statistical Association* (Vol. 85, No. 410): "The analyses conducted consistently indicate that these data provide no support for the hypothesis that executions deter murder in the short term." This is a major social policy issue, and the efforts of people such as Professor Grogger help to dispel misconceptions so that we have accurate information with which to address such issues.

4. You will see a four-step dialog box. Start with Step 1 by clicking on the bar labeled **Set $\mu0$.** You will get a smaller dialog box with a position for entering the value of the claimed population mean.
   - Enter the claimed value of the population mean.
   - Click **OK.**

5. Proceed to Step 2 where you select the significance level.

6. In Step 3, select the form of the alternative hypothesis that is appropriate for the test.

7. In Step 4, click on the **Compute** bar. Shown below is the Excel display for the example of this section. (The discrepancy between the test statistic value of 6.393 and the value of 6.440 in the example is due to a rounding error in the example. If the example had used more precise values of the sample mean and standard deviation, the test statistic would become 6.393 and would agree with the test statistic in the Excel display.) Note that the display includes the P-value, but not the critical values.

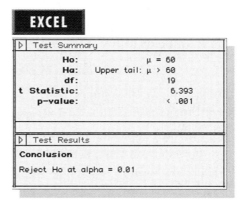

7-4 Basic Skills and Concepts

Traditional Method.   In Exercises 1–4, assume that the sample is a simple random sample selected from a normally distributed population. Test the given claim by using the traditional method of testing hypotheses. Include the test statistic, critical value(s), and conclusions.

1. Exam Scores Test the claim that for the population of statistics final exams, the mean score is given by $\mu = 75$. Sample data are summarized as $n = 16$, $\overline{x} = 77$, and $s = 15$. Test at the $\alpha = 0.05$ significance level.

2. Heights of Males Test the claim that for the population of male college students, the mean height is given by $\mu = 68.0$ in. Sample statistics include $n = 20$, $\overline{x} = 70.0$ in., and $s = 2.6$ in. Use a significance level of $\alpha = 0.01$.

3. Grade-Point Average Test the claim that the population of sophomore college students has a mean grade-point average greater than 2.00. Sample statistics include $n = 24$, $\overline{x} = 2.35$, and $s = 0.70$. Use a significance level of $\alpha = 0.01$.

4. Ages of Jets Test the claim that for the population of U.S. commercial jets, the mean age is less than 18 years. Sample statistics include $n = 8$, $\overline{x} = 15.8$ years, and $s = 8.3$ years. Use a significance level of $\alpha = 0.025$.

P-Value Method.   In Exercises 5–8, refer to the indicated exercise and Table A-3 to describe the P-value. (See the example in this section.)

5. Exercise 1

6. Exercise 2

7. Exercise 3

8. Exercise 4

Traditional Method.   For hypothesis tests in Exercises 9–24, draw a graph showing the test statistic and critical values. In each case, assume that the population has a distribution that is approximately normal and that the sample is a simple random sample. Caution: Some of the exercises require the use of the normal distribution (as described in the preceding section) instead of the Student t distribution (as described in this section); be sure to check the conditions to determine which distribution is appropriate.

EXCEL 9. Body Temperatures Refer to Data Set 6 in Appendix B. Using only the first 25 body temperatures listed for 12 A.M. of Day 2, test the claim that the mean body temperature of all healthy adults is equal to 98.6°F. For the significance level, use $\alpha = 0.05$. For the first 25 scores, $\bar{x} = 98.24$ and $s = 0.56$.

EXCEL 10. Body Temperatures Refer to Data Set 6 in Appendix B. Using only the first 35 body temperatures listed for 12 A.M. of Day 2, test the claim that the mean body temperature of all healthy adults is equal to 98.6°F. Use a 0.05 significance level. For the first 35 scores, $\bar{x} = 98.27$ and $s = 0.65$.

EXCEL 11. BMW Crash Tests Because of the expense involved, car crash tests often use small samples. When five BMW cars are crashed under standard conditions, the repair costs (in dollars) are as shown in the accompanying list. Use a 0.05 significance level to test the claim that the mean for all BMW cars is less than $1000. Would BMW be justified in advertising that under the standard conditions, the repair costs are less than $1000?

$$797 \quad 571 \quad 904 \quad 1147 \quad 418$$

12. Effectiveness of SAT Training Course Tom Wolfson is absolutely convinced that his Tom Wolfson School for Standard Tests is effective in helping test subjects raise their SAT scores. Tom has never been accused of being too humble. Test the claim that the Wolfson students have a mean score greater than 1017, which is the mean for the general population. Use these results from 20 subjects, who took the Wolfson course prior to taking the SAT: $\bar{x} = 1040$, $s = 207$. Is there a better way to design the experiment for testing the effectiveness of the Wolfson course?

13. Testing via Video A study was conducted to determine whether a standard clerical test would need revision for use on video display terminals (VDTs). The VDT scores of 22 subjects have a mean of 170.2 and a standard deviation of 35.3 (based on data from "Modification of the Minnesota Clerical Test to Predict Performance on Video Display Terminals," by Silver and Bennett, *Journal of Applied Psychology,* Vol. 72, No. 1). At the 0.05 significance level, test the claim that the mean for all subjects taking the VDT test differs from the mean of 243.5 for the standard printed version of the test. Based on the result, should the VDT test be revised?

14. Effects of Hypnotism In a study of factors affected by hypnotism, visual analogue scale (VAS) sensory ratings were obtained for 16 subjects. For these sample ratings, the mean is 8.33 and the standard deviation is 1.96 (based on data from "An Analysis of Factors That Contribute to the Efficacy of Hypnotic Analgesia," by Price and Barber, *Journal of Abnormal Psychology,* Vol. 96, No. 1). At the 0.01 significance level, test the claim that this sample comes from a population with a mean rating of less than 10.00.

15. Reading Test Scores The following reading test results were obtained for a sample of 15 third-grade students: $\bar{x} = 31.0$, $s = 10.5$. (The data are based on "A Longitudinal Study of the Effects of Retention/Promotion on Academic Achievement," by Peterson et al., *American Educational Research Journal,* Vol. 24, No. 1.) Does this third-grade sample mean differ significantly from a first-grade population mean of 41.9? Assume a 0.01 significance level.

16. Mean Time for Four-Year Degree Kim Greco is a high school senior who is concerned about attending college because she knows that many college students require more than four years to earn a bachelor's degree. At the 0.10 significance level, test the claim of her guidance counselor, who states that the mean time is greater than 5 years. Sample data consist of a mean of 5.15 years and a standard deviation of 1.68 years for 80 randomly selected college graduates (based on data from the National Center for Education Statistics).

17. Weights of Brown M&Ms Using the weights of only the *brown* M&Ms listed in Data Set 10 in Appendix B, test the claim that the mean is greater than 0.9085 g, the mean value necessary for the 1498 M&Ms to produce a total of 1361 g as the package indicates. The Excel workbook is M&M.xls. Use a 0.05 significance level. For the brown M&Ms, $\bar{x} = 0.9128$ g and $s = 0.0395$ g. Based on the result, can we conclude that the packages contain more than the claimed weight printed on the label?

18. Weights of Blue M&Ms Using the weights of only the *blue* M&Ms listed in Data Set 10 in Appendix B, test the claim that the mean is at least 0.9085 g, the mean value necessary for the 1498 M&Ms to produce a total of 1361 g as the package indicates. Use a 0.05 significance level. The Excel workbook is M&M.xls. For the blue M&Ms, $\bar{x} = 0.9014$ g and $s = 0.0573$ g. Based on the result, can we conclude that the package contents do not agree with the claimed weight printed on the label?

19. Effectiveness of New Chicken Feed When a poultry farmer uses her regular feed, the newborn chickens have normally distributed weights with a mean of 62.2 oz. In an experiment with an enriched feed mixture, nine chickens are born with the weights (in ounces) given below. Use a 0.01 significance level to test the claim that the mean weight is higher with the enriched feed.

61.4   62.2   66.9   63.3   66.2   66.0   63.1   63.7   66.6

20. Olympic Winners The accompanying list shows the winning times (in seconds) of men in the 100-meter dash for consecutive summer Olympic games, listed in order by row. Assuming that these results are sample data randomly selected from the population of all past and future Olympic games, test the claim that the mean time is equal to 10.5 sec. What do you observe about the precision of the numbers? What extremely important characteristic of the data set is not considered in this hypothesis test? Do the results from the hypothesis test suggest that future winning times should be around 10.5 sec, and is such a conclusion valid?

12.0   11.0   11.0   11.2   10.8   10.8   10.8   10.6   10.8   10.3   10.3   10.3
10.4   10.5   10.2   10.0   9.95   10.14   10.06   10.25   9.99   9.92   9.96

EXCEL **21.** Mean Pepsi Volumes Refer to Data Set 1 in Appendix B for the volumes (in ounces) of cans of diet Pepsi. The Excel workbook is COLA.xls. Is there sufficient evidence to conclude that the mean is greater than the 12 oz that is printed on the label?

EXCEL **22.** Nicotine in Cigarettes Refer to the nicotine contents of cigarettes listed in Data Set 8 in Appendix B. The Excel workbook is CIGARET.xls. Use the sample data to test the claim that the population mean is less than 1.0 mg. Does the conclusion apply to the mean nicotine level of all 100 mm cigarettes (not menthol or light) smoked by consumers? Why or why not?

EXCEL **23.** Effect of Vitamin Supplement on Birth Weight The accompanying list shows the birth weights (in kilograms) of male babies born to mothers taking a special vitamin supplement (based on data from the New York State Department of Health). At the 0.05 significance level, test the claim that the mean birth weight for all male babies of mothers given vitamins is equal to 3.39 kg, which is the mean for the population of all males. Based on the result, does the vitamin supplement appear to have an effect on birth weight?

$$3.73 \quad 4.37 \quad 3.73 \quad 4.33 \quad 3.39 \quad 3.68 \quad 4.68 \quad 3.52$$

$$3.02 \quad 4.09 \quad 2.47 \quad 4.13 \quad 4.47 \quad 3.22 \quad 3.43 \quad 2.54$$

EXCEL **24.** Nicotine in Cigarettes The Carolina Tobacco Company advertised that its best-selling nonfiltered cigarettes contain at most 40 mg of nicotine, but *Consumer Advocate* magazine ran tests of 10 randomly selected cigarettes and found the amounts shown in the accompanying list. The sample is small because the laboratory work required to extract the nicotine is time-consuming and expensive. It's a serious matter to charge that the company advertising is wrong, so the magazine editor chooses a significance level of $\alpha = 0.01$ in testing her belief that the mean nicotine content is greater than 40 mg. Using a 0.01 significance level, test the editor's belief that the mean is greater than 40 mg.

$$47.3 \quad 39.3 \quad 40.3 \quad 38.3 \quad 46.3 \quad 43.3 \quad 42.3 \quad 49.3 \quad 40.3 \quad 46.3$$

EXCEL **25.** Calcium Carbonate in Antacid Tablets Bottles of antacid tables include a printed label claiming that they contain 1000 mg of calcium carbonate. In a simple random sample of tablets made by the Medassist Pharmaceutical Company, the amounts of calcium carbonate are measured. When testing the claim that consumers are being cheated because the mean is less than 1000 mg, Excel displays the results shown. Interpret those results and determine whether there is sufficient evidence to support the claim that consumers are being cheated.

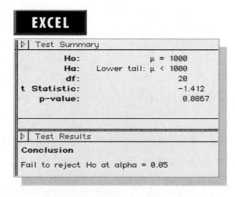

EXCEL

▷	Test Summary		
	Ho:		μ = 1000
	Ha:	Lower tail:	μ < 1000
	df:		20
t Statistic:			−1.412
	p−value:		0.0867

▷	Test Results
Conclusion	
Fail to reject Ho at alpha = 0.05	

26. **Time for Laughing** Rita Gibbons is a stand-up comedian who videotapes her performances and records the total of the times she must wait for audience laughter to subside. She recorded the times (in seconds) for 15 different shows in which she used a new routine. When testing the claim that her new routine is better than the old one, she obtains the Excel results shown here. Interpret those results. Does her new routine seem to be better than the old one?

27. **Acid Rain** The results shown below result from using Excel to test the claim that the amounts of acid rain sulfate deposits in Pennsylvania come from a population with a mean greater than 12.00 kg/hectare (based on data from the U.S. Department of Agriculture). Use a 0.01 significance level and interpret the Excel results from the hypothesis test.

28. **Reliability of Aircraft Radios** The mean time between failures (in hours) for a Telektronic Company radio used in light aircraft is 420 h. After 15 new radios were modified in an attempt to improve reliability, tests were conducted to measure the times between failures. When Excel is used to test the claim that the modified radios have a mean greater than 420 h, the results are as shown here. Does it appear that the modifications improved reliability?

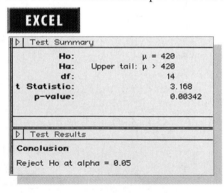

7-4 Beyond the Basics

 29. *Using Computer Results* Refer to the Excel display in Exercise 28. If the claim is changed from "greater than 420 h" to "not equal to 420 h," how is the *P*-value affected?

30. *Using the Wrong Distribution* Because of certain conditions, a hypothesis test requires the Student *t* distribution, as described in this section. Assume that the standard normal distribution was incorrectly used instead. Does using the standard normal distribution make you more or less likely to reject the null hypothesis, or does it not make a difference? Explain.

31. *Finding Critical t Values* When finding critical values, we sometimes need significance levels other than those available in Table A-3. Some computer programs approximate critical *t* values by calculating

$$t = \sqrt{\mathrm{df} \cdot (e^{A^2/\mathrm{df}} - 1)}$$

where

$$\mathrm{df} = n - 1$$

$$e = 2.718$$

$$A = z\left(\frac{8 \cdot \mathrm{df} + 3}{8 \cdot \mathrm{df} + 1}\right)$$

and *z* is the critical *z* score. Use this approximation to find the critical *t* score corresponding to *n* = 10 and a significance level of 0.05 in a right-tailed case. Compare the results to the critical *t* value found in Table A-3.

32. *Probability of Type II Error* Refer to Exercise 24 and assume that you're testing the claim that $\mu > 40$ mg. Find β (the probability of a type II error), given that the actual value of the population mean is $\mu = 45.0518$ mg. (See Exercise 31 in Section 7-3.)

 ## 7-5 Testing a Claim about a Proportion

In the preceding sections we introduced the basic methods of hypothesis testing, but they were used to address claims made about population *means* only. In this section we see how to apply those same basic methods to claims made about population *proportions*. The proportions can also represent probabilities or the decimal equivalents of percents. The following are examples of the types of claims we will be able to test.

- Fewer than 1/4 of all college graduates smoke.
- The percentage of late-night television viewers who watch *The Late Show with David Letterman* is equal to 18%.
- If a fatal car crash occurs, there is a 0.44 probability that it involves a driver who had been drinking.

Assumptions

1. The sample observations are a simple random sample. (Let us never forget the critical importance of sound sampling methods.)

Polls and Psychologists

Poll results can be dramatically affected by the wording of questions. A phrase such as "over the last few years" is interpreted differently by different people. Over the last few years (actually, since 1980), survey researchers and psychologists have been working together to improve surveys by decreasing bias and increasing accuracy. In one case, psychologists studied the finding that 10 to 15 percent of those surveyed say they voted in the last election when they did not. They experimented with theories of faulty memory, a desire to be viewed as responsible, and a tendency of those who usually vote to say that they voted in the most recent election, even if they did not. Only the last theory was actually found to be part of the problem.

2. The conditions for a *binomial distribution* are satisfied. (There are a fixed number of independent trials having constant probabilities, and each trial has two outcome categories, called "success" and "failure.")

3. The conditions $np \geq 5$ and $nq \geq 5$ are both satisfied, so **the binomial distribution of sample proportions can be approximated by a normal distribution with $\mu = np$ and $\sigma = \sqrt{npq}$** (as described in Section 5-6).

If these assumptions are not all satisfied, we may be able to use other methods not described in this section. In this section, all examples and exercises involve cases in which the assumptions are satisfied, so the sampling distribution of sample proportions can be approximated by the normal distribution. We use the following notation and test statistic.

Notation

n = number of trials

$\hat{p} = \dfrac{x}{n}$ (*sample* proportion)

p = population proportion (used in the null hypothesis)

$q = 1 - p$

Test Statistic for Testing a Claim about a Proportion

$$z = \frac{\hat{p} - p}{\sqrt{\dfrac{pq}{n}}}$$

The Traditional Method

If you are using the traditional method of testing hypotheses as outlined in Figure 7-5, find critical z values from Table A-2 (standard normal distribution) by using the procedures described in Section 7-2. For example, in a two-tailed test with significance level $\alpha = 0.05$, divide α equally between the two tails, then refer to Table A-2 for the z score corresponding to an area of $0.5 - 0.025 = 0.475$; the result is $z = 1.96$, so the critical values are $z = -1.96$ and $z = 1.96$.

The *P*-Value Method

To find the *P*-value corresponding to a z test statistic, use the following procedure, which was shown in Figure 7-8 and described in Section 7-3.

Right-tailed test: *P*-value = area to right of test statistic *z*

Left-tailed test: *P*-value = area to left of test statistic *z*

Two-tailed test: *P*-value = *twice* the area of the extreme region bounded by the test statistic *z*

As in Section 7-3, we use this decision criterion:

Reject the null hypothesis if the *P*-value is less than or equal to the significance level α.

When testing a claim about a population proportion *p*, be careful to identify correctly the sample proportion $\hat{p}$. The sample proportion $\hat{p}$ is sometimes given directly, but in other cases it must be calculated. See the examples below.

Given Statement	Finding $\hat{p}$
10% of the observed sports cars are red.	$\hat{p}$ is given directly: $\hat{p} = 0.10$
96 surveyed households have cable TV and 54 do not.	$\hat{p}$ must be calculated using $\hat{p} = x/n$: $$\hat{p} = \frac{x}{n} = \frac{96}{(96 + 54)} = 0.64$$

Caution: When a calculator or computer display of $\hat{p}$ results in many decimal places, use all of those decimal places when evaluating the *z* test statistic. Large errors can result from rounding $\hat{p}$ too much.

▷ **EXAMPLE   Survey of Voters** In a survey of 1002 people, 701 said that they voted in the recent presidential election (based on data from the ICR Research Group). Test the claim that when surveyed, the proportion of people who say that they voted is equal to 0.61, which is the proportion of people who actually did vote. Use a 0.05 significance level with the traditional method of testing hypotheses.

SOLUTION   We will use the traditional method of testing hypotheses as outlined in Figure 7-5. We use the sample proportion of $\hat{p} = 701/1002 = 0.6996007984$ and the claimed population proportion of $p = 0.61$.

Step 1:   The original claim is that the proportion of people who say that they voted is 0.61. We express this in symbolic form as $p = 0.61$.

Step 2:   The opposite of the original claim is $p \neq 0.61$.

Step 3:   Because $p = 0.61$ contains equality, we have

H_0: $p = 0.61$ (null hypothesis and original claim)

H_1: $p \neq 0.61$ (alternative hypothesis)

Step 4:   The significance level is α 5 0.05.

Step 5:   The statistic relevant to this test is $\hat{p} = 701/1002 = 0.6996007984$. The sampling distribution of sample proportions is approximated by the normal distribution. (The requirements $np \geq 5$ and $nq \geq 5$ are both satisfied with $n = 1002$, $p = 0.61$, and $q = 0.39$.)

Step 6:   The test statistic of $z = 5.81$ is found as follows:

$$z = \frac{\hat{p} - p}{\sqrt{\dfrac{pq}{n}}} = \frac{0.6996007984 - 0.61}{\sqrt{\dfrac{(0.61)(0.39)}{1002}}} = 5.81$$

The critical values of $z = -1.96$ and $z = 1.96$ are found from Table A-2. The test statistic of $z = 5.81$ is greater than the right-tail critical value of $z = 1.96$, so the test statistic is in the critical region.

Step 7:   Because the test statistic does fall within the critical region, reject the null hypothesis.

INTERPRETATION   There is sufficient evidence to warrant rejection of the claim that when surveyed, the proportion of people who say they voted is equal to 0.61. It appears that people do not accurately report whether they voted.

◀

> **EXAMPLE**   *P*-Value Approach   Repeat the preceding example, but use the *P*-value approach.

SOLUTION   Steps 1–4 will be the same as in the preceding solution. In Step 6, the test statistic $z = 5.81$ is computed as shown in the preceding example. Instead of finding the critical values, we proceed to find the *P*-value as follows (see Figure 7-8).

The test is two-tailed and the test statistic is to the right of center, so the *P*-value is twice the area to the right of the test statistic. Table A-2 indicates that the area between $z = 0$ and $z = 5.81$ is 0.4999, so the area to the right of the test statistic is $0.5 - 0.4999 = 0.0001$. The *P*-value is $2 \times 0.0001 = 0.0002$. (More precise tables would show that the exact *P*-value is actually much smaller.)

INTERPRETATION   Because the *P*-value of 0.0002 is less than or equal to the significance level of 0.05, we reject the null hypothesis. We again conclude that there is sufficient evidence to warrant rejection of the claim that when surveyed, the proportion of people who say they voted is equal to 0.61. Again, the *P*-value method is simply another way of arriving at the same conclusion reached using the traditional approach.

◀

Rationale for Test Statistic: The test statistic used in this section is justified by noting that when using the normal distribution to approximate a binomial distribution, we substitute $\mu = np$ and $\sigma = \sqrt{npq}$ to get

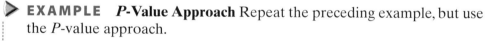

$$z = \frac{x - \mu}{\sigma} = \frac{x - np}{\sqrt{npq}}$$

In this expression, x is the number of successes among n trials. Divide the numerator and denominator of this expression by n, then replace x/n by the symbol $\hat{p}$, and you get the test statistic we are using. In other words, the test statistic is simply the same standard score (from Section 2-5) of $z = (x - \mu)/\sigma$, but modified for the binomial notation.

Using Excel

To conduct a hypothesis test for a claim about a population proportion, you must know the number of successes x and the total number of trials n. You must use the DDXL add-in, as described below.

1. First enter the number of successes in cell A1, and enter the total number of trials in cell B1.

2. Click on **DDXL** to use the Data Desk / XL add-in.

3. Select the menu item of **Hypothesis Tests,** then select **Summ 1 Var Prop Test** (for testing a claimed proportion using summary data for one variable).

4. Click on the pencil icon for the number of successes and enter A1, then click on the pencil icon for the number of trials and enter B1. Click **OK.**

5. You should now see a four-step dialog box. Begin with Step 1 by clicking on **Set p0.** In the smaller dialog box that appears, enter the value of the claimed population proportion, then click **OK.**

6. Proceed to Step 2 and select a significance level, then go to Step 3 and select the form of the alternative hypothesis. Finally, go to Step 4 and click on **Compute.** Shown below is the display resulting from the example of this section. Note that the P-value is displayed, but critical values are not shown. Correct interpretation of such displays requires an understanding of the P-value approach to hypothesis testing.

7-5   Basic Skills and Concepts

Testing Claims about Proportions.   *In Exercises 1–20, assume that the sample is a simple random sample and test the given claim. Use the traditional method or the P-value method of testing hypotheses. (Your instructor may specify which method should be used for some or all of the assigned exercises.)*

EXCEL **1.** Percentage of E-Mail Users Technology is dramatically changing the way we communicate. In 1997, a survey of 880 U.S. households showed that 149 of them use e-mail (based on data from *The World Almanac and Book of Facts*). Use those sample results to test the claim that less than 20% of U.S. households use e-mail. Use a 0.05 significance level. Is the conclusion valid today? Why or why not?

EXCEL **2.** Percentage of Telephone Users A recent survey of 4276 randomly selected households showed that 4019 of them had telephones (based on data from the U.S. Census Bureau). Use those survey results to test the claim that the percentage of households is now greater than the 35% rate that was found in 1920. Use a 0.01 significance level. The current rate of 4019/4276 (or 94%) appears to be significantly greater than the 1920 rate of 35%, but is there sufficient evidence to support that claim?

EXCEL **3.** Driving Distractions The Hartford Insurance Company obtains a simple random sample of 850 drivers and finds that 544 of them change tapes or CDs while driving (based on data from *Prevention* magazine). Use a 0.01 significance level to test the claim that most drivers change tapes or CDs while driving. Does changing tapes or CDs appear to be a driving hazard?

EXCEL **4.** Percentage of Theme Park Visits Each year, billions of dollars are spent at theme parks owned by Disney, Universal Studios, Sea World, Busch Gardens, and others. A survey of 1233 people who took trips revealed that 111 of them included a visit to a theme park (based on data from the Travel Industry Association of America). Based on those survey results, management consultant Laura Croft claims that less than 10% of trips include a theme park visit. Use a 0.05 significance level to test her claim. Would it be wise for her to use that claim in trying to convince theme park management to increase advertising spending?

EXCEL **5.** Voting for the Winning Candidate In a recent presidential election, 611 voters were surveyed, and 308 of them said that they voted for the candidate who won (based on data from the ICR Survey Research Group). Use a 0.05 significance level to test the claim that among all voters, 43% say that they voted for the winning candidate. (Voting records showed that the actual percentage who voted for the winning candidate was equal to 43%.) What does the result suggest about voter perceptions?

EXCEL **6.** Aviation Fatalities Researchers studied crashes of general aviation (noncommercial and nonmilitary) airplanes and found that pilots died in 437 of 8411 crash landings (based on data from "Risk Factors for Pilot Fatalities in General Aviation Airplane Crash Landings," by Rostykus, Cummings, and Mueller, *Journal of the American Medical Association*, Vol. 280, No. 11). Use a 0.05 significance level to test the claim that pilots die in more than 5% of such crashes. From the result, can we conclude that the pilot fatality rate does *not* exceed 5%? Why or why not?

EXCEL **7.** Store Checkout-Scanner Accuracy In a study of store checkout-scanners, 1234 items were checked and 20 of them were found to be overcharges (based on data from "UPC Scanner Pricing Systems: Are They Accurate?" by Goodstein, *Journal of Marketing,* Vol. 58). Use a 0.05 significance level to test the claim that with scanners, 1% of sales are overcharges. (Before scanners were used, the overcharge rate was estimated to be about 1%.) Based on these results, do scanners appear to help consumers avoid overcharges?

EXCEL **8.** Drug Testing of Job Applicants In 1990, 5.8% of job applicants who were tested for drugs failed the test. At the 0.01 significance level, test the claim that the failure rate is now lower if a simple random sample of 1520 current job applicants results in 58 failures (based on data from the American Management Association). Does the result suggest that fewer job applicants now use drugs?

9. *Awareness of Holocaust* Ralph Carter is a high school history teacher who says that if students are not aware of the Holocaust, then the curriculum should be revised to correct that deficiency. A Roper Organization survey of 506 high school students showed that 268 of them did not know that the term "Holocaust" refers to the Nazi killing of about 6 million Jews during World War II. Using the sample data and a 0.05 significance level, test the claim that most (more than 50%) students don't know what "Holocaust" refers to. Based on these results, should Ralph Carter seek revisions to the curriculum?

10. *Drug Testing for Adverse Reactions* The drug Ziac is used to treat hypertension. In a clinical test, 3.2% of 221 Ziac users experienced dizziness (based on data from Lederle Laboratories). Further tests must be conducted for adverse reactions that occur in at least 5% of treated subjects. Using a 0.01 significance level, test the claim that fewer than 5% of all Ziac users experience dizziness.

11. *Testing Effectiveness of Nicotine Patches* In one study of 71 smokers who tried to quit smoking with nicotine patch therapy, 32 were not smoking one year after the treatment (based on data from "High-Dose Nicotine Patch Therapy," by Dale et al., *Journal of the American Medical Association,* Vol. 274, No. 17). Use a 0.10 significance level to test the claim that among smokers who try to quit with nicotine patch therapy, the majority are smoking a year after the treatment. Do these results suggest that the nicotine patch therapy is not effective?

12. *Smoking and College Education* One survey showed that among 785 randomly selected subjects who completed four years of college, 18.3% smoke (based on data from the American Medical Association). Use a 0.01 significance level to test the claim that the rate of smoking among those with four years of college is less than the 27% rate for the general population.

13. *Computer Usage in Schools* In a survey of 500 students aged 12–18, it is found that 82% use computers in school (based on data from the Consumer Electronics Manufacturers Association). Test the claim that the proportion of students aged 12–18 who use computers in school is equal to 3/4.

14. *Percentage of Left-Handed Golfers* Arnold Handly is a production supervisor for the Spalding Corporation, and he claims that the percentage of left-handed golfers is greater than the 10% rate of people from the general population who are left-handed. In a simple random sample of 250 golfers, 15.2% are left-handed (based on a *USA Today* report). Is there sufficient evidence to support Arnold Handly's claim?

15. *Percentage of Weapons on Campus* Concerned about campus safety, the president of the College of Little Falls commissions a study. Among 400 students surveyed at this college, it is found that 7% carry a gun, knife, or other such weapon. Use a 0.01 significance level to test the claim that less than 10% of all U.S. college students carry such weapons. What is wrong with this conclusion?

16. *TV Ratings* A simple random sample of 1200 households with TV sets in use shows that 22% of them were tuned to *60 Minutes.* Use a 0.025 significance level to test the claim of a CBS executive that "*60 Minutes* gets more than a 20 share," which means that more than 20% of the sets in use are tuned to *60 Minutes*. If you are a commercial advertiser and you are trying to negotiate lower costs, what would you argue?

EXCEL **17.** Interpreting Calculator Display The Federal Aviation Administration will fund research on spatial disorientation of pilots if there is sufficient sample evidence (at the 0.01 significance level) to conclude that among aircraft accidents involving such disorientation, more than three-fourths result in fatalities. A study of 500 aircraft accidents involving spatial disorientation of the pilot found that 91% of those accidents resulted in fatalities (based on data from the U. S. Department of Transportation). The accompanying Excel display is obtained. Interpret that display. Based on these sample results, will the funding be approved?

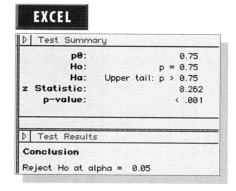

EXCEL **18.** Interpreting Calculator Display A television executive claims that "fewer than half of all adults are annoyed by the violence shown on television." Sample data from a Roper poll showed that 48% of 1,998 surveyed adults indicated their annoyance with television violence. The accompanying Excel display is obtained. Use a 0.05 significance level and interpret that display. Is the executive's claim supported by the sample data?

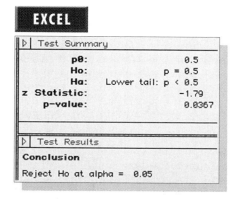

EXCEL **19.** Using M&M Data Refer to Data Set 10 in Appendix B and find the sample proportion of M&Ms that are blue. The Excel workbook is M&M.xls. Use that result to test the claim of Mars, Inc., that 10% of its plain M&M candies are blue.

EXCEL **20.** Using Survey Data Refer to Data Set 14 (Excel workbook STATSURV.xls) in Appendix B and consider only those statistics students who are 21 or older. Find the sample percentage of smokers in that age group, then test the claim that statistics students aged 21 or over smoke at a rate that is less than 32%, which is the smoking rate for the general population of persons aged 21 and over (based on data from the U.S. National Institute on Drug Abuse). Is there a reason that statistics students 21 and over would smoke at a rate lower than the rate for the general population in that age group?

7-5 Beyond the Basics

21. *Using a Different Method* In a study of perception, 80 men are tested and 7 are found to have red/green color blindness (based on data from *USA Today*). We want to use a 0.01 significance level to test the claim that men have a red/green color-blindness rate that is greater than the 0.25% rate for women.
   a. Why can't we use the methods of this section?
   b. Assuming that the red/green color-blindness rate for men is equal to the 0.25% rate for women, find the probability that among 80 randomly selected men, at least 7 will have that type of color blindness. Describe the method used to find that probability.
   c. Based on the result from part (b), what do you conclude?

22. *Dealing with No Successes* In a simple random sample of 50 plain M&M candies, it is found that none of them are blue. We want to use a 0.01 significance level to test the claim of Mars, Inc., that the proportion of M&M candies that are blue is equal to 0.10. Can the methods of this section be used? If so, test the claim. If not, explain why not.

23. *Misleading with Statistics* Chemco, a supplier of chemical-waste containers, finds that 3% of a sample of 500 units are defective. Being fundamentally dishonest, the Chemco production manager wants to make a claim that the rate of defective units is no more than some specified percentage, and he doesn't want that claim rejected at the 0.05 significance level if the sample data are used. What is the *lowest* defective rate he can claim under these conditions?

24. *Probability of Type II Error* Refer to Exercise 18. If the true value of p is 0.45, find β, the probability of a type II error (see Exercise 31 in Section 7-3). [*Hint:* In Step 3, use the values $p = 0.45$ and $pq/n = (0.45)(0.55)/1998$.]

25. *False Claim* A researcher claimed that when 20 mice were treated, the success rate was equal to 47%. What is the basis for rejecting that claim?

7-6 Testing a Claim about a Standard Deviation or Variance

Quality-control engineers want to ensure that a product is, on the average, acceptable, but they also want to produce items of *consistent* quality so that there will be few defective products. For example, the consistency of aircraft altimeters is governed by Federal Aviation Regulation 91.36, which requires that aircraft altimeters be tested and calibrated to give a reading "within 125 feet (on a 95-percent probability basis)." Even if the mean altitude reading is exactly correct, an excessively large standard deviation will result in individual readings that are dangerously low or high. Consistency is improved by reducing the standard deviation or variance. In the preceding sections of this chapter we described methods for testing claims made about

Ethics in Experiments

Sample data can often be obtained by simply observing or surveying members selected from the population. Many other situations require that we somehow manipulate circumstances to obtain sample data. In both cases ethical questions may arise. Researchers in Tuskegee, Alabama, withheld the effective penicillin treatment to syphilis victims so that the disease could be studied. This experiment continued for a period of 27 years!

population means and proportions. This section focuses on variation, which is critically important in many applications, including quality control. The main objective of this section is to present methods to test claims made about a population standard deviation σ or variance σ^2. We begin by identifying the assumptions that apply to such tests.

Assumptions

1. The samples are simple random samples. (Remember the importance of good sampling methods.)

2. The population has values that are normally distributed.

Other methods of testing hypotheses also require a normally distributed population, but tests of claims about standard deviations or variances are not as *robust,* meaning that the inferences can be very misleading if the population does not have a normal distribution. Therefore, the condition of a normally distributed population is a much stricter requirement in this section.

When testing a claim made about a population standard deviation σ or variance σ^2, we use the chi-square test statistic and critical values described as follows.

Test Statistic for Testing Hypotheses about σ or σ^2

$$\chi^2 = \frac{(n-1)s^2}{\sigma^2}$$

where
n = sample size
s^2 = sample variance
σ^2 = population variance (given in the null hypothesis)

Critical Values

1. Critical values are found in Table A-4.

2. Degrees of freedom = $n - 1$.

Don't be confused by the use of both the normal and the chi-square distributions. After verifying that the sample data appear to come from a normally distributed population, we should then shift gears and think "chi-square." The chi-square distribution was introduced in Section 6-6, where we noted the following important properties.

Properties of the Chi-Square Distribution

1. All values of χ^2 are nonnegative, and the distribution is not symmetric (see Figure 7-12).

2. There is a different distribution for each number of degrees of freedom (see Figure 7-13).

3. The critical values are found in Table A-4 using

degrees of freedom = $n - 1$

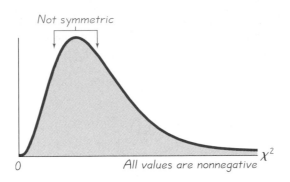

FIGURE 7-12
Properties of the Chi-Square Distribution

FIGURE 7-13
Chi-Square Distributions for 10 and 20 Degrees of Freedom

There is a different distribution for each number of degrees of freedom.

Critical values are found in Table A-4 by first locating the row corresponding to the appropriate number of degrees of freedom (where df = $n - 1$). Next, the significance level α is used to determine the correct column as described in the following list.

Right-tailed test: Locate the area at the top of Table A-4 that is equal to the significance level α.

Left-tailed test: Calculate $1 - \alpha$, then locate the area at the top of Table A-4 that is equal to $1 - \alpha$.

Two-tailed test: Calculate $1 - \alpha/2$ and $\alpha/2$, then locate the area at the top of Table A-4 that equals $1 - \alpha/2$ (for the left critical value) and locate the area at the top equal to $\alpha/2$ (for the right critical value). (See Figure 6-9 and the example on pages 361–362.)

▶ **EXAMPLE   IQ Scores of Statistics Professors** For a simple random sample of adults, IQ scores are normally distributed with a mean of 100 and a standard deviation of 15. A simple random sample of 13 statistics professors yields a standard deviation of $s = 7.2$. A psychologist is quite sure that statistics professors have IQ scores that have a mean greater than 100. He doesn't understand the concept of standard deviation very well and does not realize that the standard deviation should be lower than 15 (because statistics professors have less variation than the general population). Instead, he claims that statistics professors have IQ scores with a standard deviation equal to 15, the same standard deviation for the general population. Use a 0.05 significance level to test the claim that $\sigma = 15$. Based on the result, what do you conclude about the standard deviation of IQ scores for statistics professors?

SOLUTION   We will use the traditional method of testing hypotheses as outlined in Figure 7-5.

Step 1:   The claim is expressed in symbolic form as $\sigma = 15$.

Step 2:   If the original claim is false, then $\sigma \neq 15$.

Step 3:   The null hypothesis must contain equality, so we have

$$H_0: \sigma = 15 \text{ (original claim)} \qquad H_1: \sigma \neq 15$$

Step 4:   The significance level is $\alpha = 0.05$.

Step 5:   Because the claim is made about σ, we use the chi-square distribution.

Step 6:   The test statistic is

$$\chi^2 = \frac{(n-1)s^2}{\sigma^2} = \frac{(13-1)(7.2)^2}{15^2} = 2.765$$

The critical values of 4.404 and 23.337 are found in Table A-4, in the 12th row (degrees of freedom = $n - 1 = 12$) in the columns corresponding to 0.975 and 0.025. See the test statistic and critical values shown in Figure 7-14.

Step 7:   Because the test statistic is in the critical region, we reject the null hypothesis.

INTERPRETATION   There is sufficient evidence to warrant rejection of the claim that the standard deviation is equal to 15. It appears that statistics professors have IQ scores with a standard deviation that is significantly different than the standard deviation of 15 for the general population. ◀

FIGURE 7-14
Hypothesis Test of Claim that $\sigma = 15$

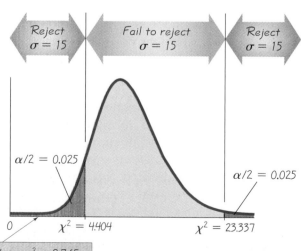

Reject $\sigma = 15$ Fail to reject $\sigma = 15$ Reject $\sigma = 15$

$\alpha/2 = 0.025$

$\alpha/2 = 0.025$

0 $\chi^2 = 4.404$ $\chi^2 = 23.337$

Sample data: $\chi^2 = 2.765$

The *P*-Value Method

Instead of the traditional approach to hypothesis testing, we can also use the *P*-value approach summarized in Figures 7-8 and 7-9. If we use Table A-4, we usually cannot find *exact* *P*-values because that chi-square distribution table includes only selected values of α. If using Table A-4, we can identify limits that contain the *P*-value. The test statistic from the last example is $\chi^2 = 2.765$, and we know that the test is two-tailed with 12 degrees of freedom. Refer to the 12th row of Table A-4 and see that the test statistic of 2.765 is less than every entry in that row, which means that the area to the left of the test statistic is less than 0.005. The *P*-value is therefore less than 0.01. (Figure 7-8 shows that in a two-tailed test with a test statistic to the left of the center, the *P*-value is *twice* the tail area beyond the test statistic.) Because the *P*-value is less than the significance level of $\alpha = 0.05$, we reject the null hypothesis. Again, the traditional method and the *P*-value method are equivalent in the sense that they always lead to the same conclusion.

Using Excel

Excel is not designed for tests of claims made about a population standard deviation or variance. The test statistic is quite easy to compute, and you could either refer to Table A-4 for the critical value(s), or you could use Excel to help find the *P*-value. Here is the Excel procedure for finding the *P*-value that corresponds to a computed test statistic.

1. Click on *fx*, select the function category of **Statistical,** then select the function name of **CHIDIST** and click **OK.**

2. In the dialog box, enter the value of the χ^2 test statistic in the box labeled x, and enter the number of degrees of freedom (equal to $n - 1$), then click OK.

3. Excel will return the area to the *right* of the test statistic. Let's refer to this area that Excel displays as the "Excel area." The *P*-value can be found as follows:

Right-tailed test: *P*-value = Excel area

Left-tailed test: *P*-value = 1 − (Excel area)

Two-tailed test: *P*-value = smaller of these two values:

$$2 \cdot (\text{Excel area})$$

$$2 \cdot (1 - \text{Excel area})$$

For the example of this section, the test is two-tailed, the test statistic is $\chi^2 = 2.765$, and the sample size is 13 so that the number of degrees of freedom is 12. If you follow the above procedure using a test statistic of 2.765 and 12 degrees of freedom, Excel will return an area of 0.996987952. Because the test is two-tailed, the *P*-value is the smaller of 2(0.996987952) = 1.993975904 and 2(1 − 0.996987952) = 0.006024096. The *P*-value is therefore 0.006024096. (The *P*-value cannot exceed 1 so that the value of 1.993975904 could not be used.) Because this *P*-value is less than the 0.05 significance level, we reject the null hypothesis, as in the example of this section.

7-6 Basic Skills and Concepts

Finding Critical Values.   *In Exercises 1 and 2, use Table A-4 to find the critical values of χ^2 based on the given information.*

1.   a. $H_0: \sigma = 15$ b. $H_1: \sigma > 0.62$ c. $H_1: \sigma < 14.4$

$n = 10$ $n = 27$ $n = 21$

$\alpha = 0.01$ $\alpha = 0.01$ $\alpha = 0.05$

2.   a. $H_1: \sigma < 1.22$ b. $H_1: \sigma > 92.5$ c. $H_0: \sigma = 0.237$

$n = 23$ $n = 12$ $n = 16$

$\alpha = 0.025$ $\alpha = 0.10$ $\alpha = 0.05$

Testing Claims about Variation.   *In Exercises 3–14, test the given hypotheses. Follow the steps outlined in Figure 7-5, and draw the appropriate graph. In all cases, assume that the population is normally distributed and that the sample is a simple random sample.*

3.   Use a 0.05 significance level to test the claim that $\sigma = 15$. The sample statistics are as follows: $n = 10$, $\overline{x} = 103.5$, $s = 18.0$.

4.   Use a 0.01 significance level to test the claim that $\sigma > 5$. The sample statistics are as follows: $n = 16$, $\overline{x} = 47.3$, $s = 8.0$.

5.   Variation in Peanut M&Ms Use a 0.01 significance level to test the claim that peanut M&M candies have weights that vary more than the weights of plain M&M candies. The standard deviation for the weights of plain M&M candies is 0.04 g. A sample of 40 peanut M&Ms has weights with a standard deviation of 0.31 g. Why should peanut M&M candies have weights that vary more than the weights of plain M&M candies?

6.   Random Generation of Data Excel can be used to generate random data from a normally distributed population. One such generated sample of 50 values has a mean of 98.4 and a standard deviation of 16.3. Use a 0.10 significance level to test the claim that this sample actually does come from a population with a standard deviation equal to 15. What does the result say about the variation among the generated sample values?

7.   Manufacturing Aircraft Altimeters The Stewart Aviation Products Company uses a new production method to manufacture aircraft altimeters. A simple random sample of 81 altimeters is tested in a pressure chamber, and the errors in altitude are recorded as positive values (for readings that are too high) or negative values (for readings that are too low). The sample has a standard deviation of $s = 52.3$ ft. At the 0.05 significance level, test the claim that the new production line has errors with a standard deviation different from 43.7 ft, which was the standard deviation for the old production method. If it appears that the standard deviation has changed, does the new production method appear to be better or worse than the old method?

8.   Is the New Machine Better? The Medassist Pharmaceutical Company uses a machine to pour cold medicine into bottles in such a way that the standard deviation of the weights is 0.15 oz. A new machine is tested on 71 bottles, and the standard deviation for this sample is 0.12 oz. The Dayton Machine Company, which manufactures the new machine, claims that it fills bottles with a lower

variation. At the 0.05 significance level, test the claim made by the Dayton Machine Company. If Dayton's machine is being used on a trial basis, should its purchase be considered?

9. Statistics Test Scores Tests in the author's past statistics classes have scores with a standard deviation equal to 14.1. One of his current classes now has 27 test scores with a standard deviation of 9.3. Use a 0.01 significance level to test the claim that this current class has less variation than past classes. Does a lower standard deviation suggest that the current class is doing better?

10. Bank Customer Waiting Times With individual lines at its various windows, the Jefferson Valley Bank found that the standard deviation for normally distributed waiting times on Friday afternoons was 6.2 min. The bank experimented with a single main waiting line and found that for a simple random sample of 25 customers, the waiting times have a standard deviation of 3.8 min. On the basis of previous studies, we can assume that the waiting times are normally distributed. Use a 0.05 significance level to test the claim that a single line causes lower variation among the waiting times. Why would customers prefer waiting times with less variation? Does the use of a single line result in a shorter wait?

11. Blood Pressure Measurements Systolic blood pressure results from contraction of the heart. In comparing systolic blood pressure levels of men and women, Dr. Jane Taylor obtained readings for a simple random sample of 50 women. The sample mean and standard deviation were found to be 130.7 and 23.4, respectively. If systolic blood pressure levels for men are known to have a mean and standard deviation of 133.4 and 19.7, respectively, test the claim that women have more variation. Use a 0.05 significance level. (All readings are in millimeters of mercury, and data are based on the National Health Survey.)

12. Body Temperatures In Section 7-3, we used the sample data given at the beginning of the chapter ($n = 106$, $\bar{x} = 98.20$, $s = 0.62$) to test the claim that the mean body temperature of healthy adults is equal to 98.6°F, and we rejected that claim. The test statistic will cause rejection of $\mu = 98.6$°F as long as the standard deviation is less than 2.11°F. Use the sample statistics and a 0.005 significance level to test the claim that $\sigma < 2.11$°F.

EXCEL 13. Using Data of Heights of Men Based on data from the National Health Survey, men aged 25–34 have heights with a standard deviation of 2.9 in. Test the claim that men aged 45–54 have heights with a different standard deviation. The heights of 25 randomly selected men in the 45–54 age bracket are listed below.

66.80   71.22   65.80   66.24   69.62   70.49   70.00   71.46   65.72

68.10   72.14   71.58   66.85   69.88   68.69   72.77   67.34   68.40

68.96   68.70   72.69   68.67   67.79   63.97   67.19

EXCEL 14. Using Birth Weight Data Shown below are birth weights (in kilograms) of male babies born to mothers taking a special vitamin supplement (based on data from the New York State Department of Health). Test the claim that this sample comes from a population with a standard deviation equal to 0.470 kg, which is the standard deviation for male birth weights in general. Does the vitamin supplement appear to affect the variation among birth weights?

3.73   4.37   3.73   4.33   3.39   3.68   4.68   3.52

3.02   4.09   2.47   4.13   4.47   3.22   3.43   2.54

EXCEL **15.** Supermodel Weights Use a 0.01 significance level to test the claim that weights of female supermodels vary less than the weights of women in general. The standard deviation of weights of the population of women is 29 lb. Listed below are the weights (in pounds) of nine randomly selected supermodels.

125 (Taylor)	119 (Auermann)	128 (Schiffer)	128 (MacPherson)
119 (Turlington)	127 (Hall)	105 (Moss)	123 (Mazza)
115 (Hume)			

EXCEL **16.** Supermodel Heights Use a 0.05 significance level to test the claim that heights of female supermodels vary less than the heights of women in general. The standard deviation of heights of the population of women is 2.5 in. Listed below are the heights (in inches) of randomly selected supermodels (Taylor, Harlow, Mulder, Goff, Evangelista, Avermann, Schiffer, MacPherson, Turlington, Hall, Crawford, Campbell, Herzigova, Seymour, Banks, Moss, Mazza, Hume).

71 71 70 69 69.5 70.5 71 72 70 70 69 69.5 69 70 70 66.5 70 71

7-6 Beyond the Basics

17. Finding P-Values Use Table A-4 to find the range of possible P-values in the given exercises.
a. Exercise 3   **b.** Exercise 5   **c.** Exercise 15

18. Finding Critical Values of χ^2 For large numbers of degrees of freedom, we can approximate critical values of χ^2 as follows:

$$\chi^2 = \frac{1}{2}(z + \sqrt{2k-1})^2$$

Here k is the number of degrees of freedom and z is the critical value, found in Table A-2. For example, if we want to approximate the two critical values of χ^2 in a two-tailed hypothesis test with $\alpha = 0.05$ and a sample size of 150, we let $k = 149$ with $z = -1.96$, followed by $k = 149$ and $z = 1.96$.
a. Use this approximation to estimate the critical values of χ^2 in a two-tailed hypothesis test with $n = 101$ and $\alpha = 0.05$. Compare the results to those found in Table A-4.
b. Use this approximation to estimate the critical values of χ^2 in a two-tailed hypothesis test with $n = 150$ and $\alpha = 0.05$.

19. Finding Critical Values of χ^2 Repeat Exercise 18 using this approximation (with k and z as described in Exercise 18):

$$\chi^2 = k\left(1 - \frac{2}{9k} + z\sqrt{\frac{2}{9k}}\right)^3$$

20. Effect of Outlier When using the hypothesis testing procedure of this section, will the result be dramatically affected by the presence of an outlier? Describe how you arrived at your response.

21. Last-Digit Analysis The last digits of sample data are sometimes used in an attempt to determine whether the data have been measured or simply reported by the subject. Reported data often have last digits with disproportionately more 0s and 5s. Measured data tend to have last digits with a mean of 4.5, a standard deviation of about 3, and the digits should occur with roughly the same frequency.

a. How is the standard deviation of the data affected if there are disproportionately more 0s and 5s?

b. Why can't we use the methods of this section to test that the last digits of the sample data have a standard deviation equal to 3?

22. Probability of a Type II Error Refer to Exercise 10. Assuming that σ is actually 4.0, find β (the probability of a type II error). See Exercise 31 in Section 7-3, and modify the procedure so that it applies to a hypothesis test involving σ instead of μ.

"Freshman 15": Fact or Fantasy?

BOSTON Along with all of the typical "back-to-school" hype about lunch boxes and school buses, each September is typically greeted with media reports and advice about the "freshman 15," which is the popular name given to the phenomenon of first-year college students gaining 15 lb during their freshman year. But does this 15-lb weight gain actually occur, or is it simply a myth? Carole Nhu'y Hodge, Linda Jackson, and Linda Sullivan are Michigan State University researchers who conducted their own investigation. They studied 61 Michigan State female students who took an introductory psychology course. The volunteers, who were given extra credit for participation in the experiment, were weighed at the beginning of their freshman year and at a point in time six months later. Among their findings reported in *Psychology of Women Quarterly*: "Body weight at the beginning of the first college year (Time 1) was compared with weight approximately 6 months later (Time 2).

Average weight at Time 2, 131.45 lb (59.62 kg), was no different from average weight at Time 1, 130.57 lb (59.23 kg)." They also state that "Our findings suggest it (the 15-lb weight gain) is fantasy, although additional research is needed before drawing firm conclusions."

1. What do the researchers infer when they say that there is "no difference" between the mean weight at Time 1 (130.57 lb) and the mean weight at Time 2 (131.45 lb), when there is an apparent difference of 0.88 lb?

2. What are the limitations of this particular study? That is, if the sample data are used to make inferences about a population, identify the specific population in question.

3. Identify any aspects of the experiment that could potentially threaten the validity of the results.

4. Identify a possible null hypothesis and alternative hypothesis for this experiment.

VOCABULARY LIST

hypothesis

hypothesis test

test of significance

null hypothesis

alternative hypothesis

test statistic

critical region

significance level

critical value

two-tailed test

right-tailed test

left-tailed test

type I error

type II error

alpha (α)

beta (β)

power

traditional method

classical method

P-value

probability value

REVIEW

In Chapters 6 and 7 we described basic procedures for two extremely important methods of inferential statistics: estimating values of population parameters and testing claims made about population parameters. In this chapter we presented methods for testing claims about a population mean, proportion, standard deviation, and variance.

In Section 7-2 we presented the fundamental concepts of a hypothesis test: null hypothesis, alternative hypothesis, test statistic, critical region, significance level, critical value, type I error, and type II error. We also discussed two-tailed tests, left-tailed tests, right-tailed tests, and the statement of conclusions. In Section 7-3 we used those components in identifying three different methods for testing hypotheses:

1. The traditional method (summarized in Figure 7-5)

2. The *P*-value method (summarized in Figure 7-9)

3. Confidence intervals (discussed in Chapter 6)

In Sections 7-3 through 7-6 we discussed specific methods for dealing with the different parameters. Because it is so important to be correct in selecting the distribution and test statistic, we provide Figure 7-15 on the next page, which summarizes the key decisions to be made.

REVIEW EXERCISES

1. **a.** You have just developed a new cure for the common cold and you plan to conduct a formal test to justify its effectiveness. Which *P*-value would you most prefer: 0.99, 0.05, 0.5, 0.01, or 0.001?

   b. For healthy women aged 18–24, systolic blood pressure readings (in mm Hg) are normally distributed with a mean of 114.8 and a standard deviation of 13.1 (based on data from the National Health Survey). If you want to use a formal hypothesis test in an experiment to show that the drug Xynon lowers blood pressure levels for women in that age group, how should your claim be stated?

   c. In testing the claim that the mean IQ score of convicted felons is less than 90, what is the null hypothesis?

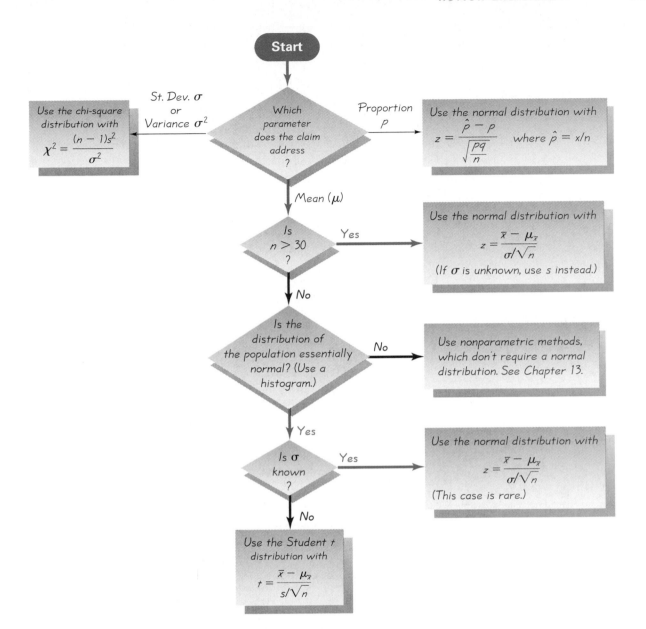

Start

Which parameter does the claim address ?

St. Dev. σ or Variance σ²

Use the chi-square distribution with
$$\chi^2 = \frac{(n-1)s^2}{\sigma^2}$$

Proportion p

Use the normal distribution with
$$z = \frac{\hat{p} - p}{\sqrt{\frac{pq}{n}}} \quad \text{where } \hat{p} = x/n$$

Mean (μ)

Is n > 30 ?

Yes

Use the normal distribution with
$$z = \frac{\bar{x} - \mu_{\bar{x}}}{\sigma/\sqrt{n}}$$
(If σ is unknown, use s instead.)

No

Is the distribution of the population essentially normal? (Use a histogram.)

No

Use nonparametric methods, which don't require a normal distribution. See Chapter 13.

Yes

Is σ known ?

Yes

Use the normal distribution with
$$z = \frac{\bar{x} - \mu_{\bar{x}}}{\sigma/\sqrt{n}}$$
(This case is rare.)

No

Use the Student t distribution with
$$t = \frac{\bar{x} - \mu_{\bar{x}}}{s/\sqrt{n}}$$

FIGURE 7-15
Testing a Claim about a Mean, Proportion, Standard Deviation, or Variance

d. In testing the claim that the mean amount of cola in cans is greater than 12 oz, you fail to reject the null hypothesis. State the final conclusion that addresses the original claim.

e. Complete the statement: "A type I error is the mistake of ... "

2. Finding Critical Values Find the appropriate critical values. In all cases, assume that the population standard deviation σ is unknown.

a. $H_1: \mu < 27.3$
   $n = 10$
   $\alpha = 0.05$

b. $H_1: p \neq 0.5$
   $n = 150$
   $\alpha = 0.01$

c. $H_1: \sigma \neq 15$
   $n = 20$
   $\alpha = 0.05$

(continued)

d. $H_0: \mu = 19.9$

$n = 40$

$\alpha = 0.10$

e. $H_0: p \geq 0.25$

$n = 540$

$\alpha = 0.01$

EXCEL 3. Does the Majority Favor Gun Control? A *USA Today*/CNN/Gallup survey found that of the gun owners surveyed, 57% favor stricter gun laws. Assume that the survey group consisted of 504 randomly selected gun owners. Test the claim that the majority (more than 50%) of gun owners favor stricter gun laws. Use a 0.01 significance level.

4. Are Consumers Being Cheated? The Orange County Bureau of Weights and Measures received complaints that the Windsor Bottling Company was cheating consumers by putting less than 12 oz of root beer in its cans. When 24 cans are randomly selected and measured, the amounts are found to have a mean of 11.4 oz and a standard deviation of 0.62 oz. The company president, Harry Windsor, claims that the sample is too small to be meaningful. Use the sample data to test the claim that consumers are being cheated. Does Harry Windsor's argument have any validity?

5. Testing Consistency of Filled Bags The Kansas Farm Products Company uses a machine that fills 50-lb corn-seed bags. In the past, the machine has had a standard deviation of 0.75 lb. In an attempt to get more consistent weights, mechanics have replaced some worn machine parts. A simple random sample of 61 bags taken from the repaired machinery produced a sample mean of 50.13 lb and a sample standard deviation of 0.48 lb. At the 0.05 significance level, test the claim that the weights are "more consistent" with the repaired machinery than in the past. If the weights are more consistent, how is the standard deviation affected?

EXCEL 6. Percentage Believing That Elvis Is Alive *USA Today* ran a report about a University of North Carolina poll of 1248 adults from the southern United States. It was reported that 8% of those surveyed believe that Elvis Presley still lives. The article began with the claim that "almost 1 out of 10" Southerners still thinks Elvis is alive. At the 0.01 significance level, test the claim that the true percentage is less than 10%. Based on the result, determine whether the 8% sample result justifies the phrase "almost 1 out of 10."

7. Driving Distances of Women The New England Insurance Company is reviewing the driving habits of women aged 16–24 to determine whether they should continue to pay higher premiums than women in a higher age bracket. In a study of 750 randomly selected women drivers aged 16–24, the mean driving distance for one year is 6047 mi and the standard deviation is 2944 mi (based on data from the Federal Highway Administration). Use a 0.01 significance level to test the claim that the population mean for women in the 16–24 age bracket is less than 7124 mi, which is the known mean for women in the higher age bracket. If women in the 16–24 age bracket drive less, should they be charged lower insurance premiums?

8. Uses of Fruitcakes A Bruskin-Goldring Research poll of 1012 adults showed that among those who used fruitcakes, 28% ate them and 72% used them for other purposes, such as doorstops, birdfeed, and landfill. This surprised fruitcake producers, who believe that a fruitcake is an appealing food. The president of the Kansas Food Products Company claims that the poll results are a fluke and, in reality, at least half of all adults eat their fruitcakes. Use a 0.01 significance level to test that claim. Based on the result, does it appear that fruitcake producers should consider changes to make their product more appealing as a food or better suited for its uses as a doorstop, birdfeed, and so on?

EXCEL **9.** Seizure Reduction Drug The Medassist Pharmaceutical Company makes a pill intended to reduce the chance of seizures among children. The pill is supposed to contain 20.0 mg of phenobarbital. A simple random sample of 20 pills yielded the amounts (in mg) listed here. Are these pills acceptable at the $\alpha = 0.01$ significance level?

27.5	26.0	22.9	23.4	23.0
23.9	32.6	20.9	22.9	24.3
24.8	16.1	24.3	17.3	18.9
20.7	33.0	15.6	24.3	23.3

10. Claim about Mean Annual Income The Orange County Supervisor boasts that full-time salaried employees in her county have a mean annual income that is greater than $30,000. If she based that claim on a sample of 150 such employees who had a mean of $30,122 and a standard deviation of $14,276, is her claim justified? Use a 0.01 significance level.

CUMULATIVE REVIEW EXERCISES

EXCEL **1.** Analyzing SAT Scores The College of Newport randomly selects female applicants for admission and records their scores on the math portion of the SAT. Here are the results: 490, 570, 630, 440, 460, 720, 410, 780.
  a. Find the mean of this sample.
  b. Find the median.
  c. Find the standard deviation.
  d. Find the variance.
  e. Find the range.
  f. Construct a 95% confidence interval estimate of the population mean.
  g. Use a 0.05 significance level to test the claim that this sample comes from a population with a mean equal to 496, which is the reported mean of the SAT math scores for all women.
  h. Based on the preceding results, what do you conclude about the female applicants to the College of Newport? Are their SAT math scores about the same as the general population, or are they better or worse? (The population of all women who take the math portion of the SAT has a mean score of 496.)

EXCEL **2.** SAT Math Scores of Women The math SAT scores for women are normally distributed with a mean of 496 and a standard deviation of 108.
  a. If a woman who takes the math portion of the SAT is randomly selected, find the probability that her score is above 500.
  b. If five math SAT scores are randomly selected from the population of women who take the test, find the probability that all five of the scores are above 500.
  c. If five women who take the math portion of the SAT are randomly selected, find the probability that their mean is above 500.
  d. Find P_{90}, the score separating the bottom 90% from the top 10%.

EXCEL **3.** Blood Pressure Readings A medical researcher obtains the systolic blood pressure readings (in mm Hg) in the accompanying list from a sample of women aged 18–24 who have a new strain of viral infection. (Healthy women in that age group have readings that are normally distributed with a mean of 114.8 and a standard deviation of 13.1.)

| 134.9 | 78.7 | 108.9 | 133.0 | 123.7 | 96.1 | 126.9 | 89.8 |
| 132.0 | 134.7 | 132.1 | 121.7 | 112.3 | 150.2 | 158.3 | 154.4 |

a. Find the sample mean $\bar{x}$ and standard deviation s.

b. Use a 0.05 significance level to test the claim that the sample comes from a population with a mean blood pressure equal to 114.8.

c. Use the sample data to construct a 95% confidence interval for the population mean μ. Do the confidence interval limits contain the value of 114.8, which is the mean for healthy women aged 18–24?

d. Use a 0.05 significance level to test the claim that the sample comes from a population with a standard deviation equal to 13.1, which is the standard deviation for healthy women aged 18–24.

e. Based on the preceding results, does it seem that the new strain of viral infection affects systolic blood pressure?

EXCEL **4.** ESP A student majoring in psychology designs an experiment to test for extrasensory perception (ESP). In this experiment, a card is randomly selected from a shuffled deck, and the blindfolded subject must guess the suit (clubs, diamonds, hearts, spades) of the card selected. The experiment is repeated 25 times, with the card replaced and the deck reshuffled each time.

a. For subjects who make random guesses with no ESP, find the mean number of correct responses.

b. For subjects who make random guesses with no ESP, find the standard deviation for the numbers of correct responses.

c. For subjects who make random guesses with no ESP, find the probability of getting more than 12 correct responses.

d. If a subject gets more than 12 correct responses, test the claim that they made random guesses. Use a 0.05 significance level.

e. You want to conduct a survey to estimate the percentage of adult Americans who believe that some people have ESP. How many people must you survey if you want 90% confidence that your sample percentage is in error by no more than four percentage points?

COOPERATIVE GROUP ACTIVITIES

1. *In-class activity:* Each student should estimate the length of the classroom. The values should be based on visual estimates, with no actual measurements being taken. After the estimates have been collected, measure the length of the room, then test the claim that the sample mean is equal to the actual length of the classroom. Is there a "collective wisdom," whereby the class mean is approximately equal to the actual room length?

2. *Out-of-class activity:* Using a wristwatch that is reasonably accurate, set the time to be exact. Use a radio station or telephone time report which states

that "at the tone, the time is . . . ." If you cannot set the time to the nearest second, record the error for the watch you are using. Now compare the time on
your watch to the time on others. Record the errors with positive signs for
watches that are ahead of the actual time and negative signs for those watches
that are behind the actual time. Use the data to test the claim that the mean
error of all wristwatches is equal to 0. Do we collectively run on time, or are
we early or late? Also test the claim that the standard deviation of errors is
less than 1 min. What are the practical implications of a standard deviation
that is excessively large?

3. *In-class activity:* In a group of three or four people, conduct an ESP experiment
by selecting one of the group members as the subject. Draw a circle on one
small piece of paper and draw a square on another sheet of the same size.
Repeat this experiment 20 times: Randomly select the circle or the square and
place it in the subject's hand behind his or her back so that it cannot be seen,
then ask the subject to identify the shape (without looking at it); record whether
the response is correct. Test the claim that the subject has ESP because the proportion of correct responses is greater than 0.5.

4. *In-class activity:* After dividing into groups with sizes between 10 and 20 people,
each group member should record the number of heartbeats in a minute. After
calculating $\overline{x}$ and s, each group should proceed to test the claim that the mean is
greater than 60, which is the author's result. (When people exercise, they tend to
have lower pulse rates, and the author runs five miles a few times each week.
What a guy.)

EXCEL PROJECT

Excel can be used to randomly generate data from a normally distributed population with a given mean and standard deviation.

a. Use Excel to randomly generate five values from a normally distributed population with a mean of 100 and a standard deviation of 15 (the parameters for
typical IQ tests). Here is the Excel procedure for generating five such values:
Click on **Tools,** select **Data Analysis,** then select **Random Number Generation**
and click **OK.** In the dialog box, enter 1 for the number of variables, enter 5 for
the number of random numbers, select the distribution option of **Normal,** enter
a mean of 100, enter a standard deviation of 15, then click **OK.**

b. Using the five sample values generated in part (a), test the claim that the sample
is from a population with a mean equal to 100. Use a 0.10 level of significance.
Here is the Excel procedure for testing that claim: Click **DDXL** and select
Hypothesis Tests. Select the item of **1 Var t Test.** Click on the pencil icon and
enter the range of cells containing the generated sample data. For example,
enter A1:A5 for five values in rows 1 through 5 of column A. Click **OK.** In the
dialog box, click on the bar in Step 1 and proceed to enter a claimed mean of
100. Proceed to click on the bars in the remaining steps. After clicking **Compute,**
record the conclusion (reject the null hypothesis or fail to reject the null
hypothesis).

c. Repeat parts (a) and (b) nine more times, so that a total of 10 different samples
have been generated, and 10 different hypothesis tests have been executed.

d. With a 0.10 significance level, there is 0.10 probability of making a type I error (rejecting the null hypothesis when it is true). Because of the way that the sample data are generated, we know that 100 is the true population mean, so we do make a type I error in this experiment anytime that we reject the null hypothesis of $\mu = 100$. For the 10 trials of this experiment, how often was the null hypothesis actually rejected? When we conduct 10 such trials, how many times do we expect to reject the null hypothesis? Are the actual results consistent with the theoretical results? Explain.

from **DATA** *to* **DECISION**

Critical Thinking: Should this drug be approved?

Standard methods of hypothesis testing are routinely used to decide whether newly developed drugs are actually effective. Pharmaceutical companies routinely research and develop drugs that accomplish objectives such as lowering blood pressure, lowering cholesterol levels, and growing hair on the bald pates of men. Men aged 18–24 have a mean cholesterol level of 178.1 (based on data from the National Health Survey), and one criterion for identifying risk of coronary disease is a cholesterol level above 300.

Analyzing the results

The accompanying list shows the serum cholesterol levels (in mg/100 mL) of randomly selected men aged 18–24 who took the drug Sinatrastatin.

a. Use the given data to test the claim that this treatment group has a mean level less than 178.1.

b. Based on the result, does it appear that the drug is effective in lowering cholesterol?

c. This experiment involves treating randomly selected subjects, then comparing the sample mean to the population mean. Is there a more effective way that the experiment could be designed? If so, describe the better design.

131	159	174	227	232	204
111	194	159	148	233	112
169	111	168	173	180	257
138	192	227	147	153	178
164	150	147	151	206	213
225	189	158	123	198	215

Internet Project

Hypothesis Testing

This chapter discussed the important method of inferential statistics known as hypothesis testing. This Internet Project will have you conduct tests using a variety of data sets. For each data set, you will be asked to

- Collect data available on the Internet.
- Formulate a hypothesis based on a given question.
- Conduct the test at a specified level of significance.

- Summarize your conclusions

Go to the Web site for this text:

http://www.awlonline.com/triola

There you will find instructions outlining tasks using data from business and economics, education, and sports, as well as a classic example from physical sciences.

Statistics
at work

"If I didn't have any background in statistics, I would not be able to fully understand the data my company produces… help protect our workers and customers."

Jeffrey Foy

Jeffrey Foy is a toxicologist working for the Cabot Corporation, a chemical company.

Jeffrey Foy is also responsible for the hazard evaluation of the chemicals Cabot Corporation produces. It is his job to understand how the company's products may affect humans or the environment and help decide on the best ways to protect both.

What do you do in your job?

My responsibilities include arranging and evaluating toxicological studies, writing material safety data sheets, and helping our research and development groups produce materials that are safe for both people and the environment or to understand what potential hazards the materials might have.

What concepts of statistics do you use?

The primary concept I use is hypothesis testing (probability testing).

How do you use statistics on the job?

I use statistics daily. Statistical methods have been and are used in two ways in my work. First, statistics is used to help determine how I design my experiments. Second, statistics is used to determine if the data generated are significant or sometimes if they're even good enough to use. If you start an experiment without thinking of how you are going to use or evaluate the data, you may not sufficiently account for all the endpoints you need to examine. If you start your work and realize that you didn't include enough subjects, time points, or dose ranges, then you may have just wasted time. If you're lucky, then that's all you've wasted. Studies that I am involved in can cost from as little as $1,000, to as much as

$500,000 or more, and if you don't properly determine how you are going to evaluate the data, you could cost your company a great deal of time and money. If the experiment is done properly, then we move on to analyze the data. The data from the studies we perform are used to assess any potential health effects our products may have on our workers, customers or the environment. The results are used to determine how chemicals can be sold or handled. When performing experiments at a testing laboratory or drug company, you want to determine if your materials have an effect, whether desired (a drug curing a disease) or undesired (that same drug being toxic). Statistics plays an enormous role in our evaluation of the significance of the effects.

Please describe one specific example illustrating how the use of statistics was successful in improving a product or service.

A toxicology study costing about $300,000 dollars was recently conducted. The data from the study were to be used to help determine if a particular chemical caused any effects in the subjects studied. After the study was performed, flaws were found in both the data and statistics used. It took an additional 2 years to properly review the data and finish the health evaluation. If the proper methods and endpoints had been chosen, then the additional time and money may not have been necessary. It was the understanding of the data and correct statistical evaluation that helped prevent the failure and potential repeat of the study.

How do you ensure objectivity in your studies?

Objectivity is attained by having an external source do our studies and perform the initial statistical analysis of the data.

How beneficial is your knowledge of statistics in your job?

If I didn't have any background in statistics, I would not be able to fully understand the data my company produces. I would not be able to use the data to help protect our workers and customers.

In terms of statistics, what would you recommend for prospective employees?

If you are doing scientific research on a daily basis, then you must thoroughly understand which methods of statistics are needed to evaluate your data. Not all data are evaluated the same way, and a researcher needs to know the correct ways of testing his or her hypothesis. It could make the difference between the commercial success or failure of a product. If you are not doing as much with data evaluation, then you should be able to look at data and statistics com-

piled by others and be able to evaluate their results.

Do you recommend statistics for today's college student? Why?

Statistics is essential.

Which other skills are important for today's college students?

Public speaking and a working knowledge of a second language (2 years of study), since so many companies are now global enterprises.

Inferences from Two Samples

Do male statistics students exaggerate their heights?

Male statistics students were given a survey that included a question asking them to report their height in inches. They weren't told that their height would be measured, but heights were accurately measured after the survey was completed. Anonymity was maintained, with code numbers used instead of names, so that no personal information would be publicly announced and nobody would be embarrassed by the results. Sample data are listed in Table 8-1. Is there a significant difference between the reported heights and the measured heights?

TABLE 8-1	Reported and Measured Heights in inches of Male Statistics Students											
Student	A	B	C	D	E	F	G	H	I	J	K	L
Reported height	68	74	82.25	66.5	69	68	71	70	70	67	68	70
Measured height	66.8	73.9	74.3	66.1	67.2	67.9	69.4	69.9	68.6	67.9	67.6	68.8
Difference	1.2	0.1	**7.95**	0.4	1.8	0.1	1.6	0.1	1.4	**−0.9**	0.4	1.2

8-1 Overview

Chapter 6 introduced an important activity of inferential statistics: Sample data were used to estimate values of population parameters. Chapter 7 introduced a second important activity of inferential statistics: Sample data were used to test hypotheses about population parameters. In both of those chapters, all examples and exercises involved the use of *one* sample to form an inference about *one* population. In reality, however, there are many important and meaningful situations in which it becomes necessary to compare *two* sets of sample data. The following are examples typical of those found in this chapter, which presents methods for using data from two samples so that inferences can be made about the populations from which they came.

- When testing the effectiveness of a drug, determine whether there is a significant difference between a group of subjects who were treated with the drug and a second group of subjects who were given a placebo.

- When surveying car drivers, determine whether the percentage of women who use cell phones is different from the percentage of men who use cell phones.

8-2 Inferences about Two Means: Independent and Large Samples

In this section we consider methods for using sample data from two independent and large ($n > 30$) samples to test hypotheses made about two population means or to construct confidence interval estimates of the difference between two population means. We begin by formally defining *independent* and *dependent*.

> ### DEFINITIONS
>
> Two samples are **independent** if the sample values selected from one population are not related to or somehow paired with the sample values selected from the other population. If the values in one sample are related to the values in the other sample, the samples are **dependent**. Such samples are often referred to as **matched pairs**, or **paired samples**. (We will use *matched pairs*.)

▷ **EXAMPLE**   **Drug Testing**

Independent samples: One group of subjects is treated with a drug while a separate group of subjects is given a placebo. These two sample groups are independent because there is no direct relationship between the treatment group and the placebo group.

Matched pairs: Blood pressure levels of subjects are measured before and after taking a drug. Each "before" value is matched with the "after" value because each before/after pair of measurements comes from the same person. ◀

For the hypothesis tests and confidence intervals described in this section, we make the following assumptions.

Assumptions

1. The two samples are *independent.*
2. The two sample sizes are *large.* That is, $n_1 > 30$ and $n_2 > 30$.
3. Both samples are *simple random samples.*

Exploring the Data Sets

Let's suppose that you want to investigate the difference between two simple random samples selected from two populations. Instead of immediately running a hypothesis test or constructing a confidence interval, you might first consider *exploring* the two samples using the methods described in Chapter 2. It could be very helpful to do the following:

- Find descriptive statistics for both data sets, including n, $\bar{x}$, and s.
- Create boxplots of both data sets, drawn on the same scale so that they can be compared.
- Create histograms of both data sets, so that their distributions can be compared.
- Identify any outliers.

Hypothesis Tests

In testing a claim about the difference $\mu_1 - \mu_2$, we use the following test statistic.

Test Statistic for Two Means: Independent and Large Samples

$$z = \frac{(\bar{x}_1 - \bar{x}_2) - (\mu_1 - \mu_2)}{\sqrt{\dfrac{\sigma_1^2}{n_1} + \dfrac{\sigma_2^2}{n_2}}}$$

σ_1 and σ_2:   If σ_1 and σ_2, are not known, use s_1 and s_2 in their places, provided that both samples are large.

P-value: Use the computed value of the test statistic z, and find the P-value by following the procedure summarized in Figure 7-8.

Critical values: Based on the significance level α, find critical values by using the procedures introduced in Section 7-2.

Except for the form of this test statistic, the method of hypothesis testing is essentially the same as that described in Chapter 7. (Later in this section, we will discuss the rationale for the form of the given test statistic.) As in Chapter 7, if the values of σ_1 and σ_2 are not known, we can use s_1 and s_2 in their places, provided that both samples are large. If σ_1 and σ_2 are known, we use those values in calculating the test statistic, but in reality they are rarely known.

▶ **EXAMPLE**   **Coke versus Pepsi** Data Set 1 in Appendix B includes the weights (in pounds) of samples of regular Coke and regular Pepsi. Sample statistics and boxplots are shown below. Use a 0.01 significance level to test the claim that the mean weight of regular Coke is different from the mean weight of regular Pepsi.

	Regular Coke	Regular Pepsi
n	36	36
$\overline{x}$	0.81682	0.82410
s	0.007507	0.005701

SOLUTION   In the spirit of exploring the two data sets, we see that the means do not appear to be dramatically different, but the boxplots seem to display a difference. Let's proceed with a formal hypothesis test to determine whether the difference between the two sample means is really significant. (The outlier shown in the boxplot does not have a substantial effect on these results.)

Step 1:   The claim of different means can be expressed symbolically as $\mu_1 \neq \mu_2$.

Step 2:   If the original claim is false, then $\mu_1 = \mu_2$.

Step 3:   The null hypothesis must contain equality, so we have

$$H_0: \mu_1 = \mu_2 \qquad H_1: \mu_1 \neq \mu_2 \text{ (original claim)}$$

We now proceed with the assumption that $\mu_1 = \mu_2$, or $\mu_1 - \mu_2 = 0$.

Step 4:   The significance level is $\alpha = 0.01$.

Step 5:   Because we have two independent and large samples and we are testing a claim about the two population means, we use a normal distribution with the test statistic given earlier in this section.

Step 6:   The values of σ_1 and σ_2 are unknown, but the samples are both large, so we can use the sample standard deviations as esti-

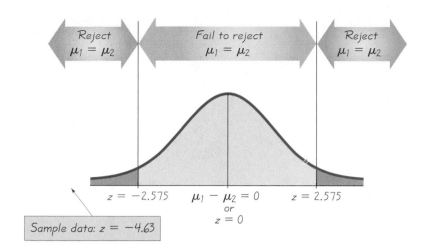

FIGURE 8-1
**Distribution of Differences
Between Means of Weights of
Regular Coke and Regular
Pepsi**

mates of the population standard deviations, and the test statistic is calculated as follows:

$$z = \frac{(\bar{x}_1 - \bar{x}_2) - (\mu_1 - \mu_2)}{\sqrt{\dfrac{\sigma_1^2}{n_1} + \dfrac{\sigma_2^2}{n_2}}} = \frac{(0.81682 - 0.82410) - 0}{\sqrt{\dfrac{0.007507^2}{36} + \dfrac{0.005701^2}{36}}} = -4.63$$

Because we are using a normal distribution, the critical values of $z = \pm 2.575$ are found from Table A-2. (With $\alpha = 0.01$ divided equally between the two tails, find the z score corresponding to an area of $0.5 - 0.01 \div 2$, or 0.4950.) The P-value is twice the area to the left of the test statistic (see Figure 7-8); using Table A-2, the P-value is 0.0002. (Using more accurate software, the P-value is smaller.) The test statistic, critical value, and critical region are shown in Figure 8-1.

Step 7:   Because the test statistic falls within the critical region, reject the null hypothesis $\mu_1 = \mu_2$ (or $\mu_1 - \mu_2 = 0$).

INTERPRETATION   There is sufficient evidence to support the claim that there is a difference between the mean weight of Coke and the mean weight of Pepsi. The *practical* significance of that difference is another issue not addressed by this hypothesis test, but the magnitude of the difference does not appear to be anything that a consumer would ever notice. Also, note that we are dealing with weights, not volumes, so the apparent lower mean for Coke simply indicates that the ingredients weigh less; this does not suggest that Coke is providing less of its drink than Pepsi.   ◀

Confidence Intervals

We can construct confidence interval estimates of the difference between two population means, which we denote as $\mu_1 - \mu_2$.

Commercials

Television networks have their own clearance departments for screening commercials and verifying claims. The National Advertising Division, a branch of the Council of Better Business Bureaus, investigates advertising claims. The Federal Trade Commission and local district attorneys also become involved. In the past, Firestone had to drop a claim that its tires resulted in 25% faster stops, and Warner Lambert had to spend $10 million informing customers that Listerine doesn't prevent or cure colds. Many deceptive ads are voluntarily dropped, and many others escape scrutiny simply because the regulatory mechanisms can't keep up with the flood of commercials.

> **Confidence Interval Estimate of $\mu_1 - \mu_2$ (for Independent and Large Samples)**
>
> The confidence interval estimate of the difference $\mu_1 - \mu_2$ is
>
> $$(\bar{x}_1 - \bar{x}_2) - E < (\mu_1 - \mu_2) < (\bar{x}_1 - \bar{x}_2) + E$$
>
> where
> $$E = z_{\alpha/2}\sqrt{\frac{\sigma_1^2}{n_1} + \frac{\sigma_2^2}{n_2}}$$

As with hypothesis testing, if σ_1 and σ_2 are not known, use s_1 and s_2 in their places, provided that both samples are large.

> **EXAMPLE**   **Coke and Pepsi** Using the sample data given in the preceding example, construct a 99% confidence interval estimate of the difference between the mean weights of regular Coke and regular Pepsi.
>
> **SOLUTION**   We first find the value of the margin of error E.
>
> $$E = z_{\alpha/2}\sqrt{\frac{\sigma_1^2}{n_1} + \frac{\sigma_2^2}{n_2}} = 2.575\sqrt{\frac{0.007507^2}{36} + \frac{0.005701^2}{36}} = 0.004045$$
>
> We now find the desired confidence interval as follows:
>
> $$(\bar{x}_1 - \bar{x}_2) - E < (\mu_1 - \mu_2) < (\bar{x}_1 - \bar{x}_2) + E$$
> $$(0.81682 - 0.82410) - 0.004045 < (\mu_1 - \mu_2) < (0.81682 - 0.82410) + 0.004045$$
> $$-0.01133 < (\mu_1 - \mu_2) < -0.00324$$
>
> **INTERPRETATION**   We are 99% confident that the limits of -0.01133 and -0.00324 actually do contain the difference between the two population means. This result could be more clearly presented by stating that μ_2 exceeds μ_1 by an amount that is between 0.00324 lb and 0.01133 lb. Because those limits do not contain 0, this confidence interval suggests that it is very unlikely that the two population means are equal.

Rationale: Why Do the Test Statistic and Confidence Interval Have the Particular Forms We Have Presented? Recall from the central limit theorem that sample means tend to be normally distributed. When applied to differences between sample means, the central limit theorem leads to the following key properties of the values that are the differences $\bar{x}_1 - \bar{x}_2$:

- The *differences* between sample means $(\bar{x}_1 - \bar{x}_2)$ tend to be normally distributed.
- The population mean of the sample $\bar{x}_1 - \bar{x}_2$ values is equal to $\mu_1 - \mu_2$.
- The population standard deviation of the $\bar{x}_1 - \bar{x}_2$ values is $\sqrt{\sigma_1^2/n_1 + \sigma_2^2/n_2}$.

To clarify the third statement above: The variance of the *differences* between two independent random variables equals the variance of the first random variable *plus* the variance of the second random variable. That is, the variance of sample values $\bar{x}_1 - \bar{x}_2$ will tend to equal $\sigma^2_{\bar{x}_1} + \sigma^2_{\bar{x}_2}$, provided that $\bar{x}_1$ and $\bar{x}_2$ are independent. (See Exercise 23.) For large simple random samples, the standard deviation of sample means is $\sigma/\sqrt{n}$, so the variance of sample means is σ^2/n. If we use this expression, σ^2/n, for the variance of sample means and use the property that the variance of the *differences* between two independent random variables equals the variance of the first random variable *plus* the variance of the second random variable, we get the following:

$$\sigma^2_{\bar{x}_1 - \bar{x}_2} = \sigma^2_{\bar{x}_1} + \sigma^2_{\bar{x}_2} = \frac{\sigma^2_1}{n_1} + \frac{\sigma^2_2}{n_2}$$

Taking the square root, we see that the standard deviation of the differences $\bar{x}_1 - \bar{x}_2$ can be expressed as follows:

$$\text{standard deviation of the } \bar{x}_1 - \bar{x}_2 \text{ sample values} = \sqrt{\frac{\sigma^2_1}{n_1} + \frac{\sigma^2_2}{n_2}}$$

Using the three properties listed above, and assuming that the samples are independent and large, the test statistic used in hypothesis tests has the same basic format we have already used:

$$\frac{(\text{sample statistic}) - (\text{claimed population parameter})}{(\text{standard deviation of sample statistics})} = \frac{(\bar{x}_1 - \bar{x}_2) - (\mu_1 - \mu_2)}{\sqrt{\dfrac{\sigma^2_1}{n_1} + \dfrac{\sigma^2_2}{n_2}}}$$

The format of the confidence interval given in this section is really the same format introduced in Chapter 6, except that the sample mean $\bar{x}$ is replaced by $\bar{x}_1 - \bar{x}_2$, the population mean μ is replaced by $\mu_1 - \mu_2$, and the standard deviation $\sigma/\sqrt{n}$ is replaced by $\sqrt{\sigma^2_1/n_1 + \sigma^2_2/n_2}$.

Using Excel

Notes: 1.   The DDXL add-in requires that you have the original lists of sample values. If you have only summary statistics, use either one of the following two approaches:

    a.   Create lists of data by using the procedure given in the Chapter 5 Excel Project.

    b.   Calculate the value of the z test statistic as described in this section, then use that value in the Excel command **=NORMSDIST(z),** then find the *P*-value as follows:

        Left-tailed test: *P*-value = result of **NORMSDIST(z).**
        Right-tailed test: *P*-value = 1 – result of **NORMSDIST(z).**
        Two-tailed test: *P*-value = smaller of the following two values:

$$2 \times \text{result of } \textbf{NORMSDIST}(z)$$
$$2 \times [1 - \text{result of } \textbf{NORMSDIST}(z)]$$

  2.   The DDXL program uses a t distribution, so the results may differ slightly from those given in the text. For example, the confidence interval limits in the Coke and Pepsi example are found using a critical value of $z = 2.575$, and the resulting confidence interval is

The Placebo Effect

It has been a common belief that when patients are given a placebo (a treatment with no medicinal value), about one-third of them show some improvement. However, a more recent study of 6000 patients showed that for those with mild medical problems, the placebos seemed to result in improvement in about two-thirds of the cases. The placebo effect seems to be strongest when patients are very anxious and they like their physicians. Because it could cloud studies of new treatments, the placebo effect is minimized by using a *double-blind* experiment in which neither the patient nor the physician knows whether the treatment is a placebo or a real medicine.

9. Weights of Men in Different Age Brackets As part of the National Health Survey, data were collected on the weights of men in two different age brackets. For 804 men aged 25–34, the mean is 176 lb and the standard deviation is 35.0 lb. For 1657 men aged 65–74, the mean and standard deviation are 164 lb and 27.0 lb, respectively.

    a. Test the claim that the older men come from a population with a mean that is less than the mean for men in the 25–34 age bracket. Use a 0.01 significance level.

    b. Construct a 99% confidence interval for the difference between the means of the men in the two age brackets. Do the confidence interval limits contain 0? Does this indicate that there is or is not a significant difference between the two means?

10. Seat Belts and Hospital Time A study of seat belt use involved children who were hospitalized as a result of motor vehicle crashes. For a group of 290 children who were not wearing seat belts, the numbers of days spent in intensive care units (ICUs) have a mean of 1.39 and a standard deviation of 3.06. For a group of 123 children who were wearing seat belts, the numbers of days in ICU have a mean of 0.83 and a standard deviation of 1.77 (based on data from "Morbidity Among Pediatric Motor Vehicle Crash Victims: The Effectiveness of Seat Belts," by Osberg and Di Scala, *American Journal of Public Health,* Vol. 82, No. 3). Using a 0.01 significance level, is there sufficient evidence to support the claim that the population of children not wearing seat belts has a higher mean number of days spent in ICU? Based on the result, is there significant evidence in favor of seat belt use among children?

Zinc Supplement Group	Placebo Group
$n_1 = 294$	$n_2 = 286$
$\bar{x}_1 = 3214$	$\bar{x}_2 = 3088$
$s_1 = 669$	$s_2 = 728$

11. Zinc for Mothers A study of zinc-deficient mothers was conducted to determine whether zinc supplementation during pregnancy results in babies with increased weights at birth. Sample data are listed in the margin (based on data from "The Effect of Zinc Supplementation on Pregnancy Outcome," by Goldenberg et al., *Journal of the American Medical Association,* Vol. 274, No. 6). The weights were measured in grams. Using a 0.05 significance level, is there sufficient evidence to support the claim that zinc supplementation does result in increased birth weights?

16–19 Years	65 Years and Older
$n_1 = 50$	$n_2 = 60$
$\bar{x}_1 = 58.2$	$\bar{x}_2 = 72.8$
$s_1 = 33.4$	$s_2 = 46.7$

12. Driving Times for Different Age Groups In a study of times (in minutes) that people drive each day, the sample results in the margin were obtained for men aged 16–19 and men aged 65 and older (based on data from the Federal Highway Administration).

    a. Is there sufficient evidence to support a claim that the two populations have different means?

    b. Construct a 99% confidence interval estimate of the difference between the means of the two populations. Do the confidence interval limits contain 0? How do you interpret the results, given that the confidence interval limits do or do not contain 0?

13. Textbook Prices Refer to Data Set 2 (TEXTBOOK.xls) in Appendix B. Is there sufficient evidence to reject the claim that the mean price of new textbooks at the University of Massachusetts is equal to the mean price of new textbooks at Dutchess Community College?

14. Means of Diet Coke and Diet Pepsi Refer to Data Set 1 (COLA.xls) in Appendix B. Is there sufficient evidence to reject a claim that the mean weight of cans of diet Coke is equal to the mean weight of cans of diet Pepsi?

EXCEL **15.** Effects of Second-hand Smoke Refer to Data Set 9 (COTININE.xls) in Appendix B. Test the hypothesis that the mean serum cotinine level of non-smokers who are not exposed to smoke (see the NOETS sample) is equal to the mean level of nonsmokers who are exposed to smoke (see the ETS sample).

EXCEL **16.** Male and Female Student Pulse Rates Refer to the pulse rates in Data Set 14 (STATSURV.xls) in Appendix B. First identify and delete the two out-liers that are unrealistic pulse rates of living and reasonably healthy statis-tics students.
  a. Use a 0.05 significance level and the traditional method of testing hypotheses to test the claim that male and female statistics students have the same mean pulse rate.
  b. Construct a 95% confidence interval for the difference $\mu_1 - \mu_2$, where μ_1 and μ_2 are the mean pulse rates for male and female statistics students, respectively.

EXCEL **17.** Treatment/Placebo: Interpreting Excel Display When testing for a difference between the means of a treatment group and a placebo group, the accompany-ing Excel display is obtained. Using a 0.05 significance level, is there sufficient evidence to support the claim that the treatment group (Variable 1) comes from a population with a mean that is less than the mean for the placebo population? Explain.

EXCEL

	A	B	C
1	z-Test: Two Sample for Means		
2		Variable 1	Variable 2
3	Mean	152.0739	154.9669
4	Known Variance	438.5388	239.1461
5	Observations	50	50
6	Hypothesized	0	
7	Mean Difference		
8	z	-0.7858	
9	P(Z<=z) one-tail	0.215991	
10	z Critical one-tail	1.644853	
11	P(Z<=z) two-tail	0.431982	
12	z Critical two-tail	1.959961	

EXCEL **18.** Treatment/Placebo: Interpreting Excel Display When testing for a difference between the means of a treatment group and a placebo group, the accompany-ing Excel display is obtained. Using a 0.02 significance level, is there sufficient evidence to support the claim that the treatment group (Variable 1) comes from a population with a mean that is different from the mean for the placebo population? Explain. (See the Excel display on the next page.)

EXCEL

	A	B	C
1	z-Test: Two Sample for Means		
2		Variable 1	Variable 2
3	Mean	75.34798	73.81763
4	Known Variance	12.55383	11.65292
5	Observations	50	50
6	Hypothesized Mean Difference	0	
7	z	2.199422	
8	P(Z<=z) one-tail	0.013924	
9	z Critical one-tail	1.644853	
10	P(Z<=z) two-tail	0.027848	
11	z Critical two-tail	1.959961	

EXCEL 19. **Heights of Men and Women: Interpreting Excel Display** As part of the National Health Survey, sample data were collected on the heights of men and women. When testing the claim that the mean height of men is greater than the mean height of women, the results from the Excel DDXL add-in are as shown. Is there sufficient evidence to support the claim? Explain.

EXCEL 20. **Interpreting Excel Display** If you are using the same sample data as in Exercise 19 in a test of the claim that the mean height of men is *equal* to the mean height of women, what is the *P*-value? What do you conclude?

8-2 Beyond the Basics

EXCEL 21. **Effects of an Outlier** Refer to Data Set 12 in Appendix B and determine whether there is sufficient evidence to support the claim that the cans 0.0111 in. thick have a mean axial load that is greater than the mean axial load for cans that are 0.0109 in. thick. The sample of cans 0.0111 in. thick includes an outlier of 504 lb. When investigating the difference between the two population means, how are the hypothesis test and the confidence interval affected if we decide that the outlier is actually an error that should be deleted?

EXCEL 22. **Effects of Units of Measurement** How are the results of Exercise 21 affected if the axial loads are all converted from pounds to kilograms? In general, does the choice of the scale affect the conclusions about equality of the two population means?

EXCEL 23. **a.** Find the variance for this *population* of *x* values: 5, 10, 15. (See Section 2-5 for the variance σ^2 of a population.)
b. Find the variance for this *population* of *y* values: 1, 2, 3.
c. List the *population* of all possible differences $x - y$, and find the variance of this population.

d. Use the results from parts (a), (b), and (c) to verify that $\sigma^2_{x-y} = \sigma^2_x + \sigma^2_y$. (This principle is used to derive the test statistic and confidence interval for this section.)

e. How is the *range* of the differences $x - y$ related to the range of the x values and the range of the y values?

24. Finding Type II Error Refer to Exercise 9. If the actual difference between the population means is 6.0, find β, the probability of a type II error. (See Exercise 31 in Section 7-3.) [*Hint:* In Step 1, replace $\overline{x}$ by $(\overline{x}_1 - \overline{x}_2)$, replace $\mu_{\overline{x}}$ by 0, and replace $\sigma_{\overline{x}}$ by $\sqrt{\sigma^2_1/n_1 + \sigma^2_2/n_2}$.]

8-3 Inferences about Two Means: Matched Pairs

In this section we develop methods for testing hypotheses and constructing confidence interval estimates of the differences between samples consisting of *matched pairs*. For example, the data in Table 8-1 (reproduced here) are matched pairs because each pair of reported/measured heights is obtained from the same person.

For the hypothesis tests and confidence intervals described in this section, we make the following assumptions.

Assumptions

1. The sample data consist of matched pairs.

2. The samples are simple random samples.

3. If the number of pairs of sample data is small ($n \leq 30$), then the population of differences in the paired values must be approximately *normally distributed*. (If there is a radical departure from a normal distribution, we should not use the methods given in this section, in which case we may be able to use nonparametric methods, discussed in Chapter 13.)

Exploring the Data Sets

As always, we suggest that you begin by exploring the data to see what might be learned. A simple examination of Table 8-1 should immediately reveal one notable characteristic: The third statistics student constitutes an

TABLE 8-1	Reported and Measured Heights (in inches) of Male Statistics Students											
Student	A	B	C	D	E	F	G	H	I	J	K	L
Reported height	68	74	82.25	66.5	69	68	71	70	70	67	68	70
Measured height	66.8	73.9	74.3	66.1	67.2	67.9	69.4	69.9	68.6	67.9	67.6	68.8
Difference	1.2	0.1	**7.95**	0.4	1.8	0.1	1.6	0.1	1.4	−0.9	0.4	1.2

outlier, because the difference between his reported height and his measured height is 7.95 in., a value very far away from all of the other differences. One possible explanation is that this student made an arithmetic error when he converted his height to inches. It is unlikely that he really believed that he was nearly 8 in. taller than he really is. Because this outlier appears to be an error, we will discard it, so that it will not have a dramatic effect on our analysis. We list this outlier along with some other notable characteristics.

- *Outlier:* The third student's reported height is an outlier, which will be discarded because it appears to be an error.
- *Pattern:* Only one student had a measured height greater than his reported height.
- *Rounding:* The reported heights are obviously rounded to the nearest inch in most of the sample values.
- *Mean:* After discarding the values for the third student, the mean of the differences is 0.67. (A mean of 0 would suggest agreement between the reported and measured heights.)
- *Distribution of differences:* See the Excel-generated histogram of the differences. The differences appear to be approximately normally distributed.

Hypothesis Tests

In dealing with sample data consisting of matched pairs, we base our calculations on the differences (d) between the pairs of data, because comparing two individual sample means would waste important information about the paired data. The following notation is based on the differences.

> **Notation for Matched Pairs**
>
> μ_d = mean value of the differences d for the *population* of paired data
>
> $\bar{d}$ = mean value of the differences d for the paired *sample* data (equal to the mean of the $x - y$ values)
>
> s_d = standard deviation of the differences d for the paired *sample* data
>
> n = number of *pairs* of data

Research in Twins

Identical twins occur when a single fertilized egg splits in two, so that both twins share the same genetic makeup. There is now an explosion in research focused on those twins. Speaking for the Center for Study of Multiple Birth, Louis Keith notes that now "we have far more ability to analyze the data on twins using computers with new, built-in statistical packages." A common goal of such studies is to explore the classic issue of "nature versus nurture." For example, Thomas Bouchard, who runs the Minnesota Study of Twins Reared Apart, has found that IQ is 50%-60% inherited, while the remainder is the result of external forces.

Identical twins are matched pairs that provide better results by allowing us to reduce the genetic variation that is inevitable with unrelated pairs of people.

We now use this notation to describe the test statistic to be used in hypothesis tests of claims made about the difference between the means of two populations, given that the sample data consist of matched pairs. If we have simple random paired sample data from populations in which the paired differences have a normal distribution, we use the sample differences and proceed with a single test. We use the following test statistic with a Student t distribution.

Test Statistic for Matched Pairs of Sample Data

$$t = \frac{\bar{d} - \mu_d}{\frac{s_d}{\sqrt{n}}}$$

where degrees of freedom $= n - 1$.

Critical values: If $n \leq 30$, critical values are found in Table A-3 (t distribution).

If $n > 30$, critical values are found in Table A-2 (normal distribution).

EXAMPLE **Do Male Statistics Students Exaggerate Their Heights?** Using the sample data in Table 8-1 with the outlier (third pair) deleted, test the claim that male statistics students do exaggerate by reporting heights that are greater than their actual measured heights. Use a 0.05 significance level.

SOLUTION We will follow the same basic method of hypothesis testing that was introduced in Chapter 7, but we will use the above test statistic.

Step 1: The claim that reported heights are greater than measured heights can be expressed in symbolic form as $\mu_d > 0$. (Refer to the format of Table 8-1 and note that the differences are values of "reported height minus measured height.")

Step 2: If the original claim is not true, we have $\mu_d \leq 0$.

Step 3: The null hypothesis must contain equality, so we have

$$H_0 : \mu_d \leq 0 \qquad H_1 : \mu_d > 0 \text{ (original claim)}$$

Step 4: The significance level is $\alpha = 0.05$.

Step 5: Because we are testing a claim about matched pairs of data, we use the Student t distribution.

Step 6: Before finding the value of the test statistic, we must first find the values of $\bar{d}$ and s_d. Refer to Table 8-1 and, excluding the outlier of 7.95, use the differences of 1.2, 0.1, 0.4, 1.8, 0.1, 1.6, 0.1, 1.4, -0.9, 0.4, and 1.2 to find these sample statistics:

$$\bar{d} = \frac{\Sigma d}{n} = \frac{7.4}{11} = 0.672727$$

$$s_d = \sqrt{\frac{n(\Sigma d^2) - (\Sigma d)^2}{n(n-1)}} = \sqrt{\frac{11(11.8) - (7.4)^2}{11(11-1)}} = 0.825943$$

Crest and Dependent Samples

In the late 1950s, Procter & Gamble introduced Crest toothpaste as the first such product with fluoride. To test the effectiveness of Crest in reducing cavities, researchers conducted experiments with several sets of twins. One of the twins in each set was given Crest with fluoride, while the other twin continued to use ordinary toothpaste without fluoride. It was believed that each pair of twins would have similar eating, brushing, and genetic characteristics. Results showed that the twins who used Crest had significantly fewer cavities than those who did not. This use of twins as dependent samples allowed the researchers to control many of the different variables affecting cavities.

FIGURE 8-2
Distribution of Differences Between Reported Heights and Measured Heights

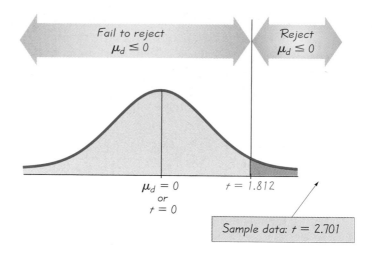

Using these sample statistics and the assumption of the hypothesis test that $\mu_d = 0$, we can now find the value of the test statistic.

$$t = \frac{\overline{d} - \mu_d}{\dfrac{s_d}{\sqrt{n}}} = \frac{0.672727 - 0}{\dfrac{0.825943}{\sqrt{11}}} = 2.701$$

The critical value of $t = 1.812$ is found from Table A-3; use the column for 0.05 (one tail), and use the row with degrees of freedom of $n - 1 = 10$. Figure 8-2 shows the test statistic, critical values, and critical region.

Step 7:   Because the test statistic does fall in the critical region, we reject the null hypothesis of $\mu_d \le 0$.

INTERPRETATION   There is sufficient evidence to support the claim that male statistics students exaggerate their heights by reporting values that are greater than their measured heights.   ◄

The preceding example used the traditional method, but the *P*-value approach could be used by modifying Steps 6 and 7. In Step 6, use the test statistic of $t = 2.701$ and refer to the 10th row of Table A-3 to find that the test statistic is between 2.764 and 2.228, indicating that the *P*-value is between the one-tailed values of 0.01 and 0.025. We again reject the null hypothesis, because the *P*-value is less than the significance level of $\alpha = 0.05$.

Confidence Intervals

We can develop a confidence interval estimate of the population mean difference μ_d by using the sample mean $\overline{d}$, the standard deviation of sample means d (which is $s_d/\sqrt{n}$), and the critical value $t_{a/2}$.

Confidence Intervals

The confidence interval estimate of the mean difference μ_d is

$$\overline{d} - E < \mu_d < \overline{d} + E$$

where $\qquad E = t_{\alpha/2}\dfrac{s_d}{\sqrt{n}}$ and   degrees of freedom $= n - 1$

EXAMPLE   How Much Do Male Statistics Students Exaggerate Their Heights? Using the sample data from Table 8-1 with the outlier excluded, construct a 95% confidence interval estimate of μ_d, which is the mean of the differences between reported heights and measured heights of male statistics students.

SOLUTION   We use the values of $\overline{d} = 0.672727$, $s_d = 0.825943$, $n = 11$, and $t_{\alpha/2} = 2.228$ (found from Table A-3 with 10 degrees of freedom and 0.05 in two tails). We first find the value of the margin of error E.

$$E = t_{\alpha/2}\frac{s_d}{\sqrt{n}} = (2.228)\left(\frac{0.825943}{\sqrt{11}}\right) = 0.554841$$

The confidence interval can now be found.

$$\overline{d} - E < \mu_d < \overline{d} + E$$
$$0.672727 - 0.554841 < \mu_d < 0.672727 + 0.554841$$
$$0.12 < \mu_d < 1.23$$

INTERPRETATION   The result is sometimes expressed as $\mu_d = 0.67 \pm 0.55$ or as (0.12, 1.23). In the long run, 95% of such samples will lead to confidence interval limits that actually do contain the true population mean of the differences. Note that the confidence interval limits do not contain 0, indicating that the true value of μ_d is significantly different from 0. That is, the mean value of the "reported − measured" differences is different from 0. On the basis of the confidence interval, we conclude that there is sufficient evidence to support the claim that there is a difference between the reported heights and the measured heights of male statistics students.

◀

Using Excel

Hypothesis Tests

You can use the DDXL add-in or the Excel Data Analysis add-in.

1. Begin by entering the values for the first sample in column A, then enter the matched values for the second sample in column B.

2. To use the Data Desk/XL add-in, click on **DDXL** and select **Hypothesis Tests,** then select **Paired t Test.** (You could also use Excel's Data Analysis add-in by selecting **Tools, Data Analysis,** then **t-Test Paired Two Sample for Means.**)

3. In the dialog box, click on the pencil icon for the first quantitative column, then enter the range of values for the first sample, such as A1:A11. Click on the pencil icon for the second quantitative column and enter the range of values for the second sample. Click **OK.**

4. Now complete the new dialog box by following the indicated steps. Click on **Compute** when finished.

Confidence Intervals

Excel alone does not have the ability of generate confidence intervals for the mean difference between matched pairs, so use the DDXL add-in as follows.

1. Begin by entering the values for the first sample in column A, then enter the matched values for the second sample in column B.

2. Click on **DDXL** and select **Confidence Intervals,** then select **Paired t Interval.**

3. In the dialog box, click on the pencil icon for the first quantitative column, then enter the range of values for the first sample, such as A1:A11. Click on the pencil icon for the second quantitative column and enter the range of values for the second sample. Click **OK.**

4. Now complete the new dialog box by following the indicated steps.

8-3 Basic Skills and Concepts

Calculations for Matched Pairs.   In Exercises 1 and 2, assume that you want to test the claim that the paired sample data come from a population for which the mean difference is $\mu_d = 0$. Assuming a 0.05 significance level, find (a) $\overline{d}$, (b) s_d, (c) the t test statistic, and (d) the critical values.

1.

x	8	8	6	9	7
y	3	2	6	4	9

2.

x	20	25	27	27	23	29	30	26
y	20	24	25	29	20	29	32	29

3. Using the sample paired data in Exercise 1, construct a 95% confidence interval for the population mean of all differences $x - y$.

4. Using the sample paired data in Exercise 2, construct a 99% confidence interval for the population mean of all differences $x - y$.

5. Reported and Measured Female Heights This section used reported heights and measured heights for a sample of *male* statistics students. Listed here are data collected from a sample of *female* statistics students.
    a. Is there sufficient evidence to support the claim that female statistics students exaggerate by reporting heights that are greater than their actual measured heights? Use a 0.05 significance level.

b. Construct a 95% confidence interval estimate of the mean difference between reported heights and measured heights. Interpret the resulting confidence interval, and comment on the implications of whether the confidence interval limits contain 0.

Reported height	64	63	64	65	64	64	63	59	66	64
Measured height	63.5	63.1	63.8	63.4	62.1	64.4	62.7	59.3	65.4	62.2

EXCEL **6.** Crash Test Repair Costs In low-speed crash tests of five BMW cars, the repair costs were computed for a factory-authorized repair center and an independent repair facility. The results are listed in the accompanying table.
   a. Is there sufficient evidence to support the claim that the independent center has lower repair costs? Use a 0.01 significance level.
   b. Construct a 99% confidence interval estimate of the mean difference between the repair costs of the factory-authorized repair center and the independent repair center. Do the confidence interval limits contain 0?
   c. Apart from any monetary difference in repair costs, what other key factor is critical in choosing between the authorized repair center and the independent repair center?

Authorized repair center	$797	$571	$904	$1147	$418
Independent repair center	$523	$488	$875	$911	$297

EXCEL **7.** Effectiveness of SAT Course Refer to the data in the table that lists SAT scores before and after the sample of 10 students took a preparatory course.
   a. Is there sufficient evidence to conclude that the preparatory course is effective in raising scores? Use a 0.05 significance level.
   b. Construct a 95% confidence interval estimate of the mean difference between the before and after scores. Write a statement that interprets the resulting confidence interval.

Student	A	B	C	D	E	F	G	H	I	J
SAT score before course (x)	700	840	830	860	840	690	830	1180	930	1070
SAT score after course (y)	720	840	820	900	870	700	800	1200	950	1080

Based on data from the College Board and "An Analysis of the Impact of Commercial Test Preparation Courses on SAT Scores," by Sesnowitz, Bernhardt, and Knain, *American Educational Research Journal*, Vol. 19, No. 3.

EXCEL **8.** Effectiveness of Drug Captopril is a drug designed to lower systolic blood pressure. When subjects were tested with this drug, their systolic blood pressure readings (in mm of mercury) were measured before and after the drug was taken, with the results given in the accompanying table.
   a. Use the sample data to construct a 99% confidence interval for the mean difference between the before and after readings.
   b. Is there sufficient evidence to support the claim that captopril is effective in lowering systolic blood pressure?

Subject	A	B	C	D	E	F	G	H	I	J	K	L
Before	200	174	198	170	179	182	193	209	185	155	169	210
After	191	170	177	167	159	151	176	183	159	145	146	177

Based on data from "Essential Hypertension: Effect of an Oral Inhibitor of Angiotensin-Converting Enzyme," by MacGregor et al., *British Medical Journal*, Vol. 2.

9. Effectiveness of Hypnotism in Reducing Pain A study was conducted to investigate the effectiveness of hypnotism in reducing pain. Results for randomly selected subjects are given in the accompanying table. The values are before and after hypnosis; the measurements are in centimeters on a pain scale.

    a. Construct a 95% confidence interval for the mean of the "before − after" differences.

    b. At the 0.05 significance level, test the claim that the sensory measurements are lower after hypnotism.

    c. Does hypnotism appear to be effective in reducing pain?

Subject	A	B	C	D	E	F	G	H
Before	6.6	6.5	9.0	10.3	11.3	8.1	6.3	11.6
After	6.8	2.4	7.4	8.5	8.1	6.1	3.4	2.0

Based on "An Analysis of Factors That Contribute to the Efficacy of Hypnotic Analgesia," by Price and Barber, *Journal of Abnormal Psychology,* Vol. 96, No. 1.

10. Measuring Intelligence in Children Mental measurements of young children are often made by giving them blocks and telling them to build a tower as tall as possible. One experiment of block building was repeated a month later, with the times (in seconds) listed in the accompanying table.

    a. Is there sufficient evidence to support the claim that there is a difference between the two times? Use a 0.01 significance level.

    b. Construct a 99% confidence interval for the mean of the differences. Do the confidence interval limits contain 0, indicating that there is not a significant difference between the times of the first and second trials?

Child	A	B	C	D	E	F	G	H	I	J	K	L	M	N	O
First trial	30	19	19	23	29	178	42	20	12	39	14	81	17	31	52
Second trial	30	6	14	8	14	52	14	22	17	8	11	30	14	17	15

Based on data from "Tower Building," by Johnson and Courtney, *Child Development,* Vol. 3.

11. Morning and Night Body Temperatures Refer to Data Set 6 in Appendix B. Use the paired data consisting of body temperatures of women at 8:00 A.M. and at 12:00 A.M. on Day 2.

    a. Construct a 95% confidence interval for the mean difference of 8 A.M. temperatures minus 12 A.M. temperatures.

    b. Using a 0.05 significance level, test the claim that for those temperatures, the mean difference is 0. Based on the result, do morning and night body temperatures appear to be about the same?

12. Qwerty/Dvorak Keyboards The following sentence, which contains every letter of the alphabet, was used for testing typewriters: "The quick brown fox jumped over the lazy red dog." If we use the word-rating procedure described in the Chapter Problem for Chapter 2, we find that the 10 words in that sentence have Qwerty ratings of 2, 5, 7, 3, 5, 5, 2, 3, 2, 1. The corresponding Dvorak ratings are 0, 5, 5, 3, 5, 3, 0, 4, 1, 1.

    a. Is there sufficient evidence to support the claim that the Dvorak keyboard has lower ratings?

    b. Construct a 95% confidence interval estimate of the difference. Does it contain 0, suggesting that there is not a significant difference between the Qwerty and Dvorak word ratings?

13. Treating Motion Sickness The following Excel display resulted from an experiment in which 10 subjects were tested for motion sickness before and after taking the drug astemizole. The sample values are the numbers of head movements that the subjects could endure without becoming nauseous.

a. Use a 0.05 significance level to test the claim that astemizole has an effect (for better or worse) on vulnerability to motion sickness. Based on the result, would you use astemizole if you were concerned about motion sickness while on a cruise ship?

b. Instead of testing for some effect (for better or worse), suppose we want to test the claim that astemizole is effective in *preventing* motion sickness? What is the *P*-value, and what do you conclude?

14. **Dieting: Interpreting Excel Display** Researchers obtained weight loss data from a sample of dieters using the New World Athletic Club facilities. The before/after weights are recorded. The Excel results are shown for a test of the claim that the diet is effective. Is there sufficient evidence to support the claim that the diet is effective? Explain.

8-3   Beyond the Basics

15. Effects of an Outlier In the examples in this section, we excluded the outlier in Table 8-1 and used the remaining 11 pairs of data. How are the results affected if that outlier is included? In general, can an outlier have a dramatic effect on the hypothesis test and confidence interval?

16. Effects of Units of Measurement The examples in this section used heights measured in inches. Suppose we convert all reported and measured heights from inches to centimeters. Is the hypothesis test affected by such a change in units? Is the confidence interval affected by such a change in units? How?

17. Confidence Intervals and One-Sided Tests The 95% confidence interval for a collection of paired sample data is $0.0 < \mu_d < 1.2$. Based on this confidence interval, the traditional method of hypothesis testing leads to the conclusion that the claim of $\mu_d > 0$ is supported. What is the smallest possible value of the significance level of the hypothesis test?

18. Using the Wrong Procedure Section 8-2 included an example of a hypothesis test that used the weights of 36 cans of regular Coke and 36 cans of regular Pepsi. The weights are listed in Data Set 1 (COLA.xls) in Appendix B. In Section 8-2 we tested the hypothesis of no difference between the two population means.

(continued)

Suppose we incorrectly believe that those weights are paired, so that the first regular Coke weight is paired with the first regular Pepsi weight, the second weights are paired, and so on. Compare the results obtained by using the methods of this section to the results found in Section 8-2. Are the results seriously affected if we choose the wrong procedure?

8-4 Inferences about Two Proportions

A strong argument could be made that this section is one of the most important sections in the book because this is where we describe methods for dealing with two sample proportions—a situation that is very common in real applications. Although this section is based on proportions, we can deal with percentages by using the corresponding decimal equivalents. For example, we might want to determine whether there is a difference between the percentage of adverse reactions in a placebo group and the percentage of adverse reactions in a drug treatment group. We can convert the percentages to their corresponding decimal values and proceed to use the methods of this section.

When testing a hypothesis made about two population proportions—such as the proportions of cured patients in a population given some treatment and a second population given a placebo—or when constructing a confidence interval for the difference between two population proportions, we make the following assumptions and use the following notation.

Assumptions

1. We have proportions from two *independent* simple random samples.
2. For both samples, the conditions $np \geq 5$ and $nq \geq 5$ are satisfied. That is, there are at least five successes and five failures in each of the two samples. (In many cases, we will test the claim that two populations have equal proportions so that $p_1 - p_2 = 0$. Because we assume that $p_1 - p_2 = 0$, it is not necessary to specify the particular value that p_1 and p_2 have in common. In such cases, the conditions $np \geq 5$ and $nq \geq 5$ can be checked by replacing p with the estimated pooled proportion $\overline{p}$, which will be described later.)

Notation for Two Proportions

For population 1 we let

$p_1 = population$ proportion

$n_1 = $ size of the sample

$x_1 = $ number of successes in the sample

$\hat{p}_1 = \dfrac{x_1}{n_1}$ (the *sample* proportion)

$\hat{q}_1 = 1 - \hat{p}_1$

The corresponding meanings are attached to p_2, n_2, x_2, $\hat{p}_2$, and $\hat{q}_2$, which come from population 2.

Finding the Numbers of Successes x_1 and x_2: The calculations for hypothesis tests and confidence intervals require that we have specific values for x_1, n_1, x_2, and n_2. Sometimes the available sample data include those specific numbers, but sometimes it is necessary to calculate the values of x_1 and x_2. The first example in this section reports that "when 734 men were treated with Viagra, 16% of them experienced headaches." From that statement we can see that $n_1 = 734$ and $\hat{p}_1 = 0.16$, but the actual number of successes x_1 is not given. However, from $\hat{p}_1 = x_1/n_1$, we know that

$$x_1 = n_1 \cdot \hat{p}_1$$

so that $x_1 = 734 \cdot 0.16 = 117.44$. But you cannot have 117.44 men who experienced headaches, because everyone either experiences a headache or does not, and the number of successes x_1 must therefore be a whole number. We can round 117.44 to 117. We can now use $x_1 = 117$ in the calculations that require its value. It's really quite simple: 16% of 734 means 0.16×734, which results in 117.44, which we round to 117.

Hypothesis Tests

In Section 7-5 we discussed tests of hypotheses made about a single population proportion. We will now consider tests of hypotheses made about two population proportions, but *we will be testing only claims that $p_1 = p_2$*, and we will use the following pooled (or combined) estimate of the value that p_1 and p_2 have in common. (For claims that the difference between p_1 and p_2 is equal to a nonzero constant, see Exercise 21 of this section.) You can see from the form of the pooled estimate $\bar{p}$ that it basically combines the two different samples into one big sample.

Pooled Estimate of p_1 and p_2

The **pooled estimate of p_1 and p_2** is denoted by $\bar{p}$ and is given by

$$\bar{p} = \frac{x_1 + x_2}{n_1 + n_2}$$

We denote the complement of $\bar{p}$ by $\bar{q}$, so $\bar{q} = 1 - \bar{p}$.

Test Statistic for Two Proportions

The following test statistic applies to null and alternative hypotheses that fit one of these three formats:

$$H_0: p_1 = p_2 \qquad H_0: p_1 \geq p_2 \qquad H_0: p_1 \leq p_2$$
$$H_1: p_1 \neq p_2 \qquad H_1: p_1 < p_2 \qquad H_1: p_1 > p_2$$

$$z = \frac{(\hat{p}_1 - \hat{p}_2) - (p_1 - p_2)}{\sqrt{\dfrac{\bar{p}\bar{q}}{n_1} + \dfrac{\bar{p}\bar{q}}{n_2}}}$$

(continued)

Polio Experiment

In 1954 an experiment was conducted to test the effectiveness of the Salk vaccine as protection against the devastating effects of polio. Approximately 200,000 children were injected with an ineffective salt solution, and 200,000 other children were injected with the vaccine. The experiment was "double blind" because the children being injected didn't know whether they were given the real vaccine or the placebo, and the doctors giving the injections and evaluating the results didn't know either. Only 33 of the 200,000 vaccinated children later developed paralytic polio, whereas 115 of the 200,000 injected with the salt solution later developed paralytic polio. Statistical analysis of these and other results led to the conclusion that the Salk vaccine was indeed effective against paralytic polio.

where $p_1 - p_2 = 0$ (assumed in the null hypothesis)

$$\hat{p}_1 = \frac{x_1}{n_1} \quad \text{and} \quad \hat{p}_2 = \frac{x_2}{n_2}$$

$$\overline{p} = \frac{x_1 + x_2}{n_1 + n_2}$$

$$\overline{q} = 1 - \overline{p}$$

P-value: Use the computed value of the test statistic z and find the P-value by following the procedure summarized in Figure 7-8.

Critical values: Based on the significance level α, find critical values by using the procedures introduced in Section 7-2.

Once again, the test statistic fits the common format of

$$\frac{(\text{sample statistic}) - (\text{claimed population parameter})}{(\text{standard deviation of sample statistics})}$$

The following example will illustrate one method used to compare a placebo group and a drug treatment group. It will also help clarify the roles of $x_1, n_1, \hat{p}_1, \overline{p}$, and so on. In particular, you should recognize that under the assumption of equal proportions, the best estimate of the common proportion is obtained by pooling both samples into one big sample, so that $\overline{p}$ becomes a more obvious estimate of the common population proportion.

▷ **EXAMPLE**   **Viagra Treatment and Placebo** The drug Viagra has become quite well known, and it has had a substantial economic impact on its producer, Pfizer Pharmaceuticals. In preliminary tests for adverse reactions, it was found that when 734 men were treated with Viagra, 16% of them experienced headaches. (There's some real irony there.) Among 725 men in a placebo group, 4% experienced headaches. These results were provided by Pfizer Pharmaceuticals. Is there sufficient evidence to support the claim that among those men who take Viagra, headaches occur at a rate that is greater than the rate for those who do not take Viagra? Use a 0.01 significance level.

SOLUTION   For notation purposes, we stipulate that sample 1 is the treatment group of men who took Viagra, and sample 2 is the group of men who took a placebo instead. Because 16% of the 734 men in the Viagra treatment group experienced headaches, the actual number of those men who experienced headaches is

16% of 734 = 0.16×734
                  = $117.44 = 117$ (rounded to a whole number)

A similar calculation shows that if 4% of the 725 men in the placebo group experienced headaches, then the actual number of those men who experienced headaches is 29. We can now summarize the sample data as follows.

Viagra Treatment Group	Placebo Group
$n_1 = 734$	$n_2 = 725$
$x_1 = 117$	$x_2 = 29$
$\hat{p}_1 = \dfrac{117}{734} = 0.159401$	$\hat{p}_2 = \dfrac{29}{725} = 0.04$

We will now use the traditional method of hypothesis testing, as summarized in Figure 7-5.

Step 1: The claim of a greater headache rate for Viagra users can be represented by $p_1 > p_2$.

Step 2: If $p_1 > p_2$ is false, then $p_1 \leq p_2$.

Step 3: Because our claim of $p_1 > p_2$ does not contain equality, it becomes the alternative hypothesis, and we have

$$H_0: p_1 \leq p_2 \qquad H_1: p_1 > p_2 \text{ (original claim)}$$

Step 4: The significance level is $\alpha = 0.01$.

Step 5: We will use the normal distribution (with the test statistic previously given) as an approximation to the binomial distribution. We have two independent samples, and the conditions $np \geq 5$ and $nq \geq 5$ are satisfied for each of the two samples. To check this, we note that in conducting this test, we assume that $p_1 = p_2$, where their common value is the pooled estimate $\overline{p}$, calculated as

$$\overline{p} = \frac{x_1 + x_2}{n_1 + n_2} = \frac{117 + 29}{734 + 725} = 0.100069$$

With $\overline{p} = 0.100069$, it follows that $\overline{q} = 1 - 0.100069 = 0.899931$. We verify that $np \geq 5$ and $nq \geq 5$ for both samples as follows:

Sample 1	Sample 2
$n_1 p = (734)(0.100069) = 73.45 \geq 5$	$n_2 p = (725)(0.100069) = 72.55 \geq 5$
$n_1 q = (734)(0.899931) = 660.55 \geq 5$	$n_2 q = (725)(0.899931) = 652.45 \geq 5$

Step 6: We can now find the value of the test statistic.

$$z = \frac{(\hat{p}_1 - \hat{p}_2) - (p_1 - p_2)}{\sqrt{\dfrac{\overline{p}\,\overline{q}}{n_1} + \dfrac{\overline{p}\,\overline{q}}{n_2}}}$$

$$= \frac{\left(\dfrac{117}{734} - \dfrac{29}{725}\right) - 0}{\sqrt{\dfrac{(0.100069)(0.899931)}{734} + \dfrac{(0.100069)(0.899931)}{725}}} = 7.60$$

The critical value of $z = 2.33$ is found by observing that we have a right-tailed test with $\alpha = 0.01$. The value of $z = 2.33$ is found from Table A-2 as the z score corresponding to an area of $0.5 - 0.01 = 0.4900$.

The Lead Margin of Error

Authors Stephen Ansolabehere and Thomas Belin wrote in their article "Poll Faulting" (*Chance* magazine) that "our greatest criticism of the reporting of poll results is with the margin of error of a single proportion (usually ±3%) when media attention is clearly drawn to the *lead* of one candidate." They point out that the lead is really the *difference* between two proportions $(p_1 - p_2)$ and go on to explain how they developed the following rule of thumb: The lead is approximately $\sqrt{3}$ times larger than the margin of error for any one proportion. For a typical pre-election poll, a reported ±3% margin of error translates to about ±5% for the lead of one candidate over the other. They write that the margin of error for the lead should be reported.

FIGURE 8-3
Distribution of Differences Between Proportions of Headaches in the Viagra Treatment Group and the Placebo Group

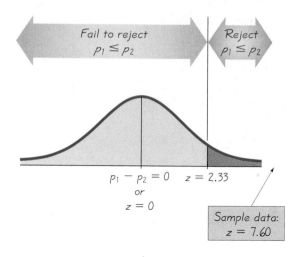

Step 7: In Figure 8-3 we see that the test statistic falls within the critical region, so we reject the null hypothesis of $p_1 \le p_2$.

INTERPRETATION We conclude that there is sufficient evidence to support the claim that the headache rate is greater for the men treated with Viagra than for the men who took a placebo.

These steps use the traditional approach to hypothesis testing, but it would be quite easy to use the *P*-value approach. In Step 6, instead of finding the critical value of *z*, we would find the *P*-value by using the procedure summarized in Figures 7-8 and 7-9. With a test statistic of $z = 7.60$ and a right-tailed test, we get

$$P\text{-value} = (\text{area to the right of } z = 7.60) = 0.0001$$

The *P*-value of 0.0001 comes from Table A-2, where all *z* scores above 3.09 yield a right-tailed area of 0.0001. If we use Excel we get a *P*-value less than 0.001. Again, we reject the null hypothesis because the *P*-value is less than the significance level of $\alpha = 0.01$. Again, we conclude that there is sufficient evidence to support the claim that the headache rate is greater for the men treated with Viagra than for the men who took a placebo ◀

Confidence Intervals

In the preceding example we found that the 16% rate of headaches among men who take Viagra is significantly greater than the 4% rate for men who do not. If we compute the difference between the two rates, we get 12%, which is a point estimate of the difference between the two population rates. But, with the same assumptions given at the beginning of this section, we can construct a confidence interval estimate of the difference between population proportions $(p_1 - p_2)$ by evaluating the following.

Confidence Interval Estimate of $p_1 - p_2$

The confidence interval estimate of the difference $p_1 - p_2$ is:

$$(\hat{p}_1 - \hat{p}_2) - E < (p_1 - p_2) < (\hat{p}_1 - \hat{p}_2) + E$$

where

$$E = z_{\alpha/2}\sqrt{\frac{\hat{p}_1\hat{q}_1}{n_1} + \frac{\hat{p}_2\hat{q}_2}{n_2}}$$

> **EXAMPLE** **Viagra Treatment and Placebo** Use the sample data given in the preceding example to construct a 99% confidence interval for the difference between the two population proportions.

SOLUTION With a 99% degree of confidence, $z_{\alpha/2} = 2.575$ (from Table A-2). We first calculate the value of the margin of error E as shown.

$$E = z_{\alpha/2}\sqrt{\frac{\hat{p}_1\hat{q}_1}{n_1} + \frac{\hat{p}_2\hat{q}_2}{n_2}} = 2.575\sqrt{\frac{\left(\frac{117}{734}\right)\left(\frac{617}{734}\right)}{734} + \frac{\left(\frac{29}{725}\right)\left(\frac{696}{725}\right)}{725}}$$

$$= 0.0395$$

With $\hat{p}_1 = 117/734 = 0.1594$, $\hat{p}_2 = 29/725 = 0.0400$, and $E = 0.0395$, the confidence interval is evaluated as follows.

$$(\hat{p}_1 - \hat{p}_2) - E < (p_1 - p_2) < (\hat{p}_1 - \hat{p}_2) + E$$
$$(0.1594 - 0.0400) - 0.0395 < (p_1 - p_2) < (0.1594 - 0.0400) + 0.0395$$
$$0.0799 < (p_1 - p_2) < 0.1589$$

INTERPRETATION The confidence interval limits do not contain 0, suggesting that there is a significant difference between the two proportions. As always, we should be careful when interpreting confidence intervals. Because p_1 and p_2 have fixed values and are not variables, it is wrong to state that there is a 99% chance that the value of $p_1 - p_2$ falls between 0.0799 and 0.1589. Instead, we should say that if we were to repeat the same sampling process and construct 99% confidence intervals, in the long run 99% of the intervals will actually contain the value of $p_1 - p_2$. ◀

Rationale: Why Do the Procedures of This Section Work? The test statistic given for hypothesis tests is justified by the following:

1. With $n_1 p_1 \geq 5$ and $n_1 q_1 \geq 5$, the distribution of $\hat{p}_1$ can be approximated by a normal distribution with mean p_1, standard deviation $\sqrt{p_1 q_1 / n_1}$, and variance $p_1 q_1 / n_1$. These conclusions are based on Sections 5-6 and 6-5, and they also apply to the second sample.

Drug Screening: False Positives

The American Management Association reports that more than half of all companies test for drugs. For a job applicant undergoing drug screening, a *false positive* is an indication of drug use when he or she does not use drugs. A *false negative* is an indication of no drug use when he or she is a user. The *test sensitivity* is the probability of a positive indication for a drug user; the *test specificity* is the probability of a negative indication for a nonuser. Suppose that 3% of job applicants use drugs, and a test has a sensitivity of 0.99 and a specificity of 0.98. The high probabilities make the test seem reliable, but 40% of the positive indications will be false positives.

Does Aspirin Help Prevent Heart Attacks?

In a recent study of 22,000 male physicians, half were given regular doses of aspirin while the other half were given placebos. The study ran for six years at a cost of $4.4 million. Among those who took the aspirin, 104 suffered heart attacks. Among those who took the placebos, 189 suffered heart attacks. (The figures are based on data from *Time* and the *New England Journal of Medicine*, Vol. 318, No. 4.) This is a classic experiment involving a treatment group (those who took the aspirin) and a placebo group (those who took pills that looked and tasted like the aspirin pills, but no aspirin was present). We can use methods presented in this chapter to address the issue of whether the results show a statistically significant lower rate of heart attacks among the sample group who took aspirin. The issue is clearly important because it can affect many lives. This particular study was criticized because only male physicians were included, and the results might not apply to women.

2. Because $\hat{p}_1$ and $\hat{p}_2$ are each approximated by a normal distribution, $\hat{p}_1 - \hat{p}_2$ will also be approximated by a normal distribution with mean $p_1 - p_2$ and variance

$$\sigma^2_{(\hat{p}_1 - \hat{p}_2)} = \sigma^2_{\hat{p}_1} + \sigma^2_{\hat{p}_2} = \frac{p_1 q_1}{n_1} + \frac{p_2 q_2}{n_2}$$

(In Section 8-2 we established that the variance of the *difference* between two independent random variables is the *sum* of their individual variances.)

3. Because the values of p_1, p_2, q_1, and q_2 are typically unknown and from the null hypothesis we assume that $p_1 = p_2$, we can pool (or combine) the sample data. The pooled estimate of the common value of p_1 and p_2 is $\overline{p} = (x_1 + x_2)/(n_1 + n_2)$. If we replace p_1 and p_2 by $\overline{p}$ and replace q_1 and q_2 by $\overline{q} = 1 - \overline{p}$, the variance from Step 2 leads to the following standard deviation.

$$\sigma_{(\hat{p}_1 - \hat{p}_2)} = \sqrt{\frac{\overline{p}\,\overline{q}}{n_1} + \frac{\overline{p}\,\overline{q}}{n_2}}$$

4. We now know that the distribution of $p_1 - p_2$ is approximately normal, with mean $p_1 - p_2$ and standard deviation as shown in Step 3, so that the z test statistic has the form given earlier.

The form of the confidence interval requires an expression for the variance different from the one given in Step 3. In Step 3 we are assuming that $p_1 = p_2$, but if we don't make that assumption (as in the construction of a confidence interval), we estimate the variance of $\hat{p}_1 - \hat{p}_2$ as

$$\sigma^2_{(\hat{p}_1 - \hat{p}_2)} = \sigma^2_{\hat{p}_1} + \sigma^2_{\hat{p}_2} = \frac{\hat{p}_1 \hat{q}_1}{n_1} + \frac{\hat{p}_2 \hat{q}_2}{n_2}$$

and the standard deviation becomes

$$\sqrt{\frac{\hat{p}_1 \hat{q}_1}{n_1} + \frac{\hat{p}_2 \hat{q}_2}{n_2}}$$

In the test statistic

$$z = \frac{(\hat{p}_1 - \hat{p}_2) - (p_1 - p_2)}{\sqrt{\dfrac{\hat{p}_1 \hat{q}_1}{n_1} + \dfrac{\hat{p}_2 \hat{q}_2}{n_2}}}$$

let z be positive and negative (for two tails) and solve for $p_1 - p_2$. The results are the limits of the confidence interval given earlier.

Using Excel

You must use the Data Desk/XL add-in.

1. First make these entries:

    Cell A1: Enter the number of successes for sample 1.

    Cell B1: Enter the number of trials for sample 1.

    Cell C1: Enter the number of successes for sample 2.

    Cell D1: Enter the number of trials for sample 2.

2. Click on **DDXL.** Depending on whether you want to conduct a hypothesis test or construct a confidence interval, choose one of the following paths.

> **Hypothesis Tests, Summ 2 Var Prop Test**
> **Confidence Intervals, Summ 2 Var Prop Interval**

3. In the dialog box, click on the four pencil icons and enter A1, B1, C1, and D1 in the four input boxes. Click **OK.**

4. Proceed to complete the new dialog box.

8-4 Basic Skills and Concepts

Finding Number of Successes.   *In Exercises 1–4, find the number of successes x suggested by the given statement.*

1. From the *Uniform Crime Reports*: Of 2750 randomly selected arrests of criminals younger than 21 years of age, 4.25% involve violent crimes.

2. From the *New York Times*: Among 240 vinyl gloves subjected to stress tests, 63% leaked viruses.

3. From *Sociological Methods and Research*: When 294 central-city residents were surveyed, 28.9% refused to respond.

4. From *Science* magazine: Among 850 teens surveyed with an anonymous computer program, 12.4% said that they carried a gun within the last 30 days.

Calculations for Testing Claims.   *In Exercises 5 and 6, assume that you plan to use a significance level of $\alpha = 0.05$ to test the claim that $p_1 = p_2$. Use the given sample sizes and numbers of successes to find (a) the pooled estimate $\overline{p}$, (b) the z test statistic, (c) the critical z values, and (d) the P-value.*

5. Treatment Group   Placebo Group

$$n_1 = 20 \qquad n_2 = 25$$
$$x_1 = 10 \qquad x_2 = 15$$

6. Students   Faculty

$$n_1 = 150 \qquad n_2 = 50$$
$$x_1 = 30 \qquad x_2 = 20$$

7. Testing Effectiveness of Vaccine In a *USA Today* article about an experimental nasal spray vaccine for children, the following statement was presented: "In a trial involving 1602 children only 14 (1%) of the 1070 who received the vaccine developed the flu, compared with 95 (18%) of the 532 who got a placebo." The article also referred to a study claiming that the experimental nasal spray "cuts children's chances of getting the flu." Is there sufficient sample evidence to support the stated claim?

8. Color Blindness in Men and Women In a study of red/green color blindness, 500 men and 2100 women are randomly selected and tested. Among the men, 45 have red/green color blindness. Among the women, 6 have red/green color blindness (based on data from *USA Today*).
   a. Is there sufficient evidence to support the claim that men have a higher rate of red/green color blindness than women?
   b. Construct the 99% confidence interval for the difference between the color blindness rates of men and women. Do the confidence interval limits contain 0?
   c. Why would the sample size for women be so much larger than the sample size for men?

EXCEL **9.** Seat Belts and Hospital Time A study was made of 413 children who were hospitalized as a result of motor vehicle crashes. Among 290 children who were not using seat belts, 50 were injured severely. Among 123 children using seat belts, 16 were injured severely (based on data from "Morbidity Among Pediatric Motor Vehicle Crash Victims: The Effectiveness of Seat Belts," by Osberg and Di Scala, *American Journal of Public Health,* Vol. 82, No. 3). Is there sufficient sample evidence to conclude, at the 0.05 significance level, that the rate of severe injuries is lower for children wearing seat belts? Based on these results, what action should be taken?

EXCEL **10.** Interpreting a Computer Display A U.S. Department of Justice report (NCJ-156831) included the claim that "in spouse murder cases, wife defendants were less likely to be convicted than husband defendants." Sample data consisted of 277 convictions among 318 husband defendants, and 155 convictions among 222 wife defendants. Test the stated claim and identify one possible explanation for the result. The Excel results are shown here.

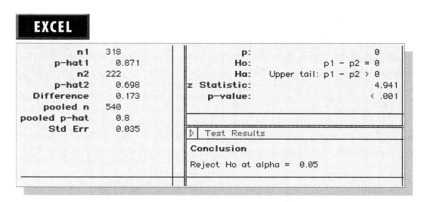

EXCEL **11.** Effectiveness of Salk Vaccine for Polio In initial tests of the Salk vaccine, 33 of 200,000 vaccinated children later developed polio. Of 200,000 children vaccinated with a placebo, 115 later developed polio. The Excel display is shown here. At the 0.01 significance level, test the claim that the Salk vaccine is effective in lowering the polio rate. Does it appear that the vaccine is effective?

EXCEL **12.** Airline Load Factor In a recent year, Southwest Airlines had 3,131,727 aircraft seats available on all of its flights, and 2,181,604 of them were occupied by passengers. America West had 2,091,859 seats available, and 1,448,255 of them were occupied. The percentage of seats occupied is called the *load factor,* so these results show that the load factor is 69.7% (rounded) for Southwest Airlines and 69.2% (rounded) for America West. (The data are from the U. S. Department of Transportation.) Answer the following by assuming that the results are from randomly selected samples.
a. Test the claim that both airlines have the same load factor.
b. Given that 69.7% and 69.2% appear to be so obviously close, how do you explain the results from part (a)?

c. Generalize the key point of this example by completing the following statement: "If two sample sizes are extremely large, even seemingly small differences in sample proportions . . . "

EXCEL **13.** Violent Crime and Age Group The newly appointed head of the state mental health agency claims that a smaller proportion of the crimes committed by persons younger than 21 years of age are violent crimes (when compared to the crimes committed by persons 21 years of age or older). Of 2750 randomly selected arrests of criminals younger than 21 years of age, 4.25% involve violent crimes. Of 2200 randomly selected arrests of criminals 21 years of age or older, 4.55% involve violent crimes (based on data from the Uniform Crime Reports). Construct a 95% confidence interval for the difference between the two proportions of violent crimes. Do the confidence interval limits contain 0? Does this indicate that there isn't a significant difference between the two rates of violent crimes?

EXCEL **14.** Testing Laboratory Gloves The *New York Times* ran an article about a study in which Professor Denise Korniewicz and other Johns Hopkins researchers subjected laboratory gloves to stress. Among 240 vinyl gloves, 63% leaked viruses. Among 240 latex gloves, 7% leaked viruses. At the 0.005 significance level, test the claim that vinyl gloves have a larger virus leak rate than latex gloves.

EXCEL **15.** Written Survey and Computer Survey In a study of 1700 teens aged 15–19, half were given written surveys and half were given surveys using an anonymous computer program. Among those given the written surveys, 7.9% say that they carried a gun within the last 30 days. Among those given the computer surveys, 12.4% say that they carried a gun within the last 30 days (based on data from the Urban Institute).

a. The sample percentages of 7.9% and 12.4% are obviously not equal, but is the difference significant? Explain.

b. Construct a 99% confidence interval estimate of the difference between the two population percentages, and interpret the result.

EXCEL **16.** Adverse Drug Reactions In clinical tests of adverse reactions to the drug Viagra, 4.0% of the 734 subjects in the treatment group experienced nasal congestion, but 2.1% of the 725 subjects in the placebo group experienced nasal congestion (based on data from Pfizer Pharmaceuticals).

a. Using a 0.01 significance level, is there sufficient evidence to support the claim that nasal congestion occurs at a higher rate among Viagra users than those who do not use Viagra?

b. Construct the 99% confidence interval estimate of the difference between the rate of nasal congestion among Viagra users and the nasal congestion rate for those who use a placebo. Do the confidence interval limits contain 0, and what does this suggest about the two rates?

EXCEL **17.** Drinking and Crime Karl Pearson, who developed many important concepts in statistics, collected crime data in 1909. Of those convicted of arson, 50 were drinkers and 43 abstained. Of those convicted of fraud, 63 were drinkers and 144 abstained. Use a 0.01 significance level to test the claim that the proportion of drinkers among convicted arsonists is greater than the proportion of drinkers convicted of fraud. Does it seem reasonable that drinking might have had an effect on the type of crime? Why?

EXCEL **18.** Poll Refusal Rate Professional pollsters are becoming concerned about the growing rate of refusals among potential survey subjects. In analyzing the problem, there is a need to know if the refusal rate is universal or if there is a difference

(continued)

between the rates for central-city residents and those not living in central cities. Specifically, it was found that when 294 central-city residents were surveyed, 28.9% refused to respond. A survey of 1015 residents not in a central city resulted in a 17.1% refusal rate (based on data from "I Hear You Knocking But You Can't Come In," by Fitzgerald and Fuller, *Sociological Methods and Research,* Vol. 11, No. 1). At the 0.01 significance level, test the claim that the central-city refusal rate is the same as the refusal rate in other areas.

19. Warmer Surgical Patients Recover Better? An article published in *USA Today* stated that "in a study of 200 colorectal surgery patients, 104 were kept warm with blankets and intravenous fluids; 96 were kept cool. The results show: Only 6 of those warmed developed wound infections vs. 18 who were kept cool." Use a 0.05 significance level to test the claim of the article's headline: "Warmer surgical patients recover better." If these results are verified, should surgical patients be routinely warmed?

20. Home Field Advantage When games were sampled from throughout a season, it was found that the home team won 127 of 198 professional *basketball* games, and the home team won 57 of 99 professional *football* games (based on data from "Predicting Professional Sports Game Outcomes from Intermediate Game Scores," by Cooper et al., *Chance,* Vol. 5, No. 3–4). Construct a 95% confidence interval for the difference between the proportions of home wins. Do the confidence interval limits contain 0? Based on the results, is there a significant difference between the proportions of home wins? What do you conclude about the home field advantage?

8-4 Beyond the Basics

21. Testing for Constant Difference To test the null hypothesis that the difference between two population proportions is equal to a nonzero constant c, use the test statistic

$$z = \frac{(\hat{p}_1 - \hat{p}_2) - c}{\sqrt{\dfrac{\hat{p}_1(1 - \hat{p}_1)}{n_1} + \dfrac{\hat{p}_2(1 - \hat{p}_2)}{n_2}}}$$

As long as n_1 and n_2 are both large, the sampling distribution of the test statistic z will be approximately the standard normal distribution. Refer to the sample data included with the example presented in this section, and use a 0.05 significance level to test the claim that the headache rate of Viagra users is 10 percentage points more than the percentage for those who use a placebo.

22. Transitivity of Hypothesis Tests Sample data are randomly drawn from three independent populations, each of size 100. The sample proportions are $\hat{p}_1 = 40/100$, $\hat{p}_2 = 30/100$, and $\hat{p}_3 = 20/100$.
   a.   At the 0.05 significance level, test H_0: $p_1 = p_2$.
   b.   At the 0.05 significance level, test H_0: $p_2 = p_3$.
   c.   At the 0.05 significance level, test H_0: $p_1 = p_3$.
   d.   In general, if hypothesis tests lead to the conclusions that $p_1 = p_2$ and $p_2 = p_3$ are reasonable, does it follow that $p_1 = p_3$ is also reasonable? Why or why not?

23. Determining Sample Size The sample size needed to estimate the difference between two population proportions to within a margin of error E with a confidence level of $1 - \alpha$ can be found as follows. In the expression

$$E = z_{\alpha/2} \sqrt{\frac{p_1 q_1}{n_1} + \frac{p_2 q_2}{n_2}}$$

replace n_1 and n_2 by n (assuming that both samples have the same size) and replace each of $p_1, q_1, p_2,$ and q_2 by 0.5 (because their values are not known). Then solve for n.

   Use this approach to find the size of each sample if you want to estimate the difference between the proportions of men and women who own cars. Assume that you want 95% confidence that your error is no more than 0.03.

24. Interpreting Drug Test Results Ziac is a Lederle Laboratories drug developed to treat hypertension. Lederle Laboratories reported that when 221 people were treated with Ziac, 3.2% of them experienced dizziness. It was also reported that among the 144 people in the placebo group, 1.8% experienced dizziness.
   a. Can you use the methods of this section to test the claim that there is a significant difference between the two rates of dizziness? Why or why not?
   b. Can the given information be correct? Why or why not?

8-5 Comparing Variation in Two Samples

Because the characteristic of variation among data is extremely important, this section presents a method for using two samples to compare the variances of the populations from which the samples are drawn. In Section 2-5 we saw that variation in a sample could be described by measures that include standard deviation and variance. Because standard deviation is a very effective measure of variation, and because it is easier to understand than variance, we have tended to stress the use of standard deviation instead of variance. Let's briefly review the relationship between standard deviation and variance by noting that the variance is the square of the standard deviation.

Measures of Variation	
s = standard deviation of *sample*	s^2 = variance of *sample* (sample standard deviation squared)
σ = standard deviation of *population*	σ^2 = variance of *population* (population standard deviation squared)

Although the basic procedure of this section is designed for variances, we can also use it for standard deviations. The method we use requires the following assumptions.

Assumptions

1. The two populations are *independent* of each other. (Recall from Section 8-2 that two samples are independent if the sample selected from one population is not related to the sample selected from the other population.)

Do Air Bags Save Lives?

The National Highway Transportation Safety Administration reported that for a recent year, 3,448 lives were saved because of air bags. It was reported that for car drivers involved in frontal crashes, the fatality rate was reduced 31%; for passengers, there was a 27% reduction. It was noted that "calculating lives saved is done with a mathematical analysis of the real-world fatality experience of vehicles with air bags compared with vehicles without air bags. These are called double-pair comparison studies, and are widely accepted methods of statistical analysis."

2. The two populations are each *normally distributed*. (This assumption is important because the methods of this section are extremely sensitive to departures from normality.)

Exploring the Data

Because the requirement of normal distributions is so important, we should begin by comparing the two sets of sample data by using tools such as histograms, boxplots, and normal quantile plots (see Section 5-7). We should search for outliers, and we should find the values of the sample statistics, especially the standard deviations. For example, consider the 36 weights of regular Coke in 36 different cans. Section 8-2 included a boxplot of that data set. Shown here are a histogram and a normal probability plot from Excel. The histogram shows that the data have a distribution that is approximately normal and that there is one value that is a potential outlier. The normal probability plot, which can be interpreted as if it were a normal quantile plot, shows that the points are reasonably close to a straight line, but they don't fit the straight line perfectly. This data set clearly satisfies a requirement of a distribution that is *approximately* normal, but it isn't so clear that this data set satisfies the stricter requirements of normality that apply to the methods of this section.

Hypothesis Tests

The computations of this section will be greatly simplified if we stipulate that s_1^2 represents the *larger* of the two sample variances. It doesn't really matter which sample is designated as sample 1, so we choose to use the following notation.

Notation for Hypothesis Tests with Two Variances or Standard Deviations

$s_1^2 = larger$ of the two sample variances

$n_1 = $ size of the sample with the *larger* variance

$\sigma_1^2 = $ variance of the population from which the sample with the *larger* variance was drawn

The symbols s_2^2, n_2, and σ_2^2 are used for the other sample and population.

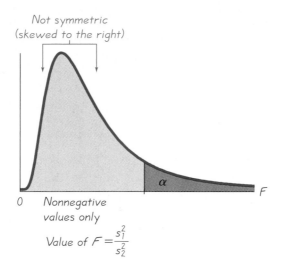

FIGURE 8-4
F Distribution

**There is a different F distri-
bution for each different
pair of degrees of freedom
for the numerator and the
denominator.**

For two normally distributed populations with equal variances (that is,
$\sigma_1^2 = \sigma_2^2$), the sampling distribution of the following test statistic is the **F distribution** shown in Figure 8-4 with critical values listed in Table A-5. If you
continue to repeat an experiment of randomly selecting samples from two
normally distributed populations with equal variances, the distribution of
the ratio s_1^2/s_2^2 of the sample variances is the *F* distribution.

In Figure 8-4, note these properties of the *F* distribution:

- The *F* distribution is not symmetric.
- Values of the *F* distribution cannot be negative.
- The exact shape of the *F* distribution depends on two different degrees
  of freedom.

Test Statistic for Hypothesis Tests with Two Variances

$$F = \frac{s_1^2}{s_2^2}$$

Critical values: Using Table A-5, we obtain critical *F* values that are determined by the following three values:

1. The significance level α (Table A-5 has six pages of critical values for
    $\alpha = 0.01, 0.025,$ and 0.05.)
2. **Numerator degrees of freedom** $= n_1 - 1$
3. **Denominator degrees of freedom** $= n_2 - 1$

To find a critical value, first refer to the part of Table A-5 corresponding
to α (for a one-tailed test) or $\alpha/2$ (for a two-tailed test), then intersect the
column representing the degrees of freedom for s_1^2 with the row representing the degrees of freedom for s_2^2. Because we are stipulating that the larger
sample variance is s_1^2, all one-tailed tests will be right-tailed and all two-

tailed tests will require that we find only the critical value located to the right. Good news: We have no need to find a critical value separating a left-tailed critical region. (Because the F distribution is not symmetric and has only nonnegative values, a left-tailed critical value cannot be found by using the negative of the right-tailed critical value; instead, a left-tailed critical value is found by using the reciprocal of the right-tailed value with the numbers of degrees of freedom reversed. See Exercise 14.)

We often have numbers of degrees of freedom that are not included in Table A-5. We could use linear interpolation to approximate the missing values, but in most cases that's not necessary because the F test statistic is either less than the lowest possible critical value or greater than the largest possible critical value. For example, Table A-5 shows that for $\alpha = 0.025$ in the right tail, 20 degrees of freedom for the numerator, and 34 degrees of freedom for the denominator, the critical F value is between 2.0677 and 2.1952. Any F test statistic below 2.0677 will result in failure to reject the null hypothesis, any F test statistic above 2.1952 will result in rejection of the null hypothesis, and interpolation is necessary only if the F test statistic happens to fall between 2.0677 and 2.1952.

Interpreting the F Test Statistic: If the two populations really do have equal variances, then s_1^2/s_2^2 tends to be close to 1 because s_1^2 and s_2^2 tend to be close in value. But if the two populations have radically different variances, s_1^2 and s_2^2 tend to be very different numbers. Denoting the larger of the sample variances by s_1^2, we see that the ratio s_1^2/s_2^2 will be a large number whenever s_1^2 and s_2^2 are far apart in value. Consequently, a value of F near 1 will be evidence in favor of the conclusion that $\sigma_1^2 = \sigma_2^2$, but a large value of F will be evidence against the conclusion of equality of the population variances.

Large **values of F are evidence against $\sigma_1^2 = \sigma_2^2$.**

Claims about Standard Deviations: The F test statistic applies to a claim made about two variances, but we can also use it for claims about two population standard deviations. Any claim about two population standard deviations can be restated in terms of the corresponding variances.

▷ **EXAMPLE** **Coke versus Pepsi** Data Set 1 in Appendix B includes the weights (in pounds) of samples of regular Coke and regular Pepsi. The sample statistics are summarized in the accompanying table. Use a 0.05 significance level to test the claim that the weights of regular Coke and the weights of regular Pepsi have the same standard deviation.

	Regular Coke	Regular Pepsi
n	36	36
$\bar{x}$	0.81682	0.82410
s	0.007507	0.005701

SOLUTION Instead of using the sample standard deviations to test the claim of equal standard deviations, we will use the sample variances to test the claim of equal variances. Because we stipulate in this section that the larger variance is denoted by s_1^2, we let $s_1^2 = 0.007507^2$, $n_1 = 36$, $s_2^2 = 0.005701^2$, and $n_2 = 36$. We now proceed to use the traditional method of testing hypotheses as outlined in Figure 7-5.

Step 1:   The claim of equal standard deviations is equivalent to a claim of equal variances, which we express symbolically as $\sigma_1^2 = \sigma_2^2$.

Step 2:   If the original claim is false, then $\sigma_1^2 \neq \sigma_2^2$.

Step 3:   Because the null hypothesis must contain equality, we have

$$H_0: \sigma_1^2 = \sigma_2^2 \text{ (original claim)} \qquad H_1: \sigma_1^2 \neq \sigma_2^2$$

Step 4:   The significance level is $\alpha = 0.05$.

Step 5:   Because this test involves two population variances, we use the F distribution.

Step 6:   The test statistic is

$$F = \frac{s_1^2}{s_2^2} = \frac{0.007507^2}{0.005701^2} = 1.7339$$

For the critical values, first note that this is a two-tailed test with 0.025 in each tail. As long as we are stipulating that the larger variance is placed in the numerator of the F test statistic, we need to find only the right-tailed critical value. From Table A-5 we see that the critical value of F is between 1.8752 and 2.0739, which we find by referring to 0.025 in the right tail, with 35 degrees of freedom for the numerator and 35 degrees of freedom for the denominator. (Excel provides a critical value of 1.9611. To get a critical F value, use *fx*, **Statistical, FINV,** and enter the area to the *right* of the critical value.)

Step 7:   Figure 8-5 shows that the test statistic $F = 1.7339$ does not fall within the critical region, so we fail to reject the null hypothesis of equal variances.

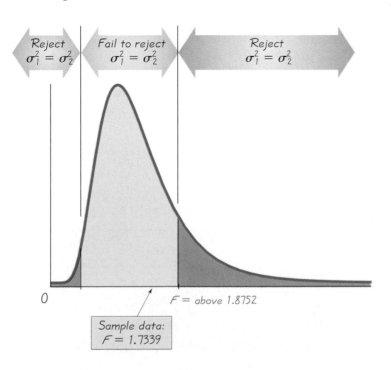

FIGURE 8-5
Distribution of s_1^2/s_2^2 for Weights of Regular Coke and Regular Pepsi.

Lower Variation, Higher Quality

Ford and Mazda were producing similar transmissions that were supposed to be made with the same specifications. But the American-made transmissions required more warranty repairs than the Japanese-made transmissions. When investigators inspected samples of the Japanese transmission gearboxes, they first thought that their measuring instruments were defective because they weren't detecting any variability among the Mazda transmission gearboxes. They realized that although the American transmissions were within the specifications, the Mazda transmissions were not only within the specifications, but consistently close to the desired value. By reducing variability among transmission gearboxes, Mazda reduced the costs of inspection, scrap, rework, and warranty repair.

INTERPRETATION   There is not sufficient evidence to warrant rejection of the claim that the two variances are equal. However, we should recognize that the F test is extremely sensitive to distributions that are not normally distributed, so this conclusion might make it appear that there is no significant difference between the population variances when there really is a difference that was hidden by nonnormal distributions.

In the preceding example we used a two-tailed test for the claim of equal variances. A right-tailed test would yield the same test statistic of $F = 1.7339$, but a different critical value of F.

We have described the traditional method of testing hypotheses made about two population variances. Exercise 13 deals with the P-value approach, and Exercise 15 deals with the construction of confidence intervals.

Using Excel

Note that Excel requires the original lists of sample values. If only the summary statistics are known, see the Excel Project at the end of Chapter 5 for a way to create lists of sample data with the desired statistics.

1. First enter the data from the first sample in the first column A, then enter the values of the second sample in column B.

2. Click on **Tools, Data Analysis,** then **F-Test Two-Sample for Variances.**

3. In the dialog box, enter the range of values for the first sample (such as A1:A36) and enter the range of values for the second sample. Enter the value of the significance level in the "Alpha" box. Click **OK.** Excel will provide the F test statistic, the P-value for the one-tailed case, and the critical F value for the one-tailed case. For a two-tailed test, double the P-value given by Excel. (To get a critical value for a one-tailed test, use ***fx*, Statistical, FINV,** and enter the area to the *right* of the desired critical value.) Shown below is the Excel display for the example of this section.

EXCEL

F-Test Two-Sample for Variances

	Variable 1	Variable 2
Mean	0.816822	0.824103
Variance	5.64E-05	3.25E-05
Observations	36	36
df	35	35
F	1.734067	
P(F<=f) one-tail	0.054054	
F Critical one-tail	1.75714	

8-5 Basic Skills and Concepts

Hypothesis Test of Equal Variances.   *In Exercises 1 and 2, test the given claim. Use a significance level of $\alpha = 0.05$, and assume that all populations are normally distributed.*

EXCEL
1. Claim: The treatment population and the placebo population have different variances ($\sigma_1^2 \neq \sigma_2^2$).

Treatment group: $n = 16$, $\bar{x} = 482.8$, $s = 150.0$
Placebo group: $n = 30$, $\bar{x} = 503.7$, $s = 75.0$

EXCEL
2. Claim: Women have a larger variance than men.

Men: $n = 15$, $\bar{x} = 107$, $s = 12.4$
Women: $n = 25$, $\bar{x} = 75.7$, $s = 16.8$

EXCEL
3. Weights of Regular Coke and Diet Coke This section included an example about a hypothesis test of the claim that weights of regular Coke and regular Pepsi have the same standard deviation. Test the claim that regular Coke and diet Coke have weights with different standard deviations. Sample weights are found in Data Set 1 (COLA.xls) in Appendix B, but here are the summary statistics: The sample of 36 weights of regular Coke have a standard deviation of 0.007507 lb, and the sample of 36 weights of diet Coke have a standard deviation of 0.004391 lb. Use a 0.05 significance level. If the results were to show that the standard deviations are significantly different, what would be an important factor that might explain the difference?

EXCEL
4. Axial Loads of Aluminum Cans Data Set 12 (CANS.xls) in Appendix B includes axial loads (in pounds) of a sample of 175 aluminum cans that are 0.0109 in. thick and another sample of 175 aluminum cans that are 0.0111 in. thick. (An axial load is the maximum weight that the sides can support. It is measured by using a plate to apply increasing pressure to the top of the can until it collapses.) The sample of 0.0109 in. cans has axial loads with a mean of 267.1 lb and a standard deviation of 22.1 lb. The sample of 0.0111 in. cans has axial loads with a mean of 281.8 lb and a standard deviation of 27.8 lb. Use a 0.05 significance level to test the claim that the samples come from populations with the same standard deviation.

Nonstress	Stress
$n_1 = 40$	$n_2 = 40$
$\bar{x}_1 = 53.3$	$\bar{x}_2 = 45.3$
$s_1 = 11.6$	$s_2 = 13.2$

5. Police as Eyewitnesses An experiment was conducted to investigate the effect of stress on the recall ability of police eyewitnesses. The experiment involved a non-stressful interrogation of a cooperative suspect and a stressful interrogation of an uncooperative and belligerent suspect. The numbers of details recalled a week after the incident are summarized in the margin (based on data from "Eyewitness Memory of Police Trainees for Realistic Role Plays," by Yuille et al., *Journal of Applied Psychology,* Vol. 79, No. 6). Use a 0.10 significance level to test the claim that the samples come from populations with different standard deviations.

6. Weights of Men in Different Age Groups As part of the National Health Survey, data were collected on the weights of men. For 804 men aged 25 to 34, the mean is 176 lb and the standard deviation is 35.0 lb. For 1657 men aged 65 to 74, the mean and standard deviation are 164 lb and 27.0 lb, respectively. At the 0.01 significance level, test the claim that the older men come
(continued)

from a population with a standard deviation less than that for men in the 25 to 34 age bracket.

7. Ages of Faculty and Student Cars Students at the author's college randomly selected 217 student cars and found that they had ages with a mean of 7.89 years and a standard deviation of 3.67 years. They also randomly selected 152 faculty cars and found that they had ages with a mean of 5.99 years and a standard deviation of 3.65 years. Is there sufficient evidence to support the claim that the ages of faculty cars vary less than the ages of student cars?

Zinc Supplement Group	Placebo Group
$n_1 = 294$	$n_2 = 286$
$\bar{x}_1 = 3214$	$\bar{x}_2 = 3088$
$s_1 = 669$	$s_2 = 728$

8. Testing Effects of Zinc A study of zinc-deficient mothers was conducted to determine effects of zinc supplementation during pregnancy. Sample statistics are listed in the margin (based on data from "The Effect of Zinc Supplementation on Pregnancy Outcome," by Goldenberg et al., *Journal of the American Medical Association*, Vol. 274, No. 6). The weights were measured in grams. Using a 0.05 significance level, is there sufficient evidence to support the claim that the variation of birth weights for the placebo population is greater than the variation for the population treated with zinc supplements?

EXCEL **9.** Blanking Out on Tests Many students have had the unpleasant experience of panicking on a test because the first question was exceptionally difficult. The arrangement of test items was studied for its effect on anxiety. The following scores are measures of "debilitating test anxiety" (which most of us call panic or blanking out). The Excel display for the given values is also shown. At the 0.05 significance level, test the claim that the two given samples come from populations with the same variance.

Questions Arranged from Easy to Difficult					Questions Arranged from Difficult to Easy			
24.64	39.29	16.32	32.83	28.02	33.62	34.02	26.63	30.26
33.31	20.60	21.13	26.69	28.90	35.91	26.68	29.49	35.32
26.43	24.23	7.10	32.86	21.06	27.24	32.34	29.34	33.53
28.89	28.71	31.73	30.02	21.96	27.62	42.91	30.20	32.54
25.49	38.81	27.85	30.29	30.72				

Based on data from "Item Arrangement, Cognitive Entry Characteristics, Sex and Test Anxiety as Predictors of Achievement in Examination Performance," by Klimko, *Journal of Experimental Education*, Vol. 52, No. 4.

EXCEL

	A	B	C
1	F-Test Two-Sample for Variances		
2		Variable 1	Variable 2
3	Mean	27.1152	31.72813
4	Variance	47.01983	18.1489
5	Observations	25	16
6	df	24	15
7	F	2.590782	
8	P(F<=f) one-tail	0.029928	
9	F Critical one-tail	2.287827	

10. Calcium and Blood Pressure Sample data were collected in a study of calcium supplements and their effects on blood pressure. A placebo group and a calcium group began the study with blood pressure measurements (based on data from "Blood Pressure and Metabolic Effects of Calcium Supplementation in Normotensive White and Black Men," by Lyle et al., *Journal of the American Medical Association,* Vol. 257, No. 13). Sample values are listed and an Excel display is shown. At the 0.10 significance level, test the claim that the two sample groups come from populations with the same standard deviation. If the experiment requires groups with equal standard deviations, are these two groups acceptable?

Placebo: 124.6   104.8   96.5   116.3   106.1   128.8   107.2   123.1
           118.1   108.5   120.4   122.5   113.6

Calcium: 129.1   123.4   102.7   118.1   114.7   120.9   104.4   116.3
           109.6   127.7   108.0   124.3   106.6   121.4   113.2

EXCEL

F-Test Two-Sample for Variances		
	Variable 1	*Variable 2*
Mean	114.6538	116.0267
Variance	89.49269	71.7221
Observations	13	15
df	12	14
F	1.24777	
P(F<=f) one-tail	0.342588	
F Critical one-tail	2.053714	

11. Comparing Diet Coke and Diet Pepsi Refer to Data Set 1 (COLA.xls) in Appendix B, and use a 0.05 significance level to test the claim that diet Coke and diet Pepsi have weights with different standard deviations.

12. Comparing Textbook Prices Refer to Data Set 2 (TEXTBOOK.xls) in Appendix B. Is there sufficient evidence to support the claim that the standard deviations of new textbook prices at the two colleges are different? Do the sample values appear to satisfy the requirement of coming from populations with normal distributions?

8-5   Beyond the Basics

13. Determining *P*-Values To test a claim about two population variances by using the *P*-value approach, first find the *F* test statistic, then refer to Table A-5 to determine how it compares to the critical values listed for $\alpha = 0.01$, $\alpha = 0.025$, and $\alpha = 0.05$. Referring to Exercise 3, what can be concluded about the *P*-value?

14. Finding Lower Critical *F* Values In this section, for hypothesis tests that were two-tailed, we found only the upper critical value. Let's denote that value by F_R, where the subscript suggests the critical value for the right tail. The lower critical value F_L (for the left tail) can be found as follows: First interchange the degrees of freedom, and then take the reciprocal of the resulting *F* value found in Table A-5. (F_R is often denoted by $F_{\alpha/2}$, and F_L is often denoted by $F_{1-\alpha/2}$.) Find the critical values F_L and F_R for two-tailed hypothesis tests based on the following values.   (*continued*)

a. $n_1 = 10, n_2 = 10, \alpha = 0.05$ b. $n_1 = 10, n_2 = 7, \alpha = 0.05$
c. $n_1 = 7, n_2 = 10, \alpha = 0.05$ d. $n_1 = 25, n_2 = 10, \alpha = 0.02$
e. $n_1 = 10, n_2 = 25, \alpha = 0.02$

15. Contructing Confidence Intervals In addition to testing claims involving σ_1^2 and σ_2^2, we can also construct confidence interval estimates of the ratio σ_1^2/σ_2^2 using the following expression.

$$\left(\frac{s_1^2}{s_2^2} \cdot \frac{1}{F_R}\right) < \frac{\sigma_1^2}{\sigma_2^2} < \left(\frac{s_1^2}{s_2^2} \cdot \frac{1}{F_L}\right)$$

Here F_L and F_R are as described in Exercise 14. Refer to the data in Exercise 10 and construct a 95% confidence interval estimate for the ratio of the placebo group variance to the calcium-supplement group variance.

16. Effects of Transformations on Data Sample data consist of temperatures recorded for two different groups of items that were produced by two different production techniques. A quality-control specialist plans to analyze the results. She begins by testing for equality of the two population standard deviations.
   a. If she adds the same constant to every temperature from both groups, how is the value of the test statistic F affected?
   b. If she multiplies every value from both groups by the same constant, how is the value of the test statistic F affected?
   c. If she converts all temperatures from the Fahrenheit scale to the Celsius scale, how is the value of the test statistic F affected?

8-6 Inferences about Two Means: Independent and Small Samples

In this section we present methods of inferential statistics (hypothesis tests and confidence intervals) for situations involving the means of two independent populations, when (unlike Section 8-2) at least one of the two samples is small (with $n \leq 30$). We describe methods for testing hypotheses and constructing confidence intervals for situations in which the following assumptions apply.

Assumptions

1. The two samples are *independent*.
2. The two samples are simple random samples *selected from normally distributed* populations.
3. At least one of the two samples is small ($n \leq 30$).

   When these conditions are satisfied, we use one of three different procedures corresponding to the following cases:

Case 1: The values of both population variances are known. (In reality, this case seldom occurs.)

Case 2: The two populations have equal variances. (That is, $\sigma_1^2 = \sigma_2^2$.)

Case 3: The two populations have unequal variances. (That is, $\sigma_1^2 \neq \sigma_1^2$.)

Case 1: Both Population Variances Are Known

In reality, Case 1 almost never occurs. Finding σ_1^2 and σ_2^2 typically requires that we know all of the values of both populations, and we can therefore find the values of μ_1 and μ_2, so there is no need to make inferences about μ_1 and μ_2 by testing claims or constructing confidence intervals. If some strange set of circumstances allows us to know the values of σ_1^2 and σ_2^2 but not μ_1 and μ_2, then we use the methods described in Section 8-2 for large samples. Because this case of known population variances is so unlikely to occur, we will not pursue it.

Choosing Between Cases 2 and 3: How Do We Decide That Two Populations Have Equal or Unequal Variances?

If the three assumptions listed at the beginning of this section are satisfied, inferences about the two population means are made with the procedures we will describe in Cases 2 and 3. Case 2 applies when the two populations have *equal* variances, and Case 3 applies when the two populations have *different* variances. But how do we determine whether the two population variances are equal? In Section 8-5 we described a procedure for testing the claim that the two populations have equal variances, and this approach, which we will now call the *preliminary F test approach*, is used by some statisticians and is designed into some software. It is somewhat controversial, however. Some argue that if we apply the F test with a certain significance level and then do a t test at the same level, the overall result will not be at the same level of significance. Also, in addition to being sensitive to differences in population variances, the F statistic is so sensitive to departures from normal distributions that we might choose a method for the wrong reason. (For example, the F test might make us think that the two populations have unequal variances, when the variances are actually equal but the test was misleading because the two population distributions were not both normal.)

Some textbooks completely dodge the issue by presenting only those methods that assume equal population variances. Other textbooks describe both methods but ignore any discussion of criteria for choosing between them. (The exercises include notes, such as "assume equal population variances.") Moser and Stevens, in "Homogeneity of Variance in the Two-Sample Means Test" (*The American Statistician,* Vol. 46, No. 1), argue against the preliminary F test and recommend instead that we always assume that the two population variances are *not* equal (as in Case 3 below). Here's another strategy: Make the inferences using *both* methods (for equal variances and for unequal variances). If the two sets of results agree, your conclusion is probably OK. Here is a summary of two different schools of thought:

1. *Preliminary F test approach:* Apply the F test described in Section 8-5 to test the null hypothesis that $\sigma_1^2 = \sigma_2^2$. Use the conclusion of that test as follows.

   • *Fail to reject $\sigma_1^2 = \sigma_2^2$:* Assume *equal* population variances. (Case 2)
   • *Reject $\sigma_1^2 = \sigma_2^2$:* Assume *unequal* population variances. (Case 3)

2. *Assume unequal variances:* Assume that the two populations have *unequal* variances ($\sigma_1^2 \neq \sigma_2^2$) and use the methods of Case 3. (Remember, this

assumption applies only to cases in which we have two independent samples, at least one of which is small, and the population variances are not known.)

Case 2: Equal Population Variances: Pool the Two Sample Variances

If the two populations have unknown but equal variances, we calculate a **pooled estimate of σ^2** that is common to both populations. The pooled estimate is denoted by s_p^2 and is a weighted average of s_1^2 and s_2^2, as shown in the box. Hypotheses about μ_1 and μ_2 can be tested and confidence interval estimates of $\mu_1 - \mu_2$ can be constructed by using the following methods.

Test Statistic (Small Independent Samples and Equal Variances)

$$t = \frac{(\bar{x}_1 - \bar{x}_2) - (\mu_1 - \mu_2)}{\sqrt{\dfrac{s_p^2}{n_1} + \dfrac{s_p^2}{n_2}}}$$

where

$$s_p^2 = \frac{(n_1 - 1)s_1^2 + (n_2 - 1)s_2^2}{(n_1 - 1) + (n_2 - 1)}$$

and the number of degrees of freedom is given by df $= n_1 + n_2 - 2$.

Confidence Interval (Small Independent Samples and Equal Variances)

$$(\bar{x}_1 - \bar{x}_2) - E < (\mu_1 - \mu_2) < (\bar{x}_1 - \bar{x}_2) + E$$

where

$$E = t_{\alpha/2}\sqrt{\frac{s_p^2}{n_1} + \frac{s_p^2}{n_2}}$$

and s_p^2 is as given in the test statistic.

▷ **EXAMPLE**   **Cigarette Filters and Nicotine** Refer to Table 8-2 for the measured nicotine contents of randomly selected filtered and nonfiltered king-size cigarettes. All measurements are in milligrams, and the data are from the Federal Trade Commission. Use a 0.05 significance level to test

TABLE 8-2	Nicotine Contents (in mg) of Cigarettes
Filtered	1.2 1.3 1.1 1.1 1.0 0.1 1.1 1.0 0.8 1.0 0.9 0.8 1.0 1.0 1.0 0.9 1.2 1.1 0.1 1.2 0.9
Nonfiltered	1.6 1.9 1.6 1.8 1.7 1.7 1.4 1.5

the claim that the mean amount of nicotine in filtered king-size cigarettes is equal to the mean amount of nicotine in nonfiltered king-size cigarettes.

SOLUTION   Before performing the hypothesis test, first explore the data sets. The sample statistics are listed in the margin, and the EXCEL-generated boxplots are also shown.

	Nicotine (mg)	
	Filtered Kings	Nonfiltered Kings
	$n_1 = 21$	$n_2 = 8$
	$\bar{x}_1 = 0.94$	$\bar{x}_2 = 1.65$
	$s_1 = 0.31$	$s_2 = 0.16$

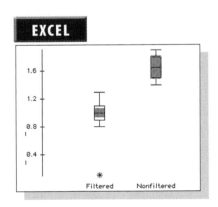

The sample statistics and boxplots seem to indicate a clear difference between the two means, but those results are based on small samples, so are the differences really significant? The two samples are independent (because they are separate and are not matched), the sample sizes are small, and we don't know the values of σ_1 and σ_2, so we are dealing with either Case 2 or Case 3. Using a preliminary F test (described in Section 8-5) to choose between Case 2 and Case 3, we test H_0: $\sigma_1^2 = \sigma_2^2$ with the test statistic

$$F = \frac{s_1^2}{s_2^2} = \frac{0.31^2}{0.16^2} = 3.7539$$

With $\alpha = 0.05$ in a two-tailed F test and with 20 and 7 degrees of freedom for the numerator and denominator, respectively, we use Table A-5 to find the critical F value of 4.4667. Because the computed test statistic of $F = 3.7539$ does *not* fall within the critical region, we fail to reject the null hypothesis of equal variances and proceed by using the approach outlined in Case 2.

In using the Case 2 approach, we test the claim that $\mu_1 = \mu_2$, with the following null and alternative hypotheses:

$$H_0: \mu_1 = \mu_2 \text{ (or } \mu_1 - \mu_2 = 0) \qquad H_1: \mu_1 \neq \mu_2$$

The test statistic for this case of equal variances requires the value of the pooled variance s_p^2 so we find that value first.

$$
\begin{aligned}
s_p^2 &= \frac{(n_1 - 1)s_1^2 + (n_2 - 1)s_2^2}{(n_1 - 1) + (n_2 - 1)} \\
&= \frac{(21 - 1) \cdot 0.31^2 + (8 - 1) \cdot 0.16^2}{(21 - 1) + (8 - 1)} \\
&= 0.078
\end{aligned}
$$

We can now find the value of the test statistic:

$$t = \frac{(\bar{x}_1 - \bar{x}_2) - (\mu_1 - \mu_2)}{\sqrt{\dfrac{s_p^2}{n_1} + \dfrac{s_p^2}{n_2}}} = \frac{(0.94 - 1.65) - 0}{\sqrt{\dfrac{0.078}{21} + \dfrac{0.078}{8}}} = -6.119$$

The critical values of $t = -2.052$ and $t = 2.052$ are found from Table A-3 by referring to the column for $\alpha = 0.05$ (two tails) and to the row for df = 27 (the value of $21 + 8 - 2$). Figure 8-6 shows the test statistic, critical values, and critical region. We see that the test statistic does fall within the critical region, so we reject the null hypothesis that $\mu_1 = \mu_2$.

INTERPRETATION   There is sufficient evidence to warrant rejection of the claim that filtered king cigarettes and nonfiltered kings have the same mean amounts of nicotine. The subjective conclusion we formed by intuitively analyzing the boxplots and the sample statistics is now upheld by a formal hypothesis test.

Shown below is the Excel display for this example. Compare these results to those shown in the example

EXCEL

t-Test: Two-Sample Assuming Equal Variances

	Variable 1	Variable 2
Mean	0.942857	1.65
Variance	0.095571	0.025714
Observations	21	8
Pooled Variance	0.07746	
Hypothesized Mean Difference	0	
df	27	
t Stat	-6.11538	
P(T<=t) one-tail	7.81E-07	
t Critical one-tail	1.703288	
P(T<=t) two-tail	1.56E-06	
t Critical two-tail	2.051829	

FIGURE 8-6
Distribution of Differences Between Means of Nicotine Amounts in Filtered and Nonfiltered Cigarettes

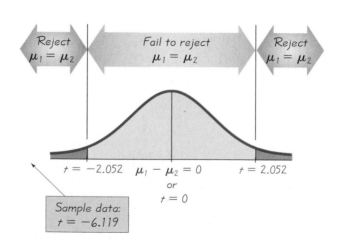

Instead of using the preliminary F test approach, we could use the same data with the methods to be described in Case 3, and we would again reject the null hypothesis that $\mu_1 = \mu_2$.

▷ **EXAMPLE   Cigarette Filters and Nicotine** Using the data in the preceding example, construct a 95% confidence interval estimate of $\mu_1 - \mu_2$.

SOLUTION It is easy to find that $\bar{x}_1 - \bar{x}_2 = 0.94 - 1.65 = -0.71$. Next, we find the value of the margin of error E.

$$E = t_{\alpha/2}\sqrt{\frac{s_p^2}{n_1} + \frac{s_p^2}{n_2}} = 2.052\sqrt{\frac{0.078}{21} + \frac{0.078}{8}} = 0.238$$

With $\bar{x}_1 - \bar{x}_2 = -0.71$ and with $E = 0.238$, we proceed to construct the confidence interval as follows.

$$(\bar{x}_1 - \bar{x}_2) - E < (\mu_1 - \mu_2) < (\bar{x}_1 - \bar{x}_2) + E$$
$$-0.71 - 0.238 < (\mu_1 - \mu_2) < -0.71 + 0.238$$
$$-0.95 < (\mu_1 - \mu_2) < -0.47$$

INTERPRETATION This final format is somewhat awkward with its negative signs, so we might interpret it by saying that we are 95% confident that nonfiltered king-size cigarettes have a mean nicotine content level that exceeds the mean for filtered kings by an amount that is between 0.47 mg and 0.95 mg.   ◁

Case 3: Unequal Population Variances

If the two populations have unequal variances, there is no exact method for testing equality of means and constructing confidence intervals. An *approximate* method is to use the following test statistic and confidence interval.

Test Statistic (Small Independent Samples and Unequal Variances)

$$t = \frac{(\bar{x}_1 - \bar{x}_2) - (\mu_1 - \mu_2)}{\sqrt{\frac{s_1^2}{n_1} + \frac{s_2^2}{n_2}}}$$

where   df = smaller of $n_1 - 1$ and $n_2 - 1$

Confidence Interval (Small Independent Samples and Unequal Variances)

$$(\bar{x}_1 - \bar{x}_2) - E < (\mu_1 - \mu_2) < (\bar{x}_1 - \bar{x}_2) + E$$

where $$E = t_{\alpha/2}\sqrt{\frac{s_1^2}{n_1} + \frac{s_2^2}{n_2}}$$

and   df = smaller of $n_1 - 1$ and $n_2 - 1$

This test statistic and confidence interval give the number of degrees of freedom as the smaller of $n_1 - 1$ and $n_2 - 1$, but this is a more conservative and simpler alternative to computing the number of degrees of freedom by using Formula 8-1. (Excel uses Formula 8-1.)

Formula 8-1

$$df = \frac{(A + B)^2}{\dfrac{A^2}{n_1 - 1} + \dfrac{B^2}{n_2 - 1}}$$

where

$$A = \frac{s_1^2}{n_1} \quad \text{and} \quad B = \frac{s_2^2}{n_2}$$

More exact results are obtained by using Formula 8-1, but they continue to be only approximate.

AS

▶ **EXAMPLE**   **Cigarette Filters and Tar** Refer to the data listed in Table 8-3 and use a 0.05 significance level to test the claim that the mean amount of tar in filtered king-size cigarettes is *less than* the mean amount of tar in nonfiltered king-size cigarettes. (All measurements are in milligrams, and the data are from the Federal Trade Commission.)

SOLUTION   The sample statistics are listed in the margin, and they appear to suggest a dramatic difference between the two means. But are those differences really significant?

The two samples are independent (because they are separate and are not matched), the sample sizes are small, and we don't know the values of σ_1 and σ_2, so we are dealing with either Case 2 or Case 3. If we use a preliminary F test to choose between Case 2 and Case 3, we test $H_0: \sigma_1^2 = \sigma_2^2$ with the test statistic

$$F = \frac{s_1^2}{s_2^2} = \frac{3.7^2}{1.7^2} = 4.7370$$

With $\alpha = 0.05$ in a two-tailed F test and with 20 and 7 degrees of freedom for the numerator and denominator, respectively, we use Table A-5 to find the critical F value of 4.4667. Because the computed test statistic of $F = 4.7370$ does fall within the critical region, we reject the null hypothesis of equal variances and proceed by using the approach outlined in Case 3.

In using the Case 3 approach, we test the claim that $\mu_1 < \mu_2$ with the following null and alternative hypotheses:

$$H_0: \mu_1 \geq \mu_2 \text{ (or } \mu_1 - \mu_2 \geq 0)$$
$$H_1: \mu_1 < \mu_2 \quad \text{(original claim)}$$

The test statistic for this case of unequal variances is

$$t = \frac{(\bar{x}_1 - \bar{x}_2) - (\mu_1 - \mu_2)}{\sqrt{\dfrac{s_1^2}{n_1} + \dfrac{s_2^2}{n_2}}} = \frac{(13.3 - 24.0) - 0}{\sqrt{\dfrac{3.7^2}{21} + \dfrac{1.7^2}{8}}} = -10.630$$

Tar (mg)	
Filtered Kings	Nonfiltered Kings
$n_1 = 21$	$n_2 = 8$
$\bar{x}_1 = 13.3$	$\bar{x}_2 = 24.0$
$s_1 = 3.7$	$s_2 = 1.7$

TABLE 8-3	Tar Contents (in mg) of Cigarettes																			
Filtered	16	15	16	14	16	1	16	18	10	14	12	11	14	13	13	13	16	16	8	16	11
Nonfiltered	23	23	24	26	25	26	21	24													

INTERPRETATION   The critical value of $t = -1.895$ is found from Table A-3 by referring to the column for $\alpha = 0.05$ (one tail) and to the row for df $= 7$ (the smaller of $21 - 1$ and $8 - 1$). (If Formula 8-1 is used to compute the number of degrees of freedom, you get df $= 26$, so the critical value is $t = -1.706$.) Because this left-tailed test has a test statistic of $t = -10.630$ and a critical value of $t = -1.895$, the test statistic does fall within the critical region, so we reject the null hypothesis that $\mu_1 \geq \mu_2$. There is sufficient evidence to support the claim that filtered king-size cigarettes have a mean tar level that is less than the mean for nonfiltered kings. (If we use the same data with the methods of Case 2 for equal variances, we again reject the null hypothesis that $\mu_1 \geq \mu_2$, so the conclusion is the same.)

EXAMPLE   **Cigarette Filters and Tar** Using the data in the preceding example, construct a 95% confidence interval estimate of $\mu_1 - \mu_2$.

SOLUTION   With $\bar{x}_1 - \bar{x}_2 = 13.3 - 24.0 = -10.7$, we proceed to find the value of the margin of error E.

$$E = t_{\alpha/2} \sqrt{\frac{s_1^2}{n_1} + \frac{s_2^2}{n_2}} = 2.365 \sqrt{\frac{3.7^2}{21} + \frac{1.7^2}{8}} = 2.381$$

With $\bar{x}_1 - \bar{x}_2 = -10.7$ and with $E = 2.381$, we construct the confidence interval as follows.

$$(\bar{x}_1 - \bar{x}_2) - E < (\mu_1 - \mu_2) < (\bar{x}_1 - \bar{x}_2) + E$$
$$-10.7 - 2.381 < (\mu_1 - \mu_2) < -10.7 + 2.381$$
$$-13.1 < (\mu_1 - \mu_2) < -8.3$$

INTERPRETATION   Again, this final format is somewhat awkward with its negative signs, so we might interpret it by saying that we are 95% confident that the mean tar content for nonfiltered kings exceeds the mean tar content for filtered kings by an amount that is between 8.3 mg and 13.1 mg.

This section, Section 8-2, and Section 8-3 all deal with inferences about the means of two populations. Determining the correct procedure can be difficult because we must consider issues such as the independence of the samples, the sizes of the samples, and whether the two populations appear to have equal variances. Figure 8-7 is designed to simplify the decisions that lead to the correct procedure.

Using Statistics to Identify Thieves

Methods of statistics can be used to determine that an employee is stealing, and they can also be used to estimate the amount stolen. The following are some of the indicators that have been used. For comparable time periods, samples of sales have means that are significantly different. The mean sale amount decreases significantly. There is a significant increase in the proportion of "no sale" register openings. There is a significant decrease in the ratio of cash receipts to checks. Methods of hypothesis testing can be used to identify such indicators. (See "How To Catch a Thief" by Manly and Thomson, *Chance*, Vol. 11, No. 4.)

FIGURE 8-7
Inferences about the Means of Two Populations

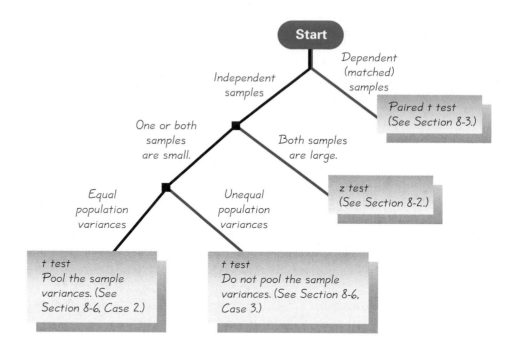

Using Excel

Notes: **1.** Excel requires the original lists of sample values. If only the summary statistics are known, create samples with the same summary statistics by using the procedure described in the Chapter 5 Excel Project.

2. You must choose between the assumption of equal variances (Case 2) or the assumption of unequal variances (Case 3).

Hypothesis Tests

1. Enter the data for the two samples in columns A and B.

2. To use Excel's Data Analysis add-in, click on **Tools** and select **Data Analysis.** (You could also use the DDXL add-in by following the same procedure described in Section 8-2 with this modification: When using the hypothesis-testing dialog box, in Step 1 click on **2-Sample** if you are assuming unequal variances, or click on **Pooled** if you are assuming equal variances.)

3. Select one of the following two items:

   t-test: Two-Sample Assuming Equal Variances

   t-test: Two-Sample Assuming Unequal Variances

4. Proceed to enter the range for the values of the first sample (such as A1:A21) and enter the range of values for the second sample. Enter a value for the claimed difference between the two population means, which will usually be 0. Enter the significance level in the "Alpha" box and click **OK.**

5. If using Excel's Data Analysis add-in, the display is condensed to fit in the spreadsheet columns. You can increase the column width to make the display more readable. Click on **Format,** select **Column,** then select **Width** and enter a larger number for the column width, then click **OK.**

Confidence Intervals

The DDXL add-in is designed to generate confidence intervals for the difference between two population means, but Excel's Data Analysis add-in cannot be used. The DDXL add-in calculates the number of degrees of freedom by using Formula 8-1 instead of the smaller of $n_1 - 1$ and $n_2 - 1$, so the DDXL results may differ somewhat from those given in this book.

1. Enter the data for the two samples in columns A and B.

2. Click on **DDXL** and select **Confidence Intervals,** then select **2 Var t Interval.**

3. In the dialog box, click on the pencil icon for the first quantitative column, then enter the range of values for the first sample, such as A1:A36. Click on the pencil icon for the second quantitative column and enter the range of values for the second sample. Click **OK.**

4. Now complete the new dialog box by following the indicated steps. Be careful to make the correct choice in Step 1:

   Assuming *unequal* population variances, select **2 Sample.**

   Assuming *equal* population variances, select **Pooled.**

8-6   Basic Skills and Concepts

Testing Equality of Means.   *In Exercises 1 and 2, test the given claim. Use a significance level of $\alpha = 0.05$, and assume that all populations are normally distributed. Use the traditional method of testing hypotheses outlined in Figure 7-5. Instead of using a preliminary F test to determine whether the two populations have equal variances, simply examine the standard deviations and make a judgment based on their obvious values.*

1. Claim: The treatment and placebo groups have the same mean ($\mu_1 = \mu_2$).

   Treatment Group: $n_1 = 5, \bar{x}_1 = 98.20, s_1 = 0.62$

   Placebo Group: $n_2 = 25, \bar{x}_2 = 104.60, s_2 = 0.61$

2. Claim: The treatment and placebo groups have the same mean ($\mu_1 = \mu_2$,).

   Treatment Group: $n_1 = 20, \bar{x}_1 = 235.3, s_1 = 5.6$

   Placebo Group: $n_2 = 10, \bar{x}_2 = 177.7, s_2 = 31.4$

3. Using the sample data in Exercise 1, construct a 95% confidence interval estimate of the difference between the means of the two populations.

4. Using the sample data in Exercise 2, construct a 95% confidence interval estimate of the difference between the means of the two populations.

5. Testing Equality of Heights of Men and Women Samples of men and women are randomly selected, and their heights are accurately measured with the results (in inches) listed below (based on data from the National Health Survey).

  a. Assuming that the heights of men and women have *equal* variances, use a 0.01 significance level to test the claim that men and women have the same mean height.

  b. Assuming that the heights of men and women have *unequal* variances, use a 0.01 significance level to test the claim that men and women have the same mean height.

  c. Compare the conclusions to parts (a) and (b). Do both procedures lead to the same result?

  d. Based on the given sample data, what do you conclude about the heights of men and women? Do the sample data suggest that men and women have the same mean height? Do the results agree with your knowledge of the heights of men and women?

Men:	73.0	71.6	68.7	65.9	70.4	$\bar{x} = 69.920,\ s = 2.747$
Women:	63.9	67.1	63.2	64.7		$\bar{x} = 64.725,\ s = 1.698$

 6. Heights of Men and Women: Confidence Intervals Refer to the sample data used in Exercise 5.

  a. Assuming that heights of men and women have *equal* variances, construct a 99% confidence interval estimate of the difference $\mu_1 - \mu_2$. Do the confidence interval limits contain 0, and what does this suggest about equality of the two population means?

  b. Assuming that heights of men and a women have *unequal* variances, construct a 99% confidence interval estimate of the difference $\mu_1 - \mu_2$. Do the confidence interval limits contain 0, and what does this suggest about equality of the two population means?

  c. Compare the results from parts (a) and (b). Does the assumption of equal variances or unequal variances have much of an effect on the confidence interval limits?

7. Brain Right Cordate and Psychiatric Disorders: Hypothesis Test Are severe psychiatric disorders related to biological factors that can be physically observed? One study used X-ray computed tomography (CT) to collect data on brain volumes for a group of patients with obsessive-compulsive disorders and a control group of healthy persons. Sample results for volumes (in mL) follow for the right cordate (based on data from "Neuroanatomical Abnormalities in Obsessive-Compulsive Disorder Detected with Quantitative X-Ray Computed Tomography," by Luxenberg et al., *American Journal of Psychiatry,* Vol. 145, No. 9). At the 0.01 significance level, test the claim that obsessive-compulsive patients and healthy persons have the same mean brain volume. Based on this result, does it seem that obsessive-compulsive disorders have a biological basis?

Obsessive-compulsive patients: $n = 10,\ \bar{x} = 0.34,\ s = 0.08$

Control group: $n = 10,\ \bar{x} = 0.45,\ s = 0.08$

8. Brain Right Cordate and Psychiatric Disorders: Confidence Interval Using the sample data from Exercise 7, construct a 99% confidence interval estimate of the difference between the mean brain volume for the disorder patient group and the mean brain volume for the healthy control group. What does the confidence interval suggest about the difference between the two population means?

9. Brain Volume and Psychiatric Disorders: Confidence Intervals The study cited in Exercise 7 resulted in the values summarized here for total brain volumes (in mL).
   a. Using these sample statistics, construct a 95% confidence interval for the difference between the mean brain volume of obsessive-compulsive patients and the mean brain volume of healthy persons. Assume that the two populations have *equal* variances.
   b. Construct a 95% confidence interval for the difference between the mean brain volume of obsessive-compulsive patients and the mean brain volume of healthy persons. Assume that the two populations have *unequal* variances.
   c. Compare the results from parts (a) and (b). Are they very different? Which results are better?

   Obsessive-compulsive patients: $n = 10, \bar{x} = 1390.03, s = 156.84$

   Control group: $n = 10, \bar{x} = 1268.41, s = 137.97$

10. Brain Volume and Psychiatric Disorders: Hypothesis Test Refer to the sample data in Exercise 9.
    a. Assuming that the population variances are *equal*, use a 0.05 significance level to test the claim that there is no difference between the mean for obsessive-compulsive patients and the mean for healthy persons.
    b. Assuming that the population variances are *unequal*, use a 0.05 significance level to test the claim that there is no difference between the mean for obsessive-compulsive patients and the mean for healthy persons.
    c. Compare the results from parts (a) and (b). Which results are better? Based on the results, does it seem that obsessive-compulsive disorders have a biological basis?

EXCEL 11. Red/Orange M&Ms: Hypothesis Test Data Set 10 (M&M.xls) in Appendix B includes a sample of 21 red M&Ms with weights having a mean of 0.9097 g and a standard deviation of 0.0275 g. The data set also includes a sample of 8 orange M&Ms with weights having a mean of 0.9251 g and a standard deviation of 0.0472 g. Use a preliminary F test to determine whether the populations have equal variances, then use a 0.05 significance level to test for equality of the two population means. If you were responsible for controlling production so that red and orange M&Ms have weights with the same mean, would you take corrective action?

EXCEL 12. Testing the Anchoring Effect Randomly selected statistics students were given five seconds to estimate the value of a product of numbers with the results given in the accompanying table. Use a preliminary F test to determine whether the two populations have equal variances, then use a 0.05 significance level to test the claim that the two samples come from populations with the same mean. Does the order of the numbers appear to have an effect on the estimate? (See the Cooperative Group Activities at the end of Chapter 2.)

Estimates from Students Given $1 \times 2 \times 3 \times 4 \times 5 \times 6 \times 7 \times 8$					Estimates from Students Given $8 \times 7 \times 6 \times 5 \times 4 \times 3 \times 2 \times 1$				
1560	169	5635	25	842	100,000	2000	42,000	1500	52,836
40,320	5000	500	1110	10,000	2050	428	372	300	225
200	1252	4000	2040	175	500	1200	400	49,000	4000
856	42,200	49,654	560	800	1876	3600	354	750	640
					64,582	23,410			

EXCEL 13. Carbon Monoxide and Cigarettes Refer to the given data for the measured amounts of carbon monoxide (CO) from samples of filtered and nonfiltered king-size cigarettes. All measurements are in milligrams, and the data are from the Federal Trade Commission. Use a 0.05 significance level to test the claim that the mean amount of carbon monoxide in filtered king-size cigarettes is equal to the mean amount of carbon monoxide for nonfiltered king-size cigarettes. Based on this result and the examples of this section, are cigarette filters effective, or are they just a sales gimmick with no real effect?

Filtered: 14 12 14 16 15 2 14 16 11 13 13 12 13 12 13 14 14 14 9 17 12

Nonfiltered: 14 15 17 17 16 16 14 16

14. Effects of Alcohol An experiment was conducted to test the effects of alcohol. The errors were recorded in a test of visual and motor skills for a treatment group of people who drank ethanol and another group given a placebo. The results are shown in the accompanying table. Use a 0.05 significance level to test the claim that the two groups come from populations with the same mean. Do these results support the common belief that drinking is hazardous for drivers, pilots, ship captains, and so on?

Treatment Group	Placebo Group
$n_1 = 22$	$n_2 = 22$
$\overline{x}_1 = 4.20$	$\overline{x}_2 = 1.71$
$s_1 = 2.20$	$s_2 = 0.72$

Based on data from "Effects of Alcohol Intoxication on Risk Taking, Strategy, and Error Rate in Visuomotor Performance," by Streufert et al., *Journal of Applied Psychology,* Vol. 77, No. 4.

15. Peanut and Plain M&Ms A simple random sample of 21 peanut M&M candies is selected, and each candy is weighed. The mean is 2.4658 g and the standard deviation is 0.3127 g. The 100 M&M plain candies listed in Data Set 10 in Appendix B have a mean of 0.9147 g and a standard deviation of 0.0369 g. Is there sufficient evidence to support the claim that there is a difference between the mean weight of peanut M&Ms and the mean weight of plain M&Ms?

EXCEL 16. Blanking Out on Tests Many students have had the unpleasant experience of panicking on a test because the first question was exceptionally difficult. The arrangement of test items was studied for its effect on anxiety. The accompanying scores are measures of "debilitating test anxiety" (which most of us call panic or blanking out). Is there sufficient evidence to support the claim that the two populations of scores have the same mean? Is there sufficient evidence to support the claim that the arrangement of the test items has an effect on the score?

Questions Arranged from Easy to Difficult

24.64	39.29	16.32	32.83	28.02
33.31	20.60	21.13	26.69	28.90
26.43	24.23	7.10	32.86	21.06
28.89	28.71	31.73	30.02	21.96
25.49	38.81	27.85	30.29	30.72

Questions Arranged from Difficult to Easy

33.62	34.02	26.63	30.26
35.91	26.68	29.49	35.32
27.24	32.34	29.34	33.53
27.62	42.91	30.20	32.54

Based on data from "Item Arrangement, Cognitive Entry Characteristics, Sex and Test Anxiety as Predictors of Achievement in Examination Performance," by Klimko, *Journal of Experimental Education,* Vol. 52, No. 4.

8-6 Beyond the Basics

17. Effect of No Variation in Sample An experiment was conducted to test the effects of alcohol. The breath alcohol levels were measured for a treatment group of people who drank ethanol and another group given a placebo. The results are given in the accompanying table. Use a 0.05 significance level to test the claim that the two groups come from populations with the same mean.

Treatment Group	Placebo Group
$n_1 = 22$	$n_2 = 22$
$\bar{x}_1 = 0.049$	$\bar{x}_2 = 0.000$
$s_1 = 0.015$	$s_2 = 0.000$

Based on data from "Effects of Alcohol Intoxication on Risk Taking, Strategy, and Error Rate in Visuomotor Performance," by Streufert et al., *Journal of Applied Psychology,* Vol. 77, No. 4.

18. Equal Sample Standard Deviations Assume that two samples have the same standard deviation and that both are independent, small, and randomly selected from normally distributed populations. Also assume that we want to test the claim that the samples come from populations with the same mean.
   a. Is it necessary to conduct a preliminary F test?
   b. If both samples have standard deviation s, what is the value of s_p^2 expressed in terms of s?

19. a. Refer to the two examples of this section that used the tar amounts in cigarettes. (See the Case 3 subsection.) How are the hypothesis test and confidence interval affected if the number of degrees of freedom is calculated with Formula 8-1, instead of by using the smaller of $n_1 - 1$ and $n_2 - 1$?
   b. Refer to the two examples in this section that used the tar amounts in cigarettes. (See the Case 3 subsection.) How are the hypothesis test and confidence interval affected if the one extreme outlier is deleted?

20. Consequences of Wrong Test Refer to Table 8-1 at the beginning of this chapter for the reported/measured heights of male statistics students. Delete the outlier values for the third student, as in Section 8-3. Because the reported/measured heights are matched pairs, we used the paired *t* test described in Section 8-3, where we obtained a test statistic of $t = 2.701$ when we tested the claim that reported heights are greater than measured heights. How are the results affected if we treat the sample data as two *independent* sets of data, instead of *matched pairs*? In general, if paired data are incorrectly treated as being independent, what is the effect on the results?

STATISTICS IN THE NEWS

The Fear of Flying

POUGHKEEPSIE The lives of many people are affected by a fear that prevents them from flying. Sports announcer John Madden gained notoriety as he crossed the country by rail or motorhome, traveling from one football stadium to another. The Marist Institute for Public Opinion conducted a poll of 1014 adults, 48% of whom were men. The results are depicted in the accompanying illustration from *USA Today*. The poll results show that 12% of the men and 33% of the women fear flying.

1. Is there sufficient evidence to conclude that there is a significant difference between the percentage of men and the percentage of women who fear flying?

2. Construct a 95% confidence interval estimate of the difference between the percentage of men and the percentage of women who fear flying. Do the confidence interval limits contain 0, and what is the significance of whether they do or do not?

3. Construct a 95% confidence interval for the percentage of men who fear flying.

4. Based on the result from Exercise 3, complete the following statement, which is typical of the statement that would be reported in a newspaper or magazine: "Based on the Marist Institute for Public Opinion poll, the percentage of men who fear flying is 22% with a margin of error of _____."

5. Examine the completed statement in Exercise 4. What important piece of information should be included, but is not included?

Just plane scared
While 47% of adults believe flying is the safest way to travel (vs. 39% who say cars and 14% for trains), about 41 million are afraid. Percent of Americans who fear flying:

All adults 22%
Men 12%
Women 33%

Source: Marist Institute for Public Opinion   By Scott Boeck and Genevieve Lynn, USA TODAY
USA Today Snapshot. "A look at statistics that shape the nation."

VOCABULARY LIST
---

independent samples
dependent samples
matched pairs
paired samples
pooled estimate of p_1 and p_2

F distribution
numerator degrees of freedom
denominator degrees of freedom
pooled estimate of σ^2

REVIEW

In Chapters 6 and 7 we introduced two major concepts of inferential statistics: the estimation of population parameters and the methods of testing hypotheses made about population parameters. Whereas those chapters considered only cases involving a single population, this chapter considered two samples drawn from two populations.

- Section 8-2 considered inferences made about the means of two independent populations, but the involved samples were both large ($n > 30$).
- Section 8-3 considered inferences made about the means of population data consisting of matched pairs.
- Section 8-4 discussed the procedures for making inferences about two population proportions.
- Section 8-5 presented methods for testing claims about the means of two population standard deviations or variances.
- Section 8-6 described procedures for making inferences about the means of two independent populations, with at least one of the samples being small ($n \leq 30$). Sections 8-2, 8-3, and 8-6 dealt with a variety of different cases for making inferences about two population means; see Figure 8-7 for help in selecting the correct procedure.

REVIEW EXERCISES
---

1. Testing Adverse Drug Reactions In clinical tests of adverse reactions to the drug Viagra, 7% of the 734 subjects in the treatment group experienced dyspepsia (indigestion), but 2% of the 725 subjects in the placebo group experienced dyspepsia (based on data from Pfizer Pharmaceuticals).
   a. Is there sufficient evidence to support the claim that dyspepsia occurs at a higher rate among Viagra users than those who do not use Viagra?
   b. Construct a 95% confidence interval estimate of the difference between the dyspepsia rate for Viagra users and the dyspepsia rate for those who use a placebo. Do the confidence interval limits contain 0, and what does this suggest about the two dyspepsia rates?

2. Testing Effectiveness of Training Program The effectiveness of a mental training program was tested in the military. In an antiaircraft artillery examination, scores for an experimental group and a control group were recorded.

(continued)

EXCEL **1.** Speeding Tickets for Men and Women The data in the accompanying table were obtained through a survey of randomly selected subjects.

   a. If one of the survey subjects is randomly selected, find the probability of getting someone ticketed for speeding.

   b. If one of the survey subjects is randomly selected, find the probability of getting a man or someone ticketed for speeding.

   c. Find the probability of getting someone ticketed for speeding, given that the selected person is a man.

   d. Find the probability of getting someone ticketed for speeding, given that the selected person is a woman.

   e. Use a 0.05 significance level to test the claim that the percentage of women ticketed for speeding is less than the percentage of men. Can we conclude that men generally speed more than women?

	Ticketed for Speeding Within the Last Year?	
	Yes	No
Men	26	224
Women	27	473

Based on data from R.H. Bruskin Associates.

2. Cell Phones and Crashes: Analyzing Newspaper Report In an article from the Associated Press, it was reported that researchers "randomly selected 100 New York motorists who had been in an accident and 100 who had not. Of those in accidents, 13.7 percent owned a cellular phone, while just 10.6 percent of the accident-free drivers had a phone in the car." Analyze these results.

EXCEL **3.** Good Test Question A test question is considered good if it discriminates between prepared and unprepared students. The first question on a test was answered correctly by 62 of 80 prepared students and by 23 of 50 unprepared students.

   a. At the 0.05 significance level, test the claim that the majority of the prepared students answer this question correctly.

   b. At the 0.05 significance level, test the claim that the proportion of prepared students who answer this question correctly is greater than the proportion of unprepared students who answer it correctly.

4. Gender and GRE Test Scores A study was conducted to investigate relationships between different types of standard test scores. On the Graduate Record Examination verbal test, 68 women had a mean of 538.82 and a standard deviation of 114.16, and 86 men had a mean of 525.23 and a standard deviation of 97.23. (See "Equivalencing MAT and GRE Scores Using Simple Linear Transformation and Regression Methods," by Kagan and Stock, *Journal of Experimental Education*, Vol. 49, No. 1.)

   a. At the 0.02 significance level, test the claim that the two groups come from populations with the same standard deviation.

   b. At the 0.02 significance level, test the claim that the two groups come from populations with the same mean.

    c. Construct a 98% confidence interval estimate of the mean score for all women who take the test.

COOPERATIVE GROUP ACTIVITIES

1. *Out-of-class activity:* Are estimates influenced by anchoring numbers? Refer to the related Chapter 2 Cooperative Group Activity. In Chapter 2 we noted that, according to author John Rubin, when people must estimate a value, their estimate is often "anchored" to (or influenced by) a preceding number. In that Chapter 2 activity, some subjects were asked to quickly estimate the value of $8 \times 7 \times 6 \times 5 \times 4 \times 3 \times 2 \times 1$, and others were asked to quickly estimate the value of $1 \times 2 \times 3 \times 4 \times 5 \times 6 \times 7 \times 8$. In Chapter 2, we could compare the two sets of results by using statistics (such as the mean) and graphs (such as boxplots). The methods of Chapter 8 now allow us to compare the results with a formal hypothesis test. (See Exercise 12 in Section 8-6.) Specifically, collect your own sample data and test the claim that when we begin with larger numbers (as in $8 \times 7 \times 6$), our estimates tend to be larger.

2. *In-class activity:* Collect your own data to test the claim that male and female statistics students have the same mean pulse rate. Divide into groups according to gender, with about 10 or 12 students in each group. Each group member should record his or her pulse rate by counting the number of heartbeats in one minute, and the group statistics (n, $\bar{x}$, s) should be calculated. The groups should exchange their results and test the null hypothesis of no difference between the two population means. Is there a difference? Explain.

3. *In-class activity:* Divide into groups of about 10 or 12 students and use the reaction timer included with the Chapter 5 Cooperative Group Activities. Each group member should be tested for right-hand reaction time and left-hand reaction time. Using the group results, test the claim that there is no difference between the right-hand and left-hand reaction times. Compare the conclusion to the conclusions reached by other groups. Is there a difference?

4. *In-class activity:* Divide into groups of three or four to test the claim that the Dvorak keyboard is easier to use than the Qwerty keyboard. See the Chapter Problem for Chapter 2, and also see the exercise in this chapter in the feature titled "From Data to Decision: Critical Thinking." Instead of using the given data sets, collect your own data. Determine whether it is better to select a sample of text that will be used to rate the words by both methods, or whether it is better to select two separate samples of text so that the Qwerty rating scheme can be applied to one while the Dvorak rating scheme can be applied to the other. Collect your own data, then use the appropriate method to test the hypothesis that the Dvorak keyboard configuration is better.

EXCEL PROJECT

Excel can be used to generate normally distributed data drawn from a population with a specified mean and standard deviation, as follows:

1. Click on **Tools, Data Analysis, Random Number Generation,** then **OK.**
2. In the dialog box, enter 1 for the number of variables, enter the desired sample size for the number of random numbers, select a **Normal** distribution, and enter values for the mean and standard deviation.

Use this procedure to generate two sets of sample data that represent simulated IQ scores, as shown below.

IQ Scores of Treatment Group: Generate 10 sample values from a normally distributed population with mean 100 and standard deviation 15.

IQ Scores of Placebo Group: Generate 12 sample values from a normally distributed population with mean 100 and standard deviation 15.

You can see from the way the data are generated that both data sets really come from the same population, so there should be no difference between the two sample means.

a. After generating the two data sets, use a 0.10 significance level to test the claim that the two samples come from populations with the same mean.

b. If this experiment is repeated, what percentage of trials will lead to the conclusion that the two population means are different? How does this relate to a type I error?

c. If your generated data should lead to the conclusion that the two population means are different, would this conclusion be correct or incorrect in reality? How do you know?

d. Repeat the experiment until you get two samples that cause you to reject the null hypothesis of equal means. Based on that number, how would you describe the likelihood of rejecting equality of the population means?

from DATA to DECISION

Critical Thinking:
Is there a real difference between typing on a Qwerty keyboard and typing on a Dvorak keyboard?

Is there a real difference in ease of typing between using the Qwerty keyboard configuration and the Dvorak keyboard configuration? (See the Chapter Problem for Chapter 2 for illustrations of the actual keyboard configurations.) The Chapter 2 introductory problem presented two data sets obtained by using the 52 words in the Preamble to the Constitution. The first data set consisted of the point rating score for each word typed on the Qwerty keyboard, and the second data set listed the word ratings for the Dvorak keyboard. Using the methods of Chapter 2, we can compare the data sets using tools such as the sample statistics $(n, \bar{x}, s)$ and boxplots. The sample statistics are summarized in the accompanying table. Comparing the sample means in that table, we see that the Qwerty mean of 4.4 appears to be considerably higher than the Dvorak mean of 1.7, suggesting that the Qwerty keyboard is more difficult to use (because higher

word ratings result from words that are more difficult to type). But is this apparent difference really *significant*?

Analyzing the results
Refer to the original data listed in Tables 2-1 and 2-2 at the beginning of Chapter 2, and conduct a hypothesis test to determine whether the Qwerty population mean is different from the Dvorak population mean. Identify the method you use and be sure to justify the requirements that must be satisfied for that method. Write a brief report summarizing your procedure and conclusions.

Qwerty and Dvorak Sample Statistics

	Qwerty	Dvorak
n	52	52
$\bar{x}$	4.4	1.7
s	2.8	1.8

TABLE 2-1	Qwerty Keyboard Word Ratings								
2	2	5	1	2	6	3	3	4	2
4	0	5	7	7	5	6	6	8	10
7	2	2	10	5	8	2	5	4	2
6	2	6	1	7	2	7	2	3	8
1	5	2	5	2	14	2	2	6	3
1	7								

TABLE 2-2	Dvorak Keyboard Word Ratings								
2	0	3	1	0	0	0	0	2	0
4	0	3	4	0	3	3	1	3	5
4	2	0	5	1	4	0	3	5	0
2	0	4	1	5	0	4	0	1	3
0	1	0	3	0	1	2	0	0	0
1	4								

Internet Project

Hypothesis Testing

Whereas Chapter 7 presented methods for testing hypotheses about a single population, the methods of this chapter extend those methods to two populations. Similarly, the Internet Project for this chapter difers from that of Chapter 7 through the use of two data sets in each of the hypothesis tests. The project can be found at

http://awlonline.com/triola

There you will find several hypothesis-testing problems that examine population demographics, salary fairness, and a traditional superstition. In each case you will formulate the problem as a hypothesis test using the notation of the chapter, seek out appropriate data, conduct the test, and then interpret the results with a summarizing conclusion.

Statistics *at work*

"The basic role of statistics here at Consumers Union is to ensure that the studies we do are valid, reliable, and understandable."

David Burn

Director, Technical Information Management for Consumers Union, which publishes Consumer Reports.

When they're not designing experiments or analyzing the results, statisticians (and other Consumers Union staff) have the opportunity to participate in product tests. As an avid runner and triathlete, David often volunteers for tests of fitness equipment. In this case, his assessment of various treadmills is data used to rate their quality.

Author's note: The author met with David Burn and the other statisticians at Consumers Union: M. Edna Derderian, Keith Newsom-Stewart, Martin Romm, and Jed Schneider. The author toured the testing facility and saw that Consumers Union meticulously designs, executes, and analyzes objective experiments.

What is the role of statistics in your product evaluation?

The basic role of statistics here at Consumers Union is to ensure that the studies we do are valid, reliable, and understandable. We look for significance in results in the sense that if we conclude that product A is better than product B, that conclusion is based on good, solid statistical theory. But we always remind ourselves that statistical significance does not outweigh practical significance. If there is a statistical significance of 0.01 in the difference between two speakers, but there is a practical significance of 0.1 in the way the human ear hears those speakers, then the statistical significance is irrelevant. That's the way the article would be written. We might say that the tested speakers are of such a high quality that there is really no difference, so you could buy based on price.

Who works on a typical evaluation project?

We work with different departments having people from a variety of backgrounds. Typically, a project leader is from the department where the product belongs. An electronics engineer might be the project leader for a study of CD players. Other team members typically include a statistician, a market information analyst, a writer, and an editor.

What does the statistician do?

For the statistician, the work really begins when the market survey report comes in and we can see what consumers want to know about. The decision on what to test is made by the project team with input from all members, including the statistician. We design an experiment and method of sampling. For a CD player, we might want to test sound quality and resistance to skipping when the unit is jarred. We work with the project leader on the test protocol, sampling, and other experimental design issues. The statistician begins analyzing the test data by using general exploratory methods with graphs and numerical measurements. We explore the data because we don't necessarily know what we are going to get. At this stage we are using tools like scatterplots, boxplots, and histograms to look for features such as outliers. Then we do analyses that might make some products stand out or appear to be different from the others. Ultimately, we use statistical methods to develop overall scores we call ratings. A statistical report is written and checked by one of the other statisticians. The type of analysis is questioned to be sure that it is the best choice, and we check that everything was done correctly and that the results are valid.

How do you ensure objectivity?

We don't accept any type of outside advertising. We don't represent any organizations or companies: We represent the consumer. We buy all of our products on the open market, just the way any consumer would. We have anonymous shoppers throughout the United States in all major areas, and the statistical design for the sampling plan might require shopping in San Francisco, Florida, New York, and Texas—all at the same time for the same product. That way we get a representative sample of the product that the manufacturer is making.

Do you use computers in your work?

We all use state-of-the-art computers, and a variety of statistical and other software packages.

Would you recommend a statistics course to college students?

I would recommend several statistics courses. Certainly, elementary statistics courses are important for an overview of parametric and nonparametric methods. In addition, linear models, sampling, and design of experiment provide the depth and appreciation needed for critical review of many issues involving statistical inquiries. Beyond this, the ability to communicate your ideas is essential. Because we work on several projects simultaneously, personal organization and project management are critical.

Correlation and Regression

How much of a tip should I leave?

All of us are confronted with the problem of determining how much of a tip we should leave a waiter or waitress in a restaurant. Although unofficial, many of us have heard that the tip should be 15% of the bill. Some of us try to walk a fine line between giving away too much money and looking too cheap. Others take genuine pleasure in rewarding service with a very generous tip, and there are those who have a firm policy of giving either nothing or as little as possible. Almost all of us believe that there is a relationship between the amount of the bill and the amount of the tip: Larger bills result in larger tips. Let's consider Table 9-1, which lists some sample data collected from the author's students. (Much better results could be obtained by using a sample that is larger than the six pairs of data included in Table 9-1, but we are using a sample of six pairs so that it will be easy to follow the calculations in the sections that follow.)

In Table 9-1, there appears to be a pattern of larger tips for larger bills. We will analyze the data in Table 9-1 as we address two questions:

1. Is there sufficient evidence to conclude that there is a relationship between the amount of the bill and the amount of the tip?

2. If there is a relationship, how do we use it to determine how much of a tip should be left?

TABLE 9-1	Paired Data for Six Dining Parties					
Bill ($)	33.46	50.68	87.92	98.84	63.60	107.34
Tip ($)	5.50	5.00	8.08	17.00	12.00	16.00

9-1 Overview

The main focus of this chapter is to form inferences based on sample data that come in *pairs*. Given such paired sample data, we want to determine whether there is a relationship between the two variables and, if so, to identify what the relationship is. For example, using the data in Table 9-1, we want to determine whether there is a relationship between the amount of a restaurant bill and the amount of the tip. If such a relationship exists, we want to describe it with an equation so that we can identify the rule that people follow when they leave tips.

We begin in Section 9-2 by considering the concept of correlation, which is used to determine whether there is a statistically significant relationship between two variables. We investigate correlation using the scatterplot (a graph) and the linear correlation coefficient (a measure of the strength of linear association between two variables). In Section 9-3 we investigate regression analysis; we describe the relationship between two variables with an equation that relates them and show how to use that equation to predict values of one variable when we know values of the other variable.

In Section 9-4 we analyze the differences between predicted values and actual observed values of a variable. In Sections 9-2 through 9-4 we consider relationships between two variables, and in Section 9-5 we use concepts of multiple regression to describe the relationship among three or more variables. Finally, in Section 9-6 we describe some basic methods for developing a mathematical model that can be used to describe the relationship between two variables. Although Section 9-3 is limited to linear relationships, Section 9-6 includes some common nonlinear relationships.

 ## 9-2 Correlation

The main objective of this section is to take a collection of paired sample data (sometimes called **bivariate data**) and determine whether there appears to be a relationship between the two variables. In statistics, we refer to such a relationship as a *correlation*. (We will consider only *linear* relationships, which means that when graphed, the points approximate a straight-line pattern.)

> **DEFINITION**
>
> A **correlation** exists between two variables when one of them is related to the other in some way.

Table 9-1, for example, consists of paired bill/tip data. We will determine whether there is a correlation between the variable x (amount of bill) and the variable y (amount of tip).

As we work with sample data and develop methods of forming inferences about populations, we make the following assumptions.

Assumptions

1. The sample of paired (x, y) data is a *random* sample.

2. The pairs of (x, y) data have a **bivariate normal distribution.** (Normal distributions are discussed in Chapter 5, but this assumption basically requires that for any fixed value of x, the corresponding values of y have a distribution that is bell-shaped, and for any fixed value of y, the values of x have a distribution that is bell-shaped.)

The second assumption is usually difficult to check, but a partial check can be made by determining whether the values of both x and y have distributions that are basically bell-shaped.

Exploring the Data

Before working with the more formal computational methods of this section, we should first explore the data set to see what we can learn. We can often see a relationship between two variables by constructing a graph called a *scatterplot*, or *scatter diagram*.

DEFINITION

A **scatterplot** (or **scatter diagram**) is a graph in which the paired (x, y) sample data are plotted with a horizontal x-axis and a vertical y-axis. Each individual (x, y) pair is plotted as a single point.

As an example, see the Excel display of the six pairs of data listed in Table 9-1. Such scatterplots can be generated using Excel's Chart Wizard or the DDXL add-in. See the directions at the end of this section.

Predicting the Stock Market

Forecasting and predicting are important goals of statistics. Investment advisors seek indicators that can be used to forecast stock market performance. Some indicators are quite colorful, such as the hemline index: Rising skirt hemlines supposedly precede a rise in the Dow Jones Industrial Average. According to the Super Bowl omen, a Super Bowl victory by a team with NFL origins is followed by a year in which the New York Stock Exchange index rises; otherwise, it falls. (In 1970, the NFL and AFL merged into the current NFL.) This indicator has been correct in 28 of the past 32 years, largely due to the fact that NFL teams win much more often, and the stock market tends to follow an upward trend. Other indicators: aspirin sales, limousines on Wall Street, and elevator traffic at the New York Stock Exchange.

When we examine such a scatterplot, we should study the overall pattern of the plotted points. If there is a pattern, we should note its direction. That is, as one variable increases, does the other seem to increase or decrease? We should observe whether there are any outliers, which are points that lie very far away from all of the other points. The Excel-generated scatterplot does seem to reveal a pattern showing that larger bills tend to go along with larger tips. There do not appear to be any outliers.

Other examples of scatterplots are shown in Figure 9-1. The graphs in Figure 9-1(a), (b), and (c) depict a pattern of increasing values of y that correspond to increasing values of x. As you proceed from (a) to (c), the

FIGURE 9-1
Scatterplots

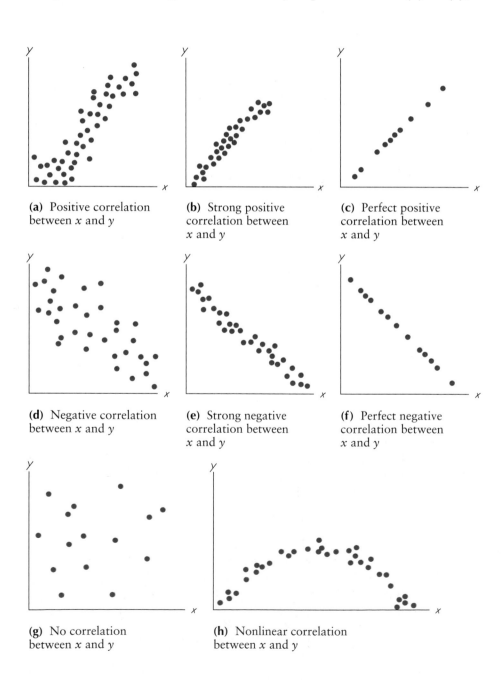

(a) Positive correlation between x and y

(b) Strong positive correlation between x and y

(c) Perfect positive correlation between x and y

(d) Negative correlation between x and y

(e) Strong negative correlation between x and y

(f) Perfect negative correlation between x and y

(g) No correlation between x and y

(h) Nonlinear correlation between x and y

dot pattern becomes closer to a straight line, suggesting that the relationship between x and y becomes stronger. The scatterplots in (d), (e), and (f) depict patterns in which the y-values decrease as the x-values increase. Again, as you proceed from (d) to (f), the relationship becomes stronger. In contrast to the first six graphs, the scatterplot of (g) shows no pattern and suggests that there is no correlation (or relationship) between x and y. Finally, the scatterplot of (h) shows a pattern, but it is not a straight-line pattern.

Linear Correlation Coefficient

Because visual examinations of scatterplots are largely subjective, we need more precise and objective measures. We will use the linear correlation coefficient r, which is useful for detecting straight-line patterns.

DEFINITION

The **linear correlation coefficient** r measures the strength of the linear relationship between the paired x- and y-values in a *sample*. Its value is computed by using Formula 9-1, which follows. [The linear correlation coefficient is sometimes referred to as the **Pearson product moment correlation coefficient** in honor of Karl Pearson (1857–1936), who originally developed it.]

Formula 9-1
$$r = \frac{n\Sigma xy - (\Sigma x)(\Sigma y)}{\sqrt{n(\Sigma x^2) - (\Sigma x)^2}\sqrt{n(\Sigma y^2) - (\Sigma y)^2}}$$

Because r is calculated using sample data, it is a sample statistic used to measure the strength of the linear correlation between x and y. If we had every pair of population values for x and y, the result of Formula 9-1 would be a population parameter, represented by ρ (Greek rho).

We will describe how to compute and interpret the linear correlation coefficient r given a list of paired data, but let's first identify the notation relevant to Formula 9-1. Later in this section we will present the underlying theory that led to the development of Formula 9-1.

Notation for the Linear Correlation Coefficient

n	represents the number of pairs of data present.
Σ	denotes the addition of the items indicated.
Σx	denotes the sum of all x-values.
Σx^2	indicates that each x-value should be squared and then those squares added.
$(\Sigma x)^2$	indicates that the x-values should be added and the total then squared. It is extremely important to avoid confusing Σx^2 and $(\Sigma x)^2$.

(continued)

Σxy indicates that each x-value should first be multiplied by its corresponding y-value. After obtaining all such products, find their sum.

r represents the linear correlation coefficient for a *sample*.

ρ represents the linear correlation coefficient for a *population*.

Rounding the Linear Correlation Coefficient

Round the linear correlation coefficient r to three decimal places (so that its value can be directly compared to critical values in Table A-6). When calculating r and other statistics in this chapter, rounding in the middle of a calculation often creates substantial errors, so try using your calculator's memory to store intermediate results and round off only at the end. Many inexpensive calculators have Formula 9-1 built in so that you can automatically evaluate r after entering the sample data.

EXAMPLE **Tipping** Using the data in Table 9-1, find the value of the linear correlation coefficient r. (A later example will use this value to determine whether there is a relationship between the amounts of bills and the amounts of tips.)

SOLUTION For the sample paired data in Table 9-1, $n = 6$ because there are six pairs of data. The other components required in Formula 9-1 are found from the calculations in Table 9-2. Note how this vertical format makes the calculations easier.

TABLE 9-2 Finding Statistics Used to Calculate r

Bill ($)		Tip ($)			
x		y	$x \cdot y$	x^2	y^2
33.46	×	5.50 =	184.0300	1119.5716	30.2500
50.68	×	5.00 =	253.4000	2568.4624	25.0000
87.92		8.08	710.3936	7729.9264	65.2864
98.84		17.00	1680.2800	9769.3456	289.0000
63.60		12.00	763.2000	4044.9600	144.0000
107.34		16.00	1717.4400	11,521.8756	256.0000
Total 441.84		63.58	5308.7436	36,754.1416	809.5364
↑		↑	↑	↑	↑
Σx		Σy	Σxy	Σx^2	Σy^2

Using the calculated values and Formula 9-1, we can now evaluate r as follows:

$$r = \frac{n\Sigma xy - (\Sigma x)(\Sigma y)}{\sqrt{n(\Sigma x^2) - (\Sigma x)^2}\sqrt{n(\Sigma y^2) - (\Sigma y)^2}}$$

$$= \frac{6(5308.7436) - (441.84)(63.58)}{\sqrt{6(36,754.1416) - (441.84)^2}\sqrt{6(809.5364) - (63.58)^2}}$$

$$= \frac{3760.2744}{\sqrt{25,302.264}\sqrt{814.802}} = 0.828 \quad \rightarrow \text{close to } 1 \text{ is very strong}$$

These calculations get quite messy with large data sets, so it's fortunate that the linear correlation coefficient can be found automatically with many different calculators and computer programs. With Excel, use the command CORREL. Using the data in Table 9-1, enter the x values in cells A1 through A6, enter the y values in cells B1 through B6, then enter the command **=CORREL(A1: A6, B1: B6)** to get a result of 0.828159. ◄

Interpreting the Linear Correlation Coefficient

We need to interpret a calculated value of r, such as the value of 0.828 found in the preceding example. Given the way that Formula 9-1 is constructed, the value of r must always fall between -1 and $+1$ inclusive. If r is close to 0, we conclude that there is no significant linear correlation between x and y, but if r is close to -1 or $+1$, we conclude that there is a significant linear correlation between x and y. Interpretations of "close to" 0 or 1 or -1 are vague, so we use the following very specific decision criterion:

> **If the absolute value of the computed value of r exceeds the value in Table A-6, conclude that there is a significant linear correlation. Otherwise, there is not sufficient evidence to support the conclusion of a significant linear correlation.**

When there really is no linear correlation between x and y, Table A-6 lists values that are "critical" in this sense: They separate *usual* values of r from those that are *unusual*. For example, Table A-6 shows us that with $n = 6$ pairs of sample data, the critical values are 0.811 (for $\alpha = 0.05$) and 0.917 (for $\alpha = 0.01$). Critical values and the role of α are carefully described in Chapters 6 and 7. Here's how we interpret those numbers: With six pairs of data and no linear correlation between x and y, there is a 5% chance that the absolute value of the computed linear correlation coefficient r will exceed 0.811. With $n = 6$ and no linear correlation, there is a 1% chance that $|r|$ will exceed 0.917.

Student Ratings of Teachers

Many colleges equate high student ratings with good teaching—an equation often fostered by the fact that student evaluations are easy to administer and measure.

However, one study that compared student evaluations of teachers with the amount of material learned found a strong *negative* correlation between the two factors. Teachers rated highly by students seemed to induce less learning.

In a related study, an audience gave a high rating to a lecturer who conveyed very little information but was interesting and entertaining.

EXAMPLE   **Tipping** Given the sample data in Table 9-1, for which $r = 0.828$, refer to Table A-6 to determine whether there is a significant linear correlation between the amounts of bills and the amounts of tips. In Table A-6, use the critical value for $\alpha = 0.05$. (With $\alpha = 0.05$, we will

Palm Reading

Some people believe that the length of their palm's lifeline can be used to predict longevity. In a letter published in the *Journal of the American Medical Association*, authors M. E. Wilson and L. E. Mather refuted that belief with a study of cadavers. Ages at death were recorded, along with the lengths of palm lifelines. The authors concluded that there is no significant correlation between age at death and length of lifeline. Palmistry lost, hands down.

conclude that there is a significant linear correlation only if the sample is unlikely in the sense that such a value of r occurs 5% of the time or less.)

SOLUTION   Referring to Table A-6, we locate the row for which $n = 6$ (because there are six pairs of data). That row contains the critical values of 0.811 (for $\alpha = 0.05$) and 0.917 (for $\alpha = 0.01$). Using the critical value for $\alpha = 0.05$, we see that there is less than a 5% chance that with no linear correlation, the absolute value of the computed r will exceed 0.811. Because $r = 0.828$, its absolute value does exceed 0.811, so we conclude that there is a significant linear correlation between bills and tips. ◄

We have already noted that the format of Formula 9-1 requires that the calculated value of r always fall between -1 and $+1$ inclusive. We list that property along with other important properties.

Properties of the Linear Correlation Coefficient r

1. The value of r is always between -1 and 1 inclusive. That is,
$$-1 \leq r \leq 1$$

2. *The value of r does not change if all values of either variable are converted to a different scale.* For example, if the U.S. dollar amounts in Table 9-1 are converted to French francs, the value of r will not change.

3. *The value of r is not affected by the choice of x or y.* Interchange all x- and y-values and the value of r will not change.

4. *r measures the strength of a linear relationship.* It is not designed to measure the strength of a relationship that is not linear.

Interpreting r: Explained Variation

If we conclude that there is a significant linear correlation between x and y, we can find a linear equation that expresses y in terms of x, and that equation can be used to predict values of y for given values of x. In Section 9-3 we will describe a procedure for finding such equations and show how to predict values of y when given values of x. But a predicted value of y will not necessarily be the exact result, because in addition to x, there are other factors affecting y, such as random variation and other characteristics not included in the study. In Section 9-4 we will present a rationale and more details about this important principle:

The value of r^2 is the proportion of the variation in y that is explained by the linear relationship between x and y.

EXAMPLE   **Tipping** Using the bill/tip data in Table 9-1, we have found that the linear correlation coefficient is $r = 0.828$. What proportion of the variation in the tip can be explained by the variation in the bill?

SOLUTION   With $r = 0.828$, we get $r^2 = 0.686$, and we conclude that 0.686 (or about 69%) of the variation in tips can be explained by the linear relationship between the bill and the tip. This implies that about 31% of the variation in tips can be explained by factors other than the amount of the bill. Such other factors might include the friendliness of the waiter or waitress, the state of the economy, the wealth of the person giving the tip, the quality of the food, the ambience, and so on.   ◀

Common Errors Involving Correlation

We now identify three of the most common sources of errors made in interpreting results involving correlation:

1. *A common source of error involves concluding that correlation implies causality.* One study showed a strong correlation between the salaries of statistics professors and per capita beer consumption, but those two variables are affected by the state of the economy, a third variable lurking in the background. (A **lurking variable** is one that affects the variables being studied, but is not included in the study.)

2. *Another source of error arises with data based on averages.* Averages suppress individual variation and may inflate the correlation coefficient. One study produced a 0.4 linear correlation coefficient for paired data relating income and education among individuals, but the linear correlation coefficient became 0.7 when regional averages were used.

3. *A third source of error involves the property of linearity.* A relationship may exist between x and y even when there is no significant linear correlation. The data depicted in Figure 9-2 result in a value of $r = 0$, which is an indication of no *linear* correlation between the two variables. However, we can easily see from looking at the figure that there is a pattern reflecting a very strong *nonlinear* relationship. (Figure 9-2 is a scatterplot that depicts the relationship between distance above ground and time elapsed for an object thrown upward.)

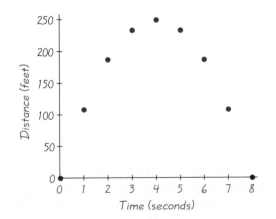

FIGURE 9-2
Scatterplot of Distance above Ground and Time for Object Thrown Upward

Formal Hypothesis Test (Requires Coverage of Chapter 7)

We present two methods (summarized in Figure 9-3) for using a formal hypothesis test to determine whether there is a significant linear correlation between two variables. Some instructors prefer Method 1 because it reinforces concepts introduced in earlier chapters. Others prefer Method 2 because it involves easier calculations.

Figure 9-3 shows that the null and alternative hypotheses will be expressed as follows:

$$H_0: \rho = 0 \qquad \text{(No linear correlation)}$$

$$H_1: \rho \neq 0 \qquad \text{(Linear correlation)}$$

For the test statistic, we use one of the following methods.

Method 1: Test Statistic Is t This method follows the format presented in earlier chapters. It uses the Student t distribution with a test statistic having the form $t = (r - \mu_r)/s_r$, where μ_r and s_r denote the claimed value of the mean and the sample standard deviation of r values. Because we assume that $\rho = 0$, it follows that $\mu_r = 0$. Also, it can be shown that s_r, the standard deviation of linear correlation coefficients, can be expressed as $\sqrt{(1 - r^2)/(n - 2)}$. We can therefore use the following test statistic.

Test Statistic t for Linear Correlation

$$t = \frac{r}{\sqrt{\dfrac{1 - r^2}{n - 2}}}$$

Critical values: Use Table A-3 with degrees of freedom $= n - 2$.

Method 2: Test Statistic Is r This method requires fewer calculations. Instead of calculating the test statistic just given, we use the computed value of r as the test statistic. Critical values are found in Table A-6.

Test statistic r for Linear Correlation

Test statistic: r
Critical values: Refer to Table A-6.

Figure 9-3 shows that the decision criterion is to reject the null hypothesis of $\rho = 0$ if the absolute value of the test statistic exceeds the critical values; rejection of $\rho = 0$ means that there is sufficient evidence to support a claim of a linear correlation between the two variables. If the absolute value of the test statistic does not exceed the critical values, then we fail to reject

FIGURE 9-3
Testing for a Linear
Correlation

Start

Let $H_0: \rho = 0$
$H_1: \rho \neq 0$

Select a
significance
level α.

Calculate r using
Formula 9-1.

Method 1
(Follows format of
earlier chapters)

Method 2
(Uses fewer
calculations)

The test statistic is

$$t = \frac{r}{\sqrt{\dfrac{1 - r^2}{n - 2}}}$$

Critical values of t are from
Table A-3 with $n - 2$ degrees
of freedom.

The test statistic is r.

Critical values of r are from
Table A-6.

If the absolute value of the
test statistic exceeds the
critical values, reject $H_0: \rho = 0$.
Otherwise, fail to reject H_0.

If H_0 is rejected, conclude that there
is a significant linear correlation.
If you fail to reject H_0, then there is
not sufficient evidence to conclude
that there is a linear correlation.

$\rho = 0$; that is, there is not sufficient evidence to conclude that there is a linear correlation between the two variables.

EXAMPLE   Tipping Using the sample data in Table 9-1, test the claim that there is a linear correlation between amounts of bills and amounts of tips. For the test statistic, use both (a) Method 1 and (b) Method 2.

SOLUTION   Refer to Figure 9-3. To claim that there is a significant linear correlation is to claim that the population linear correlation coefficient ρ is different from 0. We therefore have the following hypotheses.

$$H_0: \rho = 0 \qquad \text{(No linear correlation)}$$
$$H_1: \rho \neq 0 \qquad \text{(Linear correlation)}$$

No significance level α was specified, so use $\alpha = 0.05$.

In a preceding example we already found that $r = 0.828$. With that value, we now find the test statistic and critical value, using each of the two methods just described.

a.   *Method 1:* The test statistic is

$$t = \frac{r}{\sqrt{\dfrac{1 - r^2}{n - 2}}} = \frac{0.828}{\sqrt{\dfrac{1 - 0.828^2}{6 - 2}}} = 2.953$$

The critical values of $t = -2.776$ and $t = 2.776$ are found in Table A-3, where 2.776 corresponds to 0.05 divided between two tails (with 0.025 in each tail) and the number of degrees of freedom is $n - 2 = 4$. See Figure 9-4 for the graph that includes the test statistic and critical values.

b.   *Method 2:* The test statistic is $r = 0.828$. The critical values of $r = -0.811$ and $r = 0.811$ are found in Table A-6 with $n = 6$ and $\alpha = 0.05$. See Figure 9-5 for a graph that includes this test statistic and critical values.

Using either of the two methods, we find that the absolute value of the test statistic does exceed the critical value (Method 1: 2.953 > 2.776; Method 2: 0.828 > 0.811); that is, the test statistic does fall within the critical region. We therefore reject $H_0: \rho = 0$. There is sufficient evidence to support the claim of a linear correlation between bills and tips. The

FIGURE 9-4
Testing $H_0: \rho = 0$ with Method 1

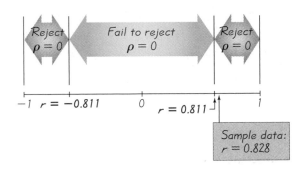

FIGURE 9-5
Testing H_0: $\rho = 0$ with
Method 2

amount of the tip does seem to correspond to the amount of the bill, as we would expect. ◀

One-tailed Tests: The preceding example and Figures 9-4 and 9-5 illustrate a two-tailed hypothesis test. The examples and exercises in this section will generally involve only two-tailed tests, but one-tailed tests can occur with a claim of a positive linear correlation or a claim of a negative linear correlation. In such cases, the hypotheses will be as shown here.

Claim of Negative Correlation **(Left-tailed test)**	**Claim of Positive Correlation** **(Right-tailed test)**
H_0: $\rho \geq 0$	H_0: $\rho \leq 0$
H_1: $\rho < 0$	H_1: $\rho > 0$

For these one-tailed tests, Method 1 can be handled as in earlier chapters. For Method 2, either calculate the critical value as described in Exercise 23 or modify Table A-6 by replacing the column headings of $\alpha = 0.05$ and $\alpha = 0.01$ by the one-sided critical values of $\alpha = 0.025$ and $\alpha = 0.005$, respectively.

Rationale: We have presented Formula 9-1 for calculating r and have illustrated its use; we will now give a justification for it. Formula 9-1 simplifies the calculations used in this equivalent formula:

$$r = \frac{\Sigma(x - \bar{x})(y - \bar{y})}{(n - 1)s_x s_y}$$

We will temporarily use this latter version of Formula 9-1 because its form relates more directly to the underlying theory. We will consider the following paired data, which are depicted in the scatterplot shown in Figure 9-6.

x	1	1	2	4	7
y	4	5	8	15	23

Figure 9-6 on the next page includes the point $(\bar{x}, \bar{y}) = (3, 11)$, which is called the *centroid* of the sample points.

DEFINITION

Given a collection of paired (x, y) data, the point $(\bar{x}, \bar{y})$ is called the **centroid**.

FIGURE 9-6
Scatterplot Partitioned into Quadrants

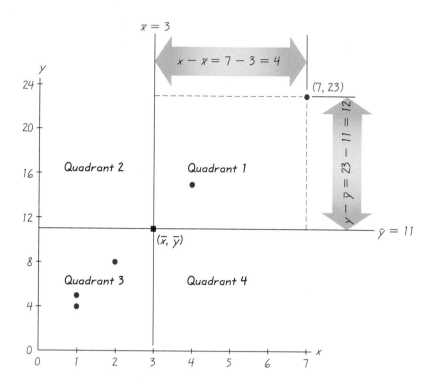

The statistic *r*, sometimes called the *Pearson product moment,* was first developed by Karl Pearson. It is based on the sum of the product of the moments $(x - \overline{x})$ and $(y - \overline{y})$; that is, on the statistic $\Sigma(x - \overline{x})(y - \overline{y})$. In any scatterplot, vertical and horizontal lines through the centroid $(\overline{x}, \overline{y})$ divide the diagram into four quadrants, as in Figure 9-6. If the points of the scatterplot tend to approximate an uphill line (as in the figure), individual values of the product $(x - \overline{x})(y - \overline{y})$ tend to be positive because most of the points are found in the first and third quadrants, where the products of $(x - \overline{x})$ and $(y - \overline{y})$ are positive. If the points of the scatterplot approximate a downhill line, most of the points are in the second and fourth quadrants, where $(x - \overline{x})$ and $(y - \overline{y})$ are opposite in sign, so $\Sigma(x - \overline{x})(y - \overline{y})$ is negative. Points that follow no linear pattern tend to be scattered among the four quadrants, so the value of $\Sigma(x - \overline{x})(y - \overline{y})$ tends to be close to 0.

The sum $\Sigma(x - \overline{x})(y - \overline{y})$ depends on the magnitude of the numbers used. For example, if you change *x* from inches to feet, that sum will change. To make *r* independent of the particular scale used, we include the sample standard deviations as follows:

$$r = \frac{\Sigma(x - \overline{x})(y - \overline{y})}{(n - 1)s_x s_y}$$

This expression can be algebraically manipulated into the equivalent form of Formula 9-1.

In preceding chapters we discussed methods of inferential statistics by addressing methods of hypothesis testing, as well as methods for constructing confidence interval estimates. A similar procedure may be used to find confidence intervals for ρ. However, because the construction of such confidence intervals involves somewhat complicated transformations, that process is presented in Exercise 24 (Beyond the Basics).

We can use the linear correlation coefficient to determine whether there is a linear relationship between two variables. Using the data in Table 9-1, we have concluded that there is a linear correlation between amounts of restaurant bills and amounts of tips left by the customers. Having concluded that a relationship exists, we would like to determine what that relationship is so that we can calculate the amount of the tip once we know the amount of the bill. This next stage of analysis is addressed in the following section.

Using Excel

Begin by entering the paired sample data in columns A and B. You can use either of the following two procedures, but we recommend using the DDXL add-in because it lets you use one simple procedure to get both the value of the linear correlation coefficient r and a scatterplot.

DDXL for Both *r* and Scatterplot

1. Enter the paired sample data in columns A and B.
2. Click on **DDXL.**
3. Select **Regression,** then click on the Function Type box and select **Correlation.**
4. In the dialog box, click on the pencil icon for the X-Axis Variable column and enter the range of values for the first variable, such as A1:A6 for the *x*-values (bill amounts) in Table 9-1.
5. Click on the pencil icon for the Y-Axis Variable column and enter the range of values for the second variable, such as B1:B6 for the *y*-values (tip amounts) in Table 9-1.
6. Click **OK** and the value of the correlation coefficient will be displayed with the scatterplot.

Excel's CORREL Function for Finding *r*

1. Enter the paired sample data in columns A and B.
2. Click on the *f*x function key located on the main menu bar.
3. Select the function category of **Statistical** and select the function name of **Correl,** then click **OK.** In the dialog box, make these entries: In the Array 1 box, enter the cell ranges of values for *x*, such as A1:A6. In the Array 2 box, enter the cell ranges of values for *y*, such as B1:B6. The value of the linear correlation coefficient *r* will be displayed.

Excel's Chart Wizard for Scatterplot

1. With the paired sample data entered in columns A and B, click on the **Chart Wizard** (with the icon that resembles a bar graph).

2. Select the chart type identified as **XY (Scatter).**
3. In the dialog box, enter the input range of the data, such as A1:B6 for six pairs of data in columns A and B. Click **Next** and proceed to use the dialog boxes to modify the graph as desired.

9-2 Basic Skills and Concepts

In Exercises 1 and 2, use a significance level of $\alpha = 0.05$.

1. Testing for Correlation and Describing Variation Assume that 40 pairs of data result in a linear correlation coefficient of $r = 0.401$.
   a. Is there a significant linear correlation between x and y? Explain.
   b. What proportion of the variation in y can be explained by the variation in x?

2. Testing for Correlation and Describing Variation Assume that 60 pairs of data result in a linear correlation coefficient of $r = -0.222$.
   a. Is there a significant linear correlation between x and y? Explain.
   b. What proportion of the variation in y can be explained by the variation in x?

Testing for a Linear Correlation. *In Exercises 3 and 4, use a scatterplot and the linear correlation coefficient r to determine whether there is a correlation between the two variables.*

EXCEL 3.

x	1	2	3	5	5
y	19	12	4	20	18

EXCEL 4.

x	2	3	3	5	8
y	5	9	10	12	25

Testing for a Linear Correlation. *In Exercises 5–16, construct a scatterplot, find the value of the linear correlation coefficient r and use a significance level of $\alpha = 0.05$ to determine whether there is a significant linear correlation between the two variables. Save your work because the same data sets will be used in the next section.*

EXCEL 5. Supermodel Heights and Weights Listed below are heights (in inches) and weights (in pounds) for supermodels Niki Taylor, Nadia Avermann, Claudia Schiffer, Elle MacPherson, Christy Turlington, Bridget Hall, Kate Moss, Valerie Mazza, and Kristy Hume.

Height (in.)	71	70.5	71	72	70	70	66.5	70	71
Weight (lb.)	125	119	128	128	119	127	105	123	115

EXCEL 6. Bear Chest Sizes and Weights When bears were anesthetized, researchers measured the distances (in inches) around the bears' chests and weighed the bears (in pounds). The results are given below for eight male bears. Based on the results, does a bear's weight seem to be related to its chest size? Do the results change if the chest measurements are converted to feet, with each of those values divided by 12?

x Chest (in.)	26	45	54	49	41	49	44	19
y Weight (lb)	90	344	416	348	262	360	332	34

Based on data from Minitab and Gary Alt.

EXCEL **7.** Discarded Plastic and Household Size The accompanying table lists weights (in pounds) of plastic discarded by a sample of households, along with the sizes of the households. Is there a significant linear correlation? This issue is important to the Census Bureau, which provided project funding, because the presence of a correlation implies that we can predict population size by analyzing discarded garbage.

Plastic (lb)	0.27	1.41	2.19	2.83	2.19	1.81	0.85	3.05
Household size	2	3	3	6	4	2	1	5

Based on data provided by Masakazu Tani and the Garbage Project at the University of Arizona.

EXCEL **8.** Discarded Paper and Household Size The paired data below consist of weights (in pounds) of discarded paper and sizes of households.

Paper (lb)	2.41	7.57	9.55	8.82	8.72	6.96	6.83	11.42
Household size	2	3	.3	6	4	2	1	5

Based on data obtained from the Garbage Project at the University of Arizona.

EXCEL **9.** DWI and Jail A study was conducted to investigate the relationship between age (in years) and BAC (blood alcohol concentration) measured when convicted DWI (driving while intoxicated) jail inmates were first arrested. Sample data are given below for randomly selected subjects. Based on the result, does the BAC level seem to be related to the age of the person tested?

Age	17.2	43.5	30.7	53.1	37.2	21.0	27.6	46.3
BAC	0.19	0.20	0.26	0.16	0.24	0.20	0.18	0.23

Based on data from the Dutchess County STOP-DWI Program.

EXCEL **10.** Weapons and Murder Rate The accompanying table lists the number of registered automatic weapons (in thousands), along with the murder rate (in murders per 100,000 of population), for randomly selected states. Automatic weapons are guns that continue to fire repeatedly while the trigger is held back. Are firearm murders often committed with automatic weapons? Does a significant linear correlation imply that increased numbers of automatic weapons result in more murders?

Automatic weapons	11.6	8.3	3.6	0.6	6.9	2.5	2.4	2.6
Murder rate	13.1	10.6	10.1	4.4	11.5	6.6	3.6	5.3

Data provided by the FBI and the Bureau of Alcohol, Tobacco, and Firearms.

EXCEL **11.** Old Faithful Geyser Refer to Data Set 11 in Appendix B. The Excel workbook is OLDFAITH.xls.
   a. Use the paired data for durations and intervals after eruptions of the geyser. Is there a significant linear correlation, suggesting that the interval after an eruption is related to the duration of the eruption?
   b. Use the paired data for intervals after eruptions and heights of eruptions of the Old Faithful geyser. Is there a significant linear correlation, suggesting that the interval after an eruption is related to the height of the eruption?
   c. Assume that you want to develop a method for predicting the time interval to the next eruption. Based on the results from parts (a) and (b), which factor would be more relevant: eruption duration or eruption height? Why?

EXCEL **12.** Cigarette Nicotine, Tar, and Carbon Monoxide Refer to Data Set 8 in Appendix B. The Excel workbook is CIGARET.xls.

(continued)

a. Use the paired data consisting of tar and nicotine. Based on the result, does there appear to be a significant linear correlation between cigarette tar and nicotine? If so, can researchers reduce their laboratory expenses by measuring only one of these two variables?

b. Use the paired data consisting of carbon monoxide and nicotine. Based on the result, does there appear to be a significant linear correlation between cigarette carbon monoxide and nicotine? If so, can researchers reduce their laboratory expenses by measuring only one of these two variables?

c. Assume that researchers want to develop a method for predicting the amount of nicotine, and they want to measure only one other item. In choosing between tar and carbon monoxide, which is the better choice? Why?

13. Diamond Prices, Carats, and Color Refer to Data Set 3 in Appendix B. The Excel workbook is DIAMONDS.xls.

a. Use the paired data consisting of price and carat (weight). Is there a significant linear correlation between the price of a diamond and its weight in carats?

b. Use the paired price/color data. Is there a significant linear correlation between the price of a diamond and its color?

c. Assume that you are planning to buy a diamond engagement ring. In considering the value of a diamond, which characteristic should you consider to be more important: the carat weight or the color? Why?

14. Car Fuel Consumption, Weight, and Displacement Refer to Data Set 18 in Appendix B. The Excel workbook is CARS.xls.

a. Use the paired data consisting of highway fuel consumption rate (mi/gal) and weight (pounds). Is there a significant linear correlation between those two variables?

b. Use the paired data consisting of highway fuel consumption rate (mi/gal) and the engine displacement (in liters). Is there a significant linear correlation between those two variables?

c. Assume that you want to develop a way to predict a car's fuel consumption rate without actually measuring it. Which factor would be more relevant: the car's weight or its engine displacement? Why?

15. Movie Grosses, Budgets, and Viewer Ratings Refer to Data Set 15 in Appendix B. The Excel workbook is MOVIES.xls.

a. Use the paired data consisting of the amount of money grossed and the amount of money budgeted. Is there a significant linear correlation between those two variables?

b. Use the paired data consisting of the amount of money grossed and the viewer rating. Is there a significant linear correlation between those two variables?

c. Based on the preceding results, if you plan to invest in a company that makes movies, which is the better choice: a company that makes big budget movies or a company that makes movies that viewers rate highly? Why?

16. Home Selling Prices, List Prices, and Taxes Refer to Data Set 16 in Appendix B. The Excel workbook is HOME.xls.

a. Use the paired data consisting of home selling price and list price. We might expect that these variables would be related, but is there sufficient evidence to support that expectation?

b. Use the paired data consisting of home selling price and the amount of taxes. The tax bill is supposed to be based on the value of the house. Is it? Explain.

Identifying Correlation Errors.   *In Exercises 17–20, describe the error in the stated conclusion. (See the list of common sources of errors included in this section.)*

17. *Given:* The paired sample data of the ages of subjects and their scores on a test of reasoning result in a linear correlation coefficient very close to 0.

    Conclusion: Younger people tend to get higher scores.

18. *Given:* There is a significant linear correlation between personal income and years of education.

    Conclusion: More education causes a person's income to rise.

19. *Given:* Subjects take a test of verbal skills and a test of manual dexterity, and those pairs of scores result in a linear correlation coefficient very close to 0.

    Conclusion: Scores on the two tests are not related in any way.

20. *Given:* There is a significant linear correlation between state average tax burdens and state average incomes.

    Conclusion: There is a significant linear correlation between individual tax burdens and individual incomes.

9-2 Beyond the Basics

21. Using Data from Scatterplot Sometimes, instead of having numerical data, we have only graphical data. The Excel scatterplot below is similar to one that was included in "The Prevalence of Nosocomial Infection in Intensive Care Units in Europe," by Jean-Louis Vincent et al., *Journal of the American Medical Association,* Vol. 274, No. 8. Each point represents a different European country. Estimate the value of the linear correlation coefficient, and determine whether there is a significant linear correlation between the mortality rate and the rate of infections acquired in intensive care units.

EXCEL **22.** Correlations with Transformed Data In addition to testing for a linear correlation between x and y, we can often use transformations of data to explore for other relationships. For example, we might replace each x-value by x^2 and use the methods of this section to determine whether there is a linear correlation between y and x^2. Given the paired data in the accompanying table, construct the scatterplot and then test for a linear correlation between y and each of the following. Which case results in the largest value of r?

a. x **b.** x^2 **c.** $\log x$ **d.** $\sqrt{x}$ **e.** $1/x$

x	1.3	2.4	2.6	2.8	2.4	3.0	4.1
y	0.11	0.38	0.41	0.45	0.39	0.48	0.61

23. Finding Critical r-Values The critical values of r in Table A-6 are found by solving

$$t = \frac{r}{\sqrt{\dfrac{1 - r^2}{n - 2}}}$$

for r to get

$$r = \frac{t}{\sqrt{t^2 + n - 2}}$$

where the t-value is found from Table A-3 by assuming a two-tailed case with $n - 2$ degrees of freedom. Table A-6 lists the results for selected values of n and α. Use the formula for r given here and Table A-3 (with $n - 2$ degrees of freedom) to find the critical values of r for the given cases.

a. $H_0: \rho = 0, n = 50, \alpha = 0.05$
b. $H_1: \rho \neq 0, n = 75, \alpha = 0.10$
c. $H_0: \rho \geq 0, n = 20, \alpha = 0.05$
d. $H_0: \rho \leq 0, n = 10, \alpha = 0.05$
e. $H_1: \rho > 0, n = 12, \alpha = 0.01$

24. Constructing Confidence Intervals for ρ Given n pairs of data from which the linear correlation coefficient r can be found, use the following procedure to construct a confidence interval about the population parameter ρ.

Step a.   Use Table A-2 to find the $z_{\alpha/2}$ value that corresponds to the desired degree of confidence.

Step b.   Evaluate the interval limits w_L and w_R:

$$w_L = \frac{1}{2}\ln\left(\frac{1 + r}{1 - r}\right) - z_{\alpha/2} \cdot \frac{1}{\sqrt{n - 3}}$$

$$w_R = \frac{1}{2}\ln\left(\frac{1 + r}{1 - r}\right) + z_{\alpha/2} \cdot \frac{1}{\sqrt{n - 3}}$$

Step c.   Now evaluate the confidence interval limits in the expression below.

$$\frac{e^{2w_L} - 1}{e^{2w_L} + 1} < \rho < \frac{e^{2w_R} - 1}{e^{2w_R} + 1}$$

Use this procedure to construct a 95% confidence interval for ρ, given 50 pairs of data for which $r = 0.600$.

25. Is the Real Issue Correlation? Refer to Table 8-1, located with the Chapter Problem at the beginning of Chapter 8. Table 8-1 includes paired data representing the reported heights and the measured heights of 12 male statistics students. Can correlation be used to address the issue of whether male statistics students exaggerate their heights? Why or why not?

9-3 Regression

In Section 9-2 we analyzed paired data with the goal of determining whether there is a significant linear correlation between two variables. The main objective of this section is to describe the relationship between two variables by finding the graph and equation of the straight line that represents the relationship. This straight line is called the *regression line*, and its equation is called the *regression equation*. Sir Francis Galton (1822–1911) studied the phenomenon of heredity and showed that when tall or short couples have children, the heights of those children tend to *regress*, or revert to the more typical mean height for people of the same gender. We continue to use Galton's "regression" terminology, even though our data might not involve the same height phenomena.

DEFINITIONS

Given a collection of paired sample data, the **regression equation**

$$\hat{y} = b_0 + b_1 x$$

algebraically describes the relationship between the two variables. The graph of the regression equation is called the **regression line** (or *line of best fit,* or *least-squares line*).

This definition expresses a relationship between x (called the **independent variable,** or **predictor variable**) and $\hat{y}$ (called the **dependent variable,** or **response variable**). In the preceding definition, the typical equation of a straight line $y = mx + b$ is expressed in the form $\hat{y} = b_0 + b_1 x$, where b_0 is the y-intercept and b_1 is the slope. The following notation box shows that b_0 and b_1 are sample statistics used to estimate the population parameters β_0 and β_1.

Notation for Regression Equation		
	Population Parameter	**Sample Statistic**
y-<u>intercept</u> of regression equation	β_0	b_0
<u>Slope</u> of regression equation	β_1	b_1
Equation of the regression line	$y = \beta_0 + \beta_1 x$	$\hat{y} = b_0 + b_1 x$

For the regression methods given in this section, we make the following assumptions.

Assumptions

1. We are investigating only *linear* relationships.

2. For each *x*-value, *y* is a random variable having a normal (bell-shaped) distribution. All of these *y* distributions have the same variance. Also, for a given value of *x,* the distribution of *y*-values has a mean that lies on the regression line. (Results are not seriously affected if departures from normal distributions and equal variances are not too extreme.)

An important goal of this section is to use paired sample data to estimate the regression equation. Using only sample data, we can't find the exact values of the population parameters β_0 and β_1, but we can use the sample data to estimate them with b_0 and b_1, which are found by using Formulas 9-2 and 9-3.

Formula 9-2
$$b_0 = \frac{(\Sigma y)(\Sigma x^2) - (\Sigma x)(\Sigma xy)}{n(\Sigma x^2) - (\Sigma x)^2} \qquad \textit{y-intercept}$$

Formula 9-3
$$b_1 = \frac{n(\Sigma xy) - (\Sigma x)(\Sigma y)}{n(\Sigma x^2) - (\Sigma x)^2} \qquad \text{(slope)}$$

These formulas might look intimidating, but they are programmed into many calculators and computer programs, so the values of b_0 and b_1 can be easily found. (See "Using Excel" at the end of this section.) In those cases when we must use formulas instead of a calculator or computer, the required computations will be much easier if we keep in mind the following facts:

1. If the linear correlation coefficient *r* has been computed using Formula 9-1, the values of Σx, Σy, Σx^2, and Σxy have already been found, and they can be used again in Formula 9-3. (Also, the numerator for *r* in Formula 9-1 is the same numerator for b_1 in Formula 9-3; the denominator for *r* includes the denominator for b_1. If the calculation for *r* is set up carefully, the calculation for b_1 requires the simple division of one known number by another.)

2. If you find the slope b_1 first, you can use Formula 9-4 to find the *y*-intercept b_0. [The regression line always passes through the centroid $(\overline{x}, \overline{y})$ so that $\overline{y} = b_0 + b_1\overline{x}$ must be true, and this equation can be expressed as Formula 9-4. It is usually easier to find the *y*-intercept b_0 by using Formula 9-4 than by using Formula 9-2.]

Formula 9-4
$$b_0 = \overline{y} - b_1\overline{x}$$

Once we have evaluated b_0 and b_1, we can identify the estimated regression equation, which has the following special property: *The regression line fits the sample points best.* (The specific criterion used to determine which line fits "best" is the least-squares property, which will be described later.)

We will now briefly discuss rounding and then illustrate the procedure for finding and applying the regression equation.

Rounding the y-Intercept b_0 and the Slope b_1

It's difficult to provide a simple universal rule for rounding values of b_0 and b_1, but we usually try to round each of these values to *three significant digits* or use the values provided by Excel. Because these values are very sensitive to rounding at intermediate steps of calculations, try to carry at least six significant digits (or use exact values) in the intermediate steps. Depending on how you round, this book's answers to examples and exercises may be slightly different from your answers.

> **EXAMPLE**   **Tipping** In Section 9-2 we used the Table 9-1 data ($x =$ amount of bill; $y =$ amount of tip) to find that the linear correlation coefficient of $r = 0.828$ indicates that there is a significant linear correlation. Now find the regression equation of the straight line that relates x and y.

SOLUTION   We will find the regression equation by using Formulas 9-3 and 9-4 and these values already found in Table 9-2 in Section 9-2:

$$n = 6 \qquad \Sigma x = 441.84 \qquad \Sigma y = 63.58$$
$$\Sigma x^2 = 36{,}754.1416 \qquad \Sigma y^2 = 809.5364 \qquad \Sigma xy = 5308.7436$$

First find the slope b_1 by using Formula 9-3:

$$b_1 = \frac{n(\Sigma xy) - (\Sigma x)(\Sigma y)}{n(\Sigma x^2) - (\Sigma x)^2}$$

$$= \frac{6(5308.7436) - (441.84)(63.58)}{6(36{,}754.1416) - (441.84)^2} = \frac{3760.2744}{25{,}302.264} = 0.148614$$

$$= 0.149 \quad \text{(rounded)}$$

Next, find the y-intercept b_0 by using Formula 9-4:

$$b_0 = \overline{y} - b_1 \overline{x}$$
$$= \frac{63.58}{6} - (0.148614)\frac{441.84}{6} = -0.347 \quad \text{(rounded)}$$

Knowing the slope b_1 and y-intercept b_0, we can now express the estimated equation of the regression line as

$$\hat{y} = -0.347 + 0.149x$$

We should realize that this equation is an *estimate* of the true regression equation $y = \beta_0 + \beta_1 x$. This estimate is based on one particular set of sample data listed in Table 9-1, but another sample drawn from the same population would probably lead to a slightly different equation.

The Excel display on the next page shows the regression line plotted on the scatterplot.

Cell Phones and Crashes

Because some countries have banned the use of cell phones in cars while other countries are considering such a ban, researchers studied the issue of whether the use of cell phones while driving increases the chance of a crash. A sample of 699 drivers was obtained. Members of the sample group used cell phones and were involved in crashes. Subjects completed questionnaires and their telephone records were checked. Telephone usage was compared to the time interval immediately preceding a crash to a comparable time period the day before. Conclusion: Use of a cell phone was associated with a crash risk that was about four times as high as the risk when a cell phone was not used. (See "Association between Cellular-Telephone Calls and Motor Vehicle Collisions," by Redelmeier and Tibshirani, *New England Journal of Medicine,* Vol. 336, No. 7.)

Pizza Correlates with Crisis

When former President Clinton was threatened with impeachment by Congress, government employees worked late and ordered record numbers of pizzas. Frank Meeks, owner of 59 Domino's Pizza outlets in Washington, D. C., reported that on the Saturday during the height of the impeachment crisis, Capitol Hill pizza deliveries exceeded $10,000 while White House pizza deliveries totaled $3,000. Meeks noted that pizza sales also peaked during the Persian Gulf War, and they peak annually during budget debates.

We can see that this line fits the data well. Note that even though the regression line fits the data points well, it does not actually pass through any of the original data points. In some other cases, the regression line might pass through one or more of the sample data points, but it is common for the regression line to miss all of them.

Using the Regression Equation for Predictions

Regression equations can be helpful when used in *predicting* the value of one variable, given some particular value for the other variable. If the regression line fits the data quite well, then it makes sense to use its equation for predictions, provided that we don't go beyond the scope of the available values. However, *we should use the equation of the regression line only if r indicates that there is a significant linear correlation. In the absence of a significant linear correlation, we should not use the regression equation for projecting or predicting; instead, our best estimate of the second variable is simply its sample mean.*

In predicting a value of *y* based on some given value of *x* . . .

1. If there is *not* a significant linear correlation, the best predicted *y*-value is $\bar{y}$.

2. If there is a significant linear correlation, the best predicted *y*-value is found by substituting the *x*-value into the regression equation.

Figure 9-7 summarizes this process, which is easier to understand if we think of *r* as a measure of how well the regression line fits the sample data. If *r* is near -1 or $+1$, then the regression line fits the data well, but if *r* is near 0, then the regression line fits poorly (and should not be used for predictions).

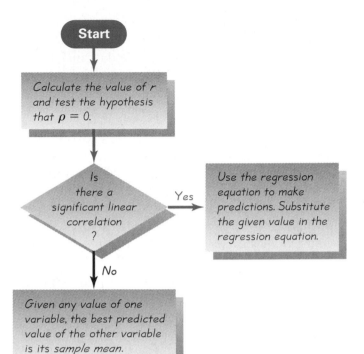

FIGURE 9-7
**Predicting the Value of a
Variable**

EXAMPLE   Predicting Tip Using the sample data in Table 9-1, we found that there is a significant linear correlation between amounts of bills and amounts of tips, and we also found that the regression equation is $\hat{y} = -0.347 + 0.149x$. If the amount of a bill is \$100, predict the amount of the tip that will be left.

SOLUTION   There's a strong temptation to jump in and substitute 100 for x in the regression equation, but we should first consider whether there is a significant linear correlation that justifies the use of that equation. In this example, we do have a significant linear correlation (with $r = 0.828$), so our predicted value is found as follows:

$$\hat{y} = -0.347 + 0.149x$$
$$= -0.347 + 0.149(100) = \$14.55$$

The predicted tip for a \$100 bill is \$14.55. (If there had not been a significant linear correlation, our best predicted tip would have been $\overline{y} = \$63.58/6 = \$10.60$.)

EXAMPLE   Shoe Sizes and Tips There is obviously no significant linear correlation between shoe sizes of adults and the amounts that they tip waiters and waitresses. Given that an adult has a shoe size of 9, what is the best prediction of the amount of tip this person would leave?

SOLUTION   Because there is no significant linear correlation, we do not use a regression equation. Instead, the best predicted tip is simply the *mean* tip that is left. Based on the sample data in Table 9-1, the mean tip is $\overline{y} = \$63.58/6 = \$10.60$, so that is the best predicted value.   ◄

Carefully compare the solutions to the preceding two examples and note that we used the regression equation when there was a significant linear correlation, but in the absence of such a correlation, the best predicted value of y is simply the value of the sample mean $\overline{y}$. Also, common sense suggests that if someone is given a restaurant bill of $100 and that person has a shoe size of 9, the tip value of $14.55 (based on the regression equation) is a better prediction than $10.60 (based on the mean tip amount for everyone).

Interpreting Predicted Values   We should understand that when we determine a predicted value of y based on some given value of x, the predicted value will not necessarily be the exact result. There are other factors, such as random variation and characteristics not included in the study, that can also affect y. Based on the sample data in Table 9-1, we see that for a bill of $100, the best predicted tip is $14.55. This does not mean that every bill of $100 will result in a tip of exactly $14.55; some will be less and some will be more. Based on the regression equation we have found, many different bills of $100 will result in many different tips, but the single best estimate of the tip is $14.55.

A common error is to use the regression equation for making a prediction when there is no significant linear correlation. That error violates the first of the following guidelines.

Guidelines for Using the Regression Equation

1. *If there is no significant linear correlation, don't use the regression equation to make predictions.*

2. *When using the regression equation for predictions, stay within the scope of the available sample data.* If you find a regression equation that relates women's heights and shoe sizes, it's absurd to predict the shoe size of a woman who is 10 ft tall. (The regression equation is like a function whose domain is the scope of the sample x-values.)

3. *A regression equation based on old data is not necessarily valid now.* The regression equation relating used-car prices and ages of cars is no longer usable if it's based on data from the 1970s.

4. *Don't make predictions about a population that is different from the population from which the sample data were drawn.* If we collect sample data from men and develop a regression equation relating age and TV remote-control usage, the results don't necessarily apply to women. If we use state *averages* to develop a regression equation relating SAT math scores and SAT verbal scores, the results don't necessarily apply to *individuals*.

Interpreting the Regression Equation: Marginal Change

We can use the regression equation to see the effect on one variable when the other variable changes by some specific amount.

> ### DEFINITION
>
> In working with two variables related by a regression equation, the **marginal change** in a variable is the amount that it changes when the other variable changes by exactly one unit.

The slope b_1 in the regression equation represents the marginal change in y that occurs when x changes by one unit. For the bill/tip data of Table 9-1, the regression line has a slope of 0.149, which shows that if we increase x (the amount of the bill) by 1 dollar, the tip will increase by 0.149 dollars, or about 15 cents. Marginal change in this case can be simply described by stating that for every dollar the bill goes up, the tip goes up by about 15 cents.

Outliers and Influential Points

A correlation/regression analysis of bivariate (paired) data should include an investigation of *outliers* and *influential points,* defined as follows.

> ### DEFINITIONS
>
> In a scatterplot, an **outlier** is a point lying far away from the other data points.
>
> Paired sample data may include one or more **influential points,** which are points that strongly affect the graph of the regression line.

An outlier is easy to identify: Examine the scatterplot and identify a point that is far away from the others. Here's how to determine whether a point is an influential point: Graph the regression line resulting from the data with the point included, then graph the regression line resulting from the data with the point excluded. If the graph changes by a considerable amount, the point is influential. Influential points are often found by identifying those outliers that are *horizontally* far away from the other points.

For example, refer to the preceding Excel display. Suppose that we include one more pair of data coming from a really hungry couple who ran up a dinner bill of $500, then left a tip of $75. This additional point would be an outlier because it would be far away from the other points, but it would not be an influential point because it would not change the position of the regression line very much. Now suppose the same couple ran up a $500 bill but left no tip ($0). This additional point would now become an influential point because the graph of the regression line would change considerably, as shown by the Excel display here. Compare this regression line to the one shown in the last Excel display, and you will see clearly that the addition of that one hungry and stingy couple has a dramatic effect on the regression line.

Residuals and the Least-Squares Property

We have stated that the regression equation represents the straight line that fits the data "best," and we will now describe the criterion used in determining the line that is better than all others. This criterion is based on the vertical distances between the original data points and the regression line. Such distances are called *residuals*.

DEFINITION

For a sample of paired (x, y) data, a **residual** is the difference $(y - \hat{y})$ between an observed sample y-value and the value of $\hat{y}$, which is the value of y that is predicted by using the regression equation.

This definition might seem as clear as tax-form instructions, but you can easily understand residuals by referring to Figure 9-8, which corresponds to the paired sample data in the margin. In Figure 9-8, the residuals are represented by the dashed lines. For a specific example, see the residual indicated as 7, which is directly above $x = 5$. If we substitute $x = 5$ into the regression equation $\hat{y} = 5 + 4x$, we get a predicted value of $\hat{y} = 25$. When $x = 5$, the *predicted* value of y is $\hat{y} = 25$, but the actual *observed* sample value is $y = 32$. The difference $y - \hat{y} = 32 - 25 = 7$ is a residual.

The regression equation represents the line that fits the points "best" according to the *least-squares property*.

DEFINITION

A straight line satisfies the **least-squares property** if the sum of the squares of the residuals is the smallest sum possible.

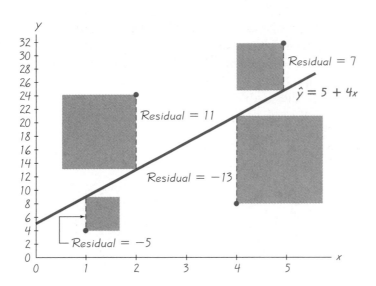

From Figure 9-8, we see that the residuals are $-5, 11, -13$, and 7, so the sum of their squares is

$$(-5)^2 + 11^2 + (-13)^2 + 7^2 = 364$$

We can visualize the least-squares property by referring to Figure 9-8, where the squares of the residuals are represented by the red square areas. The sum of the red square areas is 364, which is the smallest sum possible. Use any other straight line, and the red squares will combine to produce an area larger than the combined red area of 364.

Fortunately, we need not deal directly with the least-squares property when we want to find the equation of the regression line. Calculus has been used to build the least-squares property into Formulas 9-2 and 9-3. Because the derivations of these formulas require calculus, we don't include them in this text.

Using Excel for Regression

The equation of the regression line and a graph of the regression line can be obtained with the DDXL add-in or Excel's Data Analysis add-in. See the following procedures.

DDXL for Regression Line

1. Enter the paired data in columns A and B.
2. Click on **DDXL.**
3. Select **Regression,** then select **Simple Regression.**
4. Click on the pencil icon for the Response Variable and enter the range of cells containing the values for the *y*-variable, such as B1:B6.
5. Click on the pencil icon for the Explanatory Variable and enter the range of cells containing the values for the *x*-variable, such as A1:A6, then click **OK.**
6. The display will include several different items, including a scatterplot with the graph of the regression line. Shown on the next page is

a portion of the display that contains the y-intercept and slope of the regression line. This portion of the display results from using the data in Table 9-1. See the *coefficient* column for the values that lead to this regression equation: $y = -0.347 + 0.149x$. (The "constant" is the y-intercept, and the second value is the slope.)

EXCEL

```
Dependent variable is:    B1:B6
No Selector
R squared = 68.6%       R squared (adjusted) = 60.7%
s =  3.266  with  6 - 2 = 4  degrees of freedom

Source        Sum of Squares   df   Mean Square   F-ratio
Regression    93.1383          1    93.1383       8.73
Residual      42.662           4    10.6655

Variable    Coefficient   s.e. of Coeff   t-ratio   prob
Constant    -0.347279     3.936           -0.0882   0.9339
A1:A6       0.148614      0.05029         2.96      0.0418
```

Excel's Data Analysis

1. Enter the paired data in columns A and B.

2. Click on **Tools.**

3. Select **Data Analysis,** then select **Regression.**

4. A dialog box should appear. In the box for "Input Y Range" enter the range of cells for the y-variable, such as B1:B6.

5. In the box for "Input X Range" enter the range of cells for the x-variable, such as A1:A6.

6. Click on the box for **Line Fit Plots.** Shown below is a portion of the display. This portion of the display results from using the data in Table 9-1. See the *coefficients* column for the values that lead to this regression equation: $y = -0.347 + 0.149x$.

EXCEL

	Coefficients	Standard Error	t Stat	P-value	Lower 95%	Upper 95%
Intercept	-0.3472792	3.936081495	-0.08823	0.933935	-11.2756	10.58106
X Variable	0.14861415	0.050290571	2.95511	0.041757	0.008985	0.288243

7. The original Excel graph will take considerable editing to achieve an effective appearance. Edit the graph by right-clicking or double-clicking on different parts of the display. For example, position the mouse pointer on one of the plotted points and right-click, then select the option of **Add Trendline.** Among the variety of models that will appear, select **Linear.** Also, click on the **Options** tab and place check marks in the boxes to display the equation of the regression line and the value of R^2. Make the graph larger by clicking inside the border,

then dragging border boxes to the right or left or up or down. Shown below are the original Excel graph and the same graph after it was edited for improvements.

Original Excel Graph

Edited Excel Graph

9-3 Basic Skills and Concepts

Finding the Equation of the Regression Line.   *In Exercises 1–4, use the given data to find the equation of the regression line.*

 1.

x	1	3	3	5
y	6	8	8	10

2.

x	2	4	5	6	8
y	3	1	0	−1	−3

3.

x	1	2	3	5	5
y	19	12	4	20	18

4.

x	2	3	3	5	8
y	5	9	10	12	25

Finding the Equation of the Regression Line and Making Predictions.   *Exercises 5–16 use the same data sets as the exercises in Section 9-2. In each case, find the regression equation, letting the first variable be the independent (x) variable. Find predicted values where they are requested. Caution: When finding predicted values, be sure to follow the prediction procedure described in this section.*

 5. Supermodel Heights and Weights

Height (in.)	71	70.5	71	72	70	70	66.5	70	71
Weight (lb)	125	119	128	128	119	127	105	123	115

Find the best predicted weight of a supermodel who is 69 in. tall.

6. Bear Chest Sizes and Weights

x Chest (in.)	26	45	54	49	41	49	44	19
y Weight (lb)	90	344	416	348	262	360	332	34

Find the best predicted weight of a bear with a chest size of 52 in.

7. Discarded Plastic and Household Size

Plastic (lb)	0.27	1.41	2.19	2.83	2.19	1.81	0.85	3.05
Household size	2	3	3	6	4	2	1	5

What is the best predicted size of a household that discards 0.50 lb of plastic?

8. Discarded Paper and Household Size

Paper (lb)	2.41	7.57	9.55	8.82	8.72	6.96	6.83	11.42
Household size	2	3	3	6	4	2	1	5

What is the best predicted size of a household that discards 10.00 lb of paper?

9. DWI and Jail

Age	17.2	43.5	30.7	53.1	37.2	21.0	27.6	46.3
BAC	0.19	0.20	0.26	0.16	0.24	0.20	0.18	0.23

What is the best predicted blood alcohol level of a person 21.0 years of age who has been convicted and jailed for DWI? (The BAC level is measured at the time of arrest.)

10. Weapons and Murder Rate

Automatic weapons	11.6	8.3	3.6	0.6	6.9	2.5	2.4	2.6
Murder rate	13.1	10.6	10.1	4.4	11.5	6.6	3.6	5.3

What is the best predicted murder rate for a state with 10,000 registered automatic weapons? (In this table, the numbers of automatic weapons are in thousands, and the murder rate is number of murders per 100,000 people.)

11. Old Faithful Geyser Refer to Data Set 11 (OLDFAITH.xls) in Appendix B.
   a. Use the paired data for durations and intervals after eruptions of the Old Faithful geyser. What is the best predicted time before the next eruption if the previous eruption lasted for 210 sec?
   b. Use the paired height and interval data for eruptions of the Old Faithful geyser. What is the best predicted time before the next eruption if the previous eruption had a height of 275 ft?
   c. Which predicted time is better: the result from part (a) or the result from part (b)? Why?

12. Cigarette Nicotine, Tar, and Carbon Monoxide Refer to Data Set 8 (CIGARET.xls) in Appendix B.
   a. Use the paired data consisting of tar and nicotine. What is the best predicted amount of nicotine for a cigarette with 3 mg of tar?
   b. Use the paired data consisting of carbon monoxide and nicotine. What is the best predicted amount of nicotine for a cigarette with 18 mg of carbon monoxide?
   c. Which prediction is better: the result from part (a) or the result from part (b)? Why?

13. Diamond Prices, Carats, and Color Refer to Data Set 3 (DIAMONDS.xls) in Appendix B.
   a. Use the paired data consisting of price and carat (weight). What is the best predicted price of a diamond with a weight of 1.5 carats?
   b. Use the paired price/color data. What is the best predicted price of a diamond with a color rating of 3?

c. Which prediction is better: the result from part (a) or the result from part (b)? Why?

EXCEL 14. Car Fuel Consumption, Weight, and Displacement Refer to Data Set 18 (CARS.xls) in Appendix B.

a. Use the paired data consisting of highway fuel consumption rate (mi/gal) and weight (pounds). What is the best predicted fuel consumption rate for a car that weighs 3000 lb?

b. Use the paired data consisting of highway fuel consumption rate (mi/gal) and the engine displacement (in liters). What is the best predicted fuel consumption rate for a car with an engine displacement of 4.0 liters?

c. Which prediction is better: the result from part (a) or the result from part (b)? Why?

EXCEL 15. Movie Grosses, Budgets, and Viewer Ratings Refer to Data Set 15 (MOVIES.xls) in Appendix B.

a. Use the paired data consisting of the amount of money grossed and the amount of money budgeted. What is the best predicted gross amount if a movie has a budgeted amount of $15 million?

b. Use the paired data consisting of the amount of money grossed and the viewer rating. What is the best predicted gross amount if a movie has a viewer rating of 7?

c. Is the prediction from part (a) better than the prediction from part (b)? Why or why not?

EXCEL 16. Home Selling Prices, List Prices, and Taxes Refer to Data Set 16 (HOMES.xls) in Appendix B.

a. Use the paired data consisting of home selling price and list price. What is the best predicted selling price of a home with a list price of $200,000?

b. Use the paired data consisting of home selling price and the amount of taxes. The tax bill is supposed to be based on the value of the house. Is it?

17. Predicting Values In each of the following cases, find the best predicted value of y given that $x = 3.00$. The given statistics are summarized from paired sample data.

a. $r = 0.931$, $\bar{y} = 7.00$, $n = 10$, and the equation of the regression line is $\hat{y} = 4.00 + 2.00x$.

b. $r = -0.033$, $\bar{y} = 2.50$, $n = 80$, and the equation of the regression line is $\hat{y} = 5.00 - 2.00x$.

18. Predicting Values In each of the following cases, find the best predicted value of y given that $x = 2.00$. The given statistics are summarized from paired sample data.

a. $r = -0.882$, $\bar{y} = 3.57$, $n = 15$, and the equation of the regression line is $\hat{y} = 23.00 - 8.00x$.

b. $r = 0.187$, $\bar{y} = 9.33$, $n = 60$, and the equation of the regression line is $\hat{y} = 4.00 + 8.00x$.

19. Identifying Outliers and Influential Points Refer to the six pairs of data listed in Table 9-1. If we include another dining party with a bill of $400 and a tip of $50, is the new point an outlier? Is it an influential point?

20. Identifying Outliers and Influential Points Refer to the six pairs of data listed in Table 9-1. If we include another dining party with a bill of $600 and a tip of $5, is the new point an outlier? Is it an influential point?

9-3 Beyond the Basics

21. *How Is Regression Equation Affected by Change in Scale?* Large numbers, such as those in the accompanying table, often cause computational problems. First use the given data to find the equation of the regression line, then find the equation of the regression line after each x-value has been divided by 1000. How are the results affected by the change in x? How would the results be affected if each y-value were divided by 1000?

x	924,736	832,985	825,664	793,427	857,366
y	142	111	109	95	119

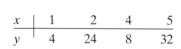

x	1	2	4	5
y	4	24	8	32

22. *Testing Least-Squares Property* According to the least-squares property, the regression line minimizes the sum of the squares of the residuals. We noted that with the paired data in the margin, the regression equation is $\hat{y} = 5 + 4x$ and the sum of the squares of the residuals is 364. Show that the equation $\hat{y} = 8 + 3x$ results in a sum of squares of residuals that is greater than 364.

23. *Using Logarithms to Transform Data* If the scatterplot reveals a nonlinear (not a straight line) pattern that you recognize as another type of curve, you may be able to apply the methods of this section. For the data given in the margin, find the linear equation ($y = b_0 + b_1 x$) that best fits the sample data, and find the logarithmic equation ($y = a + b \ln x$) that best fits the sample data. (*Hint:* Begin by replacing each x-value with $\ln x$.) Which of these two equations fits the data better? Why?

x	2.0	2.5	4.2	10.0
y	12.0	18.7	53.0	225.0

24. *Equivalent Hypothesis Tests* Explain why a test of the null hypothesis $H_0: \rho = 0$ is equivalent to a test of the null hypothesis $H_0: \beta_1 = 0$, where ρ is the linear correlation coefficient for a population of paired data, and β_1 is the slope of the regression line for that same population.

9-4 Variation and Prediction Intervals

The main objective of this section is to describe different types of variation that can be used for two major applications:

1. To determine the proportion of the variation in y that can be explained by the linear relationship between x and y.

2. To construct confidence interval estimates of predicted y-values. Such confidence intervals are called *prediction intervals*, which are formally defined later in the section.

Explained and Unexplained Variation

In Section 9-2 we introduced the concept of correlation and used the linear correlation coefficient r in determining whether there is a significant linear correlation between two variables, denoted by x and y. In addition to serving as a measure of the linear correlation between two variables, the value of r can also provide us with additional information about the variation of sample points about the regression line. We begin with a sample case, which leads to an important definition (coefficient of determination).

Suppose we have a large collection of paired data, which yields the following results:

- There is a significant linear correlation.
- The equation of the regression line is $\hat{y} = 3 + 2x$.
- The mean of the y-values is given by $\bar{y} = 9$.
- One of the pairs of sample data is $x = 5$ and $y = 19$.
- The point $(5, 13)$ is one of the points on the regression line, because substituting $x = 5$ into the regression equation yields $\hat{y} = 13$.

$$\hat{y} = 3 + 2x = 3 + 2(5) = 13$$

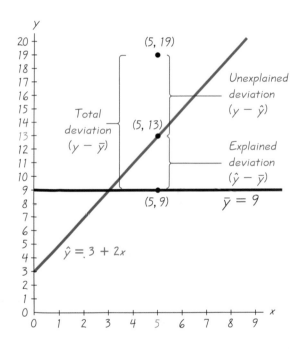

FIGURE 9-9
Unexplained, Explained, and Total Deviation

Figure 9-9 shows that the point $(5, 13)$ lies on the regression line, but the point $(5, 19)$ is from the original data set and does not lie on the regression line because it does not satisfy the regression equation. Take time to examine Figure 9-9 carefully and note the differences defined as follows.

DEFINITIONS

Assume that we have a collection of paired data containing the sample point (x, y), that $\hat{y}$ is the predicted value of y (obtained by using the regression equation), and that the mean of the sample y-values is $\bar{y}$.

The **total deviation** (from the mean) of the particular point (x, y) is the vertical distance $y - \bar{y}$, which is the distance between the point (x, y) and the horizontal line passing through the sample mean $\bar{y}$.   *(continued)*

The **explained deviation** is the vertical distance $\hat{y} - \bar{y}$, which is the distance between the predicted y-value and the horizontal line passing through the sample mean $\bar{y}$.

The **unexplained deviation** is the vertical distance $y - \hat{y}$, which is the vertical distance between the point (x, y) and the regression line. (The distance $y - \hat{y}$ is also called a *residual,* as defined in Section 9-3.)

Wage Gender Gap

Many articles note that, on average, full-time female workers earn about 70¢ for each $1 earned by full-time male workers. Researchers at the Institute for Social Research at the University of Michigan analyzed the effects of various key factors and found that about one-third of the discrepancy can be explained by differences in education, seniority, work interruptions, and job choices. The other two-thirds remains unexplained by such labor factors.

For the specific data under consideration, we get these results:

Total deviation of $(5, 19) = y - \bar{y} = 19 - 9 = 10$

Explained deviation of $(5, 19) = \hat{y} - \bar{y} = 13 - 9 = 4$

Unexplained deviation of $(5, 19) = y - \hat{y} = 19 - 13 = 6$

If we were totally ignorant of correlation and regression concepts and wanted to predict a value of y given a value of x and a collection of paired (x, y) data, our best guess would be $\bar{y}$. But we are not totally ignorant of correlation and regression concepts: We know that in this case (with a significant linear correlation) the way to predict the value of y when $x = 5$ is to use the regression equation, which yields $\hat{y} = 13$, as calculated earlier. We can explain the discrepancy between $\bar{y} = 9$ and $\hat{y} = 13$ by simply noting that there is a significant linear correlation best described by the regression line. Consequently, when $x = 5$, y *should be* 13, not the mean value of 9. But whereas y should be 13, *it is* 19. The discrepancy between 13 and 19 cannot be explained by the regression line, and it is called an *unexplained deviation*, or a *residual.* The specific case illustrated in Figure 9-9 can be generalized as follows:

(total deviation) = (explained deviation) + (unexplained deviation)

or $(y - \bar{y})$ = $(\hat{y} - \bar{y})$ + $(y - \hat{y})$

This last expression applies to a particular point (x, y), but it can be further generalized and modified to include all of the pairs of sample data, as shown in Formula 9-5. In that formula, the **total variation** is expressed as the sum of the squares of the total deviation values, the **explained variation** is the sum of the squares of the explained deviation values, and the **unexplained variation** is the sum of the squares of the unexplained deviation values. Excel provides the value of explained variation under the labels of SS (Sum of Squares) and Regression; the value of unexplained variation can be found from the labels of SS (Sum of Squares) and Residuals. See the Excel directions at the end of this section.

Formula 9-5

(total variation) = (explained variation) + (unexplained variation)

or $\Sigma(y - \bar{y})^2$ = $\Sigma(\hat{y} - \bar{y})^2$ + $\Sigma(y - \hat{y})^2$

Coefficient of Determination

The components of Formula 9-5 are used in the following important definition.

> ### DEFINITION
>
> The **coefficient of determination** is the amount of the variation in y that is explained by the regression line. It is computed as
>
> $$r^2 = \frac{\text{explained variation}}{\text{total variation}}$$

We can compute r^2 by using the definition just given with Formula 9-5, or we can simply square the linear correlation coefficient r, which is found by using the methods described in Section 9-2. For example, in Section 9-2 we noted that if $r = 0.828$, then $r^2 = 0.686$, which means that _68.6% of the total variation in y can be explained by the linear relationship between x and y (as described by the regression equation). It follows that 31.4% of the total variation in y remains unexplained._

 EXAMPLE   Diamonds In Exercise 13(a) in Section 9-2, we found that for the paired data consisting of weights (in carats) of diamonds and the prices of the diamonds, the linear correlation coefficient is given by $r = 0.767$. Find the percentage of the variation in y (price) that can be explained by the linear relationship between the weight and the price.

SOLUTION   The coefficient of determination is $r^2 = 0.767^2 = 0.588$, indicating that the ratio of explained variation in y to total variation in y is 0.588. We can now state that 58.8% of the total variation in y can be explained by the regression equation. We interpret this to mean that 58.8% of the total variation in diamond prices can be explained by the variation in their weights; the other 41.2% is attributable to other factors, such as color, clarity, and random chance. But remember that these results are estimates based on the given sample data. Other sample data will likely result in different estimates.   ◀

Prediction Intervals

In Section 9-3 we used the Table 9-1 sample data to find the regression equation $\hat{y} = -0.347 + 0.149x$, where $\hat{y}$ represents the predicted amount of tip and x represents the amount of the bill. We then used that equation to predict the y-value, given that $x = \$100$. We found that the best predicted tip amount for a $100 restaurant bill is $14.55. If we use the unrounded

from **DATA** *to* **DECISION**

Critical Thinking:
Which home should you buy?

Refer to Data Set 16 in Appendix B. Construct a scatterplot and test for a correlation between selling price and list price. Describe a way to visually examine the graph of the scatterplot and identify homes that represent "good buys" in the sense that they sell for disproportionately less than the amount predicted by the regression equation.

Analyze the Results
What is the predicted selling price of a home that lists for $200,000? Is that predicted selling price likely to be reasonably accurate or a very rough approximation? Explain. Test for a correlation between selling price and

tax bill. Assuming that the tax bill is supposed to be a fair assessment of a home's value, identify a home that appears to be assessed too low and identify another home that appears to be assessed too high.

Internet Project

Linear Regression

The linear correlation coefficient is a tool that can be used to measure the strength of the linear relationship between two variables. We can evaluate the linear correlation coefficient for any two sets of paired data, and then we can proceed to analyze the resulting values. Does it make sense for the two variables to be linearly correlated? Could a high correlation be caused by a third variable that is correlated with each of the two original variables? Go to the Web site for this book:

http://www.awlonline.com/triola

There you will find the Internet Project for this chapter. There you will be told how to find several paired data sets in the fields of sports, medicine, and economics. You will then apply the correlation and regression methods of this chapter to answer some of the questions that arise in this type of analysis.

Statistics
at work

Angela Gillespie

Traffic Analyst, Lycos.com

As a Traffic Analyst for Lycos, Inc., Angela reports on major and minor traffic metrics. She monitors changes in trends and behavior patterns for use, enhancing the site to increase reach and stickiness (the amount of time people spend online at any particular web site).

What is your job at Lycos?

I produce traffic reports of our site's activities each week. These are reviewed by our product group teams and senior management. They see what is increasing, what is decreasing, and make decisions about where our resources are spent.

My reports basically analyze trends on the sites and give projections for where we will be in a year or in any given time frame.

What concepts of statistics do you use?

Regression analysis and R-squared values.

How do you use statistics on the job?

To determine what is working and what is not working for our users. To determine the effectiveness of advertising campaigns, and to project future growth.

Please describe one specific example illustrating how the use of statistics was successful in improving a product or service.

At the end of our last fiscal year our CEO, Bob Davis, presented the company with an average daily pageview goal to be reached by the end of the next fiscal year. Using two years worth of pageview data, I put together a projection showing where we would be at the end of the next fiscal year if things remained static. Using an R-squared value gave these charts the oomph I needed to be effective. I updated the charts each week and presented them to the Product Management team. The data helped them understand what adjustments to make to their products and each week they got closer and closer to their goals. When Bob first presented the pageview goal, we all thought he had gone mad, but I am happy to say that at the end of the next fiscal year, we will have either reached our goal or be within 98% of it. Without the representation I supplied, product management would not have known where to focus their energy and resources. Because they were an efficient team, we have reached our unreachable goal.

Is your use of probability and statistics increasing, decreasing, or remaining stable?

As Lycos gets more sophisticated, they (management) expect more and more sophisticated reporting. It is increasing.

Do you feel job applicants are viewed more favorably if they have studied some statistics?

Absolutely, and not just within Lycos Reporting, but also in Product Marketing and Finance.

10

Multinomial Experiments and Contingency Tables

Titanic revisited: Women and children first?

The Chapter Problem for Chapter 3 included the table reproduced here as Table 10-1. This table summarizes the fate of the passengers and crew when the *Titanic* sank on Monday, April 15, 1912. If we examine the data, we see that 19.6% of the men (332 out of 1692) survived, 75.4% of the women (318 out of 422) survived, 45.3% of the boys (29 out of 64) survived, and 60% of the girls (27 out of 45) survived. These percentages are graphed in the accompanying Excel-generated displays. The first graph is a bar chart showing the percentage of survivors for each group. The second graph is a stacked bar chart; each bar represents the number who died stacked above the number who survived. The Excel graphs make it appear that there are differences, but are the differences really *significant*?

We will proceed to analyze the data in Table 10-1, and we will treat the 2223 people aboard the *Titanic* as a *sample*. We could take the position that the *Titanic* data in Table 10-1 constitute a *population* and therefore should not be treated as a sample, so that methods of inferential statistics do not apply. Let's stipulate that the Table 10-1 data are sample data randomly selected from the population of all theoretical people who would find themselves in the same conditions. Realistically, no other people will actually find themselves in the same conditions, but we will make that assumption for the purposes of this discussion and analysis. We can then determine whether the observed differences have statistical significance. (See also Paul Velleman's *ActivStats* software for the example involving the *Titanic*.)

TABLE 10-1	*Titanic* Mortality				
	Men	Women	Boys	Girls	**Total**
Survived	332	318	29	27	**706**
Died	1360	104	35	18	**1517**
Total	**1692**	**422**	**64**	**45**	**2223**

EXCEL

10-1 Overview

In this chapter we continue the pattern of applying inferential methods to different configurations of data. Recall from Chapter 1 that categorical (or qualitative, or attribute) data are those data that can be separated into different categories (often called **cells**) that are distinguished by some nonnumeric characteristic. For example, we might separate a sample of M&Ms into the color categories of red, orange, yellow, brown, blue, and green. After finding the frequency count for each category, we might proceed to test the claim that the frequencies fit (or agree with) the color distribution claimed by the manufacturer (Mars, Inc.). The main objective of this chapter is to test claims about categorical data consisting of frequency counts for different categories. In Section 10-2 we will consider multinomial experiments, which consist of observed frequency counts arranged in a single row or column (called a one-way frequency table), and we will test the claim that the observed frequency counts agree with some claimed distribution. In Section 10-3 we will consider contingency tables (or two-way frequency tables), which consist of frequency counts arranged in a table such as Table 10-1. We will use contingency tables for two types of very similar tests: (1) tests of independence, which test the claim that the row and column variables are independent; and (2) tests of homogeneity, which test the claim that different populations have the same proportion of some specified characteristic.

We will see that the methods of this chapter use the same χ^2 (chi-square) distribution that was first introduced in Section 6-6. Recall the following important properties of the chi-square distribution:

1. Unlike the normal and Student t distributions, the chi-square distribution is not symmetric. (See Figure 10-1.)

2. The values of the chi-square distribution can be 0 or positive, but they cannot be negative. (See Figure 10-1.)

3. The chi-square distribution is different for each number of degrees of freedom. (See Figure 10-2.)

Critical values of the chi-square distribution are found in Table A-4.

FIGURE 10-1
The Chi-Square Distribution

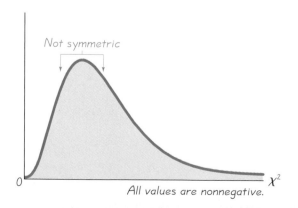

Not symmetric

0 All values are nonnegative. χ^2

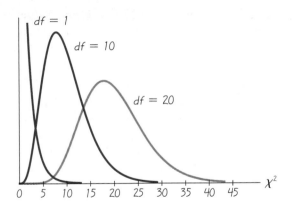

FIGURE 10-2
Chi-Square Distributions for 1, 10, and 20 Degrees of Freedom

 # 10-2 Multinomial Experiments: Goodness-of-Fit

Each data set in this section consists of qualitative data that have been separated into different categories. The main objective is to determine whether the distribution agrees with or "fits" some claimed distribution.

The following assumptions apply when we test a hypothesis that the population proportion for each of the categories is as claimed.

Assumptions

1. The data have been randomly selected.

2. The sample data consist of frequency counts for each of the different categories.

3. For each category, the *expected* frequency is at least 5. (The expected frequency for a category is the frequency that would occur if the assumed distribution works exactly as it should. There is no requirement that the *observed* frequency for each category must be at least 5.)

We define a *multinomial experiment* the same way we defined a binomial experiment (Section 4-3), except that a multinomial experiment has more than two categories (unlike a binomial experiment, which has exactly two categories).

> ### DEFINITION
>
> A **multinomial experiment** is an experiment that meets the following conditions.
>
> 1. The number of trials is fixed.
> 2. The trials are independent.
> 3. All outcomes of each trial must be classified into exactly one of several different categories.
> 4. The probabilities for the different categories remain constant for each trial.

TABLE 10-2

Last-Digits of Mark McGwire's Home-Run Distances

Last Digit	Frequency
0	55
1	2
2	1
3	1
4	0
5	3
6	0
7	2
8	4
9	2

> **EXAMPLE**   **Last-Digit Analysis of Home-Run Distances** In 1998, Mark McGwire hit 70 home runs and broke a record that had been intact for about 30 years. *USA Today* published the distances of those home runs, and Table 10-2 summarizes the *last digits* of those distances. If such distances are actually measured, we usually expect that the last digits will occur with relative frequencies (or probabilities) that are roughly the same. In contrast, estimated values tend to have 0 or 5 occurring much more often as last digits. In Table 10-2, it appears that there is a disproportionate occurrence of 0s. Later, we will analyze the data, but at this point simply verify that the four conditions of a multinomial experiment are satisfied.

SOLUTION   Here is the verification that the four conditions of a multinomial experiment are all satisfied:

1. The number of trials (last digits) is the fixed number 70.
2. The trials are independent, because the last digit of the length of a home run does not affect the last digit of the length of any other home run.
3. Each outcome (last digit) is classified into exactly 1 of 10 different categories. The categories are identified as 0, 1, 2, . . . , 9.
4. Finally, if we assume that the home-run distances are measured, the last digits should be equally likely, so that each possible digit has a probability of 1/10.

In this section we are presenting a method for testing a claim that in a multinomial experiment, the frequencies observed in the different categories fit a particular distribution. Because we test for how well an observed frequency distribution fits some specified theoretical distribution, this method is often called a *goodness-of-fit test.*

> **DEFINITION**
>
> A **goodness-of-fit test** is used to test the hypothesis that an observed frequency distribution fits (or conforms to) some claimed distribution.

For example, using the data in Table 10-2, we can test the hypothesis that the data fit a uniform distribution, with all of the digits being equally likely. Our goodness-of-fit tests will incorporate the following notation.

> **Notation**
>
> O represents the *observed frequency* of an outcome.
> E represents the *expected frequency* of an outcome.
> k represents the *number of different categories* or outcomes.
> n represents the total *number of trials*.

Finding Expected Frequencies

In Table 10-2 we see that the observed frequencies are denoted by $O = 55$, $O = 2, O = 1$, and so on. The sum of the observed frequencies is 70, so $n = 70$. If we assume that the 70 digits were obtained from a population in which all digits are equally likely, then we *expect* that each digit should occur in 1/10 of the 70 trials, so each of the 10 expected frequencies is given by $E = 7$. If we generalize this result, we get an easy procedure for finding expected frequencies whenever we are assuming that all of the expected frequencies are equal: Simply divide the total number of observations by the number of different categories ($E = n/k$). In other cases where the expected frequencies are not all equal, we can often find the expected frequency for each category by multiplying the sum of all observed frequencies and the probability p for the category, so $E = np$. We summarize these two procedures here.

- **If all expected frequencies are equal, then each expected frequency is the sum of all observed frequencies divided by the number of categories, so that $E = n/k$.**

- **If the expected frequencies are not all equal, then each expected frequency is found by multiplying the sum of all observed frequencies by the probability for the category, so $E = np$ for each category.**

As good as these two formulas for E might be, it would be better to use an informal approach based on an understanding of the circumstances. Just ask yourself, "How can the observed frequencies be split up among the different categories so that there is perfect agreement with the claimed distribution?" See the two examples given later in this section.

We know that sample frequencies typically deviate somewhat from the values we theoretically expect, so we now present the key question: Are the differences between the actual *observed* values O and the theoretically *expected* values E statistically significant? We need a measure of the discrepancy between the O and E values, so we use the following test statistic. (Later, we will explain how this test statistic was developed, but you can see that it has differences of $O - E$ as a key component.)

Test Statistic for Goodness-of-Fit Tests in Multinomial Experiments

$$\chi^2 = \sum \frac{(O - E)^2}{E}$$

Critical Values

1. Critical values are found in Table A-4 by using $k - 1$ degrees of freedom, where k = number of categories.

2. Goodness-of-fit hypothesis tests are always *right-tailed*.

The form of the χ^2 test statistic is such that *close agreement* between observed and expected values will lead to a *small* value of χ^2 and a *large* P-value. A large discrepancy between observed and expected values will lead to a *large* value of χ^2 and a *small* P-value. The hypothesis tests of this section are therefore always right-tailed, because the critical value and critical region are located at the extreme right of the distribution. These relationships are summarized and illustrated in Figure 10-3.

Once we know how to find the value of the test statistic and the critical value, we can test hypotheses by using the procedure introduced in Chapter 7 and summarized in Figure 7-5.

> **EXAMPLE**   **Last Digit Analysis of Home Runs: Equal Expected Frequencies**   Let's again refer to Table 10-2 for the last digits of Mark McGwire's home-run distances. The value of 0 seems to occur considerably more often, but is that really significant? Test the claim that the digits do not occur with the same frequency.

SOLUTION   The claim that the digits do not occur with the same frequency is equivalent to the claim that the relative frequencies or probabilities of the 10 cells $(p_0, p_1, \ldots, p_9)$ are not all equal. We will apply our standard procedure for testing hypotheses.

FIGURE 10-3
Relationships Among Components in Goodness-of-Fit Hypothesis Test

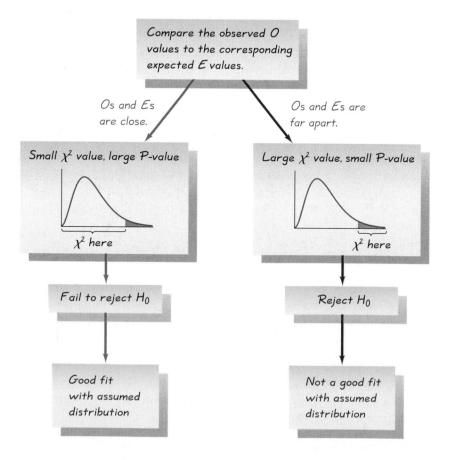

Step 1:   The original claim is that the digits do not occur with the same frequency. That is, at least one of the probabilities p_0, p_1, . . . , p_9 is different from the others.

Step 2:   If the original claim is false, then all of the probabilities are the same. That is, $p_0 = p_1 = \cdots = p_9$.

Step 3:   The null hypothesis must contain the condition of equality, so we have

H_0: $p_0 = p_1 = p_2 = p_3 = p_4 = p_5 = p_6 = p_7 = p_8 = p_9$
H_1: At least one of the probabilities is different from the others.

Step 4:   No significance level was specified, so we select $\alpha = 0.05$, a very common choice.

Step 5:   Because we are testing a claim about the distribution of the last digits being a uniform distribution, we use the goodness-of-fit test described in this section. The χ^2 distribution is used with the test statistic given earlier.

Step 6:   The observed frequencies O are listed in Table 10-2, and each corresponding expected frequency E is equal to 7 (because the 70 digits would be uniformly distributed through the 10 categories). Table 10-3 on the next page shows the computation of the χ^2 test statistic. The test statistic is $\chi^2 = 367.714$ (rounded). The critical value is $\chi^2 = 16.919$ (found in Table A-4 with $\alpha = 0.05$ in the right tail and degrees of freedom equal to $k - 1 = 9$). The test statistic and critical value are shown in Figure 10-4 on the next page.

Step 7:   Because the test statistic falls within the critical region, there is sufficient evidence to reject the null hypothesis.

Step 8:   There is sufficient evidence to support the claim that the last digits do not occur with the same relative frequency. We now have very strong evidence indicating that the home-run distances were not actually measured. It is reasonable to speculate that the distances are estimates instead of actual measurements.   ◀

The techniques in this section can be used to test whether an observed frequency distribution is a good fit with some theoretical frequency distribution. The preceding example tested for goodness-of-fit with a uniform distribution. Because many statistical analyses require a normally distributed population, we can use the chi-square test in this section to help determine whether given samples are drawn from normally distributed populations (see Exercise 22).

Safest Airplane Seats

Many of us believe that the rear seats are safest in an airplane crash. Safety experts do not agree that any particular part of an airplane is safer than others. Some planes crash nose first when they come down, but others crash tail first on takeoff. Matt McCormick, a survival expert for the National Transportation Safety Board, told *Travel* magazine that "There is no one safe place to sit." Goodness-of-fit tests can be used with a null hypothesis that all sections of an airplane are equally safe. Crashed airplanes could be divided into the front, middle, and rear sections. The observed frequencies of fatalities could then be compared to the frequencies that would be expected with a uniform distribution of fatalities. The χ^2 test statistic reflects the size of the discrepancies between observed and expected frequencies, and it would reveal whether some sections are safer than others.

TABLE 10-3	Calculating the χ^2 Test Statistic for the Last Digits of Home-run Distances				
Last Digit	Observed Frequency O	Expected Frequency E	$O - E$	$(O - E)^2$	$\dfrac{(O - E)^2}{E}$
0	55	7	48	2304	329.1429
1	2	7	−5	25	3.5714
2	1	7	−6	36	5.1429
3	1	7	−6	36	5.1429
4	0	7	−7	49	7.0000
5	3	7	−4	16	2.2857
6	0	7	−7	49	7.0000
7	2	7	−5	25	3.5714
8	4	7	−3	9	1.2857
9	2	7	−5	25	3.5714
	70	70			↑

↑ ↑
(These two totals must agree.)

$$\chi^2 = \sum \frac{(O - E)^2}{E} = 367.7143$$

FIGURE 10-4
Goodness-of-Fit Test of
$p_0 = p_1 = p_2 = p_3 = p_4 = p_5 = p_6 = p_7 = p_8 = p_9$

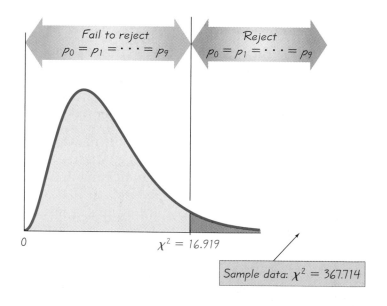

Fail to reject
$p_0 = p_1 = \cdots = p_9$

Reject
$p_0 = p_1 = \cdots = p_9$

0 $\chi^2 = 16.919$

Sample data: $\chi^2 = 367.714$

The preceding example dealt with the null hypothesis that the probabilities for the different categories are all equal. The methods of this section can also be used when the hypothesized probabilities (or frequencies) are different, as shown in the next example.

▷ **EXAMPLE**   **M&Ms: Unequal Expected Frequencies** Mars, Inc., claims that its M&M plain candies are distributed with the following color percentages: 30% brown, 20% yellow, 20% red, 10% orange, 10% green, and 10% blue. The colors of the M&Ms listed in Data Set 10 in Appendix B are summarized in Table 10-4. Using the sample data and a 0.05 significance level, test the claim that the color distribution is as claimed by Mars, Inc.

SOLUTION   We extended Table 10-4 to include the expected frequencies, which are calculated as follows. For n, we use the total number of trials (100), which is the total number of M&Ms observed in the sample. For the probabilities, we use the decimal equivalents of the claimed percentages (30%, 20%, ..., 10%).

Brown: $E = np = (100)(0.30) = 30$

Yellow: $E = np = (100)(0.20) = 20$

$\vdots$

Blue: $E = np = (100)(0.10) = 10$

In testing the given claim, Steps 1, 2, and 3 result in the following hypotheses:

H_0: $p_{br} = 0.3$ and $p_y = 0.2$ and $p_r = 0.2$ and $p_o = 0.1$ and $p_g = 0.1$ and $p_{bl} = 0.1$

H_1: At least one of the above proportions is different from the claimed value.

Steps 4, 5, and 6 lead us to use the goodness-of-fit test with a 0.05 significance level and a test statistic calculated from Table 10-5.

The test statistic is $x^2 = 5.950$. The critical value of x^2 is 11.071, and it is found in Table A-4 (using $\alpha = 0.05$ in the right tail with $k - 1 = 5$ degrees of freedom). The test statistic and critical value are shown in Figure 10-5 on the next page. Because the test statistic does not fall within the critical region, there is not sufficient evidence to warrant rejection of the null hypothesis. There is not sufficient evidence to warrant rejection of the claim that the colors are distributed with the percentages given by Mars, Inc.

In Figure 10-6 on page 599 we graph the claimed proportions of 0.30, 0.20, 0.20, 0.10, 0.10, 0.10, along with the observed proportions of 0.33,

TABLE 10-4	Frequencies of M&M Plain Candies					
	Brown	Yellow	Red	Orange	Green	Blue
Observed frequency	33	26	21	8	7	5
Expected frequency	30	20	20	10	10	10

to $k - 1$ categories, we cannot have negative frequencies nor can we have frequencies so large that their sum exceeds the total of the observed frequencies for all categories combined.

P-Values

The examples in this section used the traditional approach to hypothesis testing, but the *P*-value approach can also be used. *P*-values can be found using Excel, or they can be obtained by using the methods described in Sections 7-3 and 7-6. For instance, the preceding example resulted in a test statistic of $\chi^2 = 5.950$. That example had $k = 6$ categories, so there were $k - 1 = 5$ degrees of freedom. Referring to Table A-4, we see that for the row with 5 degrees of freedom, the test statistic of 5.950 is less than the lowest right-tailed critical value of 9.236, so the *P*-value is greater than 0.10. If Excel is used as described below, the display will include a *P*-value of 0.31111. The high *P*-value suggests that the null hypothesis should not be rejected. Remember, we reject the null hypothesis only when the *P*-value is equal to or less than the significance level.

Using Excel for Testing Goodness-of-Fit

Excel can be used to find the *P*-value in a goodness-of-fit test.
1. Enter the observed frequencies *O* in column A.
2. Enter the corresponding expected frequencies *E* in column B.
3. Click on *fx*, then select the function category of **Statistical,** then select the function name of **CHITEST.**
4. In the dialog box, make these entries:

   Actual range: Enter the cell range for the *observed* frequencies, such as A1:A6.

   Expected range: Enter the cell range for the *expected* frequencies, such as B1:B6.

   The *P*-value will be displayed, as shown in the following Excel display for the preceeding example.

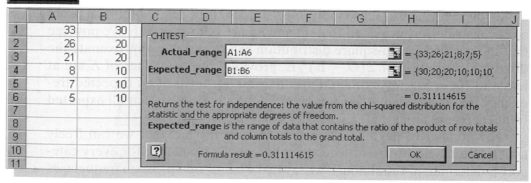

10-2 Basic Skills and Concepts

1. Testing Fairness of Roulette Wheel The author observed 500 spins of a roulette wheel at the Mirage Resort and Casino. (To the IRS: Isn't that Las Vegas trip now a tax deduction?) For each spin, the ball can land in any one of 38 different slots that are supposed to be equally likely. When the results were used to test the claim that the slots are in fact equally likely, the test statistic $\chi^2 = 38.232$ was obtained.
   a. Find the critical value assuming that the significance level is 0.10.
   b. Excel displayed a P-value of 0.41331, but what do you know about the P-value if you must use only Table A-4 along with the given test statistic of 38.232, which results from the 38 spins?
   c. Write a conclusion about the claim that the 38 results are equally likely.

2. Testing Excel's Random Number Generator Excel can generate random numbers by using the Tools/Data Analysis/Random Number Generation options. The author used Excel to generate 1000 random digits (0, 1, . . . 9) by specifying that the results should be equally likely. Testing the claim of equally likely outcomes resulted in a test statistic of $\chi^2 = 10.400$ and a P-value of 0.31908.
   a. Find the critical value assuming that the significance level is 0.025.
   b. What can you conclude about the P-value from Table A-4 if you know only the value of the test statistic and the number of categories (10)?
   c. State a conclusion about the claim that Excel generates random digits in such a way that they are equally likely.

3. Flat Tire and Missed Class A classic tale involves four car-pooling students who missed a test and gave as an excuse a flat tire. On the makeup test, the instructor asked the students to identify the particular tire that went flat. If they really didn't have a flat tire, would they be able to identify the same tire? The author asked 41 other students to identify the tire they would select. The results are listed in the following table (except for one student who selected the spare). Use a 0.05 significance level to test the author's claim that the results fit a uniform distribution. What does the result suggest about the ability of the four students to select the same tire when they really didn't have a flat?

Tire	Left front	Right front	Left rear	Right rear
Number selected	11	15	8	6

4. Do Car Crashes Occur on Different Days with the Same Frequency? It is a common belief that more fatal car crashes occur on certain days of the week, such as Friday or Saturday. A sample of motor vehicle deaths for a recent year in Montana is randomly selected. The numbers of fatalities for the different days of the week are listed in the accompanying table. At the 0.05 significance level, test the claim that accidents occur with equal frequency on the different days.

Day	Sun	Mon	Tues	Wed	Thurs	Fri	Sat
Number of fatalities	31	20	20	22	22	29	36

Based on data from the Insurance Institute for Highway Safety.

5. Are DWI Fatalities the Result of Weekend Drinking? Many people believe that fatal DWI crashes occur because of casual drinkers who tend to binge on Friday

and Saturday nights, whereas others believe that fatal DWI crashes are caused by people who drink every day of the week. In a study of fatal car crashes, 216 cases are randomly selected from the pool in which the driver was found to have a blood alcohol content over 0.10. These cases are broken down according to the day of the week, with the results listed in the accompanying table. At the 0.05 significance level, test the claim that such fatal crashes occur on the different days of the week with equal frequency. Does the evidence support the theory that fatal DWI car crashes are due to casual drinkers or that they are caused by those who drink daily?

Day	Sun	Mon	Tues	Wed	Thurs	Fri	Sat
Number	40	24	25	28	29	32	38

Based on data from the Dutchess County STOP-DWI Program.

6. Testing for Uniformly Distributed Industrial Accidents A study was made of 147 industrial accidents that required medical attention. Among those accidents, 31 occurred on Monday, 42 on Tuesday, 18 on Wednesday, 25 on Thursday, and 31 on Friday (based on results from "Counted Data CUSUM's," by Lucas, *Technometrics*, Vol. 27, No. 2). Test the claim that accidents occur with equal proportions on the five workdays. If the proportions are not the same, what factors might explain the differences?

7. Do Industrial Accidents Fit the Claimed Distribution? Use a 0.05 significance level and the industrial accident data from Exercise 6 to test the claim of a safety expert that accidents are distributed on workdays as follows: 30% on Monday, 15% on Tuesday, 15% on Wednesday, 20% on Thursday, and 20% on Friday. Does rejection of that claim provide any help in correcting the industrial accident problem?

8. Do Trucks Have the Same Color Distribution as Cars? A study of sport and compact car colors shows that 15.2% are green, 14.4% are white, 19.8% are red, 11.2% are black, and 39.4% are other colors (based on data from DuPont Automotive as reported in *USA Today*). When 100 trucks and vans are randomly selected, it is found that 16 are green, 24 are white, 16 are red, none are black, and 44 are other colors. Is there sufficient evidence to support the claim that the color distribution for trucks and vans is different from the color distribution for sport and compact cars?

9. Distribution of Digits in the Irrational Number Pi The number π is an irrational number with the property that when we try to express it in decimal form, it requires an infinite number of decimal places and there is no pattern of repetition. In the decimal representation of π, the first 100 digits occur with the frequencies described in the accompanying table. At the 0.05 significance level, test the claim that the digits are uniformly distributed.

Digit	0	1	2	3	4	5	6	7	8	9
Frequency	8	8	12	11	10	8	9	8	12	14

10. Distribution of Digits in the Rational Number 22/7 The number 22/7 is similar to π in the sense that they both require an infinite number of decimal places. However, 22/7 is a rational number because it can be expressed as the ratio of two integers, whereas π cannot. When rational numbers such as 22/7 are expressed in decimal form, there is a pattern of repetition. In the decimal representation of 22/7, the first 100 digits occur with the frequencies described in the accompanying table. At the 0.05 significance level, test the claim that the digits are uniformly distributed. How does the result differ from that found in Exercise 9?

Digit	0	1	2	3	4	5	6	7	8	9
Frequency	0	17	17	1	17	16	0	16	16	0

EXCEL 11. Car Crashes and Age Brackets Among drivers who have had a car crash in the last year, 88 are randomly selected and categorized by age, with the results listed in the accompanying table. If all ages have the same crash rate, we would expect (because of the age distribution of licensed drivers) the given categories to have 16%, 44%, 27%, and 13% of the subjects, respectively. At the 0.05 significance level, test the claim that the distribution of crashes conforms to the distribution of ages. Does any age group appear to have a disproportionate number of crashes?

Age	Under 25	25–44	45–64	Over 64
Drivers	36	21	12	19

Based on data from the Insurance Information Institute.

EXCEL 12. Do World War II Bomb Hits Fit a Poisson Distribution? In analyzing hits by V-1 buzz bombs in World War II, South London was subdivided into regions, each with an area of 0.25 km². In Section 4-5 we presented an example and included a table of actual frequencies of hits and the frequencies expected with the Poisson distribution. Use the values listed here and test the claim that the actual frequencies fit a Poisson distribution. Use a 0.05 significance level.

Number of bomb hits	0	1	2	3	4 or more
Actual number of regions	229	211	93	35	8
Expected number of regions (from Poisson distribution)	227.5	211.4	97.9	30.5	8.7

EXCEL 13. Post Position and Winning Horse Races Many people believe that when a horse races, it has a better chance of winning if its starting line-up position is closer to the rail on the inside of the track. The starting position of 1 is closest to the inside rail, followed by position 2, and so on. The accompanying table lists the numbers of wins for horses in the different starting positions. Test the claim that the probabilities of winning in the different post positions are not all the same.

Starting Position	1	2	3	4	5	6	7	8
Number of wins	29	19	18	25	17	10	15	11

Based on data from the *New York Post*.

EXCEL 14. Grade and Seating Location Do "A" students tend to sit in a particular part of the classroom? The author recorded the locations of the students who received grades of A, with these results: 17 sat in the front, 9 sat in the middle, and 5 sat in the back of the classroom. Is there sufficient evidence to support the claim that the "A" students are not evenly distributed throughout the classroom? If so, does that mean you can increase your likelihood of getting an A by sitting in the front?

EXCEL **15.** Analyzing Rainfall Refer to Data Set 17 (BOSTRAIN.xls) in Appendix B and use the last digits of the Boston rainfall amounts for Monday only. Is there sufficient evidence to support the claim that the digits do not occur with the same frequency? In the first example of this section, we saw that in some cases, the analysis of last digits can reveal whether the sample data have been obtained through actual measurements. Does that analysis apply to the last digits of the Monday rainfall amounts? Why or why not?

EXCEL **16.** Analyzing Last Digits of Sammy Sosa Home-Run Distances Refer to Data Set 19 (HOMERUNS.xls) in Appendix B and use the last digits of the distances of home runs hit by Sammy Sosa. Is there sufficient evidence to conclude that the last digits do not occur with the same frequency? What does that conclusion suggest about how the home-run distances were obtained?

10-2 Beyond the Basics

EXCEL **17.** Testing Effects of Outliers In doing a test for the goodness-of-fit as described in this section, does an outlier have much of an effect on the value of the χ^2 test statistic? Test for the effect of an outlier by repeating Exercise 3 after changing the frequency for the right rear tire from 6 to 60. Describe the general effect of an outlier.

18. Detecting Altered Experimental Data When Gregor Mendel conducted his famous hybridization experiments with peas, it appears that his gardening assistant knew the results that Mendel expected, and he altered the results to fit Mendel's expectations. Subsequent analysis of the results led to the conclusion that there is a probability of only 0.00004 that the expected results and reported results would agree so closely. How could the methods of this section be used to detect such results that are just too perfect to be realistic?

19. Equivalent Test In this exercise we will show that a hypothesis test involving a multinomial experiment with only two categories is equivalent to a hypothesis test for a proportion (Section 7-5). Assume that a particular multinomial experiment has only two possible outcomes, A and B, with observed frequencies of f_1 and f_2, respectively.
   a. Find an expression for the χ^2 test statistic, and find the critical value for a 0.05 significance level. Assume that we are testing the claim that both categories have the same frequency, $(f_1 + f_2)/2$.
   b. The test statistic

$$z = \frac{\hat{p} - p}{\sqrt{\dfrac{pq}{n}}}$$

   is used to test the claim that a population proportion is equal to some value p. With the claim that $p = 0.5$, $\alpha = 0.05$, and

$$\hat{p} = \frac{f_1}{f_1 + f_2}$$

   show that z^2 is equivalent to χ^2 [from part (a)]. Also show that the square of the critical z score is equal to the critical χ^2 value from part (a).

EXCEL **20.** Testing Goodness-of-Fit with a Binomial Distribution An observed frequency distribution is as follows:

Number of successes	0	1	2	3
Frequency	89	133	52	26

   a. Assuming a binomial distribution with $n = 3$ and $p = 1/3$, use the binomial probability formula to find the probability corresponding to each category of the table.

   b. Using the probabilities found in part (a), find the expected frequency for each category.

   c. Use a 0.05 significance level to test the claim that the observed frequencies fit a binomial distribution for which $n = 3$ and $p = 1/3$.

EXCEL **21.** Testing Goodness-of-Fit with a Poisson Distribution In a recent year, there were 116 homicide deaths in Richmond, Virginia (based on "A Classroom Note On the Poisson Distribution: A Model for Homicidal Deaths In Richmond, VA for 1991," *Mathematics and Computer Education,* by Winston A. Richards). If the frequencies of deaths on different days conform to a Poisson distribution, they will be as shown in the accompanying table. Use a 0.05 significance level to test the claim that the actual frequencies fit a Poisson distribution. (*Caution:* Not all of the expected frequencies are at least 5.)

	Number of Homicides				
	0	1	2	3	4
Actual number of days	268	79	17	1	0
Expected number of days	265.6	84.4	13.4	1.4	0.1

EXCEL **22.** Testing Goodness-of-Fit with a Normal Distribution An observed frequency distribution of sample IQ scores is as follows:

IQ score	Less than 80	80–95	96–110	111–120	More than 120
Frequency	20	20	80	40	40

   a. Assuming a normal distribution with $\mu = 100$ and $\sigma = 15$, use the methods given in Chapter 5 to find the probability of a randomly selected subject belonging to each class. (Use class boundaries of 79.5, 95.5, 110.5, and 120.5.)

   b. Using the probabilities found in part (a), find the expected frequency for each category.

   c. Use a 0.01 significance level to test the claim that the IQ scores were randomly selected from a normally distributed population with $\mu = 100$ and $\sigma = 15$.

10-3 Contingency Tables: Independence and Homogeneity

The main objective of this section is to test claims about tables, such as Table 10-1 in the Chapter Problem. Tables similar to Table 10-1 are generally called *contingency tables,* or *two-way frequency tables.*

the total of all frequencies in a contingency table, we can freely assign frequencies to only $r - 1$ rows and $c - 1$ columns before the frequency for every cell is determined. [However, we cannot have negative frequencies or frequencies so large that any row (or column) sum exceeds the total of the observed frequencies for that row (or column).]

In the preceding section we knew the corresponding probabilities and could easily determine the expected values, but the typical contingency table does not come with the relevant probabilities. For each cell in the frequency table, the expected frequency E can be calculated by applying the multiplication rule of probability for independent events. Assuming that the row and column variables are independent (which is the null hypothesis), the probability of a value being in a particular cell is the probability of being in the row containing the cell (namely, the row total divided by the sum of all frequencies) multiplied by the probability of being in the column containing the cell (namely, the column total divided by the sum of all frequencies) multiplied by the sum of all frequencies. Sound too complicated? The expected frequency for a cell can be simplified to the following equation.

Expected Frequency for a Contingency Table

$$\text{expected frequency} = \frac{(\text{row total})(\text{column total})}{(\text{grand total})}$$

Here *grand total* refers to the total of all observed frequencies in the table. For example, the expected frequency for the upper left cell of Table 10-6 (a duplicate of Table 10-1 with expected frequencies inserted in parentheses) is 537.360, which is found by noting that the total of all frequencies for the first row is 706, the total of the column frequencies is 1692, and the sum of all frequencies in the table is 2223, so we get an expected frequency of

$$E = \frac{(\text{row total})(\text{column total})}{(\text{grand total})} = \frac{(706)(1692)}{2223} = 537.360$$

TABLE 10-6 Observed Frequencies (and Expected Frequencies)

	Gender/Age Category				
	Men	Women	Boys	Girls	Row totals
Survived	332	318	29	27	706
	(537.360)	(134.022)	(20.326)	(14.291)	
Died	1360	104	35	18	1517
	(1154.640)	(287.978)	(43.674)	(30.709)	
Column totals:	1692	422	64	45	Grand total: 2223

EXAMPLE **Finding Expected Frequency** The expected frequency for the upper left cell of Table 10-6 is 537.360. Find the expected frequency for the lower left cell, assuming independence between the row variable (whether the person survived) and the column variable (whether the person is a man, woman, boy, or girl).

SOLUTION   The lower left cell lies in the second row (with total 1517) and the first column (with total 1692). The expected frequency is

$$E = \frac{(\text{row total})(\text{column total})}{(\text{grand total})} = \frac{(1517)(1692)}{2223} = 1154.640$$

INTERPRETATION   To interpret this result for the lower left cell, we can say that although 1360 men actually died, we would have expected 1154.640 men to die if survivability is independent of whether the person is a man, woman, boy, or girl. There is a discrepancy between $O = 1360$ and $E = 1154.640$, and such discrepancies are key components of the test statistic.   ◄

Home Field Advantage

In the *Chance* magazine article "Predicting Professional Sports Game Outcomes from Intermediate Game Scores," authors Harris Cooper, Kristina DeNeve, and Frederick Mosteller used statistics to analyze two common beliefs: Teams have an advantage when they play at home, and only the last quarter of professional basketball games really counts. Using a random sample of hundreds of games, they found that for the four top sports, the home team wins about 58.6% of games. Also, basketball teams ahead after 3 quarters go on to win about 4 out of 5 times, but baseball teams ahead after 7 innings go on to win about 19 out of 20 times. The statistical methods of analysis included the chi-square distribution applied to a contingency table.

To better understand the rationale for finding expected frequencies with this procedure, let's pretend that we know only the row and column totals and that we must fill in the cell expected frequencies by assuming independence (or no relationship) between the two variables involved—that is, we pretend that we know only the row and column totals shown in Table 10-6. Let's begin with the cell in the upper left corner. Because 706 of the 2223 persons survived, we have $P(\text{survived}) = 706/2223$. Similarly, 1692 of the people were men, so $P(\text{man}) = 1692/2223$. Because we are assuming independence between survivability and the column gender/age category, we can use the multiplication rule of probability to get

$$P(\text{survived and man}) = P(\text{survived}) \cdot P(\text{man}) = \frac{706}{2223} \cdot \frac{1692}{2223}$$

This equation is an application of the multiplication rule for independent events, which is expressed in general as follows: $P(A \text{ and } B) = P(A) \cdot P(B)$. Knowing the probability of being in the upper left cell, we can now find the *expected value* for that cell, which we get by multiplying the probability for that cell by the total number of people, as shown in the following equation.

$$E = n \cdot p = 2223\left(\frac{706}{2223} \cdot \frac{1692}{2223}\right) = 537.360$$

The form of this product suggests a general way to obtain the expected frequency of a cell:

$$\text{expected frequency } E = (\text{grand total}) \cdot \frac{(\text{row total})}{(\text{grand total})} \cdot \frac{(\text{column total})}{(\text{grand total})}$$

This expression can be simplified to

$$E = \frac{(\text{row total})(\text{column total})}{(\text{grand total})}$$

We can now proceed to use contingency table data for testing hypotheses, as in the following example, which uses the data given in the Chapter Problem.

EXAMPLE   *Titanic* **Sinking** At the 0.05 significance level, use the data in Table 10-1 to test the claim that when the *Titanic* sank, whether someone survived or died is independent of whether the person is a man, woman, boy, or girl.

SOLUTION   The null hypothesis and alternative hypothesis are as follows:

H_0: Whether a person survived is independent of whether the person is a man, woman, boy, or girl.

H_1: Surviving the *Titanic* sinking and being a man, woman, boy, or girl are dependent.

The significance level is $\alpha = 0.05$.

Because the data are in the form of a contingency table, we use the χ^2 distribution with this test statistic:

$$\chi^2 = \sum \frac{(O - E)^2}{E}$$

$$= \frac{(332 - 537.360)^2}{537.360} + \frac{(318 - 134.022)^2}{134.022} + \frac{(29 - 20.326)^2}{20.326}$$

$$+ \frac{(27 - 14.291)^2}{14.291} + \frac{(1360 - 1154.640)^2}{1154.640} + \frac{(104 - 287.978)^2}{287.978}$$

$$+ \frac{(35 - 43.674)^2}{43.674} + \frac{(18 - 30.709)^2}{30.709}$$

$$= 78.481 + 252.555 + 3.702 + 11.302 + 36.525 + 117.536$$

$$+ 1.723 + 5.260$$

$$= 507.084$$

The critical value is $\chi^2 = 7.815$, and it is found from Table A-4 by noting that $\alpha = 0.05$ in the right tail and the number of degrees of freedom is given by $(r - 1)(c - 1) = (2 - 1)(4 - 1) = 3$. The test statistic and critical value are shown in Figure 10-7. Because the test statistic falls within the critical region, we reject the null hypothesis that whether a person survived is independent of whether the person is a man, woman, boy, or girl. It appears that whether a person survived the *Titanic* and whether that person is a man, woman, boy, or girl are dependent variables.

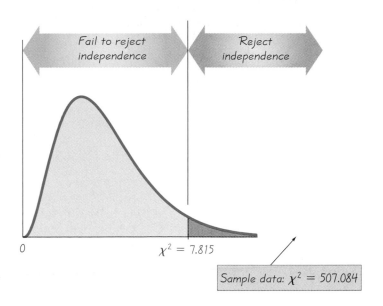

FIGURE 10-7
Test of Independence for
Table 10-1 Data

P-Values

The preceding example used the traditional approach to hypothesis testing, but we can easily use the *P*-value approach. Excel can be used to find the *P*-value. See the Excel directions at the end of this section. Also, you can estimate *P*-values from Table A-4 in Appendix A. Locate the appropriate number of degrees of freedom to isolate a particular row in that table. Find where the test statistic falls in that row, and you can identify a range of possible *P*-values by referring to the areas given at the top of each column. In the preceding example, there are 3 degrees of freedom, so go to the third row of Table A-4. Now use the test statistic of $\chi^2 = 507.084$ to see that this test statistic is greater than (and farther to the right of) every critical value of χ^2 that is in the third row, so the *P*-value is less than 0.005. On the basis of this small *P*-value, we again reject the null hypothesis and conclude that there is sufficient sample evidence to warrant rejection of the null hypothesis of independence.

As in Section 10-2, if observed and expected frequencies are close, the χ^2 test statistic will be small and the *P*-value will be large. If observed and expected frequencies are far apart, the χ^2 test statistic will be large and the *P*-value will be small. These relationships are summarized and illustrated in Figure 10-8.

Test of Homogeneity

In the preceding example, we illustrated a test of independence by using a sample of 2223 people who were aboard the *Titanic*. We were treating those 2223 people as a random sample drawn from *one* hypothetical population of all people who would find themselves in similar circumstances. However, some other samples are drawn from *different* populations, and we want to determine whether those populations have the same proportions of the characteristics being considered. The *test of homogeneity* can be used in such

FIGURE 10-8
Relationships Among Components in χ^2 Test of Independence

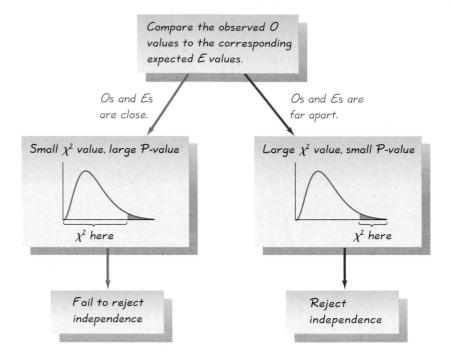

cases. (The word *homogeneous* means "having the same quality," and in this context, we are testing to determine whether the proportions are the same.)

> ## DEFINITION
>
> In a **test of homogeneity,** we test the claim that *different populations* have the same proportions of some characteristics.

Because a test of homogeneity uses data sampled from different populations, we have predetermined totals for either the rows or the columns in the contingency table. Consequently, a test of homogeneity involves random selections made in such a way that either the row totals are predetermined or the column totals are predetermined. In trying to distinguish between a test for homogeneity and a test for independence, we can therefore pose the following question:

> **Were *predetermined* sample sizes used for different populations (test of homogeneity), or was one big sample drawn so both row and column totals were determined randomly (test of independence)?**

As an example of a test of homogeneity, suppose we want to test the claim that the proportion of college students who take a statistics course is the same in public four-year colleges, public two-year colleges, private four-year colleges, and private two-year colleges. Suppose we decide to sample 200 people from public four-year colleges, 300 people from public two-year colleges, 400 people from private four-year colleges, and 500 people from private two-year colleges. Then the contingency table summarizing the results will have either the row totals or the column totals (whichever represent the different colleges) predetermined as 200, 300, 400, and 500.

In conducting a test of homogeneity, we can use the same procedures already presented in this section, as illustrated in the following example.

▷ **EXAMPLE**   **Influence of Gender** Does a pollster's gender have an effect on poll responses by men? A *U.S. News & World Report* article about polls stated: "On sensitive issues, people tend to give 'acceptable' rather than honest responses; their answers may depend on the gender or race of the interviewer." To support that claim, data were provided for an Eagleton Institute poll in which surveyed men were asked if they agreed with this statement: "Abortion is a private matter that should be left to the woman to decide without government intervention." We will analyze the effect of gender on male survey subjects only. Table 10-7 is based on the responses of surveyed men. Assume that the survey was designed so that male interviewers were instructed to obtain 800 responses from male subjects, and female interviewers were instructed to obtain 400 responses from male subjects. Using a 0.05 significance level, test the claim that the proportions of agree/disagree responses are the same for the subjects interviewed by men and the subjects interviewed by women.

SOLUTION   Because we have predetermined column totals of 800 subjects interviewed by men and 400 subjects interviewed by women, we test for homogeneity with these hypotheses:

H_0: The proportions of agree/disagree responses are the same for the subjects interviewed by men and the subjects interviewed by women.

H_1: The proportions are different.

The significance level is $\alpha = 0.05$. We use the same χ^2 test statistic described earlier, and it is calculated by using the same procedure. Instead of listing the details of that calculation, we provide the Excel display that results from the data in Table 10-7.

The Excel display shows the test statistic of $\chi^2 = 6.529$ and the *P*-value of 0.010611. Using the *P*-value approach to hypothesis testing, we reject the null hypothesis of equal (homogeneous) proportions. There is sufficient evidence to warrant rejection of the claim that the proportions are the

TABLE 10-7	Gender and Survey Responses	
	Gender of Interviewer	
	Man	Woman
Men who agree	560	308
Men who disagree	240	92

EXCEL

	A	B	C
1	560	578.6667	0.602151
2	308	289.3333	1.204306
3	240	221.3333	1.574303
4	92	110.6667	3.148605
5			
6		Chi-sq:	6.529364
7		P-value:	0.010611

same. It appears that response and the gender of the interviewer are dependent. It seems that men are influenced by the gender of the interviewer, although this conclusion is a statement of causality that is not justified by the statistical analysis.   ◀

Using Excel for Contingency Tables

Excel does not have a function for using a contingency table in a test for independence or homogeneity. Although there is a procedure for entering a table and doing all of the required calculations directly, the following procedure is much easier, but it does require that you calculate the expected value for each cell.

1. Use $E = \dfrac{(\text{row total})(\text{column total})}{(\text{grand total})}$ to compute the expected frequency for each cell.

2. Enter all of the *observed* frequencies in column A and enter all of the corresponding *expected* frequencies in column B.

3. Compute the number of degrees of freedom using $df = (r - 1)(c - 1)$.

4. Do not use the **CHITEST** function, because it would be based on the wrong number for degrees of freedom. Instead, click on cell C1 and enter this expression: **=(A1 − B1)^2/B1**

5. Press the **Enter** key and the value of $\dfrac{(O - E)^2}{E}$ will appear in the first row of column C.

6. Click on cell C1. By holding the mouse button down, click and drag the lower right corner of cell C1 downward until all of the cells adjacent to values in column B are highlighted. When the mouse button is released, the other values of $\dfrac{(O - E)^2}{E}$ should also appear.

7. Find the sum of the values in column C by using Excel's **SUM** function. Let's assume that column C has eight values. Go to an empty cell and enter **=SUM(C1:C8)**

8. Press the **Enter** key and the sum of the column C entries will appear. This is the value of $\chi^2 = \sum \dfrac{(O - E)^2}{E}$, which is the desired test statistic.

9. Now proceed to find the *P*-value. Click on *fx,* then select the function category of **Statistical,** then select the function name of **CHIDIST.**

10. In the dialog box, enter the value of the χ^2 test statistic and enter the number of degrees of freedom. The *P*-value will be generated.

If you apply the above procedure to the data of Table 10-1, the spreadsheet should appear as shown below. The small *P*-value suggests that we reject the null hypothesis of independence between the row and column variables.

EXCEL

	A	B	C
1	332	537.36	78.48133393
2	318	134.022	252.5548379
3	29	20.326	3.701578077
4	27	14.291	11.30212588
5	1360	1154.64	36.52457008
6	104	287.978	117.5364246
7	35	43.674	1.722724642
8	18	30.709	5.259652903
9			
10		Chi-sq:	507.083248
11		P-value:	1.3928E-109

10-3 Basic Skills and Concepts

EXCEL

1. Drug Testing: Is Treatment Independent of Reaction? Nicorette is a chewing gum designed to help people stop smoking cigarettes. Tests for adverse reactions yielded the results given in the accompanying table. Using this table, the Excel display includes the information shown here. At the 0.05 significance level, test the claim that the treatment (drug or placebo) is independent of the reaction (whether or not mouth or throat soreness was experienced). If you are thinking about using Nicorette as an aid to stop smoking, should you be concerned about mouth or throat soreness?

	Drug	Placebo
Mouth or throat soreness	43	35
No mouth or throat soreness	109	118

Based on data from Merrell Dow Pharmaceuticals, Inc.

EXCEL

	A	B	C
1	43	38.8721	0.438349
2	35	39.1279	0.435484
3	109	113.1279	0.150622
4	118	113.8721	0.149638
5			
6		Chi-sq:	1.174093
7		P-value:	0.278562

2. Testing Effectiveness of Bicycle Helmets A study was conducted of 531 persons injured in bicycle crashes, and randomly selected sample results are summarized in the accompanying table. The Excel results also are shown. At the 0.05 significance level, test the claim that wearing a helmet has no effect on whether facial injuries are received. Based on these results, does a helmet seem to be effective in helping to prevent facial injuries in a crash?

	Helmet Worn	No Helmet
Facial injuries received	30	182
All injuries nonfacial	83	236

Based on data from "A Case-Control Study of the Effectiveness of Bicycle Safety Helmets in Preventing Facial Injury," by Thompson, Thompson, Rivara, and Wolf, *American Journal of Public Health,* Vol. 80, No. 12.

EXCEL

	A	B	C
1	30	45.1149	5.063963
2	182	166.8851	1.368967
3	83	67.8851	3.365395
4	236	251.1149	0.909784
5			
6		Chi-sq:	10.70811
7		P-value:	0.001067

3. Is Gender Independent of Opinion about Car Seizure? Considerable controversy arose when New York City introduced a program of keeping the cars belonging to people charged with drunk driving. The Associated Press conducted a poll, and the accompanying table is based on the results. Use a 0.01 significance level to test the claim that the opinion about car seizure is independent of the gender of the person being surveyed. Identify at least one key factor that might help explain the result.

	Should Car Be Seized?	
	Yes	No
Men	391	425
Women	480	256

4. Testing for Independence Between Early Discharge and Rehospitalization of Newborn Is it safe to discharge newborns from the hospital early after their births? The accompanying table shows results from a study of this issue. Use a 0.05 significance level to test the claim that whether the newborn was discharged early or late is independent of whether the newborn was rehospitalized within a week of discharge. Does the conclusion change if the significance level is changed to 0.01?

	Rehospitalized within Week of Discharge?	
	Yes	No
Early discharge (less than 30 hours)	622	3997
Late discharge (30–78 hours)	631	4660

Based on data from "The Safety of Newborn Early Discharge," by Liu and others, *Journal of the American Medical Association,* Vol. 278, No. 4.

EXCEL **5.** Testing Influence of Gender Table 10-7 summarizes data for male survey subjects, but the accompanying table summarizes data for a sample of women. Using a 0.01 significance level, and assuming that the sample sizes of 800 men and 400 women are predetermined, test the claim that the proportions of agree/disagree responses are the same for the subjects interviewed by men and the subjects interviewed by women.

	Gender of Interviewer	
	Man	Woman
Women who agree	512	336
Women who disagree	288	64

Based on data from the Eagleton Institute.

EXCEL **6.** Testing for Discrimination In the judicial case *United States* v. *City of Chicago*, fair employment practices were challenged. A minority group (group A) and a majority group (group B) took the Fire Captain Examination. Assume that the study began with predetermined sample sizes of 24 minority candidates (Group A) and 562 majority candidates (Group B), with the results as shown in the table. At the 0.05 significance level, test the claim that the proportion of minority candidates who pass is the same as the proportion of majority candidates who pass. Based on the results, does the test appear to discriminate?

	Pass	Fail
Group A	10	14
Group B	417	145

EXCEL **7.** Is Gender Independent of Confidence in Police? A survey was conducted to determine whether there is a gender gap in the confidence people have in police. The sample results are listed in the accompanying table. Use a 0.05 significance level to test the claim that there is such a gender gap.

	Confidence in Police		
	Great Deal	Some	Very Little or None
Men	115	56	29
Women	175	94	31

Based on data from the U.S. Department of Justice and the Gallup Organization.

EXCEL **8.** Is Scanner Accuracy the Same for Specials? In a study of store checkout scanning systems, samples of purchases were used to compare the scanned prices to the posted prices. The accompanying table summarizes results for a sample of 819 items. When stores use scanners to check out items, are the error rates the same for regular-priced items as they are for advertised-special items? How might the behavior of consumers change if they believe that disproportionately more overcharges occur with advertised-special items?

	Regular-Priced Items	Advertised-Special Items
Undercharge	20	7
Overcharge	15	29
Correct price	384	364

Based on data from "UPC Scanner Pricing Systems: Are They Accurate?" by Ronald Goodstein, *Journal of Marketing*, Vol. 58.

9. Boeing 777s and Airline Companies The Boeing Commercial Airplane Group conducted a survey of 133 planes being flown or on order from three airline companies based in different countries. The sample data are summarized in the accompanying table. Is there sufficient evidence to support the claim that the distribution of planes flying and planes ordered is independent of the airline company? Use a 0.01 significance level.

	United	British Air	Singapore
Boeing 777s flying	36	22	14
Boeing 777s ordered	16	23	22

10. Firearm Training and Safety Does firearm training result in safer practices by gun owners? In one study, randomly selected subjects were surveyed with the results given in the accompanying table. Use a 0.05 significance level to test the claim that formal firearm training is independent of how firearms are stored. Does the formal training appear to have a positive effect?

	Guns Stored Loaded and Unlocked?	
	Yes	No
Had formal firearm training	122	329
Had no formal firearm training	49	299

Based on data from "Firearm Training and Storage," by Hemenway, Solnick, Azrael, *Journal of the American Medical Association,* Vol. 273, No. 1.

11. Crime and Strangers The accompanying table lists survey results obtained from a random sample of different crime victims. At the 0.05 significance level, test the claim that the type of crime is independent of whether the criminal is a stranger. How might the results affect the strategy police officers use when they investigate crimes?

	Homicide	Robbery	Assault
Criminal was a stranger	12	379	727
Criminal was acquaintance or relative	39	106	642

Based on data from the U.S. Department of Justice.

12. Is Seat-Belt Use Independent of Cigarette Smoking? A study of seat-belt users and nonusers yielded the randomly selected sample data summarized in the table on the next page. Test the claim that the amount of smoking is independent of seat-belt use. A plausible theory is that people who smoke more are less concerned about their health and safety and are therefore less inclined to wear seat belts. Is this theory supported by the sample data?

	Number of Cigarettes Smoked per Day			
	0	1–14	15–34	35 and over
Wear seat belts	175	20	42	6
Don't wear seat belts	149	17	41	9

Based on data from "What Kinds of People Do Not Use Seat Belts?" by Helsing and Comstock, *American Journal of Public Health,* Vol. 67, No. 11.

13. Is Sentence Independent of Plea? Many people believe that criminals who plead guilty tend to get lighter sentences than those who are convicted in trials.

The accompanying table summarizes randomly selected sample data for San Francisco defendants in burglary cases. All of the subjects had prior prison sentences. At the 0.05 significance level, test the claim that the sentence (sent to prison or not sent to prison) is independent of the plea. If you were an attorney defending a guilty defendant, would these results suggest that you should encourage a guilty plea?

	Guilty Plea	Not-Guilty Plea
Sent to prison	392	58
Not sent to prison	564	14

Based on data from "Does It Pay to Plead Guilty? Differential Sentencing and the Functioning of the Criminal Courts," by Brereton and Casper, *Law and Society Review,* Vol. 16, No. 1.

EXCEL 14. Is the Home Field Advantage Independent of the Sport? Winning team data were collected for teams in different sports, with the results given in the accompanying table. Use a 0.10 significance level to test the claim that home/visitor wins are independent of the sport. Given that among the four sports included here, baseball is the only sport in which the home team can modify field dimensions to favor its own players, does it appear that baseball teams are effective in using this advantage?

	Basketball	Baseball	Hockey	Football
Home-team wins	127	53	50	57
Visiting-team wins	71	47	43	42

Based on data from "Predicting Professional Sports Game Outcomes from Intermediate Game Scores," by Copper, DeNeve, and Mosteller, *Chance,* Vol. 5, No. 3–4.

EXCEL 15. Is Drinking Independent of Type of Crime? The accompanying table lists sample data that statistician Karl Pearson used in 1909. Does the type of crime appear to be related to whether the criminal drinks or abstains? Are there any crimes that appear to be associated with drinking?

	Arson	Rape	Violence	Stealing	Coining (Counterfeiting)	Fraud
Drinker	50	88	155	379	18	63
Abstainer	43	62	110	300	14	144

EXCEL 16. Survey Refusals and Age Bracket A study of people who refused to answer survey questions provided the randomly selected sample data shown in the table on the next page. At the 0.01 significance level, test the claim that the cooperation of the subject (response or refusal) is independent of the age category. Does any particular age group appear to be particularly uncooperative?

	Age					
	18–21	22–29	30–39	40–49	50–59	60 and over
Responded	73	255	245	136	138	202
Refused	11	20	33	16	27	49

Based on data from "I Hear You Knocking But You Can't Come In," by Fitzgerald and Fuller, *Sociological Methods and Research,* Vol. 11, No. 1.

EXCEL 17. Is Smoking Independent of Gender? Refer to Data Set 14 (STATSURV.xls) in Appendix B and test the claim that the gender of statistics students is independent of whether they smoke.

18. Is Exercise Independent of Gender? Refer to Data Set 14 (STATSURV.xls) in Appendix B and test the claim that whether statistics students exercise is independent of gender.

10-3 Beyond the Basics

19. Using Yates' Correction for Continuity The chi-square distribution is continuous, whereas the test statistic used in this section is discrete. Some statisticians use *Yates' correction for continuity* in cells with an expected frequency of less than 10 or in all cells of a contingency table with two rows and two columns. With Yates' correction, we replace

$$\sum \frac{(O - E)^2}{E} \quad \text{with} \quad \sum \frac{(|O - E| - 0.5)^2}{E}$$

Given the contingency table in Exercise 1, find the value of the χ^2 test statistic with and without Yates' correction. In general, what effect does Yates' correction have on the value of the test statistic?

20. Equivalent Tests Assume that a contingency table has two rows and two columns with frequencies of a and b in the first row and frequencies of c and d in the second row.

a. Verify that the test statistic can be expressed as

$$\chi^2 = \frac{(a + b + c + d)(ad - bc)^2}{(a + b)(c + d)(b + d)(a + c)}$$

b. Let $\hat{p}_1 = a/(a + c)$ and let $\hat{p}_2 = b/(b + d)$. Show that the test statistic

$$z = \frac{(\hat{p}_1 - \hat{p}_2) - 0}{\sqrt{\dfrac{\overline{p}\,\overline{q}}{n_1} + \dfrac{\overline{p}\,\overline{q}}{n_2}}}$$

where

$$\overline{p} = \frac{a + b}{a + b + c + d}$$

and

$$\overline{q} = 1 - \overline{p}$$

is such that $z^2 = \chi^2$ [the same result as in part (a)]. This result shows that the chi-square test involving a 2×2 table is equivalent to the test for the difference between two proportions, as described in Section 8-4.

Race and the Future

NEW YORK Among the many uses of polls, some are designed to help companies improve marketing strategy, some are designed to help political candidates identify areas of strength and weakness, and some are designed to measure changing perceptions about culture, society, and the future. A *Newsweek*-sponsored poll included a question in which subjects were asked to look ahead 10 years into the future and identify how they think they will fare (better, worse, same) with respect to different factors. The table includes sample results for those who responded with "better." *Newsweek* noted that "Princeton Research Associates conducted telephone interviews with 751 adults including 264 African-Americans on April 16–19."

Responses of "Better"

	Blacks	Whites
Family income	71%	59%
Quality of education in local schools	64%	55%
Job opportunities for your family	57%	48%

1. Use the reported sample sizes to convert the given percentages to frequency counts.

2. The *Newsweek* article included this statement: "Though blacks see some serious problems yet to be resolved, they are even more optimistic about the future than white Americans." Test that claim. Is there sufficient sample evidence to support the claim?

VOCABULARY LIST

cells
multinomial experiment
goodness-of-fit test
contingency table

two-way frequency table
test of independence
test of homogeneity

REVIEW

In this chapter we worked with data in the form of frequency counts for different categories. In Section 10-2 we described methods for testing goodness-of-fit in a multinomial experiment, which is similar to a binomial experiment except that there are more than two categories of outcomes. Multinomial experiments result in frequency counts arranged in a single row or column, and we tested to determine whether the observed sample frequencies agree with some claimed distribution.

In Section 10-3 we described methods for testing claims involving contingency tables (or two-way frequency tables), which have at least two rows and two columns. Contingency tables incorporate two variables: One variable is used for determining the row that describes a sample value, and the second variable is used for determining the column that describes a sample value. Section 10-3 included two types of hypothesis test: (1) a test of independence between the row and column variables; (2) a test of homogeneity to decide whether different populations have the same proportions of some characteristics. The following are some key components of the methods discussed in this chapter.

- *Section 10-2 (Test for goodness-of-fit):*

  Test statistic is $\chi^2 = \sum \dfrac{(O - E)^2}{E}$

  Test is right-tailed with $k - 1$ degrees of freedom. All expected frequencies must be at least 5.

- *Section 10-3 (Contingency table test of independence or homogeneity):*

  Test statistic is $\chi^2 = \sum \dfrac{(O - E)^2}{E}$

  Test is right-tailed with $(r - 1)(c - 1)$ degrees of freedom. All expected frequencies must be at least 5.

REVIEW EXERCISES

EXCEL

1. NRA Membership and Gun Safety The accompanying table describes sample data obtained from a survey of randomly selected gun owners. Use a 0.01 significance level to test the claim that whether gun owners store their guns loaded and unlocked is independent of whether they are a member of the National Rifle Association (NRA). If it appears that these two variables are dependent, identify at least one relevant factor that might explain the result.

	Guns Stored Loaded and Unlocked?	
	Yes	No
NRA member	40	80
Not a member of NRA	129	548

Based on data from "Firearm Training and Storage," by Hemenway, Solnick, Azrael, *Journal of the American Medical Association*, Vol. 273, No. 1

EXCEL

2. Is the Experiment Fair? The self-proclaimed amazing mentalist Bob claims that he has the power to read minds and he boldly challenges anyone to prove otherwise. If Bob wins the challenge, the challenger must pay him $100,000, and Bob agrees to disavow any mental powers if he loses the challenge. An experiment is devised whereby a neutral third party rolls a die and concentrates on the outcome for one minute; then Bob must identify the outcome while in another room. This is repeated for a total of 30 different trials, and the die results are listed in the accompanying table. Bob lost the challenge because he correctly identified only four outcomes, whereas someone making random guesses typically gets five of them correct. However, Bob charges that the experiment is unfair because the outcomes of the die are not uniform. Use a 0.05 significance level to test Bob's claim that there is sufficient evidence to conclude that the die is not fair, because all outcomes are not equally likely.

Die outcome	1	2	3	4	5	6
Frequency	2	8	4	3	7	6

3. Do Gunfire Deaths Occur More Often on Weekends? When *Time* magazine tracked U.S. deaths by gunfire during a one-week period, the results shown in the accompanying table were obtained. At the 0.05 significance level, test the claim that gunfire death rates are the same for the different days of the week. Is there any support for the theory that more gunfire deaths occur on weekends when more people are at home?

Weekday	Mon	Tues	Wed	Thurs	Fri	Sat	Sun
Number of deaths by gunfire	74	60	66	71	51	66	76

4. Are the Proportions of On-Time Flights the Same? The accompanying table summarizes results for 100 flights randomly selected from each of three different airline companies. Use a 0.05 significance level to test the claim that USAir, American, and Delta have the same proportions of on-time flights.

	USAir	American	Delta
Arrived on time	80	77	76
Arrived late	20	23	24

Based on data from the Department of Transportation.

5. Testing Adverse Reactions to Seldane Clinical tests of the allergy drug Seldane yielded results summarized in the accompanying table. At the 0.05 significance level, test the claim that the occurrence of headaches is independent of the group (Seldane, placebo, control). Based on these results, should Seldane users be concerned about getting headaches?

	Seldane Users	Placebo Users	Control
Headache	49	49	24
No headache	732	616	602

Based on data from Merrell Dow Pharmaceuticals, Inc.

CUMULATIVE REVIEW EXERCISES

1. Finding Statistics Assume that in Table 10-8, the row and column titles have no meaning so that the table contains test scores for eight randomly selected prisoners. Find the mean, median, range, variance, standard deviation, and 5-number summary.

TABLE 10-8

	A	B	C	D
x	66	80	82	75
y	77	89	94	84

2. Finding Probability Assume that in Table 10-8, the letters A, B, C, and D represent the choices on the first question of a multiple-choice quiz. Also assume that x represents men and y represents women and that the table entries are frequency counts, so 66 men chose answer A, 77 women chose answer A, 80 men chose answer B, and so on.
  a. If one response is randomly selected, find the probability that it is response C.
  b. If one response is randomly selected, find the probability that it was made by a man.
  c. If one response is randomly selected, find the probability that it is C or was made by a man.
  d. If two different responses are randomly selected, find the probability that they were both made by a woman.

EXCEL
3. Testing for Equal Proportions Using the same assumptions as in Exercise 2, test the claim that men and women choose the different answers in the same proportions.

EXCEL
4. Testing for a Relationship Assume that Table 10-8 lists test scores for four people, where the x-score is from a test of memory and the y-score is from a test of reasoning. Test the claim that there is a relationship between the x- and y-scores.

EXCEL
5. Testing for Effectiveness of Training Assume that Table 10-8 lists test scores for four people, where the x-score is from a pretest taken before a training session on memory improvement and the y-score is from a posttest taken after the training. Test the claim that the training session is effective in raising scores.

EXCEL
6. Testing for Equality of Means Assume that in Table 10-8, the letters A, B, C, and D represent different versions of the same test of reasoning. The x-scores were obtained by four randomly selected men and the y-scores were obtained by four randomly selected women. Test the claim that men and women have the same mean score.

COOPERATIVE GROUP ACTIVITIES

1. *Out-of-class activity:* Divide into groups of four or five students. Each group member should survey at least 15 male students and 15 female students at the same college by asking two questions: (1) Which political party does the subject favor most? (2) If the subject were to make up an absence excuse of a flat tire, which tire would he or she say went flat if the instructor asked? (See Exercise 3 in Section 10-2.) Ask the subject to write the two responses on an index card, and also record the gender of the subject and whether the subject wrote with the right or left hand. Use the methods of this chapter to analyze the data collected. Include these tests:

   • Political party choice is independent of the gender of the subject.
   • The tire identified as being flat is independent of the gender of the subject.
   • Political party choice is independent of whether the subject is right- or left-handed.
   • The tire identified as being flat is independent of whether the subject is right- or left-handed.
   • Gender is independent of whether the subject is right- or left-handed.
   • Political party choice is independent of the tire identified as being flat.

2. *Out-of-class activity:* Divide into groups of four or five students. Each group member should select about 15 other students and first ask them to "randomly" select four digits each. After the four digits have been recorded, ask each subject to write the last four digits of his or her social security number. Take the "random" sample results and mix them into one big sample, then mix the social security digits into a second big sample. Using the "random" sample set, test the claim that students select digits randomly. Then use the social security digits to test the claim that they come from a population of random digits. Compare the results. Does it appear that students can randomly select digits? Are they likely to select any digits more often than others? Are they likely to select any digits less often than others? Do the last digits of social security numbers appear to be randomly selected?

3. *In-class activity:* Divide into groups of three or four students. Each group should be given a die along with the instruction that it should be tested for "fairness." Is the die fair or is it biased? Describe the analysis and results.

4. *Out-of-class activity:* Divide into groups of two or three students. Some examples and exercises of this chapter were based on the analysis of last digits of values. (See the examples in Section 10-2 and Exercises 15 and 16 in Section 10-2.) It was noted that the analysis of last digits can sometimes reveal whether values are the results of actual measurements or whether they are reported estimates. Refer to an almanac and find the lengths of rivers in the world, then analyze the last digits to determine whether those lengths appear to be actual measurements or whether they appear to be reported estimates. (Instead of lengths of rivers, you could use heights of mountains, heights of the tallest buildings, lengths of bridges, and so on.)

EXCEL PROJECT

Use the following Excel procedure to randomly generate 500 digits.

1. Click on the empty cell A1.

2. Click *f*x. Select the function category of **Math & Trig,** then click on the function name of **RANDBETWEEN.** Click **OK.**

3. You should now see a dialog box. In the "Bottom" box, enter 0. In the "Top" box, enter 9. Click **OK.** (You are telling Excel to select a random number between 0 and 9. The result will be one of these numbers: 0, 1, 2, 3, 4, 5, 6, 7, 8, 9.)

4. Cell A1 should be highlighted. Use the mouse to click and drag the lower right corner of cell A1 downward so that cells A1 through A500 are highlighted, then release the mouse button. The result should be a column of random digits.

5. Record the results in the following table. Enter the bin values of 0, 1, 2, 3, 4, 5, 6, 7, 8, 9 in the first ten cells of column B, then select **Tools, Data Analysis, Histogram,** and enter A1:A500 for the "Inut Range," and enter B1:B10 for the "Bin Range." Click **OK** in the dialog box. Use the results to fill in the frequencies in the following table. (Note that the values in column A may change during this process, but they continue to be 500 random digits, which is what we want.)

Digit	0	1	2	3	4	5	6	7	8	9
Frequency										

6. Use a 0.05 significance level to test the claim that the sample digits come from a population for which all digits are equally likely. Does Excel's **RANDBETWEEN** function appear to work correctly?

Critical Thinking: Are noncombat mortality rates the same for military personnel in a combat zone as for military personnel not so deployed?

After the military operations of Desert Shield and Desert Storm in the Persian Gulf, there were claims of dramatic increases in death rates of U.S. troops who served in that region. One study compared noncombat mortality rates for U.S. military personnel who were deployed in combat situations to those not deployed. Table 10-9 summarizes sample data for deaths in three different categories. The table entries show that among those who died from unintentional injuries (from cars, aircraft, explosions, and so on), 19% were deployed; among those who died from illness, 10% were deployed; among those who died from homicide or suicide, 3% were deployed.

Analyze the Results

Are those differences significant? Analyze the sample data and form conclusions. If the cause of death and deployment appear to be dependent, describe the effect of deployment and try to identify at least one factor that might explain the dependency. Make a recommendation that might improve the situation for deployed personnel.

TABLE 10-9	Noncombat Causes of Death for Deployed and Nondeployed Military Personnel		
	Cause of Death		
	Unintentional Injury	Illness	Homicide or Suicide
Deployed	183	30	11
Not deployed	784	264	308

Table values are based on data from "Comparative Mortality Among U.S. Military Personnel in the Persian Gulf Region and Worldwide During Operations Desert Shield and Desert Storm," by Writer, DeFraites, and Brundage, *Journal of the American Medical Association,* Vol. 275, No. 2.

Internet Project

Contingency Tables

In all of the examples in Section 10-3, data were sorted into categories formed by the rows and columns of a contingency table. We used contingency tables along with the associated chi-square test to form conclusions regarding qualitative population data. Characteristics such as gender, race, and geographic location become fair game for formal hypothesis testing procedures. Go to the Internet Project for this chapter, located at

http://www.awlonline.com/triola

After collecting demographic data for a number of populations, you will apply the techniques of this chapter to form conclusions about the independence of interesting pairings of characteristics.

Analysis of Variance

Crash test dummies: Does the weight category of a car make a difference?

Table 11-1 on the following page lists recent data from car crash experiments conducted by the National Transportation Safety Administration. New cars were purchased and crashed into a fixed barrier at 35 mi/h, and measurements were recorded for the dummy in the driver's seat. (We've all seen a dummy in a driver's seat, but these dummies were not human.) Does the weight class (subcompact, compact, midsize, full size) make a difference? Let's begin by focusing on head injuries. In the spirit of exploring the data, we first list the statistics for the head injury measurements.

Examination of the head injury values shows that there are no outliers. With only five values in each sample, there isn't a real need for 5-number summaries. For visual comparisons, dotplots are shown below. The second Excel display shows the four sample means. We should usually explore the distributions of the sample data, but we can't get a good sense of their distributions with only five values in each sample.

There appear to be considerable differences, such as the difference between the means of 668.8 and 486.8. But is there sufficient sample evidence to support the conclusion that the weight category makes a difference? We will consider this problem in Section 11-2.

Head Injuries

	Subcompact	Compact	Midsize	Full Size
n	5	5	5	5
$\bar{x}$	668.8	555.8	486.8	537.8
s	242.0	91.0	167.7	154.6

Excel Dotplots

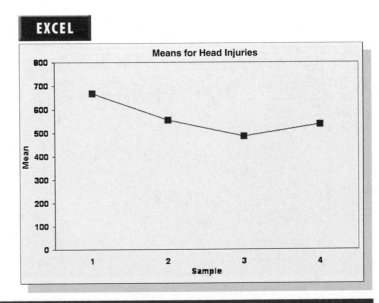

TABLE 11-1	Injuries to Car Crash Test Dummies		
	Head Injury (hic)	Chest Deceleration (g)	Left Femur Load (lb)
Subcompact Cars			
Ford Escort	681	55	595
Honda Civic	428	47	1063
Hyundai Accent	917	59	885
Nissan Sentra	898	49	519
Saturn SL4	420	42	422
Compact Cars			
Chevrolet Cavalier	643	57	1051
Dodge Neon	655	57	1193
Mazda 626 DX	442	46	946
Pontiac Sunfire	514	54	984
Subaru Legacy	525	51	584
Midsize Cars			
Chevrolet Camaro	469	45	629
Dodge Intrepid	727	53	1686
Ford Mustang	525	49	880
Honda Accord	454	51	181
Volvo S70	259	46	645
Full-Size Cars			
Audi A8	384	44	1085
Cadillac Deville	656	45	971
Ford Crown Victoria	602	39	996
Oldsmobile Aurora	687	58	804
Pontiac Bonneville	360	44	1376

11-1 Overview

In Chapter 8 we developed procedures for testing the hypothesis that two population means are equal. In Section 11-2 we introduce a procedure for testing the hypothesis that *three or more* population means are equal. (The methods used in this chapter can also be used to test for equality between two population means, but the methods used in Chapter 8 are more efficient.) A typical null hypothesis in Section 11-2 will be $H_0: \mu_1 = \mu_2 = \mu_3 = \mu_4$; the alternative hypothesis is H_1: At least one mean is different. The method we use is based on an analysis of sample variances.

> ### DEFINITION
>
> **Analysis of variance (ANOVA)** is a method of testing the equality of three or more population means by analyzing sample variances.

ANOVA is used in applications such as the following:

- When three different groups of people (such as smokers, nonsmokers exposed to environmental tobacco smoke, and nonsmokers not so exposed) are measured for cotinine (a marker of nicotine), we can test to determine whether they have different mean levels of cotinine.
- If we treat one group with two aspirin tablets each day and a second group with one aspirin tablet each day, while a third group is given a placebo each day, we can test to determine if there is sufficient evidence to support the claim that the three groups have different mean blood pressure levels.

Why Can't We Just Test Two Samples at a Time? Why do we need a new procedure when we can test for equality of two means by using the methods presented in Chapter 8? For example, if we want to use the sample data from Table 11-1 to test the claim that the four populations have the same mean, why not simply pair them off and do two at a time by testing $H_0: \mu_1 = \mu_2$, then $H_0: \mu_2 = \mu_3$, and so on? This approach (doing two at a time) requires six different hypothesis tests, so the degree of confidence could be as low as 0.95^6 (or 0.735). In general, as we increase the number of individual tests of significance, we increase the likelihood of finding a difference by chance alone (instead of a real difference in the means). The risk of a type I error—finding a difference in one of the pairs when no such difference actually exists—is far too high. The method of analysis of variance lets us avoid that particular pitfall (rejecting a true null hypothesis) by using one test for equality of several means.

tic, for categorizing the populations. This characteristic is sometimes referred to as a *treatment*, or *factor*.

> ## DEFINITION
>
> A **treatment** (or **factor**) is a property, or characteristic, that allows us to distinguish the different populations from one another.

For example, the head injury measurements listed in Table 11-1 are sample data drawn from four different populations that are distinguished according to the treatment (or factor) of weight class (subcompact, compact, midsize, full size). The term *treatment* is used because early applications of analysis of variance involved agricultural experiments in which different plots of farmland were treated with different fertilizers, seed types, insecticides, and so on.

Interpreting Excel Results

Because the calculations required for analysis of variance are extremely involved, almost everyone uses computer software or a calculator (except for one statistician living in the Nevada desert). Excel provides *P*-values, so we will use the *P*-value approach for testing hypotheses.

Procedure for Testing H_0: $\mu_1 = \mu_2 = \mu_3 = \cdots$

1. **Use Excel to obtain results.**
2. **Identify the *P*-value from the display.**
3. **Form a conclusion based on these criteria:**

   - **If *P*-value $\leq \alpha$, reject the null hypothesis of equal means.**

   - **If *P*-value $> \alpha$, fail to reject the null hypothesis of equal means.**

We will describe the nature of the *F* test statistic after the following example, but Figure 11-2 shows the relationships among the key components of the ANOVA test.

EXAMPLE   **Head Injuries in Car Crashes** Given the sample data in Table 11-1 and a significance level of $\alpha = 0.05$, use Excel to test the claim that the four samples come from populations with the same mean ($\mu_1 = \mu_2 = \mu_3 = \mu_4$).

SOLUTION   At the end of this section we will describe specific procedures for obtaining Excel results, but we will now consider the displayed

FIGURE 11-2
Relationships Among
Components of ANOVA

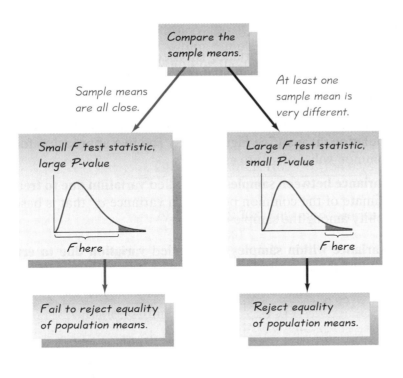

Compare the sample means.

Sample means are all close.

At least one sample mean is very different.

Small *F* test statistic, large *P*-value

F here

Large *F* test statistic, small *P*-value

F here

Fail to reject equality of population means.

Reject equality of population means.

EXCEL

	A	B	C	D	E	F	G
1	Anova: Single Factor						
2							
3	SUMMARY						
4	Groups	Count	Sum	Average	Variance		
5	Column 1	5	3344	668.8	58542.7		
6	Column 2	5	2779	555.8	8272.7		
7	Column 3	5	2434	486.8	28110.2		
8	Column 4	5	2689	537.8	23905.2		
9							
10							
11	ANOVA						
12	Source of Variation	SS	df	MS	F	P-value	F crit
13	Between Groups	88425	3	29475	0.992167	0.42157	3.238867
14	Within Groups	475323.2	16	29707.7			
15							
16	Total	563748.2	19				

results. A key element is the *P*-value, which is shown as 0.42157. Because the *P*-value is greater than the significance level of $\alpha = 0.05$, we fail to reject the null hypothesis of equal means.

INTERPRETATION   There is not sufficient evidence to warrant rejection of the claim that the four samples come from populations with the same mean. We might feel safer in larger cars, and that feeling might be justified by the reality that occupants of larger cars have less severe head injuries in crashes, but the data in Table 11-1 do not provide enough evidence to support that claim. We would need to crash test more cars, a time-consuming and obviously expensive process.

same. The same basic reasoning applies because we calculate an F test statistic that is the ratio of two different estimates of the common population variance σ^2, but those estimates involve *weighted* measures that take the sample sizes into account, as shown below.

$$F = \frac{\text{variance between samples}}{\text{variance within samples}} = \frac{\left[\dfrac{\Sigma n_i (\overline{x}_i - \overline{\overline{x}})^2}{k - 1} \right]}{\left[\dfrac{\Sigma (n_i - 1) s_i^2}{\Sigma (n_i - 1)} \right]}$$

where

$\overline{\overline{x}}$ = mean of all sample values combined

k = number of population means being compared

n_i = number of values in the ith sample

$\overline{x}_i$ = mean of values in the ith sample

s_i^2 = variance of values in the ith sample

Note that the numerator in this test statistic is really a form of the formula for variance that was given in Chapter 2:

$$s^2 = \frac{\Sigma (x - \overline{x})^2}{n - 1}$$

The factor of n_i is included so that larger samples carry more weight. The denominator of the test statistic is simply the mean of the sample variances, but it is a weighted mean with the weights based on the sample sizes.

Because calculating this test statistic can lead to large rounding errors, the various software packages typically use a different (but equivalent) expression that involves SS (for sum of squares) and MS (for mean square) notation. Although the following notation and components are complicated and involved, the basic idea is the same: The test statistic F is a ratio with a numerator reflecting variation *between* the means of the samples and a denominator reflecting variation *within* the samples. If the populations have equal means, the F ratio tends to be close to 1, but if the population means are not equal, the F ratio tends to be significantly larger than 1. Key components in our ANOVA method are identified in the following boxes.

> **SS(total),** or total sum of squares, is a measure of the total variation (around $\overline{\overline{x}}$) in all of the sample data combined.

Formula 11-1 $SS(\text{total}) = \Sigma (x - \overline{\overline{x}})^2$

SS(total) can be broken down into two components, SS(between groups) and SS(within groups), described as follows:

> **SS (between groups)** or **SS(treatment)** is a measure of the variation between the sample means. In one-way ANOVA, it is sometimes referred to as SS(factor).

Formula 11-2

SS(between groups)

$$= n_1(\overline{x}_1 - \overline{\overline{x}})^2 + n_2(\overline{x}_2 - \overline{\overline{x}})^2 + \cdots + n_k(\overline{x}_k - \overline{\overline{x}})^2$$

$$= \Sigma n_i(\overline{x} - \overline{\overline{x}})^2$$

If the population means $(\mu_1, \mu_2, \ldots, \mu_k)$ are equal, then the sample means $\overline{x}_1, \overline{x}_2, \ldots, \overline{x}_k$ will all tend to be close together and also close to $\overline{\overline{x}}$. The result will be a relatively small value of SS(between groups). If the population means are not all equal, however, then at least one of $\overline{x}_1, \overline{x}_2, \ldots, \overline{x}_k$ will tend to be far apart from the others and also far apart from $\overline{\overline{x}}$. The result will be a relatively large value of SS(between groups).

> **SS(within groups)** or **SS(error)** is a sum of squares representing the variability that is assumed to be common to all the populations being considered.

Formula 11-3

$$\text{SS(within groups)} = (n_1 - 1)s_1^2 + (n_2 - 1)s_2^2 + \cdots + (n_k - 1)s_k^2$$

$$= \Sigma(n_i - 1)s_i^2$$

Given the preceding expressions, the following relationship will always hold.

Formula 11-4   SS(total) = SS(between groups) + SS(within groups)

SS(between groups) and SS(within groups) are both sums of squares, and if we divide each by its corresponding number of degrees of freedom, we get *mean* squares. Some of the following expressions for mean squares include the notation *N*:

N = total number of values in all samples combined

> **MS(between groups)** or **MS(treatment)** is a mean square for treatment, obtained as follows:

Formula 11-5 $\text{MS(between groups)} = \dfrac{\text{SS(between groups)}}{k - 1}$

column A followed by the last column and the *highest row number*. For example if you have A, B, C, and D with 15 values, 99 values, 10 values, and 20 values respectively, enter A1:D99. Exercise 9 is the first exercise with unequal column lengths and, for that exercise, enter A1:E11.

4. Click the button to indicate that the data are grouped by *columns,* and enter a significance level for "alpha" if the default value of 0.05 is not desired. See the dialog box shown below.

5. Click **OK** and the Excel results should appear in a format as shown with the example in this section.

11-2 Basic Skills and Concepts

EXCEL

1. Interpreting Computer Display The accompanying table shows the analysis of variance results from an Excel display. The sample values are measured consumer reactions to television commercials for different products. Assume that we want to use a 0.05 significance level in testing the null hypothesis that the different products have commercials resulting in the same mean reaction score.
   a. How many different samples are included in this study?
   b. Identify the value of the test statistic.
   c. Identify the critical value.
   d. Identify the *P*-value.
   e. Based on the preceding results, what do you conclude about equality of the population means?

EXCEL

Source of Variation	SS	df	MS	F	P-value	F crit
Between Groups	50	5	10	3.33	0.032	2.9013
Within Groups	5	15	3			
Total	55	20				

2. Interpreting Computer Display Repeat Exercise 1 assuming that the Excel results are as shown here.

EXCEL

Source of Variation	SS	df	MS	F	P-value	F crit
Between Groups	120	4	30	1.5	0.289	3.8379
Within Groups	160	8	20			
Total	280	12				

3. Mean Body Temperatures for Different Age Groups Do different age groups have different mean body temperatures? The accompanying table lists the body temperatures of five randomly selected subjects from each of three different age groups.

    a. Find the variance *between* samples by evaluating $n s_{\bar{x}}^2$ (Recall that $s_{\bar{x}}^2$ is the variance of the three sample means.)

    b. Find the variance *within* samples by evaluating the pooled variance s_p^2, which is the mean of the sample variances.

    c. Use the results from parts (a) and (b) to find the F test statistic.

    d. Find the F critical value. [Recall that with equal sample sizes, the degrees of freedom are $k - 1$ for the numerator and $k(n - 1)$ for the denominator.]

    e. Based on the preceding results, what do you conclude about the claim that the three age-group populations have the same mean body temperature?

Body Temperatures (°F) Categorized by Age

18–20	21–29	30 and older
98.0	99.6	98.6
98.4	98.2	98.6
97.7	99.0	97.0
98.5	98.2	97.5
97.1	97.9	97.3
$n_1 = 5$	$n_2 = 5$	$n_3 = 5$
$\bar{x}_1 = 97.940$	$\bar{x}_2 = 98.580$	$\bar{x}_3 = 97.800$
$s_1 = 0.568$	$s_2 = 0.701$	$s_3 = 0.752$

Based on data from Dr. Philip Mackowiak, Dr. Steven Wasserman, and Dr. Myron Levine of the University of Maryland.

4. Waste Disposal of Metal, Paper, Plastic, and Glass The City Resource Recovery Company (CRRC) collects the waste discarded by households in a region. Discarded waste must be separated into categories of metal, paper, plastic, and glass. In planning for the equipment needed to collect and process the garbage, CRRC refers to the data we have summarized in Data Set 5 (GARBAGE.xls) in Appendix B (provided by the Garbage Project at the University of Arizona). The results (weights in pounds) are summarized in the accompanying table. At the 0.05 significance level, test the claim that the four specific populations of garbage have the same mean. Based on the results, does it appear that these four categories require the same collection and processing resources? Does any single category seem to be a particularly large part of the waste-management problem?

Metal	Paper
$n = 62$	$n = 62$
$\bar{x} = 2.218$	$\bar{x} = 9.428$
$s = 1.091$	$s = 4.168$

Plastic	Glass
$n = 62$	$n = 62$
$\bar{x} = 1.911$	$\bar{x} = 3.752$
$s = 1.065$	$s = 3.108$

5. Crash Test Dummies: Mean Chest Deceleration Value The example of this section used the head injury data in Table 11-1 in a test of the null hypothesis that the means are the same for the different weight categories (subcompact, compact, midsize, full size). Use a 0.05 significance level to test the null hypothesis that the different weight categories have the same mean chest deceleration value.

6. Crash Test Dummies: Mean Left Femur Load The example of this section used the head injury data in Table 11-1 in a test of the null hypothesis that the means are the same for the different weight categories (subcompact, compact, midsize, full size). Use a 0.05 significance level to test the null hypothesis that the different weight categories have the same mean left femur load.

7. Archeology: Skull Breadths from Different Epochs The values in the following table are measured maximum breadths of male Egyptian skulls from different epochs (based on data from *Ancient Races of the Thebaid,* by Thomson and Randall-Maciver). Changes in head shape over time suggest that interbreeding occurred with immigrant populations. Use a 0.05 significance level to test the claim that the mean is the same for the different epochs.

4000 B.C.	1850 B.C.	150 A.D.
131	129	128
138	134	138
125	136	136
129	137	139
132	137	141
135	129	142
132	136	137
134	138	145
138	134	137

8. Solar Energy in Different Weather A student of the author lives in a home with a solar electric system. At the same time each day, she collected voltage readings from a meter connected to the system and the results are listed in the accompanying table. Use a 0.05 significance level to test the claim that the mean voltage reading is the same under the three different types of day. Is there sufficient evidence to support a claim of different population means? We might expect that a solar system would provide more electrical energy on sunny days than on cloudy or rainy days. Can we conclude that sunny days result in greater amounts of electrical energy?

Sunny Days	Cloudy Days	Rainy Days
13.5	12.7	12.1
13.0	12.5	12.2
13.2	12.6	12.3
13.9	12.7	11.9
13.8	13.0	11.6
14.0	13.0	12.2

9. Fabric Flammability Tests in Different Laboratories Flammability tests were conducted on children's sleepwear. The Vertical Semirestrained Test was used, in which pieces of fabric were burned under controlled conditions. After the burning stopped, the length of the charred portion was measured and recorded. Results are given in the margin for the same fabric tested at different laboratories. Because the same fabric was used, the different laboratories should have obtained the same results. Is there sufficient evidence to support the claim that the means for the different laboratories are the same? (The data were provided by Minitab, Inc.)

EXCEL **10.** Mean Weights of Different Colas Refer to Data Set 1 (COLA.xls) in Appendix B. Is there sufficient evidence to support the claim that cans of regular Coke, diet Coke, regular Pepsi, and diet Pepsi have different mean weights?

EXCEL **11.** Mean Weights of M&Ms Refer to Data Set 10 (M&M.xls) in Appendix B. At the 0.05 significance level, test the claim that the mean weight of M&Ms is the same for each of the six different color populations. If it is the intent of Mars, Inc., to make the candies so that the different color populations have the same mean weight, do these results suggest that the company has a problem requiring corrective action?

EXCEL **12.** Second-Hand Smoke in Different Groups Refer to Data Set 9 (COTININE.xls) in Appendix B. Use a 0.01 significance level to test the claim that the mean cotinine level is different for these three groups: nonsmokers who are not exposed to environmental tobacco smoke, nonsmokers who are exposed to tobacco smoke, and people who smoke. What do the results suggest about second-hand smoke?

Laboratory				
1	2	3	4	5
2.9	2.7	3.3	3.3	4.1
3.1	3.4	3.3	3.2	4.1
3.1	3.6	3.5	3.4	3.7
3.7	3.2	3.5	2.7	4.2
3.1	4.0	2.8	2.7	3.1
4.2	4.1	2.8	3.3	3.5
3.7	3.8	3.2	2.9	2.8
3.9	3.8	2.8	3.2	
3.1	4.3	3.8	2.9	
3.0	3.4	3.5		
2.9	3.3			

11-2 Beyond the Basics

EXCEL **13.** Effects of Data Transformations How are the analysis of variance results from Exercise 3 affected in each of the following cases?

  a. 2°F is added to each temperature listed for the 18–20 age group.

  b. All of the original temperatures are converted from the Fahrenheit scale to the Celsius scale. In general, how are ANOVA results affected by the scale used?

  c. The same constant is added to every one of the original sample values.

  d. Each of the original sample values is multiplied by the same constant.

  e. The order of the samples is changed.

14. Using t Test Five independent samples of 50 values each are randomly drawn from populations that are normally distributed with equal variances. We wish to test the claim that $\mu_1 = \mu_2 = \mu_3 = \mu_4 = \mu_5$.

  a. If we used only the methods given in Chapter 8, we would test the individual claims $\mu_1 = \mu_2, \mu_1 = \mu_3, \ldots, \mu_4 = \mu_5$. How many ways can we pair off five means?

  b. Assume that for each test of equality between two means, there is a 0.95 probability of not making a type I error. If all possible pairs of means are tested for equality, what is the probability of making no type I errors? (Although the tests are not actually independent, assume that they are.)

  c. If we use analysis of variance to test the claim that $\mu_1 = \mu_2 = \mu_3 = \mu_4 = \mu_5$ at the 0.05 significance level, what is the probability of not making a type I error?

  d. Compare the results of parts (b) and (c). Which approach is better in the sense of giving us a greater chance of not making a type I error?

15. Equivalent Tests In this exercise you will verify that when you have two sets of sample data, the *t* test for independent samples and the ANOVA method of this section are equivalent. Refer to the head injury measurements in Table 11-1, but use only the data for subcompact and compact cars.

a. Use a 0.05 significance level and the method of Section 8-6 to test the claim that the two samples come from populations with the same mean. (Assume that both populations have the same variance.)

b. Use a 0.05 significance level and the ANOVA method of this section to test the claim made in part (a).

c. Verify that the squares of the *t* test statistic and critical value from part (a) are equal to the *F* test statistic and critical value from part (b).

16. Components in ANOVA Table Complete the ANOVA table assuming that there are three samples with sizes of 5, 7, and 7, respectively.

Source of Variation	Sum of Squares (SS)	Degrees of Freedom	Mean Square (MS)	Test Statistic
Between groups	?	?	?	$F = ?$
Within groups	100.00	?	?	
Total	123.45	?		

 ## 11-3 Two-Way ANOVA

In Section 11-2 we used analysis of variance to decide whether three or more populations have the same mean. That section used procedures referred to as *one*-way analysis of variance (or single-factor analysis of variance) because the data are categorized into groups according to a *single* factor (or treatment). Recall that a factor, or treatment, is a property that is the basis for categorizing the different groups of data. For example, in Table 11-3, 60 SAT scores are separated into three categories according to the color of the M&M candy used as a treatment. The three populations are the people who eat red M&Ms, the people who eat green M&Ms, and the people who eat

TABLE 11-3	SAT Scores with Single Factor: Treatment with M&M Candies (Red, Green, Blue)								
Red	1130	621	813	996	1030	1257	898	743	921	1179
	1092	855	896	858	1095	1133	896	1190	908	699
Green	996	630	583	828	1121	993	1025	907	1111	1147
	780	916	793	1188	499	1180	1229	1450	1071	1153
Blue	706	1068	1013	892	1370	1611	939	1004	821	915
	866	848	1408	793	1097	1244	996	1131	1039	1159

blue M&Ms. The SAT scores are based on data from the College Board, and the M&M element is a product of author whimsy. If we use one-way analysis of variance to test that the three M&M color populations have equal means, we get the Excel display shown here.

EXCEL

ANOVA						
Source of Variation	SS	df	MS	F	P-value	F crit
Between Groups	80310	2	40155	**0.864694**	**0.426631**	3.158846
Within Groups	2646987	57	46438.37			
Total	2727297	59				

This Excel display includes a *P*-value of 0.426631, so we fail to reject the null hypothesis that the three populations have the same mean. This is pretty much what we expect: The colors of the M&Ms consumed don't appear to have an effect on SAT scores.

Two-way analysis of variance involves *two* factors, such as the Table 11-4 factors of (1) color of M&M candies used as treatment, and (2) gender. Using the two factors of M&M color and gender, we separate the data into six categories, as shown in Table 11-4. Such subcategories are often called *cells,* so Table 11-4 has six cells containing 10 scores each.

The following assumptions apply when testing with two-way analysis of variance.

Assumptions

1. For each cell, the sample values come from a population with a distribution that is approximately normal.

2. The populations have the same variance σ^2 (or standard deviation σ).

3. The samples are simple random samples. (That is, samples of the same size have the same probability of being selected.)

4. The samples are independent of each other. (The samples are not matched or paired in any way.)

5. The sample values are categorized two ways. (This is the basis for the name of the method: *two-way* analysis of variance.)

6. All of the cells have the same number of sample values. (This is called a *balanced* design.)

In analyzing the sample data in Table 11-4, we have already discussed the one-way analysis of variance for the single factor of M&M color, so it might seem reasonable to simply proceed with another one-way ANOVA for the factor of gender. Unfortunately, that approach wastes information and totally ignores a very important element: the effect from an interaction between the two factors.

| TABLE 11-4 | SAT Scores with Two Factors: Gender and Treatment with M&M Candies (Red, Green, Blue) |

	Female					Male				
Red	1130	621	813	996	1030	1257	898	743	921	1179
	1092	855	896	858	1095	1133	896	1190	908	699
Green	996	630	583	828	1121	993	1025	907	1111	1147
	780	916	793	1188	499	1180	1229	1450	1071	1153
Blue	706	1068	1013	892	1370	1611	939	1004	821	915
	866	848	1408	793	1097	1244	996	1131	1039	1159

DEFINITION

There is an **interaction** between two factors if the effect of one of the factors changes for different categories of the other factor.

As an example of an *interaction* between two factors, consider the pairings of food and wine at a quality restaurant. It is known that certain foods and wines interact well while others interact poorly. There is a good interaction between Chablis wine and oysters; the limestone in the soil where Chablis is made leaves a residue in the wine that interacts well with oysters. In contrast, red wine and turkey interact in a way that results in a bad taste. Willing to make any sacrifices to improve the quality of this book, the author experimented with different food/wine pairings and recorded taste test results in Table 11-5. The scores in the body of the table reflect these effects: (1) Red wine interacts in a positive way with red meat (see the high scores of 10, 9, and 10 for the pairing of red wine and red meat); (2) white wine goes better with chicken or fish (see the high scores of 6, 5, 7 and 8, 6, 7). There appears to be an effect due to an interaction between the type of wine and the type of food.

TABLE 11-5	Taste Test Scores for Food/Wine Pairings		
	Red Meat	Chicken	Fish
Red wine	10, 9, 10	4, 4, 2	3, 2, 4
White wine	3, 3, 2	6, 5, 7	8, 6, 7

Now that we understand the effects of an interaction, let's return to the SAT scores in Table 11-4. In using two-way ANOVA for the data of Table 11-4, we consider the effects of an interaction between M&M color and gender, as well as the effects of M&M color and the effects of gender on SAT scores. The calculations are quite involved, so *we will assume that Excel is being used.* (Procedures for using Excel are described at the end of this section.)

The Excel display includes SS (sum of squares) components similar to those described in Section 11-2. Because the circumstances of Section 11-2 involved only a single factor, we used SS(between groups) as a measure of the variation due to the different treatment categories, and we used SS(within groups) as a measure of the variation due to sampling error. We now use SS(Gender) as a measure of variation among the gender means. We use SS(M&M) as a measure of variation among the M&M color means. We continue to use SS(within groups) as a measure of variation due to sampling error. Similarly, we use MS(Gender) and MS(M&M) for the two different mean squares and continue to use MS(within groups) as before. Also, we use df(Gender) and df(M&M) for the two different degrees of freedom.

EXCEL

| ANOVA | | | | | | |
Source of Variation	SS	df	MS	F	P-value	F crit
Sample	80310	2	40155	**0.99662**	**0.375812**	3.168246
Columns	289537.1	1	289537.1	**7.186116**	**0.009723**	4.01954
Interaction	181726.5	2	90863.27	**2.255166**	**0.114644**	3.168246
Within	2175723	54	40291.17			
Total	2727297	59				

Interpreting Excel displays:
"Sample" corresponds to the row variable.
"Columns" corresponds to the column variable.

Table 11-6 compares the analyses of the data in Tables 11-3 and 11-4. Because Tables 11-3 and 11-4 have the same data in the M&M color categories, the calculated values of SS(M&M), MS(M&M), and df(M&M) are also the same in both cases. But note that as we go from the one-factor case (Table 11-3) to the two-factor case (Table 11-4) by further partitioning the data according to the second factor of gender, the value of SS(within groups) is partitioned into SS(Gender), SS(interaction), and SS(within groups). Also, df(within groups) is partitioned into df(Gender), df(interaction), and df(within groups), which can be easily verified by noting that $57 = 1 + 2 + 54$. Similar partitioning does not apply to MS(within groups); the value of MS(within groups) is 46,438, which does not equal $289,537 + 90,863 + 40,291$.

When conducting a two-way analysis of variance, we consider three effects:

1. The effect due to the interaction between the two factors of M&M color and gender

2. The effect due to the row factor (M&M color)

3. The effect due to the column factor (gender)

We now summarize the basic procedure for two-way analysis of variance. This procedure is very similar to the procedure presented in Section 11-2. We form conclusions about equal means by analyzing two estimates of variance, and the test statistic F is the ratio of those two estimates. A significantly large

TABLE 11-6	Comparison of Components Used in One-Way ANOVA and Two-Way ANOVA

One-Way ANOVA	Two-Way ANOVA
Sample data: Table 11-3	Sample data: Table 11-4
One factor: M&M Color	Two factors: M&M Color and Gender
SS(M&M) = 80,310	Same
MS(M&M) = 40,155	Same
df(M&M) = 2	Same
SS(total) = 2,727,297	Same
df(total) = 59	Same
SS(within groups) = 2,646,987 ⟶	SS(Gender) = 289,537
	SS(interaction) = 181,727
	SS(within groups) = 2,175,723
df(within groups) = 57 ⟶	df(Gender) = 1
	df(interaction) = 2
	df(within groups) = 54
	MS(Gender) = 289,537
MS(within groups) = 46,438	MS(interaction) = 90,863
	MS(within groups) = 40,291

value for F indicates that there is a statistically significant difference in means. The following procedure for two-way ANOVA is summarized in Figure 11-3.

Procedure for Two-Way ANOVA

Step 1:   *Interaction Effect:* In two-way analysis of variance, begin by testing the null hypothesis that there is no interaction between the two factors. Using Excel for the data in Table 11-4, we get the following test statistic:

$$F = \frac{MS(\text{interaction})}{MS(\text{within groups})} = \frac{90,863.27}{40,291.17} = 2.255166$$

Interpretation: The corresponding *P*-value is shown in the Excel display as 0.114644, so we fail to reject the null hypothesis of no interaction between the two factors. As we might expect, it does not appear that SAT scores are affected by an interaction between the color of consumed M&Ms and the gender of the subject.

Step 2:   *Row/Column Effects:* If we do reject the null hypothesis of no interaction between factors, then we should stop now; we should not proceed with the two additional tests. (If there is an interaction between factors, we shouldn't consider the effects of either factor without considering those of the other.)

   If we fail to reject the null hypothesis of no interaction between factors, then we should proceed to test the following two hypotheses:

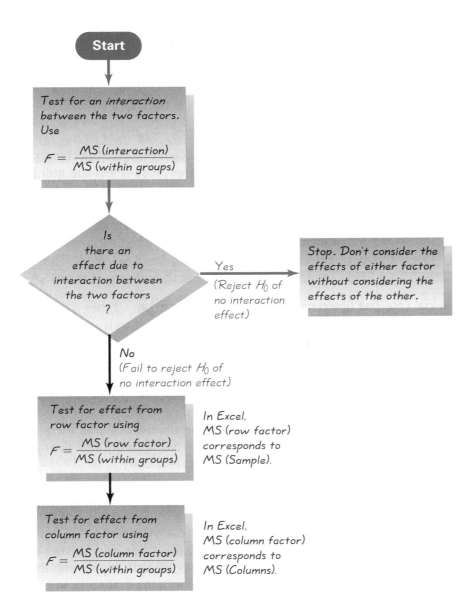

FIGURE 11-3
Procedure for Two-Way
ANOVA

H_0: There are no effects from the row factor (that is, the row means are equal).

H_0: There are no effects from the column factor (that is, the column means are equal).

In Step 1, we failed to reject the null hypothesis of no interaction between factors, so we proceed with the next two hypothesis tests identified in Step 2.

For the row factor,

$$F = \frac{MS(M\&M)}{MS(\text{within groups})} = \frac{40{,}155}{40{,}291.17} = 0.99662$$

Interpretation: This value is not significant because the corresponding *P*-value is shown in the Excel display as 0.375812. We fail to

1. Enter the sample data and labels as described above.
2. Click on **Tools,** then **Data Analysis,** then **Anova: Two Factor With Replication.**
3. In the dialog box, enter the input range. For the data in Table 11-4, enter A1:C31.
4. For "rows per sample," enter the number of values in each cell. For example, enter 10 for the data in Table 11-4.
5. Click **OK.**

Two-Way Tables with Exactly One Entry per Cell

1. Labels are not required. Enter the sample data as they appear in the table.
2. Select **Tools,** then **Data Analysis,** then **Anova: Two Factor Without Replication.**
3. In the dialog box, enter the input range of the sample values only. Do not include labels in the input range.
4. Click **OK.**

11-3 Basic Skills and Concepts

Interpreting a Computer Display. *In Exercises 1–3, use the Excel display, which results from the scores listed in the accompanying table. The sample data are SAT scores on the verbal and math portions of SAT-I and are based on reported statistics from the College Board. The "Columns" variable is SAT (verbal/math) and the "Sample" variable is gender.*

Verbal

Female	646	539	348	623	478	429	298	782	626	533
Male	562	525	512	576	570	480	571	555	519	596

Math

Female	484	489	436	396	545	504	574	352	365	350
Male	547	678	464	651	645	673	624	624	328	548

EXCEL

ANOVA						
Source of Variation	SS	df	MS	F	P-value	F crit
Sample	52635.03	1	52635.03	5.029517	0.031169	4.113161
Columns	6027.025	1	6027.025	0.57591	0.45286	4.113161
Interaction	31528.22	1	31528.22	3.012666	0.09117	4.113161
Within	376748.1	36	10465.23			
Total	466938.4	39				

EXCEL **1.** Interaction Effect Test the null hypothesis that SAT scores are not affected by an interaction between gender and test (verbal/math). What do you conclude?

EXCEL **2.** Effect of Gender Assume that SAT scores are not affected by an interaction between gender and the type of test (verbal/math). Is there sufficient evidence to support the claim that gender has an effect on SAT scores?

EXCEL **3.** Effect of Type of SAT Test Assume that SAT scores are not affected by an inter-action between gender and the type of test (verbal/math). Is there sufficient evi-dence to support the claim that the type of test (verbal/math) has an effect on SAT scores?

Interpreting a Computer Display.   In Exercises 4–6, use the Excel display, which results from the values listed in the accompanying table. The sample data are student estimates (in feet) of the length of their classroom. The actual length of the classroom is 24 ft 7.5 in. "Sample" represents gender and "Columns" represents major.

	Major		
	Math	Business	Liberal Arts
Female	28 25 30	35 25 20	40 21 30
Male	25 30 20	30 24 25	25 20 32

EXCEL

ANOVA						
Source of Variation	SS	df	MS	F	P-value	F crit
Sample	29.38889	1	29.38889	0.777941	0.395096	4.747221
Columns	10.11111	2	5.055556	0.133824	0.876032	3.88529
Interaction	14.11111	2	7.055556	0.186765	0.832005	3.88529
Within	453.3333	12	37.77778			
Total	506.9444	17				

EXCEL **4.** Interaction Effect Test the null hypothesis that the estimated lengths are not affected by an interaction between gender and major.

EXCEL **5.** Effect of Gender Assume that estimated lengths are not affected by an interac-tion between gender and major. Is there sufficient evidence to support the claim that estimated length is affected by gender?

EXCEL **6.** Effect of Major Assume that estimated lengths are not affected by an interac-tion between gender and major. Is there sufficient evidence to support the claim that estimated length is affected by major?

Interpreting a Computer Display.   In Exercises 7 and 8, refer to the given Excel dis-play. The "Rows" variable is subject, and the "Columns" variable is list. This display results from a study in which 24 subjects were given hearing tests using four different lists of words. The 24 subjects had normal hearing and the tests were conducted with no background noise. The main objective was to determine whether the four lists are equal-ly difficult to understand. In the original table of hearing test scores, each cell has one entry. The original data are from A Study of the Interlist Equivalency of the CID W-22 Word List Presented in Quiet and in Noise, *by Faith Loven, University of Iowa. The original data are available on the Internet through DASL (Data and Story Library).*

EXCEL

ANOVA						
Source of Variation	SS	df	MS	F	P-value	F crit
Rows	3231.625	23	140.5054	3.867829	6.96E-06	1.686896
Columns	920.4583	3	306.8194	8.446116	7.41E-05	2.737494
Error	2506.542	69	36.32669			
Total	6658.625	95				

7. Hearing Tests: Effect of Subject Assuming that there is no effect on hearing test scores from an interaction between subject and list, is there sufficient evidence to support the claim that the choice of subject has an effect on the hearing test score? Interpret the result by explaining why it makes practical sense.

 8. Hearing Tests: Effect of Word List Assuming that there is no effect on hearing test scores from an interaction between subject and list, is there sufficient evidence to support the claim that the choice of word list has an effect on the hearing test score?

Interpreting a Computer Display. *Exercises 9 and 10 refer to the sample data in the given table and the corresponding Excel display. The table entries are the numbers of support beams manufactured by four different operators using each of three different machines. Assume that there is no interaction effect from operator and machine.*

 9. Effect of Operator Using a 0.05 significance level, test the hypothesis that the four operators have the same mean production output. Interpret the result.

 10. Effect of Machine Using a 0.05 significance level, test the claim that the choice of machine has no effect on the production output. Interpret the result.

		Machine		
		1	2	3
	1	66	74	67
Operator	2	58	67	68
	3	65	71	65
	4	60	64	66

EXCEL

ANOVA

Source of Variation	SS	df	MS	F	P-value	F crit
Rows	59.58333	3	19.86111	2.474048	0.158993	4.757055
Columns	93.16667	2	46.58333	5.802768	0.039583	5.143249
Error	48.16667	6	8.027778			
Total	200.9167	11				

 11. Interaction Between Gender and Smoking Refer to Data Set 14 (STATSURV.xls) in Appendix B and construct a table with pulse rates categorized according to the two factors of gender and whether the individual smokes. Select nine values for each cell and test the null hypothesis of no interaction between gender and smoking.

 12. Effect of Gender on Pulse Rates Use the same data collected for Exercise 11, assume that pulse rates are not affected by an interaction between gender and smoking, and test the null hypothesis that gender has no effect on pulse rates.

 13. Effect of Smoking on Pulse Rates Use the same data collected for Exercise 11, assume that pulse rates are not affected by an interaction between gender and smoking, and test the null hypothesis that smoking has no effect on pulse rates.

11-3 Beyond the Basics

14. Transformations of Data Assume that two-way ANOVA is used to analyze sample data consisting of more than one entry per cell. How are the ANOVA results affected in each of the following cases?
  a. The same constant is added to each sample value.
  b. Each sample value is multiplied by the same nonzero constant.

c. The format of the table is transposed, so that the row and column factors are interchanged.

d. The first sample value in the first cell is changed so that it becomes an outlier.

15. Changing Interaction Effect In analyzing Table 11-4, we concluded that SAT scores are not affected by an interaction between gender and the color of M&Ms consumed, and SAT scores are not affected by color of M&Ms consumed, but they are affected by gender.

a. Change the table entries so that there is an effect from the interaction between gender and color of consumed M&Ms.

b. Change the table entries so that there is no effect from the interaction between gender and color of consumed M&Ms, but there is an effect from the color of consumed M&Ms.

c. Change the table entries so that there is no effect from the interaction between gender and color of consumed M&Ms, there is no effect from color of consumed M&Ms, and there is no effect from gender.

Birth Order Affects Height

NEW YORK *New York Times* reporter C. Claiborne Ray wrote that "many studies have found a relationship between birth order and birth weight, with the later-born children in a family tending to be larger, and birth weight has some correlation with eventual height." In referring to a German study, he noted that "the researchers found that the first-born siblings had a statistically significant deficit in average height as compared with the average of all siblings, but the gap amounted to only a fraction of an inch. That study also found that men tended to increase in height with increasing birth order, while women did not, beyond the second born."

1. Describe in detail the design of an observational study that could be used to support or refute the key point that older siblings tend to be taller. Include a sampling plan and the method of analysis. How do you control for the inherent differences in heights between men and women? How might analysis of variance be used? How do you handle children who have no brothers or sisters?

2. One approach to designing an observational study might be to collect heights only from families in which there are exactly three children, all males. Among families with exactly three children, what proportion of them are all males?

3. Consider families with exactly three male children. Claiborne noted that younger siblings tend to be shorter, "but the gap amounted to only a fraction of an inch." If the differences in mean heights of the first, second, and third males are very small, how does the smallness of those differences affect the size of the sample needed to support the statistical significance of the differences?

4. Suppose someone argues that the stated results cannot be true because he knows a family in which the three children are all males, but their heights decrease with birth order. How do you respond?

VOCABULARY LIST
---

analysis of variance (ANOVA)
one-way analysis of variance
single-factor analysis of variance
treatment
factor
variance between samples
variation due to treatment

variance within samples
variation due to error
multiple comparison procedures
completely randomized design
rigorously controlled design
two-way analysis of variance
interaction

REVIEW

In Chapter 8 we used t distributions and normal distributions to test for equality between *two* population means, but in Section 11-2 we used analysis of variance (or ANOVA) to test for equality of three or more population means. This method requires (1) normally distributed populations, (2) populations with the same standard deviation (or variance), and (3) simple random samples that are independent of each other. The methods of one-way analysis of variance are used when we have three or more samples taken from populations that are characterized according to a single factor. The following are key features of one-way analysis of variance:

- The F test statistic is based on the ratio of two different estimates of the common population variance σ^2, as shown below.

$$F = \frac{\text{variance between samples}}{\text{variance within samples}} = \frac{\text{MS(between groups)}}{\text{MS(within groups)}}$$

- Critical values of F can be found in Table A-5, but we focused on the interpretation of P-values that are included as part of a computer display.

In Section 11-3 we considered two-way analysis of variance, characterized by data categorized according to two different factors. The method for two-way analysis of variance is summarized in Figure 11-3. We also considered two-way analysis of variance with one observation per cell.

Because of the nature of the calculations required throughout this chapter, we emphasized the interpretation of computer displays.

REVIEW EXERCISES
---

1. Car Fuel Consumption: Testing for Interaction Twelve different 4-cylinder cars were tested for fuel consumption (in mi/gal) after being driven under identical highway conditions; the results are listed in the table and accompanying Excel display. At the 0.05 significance level, test the claim that fuel consumption is not

EXCEL

Source of Variation	SS	df	MS	F	P-value	F crit
Sample	40.33333	1	40.33333	3.558824	0.108178	5.987374
Columns	43.16667	2	21.58333	1.904412	0.228877	5.143249
Interaction	1.166667	2	0.583333	0.051471	0.950246	5.143249
Within	68	6	11.33333			
Total	152.6667	11				

affected by an *interaction* between engine size and transmission type. The "Sample" variable is transmission type, and the "Columns" variable is engine size.

Highway Fuel Consumption (mi/gal)
of Different 4-Cylinder Compact Cars

	Engine Size (liters)		
	1.5	2.2	2.5
Automatic transmission	31, 32	28, 26	31, 23
Manual transmission	33, 36	33, 30	27, 34

2. Effect of Engine Size Refer to the data used in Exercise 1 and assume that fuel consumption is not affected by an interaction between engine size and type of transmission. Use a 0.05 significance level to test the claim that fuel consumption is not affected by engine size.

3. Effect of Transmission Refer to the data used in Exercise 1 and assume that fuel consumption is not affected by an interaction between engine size and type of transmission. Use a 0.05 significance level to test the claim that fuel consumption is not affected by type of transmission.

4. Location, Location, Location The accompanying list shows selling prices (in thousands of dollars) for homes located on Long Beach Island in New Jersey. Different mean selling prices are expected for the different locations. Do these sample data support the claim of different mean selling prices? Use a 0.05 significance level.

Oceanside:	235	395	547	469	369	279
Oceanfront:	538	446	435	639	499	399
Bayside:	199	219	239	309	399	190
Bayfront:	695	389	489	489	599	549

5. Auto Pollution The accompanying table lists the amounts of greenhouse gases emitted by different cars in one year. (See Data Set 18 in Appendix B.) The Excel display on the top of the next page results from this table.

a. Assuming that there is no interaction effect, is there sufficient evidence to support the claim that amounts of emitted greenhouse gases are affected by the type of transmission (automatic/manual)?

b. Assuming that there is no interaction effect, is there sufficient evidence to support the claim that amounts of emitted greenhouse gases are affected by the number of cylinders?

c. Based on the results from parts (a) and (b), can we conclude that greenhouse gas emissions are not affected by the type of transmission or the number of cylinders? Why or why not?

Emission of Greenhouse Gases (tons/year)

	4 Cylinders	6 Cylinders	8 Cylinders
Automatic	10	12	14
Manual	10	12	12

Then divide into groups of three or four, and use the data from the entire class to address these questions:

- Is there a significant difference between the mean estimate for males and the mean estimate for females?

- Is there sufficient evidence to reject equality of the mean estimates for different majors? Describe how the majors were categorized.

- Does an interaction between gender and major have an effect on the estimated length?

- Does gender appear to have an effect on estimated length?

- Does major appear to have an effect on estimated length?

2. *Out-of-class activity:* Divide into groups of three or four students. Each group should survey other students at the same college by asking them to identify their major and gender. You might include other factors, such as employment (none, part-time, full-time) and age (under 21, 21–30, over 30). For each surveyed subject, determine the accuracy of the time on his or her wristwatch. First set your own watch to the correct time using an accurate and reliable source ("At the tone, the time is . . ."). For watches that are ahead of the correct time, record positive times. For watches that are behind the correct time, record negative times. Use the sample data to address questions such as these:

- Does gender appear to have an effect on the accuracy of the wristwatch?

- Does major have an effect on wristwatch accuracy?

- Does an interaction between gender and major have an effect on wristwatch accuracy?

3. *In-class activity:* Divide into groups of five or six students. This activity is a contest of reaction times to determine which group is fastest, if there is a "fastest" group. Use the reaction timer included with the Cooperative Group Activities in Chapter 5. Test and record the reaction time using the dominant hand of each group member. (Only one try per person.) Each group should calculate the values of n, $\bar{x}$, and s, and record those summary statistics on the chalkboard along with the original list of reaction times. After all groups have reported their results, identify the group with the fastest mean reaction time. But is that group actually fastest? Use ANOVA to determine whether the means are significantly different. If they are not, there is no real "winner."

4. *Out-of-class activity:* The *World Almanac and Book of Facts* includes a section called "Noted Personalities," with subsections comprised of architects, artists, business leaders, cartoonists, social scientists, military leaders, philosophers, political leaders, scientists, writers, composers, entertainers, and others. Design and conduct an observational study that begins with the selection of samples from select groups, to be followed by a comparison of life spans of people from the different categories. Do any particular groups appear to have life spans that are different from the other groups? Can you explain such differences?

EXCEL PROJECT

For U.S. presidents and the popes and British monarchs since 1690, the accompanying table lists the numbers of years that they lived after their inauguration, election, or coronation. Use boxplots and analysis of variance to determine whether the survival times for the different groups differ. Conduct the analysis of variance by using Excel. Obtain printed copies of the computer displays and write your observations and conclusions.

Presidents		Popes		Kings and Queens	
Washington	10	Alex VIII	2	James II	17
J. Adams	29	Innoc XII	9	Mary II	6
Jefferson	26	Clem XI	21	William III	13
Madison	28	Innoc XIII	3	Anne	12
Monroe	15	Ben XIII	6	George I	13
J. Q. Adams	23	Clem XII	10	George II	33
Jackson	17	Ben XIV	18	George III	59
Van Buren	25	Clem XIII	11	George IV	10
Harrison	0	Clem XIV	6	William IV	7
Tyler	20	Pius VI	25	Victoria	63
Polk	4	Pius VII	23	Edward VII	9
Taylor	1	Leo XII	6	George V	25
Fillmore	24	Pius VIII	2	Edward VIII	36
Pierce	16	Greg XVI	15	George VI	15
Buchanan	12	Pius IX	32		
Lincoln	4	Leo XIII	25		
A. Johnson	10	Pius X	11		
Grant	17	Ben XV	8		
Hayes	16	Pius XII	17		
Garfield	0	Pius XIII	19		
Arthur	7	John XXIII	5		
Cleveland	24	Paul VI	15		
Harrison	12	John Paul I	0		
McKinley	4				
T. Roosevelt	18				
Taft	21				
Wilson	11				
Harding	2				
Coolidge	9				
Hoover	36				
F. Roosevelt	12				
Truman	28				
Kennedy	3				
Eisenhower	16				
L. Johnson	9				
Nixon	25				

Based on data from *Computer-Interactive Data Analysis*, by Lunn and McNeil, John Wiley & Sons.

Intervals Between Eruptions

Year	Interval (min)						Mean	Range
1	65	72	60	69	65	67	66.3	12
2	74	65	60	69	68	59	65.8	15
3	68	66	69	64	70	73	68.3	9
4	73	65	71	77	63	77	71.0	14
5	79	67	64	61	81	77	71.5	20
6	74	76	65	69	76	64	70.7	12
7	70	73	74	77	65	73	72.0	12
8	71	68	70	79	75	82	74.2	14
9	62	63	61	48	59	77	61.7	29
10	60	74	77	57	52	78	66.3	26
11	67	73	47	81	92	57	69.5	45
12	79	84	79	72	61	80	75.8	23
13	83	78	83	74	61	68	74.5	22
14	57	68	72	75	56	79	67.8	23
15	59	76	78	86	64	72	72.5	27
16	63	63	71	77	81	65	70.0	18
17	67	84	72	75	70	70	73.0	17
18	93	83	85	79	90	74	84.0	19
19	81	74	80	65	70	84	75.7	19
20	83	67	71	67	97	88	78.8	30
21	62	61	57	86	70	77	68.8	29
22	67	75	67	89	93	81	78.7	26
23	86	65	70	74	83	74	75.3	21
24	74	67	99	75	41	83	73.2	58
25	97	93	73	81	85	90	86.5	24

Constructing Control Charts for Minted Quarters. *In Exercises 4–6, use the following information:*

> *The U.S. Mint has a goal of making quarters with a weight of 5.670 g, but any weight between 5.443 g and 5.897 g is considered acceptable. A new minting machine is placed into service and the weights are recorded for a quarter randomly selected every 12 min for 20 consecutive hours. The results are listed in the accompanying table on the next page.*

EXCEL **4.** Minting Quarters: Constructing a Run Chart Construct a run chart for the 100 values. Does there appear to be a pattern suggesting that the process is not within statistical control? What are the practical implications of the run chart?

EXCEL **5.** Minting Quarters: Constructing an R Chart Construct an R chart and determine whether the process variation is within statistical control. If it is not, identify which of the three out-of-control criteria lead to rejection of statistically stable variation.

EXCEL **6.** Minting Quarters: Constructing an $\bar{x}$ Chart Construct an $\bar{x}$ chart and determine whether the process mean is within statistical control. If it is not, identify which of the three out-of-control criteria lead to rejection of a statistically stable mean. Does this process need corrective action?

Weights (in grams) of Minted Quarters

Hour	Weight (g)					$\overline{x}$	s	Range
1	5.639	5.636	5.679	5.637	5.691	5.6564	0.0265	0.055
2	5.655	5.641	5.626	5.668	5.679	5.6538	0.0211	0.053
3	5.682	5.704	5.725	5.661	5.721	5.6986	0.0270	0.064
4	5.675	5.648	5.622	5.669	5.585	5.6398	0.0370	0.090
5	5.690	5.636	5.715	5.694	5.709	5.6888	0.0313	0.079
6	5.641	5.571	5.600	5.665	5.676	5.6306	0.0443	0.105
7	5.503	5.601	5.706	5.624	5.620	5.6108	0.0725	0.203
8	5.669	5.589	5.606	5.685	5.556	5.6210	0.0545	0.129
9	5.668	5.749	5.762	5.778	5.672	5.7258	0.0520	0.110
10	5.693	5.690	5.666	5.563	5.668	5.6560	0.0534	0.130
11	5.449	5.464	5.732	5.619	5.673	5.5874	0.1261	0.283
12	5.763	5.704	5.656	5.778	5.703	5.7208	0.0496	0.122
13	5.679	5.810	5.608	5.635	5.577	5.6618	0.0909	0.233
14	5.389	5.916	5.985	5.580	5.935	5.7610	0.2625	0.596
15	5.747	6.188	5.615	5.622	5.510	5.7364	0.2661	0.678
16	5.768	5.153	5.528	5.700	6.131	5.6560	0.3569	0.978
17	5.688	5.481	6.058	5.940	5.059	5.6452	0.3968	0.999
18	6.065	6.282	6.097	5.948	5.624	6.0032	0.2435	0.658
19	5.463	5.876	5.905	5.801	5.847	5.7784	0.1804	0.442
20	5.682	5.475	6.144	6.260	6.760	6.0642	0.5055	1.285

Constructing Control Charts for Home Energy Consumption. *In Exercises 7–9, use the following information:*

The author recorded his electrical energy consumption (in kilowatt-hours) for his home in upstate New York for two-month intervals over four years, and the results are listed in the table.

	Jan.–Feb.	Mar.–Apr.	May–June	July–Aug.	Sept.–Oct.	Nov.–Dec.
Year 1	4762	3875	2657	4358	2201	3187
Year 2	4504	3237	2198	2511	3020	2857
Year 3	3952	2785	2118	2658	2139	3071
Year 4	3863	3013	2023	2953	3456	2647

EXCEL

7. Energy Consumption: Constructing a Run Chart Construct a run chart for the 24 values. Does there appear to be a pattern suggesting that the process is not within statistical control? Is there any pattern or variation that can be explained?

EXCEL

8. Energy Consumption: Constructing an *R* Chart Use samples of size 3 by combining the first three values for each year and combining the last three values for each year. Construct an *R* chart and determine whether the process variation is within statistical control. If it is not, identify which of the three out-of-control criteria lead to rejection of statistically stable variation.

EXCEL

9. Energy Consumption: Constructing an $\overline{x}$ Chart Use samples of size 3 by combining the first three values for each year and combining the last three values

Nonparametric Statistics

It rains more on weekends!?

In an article for the Knight Ridder News Service, Usha Lee McFarling writes that "Your worst weather fears are true. It does rain more on weekends. Scientists who pored through years of rainfall data have discovered a clear and dismaying pattern. Fridays, Saturdays, and Sundays are the rainiest days of the week all along America's eastern coast, from Maine to Florida." The article refers to a study conducted by Arizona State University scientists Randall S. Cerveny and Robert C. Balling. But are their conclusions correct? Have they been reported and interpreted correctly?

Data Set 17 in Appendix B includes rainfall amounts for a recent year in Boston. In selecting a city for verifying the weekend rain phenomenon, Boston should be a good choice because it is located on the eastern coast. If we construct boxplots for the rainfall amount for each day of the week, we get the results shown on the next page. Going from top to bottom, the results represent Monday, Tuesday, . . . , Sunday. The Monday boxplot at the top is unusual because it appears to be missing the box, but the Monday rainfall amounts have so many 0s that the minimum, first quartile, and median are all 0s, with the result that the box gets crunched to the left. Greater amounts of rainfall on Friday, Saturday, and Sunday would be visible with the distributions positioned farther to the right. Is that really the case? Are the differences really significant?

We might consider using methods presented in earlier chapters to investigate this issue. Analysis of variance (Section 11-2) might come to mind, but that method requires that the samples come from populations having a normal distribution. The Excel histogram for the Monday rainfall amounts shows clearly that these values do not come from a normally distributed population. A major advantage of the methods discussed in this chapter is that they do not require a particular distribution.

Can we use the Boston data to support the claim of more rainfall on weekends? Are the rainfall amounts for the different days significantly different? We will address these questions later in the chapter.

13-1 Overview

The methods of inferential statistics presented in Chapters 6, 7, 8, 9, and 11 are called *parametric methods* because they are based on sampling from a population with specific parameters, such as the mean μ, standard deviation σ, or proportion p. Those parametric methods usually must conform to some fairly strict conditions, such as a requirement that the sample data come from a normally distributed population. This chapter introduces nonparametric methods, which do not have such strict requirements.

DEFINITIONS

Parametric tests require assumptions about the nature or shape of the populations involved; **nonparametric tests** do not require such assumptions. Consequently, nonparametric tests of hypotheses are often called **distribution-free tests.**

Although the term *nonparametric* suggests that the test is not based on a parameter, there are some nonparametric tests that do depend on a parameter such as the median. The nonparametric tests do not, however, require a particular distribution, so they are sometimes referred to as *distribution-free* tests. Although *distribution-free* is a more accurate description, the term *nonparametric* is more commonly used. The following are major advantages and disadvantages of nonparametric methods.

Advantages of Nonparametric Methods

1. Nonparametric methods can be applied to a wide variety of situations because they do not have the more rigid requirements of the corresponding parametric methods. In particular, nonparametric methods do not require normally distributed populations.

2. Unlike parametric methods, nonparametric methods can often be applied to nonnumerical data, such as the genders of survey respondents.

3. Nonparametric methods usually involve simpler computations than the corresponding parametric methods and are therefore easier to understand and apply.

Disadvantages of Nonparametric Methods

1. Nonparametric methods tend to waste information because exact numerical data are often reduced to a qualitative form. For example, in the nonparametric sign test (described in Section 13-2), weight losses by dieters are recorded simply as negative signs; the actual magnitudes of the weight losses are ignored.

2. Nonparametric tests are not as efficient as parametric tests, so with a nonparametric test we generally need stronger evidence (such as a larger sample or greater differences) before we reject a null hypothesis.

When the requirements of population distributions are satisfied, nonparametric tests are generally less efficient than their parametric counterparts, but the reduced efficiency can be compensated for by an increased sample size. For example, Section 13-6 will present a concept called *rank correlation,* which has an efficiency rating of 0.91 when compared to the linear correlation presented in Chapter 9. This means that with all other things being equal, nonparametric rank correlation requires 100 sample observations to achieve the same results as 91 sample observations analyzed through parametric linear correlation, assuming the stricter requirements for using the parametric method are met. Table 13-1 lists the nonparametric methods covered in this chapter, along with the corresponding parametric approach and **efficiency** rating. Table 13-1 shows that several nonparametric tests have

TABLE 13-1	Efficiency: Comparison of Parametric and Nonparametric Tests		
Application	Parametric Test	Nonparametric Test	Efficiency Rating of Nonparametric Test with Normal Population
Matched pairs of sample data	t test or z test	Sign test	0.63
		Wilcoxon signed-ranks test	0.95
Two independent samples	t test or z test	Wilcoxon rank-sum test	0.95
Several independent samples	Analysis of variance (F test)	Kruskal-Wallis test	0.95
Correlation	Linear correlation	Rank correlation test	0.91
Randomness	No parametric test	Runs test	No basis for comparison

efficiency ratings above 0.90, so the lower efficiency might not be a critical factor in choosing between parametric and nonparametric methods. However, because parametric tests do have higher efficiency ratings than their nonparametric counterparts, it's generally better to use the parametric tests when their required assumptions are satisfied.

Ranks

Sections 13-3 through 13-6 use methods based on ranks. Let's now briefly describe them.

> ### DEFINITION
>
> Data are *sorted* when they are arranged according to some criterion, such as smallest to largest or best to worst. A **rank** is a number assigned to an individual sample item according to its order in the sorted list. The first item is assigned a rank of 1, the second item is assigned a rank of 2, and so on.

▷ **EXAMPLE**   The numbers 5, 3, 40, 10, and 12 can be sorted (arranged from lowest to highest) as 3, 5, 10, 12, and 40, and these numbers have ranks of 1, 2, 3, 4, and 5, respectively:

5	3	40	10	12	Original values
3	5	10	12	40	Values arranged in order
↑	↑	↑	↑	↑	
1	2	3	4	5	Ranks

◁

Handling ties in ranks: If a tie in ranks occurs, the usual procedure is to find the mean of the ranks involved and then assign this mean rank to each of the tied items, as in the following example.

▷ **EXAMPLE**   The numbers 3, 5, 5, 10, and 12 are given ranks of 1, 2.5, 2.5, 4, and 5, respectively. In this case, ranks 2 and 3 were tied, so we found the mean of 2 and 3 (which is 2.5) and assigned it to the values that created the tie:

3	5	5	10	12	Original values
↑	↑	↑	↑	↑	
1	2.5	2.5	4	5	Ranks

↑_____↑

2 and 3 are tied

◁

13-2 Sign Test

The main objective of this section is to understand the *sign test* procedure, which is among the easiest for nonparametric tests.

DEFINITION

The **sign test** is a nonparametric (distribution-free) test that uses plus and minus signs to test different claims, including:

1. Claims involving matched pairs of sample data
2. Claims involving nominal data
3. Claims about the median of a single population

Assumptions

1. The sample data have been randomly selected.
2. There is *no* requirement that the sample data come from a population with a particular distribution, such as a normal distribution.

Basic Concept of the Sign Test   The basic idea underlying the sign test is to analyze the frequencies of the plus and minus signs to determine whether they are significantly different. For example, suppose that we test a treatment designed to lower blood pressure. If 100 subjects are treated and 51 of them experience lower blood pressure while the other 49 have increased blood pressure, common sense suggests that there is not sufficient evidence to say that the drug is effective, because 51 decreases out of 100 is not significant. But what about 52 decreases and 48 increases? Or 90 decreases and 10 increases? The sign test allows us to determine when such results are significant. Figure 13-1 on the next page summarizes the sign test procedure, which will be illustrated with examples that follow.

For consistency and ease, we will stipulate the following notation.

Notation for the Sign Test

$x =$ the number of times the *less frequent* sign occurs

$n =$ the total number of positive and negative signs combined tests lead to rejection of the null hypothesis, but the sign test doesn't consider the sample data to be as extreme, partly because the sign test uses only information about the *direction* of the data, ignoring the *magnitudes* of the data values. The next section introduces the Wilcoxon signed-ranks test, which largely overcomes that disadvantage.

Rationale for the test statistic used when $n > 25$: When finding critical values for the sign test, we use Table A-7 only for n up to 25. When $n > 25$, the test statistic z is based on a normal approximation to the binomial probability distribution with $p = q = 1/2$. Recall that in Section 5-6 we saw that the normal approximation to the binomial distribution is acceptable when both $np \geq 5$ and $nq \geq 5$. Recall also that in Section 4-4 we saw that $\mu = np$ and $\sigma = \sqrt{npq}$ for binomial probability distributions. Because this sign test assumes that $p = q = 1/2$, we meet the $np \geq 5$ and $nq \geq 5$ prerequisites whenever $n \geq 10$. Also, with the assumption that $p = q = 1/2$, we get $\mu = np = n/2$ and $\sigma = \sqrt{npq} = \sqrt{n/4} = \sqrt{n}/2$, so

$$z = \frac{x - \mu}{\sigma}$$

becomes

$$z = \frac{x - \left(\dfrac{n}{2}\right)}{\dfrac{\sqrt{n}}{2}}$$

Finally, we replace x by $x + 0.5$ as a correction for continuity. That is, the values of x are discrete, but because we are using a continuous probability distribution, a discrete value such as 10 is actually represented by the interval from 9.5 to 10.5. Because x represents the less frequent sign, we act conservatively by concerning ourselves only with $x + 0.5$; we thus get the test statistic z, as given in the equation and in Figure 13-1.

Using Excel for the Sign Test

Excel does not have a built-in function dedicated to the sign test, but you can use Excel's **BINOMDIST** function to find the P-value for a sign test.

1. Use the methods of this section to determine the values of x (the less frequent sign) and the value of n (the sample size).

2. Click *fx* on the main menu bar, then select the function category of **Statistical,** then select **BINOMDIST,** then click **OK.**

3. In the dialog box, first enter x, then the number of trials n, then enter a probability of 0.5, then enter **TRUE** in the box for cumulative. The resulting value is the probability of getting x or fewer successes among n trials. *Double this value for two-tailed tests.* The final result is the P-value, so reject the null hypothesis if the P-value is less than or equal to the significance level. Otherwise, fail to reject the null hypothesis.

Shown on the next page is the dialog box for the last example of this section. Using $x = 23$ and $n = 91$, the P-value for this one-sided test is displayed as 1.26108E-06, which can be written in standard form as 0.00000126108. Because this probability is less than or equal to the significance level of 0.05, we reject the null hypothesis; we support the claim that the median is less than 98.6°F.

13-2   Basic Skills and Concepts

In Exercises 1–12, use the sign test.

   1. Testing for a Difference Between Reported and Measured Female Heights This section used reported heights and measured heights for a sample of *male* statistics students. Listed here are heights (in inches) collected from a random sample of *female* statistics students. Is there sufficient evidence to support the claim that there is a difference between reported and measured heights of female statistics students? Use a 0.05 significance level.

Reported height	64	63	64	65	64	64	63	59	66	64
Measured height	63.5	63.1	63.8	63.4	62.1	64.4	62.7	59.3	65.4	62.2

   2. Testing for a Difference Between the Qwerty and Dvorak Keyboards The following sentence, which contains every letter of the alphabet, was used for testing typewriters: "The quick brown fox jumped over the lazy red dog." If we use the word rating procedure described in the Chapter Problem for Chapter 2, we find that the 10 words in that sentence have Qwerty ratings of 2, 5, 7, 3, 5, 5, 2, 3, 2, 1. The corresponding Dvorak ratings are 0, 5, 5, 3, 5, 3, 0, 4, 1, 1. Is there sufficient evidence to support the claim that the Dvorak keyboard has lower ratings?

   3. Nominal Data: Survey of Voters In a survey of 1002 people, 701 said that they voted in the recent presidential election (based on data from ICR Research Group). Is there sufficient evidence to support the claim that the majority of people say that they voted in the election?

   4. Nominal Data: Smoking and Nicotine Patches In one study of 71 smokers who tried to quit smoking with nicotine patch therapy, 41 were smoking one year after the treatment (based on data from "High-Dose Nicotine Patch Therapy," by Dale et al., *Journal of the American Medical Association*, Vol. 274, No. 17). Use a 0.05 significance level to test the claim that among smokers who try to quit with nicotine patch therapy, the majority are smoking a year after the treatment.

5. Testing for Median Volume of Coke Cans Refer to Data Set 1 in Appendix B and use the volumes of regular Coke. Test the claim that cans of regular Coke have volumes with a median greater than 12 oz. Does it appear that the Coke cans are being filled correctly?

6. Testing for Median Amount of Domino Sugar in Packets Refer to Data Set 4 in Appendix B and use the sample data to test the claim that the median amount of sugar in the packets is equal to 3.5 oz. Does it appear that the sugar packets are being filled correctly?

7. Testing for Median Weight of M&Ms Using the weights of only the *brown* M&Ms listed in Data Set 10 in Appendix B, test the claim that the median is greater than 0.9085 g. (For the 1498 M&Ms to produce a total package weight of 1361 g, the mean weight must be at least 0.9085 g.) Use a 0.05 significance level. Based on the result, does it appear that the package is labeled correctly?

8. Testing for Median Time Interval for Old Faithful Geyser Refer to Data Set 11 in Appendix B. Test the claim that the intervals between eruptions of the Old Faithful geyser have a median greater than 77 min, which was the median about 20 years ago.

9. Testing for Difference Between Right/Left Reaction Times The accompanying table lists sample data obtained when 14 subjects were tested for reaction times with their left and right hands. (Only right-handed subjects were used.) Use a 0.05 significance level to test the claim of no difference between the right- and left-hand reaction times.

Right	191	97	116	165	116	129	171	155	112	102	188	158	121	133
Left	224	171	191	207	196	165	177	165	140	188	155	219	177	174
Sign of difference (right − left)	−	−	−	−	−	−	−	−	−	−	+	−	−	−

10. Testing Effectiveness of SAT Prep Courses Does it pay to take preparatory courses for standardized tests such as the SAT? Using a 0.05 significance level, test the claim that the Allan Preparation Course has no effect on SAT scores. Use the sample data in the accompanying table.

Subject	A	B	C	D	E	F	G	H	I	J
SAT score before course	700	840	830	860	840	690	830	1180	930	1070
SAT score after course	720	840	820	900	870	700	800	1200	950	1080

Based on data from the College Board and "An Analysis of the Impact of Commercial Test Preparation Courses on SAT Scores," by Sesnowitz, Bernhardt, and Knain, *American Educational Research Journal,* Vol. 19, No. 3.

11. Testing for Effect of Drinking on Reaction Time The Life Trust Insurance Company funded a university study of drinking and driving. After 30 randomly selected drivers were tested for reaction times, they were given two drinks and tested again, with the result that 22 had slower reaction times, 6 had faster reaction times, and 2 had the same reaction times as before the drinks. At the 0.01 significance level, test the claim that the drinks had no effect on the reaction times. Based on these very limited results, does it appear that the insurance company is justified in charging higher rates for those who drink and drive?

12. Nominal Data: Testing for Gender Discrimination The Diaz Tool and Die Company claims that hiring is done without any gender bias. Among the last 54 new employees hired, 1/3 are women. Job applicants are about half men and half women, who are all qualified. Is there sufficient evidence to charge gender bias? Use a 0.01 significance level, because we don't want to make such a serious charge unless there is very strong evidence.

13-2 Beyond the Basics

 13. Procedures for Handling Ties In the sign test procedure described in this section, we excluded ties (represented by 0 instead of a sign of + or −). A second approach is to treat half of the 0s as positive signs and half as negative signs. (If the number of 0s is odd, exclude one so that they can be divided equally.) With a third approach, in two-tailed tests make half of the 0s positive and half negative; in one-tailed tests make all 0s either positive or negative, whichever supports the null hypothesis. Assume that in using the sign test on a claim that the median value is at least 100, we get 60 values below 100, 40 values above 100, and 21 values equal to 100. Identify the test statistic and conclusion for the three different ways of handling differences of 0. Assume a 0.05 significance level in all three cases.

14. IQ Scores: Agreement Between Sign Test and t Test A sample of 10 randomly selected IQ scores is obtained from the population of professional athletes. We want to test the claim that the population is centered at the IQ score of 100.
   a. Is it possible to have sample data such that the sign test leads to rejection of the null hypothesis that the median is equal to 100 while the t test (Section 7-4) conclusion is failure to reject the null hypothesis of $\mu = 100$? If so, construct such a data set.
   b. Is it possible to have sample data such that the sign test conclusion is failure to reject the null hypothesis that the median is equal to 100 while the t test (Section 7-4) conclusion is rejection of the null hypothesis that $\mu = 100$? If so, construct such a data set.

 15. Finding Critical Values Table A-7 lists critical values for limited choices of α. Use Table A-1 to add a new column in Table A-7 (down to $n = 15$) that represents a significance level of 0.03 in one tail or 0.06 in two tails. For any particular n, use $p = 0.5$, because the sign test requires the assumption that

$$P(\text{positive sign}) = P(\text{negative sign}) = 0.5$$

The probability of x or fewer like signs is the sum of the probabilities for values up to and including x.

 ## 13-3 Wilcoxon Signed-Ranks Test for Matched Pairs

In Section 13-2 we used the sign test to analyze data, including sample data consisting of matched pairs. The sign test used only the signs of the differences and did not use their actual magnitudes (how large the numbers are). This section introduces the *Wilcoxon signed-ranks test,* which is also used with sample paired data. By using ranks, this test takes the magnitudes of the differences into account. (See Section 13-1 for a description of ranks.) Because the Wilcoxon signed-ranks test incorporates and uses more information than the sign test, it tends to yield conclusions that better reflect the true nature of the data.

DEFINITION

The **Wilcoxon signed-ranks test** is a nonparametric test that uses ranks of sample data consisting of matched pairs. It is used to test for differences in the population distributions, so the null and alternative hypotheses are as follows:

H_0: The two samples come from populations with the same distribution.

H_1: The two samples come from populations with different distributions.

(The Wilcoxon signed-ranks test can also be used to test the claim that a sample comes from a population with a specified median. See Exercise 10 for this application.)

Assumptions

1. The sample data have been randomly selected
2. The population of differences (found from the pairs of data) has a distribution that is approximately *symmetric,* meaning that the left half of its histogram is roughly a mirror image of its right half. (There is *no* requirement that the data have a normal distribution.)

Procedure for Finding the Value of the Test Statistic

Step 1: For each pair of data, find the difference d by subtracting the second value from the first value. Keep the signs, but discard any pairs for which $d = 0$.

Step 2: *Ignore the signs of the differences,* then sort the differences from lowest to highest and replace the differences by the corresponding rank value (as described in Section 13-1). When differences have the same numerical value, assign to them the mean of the ranks involved in the tie.

Step 3: Attach to each rank the sign of the difference from which it came. That is, insert those signs that were ignored in Step 2.

Step 4: Find the sum of the absolute values of the negative ranks. Also find the sum of the positive ranks.

Step 5: Let T be the *smaller* of the two sums found in Step 4. Either sum could be used, but for a simplified procedure we arbitrarily select the smaller of the two sums. (See the notation for T in the Notation box.)

Step 6: Let n be the number of pairs of data for which the difference d is not 0.

Step 7: Determine the test statistic and critical values based on the sample size, as shown in the Test Statistic box.

Step 8:   When forming the conclusion, reject the null hypothesis if the sample data lead to a test statistic that is in the critical region—that is, the test statistic is less than or equal to the critical value(s). Otherwise, fail to reject the null hypothesis.

Notation

T = the smaller of the following two sums:

1.   The sum of the absolute values of the negative ranks
2.   The sum of the positive ranks

Test Statistic for the Wilcoxon Signed-Ranks Test for Matched Pairs

For $n \leq 30$: T

For $n > 30$: $z = \dfrac{T - \dfrac{n(n+1)}{4}}{\sqrt{\dfrac{n(n+1)(2n+1)}{24}}}$

Critical values:

1.   If $n \leq 30$, the critical T value is found in Table A-8.
2.   If $n > 30$, the critical z values are found in Table A-2.

> **EXAMPLE**   **Reported and Measured Heights** The data in Table 13-3 on the next page are matched pairs of heights (in inches) obtained from a random sample of 12 male statistics students. Each student reported his height, then his height was measured. Use the Wilcoxon signed-ranks test and a 0.05 significance level to test the claim that there is no difference between reported heights and measured heights.

SOLUTION   The null and alternative hypotheses are as follows:

H_0: There is no difference between reported and measured heights.

H_1: There is a difference between reported and measured heights.

The significance level is $\alpha = 0.05$. We are using the Wilcoxon signed-ranks test procedure, so the test statistic is calculated by using the eight-step procedure presented earlier in this section.

Step 1:   In Table 13-3, the row of differences is obtained by computing

$$d = \text{reported height} - \text{measured height}$$

for each pair of data.

Step 2:   Ignoring their signs, we rank the absolute differences from lowest to highest. Note that ties in ranks are handled by assigning

TABLE 13-3	Reported and Measured Heights of Male Statistics Students											
Reported height	68	74	82.25	66.5	69	68	71	70	70	67	68	70
Measured height	66.8	73.9	74.3	66.1	67.2	67.9	69.4	69.9	68.6	67.9	67.6	68.8
Differences d	1.2	0.1	7.95	0.4	1.8	0.1	1.6	0.1	1.4	−0.9	0.4	1.2
Ranks of differences	7.5	2	12	4.5	11	2	10	2	9	6	4.5	7.5
Signed ranks	+7.5	+2	+12	+4.5	+11	+2	+10	+2	+9	−6	+4.5	+7.5

the mean of the involved ranks to each of the tied values. For example, the first three values in the sorted list are each 0.1, so we find the mean of ranks 1, 2, 3 and assign that mean of 2 to each of the 0.1 values.

Step 3:   The bottom row of Table 13-3 is created by attaching to each rank the sign of the corresponding difference. If there really is no difference between reported and measured heights (as in the null hypothesis), we expect the number of positive ranks to be approximately equal to the number of negative ranks.

Step 4:   We now find the sum of the absolute values of the negative ranks, and we also find the sum of the positive ranks.

Sum of absolute values of negative ranks: 6

Sum of positive ranks: 72

(which is $7.5 + 2 + 12 + 4.5 + 11 + 2 + 10 + 2 + 9 + 4.5 + 7.5$)

Step 5:   Letting T be the smaller of the two sums found in Step 4, we find that $T = 6$.

Step 6:   Letting n be the number of pairs of data for which the difference d is not 0, we have $n = 12$.

Step 7:   Because $n = 12$, we have $n \le 30$, so we use a test statistic of $T = 6$ (and we do not calculate a z test statistic). Also, because $n \le 30$, we use Table A-8 to find the critical value of 14.

Step 8:   The test statistic $T = 6$ is less than or equal to the critical value of 14, so we reject the null hypothesis. It appears that there is a difference between reported and measured heights. ◀

If we use the sign test with the preceding example, we will arrive at the same conclusion. Although the sign test and the Wilcoxon signed-ranks test agree in this particular case, there are other cases in which they do not agree.

Rationale: In this example the unsigned ranks of 1 through 12 have a total of 78, so if there are no significant differences, each of the two signed-rank totals should be around $78 \div 2$, or 39. That is, the negative ranks and positive ranks should split up as 39-39 or something close, such as 38-40. The table of critical values shows that at the 0.05 significance level with 12 pairs of data, a 14-64 split represents a significant departure from the null hypothesis, and any split that is farther apart (such as 13-65 or 12-66 or 6-72) will also represent a significant departure from the null hypothesis. Conversely, splits like 38-40 or 15-63 do not represent significant departures from a 39-39 split, and they would not be a basis for rejecting the null hypothesis. The Wilcoxon signed-ranks test is based on the lower rank total, so instead of analyzing both numbers constituting the split, we consider only the lower number.

The sum $1 + 2 + 3 + \cdots + n$ of all the ranks is equal to $n(n + 1)/2$; if this is a rank sum to be divided equally between two categories (positive and negative), each of the two totals should be near $n(n + 1)/4$, which is half of $n(n + 1)/2$. Recognition of this principle helps us understand the test statistic used when $n > 30$. The denominator in that expression represents a standard deviation of T and is based on the principle that

$$1^2 + 2^2 + 3^3 + \cdots + n^2 = \frac{n(n + 1)(2n + 1)}{6}$$

The Wilcoxon signed-ranks test can only be used for matched pairs of data. The next section will describe a rank-sum test that can be applied to two sets of independent data that are not matched in pairs.

Using Excel for the Wilcoxon Signed-Ranks Test

Excel is not programmed for the Wilcoxon signed-ranks test, but it can be very helpful in conducting that test. See the following procedure and see the following Excel spreadsheet for the example included in this section.

1. Given a list of paired data, enter the values of the first variable in column A, and enter the values of the second variable in column B. (In the Excel spreadsheet below, the data were originally entered in the same order as Table 13-3, but subsequent sorting rearranged the order of the pairs of data.)

2. Obtain a column C of differences A − B as follows. Click on cell C1 and enter this expression: **=(A1 − B1)**, then press the **Enter** key. Cell C1 should now contain the first difference. Click on the lower right corner of cell C1 and, while holding the mouse button down, drag down the column until all of the cells adjacent to values in column B are highlighted. When the mouse button is released, column C will contain all of the differences.

3. Column C typically contains positive and negative values. Take the absolute value of all entries in column C as follows. Click on cell D1, then enter **=ABS(C1)** and press the **Enter** key.

4. Click and drag the lower corner of cell D1 down to get all of the absolute values.

5. Sort column D by clicking on **Data,** then **Sort.** Select Column D as the column to sort.

Gender Gap in Drug Testing

A study of the relationship between heart attacks and doses of aspirin involved 22,000 male physicians. This study, like many others, excluded women. The General Accounting Office recently criticized the National Institutes of Health for not including both sexes in many studies because results of medical tests on males do not necessarily apply to females. For example, women's hearts are different from men's in many important ways. When forming conclusions based on sample results, we should be wary of an inference that extends to a population larger than the one from which the sample was drawn.

6. Column D now contains the sorted list of differences, so it is easy to identify the ranks. In the Excel display below, those ranks were manually entered in column E, and the value in cell E10 was made negative, as required in Step 3 of the procedure given in this section. It is easy to see from Column E that the sum of the absolute values of the negative ranks is 6 and the sum of the positive ranks is 72. (You could use the Excel command **SORT** on column E to separate the positive and negative values, then you could use the **SUM** to find the sum of the positive ranks and the sum of the negative ranks.)

EXCEL

	A	B	C	D	E
1	74	73.9	0.1	0.1	2
2	68	67.9	0.1	0.1	2
3	70	69.9	0.1	0.1	2
4	66.5	66.1	0.4	0.4	4.5
5	68	67.6	0.4	0.4	4.5
6	67	67.9	-0.9	0.9	-6
7	68	66.8	1.2	1.2	7.5
8	70	68.8	1.2	1.2	7.5
9	70	68.6	1.4	1.4	9
10	71	69.4	1.6	1.6	10
11	69	67.2	1.8	1.8	11
12	82.25	74.3	7.95	7.95	12

13-3   Basic Skills and Concepts

Using the Wilcoxon Signed-Ranks Test.   *In Exercises 1–4, refer to the sample data for the given exercises in Section 13-2. Instead of the sign test, use the Wilcoxon signed-ranks test to test the claim that both samples come from populations having the same distribution. Use the significance level α that is given.*

EXCEL   **1.** Exercise 1; $\alpha = 0.01$ EXCEL   **2.** Exercise 2; $\alpha = 0.05$

EXCEL   **3.** Exercise 9; $\alpha = 0.05$ EXCEL   **4.** Exercise 10; $\alpha = 0.01$

In Exercises 5–8, use the Wilcoxon signed-ranks test.

EXCEL   **5.** Intelligence in Children: Testing for a Difference Between Trials Mental measurements of young children are often made by giving them blocks and telling them to build a tower as tall as possible. One experiment of block building was repeated a month later, with the times (in seconds) listed in the accompanying table. Is there sufficient evidence to support the claim that there is a difference between the two times? Use a 0.01 significance level.

Child	A	B	C	D	E	F	G	H	I	J	K	L	M	N	O
First trial	30	19	19	23	29	178	42	20	12	39	14	81	17	31	52
Second trial	30	6	14	8	14	52	14	22	17	8	11	30	14	17	15

Based on data from "Tower Building," by Johnson and Courtney, *Child Development*, Vol. 3.

   6. Testing for Difference Between Sitting Measurements and Lying Measurements In a study of techniques used to measure lung volumes, physiological data were collected for 10 subjects. The values given in the table are in liters and represent the measured functional residual capacities of the 10 subjects both in a sitting position and in a supine (lying) position. At the 0.05 significance level, test the claim that there is no significant difference between the measurements taken in the two positions.

Sitting	2.96	4.65	3.27	2.50	2.59	5.97	1.74	3.51	4.37	4.02
Supine	1.97	3.05	2.29	1.68	1.58	4.43	1.53	2.81	2.70	2.70

Based on "Validation of Esophageal Balloon Technique at Different Lung Volumes and Postures," by Baydur, Cha, and Sassoon, *Journal of Applied Physiology,* Vol. 62, No. 1.

EXCEL **7.** Testing for Drug Effectiveness Captopril is a drug designed to lower systolic blood pressure. When subjects were tested with this drug, their systolic blood pressure readings (in mm of mercury) were measured before and after the drug was taken, with the results given in the accompanying table. Is there sufficient evidence to support the claim that the drug has an effect? Does Captopril appear to lower systolic blood pressure?

Subject	A	B	C	D	E	F	G	H	I	J	K	L
Before	200	174	198	170	179	182	193	209	185	155	169	210
After	191	170	177	167	159	151	176	183	159	145	146	177

Based on data from "Essential Hypertension: Effect of an Oral Inhibitor of Angiotensin-Converting Enzyme," by MacGregor et al., *British Medical Journal,* Vol. 2.

EXCEL **8.** Testing for Difference Between Morning and Night Body Temperatures Refer to Data Set 6 in Appendix B. Use the paired data consisting of body temperatures of women at 8:00 A.M. and at 12:00 A.M. on Day 2. Is there sufficient evidence to support the claim that morning and night body temperatures are different?

13-3 Beyond the Basics

9. Finding Critical Values Assume that the Wilcoxon signed-ranks test is being used for a two-tailed hypothesis test with a 0.05 significance level.
   a. With $n = 10$ pairs of data, find the lowest and highest possible values of T.
   b. With $n = 50$ pairs of data, find the lowest and highest possible values of T.
   c. If there are $n = 100$ pairs of data with no differences of 0 and no tied ranks, find the critical value of T.

EXCEL **10.** Using the Wilcoxon Signed-Ranks Test for Claims About a Median The Wilcoxon signed-ranks test can be used to test the claim that a sample comes from a population with a specified median. The procedure used is the same as the one described in this section, except that the differences (Step 1) are obtained by subtracting the value of the hypothesized median from each value. Use the sample data consisting of the 106 body temperatures listed in Table 7-1. (See the Chapter Problem at the beginning of Chapter 7.) At the 0.05 significance level, test the claim that healthy adults have a median body temperature that is equal to 98.6°F.

13-4 Wilcoxon Rank-Sum Test for Two Independent Samples

This section introduces the *Wilcoxon rank-sum test,* which is a nonparametric test that can be applied to two independent sets of sample data. Two samples are independent if the sample values selected from one population are

TABLE 13-4

Weights (in grams) of M&Ms

Red		Yellow	
0.870	(2)	0.906	(19)
0.933	(35)	0.978	(45)
0.952	(42)	0.926	(34)
0.908	(21)	0.868	(1)
0.911	(24.5)	0.876	(5)
0.908	(21)	0.968	(44)
0.913	(27)	0.921	(30)
0.983	(46)	0.893	(15)
0.920	(29)	0.939	(38)
0.936	(37)	0.886	(10.5)
0.891	(13)	0.924	(32)
0.924	(32)	0.910	(23)
0.874	(4)	0.877	(6)
0.908	(21)	0.879	(7.5)
0.924	(32)	0.941	(40)
0.897	(16)	0.879	(7.5)
0.912	(26)	0.940	(39)
0.888	(12)	0.960	(43)
0.872	(3)	0.989	(47)
0.898	(17)	0.900	(18)
0.882	(9)	0.917	(28)
		0.911	(24.5)
		0.892	(14)
		0.886	(10.5)
		0.949	(41)
		0.934	(36)
$n_1 = 21$		$n_2 = 26$	
$R_1 = 469.5$		$R_2 = 658.5$	

The expression for μ_R is based on the following result of mathematical induction: The sum of the first n positive integers is given by $1 + 2 + 3 + \cdots + n = n(n + 1)/2$. The expression for σ_R is based on a result stating that the integers $1, 2, 3, \ldots, n$ have standard deviation $\sqrt{(n^2 - 1)/12}$.

EXAMPLE   **M&Ms** Samples of M&M plain candies are randomly selected, and the red and yellow M&Ms are weighed, with the results listed in Table 13-4 (from Data Set 10 in Appendix B). Use a 0.05 significance level to test the claim that weights of red M&Ms and yellow M&Ms have the same distribution.

SOLUTION   The null and alternative hypotheses are as follows:

H_0: Red and yellow M&M plain candies have weights with the same distribution.

H_1: The two populations have weight distributions that are different in some way.

Rank all 47 weights combined, beginning with a rank of 1 (assigned to the lowest weight of 0.868 g). Ties in ranks are handled as described in Section 13-1: Find the mean of the ranks involved and assign this mean rank to each of the tied values. The ranks corresponding to the individual sample values are shown in parentheses in Table 13-4. R denotes the sum of the ranks for the sample we choose as sample 1. If we choose the red M&Ms, we get

$$R = 2 + 35 + 42 + \cdots + 9 = 469.5$$

Because there are 21 red M&Ms, we have $n_1 = 21$. Also, $n_2 = 26$, because there are 26 yellow M&Ms. We can now determine the values of μ_R, σ_R, and the test statistic z.

$$\mu_R = \frac{n_1(n_1 + n_2 + 1)}{2} = \frac{21(21 + 26 + 1)}{2} = 504$$

$$\sigma_R = \sqrt{\frac{n_1 n_2(n_1 + n_2 + 1)}{12}} = \sqrt{\frac{(21)(26)(21 + 26 + 1)}{12}} = 46.73$$

$$z = \frac{R - \mu_R}{\sigma_R} = \frac{469.5 - 504}{46.73} = -0.74$$

The test is two-tailed because a large positive value of z would indicate that the higher ranks are found disproportionately in the first sample, and a large negative value of z would indicate that the first sample had a disproportionate share of lower ranks. In either case, we would have strong evidence against the claim that the two samples come from populations with the same distribution.

The significance of the test statistic z can be treated in the same manner as in previous chapters. We are now testing (with $\alpha = 0.05$) the hypothesis that the two populations have the same distribution, so we have a two-tailed test with critical z values of 1.96 and -1.96. The test statistic of $z = -0.74$ does not fall within the critical region, so we fail to reject the null hypothesis that red and yellow M&Ms have the same weights. It appears that red and yellow M&Ms do come from populations with the same distribution.

We can verify that if we interchange the two sets of weights and consider the sample of yellow M&Ms to be first, $R = 658.5$, $\mu_R = 624$, $\sigma_R = 46.73$, and $z = 0.74$, so the conclusion is the same.

EXAMPLE   **Wednesday and Saturday Rainfall** The Chapter Problem referred to the Boston rainfall amounts listed in Data Set 17 in Appendix B. The Chapter Problem included boxplots of the rainfall amounts for the seven days of the week, starting with Monday at the top. Comparisons of those boxplots show that Wednesday and Saturday appear to be the two days that differ most. But are those differences significant? Use the Wilcoxon rank-sum test to test the claim that the rainfall amounts for Wednesdays and Saturdays come from the same distribution.

SOLUTION   The null and alternative hypotheses are as follows:

H_0: The Wednesday and Saturday rainfall amounts come from populations with the same distribution.

H_1: The two distributions are different in some way.

Instead of manually calculating the required components, Excel was used to help find these values:

$R = 2639$ (rank sum for Wednesday)

$n_1 = 53$ $n_2 = 52$

$\mu_R = 2809$ $\sigma_R = 156.03$

Test statistic: $z = -1.09$ Critical values: $z = \pm 1.96$

Comparing the test statistic and critical values, we see that the test statistic does not lie in the critical region, so we fail to reject the null hypothesis. Bottom line: The differences between Wednesday and Saturday are not significant. This seems to contradict the media reports that it rains more on weekends, but we will consider this issue more in the following section.

Using Excel for the Wilcoxon Rank-Sum Test

Excel is not programmed for the Wilcoxon rank-sum test, but it can be helpful in conducting that test. The most tedious part of the calculations is finding the rank sum for one of the samples. Excel can be used as follows.

1. Enter all of the sample values in column A. First enter the values from the first sample, then continue to enter the values for the second sample in the same column A.

2. In column B, enter the letter A adjacent to each value from sample A, and enter B adjacent to each value from sample B.

3. Click on any cell in column A, then click on **Data,** then **Sort** and proceed to sort all of the values in column A so that all of the sample values are arranged in order. The identifying labels in column B will also follow the exact same rearrangement.

4. In column C, manually enter the ranks for the sorted data in column A. If there are no ties, enter 1, 2, 3, … If there are ties, assign the mean of the tied ranks to each of the tied values.

As an example, consider this set of data:

A	B
15	14
13	12
18	17
	12

After Steps 1 and 2, the Excel spreadsheet will appear as shown below.

EXCEL

	A	B
1	15	A
2	13	A
3	18	A
4	14	B
5	12	B
6	17	B
7	12	B

After Step 3, the Excel spreadsheet will appear as follows. (The ranks in column C were manually entered, based on the values in column A. Also, the values for Sample A are highlighted for clarity.) Column A contains a rearrangement of the original sample data, column B identifies the original locations of the sample values, and column C lists the corresponding ranks. We can see that the ranks for Sample A are 3, 5, and 7, so $R = 3 + 5 + 7 = 15$ for Sample A.

EXCEL

	A	B	C
1	12	B	1.5
2	12	B	1.5
3	13	A	3
4	14	B	4
5	15	A	5
6	17	B	6
7	18	A	7

13-4 Basic Skills and Concepts

Using the Wilcoxon Rank-Sum Test. *In Exercises 1–8, use the Wilcoxon rank-sum test.*

1. Testing Red and Brown M&Ms for Identical Populations The example in this section tested the null hypothesis that red and yellow M&M plain candies have weights with the same distribution. Refer to Data Set 10 in Appendix B and test the claim that red and brown M&M plain candies have weights with the same distribution. That is, test the claim that the populations of red and brown M&M plain candies are identical. Use a 0.05 significance level.

2. Testing the Anchoring Effect Randomly selected statistics students were given five seconds to estimate the value of a product of numbers with the results given in the accompanying table. (See the Cooperative Group Activities at the end of Chapter 2.) Is there sufficient evidence to support the claim that the two samples come from populations with different distributions?

Estimates from Students Given $1 \times 2 \times 3 \times 4 \times 5 \times 6 \times 7 \times 8$

1560	169	5635	25	842	40,320	5000	500	1110	10,000
200	1252	4000	2040	175	856	42,200	49,654	560	800

Estimates from Students Given $8 \times 7 \times 6 \times 5 \times 4 \times 3 \times 2 \times 1$

100,000	2000	42,000	1500	52,836	2050	428	372	300	225	64,582
23,410	500	1200	400	49,000	4000	1876	3600	354	750	640

3. Are Severe Psychiatric Disorders Related to Biological Factors? One study used X-ray computed tomography (CT) to collect data on brain volumes for a group of patients with obsessive-compulsive disorders and a control group of healthy persons. The accompanying list shows sample results (in milliliters) for volumes of the right cordate (based on data from "Neuroanatomical Abnormalities in Obsessive-Compulsive Disorder Detected with Quantitative X-Ray Computed Tomography," by Luxenberg et al., *American Journal of Psychiatry,* Vol. 145, No. 9).

Obsessive-compulsive patients				Control group			
0.308	0.210	0.304	0.344	0.519	0.476	0.413	0.429
0.407	0.455	0.287	0.288	0.501	0.402	0.349	0.594
0.463	0.334	0.340	0.305	0.334	0.483	0.460	0.445

At the 0.01 significance level, test the claim that obsessive-compulsive patients and healthy persons have the same brain volumes. Based on this result, can we conclude that obsessive-compulsive disorders have a biological basis?

4. Does the Arrangement of Test Items Affect the Score? The arrangement of test items was studied for its effect on anxiety. Sample results are as follows:

Easy to difficult									Difficult to easy			
24.64	39.29	16.32	32.83	28.71	31.73	30.02	21.96		33.62	34.02	26.63	30.26
28.02	33.31	20.60	21.13	25.49	38.81	27.85	30.29		35.91	26.68	29.49	35.32
26.69	28.90	26.43	24.23	30.72					27.24	32.34	29.34	33.53
7.10	32.86	21.06	28.89						27.62	42.91	30.20	32.54

(continued)

At the 0.05 significance level, test the claim that the two samples come from populations with the same scores. (The data are based on "Item Arrangement, Cognitive Entry Characteristics, Sex and Test Anxiety as Predictors of Achievement Examination Performance," by Klimko, *Journal of Experimental Education,* Vol. 52, No. 4.)

5. Are Weights of Regular Coke and Diet Coke Different? Refer to Data Set 1 (COLA.xls) in Appendix B. Is there sufficient evidence to support the claim that weights of regular Coke and diet Coke are different? How might such a difference be explained?

6. Are Weights of Regular Coke and Regular Pepsi Different? Refer to Data Set 1 (COLA.xls) in Appendix B. Is there sufficient evidence to support the claim that weights of regular Coke and regular Pepsi are different?

7. Are Textbook Prices Different at the Two Colleges? Refer to Data Set 2 (TEXTBOOK.xls) in Appendix B. Is there sufficient evidence to support the claim that new textbook prices are different at the two colleges?

8. Testing Effect of Second-Hand Smoke Refer to Data Set 9 (COTININE.xls) in Appendix B. Use a 0.01 significance level to test the claim that the population of cotinine levels is the same for these two groups: nonsmokers who are not exposed to environmental tobacco smoke (labeled as NOETS), and nonsmokers who are exposed to tobacco smoke (labeled as ETS). What do the results suggest about second-hand smoke?

13-4   Beyond the Basics

9. How Do Units of Measurement Affect Results? When using the Wilcoxon rank-sum test, describe the effect of changing the unit of measurement for each sample value. Using the M&M data from the example in this section, how are the results affected if each weight is changed from grams to ounces?

10. How Do Outliers Affect Results? When using the Wilcoxon rank-sum test, is the effect of an outlier small or large? Explain.

11. Using the Mann-Whitney U Test The Mann-Whitney U test is equivalent to the Wilcoxon rank-sum test for independent samples in the sense that they both apply to the same situations and always lead to the same conclusions. In the Mann-Whitney U test we calculate

$$z = \frac{U - \dfrac{n_1 n_2}{2}}{\sqrt{\dfrac{n_1 n_2 (n_1 + n_2 + 1)}{12}}}$$

where

$$U = n_1 n_2 + \frac{n_1(n_1 + 1)}{2} - R$$

Using the sample data from the M&M example in this section, find the z test statistic for the Mann-Whitney U test and compare it to the z test statistic of -0.74 that was found using the Wilcoxon rank-sum test.

12. Finding Critical Values Assume that we have two treatments (A and B) that produce measurable results, and we have only two observations for treatment A and two observations for treatment B. We cannot use the test statistic given in this section because both sample sizes do not exceed 10.

Rank				Rank sum for
1	2	3	4	treatment A
A	A	B	B	3

a. Complete the accompanying table by listing the five rows corresponding to the other five cases, and enter the corresponding rank sums for treatment A.

b. List the possible values of R, along with their corresponding probabilities. [Assume that the rows of the table from part (a) are equally likely.]

c. Is it possible, at the 0.10 significance level, to reject the null hypothesis that there is no difference between treatments A and B? Explain.

13-5 Kruskal-Wallis Test

This section introduces the *Kruskal-Wallis test,* which is used to test the claim that several different independent samples come from identical populations. In Section 11-2 we used one-way analysis of variance (ANOVA) to test hypotheses that differences in means among several samples are significant, but that method requires that all of the involved populations have normal distributions. The Kruskal-Wallis test does not require normal distributions.

> ### DEFINITION
>
> The **Kruskal-Wallis Test** (also called the *H* **test**) is a nonparametric test that uses ranks of sample data from three or more independent populations. It is used to test the null hypothesis that the independent samples come from populations with the same distribution; the alternative hypothesis is the claim that the population distributions are different in some way.
>
> H_0: The samples come from populations with the same distribution.
>
> H_1: The samples come from populations with different distributions.

Assumptions

1. We have at least three independent samples, all of which are randomly selected.

2. Each sample has at least five observations. (If samples have fewer than five observations, refer to special tables of critical values, such as *CRC Standard Probability and Statistics Tables and Formulae,* published by CRC Press.)

3. There is *no* requirement that the populations have a normal distribution or any other particular distribution.

In applying the Kruskal-Wallis test, we compute the *test statistic H, which has a distribution that can be approximated by the chi-square distribution as long as each sample has at least five observations.* When we use the chi-square distribution in this context, the number of degrees of freedom is $k - 1$, where k is the number of samples. (For a quick review of the key features of the chi-square distribution, see Section 6-6.)

Procedure for Finding the Value of the Test Statistic

1. Temporarily combine all samples into one big sample and assign a rank to each sample value. (Sort the values from lowest to highest, and in cases of ties, assign to each observation the mean of the ranks involved.)
2. For each sample, find the sum of the ranks and find the sample size.
3. Calculate H by using the results of Step 2 and the following notation and test statistic.

Notation for the Kruskal-Wallis Test

N = total number of observations in all samples combined
k = number of samples
R_1 = sum of ranks for sample 1
n_1 = number of observations in sample 1

For sample 2, the sum of ranks is R_2 and the number of observations is n_2, and similar notation is used for the other samples.

Test Statistic for the Kruskal-Wallis Test

$$H = \frac{12}{N(N + 1)} \left(\frac{R_1^2}{n_1} + \frac{R_2^2}{n_2} + \cdots + \frac{R_k^2}{n_k} \right) - 3(N + 1)$$

where   degrees of freedom = $k - 1$

Critical values:

1. The test is *right-tailed.*
2. Use Table A-4 (because the test statistic H can be approximated by a chi-square distribution).
3. Degrees of freedom = $k - 1$

The test statistic H is basically a measure of the variance of the rank sums $R_1, R_2, \ldots, R_k$. If the ranks are distributed evenly among the sample groups, then H should be a relatively small number. If the samples are very different, then the ranks will be excessively low in some groups and high in others, with the net effect that H will be large. Consequently, only large values of H lead to rejection of the null hypothesis that the samples come from identical populations. *The Kruskal-Wallis test is therefore a right-tailed test.*

▷ **EXAMPLE   Crash Test Dummies** We will use some of the same data that were included with the Chapter Problem for Chapter 11. See Table 13-5 for car crash results obtained from the National Transportation Safety Administration. Using the *head injury* measurements, is there sufficient

TABLE 13-5	Injuries to Car Crash Test Dummies			
	Head Injury (hic)		Chest Deceleration (g)	Left Femur Load (lb)
Subcompact Cars				
Ford Escort	681	(16)	55	595
Honda Civic	428	(5)	47	1063
Hyundai Accent	917	(20)	59	885
Nissan Sentra	898	(19)	49	519
Saturn SL4	420	(4)	42	422
Compact Cars				
Chevrolet Cavalier	643	(13)	57	1051
Dodge Neon	655	(14)	57	1193
Mazda 626 DX	442	(6)	46	946
Pontiac Sunfire	514	(9)	54	984
Subaru Legacy	525	(10.5)	51	584
Midsize Cars				
Chevrolet Camaro	469	(8)	45	629
Dodge Intrepid	727	(18)	53	1686
Ford Mustang	525	(10.5)	49	880
Honda Accord	454	(7)	51	181
Volvo S70	259	(1)	46	645
Full Size Cars				
Audi A8	384	(3)	44	1085
Cadillac Deville	656	(15)	45	971
Ford Crown Victoria	602	(12)	39	996
Oldsmobile Aurora	687	(17)	58	804
Pontiac Bonneville	360	(2)	44	1376

evidence to conclude that head injuries for the four weight categories are not all the same?

SOLUTION   We will follow the hypothesis testing procedure summarized in Figure 7-5.

Steps 1, 2, and 3: The null and alternative hypotheses are as follows.

H_0: The four weight categories have head injury measurements with identical populations.

H_1: The four populations are not identical.

Step 4:   No significance level was specified. In the absence of any overriding circumstances, we use $\alpha = 0.05$.

Step 5:   In Chapter 11 we used an F test for analysis of variance, but here we illustrate the method of the Kruskal-Wallis test.

Step 6:   In determining the value of the test statistic H, we must first rank all of the data. We begin with the lowest value of 259, which is assigned a rank of 1. Ranks are shown in parentheses with the original head injury data in Table 13-5. Next we find the sample size, n, and sum of ranks, R, for each sample. If we add the five ranks for the five subcompact cars only, we get $16 + 5 + 20 + 19 + 4 = 64$. We list this result with the others.

Subcompact	**Compact**	**Midsize**	**Full Size**
$n_1 = 5$	$n_2 = 5$	$n_3 = 5$	$n_4 = 5$
$R_1 = 64$	$R_2 = 52.5$	$R_3 = 44.5$	$R_4 = 49$

Because the total number of observations is 20, we have $N = 20$. We can now evaluate the test statistic as follows:

$$H = \frac{12}{N(N + 1)}\left(\frac{R_1^2}{n_1} + \frac{R_2^2}{n_2} + \cdots + \frac{R_k^2}{n_k}\right) - 3(N + 1)$$

$$= \frac{12}{20(20 + 1)}\left(\frac{64^2}{5} + \frac{52.5^2}{5} + \frac{44.5^2}{5} + \frac{49^2}{5}\right) - 3(20 + 1)$$

$$= 1.1914$$

Because each sample has at least five observations, the distribution of H is approximately a chi-square distribution with $k - 1$ degrees of freedom. The number of samples is $k = 4$, so we have $4 - 1 = 3$ degrees of freedom. Refer to Table A-4 to find the critical value of 7.815, which corresponds to 3 degrees of freedom and a 0.05 significance level.

Step 7:   The test statistic $H = 1.1914$ is not in the critical region bounded by 7.815, so we fail to reject the null hypothesis of identical populations. (In Section 11-2, we failed to reject the null hypothesis of equal means.)

Step 8:   There is not sufficient evidence to support the claim that head injuries for the four weight categories are not all the same. Because the double negative in that conclusion can be confusing, we might simply say that on the basis of the available evidence, the hypothesis that the four weight categories appear to have the same distribution of head injury measurements is reasonable.

EXAMPLE **Rains More on Weekends?** In the Chapter Problem we noted that media reported that it rains more on weekends all along America's eastern coast, from Maine to Florida. Data Set 17 in Appendix B includes rainfall amounts for a recent year in Boston. Using that data set, test the claim that the seven week days have distributions that are not all the same.

SOLUTION Data Set 17 seems like it could be analyzed using analysis of variance methods introduced in Section 11-2, but those methods require that the sample values come from populations having distributions that are approximately normal. The Chapter Problem includes a histogram for the Monday rainfall amounts, and it is obvious that this is not a normal distribution. The histograms for the other days of the week have the same basic shape as Monday. Because the data do not come from normal distributions, analysis of variance cannot be used and the Kruskal-Wallis test is an ideal alternative.

The calculations are tedious, but Excel can be used to help find the rank sum for each day of the week. (See "Using Excel" at the end of this section.) The rank sums for Monday through Sunday are 9607, 9059, 9195, 9915, 9857, 10,236.5, and 8925.5. Using those rank sums and $N = 365$, we find that the test statistic is $H = 2.781$. The critical value of 12.592 is found from Table A-4 using a 0.05 significance level and 6 degrees of freedom. In this right-tailed test, the test statistic of $H = 2.781$ does not exceed the critical value of 12.592, so we fail to reject the null hypothesis. There isn't enough evidence to support a claim that the rainfall amounts on the seven week days have distributions that are not all the same. The rainfall amounts appear to be the same on different days of the week.

INTERPRETATION Based on the Boston rainfall amounts, there does not appear to be evidence to support the claim that it rains more on weekends. So how did the newspaper, magazine, and television reports lead us to believe that it does rain more on weekends? The original study was conducted by Arizona State University scientists Randall S. Cerveny, and Robert C. Balling. Can they be blamed for providing false information? No. The author contacted Randall Cerveny who stated that the original paper concerned rainfall *off the coast* of the Atlantic seaboard— not the rainfall associated with any particular city. Cerveny and Balling used satellite precipitation estimates and found that *areas in the ocean and near the coast did get more rainfall on weekends*, and they explain this phenomenon by its relationship with pollution coming from coastal regions. Their findings are interesting and significant. The media misinterpreted the Cerveny/Balling conclusions with reports implying that it rains more on weekends for those of us who live along the Atlantic coast. It's an interesting case of media distortion.   ◄

Rationale: The test statistic H, as presented earlier, is the rank version of the test statistic F used in the analysis of variance discussed in Chapter 11. When we deal with ranks R instead of original values x, many components are predetermined. For example, the sum of all ranks can be expressed as $N(N + 1)/2$, where N is the total number of values in all samples combined. The expression

$$H = \frac{12}{N(N + 1)} \Sigma n_i (\overline{R}_i - \overline{\overline{R}})^2$$

where $\overline{R}_i = \dfrac{R_i}{n_i}$ $\overline{\overline{R}} = \dfrac{\Sigma R_i}{\Sigma n_i}$

combines weighted variances of ranks to produce the test statistic H given here. This expression for H is algebraically equivalent to the expression for H given earlier as the test statistic. The earlier form of H (not the one given here) is easier to work with. In comparing the procedures of the parametric F test for analysis of variance and the nonparametric Kruskal-Wallis test, we see that in the absence of computer software, the Kruskal-Wallis test is much simpler to apply. We need not compute the sample variances and sample means. We do not require normal population distributions. Life becomes so much easier. However, the Kruskal-Wallis test is not as efficient as the F test, so it might require more dramatic differences for the null hypothesis to be rejected.

Using Excel for the Kruskal-Wallis Test

Excel is not programmed for the Kruskal-Wallis test, but it can be helpful in conducting that test. The most tedious part of the calculations is finding the rank sum for each of the samples. Excel can be used as follows.

1. Enter all of the sample values in column A. First enter the values from the first sample, then continue to enter the values for the second sample in the same column A, and continue until all sample values are listed in column A. For an example, see the first Excel display that follows. This display represents the first example included in this section.

2. In column B, enter the name for each category. *Hint:* Once a particular name has been entered, you can click and drag the name down so it will be repeated without retyping. See the first Excel display that follows.

3. Click on any cell in column A, then click on **Data,** then **Sort** and proceed to sort all of the values in column A so that all of the sample val-

ues are arranged in order. The identifying labels in column B will also follow the exact same rearrangement. See the second Excel display that follows.

4. In column C, manually enter the ranks for the sorted data in column A. If there are no ties, enter 1, 2, 3, ... If there are ties, assign the mean of the tied ranks to each of the tied values. See column C in the second Excel display and note that the value of 525 occurs twice, so those common values are assigned the mean of ranks 10 and 11, which is 10.5.

5. Column A contains a rearrangement of the original sample data, column B identifies the original locations of the sample values, and column C lists the corresponding ranks. From the second Excel display we can see that the ranks for the subcompact cars produce a total given by $R_1 = 4 + 5 + 16 + 19 + 20 = 64$. The other rank sums can also be found quite easily. To further simplify this process, use **Data/Sort** to sort by column B. See the third Excel display. You can now use the **SUM** command to find the rank sum for each category.)

EXCEL

	A	B
1	681	Subcompact
2	428	Subcompact
3	917	Subcompact
4	898	Subcompact
5	420	Subcompact
6	643	Compact
7	655	Compact
8	442	Compact
9	514	Compact
10	525	Compact
11	469	Midsize
12	727	Midsize
13	525	Midsize
14	454	Midsize
15	259	Midsize
16	384	Full Size
17	656	Full Size
18	602	Full Size
19	687	Full Size
20	360	Full Size

EXCEL

	A	B	C
1	259	Midsize	1
2	360	Full Size	2
3	384	Full Size	3
4	420	Subcompact	4
5	428	Subcompact	5
6	442	Compact	6
7	454	Midsize	7
8	469	Midsize	8
9	514	Compact	9
10	525	Compact	10.5
11	525	Midsize	10.5
12	602	Full Size	12
13	643	Compact	13
14	655	Compact	14
15	656	Full Size	15
16	681	Subcompact	16
17	687	Full Size	17
18	727	Midsize	18
19	898	Subcompact	19
20	917	Subcompact	20

EXCEL

	A	B	C
1	442	Compact	6
2	514	Compact	9
3	525	Compact	10.5
4	643	Compact	14
5	655	Compact	15
6	360	Full Size	2
7	384	Full Size	3
8	565	Full Size	12
9	602	Full Size	13
10	687	Full Size	17
11	259	Midsize	1
12	454	Midsize	7
13	469	Midsize	8
14	525	Midsize	10.5
15	727	Midsize	18
16	420	Subcompact	4
17	428	Subcompact	5
18	681	Subcompact	16
19	898	Subcompact	19
20	917	Subcompact	20

13-5 Basic Skills and Concepts

Using the Kruskal-Wallis Test.   *In Exercises 1–8, use the Kruskal-Wallis test.*

1. Is Old Faithful Changing Over Time? The table on the next page lists the time intervals (in minutes) between eruptions of the Old Faithful geyser for four different years. The table also includes the ranks in parentheses and the sample sizes and rank sums at the bottom. Use the Kruskal-Wallis test to test the null hypothesis that the different years have time intervals with identical populations.

1951		1985		1995		1996	
74	(21)	89	(40)	86	(34.5)	88	(37.5)
60	(8)	90	(42)	86	(34.5)	86	(34.5)
74	(21)	60	(8)	62	(12)	85	(30.5)
42	(1)	65	(15.5)	104	(48)	89	(40)
74	(21)	82	(26)	62	(12)	83	(27)
52	(2)	84	(28)	95	(47)	85	(30.5)
65	(15.5)	54	(3)	79	(24)	91	(43.5)
68	(18.5)	85	(30.5)	62	(12)	68	(18.5)
62	(12)	58	(6)	94	(45.5)	91	(43.5)
66	(17)	79	(24)	79	(24)	56	(4)
62	(12)	57	(5)	86	(34.5)	89	(40)
60	(8)	88	(37.5)	85	(30.5)	94	(45.5)
$n_1 = 12$		$n_2 = 12$		$n_3 = 12$		$n_4 = 12$	
$R_1 = 157$		$R_2 = 265.5$		$R_3 = 358.5$		$R_4 = 395$	

Data from geologist Rick Hutchinson and the National Park Service.

EXCEL **2. Does Weight of a Car Affect Chest Injuries in a Crash?** In an example in this section, we used the head injury measurements from Table 13-5 to test for differences in the four car weight categories. Use the sample data in Table 13-5 to test for differences in chest deceleration measurements among the four weight categories. Is there sufficient evidence to conclude that chest deceleration measurements for the four car weight categories are not all the same? If not, does this show that the weight of a car appears to have no effect on chest deceleration in a crash?

EXCEL **3. Does Weight of a Car Affect Leg Injuries in a Crash?** In an example in this section, we used the head injury measurements from Table 13-5 to test for differences in the four car weight categories. Use the sample data in Table 13-5 to test for differences in left femur load measurements among the four weight categories. Is there sufficient evidence to conclude that left femur loads for the four car weight categories are not all the same? If not, does this show that the weight of a car appears to have no effect on left femur load in a crash?

EXCEL **4. Is Solar Energy the Same Every Day?** A student of the author lives in a home with a solar electric system. At the same time each day, she collected voltage readings from a meter connected to the system and the results are listed in the accompanying table. Use a 0.05 significance level to test the claim that voltage readings are the same for the three different types of day. Is there sufficient evidence to support a claim of different population distributions? We might expect that a solar system would provide more electrical energy on sunny days than on cloudy or rainy days. Can we conclude that sunny days result in greater amounts of electrical energy?

Sunny Days	Cloudy Days	Rainy Days
13.5	12.7	12.1
13.0	12.5	12.2
13.2	12.6	12.3
13.9	12.7	11.9
13.8	13.0	11.6
14.0	13.0	12.2

5. Testing for Skull-Breadth Differences in Different Times The accompanying values are measured maximum breadths of male Egyptian skulls from different epochs. Changes in head shape over time suggest that interbreeding occurred with immigrant populations. Use a 0.05 significance level to test the claim that the three samples come from identical populations.

4000 B.C.	1850 B.C.	150 A.D.
131	129	128
138	134	138
125	136	136
129	137	139
132	137	141
135	129	142
132	136	137
134	138	145
138	134	137

Based on data from *Ancient Races of the Thebaid,* by Thomson and Randall-Maciver.

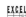

6. Laboratory Testing of Flammability of Children's Sleepwear Flammability tests were conducted on children's sleepwear. The Vertical Semirestrained Test was used, in which pieces of fabric were burned under controlled conditions. After the burning stopped, the length of the charred portion was measured and recorded. Results are given in the margin for the same fabric tested at different laboratories. Because the same fabric was used, the different laboratories should have obtained the same results. Did they?

Laboratory				
1	2	3	4	5
2.9	2.7	3.3	3.3	4.1
3.1	3.4	3.3	3.2	4.1
3.1	3.6	3.5	3.4	3.7
3.7	3.2	3.5	2.7	4.2
3.1	4.0	2.8	2.7	3.1
4.2	4.1	2.8	3.3	3.5
3.7	3.8	3.2	2.9	2.8
3.9	3.8	2.8	3.2	
3.1	4.3	3.8	2.9	
3.0	3.4	3.5		
2.9	3.3			

Data provided by Minitab, Inc.

7. Do All Colors of M&Ms Weigh the Same? Refer to Data Set 10 (M&M.xls) in Appendix B. At the 0.05 significance level, test the claim that the weights of M&Ms are the same for each of the six different color populations. If it is the intent of Mars, Inc., to make the candies so that the different color populations are the same, do your results suggest that the company has a problem that requires corrective action?

8. Testing the Effect of Second-Hand Smoke Refer to Data Set 9 (COTININE.xls) in Appendix B. Use a 0.01 significance level to test the claim that the populations of cotinine levels are identical for these three groups: nonsmokers who are not exposed to environmental tobacco smoke, nonsmokers who are exposed to tobacco smoke, and people who smoke. What do the results suggest about second-hand smoke?

13-5 Beyond the Basics

9. Testing the Effect of Transforming the Sample Data
   a. In general, how is the value of the test statistic H affected if a constant is added to (or subtracted from) each sample value?
   b. In general, how is the value of the test statistic H affected if each sample value is multiplied (or divided) by a positive constant?
   c. In general, how is the value of the test statistic H affected if a single sample value is changed to a value that causes it to become an outlier?

10. Finding Values of the Test Statistic For three samples, each of size 5, find the largest and smallest possible values of the test statistic H.

11. *Correcting the Test Statistic for Ties* In using the Kruskal-Wallis test, there is a correction factor that should be applied whenever there are many ties: Divide H by

$$1 - \frac{\Sigma T}{N^3 - N}$$

For each group of tied observations, calculate $T = t^3 - t$, where t is the number of observations that are tied within the individual group. Find t for each group of tied values, then compute the value of T for each group, then add the T values to get ΣT. The total number of observations in all samples combined is N. For Exercise 1 presented in this section, use this procedure to find the corrected value of H. Does the corrected value of H differ substantially from the value found in Exercise 1?

12. *Equivalent Tests* Show that for the case of two samples, the Kruskal-Wallis test is equivalent to the Wilcoxon rank-sum test. This can be done by showing that for the case of two samples, the test statistic H equals the square of the test statistic z used in the Wilcoxon rank-sum test. Also note that with 1 degree of freedom, the critical values of χ^2 correspond to the square of the critical z score.

13-6 Rank Correlation

In this section we describe how the nonparametric method of rank correlation is used with paired data to test for an association between two variables. In Chapter 9 we used paired sample data to compute values for the linear correlation coefficient r, but in this section we use *ranks* as the basis for measuring the strength of the correlation between two variables.

> ## DEFINITION
>
> The **rank correlation test** (or **Spearman's rank correlation test**) is a non-parametric test that uses ranks of sample data consisting of matched pairs. It is used to test for an association between two variables, so the null and alternative hypotheses are as follows (where ρ_s denotes the rank correlation coefficient for the entire population):
>
> H_0: $\rho_s = 0$ (There is *no* correlation between the two variables.)
> H_1: $\rho_s \neq 0$ (There is a correlation between the two variables.)

Assumptions

1. The sample data have been randomly selected.

2. Unlike the parametric methods of Section 9-2, there is *no* requirement that the sample pairs of data have a bivariate normal distribution (as described in Section 9-2). There is *no* requirement of a normal distribution for any population.

Advantages: Rank correlation has some distinct advantages over the parametric methods discussed in Chapter 9:

1. The nonparametric method of rank correlation can be used in a wider variety of circumstances than the parametric method of linear correlation. With rank correlation, we can analyze paired data that are ranks or can be converted to ranks. For example, if two judges rank 30 different gymnasts, we can use rank correlation, but not linear correlation.

2. Rank correlation can be used to detect some (not all) relationships that are not linear. (An example will be given later in this section.)

3. The computations for rank correlation are much simpler than the computations for linear correlation, as can be readily seen by comparing the formulas used to compute these statistics. If calculators or computers are not available, the rank correlation coefficient is easier to compute.

Disadvantage: A disadvantage of rank correlation is its efficiency rating of 0.91, as described in Section 13-1. This efficiency rating shows that with all other circumstances being equal, the nonparametric approach of rank correlation requires 100 pairs of sample data to achieve the same results as only 91 pairs of sample observations analyzed through the parametric approach, assuming the stricter requirements of the parametric approach are met.

We will use the following notation, which closely parallels the notation used in Chapter 9 for linear correlation. (Recall from Chapter 9 that r denotes the linear correlation coefficient for sample paired data, ρ denotes the linear correlation coefficient for all paired data in the entire population, and n denotes the number of pairs of data.)

Notation

r_s = rank correlation coefficient for sample paired data (r_s is a sample statistic)

ρ_s = rank correlation coefficient for all the population data (ρ_s is a population parameter)

n = number of pairs of sample data

d = difference between ranks for the two values within a pair

---

We use the notation r_s for the rank correlation coefficient so that we don't confuse it with the linear correlation coefficient r. The subscript s has nothing to do with standard deviation; it is used in honor of Charles Spearman (1863–1945), who originated the rank correlation approach. In fact, r_s is often called **Spearman's rank correlation coefficient.**

Direct Link Between Smoking and Cancer

When we find a statistical correlation between two variables, we must be extremely careful to avoid the mistake of concluding that there is a cause-effect link. The tobacco industry has consistently emphasized that correlation does not imply causality. However, Dr. David Sidransky of Johns Hopkins University now says that "we have such strong molecular proof that we can take an individual cancer and potentially, based on the patterns of genetic change, determine whether cigarette smoking was the cause of that cancer." Based on his findings, he also said that "The smoker had a much higher incidence of the mutation, but the second thing that nailed it was the very distinct pattern of mutations . . . so we had the smoking gun." Although statistical methods cannot prove that smoking *causes* cancer, such proof can be established with physical evidence of the type described by Dr. Sidransky.

ficient to test the claim that there is a relationship between corporate and graduate rankings (that is, $\rho_s \neq 0$). Use a significance level of $\alpha = 0.05$.

SOLUTION   We follow the same basic steps for testing hypotheses as outlined in Figure 7-5.

Step 1:   The claim of a correlation is expressed symbolically as $\rho_s \neq 0$.

Step 2:   The negation of the claim in Step 1 is $\rho_s = 0$.

Step 3:   Because the null hypothesis must contain the condition of equality, we have

H_0: $\rho_s = 0$

H_1: $\rho_s \neq 0$ (Original claim)

Step 4:   The significance level is $\alpha = 0.05$.

Step 5:   As we noted, we cannot use the linear correlation approach of Section 9-2 because ranks do not satisfy the requirement of a normal distribution. We will use the rank correlation approach instead.

Step 6:   We now find the value of the test statistic. Table 13-6 shows the calculation of the differences d and their squares d^2, which results in a value of $\Sigma d^2 = 148$. With $n = 10$ (for 10 pairs of data) and $\Sigma d^2 = 148$, we can find the value of the test statistic r_s as follows:

$$r_s = 1 - \frac{6\Sigma d^2}{n(n^2 - 1)} = 1 - \frac{6(148)}{10(10^2 - 1)}$$

$$= 1 - \frac{888}{990} = 0.103$$

Now we refer to Table A-9 to determine that the critical values are ± 0.648 (based on $\alpha = 0.05$ and $n = 10$). Because the test statistic $r_s = 0.103$ does not exceed the critical value of 0.648, we fail to reject the null hypothesis. There is not sufficient evidence to support the claim of a correlation between corporate and graduate rankings of business schools. It appears that corporate recruiters and business school graduates have different perceptions of the qualities of the schools.   ◀

▷ **EXAMPLE**   **Large Sample Case** Assume that the preceding example is expanded by including a total of 40 business schools and that the test statistic r_s is found to be 0.291. If the significance level is $\alpha = 0.05$, what do you conclude about the correlation between corporate and graduate rankings?

SOLUTION   Because there are 40 pairs of data, we have $n = 40$. Because n exceeds 30, we find the critical values from Formula 13-1 instead of Table A-9. With $\alpha = 0.05$ in two tails, we let $z = 1.96$ to get

$$r_s = \frac{\pm 1.96}{\sqrt{40 - 1}} = \pm 0.314$$

The test statistic of $r_s = 0.291$ does not exceed the critical value of 0.314, so we fail to reject the null hypothesis. There is not sufficient evidence to support the claim of a correlation between the corporate rankings and graduate rankings. ◄

The next example is intended to illustrate the principle that rank correlation can sometimes be used to detect relationships that are not linear.

► **EXAMPLE   Learning Tasks** A *Raiders of the Lost Ark* pinball machine (model L-7) is used to measure learning that results from repeating manual functions. Subjects were selected so that they are similar in important characteristics of age, gender, intelligence, education, and so on. Table 13-7 lists the numbers of games played and the last scores (in millions) for subjects randomly selected from the group with similar characteristics. We expect that there should be an association between the number of games played and the pinball score. Is there sufficient evidence to support the claim that there is such an association?

SOLUTION

We will test the null hypothesis of no rank correlation ($\rho_s = 0$).

H_0: $\rho_s = 0$ (no correlation)

H_1: $\rho_s \neq 0$ (correlation)

Refer to Figure 13-4, which we follow in this solution. The original scores are not ranks, so we converted them to ranks and entered the results in parentheses in Table 13-7. (Section 13-1 describes the procedure for converting scores into ranks.)

After expressing all data as ranks, we calculate the differences, d, and then square them. The sum of the d^2 values is 4. We now calculate

$$r_s = 1 - \frac{6\Sigma d^2}{n(n^2 - 1)} = 1 - \frac{6(4)}{10(10^2 - 1)}$$

$$= 1 - \frac{24}{990} = 0.976$$

Manatees Saved

Manatees are large mammals that like to float just below the water's surface, where they are in danger from powerboat propellers. A Florida study of the number of powerboat registrations and the numbers of accidental manatee deaths confirmed that there was a significant positive correlation. As a result, Florida created coastal sanctuaries where powerboats are prohibited so that manatees could thrive. (See Program 1 from the series *Against All Odds: Inside Statistics* for a discussion of this case.) This is one of many examples of the beneficial use of statistics.

TABLE 13-7	Pinball Scores (Ranks in parentheses)									
Number of games played	9 (4)	13 (6)	21 (7)	5 (1)	6 (2)	25 (8)	7 (3)	33 (9)	11 (5)	104 (10)
Score	22 (3)	62 (6)	70 (7)	2 (1)	10 (2)	82 (9)	26 (4)	78 (8)	58 (5)	86 (10)
d	1	0	0	0	0	1	1	1	0	0
d^2	1	0	0	0	0	1	1	1	0	0

Proceeding with Figure 13-4, we have $n = 10$, so we answer yes when asked if $n \leq 30$. We use Table A-9 to get the critical values of ± 0.648. Finally, the sample statistic of 0.976 exceeds 0.648, so we conclude that there is significant correlation. Higher numbers of games played appear to be associated with higher scores. Subjects appeared to better learn the game by playing more. ◀

In the preceding example, if we compute the linear correlation coefficient r (using Formula 9-1) for the original data, we get $r = 0.629$, which leads to the conclusion that there is not enough evidence to support the claim of a significant linear correlation at the 0.05 significance level. If we examine the Excel scatter diagram, we can see that the pattern of points is not a straight-line pattern. This last example illustrates these two advantages of the nonparametric approach over the parametric approach: (1) With rank correlation, we can sometimes detect relationships that are not linear; and (2) Spearman's rank correlation coefficient r_s is less sensitive to an outlier, such as the subject who played 104 games.

Using Excel for Rank Correlation

Excel does not have a function that calculates the rank correlation coefficient from original sample values, but that value can be found as follows.

1. Replace each of the original sample values by its corresponding rank. (*Hint:* Excel can be helpful in finding the ranks. Enter the values for one of the variables in column A, then use **Data** and **Sort** to sort those values. The ranks can be easily identified from the sorted list of sample values.)

2. Enter the ranks in columns A and B.

3. Click on the *f*x function key located on the main menu bar. Select the function category of **Statistical** and select the function name of **CORREL,** then click **OK.**

4. In the dialog box, enter the cell range of values for x, such as A1:A10. Also enter the cell range of values for y, such as B1:B10. Excel will display the value of the rank correlation coefficient.

13-6 Basic Skills and Concepts

EXCEL

1. Finding the Test Statistic and Critical Value For each of the following samples of paired ranks, sketch a scatter diagram, estimate the value of r_s, calculate the value of r_s, and state whether there appears to be a correlation between x and y.

   a.
x	1 3 5 4 2
y	1 3 5 4 2

   b.
x	1 2 3 4 5
y	5 4 3 2 1

   c.
x	1 2 3 4 5
y	2 5 3 1 4

2. Finding Critical Values Find the critical value(s) for r_s by using either Table A-9 or Formula 13-1, as appropriate. Assume two-tailed cases, where α represents the significance level and n represents the number of pairs of data.

   a. $n = 20$, $\alpha = 0.05$ **b.** $n = 50$, $\alpha = 0.05$
   c. $n = 40$, $\alpha = 0.02$ **d.** $n = 15$, $\alpha = 0.01$
   e. $n = 82$, $\alpha = 0.04$

Testing for Rank Correlation. In Exercises 3–14, use the rank correlation coefficient to test for a correlation between the two variables. Use a significance level of $\alpha = 0.05$.

EXCEL

3. Correlation Between Salary and Stress The accompanying table lists salary rankings and stress rankings for randomly selected jobs. Does it appear that salary increases as stress increases?

Job	Salary Rank	Stress Rank
Stockbroker	2	2
Zoologist	6	7
Electrical engineer	3	6
School principal	5	4
Hotel manager	7	5
Bank officer	10	8
Occ. safety inspector	9	9
Home economist	8	10
Psychologist	4	3
Airline pilot	1	1

Based on data from *The Jobs Rated Almanac.*

EXCEL

4. Correlation Between Salary and Physical Demand Exercise 3 includes paired salary and stress level ranks for 10 randomly selected jobs. The physical demands of the jobs were also ranked; the salary and physical demand ranks are given below. Does there appear to be a relationship between the salary of a job and its physical demands?

Salary	2	6	3	5	7	10	9	8	4	1
Physical demand	5	2	3	8	10	9	1	7	6	4

Based on data from *The Jobs Rated Almanac.*

EXCEL

5. Correlation Between Stress and Physical Demand Ten jobs were randomly selected and ranked according to stress level and physical demand, with the

results given below. Does there appear to be a relationship between the stress levels of jobs and their physical demands?

Stress level	2	7	6	4	5	8	9	10	3	1
Physical demand	5	2	3	8	10	9	1	7	6	4

Based on data from *The Jobs Rated Almanac*.

EXCEL **6.** Correlation Between Weapons and Murder Rate The accompanying table lists rankings for the number of registered automatic weapons (in thousands), along with the murder rate (in murders per 100,000), for randomly selected states. Automatic weapons are guns that continue to fire repeatedly while the trigger is held back. Are firearm murders often committed with automatic weapons? Does a significant correlation imply that increased numbers of automatic weapons result in more murders?

Automatic weapons	8	7	5	1	6	3	2	4
Murder rate	8	6	5	2	7	4	1	3

Data provided by the FBI and the Bureau of Alcohol, Tobacco, and Firearms.

EXCEL **7.** Correlation Between Restaurant Bills and Tips In Section 9-2 we included an example in which we tested for a correlation between the amount of the restaurant bill and the amount of the tip that was left. The sample data are reproduced here (based on data from students of the author). Use rank correlation to determine whether there is a correlation between the amount of the bill and the amount of the tip.

Bill (dollars)	33.46	50.68	87.92	98.84	63.60	107.34
Tip (dollars)	5.50	5.00	8.08	17.00	12.00	16.00

EXCEL **8.** Correlation Between Heights and Weights of Supermodels Listed below are heights (in inches) and weights (in pounds) for supermodels Niki Taylor, Nadia Avermann, Claudia Schiffer, Elle MacPherson, Christy Turlington, Bridget Hall, Kate Moss, Valerie Mazza, and Kristy Hume.

Height	71	70.5	71	72	70	70	66.5	70	71
Weight	125	119	128	128	119	127	105	123	115

EXCEL **9.** Correlation Between Age and BAC of DWI Prisoners A study was conducted to investigate a relationship between age (in years) and BAC (blood alcohol concentration) measured when convicted DWI jail inmates were first arrested. Sample data are given on the next page for randomly selected subjects. Based on the result, does the BAC level seem to be related to the age of the person tested?

Age	17.2	43.5	30.7	53.1	37.2	21.0	27.6	46.3
BAC	0.19	0.20	0.26	0.16	0.24	0.20	0.18	0.23

Based on data from the Dutchess County STOP-DWI Program.

EXCEL **10.** Correlation Between Car Weight and Fuel Consumption The table on the next page lists weights (in hundreds of pounds) and highway fuel consumption amounts (in mi/gal) for a sample of domestic new cars. Based on the result, can you expect to pay more for gas if you buy a heavier car? How do the results change if the weights are entered as 2900, 3500, . . . , 2400?

x Weight	29	35	28	44	25	34	30	33	28	24
y Fuel	31	27	29	25	31	29	28	28	28	33

Based on data from the EPA.

11. Old Faithful Geyser Eruption Data Refer to Data Set 11 (OLDFAITH.xls) in Appendix B.
   a. Use the paired data for durations and intervals after eruptions of the geyser. Is there a significant correlation, suggesting that the interval after an eruption is related to the duration of the eruption?
   b. Use the paired data for intervals after eruptions and heights of eruptions of the Old Faithful geyser. Is there a significant correlation, suggesting that the interval after an eruption is related to the height of the eruption?
   c. Assume that you want to develop a method for predicting the time interval to the next eruption. Based on the results from parts (a) and (b), which factor would be more relevant: eruption duration or eruption height? Why?

12. Cigarette Nicotine, Tar, Carbon Monoxide Refer to Data Set 8 (CIGARET.xls) in Appendix B.
   a. Use the paired data consisting of tar and nicotine. Based on the result, does there appear to be a significant correlation between cigarette tar and nicotine? If so, can researchers reduce their laboratory expenses by measuring only one of these two variables?
   b. Use the paired data consisting of carbon monoxide and nicotine. Based on the result, does there appear to be a significant correlation between cigarette nicotine and carbon monoxide? If so, can researchers reduce their laboratory expenses by measuring only one of these two variables?
   c. Assume that researchers want to develop a method for predicting the amount of nicotine, and they want to measure only one other item. In choosing between tar and carbon monoxide, which is the better choice? Why?

13. Diamond Prices, Weights, Color Refer to Data Set 3 (DIAMONDS.xls) in Appendix B.
   a. Use the paired data consisting of price and carat (weight). Is there a significant correlation between the price of a diamond and its weight in carats?
   b. Use the paired price/color data. Is there a significant correlation between the price of a diamond and its color?
   c. Assume that you are planning to buy a diamond engagement ring. In considering the value of a diamond, which characteristic should you consider to be more important: the carat weight or the color? Why?

14. Movie Gross, Budget, and Viewer Rating Refer to Data Set 15 (MOVIES.xls) in Appendix B.
   a. Use the paired data consisting of the amount of money grossed and the amount of money budgeted. Is there a significant correlation between those two variables?
   b. Use the paired data consisting of the amount of money grossed and the viewer rating. Is there a significant correlation between those two variables?
   c. Based on the preceding results, if you plan to invest in a company that makes movies, which is the better choice: a company that makes big budget movies or a company that makes movies that viewers rate highly? Why?

13-6 Beyond the Basics

15. Finding Critical Values One alternative to using Table A-9 to find critical values is to compute them using this approximation:

$$r_s = \pm\sqrt{\frac{t^2}{t^2 + n - 2}}$$

Here t is the t score from Table A-3 corresponding to the significance level and $n - 2$ degrees of freedom. Apply this approximation to find critical values of r_s for the following cases.

 a. $n = 8$, $\alpha = 0.05$ **b.** $n = 15$, $\alpha = 0.05$

 c. $n = 30$, $\alpha = 0.05$ **d.** $n = 30$, $\alpha = 0.01$

 e. $n = 8$, $\alpha = 0.01$

16. Effects of Scales and Rankings

 a. How is r_s affected if the scale for one of the variables is changed from feet to inches?

 b. How is r_s affected if one variable is ranked from low to high while the other variable is ranked from high to low?

 c. How is r_s affected if the two variables are interchanged?

 d. One researcher ranks both variables from low to high, while another researcher ranks both variables from high to low. How will their values of r_s compare?

 e. How is r_s affected if each value of x is replaced by log x?

 ## 13-7 Runs Test for Randomness

The main objective of this section is to understand how the runs test for randomness can be used to determine whether the sample data in a sequence are in a random order. The importance of randomness has been stressed throughout this book, and we now address one method for determining whether that characteristic is present.

> **DEFINITIONS**
>
> A **run** is a sequence of data having the same characteristic; the sequence is preceded and followed by data with a different characteristic or by no data at all.
>
> The **runs test** uses the number of runs in a sequence of sample data to test for randomness in the order of the data.

Assumptions

1. The sample data are arranged according to some ordering scheme, such as the order in which the sample values were obtained.

2. Each data value can be categorized into one of *two* separate categories.

3. The runs test for randomness is based on the *order* in which the data occur; it is *not* based on the *frequency* of the data. (For example, a sequence of 3 men and 20 women might appear to be random, but the issue of whether 3 men and 20 women constitute a *biased* sample is not addressed by the runs test.)

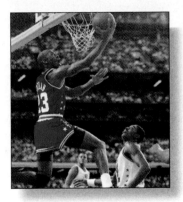

Notation

n_1 = number of elements in the sequence that have one particular characteristic (The characteristic chosen for n_1 is arbitrary.)

n_2 = number of elements in the sequence that have the other characteristic

G = number of runs

> **EXAMPLE** **Market Research** Manufacturers of cola conduct market research to determine consumer preferences. When 10 consumers are asked whether they prefer diet cola or regular cola, the responses of the first 10 subjects are listed in the order in which they were obtained. We use D to denote a consumer who prefers *diet* cola and R to denote a consumer who prefers *regular* cola. Find the number of runs G, and also identify the values of n_1 and n_2.

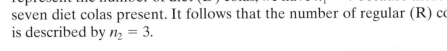

$$\underbrace{D\,D\,D\,D}_{\text{1st run}}\ \ \underbrace{R\,R}_{\text{2nd run}}\ \ \underbrace{D\,D\,D}_{\text{3rd run}}\ \ \underbrace{R}_{\text{4th run}}$$

SOLUTION   There are exactly four runs, as shown, so $G = 4$. Letting n_1 represent the number of diet (D) colas, we have $n_1 = 7$ because there are seven diet colas present. It follows that the number of regular (R) colas is described by $n_2 = 3$.

Fundamental Principle of the Runs Test

The fundamental principle of the runs test can be briefly stated as follows:

Reject randomness if the number of runs is very low or very high.

- Example: DDDDDRRRRR is not random because it has 2 runs, so the number of runs is very *low*.

- Example: DRDRDRDRDR is not random because there are 10 runs, which is very *high*.

We reject randomness if the number of runs G is too small or too large, but how do we determine exactly which values of G are too small or too large? We use the following criterion:

5% Cutoff Criterion

Reject randomness if the number of runs G is so small or so large that in repeated samplings, a value at least as extreme as G will occur 5% of the time or less.

Although this criterion wins no prizes for simplicity, it's quite easy to apply if we use Table A-10. Table A-10 identifies those values of G that are so small or so large that they belong to the category of exceptional sequences that happen 5% of the time or less. Using Table A-10, the 5% criterion can be simply restated as follows:

Simplified 5% Cutoff Criterion

Reject randomness if the number of runs G is

- **less than or equal to the smaller entry in Table A-10**
- **or greater than or equal to the larger entry in Table A-10.**

The sequence

DDDDRRRRR

results in these values: $n_1 = 4$, $n_2 = 5$, and $G = 2$ runs. With $n_1 = 4$ and $n_2 = 5$, Table A-10 indicates that we should reject randomness if the number of runs is 2 or less or 9 or greater. With $G = 2$ runs, we therefore reject randomness according to the simplified 5% cutoff criterion based on Table A-10.

Rationale: How do we find the cutoff values separating random sequences from those that are not? That is, how do we find the critical values in Table A-10? Let's consider the sequence DDDDRRRRR, with $G = 2$, $n_1 = 4$, and $n_2 = 5$. If we refer to Table A-10, we find that $n_1 = 4$ and $n_2 = 5$ correspond to cutoff values of 2 and 9. If we had an abundance of time and patience, we could list all possible sequences of 4 diet colas and 5 regular colas. Examination of that list would reveal these facts:

- There are 126 different possible sequences of 4 diet and 5 regular colas.
- Among the 126 different possible cases, 2 cases have 2 runs, 7 cases have 3 runs, and so on, as summarized in Table 13-8.

Based on the 5% cutoff criterion, and based on the 126 cases summarized in Table 13-8, we should reject randomness if the number of runs G is 2 or 9 because, with randomly selected data, we will get 2 runs or 9 runs only 2.38% of the time.

TABLE 13-8 Frequency Table of the Number of Runs

Number of runs	2	3	4	5	6	7	8	9
Frequency in 126 cases	2	7	24	30	36	18	8	1

(With only 2 cases having 2 runs and with only 1 case having 9 runs, the number of runs G is excessively low or high in 3 cases, which is 2.38% of the total number of cases. We arrived at 2.38% by converting $3 \div 126$ to 0.0238, which is equivalent to 2.38%.) It is easy to get 3, 4, 5, 6, 7, or 8 runs, because these values occur more than 95% of the time; but it is unusual to get 2 or 9 runs, because these values occur only 2.38% of the time. By analyzing the 126 cases summarized in Table 13-8, we can see how to find the cutoff values of 2 and 9 that are included in Table A-10. Although this detailed procedure for finding cutoff values is perfectly valid, it's too awkward to apply in practice, so we use Table A-10 instead.

> **EXAMPLE   Basketball Foul Shots** In the course of a game, WNBA player Cynthia Cooper shoots 12 free throws. Denoting shots made by H (for "hit") and denoting missed shots by M, her results are as follows: H, H, H, M, H, H, H, H, M, M, M, H. Use the (simplified) 5% cutoff criterion to test for randomness in the sequence of hits and misses.

SOLUTION

We must first find the values of n_1, n_2, and the number of runs G.

$$n_1 = \text{number of shots made (H)} = 8$$
$$n_2 = \text{number of shots missed (M)} = 4$$
$$G = \text{number of runs} = 5$$

With $G = 5$, we refer to Table A-10 to find the cutoff values of 3 and 10. Because $G = 5$ is not less than or equal to 3, nor is it greater than or equal to 10, *we do not reject randomness.* There is not sufficient evidence to warrant rejection of the claim that the hits and misses occur randomly. ◀

Large Sample Cases

Table A-10 applies when the following three conditions are all met:

1. We are using 5% as the cutoff for sequences that have too few or too many runs,
2. $n_1 \leq 20$, and
3. $n_2 \leq 20$.

If we wish to use the runs test for randomness but $n_1 > 20$ or $n_2 > 20$ or $\alpha \neq 0.05$, we use the property that the number of runs G has a distribution that is approximately normal with mean μ_G and standard deviation σ_G described as follows:

Formula 13-2
$$\mu_G = \frac{2n_1 n_2}{n_1 + n_2} + 1$$

Formula 13-3
$$\sigma_G = \sqrt{\frac{(2n_1 n_2)(2n_1 n_2 - n_1 - n_2)}{(n_1 + n_2)^2(n_1 + n_2 - 1)}}$$

After finding the values of μ_G and σ_G, the test statistic can be computed as $z = (G - \mu_G)/\sigma_G$. The normal approximation (with test statistic z) is quite good. If the entire table of critical values (Table A-10) had been computed using this normal approximation, no critical value would be off by more than one unit.

Test Statistic for the Runs Test for Randomness

If $\alpha = 0.05$ and $n_1 \leq 20$ and $n_2 \leq 20$, the test statistic is G.

If $\alpha \neq 0.05$ or $n_1 > 20$ or $n_2 > 20$, the test statistic is

$$z = \frac{G - \mu_G}{\sigma_G}$$

where μ_G and σ_G are from Formulas 13-2 and 13-3.

Critical values:

1. If the test statistic is G, critical values are found in Table A-10.
2. If the test statistic is z, critical values are found in Table A-2 by using the procedures introduced in Chapter 6.

Figure 13-5 on the next page summarizes the procedures for the runs test for randomness and includes cases in which the test statistic is G as well as cases in which the test statistic is z.

▶ **EXAMPLE** **Boston Rainfall On Mondays** Refer to the rainfall amounts for Boston as listed in Data Set 17 in Appendix B. Is there sufficient evidence to support the claim that rain on Mondays is not random?

SOLUTION Let D (for dry) represent Mondays with no rain (indicated by values of 0.00), and let R represent Mondays with some rain (any value greater than 0.00). The 52 consecutive Mondays are represented by this sequence:

DDDDRDRDDRDDRDDDRDDRRDDDDRDRDRRRDRDDDR
DDDRDRDDRDDDR

The null and alternative hypotheses are as follows:

H_0: The sequence is random.
H_1: The sequence is not random.

The significance level is $\alpha = 0.05$; we are using the runs test for randomness. The test statistic is obtained by first finding the number of Ds, the number of Rs, and the number of runs. It's easy to examine the sequence to find that

$$n_1 = \text{number of Ds} = 33$$
$$n_2 = \text{number of Rs} = 19$$
$$G = \text{number of runs} = 30$$

FIGURE 13-5
Runs Test for Randomness

Search for Extraterrestrial Life Continues

NEW SOUTH WALES Project Phoenix is now operational with a radio telescope located in New South Wales, Australia. This is the largest radio telescope in the Southern Hemisphere. The objective is to detect radio signals being transmitted by extraterrestrial civilizations. Radio telescopes around the world monitor signals on millions of different channels. Because of the great number of the signals received, computers are used to search for signals that seem to come from an intelligent source instead of being random noise. The ability to distinguish between an actual message and random noise is essential to such projects.

Past attempts to identify extraterrestrial intelligent life have involved efforts to send radio messages carrying information about us earthlings. Dr. Frank Drake of Cornell University developed such a radio message that could be transmitted as a series of 1271 pulses and gaps. The pulses and gaps can be thought of as 1s and 0s. If we factor 1271 into the prime numbers of 41 and 31 and then make a 41 × 31 grid and put a dot at those positions corresponding to a pulse or 1, we get the pattern shown in the accompanying figure. This pattern contains information including the location of Earth in the solar system; the symbols for hydrogen, carbon, and oxygen; and drawings of a man, woman, child, fish, and water.

1. Examine the accompanying figure and identify the man, woman, child, fish, and ocean waves.

2. Suppose we send this message and it is intercepted by extraterrestrial intelligent life. Will they think that the pattern is random? Use the methods of this chapter to determine whether the sequence appears to be random so that it would be mistaken for "random noise."

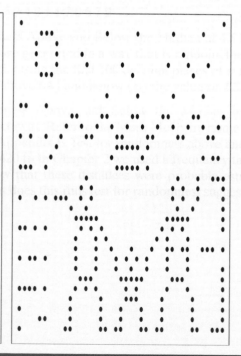

VOCABULARY LIST

parametric tests   Mann-Whitney U test
nonparametric tests   Kruskal-Wallis test
distribution-free tests H test
efficiency   rank correlation test
rank   Spearman's rank correlation test
sign test   Spearman's rank correlation coefficient
Wilcoxon signed-ranks test   run
Wilcoxon rank-sum test   runs test

REVIEW

In this chapter we examined six different nonparametric tests for analyzing sample data. Nonparametric tests are also called distribution-free tests because they do not require that the populations have a particular distribution, such as a normal distribution. However, nonparametric tests are not as efficient as parametric tests, so we generally need stronger evidence before we reject a null hypothesis.

Table 13-9 lists the nonparametric tests presented in this chapter, along with their functions. The table also lists the corresponding parametric tests.

TABLE 13-9   Summary of Nonparametric Tests

Nonparametric Test	Function	Parametric Test
Sign test (Section 13-2)	Test for claimed value of average with one sample	t test or z test (Sections 7-2, 7-3, 7-4)
	Test for differences between matched pairs	t test or z test (Section 8-3)
Wilcoxon signed-ranks test (Section 13-3)	Test for differences between matched pairs	t test or z test (Section 8-3)
Wilcoxon rank-sum test (Section 13-4)	Test for difference between two independent samples	t test or z test (Sections 8-2, 8-6)
Kruskal-Wallis test (Section 13-5)	Test that more than two independent samples come from identical populations	Analysis of variance (Section 11-2)
Rank correlation (Section 13-6)	Test for relationship between two variables	Linear correlation (Section 9-2)
Runs test (Section 13-7)	Test for randomness of sample data	(No parametric test)

Using Nonparametric Tests.   *In Exercises 1–12, use a 0.05 significance level with the indicated test. If no particular test is specified, use the appropriate nonparametric test from this chapter.*

EXCEL
1. Matched Data: Testing Effectiveness of Blood Pressure Treatment Captopril is a drug designed to lower systolic blood pressure. When subjects were tested with this drug, their systolic blood pressure readings (in mm of mercury) were measured before and after the drug was taken, with the results given in the accompanying table. Use the sign test to test the claim that the drug has no effect on systolic blood pressure readings.

Before	200	174	198	170	179	182	193	209	185	155	169	210
After	191	170	177	167	159	151	176	183	159	145	146	177

Based on data from "Essential Hypertension: Effect of an Oral Inhibitor of Angiotensin-Converting Enzyme," by MacGregor et al., *British Medical Journal*, Vol. 2.

EXCEL
2. Testing Effectiveness of Blood Pressure Treatment Do Exercise 1 using the Wilcoxon signed-ranks test.

3. Why Isn't Rank Correlation Relevant? For Exercise 1, why would the use of rank correlation *not* help to determine whether the drug has an effect?

EXCEL
4. Independent Samples: Blood Pressure Treatment Sample data were collected in a study of calcium supplements and their effects on blood pressure. A placebo group and a calcium group began the study with measures of blood pressures. Using the Wilcoxon rank-sum test, test the claim that the two sample groups come from populations with the same blood pressure levels. (The data are based on "Blood Pressure and Metabolic Effects of Calcium Supplementation in Normotensive White and Black Men," by Lyle et al., *Journal of the American Medical Association,* Vol. 257, No. 13.)

Placebo group				Calcium group			
124.6	104.8	96.5	116.3	129.1	123.4	102.7	118.1
106.1	128.8	107.2	123.1	114.7	120.9	104.4	116.3
118.1	108.5	120.4	122.5	109.6	127.7	108.0	124.3
113.6				106.6	121.4	113.2	

EXCEL
5. Is Longevity the Same for Different Classes of Leaders? The accompanying table shows samples of data from presidents, popes, and British monarchs. The values are the numbers of years the subjects lived after their inauguration, election, or coronation. Test the claim that the three populations are not all the same.

Presidents	26	15	25	1	16	17	0	12	18	21	2	36
Popes	3	18	25	6	15	11	19					
Monarchs	17	13	12	33	10	7	25					

Based on data from *Computer-Interactive Data Analysis*, by Lunn and McNeil, John Wiley & Sons.

EXCEL
6. Correlation of IQ Scores in Twins In studying the effects of heredity and environment on intelligence, scientists have learned much by analyzing the IQ scores of identical twins who were separated soon after birth. Identical twins have identical genes, which they inherited from the same fertilized egg. By studying identical twins raised apart, we can eliminate the variable of heredity

and better isolate the effects of environment. Following are the IQ scores of identical twins (first-born twins are *x*) raised apart. Is there sufficient evidence to conclude that there is a correlation between the IQs of identical twins?

x	107	96	103	90	96	113	86	99	109	105	96	89
y	111	97	116	107	99	111	85	108	102	105	100	93

Based on data from "IQ's of Identical Twins Reared Apart," by Arthur Jensen, *Behavioral Genetics.*

7. Randomness of Test Answers A statistics quiz consists of true-false questions with the answers listed as shown.

  a. Examine the sequence and try to identify a pattern showing that the answers are not random.

  b. What do you conclude from using the runs test for randomness?

  c. How do you explain the discrepancy between parts (a) and (b)?

T F T T T T F T T T T T F T T T T T F T T T T F T T T

8. Pretest/Posttest Scores The College of Portland tests the effectiveness of its writing course with a pretest taken before the course and a posttest taken after the course. Among 100 randomly selected students, it is found that on the posttest there are 62 who received higher scores, 35 who received lower scores, and 3 students who got the same scores. Is there sufficient evidence to conclude that the writing course is effective?

9. Testing Effectiveness of Employee Training Program Different employees are given a document to type into a word processing program, and their times (in minutes) are recorded. After being given a training program designed to improve their typing skills, the employees are tested a second time with the same document. A skeptical manager believes that the training program has no effect. Is there sufficient evidence to support the claim that there is a difference between the two times? Use the Wilcoxon signed-ranks test with a 0.01 significance level.

Employee	A	B	C	D	E	F	G	H	I	J	K	L	M	N	O
First trial	42	31	31	35	41	97	54	32	24	51	26	93	29	43	64
Second trial	42	18	26	20	26	64	26	34	29	20	23	42	26	29	27

10. Do All Three Age Groups Have the Same Body Temperatures? The accompanying table lists body temperatures of five randomly selected subjects from each of three different age groups. Using a significance level of $\alpha = 0.05$, test the claim that the three age-group populations of body temperatures are identical.

Body Temperatures (°F)   Categorized by Age

18–20	21–29	30 and older
98.0	99.6	98.6
98.4	98.2	98.6
97.7	99.0	97.0
98.5	98.2	97.5
97.1	97.9	97.3

Data from Dr. Philip Mackowiak, Dr. Steven Wasserman, and Dr. Myron Levine of the University of Maryland.

11. Testing for Difference Between Treatment and Placebo Groups The Medassist Pharmaceutical Company conducted a study to determine whether a drug affects eye movements. A standardized scale was developed, and the drug was administered to one group, while another group was given a placebo that produces no effects. The eye movement ratings of subjects are shown in the accompanying list. Test the claim that the drug has no effect on eye movements. Use the Wilcoxon rank-sum test.

Treatment group	Placebo group
652 512 711 621 508 603 787	674 676 821 830 565 821 837 652 549
747 516 624 627 777 729	668 772 563 703 789 800 711 598

EXCEL 12. Testing for Correlation Between Performance and Price *Consumer Reports* tested VHS tapes used in VCRs. Following are performance scores and prices (in dollars) of randomly selected tapes. Is there a correlation between performance and price? What does the conclusion suggest about buying VHS tapes?

Performance	91	92	82	85	87	80	94	97
Price	4.56	6.48	5.99	7.92	5.36	3.32	7.32	5.27

EXCEL 13. Do Beer Drinkers and Liquor Drinkers Have Different BAC Levels? The sample data in the following list show BAC (blood alcohol concentration) levels at arrest of randomly selected jail inmates who were convicted of DWI or DUI offenses. The data are categorized by the type of drink consumed (based on data from the U.S. Department of Justice).

Beer	Liquor
0.129 0.146 0.148 0.152	0.220 0.225 0.185 0.182
0.154 0.155 0.187 0.212	0.253 0.241 0.227 0.205
0.203 0.190 0.164 0.165	0.247 0.224 0.226 0.234
	0.190 0.257

Test the claim that beer drinkers and liquor drinkers have the same BAC levels. Based on these results, do both groups seem equally dangerous, or is one group more dangerous than the other?

EXCEL 14. Testing for a Correlation Between SAT Score and Cost of Education The following table ranks eight states according to SAT scores and cost per student. Is there a correlation between SAT score and cost per student?

State	NY	CA	FL	NJ	TX	NC	MD	OR
SAT score	4	3	5	6	7	8	2	1
Cost per student	1	5	6	2	7	8	3	4

Based on data from the U.S. Department of Education.

EXCEL 15. Heights of Presidential Winners and Losers The accompanying table shows the heights of presidents matched with the heights of the candidates they beat. All heights are in inches, and only the second-place candidates are included. Use the Wilcoxon signed-ranks test to test the claim that both samples come from populations with the same distribution. Is there sufficient evidence to support the claim that height is an advantage to candidates?

Winner	76	66	70	70	74	71.5	73	74
Runner-up	64	71	72	72	68	71	69.5	74

CUMULATIVE REVIEW EXERCISES

---

 1. Analyzing Poll Results A market researcher for American Airlines was instructed to randomly select passengers waiting to board. The passengers were asked several questions about the airline service. Responses were recorded along with their genders, which are listed in the order in which they were selected.

    a. At the 0.05 significance level, test the claim that the sequence is random.

    b. At the 0.05 significance level, test the claim that the proportion of women is different from 0.5.

    c. Use the sample data to construct a 95% confidence interval for the proportion of women.

    d. What do the preceding results suggest? Is the sample biased against either gender? Was the sample obtained in a random sequence? If you are the manager, do you have any problems with these results?

    M M M M M F F M M M F M M F F

    F M F M M F M M F M M F M M M

 2. Constructing a Histogram of Sunday Rainfall Amounts In Section 13-5 we used the Kruskal-Wallis test to test the null hypothesis that the rainfall amounts for Boston are the same for each of the seven days of the week. We noted that analysis-of-variance methods described in Section 11-2 cannot be used because the rainfall amounts do not appear to be normally distributed, and we illustrated this with a histogram of the Monday rainfall amounts. (See the Chapter Opening problem.) Refer to Data Set 17 (BOSTRAIN.xls) in Appendix B and construct a histogram of the 52 rainfall amounts for Sunday. Compare the result to the histogram for the Monday rainfall amounts. Do the Sunday rainfall amounts appear to come from a population having a normal distribution?

 3. Finding Descriptive Statistics for Sunday Rainfall Amounts Refer to the Sunday rainfall amounts in Data Set 17 (BOSTRAIN.xls) in Appendix B and find the following.

    a. Mean

    b. Median

    c. Mode

    d. Range

    e. Standard deviation

    f. Any outliers

 4. Proportions of Rainy Days Refer to Data Set 17 (BOSTRAIN.xls) in Appendix B. Is there sufficient evidence to support the claim that the proportions of rainy days are not the same for all seven days of the week? (There are 53 Wednesdays, but delete the last value for Wednesday and then assume that all seven days of the week occur 52 times.)

 5. Testing Effectiveness of SAT Coaching The Block Preparation Program is designed to help students achieve better scores on the SAT exam. The accompanying table lists results for randomly selected students. Use the indicated test with a 0.05 significance level to test the hypothesis that the course has no effect on SAT scores.

    a. Sign test

    b. Wilcoxon signed-ranks test

    c. t test for a claim about matched pairs of sample data (Section 8-3)

(continued)

d. How do the preceding results support the statement that nonparametric tests lack the sensitivity of parametric tests (so stronger evidence is required before a null hypothesis is rejected)?

Subject	A	B	C	D	E	F	G	H	I	J
SAT score before course	700	840	830	860	840	690	830	1180	930	1070
SAT score after course	800	840	820	980	980	800	800	1270	1080	1220

COOPERATIVE GROUP ACTIVITIES

1. *In-class activity:* Use the existing seating arrangement in your class and apply the runs test to determine whether the students are arranged randomly according to gender. After recording the seating arrangement, analysis can be done in subgroups of three or four students.

2. *In-class activity:* Divide into groups of 8 to 12 people. For each group member, *measure* his or her height and *measure* his or her arm span. For the arm span, the subject should stand with arms extended, like the wings on an airplane. It's easy to mark the height and arm span on a chalkboard, then measure the distances there. Divide the following tasks among subgroups of three or four people.
   a. Use rank correlation with the paired sample data to determine whether there is a correlation between height and arm span.
   b. Use the sign test to test for a difference between the two variables.
   c. Use the Wilcoxon signed-ranks test to test for a difference between the two variables.

3. *In-class activity:* Do activity 2 using pulse rate instead of arm span. Measure pulse rates by counting the number of heartbeats in 1 minute.

4. *In-class activity:* Divide into groups of about 10 or 12 students and use the reaction timer included with the Chapter 5 Cooperative Group Activities. Each group member should be tested for right-hand reaction time and left-hand reaction time. Analyze the results using methods from this chapter. State the methods used and the conclusions reached.

5. *In-class activity:* Divide into groups of five or six students. This activity is a contest of reaction times to determine which group is fastest, if there is a "fastest" group. Use the reaction timer included with the Cooperative Group Activities in Chapter 5. Test and record the reaction time using the dominant hand of each group member. (Only one try per person.) After all groups have reported their results, identify the group with the fastest reaction time, and identify the criterion used to determine the fastest group. Use the Kruskal-Wallis test to determine if differences among the groups are significantly different.

6. *Out-of-class activity:* Divide into groups of three or four students. Investigate the relationship between two variables by collecting your own paired sample data and using the methods of Section 13-6 to determine whether there is a significant correlation. Suggested topics:

   • Is there a relationship between taste and cost of different brands of chocolate chip cookies (or colas)? (Taste can be measured on some number scale, such as 1 to 10.)

- Is there a relationship between salaries of professional baseball (or basketball, or football) players and their season achievements?

- Rates versus weights: Is there a relationship between car fuel consumption rates and car weights?

- Is there a relationship between the lengths of men's (or women's) feet and their heights?

- Is there a relationship between student grade-point averages and the amount of television watched?

- Is there a relationship between heights of fathers (or mothers) and heights of their first sons (or daughters)?

7. *Out-of-class activity:* See the "From Data to Decision" project, which involves analysis of the 1970 lottery used for drafting men into the U.S. Army. Because the 1970 results raised concerns about the randomness of selecting draft priority numbers, design a new procedure for generating the 366 priority numbers. Use your procedure to generate the 366 numbers and test your results by using the techniques suggested in parts (a), (b), and (c) of the "From Data to Decision" project. How do your results compare to those obtained in 1970? Does your random selection process appear to be better than the one used in 1970? Write a report that clearly describes the process you designed. Also include your analyses and conclusions.

8. *Out-of-class activity:* Divide into groups of three or four. Survey students by asking them to identify their major and gender. For each surveyed subject, determine the accuracy of the time on his or her wristwatch. First set your own watch to the correct time using an accurate and reliable source ("At the tone, the time is . . . "). For watches that are ahead of the correct time, record positive times. For watches that are behind the correct time, record negative times. Use the sample data to address these questions:

- Do the errors appear to be the same for both genders?

- Do the errors appear to be the same for the different majors?

EXCEL PROJECT

Randomly generated digits are commonly used for a variety of different applications. Pollsters often select telephone numbers by using random digits, so that unlisted numbers can be included. Simulations, such as those discussed in Section 3-6, often require the random generation of digits.

1. Listed below are data sets that were generated by using the random features of different technologies. (STATDISK and Minitab are software packages, and the TI-83 Plus is a calculator.) Use the appropriate nonparametric method from this chapter to determine whether the randomly generated digits listed below come from identical populations.

Excel:	3 1 5 9 8 8 4 8 1 2 0 0 1 2 0 2 3 5 3 9
STATDISK:	1 7 8 3 0 0 4 9 7 5 6 5 9 6 1 1 5 5 8 5
Minitab:	1 2 5 1 1 3 1 8 8 5 3 5 5 3 5 3 5 2 3 3
TI-83 Plus:	5 8 4 9 5 4 0 1 4 2 4 3 1 4 9 7 5 3 7 5

2. Use Excel to generate your own data, then proceed to test for randomness. Click on the empty cell A1. Generate the data by clicking on *fx,* then **Math & Trig,** then **RANDBETWEEN.** Click **OK.** In the dialog box, enter 0 for the "bottom" value and enter 9 for the "top" value, then click **OK.** Cell A1 will contain a randomly selected digit. Now use the mouse to click on the lower right corner of cell A1; hold the mouse button down to click and drag downward until reaching cell A100, then release the mouse button. Column A should now contain 100 randomly selected digits. Test for randomness. Describe the method used and the conclusion reached.

from **DATA** *to* **DECISION**

Critical Thinking:
Was the draft lottery random?

In 1970, a lottery was used to determine who would be drafted into the U.S. Army. The 366 dates in the year were placed in individual capsules. First, the 31 January capsules were placed in a box; then the 29 February capsules were added and the two months were mixed. Then the 31 March capsules were added and the three months were mixed. This process continued until all months were included. The first capsule selected was September 14, so men born on that date were drafted first. The accompanying list shows the 366 dates in the order of selection.

Analyze the Results

a. Use the runs test to test the sequence for randomness above and below the median of 183.5.

b. Use the Kruskal-Wallis test to test the claim that the 12 months had priority numbers drawn from the same population.

c. Calculate the 12 monthly means. Then plot those 12 means on a graph. (The horizontal scale lists the 12 months, and the vertical scale ranges from 100 to 260.) Note any pattern suggesting that the original priority numbers were not randomly selected.

d. Based on the results from parts (a), (b), and (c), decide whether this particular draft lottery was fair. Write a statement either supporting your position that it was fair or explaining why you believe that it was not fair. If you decided that this lottery was unfair, describe a process for selecting lottery numbers that would have been fair.

Jan:	305	159	251	215	101	224	306	199	194	325	329	221	318	238	017	121
	235	140	058	280	186	337	118	059	052	092	355	077	349	164	211	
Feb:	086	144	297	210	214	347	091	181	338	216	150	068	152	004	089	212
	189	292	025	302	363	290	057	236	179	365	205	299	285			
Mar:	108	029	267	275	293	139	122	213	317	323	136	300	259	354	169	166
	033	332	200	239	334	265	256	258	343	170	268	223	362	217	030	
Apr:	032	271	083	081	269	253	147	312	219	218	014	346	124	231	273	148
	260	090	336	345	062	316	252	002	351	340	074	262	191	208		
May:	330	298	040	276	364	155	035	321	197	065	037	133	295	178	130	055
	112	278	075	183	250	326	319	031	361	357	296	308	226	103	313	
Jun:	249	228	301	020	028	110	085	366	335	206	134	272	069	356	180	274
	073	341	104	360	060	247	109	358	137	022	064	222	353	209		
Jul:	093	350	115	279	188	327	050	013	277	284	248	015	042	331	322	120
	098	190	227	187	027	153	172	023	067	303	289	088	270	287	193	
Aug:	111	045	261	145	054	114	168	048	106	021	324	142	307	198	102	044
	154	141	311	344	291	339	116	036	286	245	352	167	061	333	011	
Sep:	225	161	049	232	082	006	008	184	263	071	158	242	175	001	113	207
	255	246	177	063	204	160	119	195	149	018	233	257	151	315		
Oct:	359	125	244	202	024	087	234	283	342	220	237	072	138	294	171	254
	288	005	241	192	243	117	201	196	176	007	264	094	229	038	079	
Nov:	019	034	348	266	310	076	051	097	080	282	046	066	126	127	131	107
	143	146	203	185	156	009	182	230	132	309	047	281	099	174		
Dec:	129	328	157	165	056	010	012	105	043	041	039	314	163	026	320	096
	304	128	240	135	070	053	162	095	084	173	078	123	016	003	100	

Internet Project

Nonparametric Tests

This chapter introduced several different nonparametric or distribution-free tests. Such tests allow you to test hypotheses without making assumptions about the distributions underlying the associated data sets. The Internet Project for this chapter, found at

http://www.awlonline.com/triola

will ask you to revisit some of the hyotheses tested in earlier Internet Projects, but this time you will apply an appropriate nonparametric test. In the second part of the project, you will be asked to apply the runs test to determine randomness of a sequence.

Statistics
at work

"You can't effectively communicate here unless you use a common language, and that common language happens to be statistics."

Anthony DiUglio

Nuclear Analyst, Probabilistic Risk Assessment, Consolidated Edison Company of New York, Inc.

Anthony DiUglio works in the Probabilistic Risk Assessment (PRA) Group at Consolidated Edison's Indian Point Unit #2 nuclear generating facility in Buchanan, New York, In his work as a Nuclear Analyst, Tony develops probabilities that are used in quantifying various aspects of the plant-specific risk assessment. He is a former student of the author.

What is your job?

In PRA we are concerned with three basic questions about risk: What can happen, how likely is it to happen, and what are the consequences of its happening? We apply these questions about risk to the safe, reliable, and continuous operation of our power plant. When we quantify risk, we obtain numbers that are probabilities. If someone suggests a modification to a plant safety system, we analyze it from a risk perspective. Is the modification better for the system? Does it affect the operation of the plant or put the public health and safety at risk?

How do you use probability and/or statistics?

They're our primary tools. Our PRA requires that we quantify plant-specific rates for all safety-related components in our plant. In developing component-repair rates for pumps and valves, we look at industrywide data (generic) and our plant-specific data. We combine this information together, under uncertainty, and end up with component-specific repair probabilities.

How do you use probability and/or statistics in other departments at Indian Point?

Our Performance Department measures various plant parameters, such as heat rate, megawatt generation, cost per kilowatt of generation, etc. These parameters are all analyzed by use of statistics.

The statistical tools used are data trending, normal statistical curves, standard deviations, histograms, etc. Financial Planning makes extensive use of statistics in projecting budget needs and determining its constraints. Our corporate forecasters use probability theory to predict power demands at different times during the year (e.g., winter and summer, one, three, and five years down the road). We have so many people using statistics in their everyday work that statistics has now become a tool for engineers, planners, forecasters, and those of us in Risk Assessment.

In terms of statistics, what would you recommend for prospective employees?

They should have a good understanding of probability, statistics, and their applications. Because PRA is still a relatively new area, we often deal with problems that haven't been addressed before, so many of the problems we address require creative problem solving. Once you have the basic tools, your time is efficiently spent. You can't effectively communicate here unless you use a common language, and that common language happens to be statistics.

Has your work been helpful in convincing the public that your plant is safe?

Safety is always our first concern. In the early 1980s there was a series of public hearings conducted by the Nuclear

Regulatory Commission (NRC) to discuss whether or not the plant should continue operation. Con Edison maintained that the plant was safe, and we were able to help justify the continued operation of our plant through the use of our PRA. At the conclusion of those hearings the NRC agreed with our position, and we continued to operate.

Who was your best math teacher?

Professor Mario Triola.

Is your use of probability and statistics increasing, decreasing, or remaining stable?

It's increasing all the time. We are very much involved with plant performance indicators as parameters for efficient plant operation. With PRA we now have a tool that allows us to focus our attention on the more important plant components and functions. In the case where three components all need maintenance, PRA allows us to identify which component should be returned to service first. In engineering, if we have several components that should be improved, PRA allows us to identify which component should be improved first. We can quantify the effects and thereby target our resources better, thus making the plant safer.

CHAPTER 14

Projects, Procedures, Perspectives

 14-1 A Statistics Group Project

The main objective of this section is to provide some suggestions for a study that can be used as a capstone project for the introductory statistics course. One terrific advantage of this course is that it deals with skills and concepts that can be applied immediately to the real world. After only one fun semester, students are able to conduct their own studies. Some of the suggested topics can be addressed by actually conducting experiments, whereas others might be observational studies that require research of results already available. For example, testing the effectiveness of air bags by actually crashing cars is strongly discouraged, but destructive taste tests of chocolate chip cookies can be an easy and somewhat enjoyable experiment. Here is a suggested format, followed by a list of suggested topics.

Group/Individual Topics can be assigned to individuals, but group projects are particularly effective because they help develop the interpersonal skills that are so necessary in today's working environment.

Oral Report A 10- to 15-minute-long class presentation should involve all group members in a coordinated effort to clearly describe the important components of the study.

Written Report The main objective of the project is not to produce a written document equivalent to a term paper, but a written report should be submitted, and it should include the following components.

1. List of data collected
2. Description of the method of analysis
3. Relevant graphs and/or statistics, including Excel displays

4. Statement of conclusions

5. Reasons why the results might not be correct, along with a description of ways in which the study could be improved, given sufficient time and money

Suggested Topics   In addition to the topics suggested in the following list, also see the *Cooperative Group Activities* listed near the end of each chapter.

1. Ages of student cars compared to faculty/staff cars

2. Proportion of foreign cars for students compared to the proportion of foreign cars for faculty

3. Car ages in the parking lot of a discount store compared to car ages in the parking lot of an upscale department store

4. Are husbands older than their wives?

5. Are husband/wife age differences the same for young married couples as for older married couples?

6. Analysis of the ages of books in the college library

7. How do the ages of books in the college library compare to those in the library of a nearby college?

8. Comparison of the ages of science books and English books in the college library

9. How many hours do students study each week?

10. Is there a relationship between hours studied and grades earned?

11. Is there a relationship between hours worked and grades earned?

12. A study of *reported* heights compared to *measured* heights

13. A study of the accuracy of wristwatches

14. Is there a relationship between taste and cost of different brands of chocolate chip cookies?

15. Is there a relationship between taste and cost of different brands of peanut butter?

16. Is there a relationship between taste and cost of different brands of cola?

17. Is there a relationship between salaries of professional baseball (or basketball, or football) players and their season achievements?

18. Rates versus weights: Is there a relationship between car fuel-consumption rates and car weights? If so, what is it?

19. Is there a relationship between the lengths of men's (or women's) feet and their heights?

20. Are there differences in taste between ordinary tap water and different brands of bottled water?

21. Were auto fatality rates affected by laws requiring the use of seat belts?

22. Were auto fatality rates affected when the national speed limit of 55 mi/h was eliminated?

23. Were auto fatality rates affected by the presence of air bags?

24. Is there a difference in taste between Coke and Pepsi?

25. Is there a relationship between student grade-point averages and the amount of television watched? If so, what is it?

26. Is there a relationship between the selling price of a home and its living area (in square feet), lot size (in acres), number of rooms, number of baths, and the annual tax bill?

27. A comparison of the numbers of keys carried by males and females

28. A comparison of the numbers of credit cards carried by males and females

29. Are murderers now younger than they were in the past?

30. Do people who exercise vigorously tend to have lower pulse rates than those who do not?

31. Do people who exercise vigorously tend to have reaction times that are different from those of people who do not?

32. Do people who smoke tend to have higher pulse rates than those who do not?

33. For people who don't exercise, how is pulse rate affected by climbing a flight of stairs?

34. Do statistics students tend to have pulse rates that are different from those of people not studying statistics?

35. A comparison of GPAs of statistics students with those of students not taking statistics

36. Do left-handed people tend to be involved in more car crashes?

37. Do men have more car crashes than women?

38. Do young drivers have more car crashes than older drivers?

39. Are drivers who get tickets more likely to be involved in crashes?

40. Do smokers tend to be involved in more car crashes?

41. Do people with higher pulse rates tend to be involved in more/fewer car crashes?

42. A comparison of reaction times measured with right and left hands

43. Are the proportions of male and female smokers equal?

44. Do statistics students tend to smoke more (or less) than the general population?

45. Are people more likely to smoke if their parents smoked?

46. Evidence to support/refute the belief that smoking tends to stunt growth

47. Does a sports team have an advantage by playing at home instead of away?

48. Analysis of service times (in seconds) for a car drive-up window at a bank

49. A comparison of service times for car drive-up windows at two different banks

50. Analysis of times that McDonald's' patrons are seated at a table

51. Analysis of times that McDonald's' patrons wait in line

52. Analysis of times cars require for refueling

53. Is the state lottery a wise investment?

54. Comparison of casino games: craps versus roulette

55. Starting with $1, is it easier to win a million dollars by playing casino craps or by playing a state lottery?

56. Bold versus cautious strategies of gambling: When gambling with $100, does it make any difference if you bet $1 at a time or if you bet the whole $100 at once?

57. Designing and analyzing results from a test for extrasensory perception

58. Analyzing paired data consisting of heights of fathers (or mothers) and heights of their first sons (or daughters)

59. Gender differences in preferences of dinner partners among the options of Brad Pitt, the President, Nicole Kidman, Cameron Diaz, Sandra Bullock, Queen Margaret, and the Pope

60. Gender differences in preferences of activities among the options of dinner, movie, watching television, reading a book, golf, tennis, swimming, attending a baseball game, attending a football game

 # 14-2 Which Procedure Applies?

Data Collection   You can collect your own data through experiments or observational studies. It is absolutely essential to analyze the method used to collect the data, because data carelessly collected may be so completely useless that no amount of statistical torturing can salvage them. Look carefully for bias in the way data are collected, as well as bias on the part of the person or group collecting the data. Many of the procedures in this book are based on the assumption that we are working with a simple random sample, meaning that every possible sample of the same size has the same chance of being selected. If a sample is self-selected (voluntary response), it is worthless for making inferences about a population.

Exploring, Comparing, Describing   After collecting data, first consider exploring, describing, and comparing data sets using the basic tools included in Chapter 2. Be sure to address the following:

1. *Center:* Find the mean and median, which are measures of center that are representative or average values giving us an indication of where the middle of the data set is located.

2. *Variation:* Find the range and standard deviation, which are measures of the amount that the sample values vary among themselves.

3. *Distribution:* Construct a histogram to see the nature or shape of the distribution of the data, and determine if the distribution is bell-shaped, uniform, or skewed.

4. *Outliers:* Identify any sample values that lie very far away from the vast majority of the other sample values.

5. *Time:* Determine if the population is stable or if its characteristics are changing over time.

Inferences: Estimating Parameters and Hypothesis Testing   When trying to use sample data for making inferences about a population, it is often difficult to choose the particular procedure that should be applied. This text includes a wide variety of procedures that apply to many different circumstances. Here are some key questions that should be answered.

- What is the level of measurement (nominal, ordinal, interval, ratio) of the data?
- Does the study involve one, two, or more populations?
- Is there a claim to be tested or a parameter to be estimated?
- What is the relevant parameter (mean, standard deviation, proportion)?
- Is the sample large ($n > 30$) or small?
- Is there reason to believe that the population is normally distributed?
- What is the basic question or issue that you want to address?

In Figure 14-1 on the next page we list the major methods included in this book, along with a scheme for determining which of those methods should be used. To use Figure 14-1, start at the extreme left side of the figure and begin by identifying the level of measurement of the data. Proceed to follow the path suggested by the level of measurement, the number of populations, and the claim or parameter being considered.

Note: This figure applies to a fixed population. If the data are from a process that may change over time, construct a control chart (see Chapter 12) to determine whether the process is statistically stable. This figure applies to process data only if the process is statistically stable.

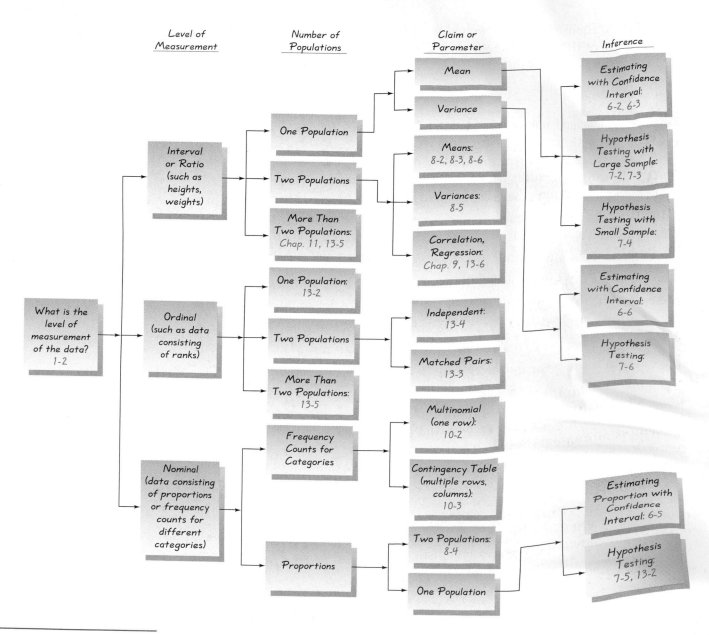

FIGURE 14-1
Inferential Statistics:
Applicable Text Sections

 14-3 A Perspective

No one expects a single introductory statistics course to transform anyone into an expert statistician. After studying several of the chapters in this book, it is natural for students to feel that they have not yet mastered the material to the extent necessary for confidently using statistics in real applications. Many important topics (such as factor analysis and discriminant analysis) are not included in this text because they are too advanced for this introductory level. Some easier topics (such as time series) have been excluded for other reasons. It is important to know that professional help is available from expert statisticians, and this introductory statistics course will help you in discussions with one of these experts.

Although this course is not designed to make you an expert statistician, it is designed to make you a better educated person with improved job marketability. You should know and understand the basic concepts of probability and chance. You should know that in attempting to gain insight into a set of data, it is important to investigate measures of center (such as mean and median), measures of variation (such as range and standard deviation), the nature of the distribution (via a frequency table or graph), the presence of outliers, and whether the population is stable or is changing over time. You should know and understand the importance of estimating population parameters (such as a mean, standard deviation, and proportion), as well as testing claims made about population parameters. You should realize that the nature and configuration of the data have a dramatic effect on the particular statistical procedures that are used.

Throughout this text we have emphasized the importance of good sampling. You should recognize that a bad sample may be beyond repair by even the most expert statisticians, using the most sophisticated techniques. There are many mail, magazine, and telephone call-in surveys that allow respondents to be "self-selected." The results of such surveys are generally worthless when judged according to the criteria of sound statistical methodology. Keep this in mind when you are exposed to self-selected surveys, so that you don't let them affect your beliefs and decisions. You should also recognize, however, that many surveys and polls obtain very good results, even though the sample sizes might seem to be relatively small. Although many people refuse to believe it, a nationwide survey of only 1700 voters can provide good results if the sampling is carefully planned and executed.

At one time a person was considered educated if he or she could read, but we are in a new millennium that is much more demanding. Today, an educated person must be able to read, write, understand the significance of the Renaissance, operate a computer, and do algebra. The larger picture combines disciplines with common goals, including the quest for truth. The study of statistics helps us see the truth that is sometimes distorted by others or concealed by data that are disorganized or perhaps not yet collected. Understanding statistics is now essential for employees, employers, and citizens. H. G. Wells once said that "statistical thinking will one day be as necessary for efficient citizenship as the ability to read and write." That day is now.

Appendices

Appendix A: Tables

								p							
n	x	.01	.05	.10	.20	.30	.40	.50	.60	.70	.80	.90	.95	.99	x
2	0	.980	.902	.810	.640	.490	.360	.250	.160	.090	.040	.010	.002	0+	0
	1	.020	.095	.180	.320	.420	.480	.500	.480	.420	.320	.180	.095	.020	1
	2	0+	.002	.010	.040	.090	.160	.250	.360	.490	.640	.810	.902	.980	2
3	0	.970	.857	.729	.512	.343	.216	.125	.064	.027	.008	.001	0+	0+	0
	1	.029	.135	.243	.384	.441	.432	.375	.288	.189	.096	.027	.007	0+	1
	2	0+	.007	.027	.096	.189	.288	.375	.432	.441	.384	.243	.135	.029	2
	3	0+	0+	.001	.008	.027	.064	.125	.216	.343	.512	.729	.857	.970	3
4	0	.961	.815	.656	.410	.240	.130	.062	.026	.008	.002	0+	0+	0+	0
	1	.039	.171	.292	.410	.412	.346	.250	.154	.076	.026	.004	0+	0+	1
	2	.001	.014	.049	.154	.265	.346	.375	.346	.265	.154	.049	.014	.001	2
	3	0+	0+	.004	.026	.076	.154	.250	.346	.412	.410	.292	.171	.039	3
	4	0+	0+	0+	.002	.008	.026	.062	.130	.240	.410	.656	.815	.961	4
5	0	.951	.774	.590	.328	.168	.078	.031	.010	.002	0+	0+	0+	0+	0
	1	.048	.204	.328	.410	.360	.259	.156	.077	.028	.006	0+	0+	0+	1
	2	.001	.021	.073	.205	.309	.346	.312	.230	.132	.051	.008	.001	0+	2
	3	0+	.001	.008	.051	.132	.230	.312	.346	.309	.205	.073	.021	.001	3
	4	0+	0+	0+	.006	.028	.077	.156	.259	.360	.410	.328	.204	.048	4
	5	0+	0+	0+	0+	.002	.010	.031	.078	.168	.328	.590	.774	.951	5
6	0	.941	.735	.531	.262	.118	.047	.016	.004	.001	0+	0+	0+	0+	0
	1	.057	.232	.354	.393	.303	.187	.094	.037	.010	.002	0+	0+	0+	1
	2	.001	.031	.098	.246	.324	.311	.234	.138	.060	.015	.001	0+	0+	2
	3	0+	.002	.015	.082	.185	.276	.312	.276	.185	.082	.015	.002	0+	3
	4	0+	0+	.001	.015	.060	.138	.234	.311	.324	.246	.098	.031	.001	4
	5	0+	0+	0+	.002	.010	.037	.094	.187	.303	.393	.354	.232	.057	5
	6	0+	0+	0+	0+	.001	.004	.016	.047	.118	.262	.531	.735	.941	6
7	0	.932	.698	.478	.210	.082	.028	.008	.002	0+	0+	0+	0+	0+	0
	1	.066	.257	.372	.367	.247	.131	.055	.017	.004	0+	0+	0+	0+	1
	2	.002	.041	.124	.275	.318	.261	.164	.077	.025	.004	0+	0+	0+	2
	3	0+	.004	.023	.115	.227	.290	.273	.194	.097	.029	.003	0+	0+	3
	4	0+	0+	.003	.029	.097	.194	.273	.290	.227	.115	.023	.004	0+	4
	5	0+	0+	0+	.004	.025	.077	.164	.261	.318	.275	.124	.041	.002	5
	6	0+	0+	0+	0+	.004	.017	.055	.131	.247	.367	.372	.257	.066	6
	7	0+	0+	0+	0+	0+	.002	.008	.028	.082	.210	.478	.698	.932	7
8	0	.923	.663	.430	.168	.058	.017	.004	.001	0+	0+	0+	0+	0+	0
	1	.075	.279	.383	.336	.198	.090	.031	.008	.001	0+	0+	0+	0+	1
	2	.003	.051	.149	.294	.296	.209	.109	.041	.010	.001	0+	0+	0+	2
	3	0+	.005	.033	.147	.254	.279	.219	.124	.047	.009	0+	0+	0+	3
	4	0+	0+	.005	.046	.136	.232	.273	.232	.136	.046	.005	0+	0+	4
	5	0+	0+	0+	.009	.047	.124	.219	.279	.254	.147	.033	.005	0+	5
	6	0+	0+	0+	.001	.010	.041	.109	.209	.296	.294	.149	.051	.003	6
	7	0+	0+	0+	0+	.001	.008	.031	.090	.198	.336	.383	.279	.075	7
	8	0+	0+	0+	0+	0+	.001	.004	.017	.058	.168	.430	.663	.923	8

NOTE: 0+ represents a positive probability less than 0.0005.

(continued)

								p								
n	x	.01	.05	.10	.20	.30	.40	.50	.60	.70	.80	.90	.95	.99	x
9	0	.914	.630	.387	.134	.040	.010	.002	0+	0+	0+	0+	0+	0+	0
	1	.083	.299	.387	.302	.156	.060	.018	.004	0+	0+	0+	0+	0+	1
	2	.003	.063	.172	.302	.267	.161	.070	.021	.004	0+	0+	0+	0+	2
	3	0+	.008	.045	.176	.267	.251	.164	.074	.021	.003	0+	0+	0+	3
	4	0+	.001	.007	.066	.172	.251	.246	.167	.074	.017	.001	0+	0+	4
	5	0+	0+	.001	.017	.074	.167	.246	.251	.172	.066	.007	.001	0+	5
	6	0+	0+	0+	.003	.021	.074	.164	.251	.267	.176	.045	.008	0+	6
	7	0+	0+	0+	0+	.004	.021	.070	.161	.267	.302	.172	.063	.003	7
	8	0+	0+	0+	0+	0+	.004	.018	.060	.156	.302	.387	.299	.083	8
	9	0+	0+	0+	0+	0+	0+	.002	.010	.040	.134	.387	.630	.914	9
10	0	.904	.599	.349	.107	.028	.006	.001	0+	0+	0+	0+	0+	0+	0
	1	.091	.315	.387	.268	.121	.040	.010	.002	0+	0+	0+	0+	0+	1
	2	.004	.075	.194	.302	.233	.121	.044	.011	.001	0+	0+	0+	0+	2
	3	0+	.010	.057	.201	.267	.215	.117	.042	.009	.001	0+	0+	0+	3
	4	0+	.001	.011	.088	.200	.251	.205	.111	.037	.006	0+	0+	0+	4
	5	0+	0+	.001	.026	.103	.201	.246	.201	.103	.026	.001	0+	0+	5
	6	0+	0+	0+	.006	.037	.111	.205	.251	.200	.088	.011	.001	0+	6
	7	0+	0+	0+	.001	.009	.042	.117	.215	.267	.201	.057	.010	0+	7
	8	0+	0+	0+	0+	.001	.011	.044	.121	.233	.302	.194	.075	.004	8
	9	0+	0+	0+	0+	0+	.002	.010	.040	.121	.268	.387	.315	.091	9
	10	0+	0+	0+	0+	0+	0+	.001	.006	.028	.107	.349	.599	.904	10
11	0	.895	.569	.314	.086	.020	.004	0+	0+	0+	0+	0+	0+	0+	0
	1	.099	.329	.384	.236	.093	.027	.005	.001	0+	0+	0+	0+	0+	1
	2	.005	.087	.213	.295	.200	.089	.027	.005	.001	0+	0+	0+	0+	2
	3	0+	.014	.071	.221	.257	.177	.081	.023	.004	0+	0+	0+	0+	3
	4	0+	.001	.016	.111	.220	.236	.161	.070	.017	.002	0+	0+	0+	4
	5	0+	0+	.002	.039	.132	.221	.226	.147	.057	.010	0+	0+	0+	5
	6	0+	0+	0+	.010	.057	.147	.226	.221	.132	.039	.002	0+	0+	6
	7	0+	0+	0+	.002	.017	.070	.161	.236	.220	.111	.016	.001	0+	7
	8	0+	0+	0+	0+	.004	.023	.081	.177	.257	.221	.071	.014	0+	8
	9	0+	0+	0+	0+	.001	.005	.027	.089	.200	.295	.213	.087	.005	9
	10	0+	0+	0+	0+	0+	.001	.005	.027	.093	.236	.384	.329	.099	10
	11	0+	0+	0+	0+	0+	0+	0+	.004	.020	.086	.314	.569	.895	11
12	0	.886	.540	.282	.069	.014	.002	0+	0+	0+	0+	0+	0+	0+	0
	1	.107	.341	.377	.206	.071	.017	.003	0+	0+	0+	0+	0+	0+	1
	2	.006	.099	.230	.283	.168	.064	.016	.002	0+	0+	0+	0+	0+	2
	3	0+	.017	.085	.236	.240	.142	.054	.012	.001	0+	0+	0+	0+	3
	4	0+	.002	.021	.133	.231	.213	.121	.042	.008	.001	0+	0+	0+	4
	5	0+	0+	.004	.053	.158	.227	.193	.101	.029	.003	0+	0+	0+	5
	6	0+	0+	0+	.016	.079	.177	.226	.177	.079	.016	0+	0+	0+	6
	7	0+	0+	0+	.003	.029	.101	.193	.227	.158	.053	.004	0+	0+	7
	8	0+	0+	0+	.001	.008	.042	.121	.213	.231	.133	.021	.002	0+	8
	9	0+	0+	0+	0+	.001	.012	.054	.142	.240	.236	.085	.017	0+	9
	10	0+	0+	0+	0+	0+	.002	.016	.064	.168	.283	.230	.099	.006	10
	11	0+	0+	0+	0+	0+	0+	.003	.017	.071	.206	.377	.341	.107	11
	12	0+	0+	0+	0+	0+	0+	0+	.002	.014	.069	.282	.540	.886	12

NOTE: 0+ represents a positive probability less than 0.0005.

(continued)

n	x	.01	.05	.10	.20	.30	.40	.50	.60	.70	.80	.90	.95	.99	x
13	0	.878	.513	.254	.055	.010	.001	0+	0+	0+	0+	0+	0+	0+	0
	1	.115	.351	.367	.179	.054	.011	.002	0+	0+	0+	0+	0+	0+	1
	2	.007	.111	.245	.268	.139	.045	.010	.001	0+	0+	0+	0+	0+	2
	3	0+	.021	.100	.246	.218	.111	.035	.006	.001	0+	0+	0+	0+	3
	4	0+	.003	.028	.154	.234	.184	.087	.024	.003	0+	0+	0+	0+	4
	5	0+	0+	.006	.069	.180	.221	.157	.066	.014	.001	0+	0+	0+	5
	6	0+	0+	.001	.023	.103	.197	.209	.131	.044	.006	0+	0+	0+	6
	7	0+	0+	0+	.006	.044	.131	.209	.197	.103	.023	.001	0+	0+	7
	8	0+	0+	0+	.001	.014	.066	.157	.221	.180	.069	.006	0+	0+	8
	9	0+	0+	0+	0+	.003	.024	.087	.184	.234	.154	.028	.003	0+	9
	10	0+	0+	0+	0+	.001	.006	.035	.111	.218	.246	.100	.021	0+	10
	11	0+	0+	0+	0+	0+	.001	.010	.045	.139	.268	.245	.111	.007	11
	12	0+	0+	0+	0+	0+	0+	.002	.011	.054	.179	.367	.351	.115	12
	13	0+	0+	0+	0+	0+	0+	0+	.001	.010	.055	.254	.513	.878	13
14	0	.869	.488	.229	.044	.007	.001	0+	0+	0+	0+	0+	0+	0+	0
	1	.123	.359	.356	.154	.041	.007	.001	0+	0+	0+	0+	0+	0+	1
	2	.008	.123	.257	.250	.113	.032	.006	.001	0+	0+	0+	0+	0+	2
	3	0+	.026	.114	.250	.194	.085	.022	.003	0+	0+	0+	0+	0+	3
	4	0+	.004	.035	.172	.229	.155	.061	.014	.001	0+	0+	0+	0+	4
	5	0+	0+	.008	.086	.196	.207	.122	.041	.007	0+	0+	0+	0+	5
	6	0+	0+	.001	.032	.126	.207	.183	.092	.023	.002	0+	0+	0+	6
	7	0+	0+	0+	.009	.062	.157	.209	.157	.062	.009	0+	0+	0+	7
	8	0+	0+	0+	.002	.023	.092	.183	.207	.126	.032	.001	0+	0+	8
	9	0+	0+	0+	0+	.007	.041	.122	.207	.196	.086	.008	0+	0+	9
	10	0+	0+	0+	0+	.001	.014	.061	.155	.229	.172	.035	.004	0+	10
	11	0+	0+	0+	0+	0+	.003	.022	.085	.194	.250	.114	.026	0+	11
	12	0+	0+	0+	0+	0+	.001	.006	.032	.113	.250	.257	.123	.008	12
	13	0+	0+	0+	0+	0+	0+	.001	.007	.041	.154	.356	.359	.123	13
	14	0+	0+	0+	0+	0+	0+	0+	.001	.007	.044	.229	.488	.869	14
15	0	.860	.463	.206	.035	.005	0+	0+	0+	0+	0+	0+	0+	0+	0
	1	.130	.366	.343	.132	.031	.005	0+	0+	0+	0+	0+	0+	0+	1
	2	.009	.135	.267	.231	.092	.022	.003	0+	0+	0+	0+	0+	0+	2
	3	0+	.031	.129	.250	.170	.063	.014	.002	0+	0+	0+	0+	0+	3
	4	0+	.005	.043	.188	.219	.127	.042	.007	.001	0+	0+	0+	0+	4
	5	0+	.001	.010	.103	.206	.186	.092	.024	.003	0+	0+	0+	0+	5
	6	0+	0+	.002	.043	.147	.207	.153	.061	.012	.001	0+	0+	0+	6
	7	0+	0+	0+	.014	.081	.177	.196	.118	.035	.003	0+	0+	0+	7
	8	0+	0+	0+	.003	.035	.118	.196	.177	.081	.014	0+	0+	0+	8
	9	0+	0+	0+	.001	.012	.061	.153	.207	.147	.043	.002	0+	0+	9
	10	0+	0+	0+	0+	.003	.024	.092	.186	.206	.103	.010	.001	0+	10
	11	0+	0+	0+	0+	.001	.007	.042	.127	.219	.188	.043	.005	0+	11
	12	0+	0+	0+	0+	0+	.002	.014	.063	.170	.250	.129	.031	0+	12
	13	0+	0+	0+	0+	0+	0+	.003	.022	.092	.231	.267	.135	.009	13
	14	0+	0+	0+	0+	0+	0+	0+	.005	.031	.132	.343	.366	.130	14
	15	0+	0+	0+	0+	0+	0+	0+	0+	.005	.035	.206	.463	.860	15

NOTE: 0+ represents a positive probability less than 0.0005.

From Frederick C. Mosteller, Robert E. K. Rourke, and George B. Thomas, Jr., *Probability with Statistical Applications*, 2nd ed., © 1970 Addison-Wesley Publishing Co., Reading, MA. Reprinted with permission.

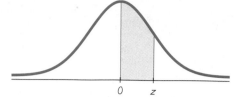

TABLE A-2	Standard Normal (z) Distribution									
z	.00	.01	.02	.03	.04	.05	.06	.07	.08	.09
0.0	.0000	.0040	.0080	.0120	.0160	.0199	.0239	.0279	.0319	.0359
0.1	.0398	.0438	.0478	.0517	.0557	.0596	.0636	.0675	.0714	.0753
0.2	.0793	.0832	.0871	.0910	.0948	.0987	.1026	.1064	.1103	.1141
0.3	.1179	.1217	.1255	.1293	.1331	.1368	.1406	.1443	.1480	.1517
0.4	.1554	.1591	.1628	.1664	.1700	.1736	.1772	.1808	.1844	.1879
0.5	.1915	.1950	.1985	.2019	.2054	.2088	.2123	.2157	.2190	.2224
0.6	.2257	.2291	.2324	.2357	.2389	.2422	.2454	.2486	.2517	.2549
0.7	.2580	.2611	.2642	.2673	.2704	.2734	.2764	.2794	.2823	.2852
0.8	.2881	.2910	.2939	.2967	.2995	.3023	.3051	.3078	.3106	.3133
0.9	.3159	.3186	.3212	.3238	.3264	.3289	.3315	.3340	.3365	.3389
1.0	.3413	.3438	.3461	.3485	.3508	.3531	.3554	.3577	.3599	.3621
1.1	.3643	.3665	.3686	.3708	.3729	.3749	.3770	.3790	.3810	.3830
1.2	.3849	.3869	.3888	.3907	.3925	.3944	.3962	.3980	.3997	.4015
1.3	.4032	.4049	.4066	.4082	.4099	.4115	.4131	.4147	.4162	.4177
1.4	.4192	.4207	.4222	.4236	.4251	.4265	.4279	.4292	.4306	.4319
1.5	.4332	.4345	.4357	.4370	.4382	.4394	.4406	.4418	.4429	.4441
1.6	.4452	.4463	.4474	.4484	.4495 *	.4505	.4515	.4525	.4535	.4545
1.7	.4554	.4564	.4573	.4582	.4591	.4599	.4608	.4616	.4625	.4633
1.8	.4641	.4649	.4656	.4664	.4671	.4678	.4686	.4693	.4699	.4706
1.9	.4713	.4719	.4726	.4732	.4738	.4744	.4750	.4756	.4761	.4767
2.0	.4772	.4778	.4783	.4788	.4793	.4798	.4803	.4808	.4812	.4817
2.1	.4821	.4826	.4830	.4834	.4838	.4842	.4846	.4850	.4854	.4857
2.2	.4861	.4864	.4868	.4871	.4875	.4878	.4881	.4884	.4887	.4890
2.3	.4893	.4896	.4898	.4901	.4904	.4906	.4909	.4911	.4913	.4916
2.4	.4918	.4920	.4922	.4925	.4927	.4929	.4931	.4932	.4934	.4936
2.5	.4938	.4940	.4941	.4943	.4945	.4946	.4948	.4949 *	.4951	.4952
2.6	.4953	.4955	.4956	.4957	.4959	.4960	.4961	.4962	.4963	.4964
2.7	.4965	.4966	.4967	.4968	.4969	.4970	.4971	.4972	.4973	.4974
2.8	.4974	.4975	.4976	.4977	.4977	.4978	.4979	.4979	.4980	.4981
2.9	.4981	.4982	.4982	.4983	.4984	.4984	.4985	.4985	.4986	.4986
3.0	.4987	.4987	.4987	.4988	.4988	.4989	.4989	.4989	.4990	.4990
3.10 and higher	.4999									

NOTE: For values of z above 3.09, use 0.4999 for the area.

*Use these common values that result from interpolation:

z score	Area
1.645	0.4500
2.575	0.4950

From Frederick C. Mosteller and Robert E. K. Rourke, *Sturdy Statistics*, 1973, Addison-Wesley Publishing Co., Reading, MA. Reprinted with permission of Frederick Mosteller.

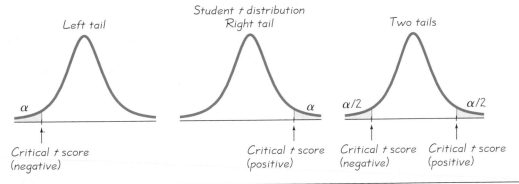

Left tail

Student *t* distribution
Right tail

Two tails

α

Critical *t* score
(negative)

α

Critical *t* score
(positive)

$\alpha/2$ $\alpha/2$

Critical *t* score   Critical *t* score
(negative)   (positive)

TABLE A-3	*t* Distribution

| | | | α | | | |

Degrees of Freedom	.005 (one tail) .01 (two tails)	.01 (one tail) .02 (two tails)	.025 (one tail) .05 (two tails)	.05 (one tail) .10 (two tails)	.10 (one tail) .20 (two tails)	.25 (one tail) .50 (two tails)
1	63.657	31.821	12.706	6.314	3.078	1.000
2	9.925	6.965	4.303	2.920	1.886	.816
3	5.841	4.541	3.182	2.353	1.638	.765
4	4.604	3.747	2.776	2.132	1.533	.741
5	4.032	3.365	2.571	2.015	1.476	.727
6	3.707	3.143	2.447	1.943	1.440	.718
7	3.500	2.998	2.365	1.895	1.415	.711
8	3.355	2.896	2.306	1.860	1.397	.706
9	3.250	2.821	2.262	1.833	1.383	.703
10	3.169	2.764	2.228	1.812	1.372	.700
11	3.106	2.718	2.201	1.796	1.363	.697
12	3.054	2.681	2.179	1.782	1.356	.696
13	3.012	2.650	2.160	1.771	1.350	.694
14	2.977	2.625	2.145	1.761	1.345	.692
15	2.947	2.602	2.132	1.753	1.341	.691
16	2.921	2.584	2.120	1.746	1.337	.690
17	2.898	2.567	2.110	1.740	1.333	.689
18	2.878	2.552	2.101	1.734	1.330	.688
19	2.861	2.540	2.093	1.729	1.328	.688
20	2.845	2.528	2.086	1.725	1.325	.687
21	2.831	2.518	2.080	1.721	1.323	.686
22	2.819	2.508	2.074	1.717	1.321	.686
23	2.807	2.500	2.069	1.714	1.320	.685
24	2.797	2.492	2.064	1.711	1.318	.685
25	2.787	2.485	2.060	1.708	1.316	.684
26	2.779	2.479	2.056	1.706	1.315	.684
27	2.771	2.473	2.052	1.703	1.314	.684
28	2.763	2.467	2.048	1.701	1.313	.683
29	2.756	2.462	2.045	1.699	1.311	.683
Large (z)	2.575	2.326	1.960	1.645	1.282	.675

TABLE A-4	Chi-Square (χ^2) Distribution

Area to the Right of the Critical Value

Degrees of Freedom	0.995	0.99	0.975	0.95	0.90	0.10	0.05	0.025	0.01	0.005
1	—	—	0.001	0.004	0.016	2.706	3.841	5.024	6.635	7.879
2	0.010	0.020	0.051	0.103	0.211	4.605	5.991	7.378	9.210	10.597
3	0.072	0.115	0.216	0.352	0.584	6.251	7.815	9.348	11.345	12.838
4	0.207	0.297	0.484	0.711	1.064	7.779	9.488	11.143	13.277	14.860
5	0.412	0.554	0.831	1.145	1.610	9.236	11.071	12.833	15.086	16.750
6	0.676	0.872	1.237	1.635	2.204	10.645	12.592	14.449	16.812	18.548
7	0.989	1.239	1.690	2.167	2.833	12.017	14.067	16.013	18.475	20.278
8	1.344	1.646	2.180	2.733	3.490	13.362	15.507	17.535	20.090	21.955
9	1.735	2.088	2.700	3.325	4.168	14.684	16.919	19.023	21.666	23.589
10	2.156	2.558	3.247	3.940	4.865	15.987	18.307	20.483	23.209	25.188
11	2.603	3.053	3.816	4.575	5.578	17.275	19.675	21.920	24.725	26.757
12	3.074	3.571	4.404	5.226	6.304	18.549	21.026	23.337	26.217	28.299
13	3.565	4.107	5.009	5.892	7.042	19.812	22.362	24.736	27.688	29.819
14	4.075	4.660	5.629	6.571	7.790	21.064	23.685	26.119	29.141	31.319
15	4.601	5.229	6.262	7.261	8.547	22.307	24.996	27.488	30.578	32.801
16	5.142	5.812	6.908	7.962	9.312	23.542	26.296	28.845	32.000	34.267
17	5.697	6.408	7.564	8.672	10.085	24.769	27.587	30.191	33.409	35.718
18	6.265	7.015	8.231	9.390	10.865	25.989	28.869	31.526	34.805	37.156
19	6.844	7.633	8.907	10.117	11.651	27.204	30.144	32.852	36.191	38.582
20	7.434	8.260	9.591	10.851	12.443	28.412	31.410	34.170	37.566	39.997
21	8.034	8.897	10.283	11.591	13.240	29.615	32.671	35.479	38.932	41.401
22	8.643	9.542	10.982	12.338	14.042	30.813	33.924	36.781	40.289	42.796
23	9.260	10.196	11.689	13.091	14.848	32.007	35.172	38.076	41.638	44.181
24	9.886	10.856	12.401	13.848	15.659	33.196	36.415	39.364	42.980	45.559
25	10.520	11.524	13.120	14.611	16.473	34.382	37.652	40.646	44.314	46.928
26	11.160	12.198	13.844	15.379	17.292	35.563	38.885	41.923	45.642	48.290
27	11.808	12.879	14.573	16.151	18.114	36.741	40.113	43.194	46.963	49.645
28	12.461	13.565	15.308	16.928	18.939	37.916	41.337	44.461	48.278	50.993
29	13.121	14.257	16.047	17.708	19.768	39.087	42.557	45.722	49.588	52.336
30	13.787	14.954	16.791	18.493	20.599	40.256	43.773	46.979	50.892	53.672
40	20.707	22.164	24.433	26.509	29.051	51.805	55.758	59.342	63.691	66.766
50	27.991	29.707	32.357	34.764	37.689	63.167	67.505	71.420	76.154	79.490
60	35.534	37.485	40.482	43.188	46.459	74.397	79.082	83.298	88.379	91.952
70	43.275	45.442	48.758	51.739	55.329	85.527	90.531	95.023	100.425	104.215
80	51.172	53.540	57.153	60.391	64.278	96.578	101.879	106.629	112.329	116.321
90	59.196	61.754	65.647	69.126	73.291	107.565	113.145	118.136	124.116	128.299
100	67.328	70.065	74.222	77.929	82.358	118.498	124.342	129.561	135.807	140.169

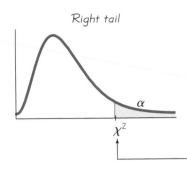

Right tail

To find this value, use the column with the area α given at the top of the table.

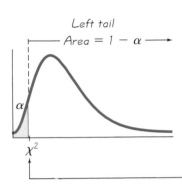

Left tail

Area $= 1 - \alpha$

To find this value, determine the area of the region to the right of this boundary (the unshaded area) and use the column with this value at the top. If the left tail has area α, use the column with the value of $1 - \alpha$ at the top of the table.

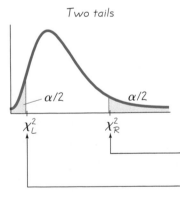

Two tails

To find this value, use the column with area $\alpha/2$ at the top of the table.

To find this value, use the column with area $1 - \alpha/2$ at the top of the table.

0.01

F

TABLE A-5 F Distribution ($\alpha = 0.01$ in the right tail)

Numerator degrees of freedom (df_1)

df_2	1	2	3	4	5	6	7	8	9
1	4052.2	4999.5	5403.4	5624.6	5763.6	5859.0	5928.4	5981.1	6022.5
2	98.503	99.000	99.166	99.249	99.299	99.333	99.356	99.374	99.388
3	34.116	30.817	29.457	28.710	28.237	27.911	27.672	27.489	27.345
4	21.198	18.000	16.694	15.977	15.522	15.207	14.976	14.799	14.659
5	16.258	13.274	12.060	11.392	10.967	10.672	10.456	10.289	10.158
6	13.745	10.925	9.7795	9.1483	8.7459	8.4661	8.2600	8.1017	7.9761
7	12.246	9.5466	8.4513	7.8466	7.4604	7.1914	6.9928	6.8400	6.7188
8	11.259	8.6491	7.5910	7.0061	6.6318	6.3707	6.1776	6.0289	5.9106
9	10.561	8.0215	6.9919	6.4221	6.0569	5.8018	5.6129	5.4671	5.3511
10	10.044	7.5594	6.5523	5.9943	5.6363	5.3858	5.2001	5.0567	4.9424
11	9.6460	7.2057	6.2167	5.6683	5.3160	5.0692	4.8861	4.7445	4.6315
12	9.3302	6.9266	5.9525	5.4120	5.0643	4.8206	4.6395	4.4994	4.3875
13	9.0738	6.7010	5.7394	5.2053	4.8616	4.6204	4.4410	4.3021	4.1911
14	8.8616	6.5149	5.5639	5.0354	4.6950	4.4558	4.2779	4.1399	4.0297
15	8.6831	6.3589	5.4170	4.8932	4.5556	4.3183	4.1415	4.0045	3.8948
16	8.5310	6.2262	5.2922	4.7726	4.4374	4.2016	4.0259	3.8896	3.7804
17	8.3997	6.1121	5.1850	4.6690	4.3359	4.1015	3.9267	3.7910	3.6822
18	8.2854	6.0129	5.0919	4.5790	4.2479	4.0146	3.8406	3.7054	3.5971
19	8.1849	5.9259	5.0103	4.5003	4.1708	3.9386	3.7653	3.6305	3.5225
20	8.0960	5.8489	4.9382	4.4307	4.1027	3.8714	3.6987	3.5644	3.4567
21	8.0166	5.7804	4.8740	4.3688	4.0421	3.8117	3.6396	3.5056	3.3981
22	7.9454	5.7190	4.8166	4.3134	3.9880	3.7583	3.5867	3.4530	3.3458
23	7.8811	5.6637	4.7649	4.2636	3.9392	3.7102	3.5390	3.4057	3.2986
24	7.8229	5.6136	4.7181	4.2184	3.8951	3.6667	3.4959	3.3629	3.2560
25	7.7698	5.5680	4.6755	4.1774	3.8550	3.6272	3.4568	3.3239	3.2172
26	7.7213	5.5263	4.6366	4.1400	3.8183	3.5911	3.4210	3.2884	3.1818
27	7.6767	5.4881	4.6009	4.1056	3.7848	3.5580	3.3882	3.2558	3.1494
28	7.6356	5.4529	4.5681	4.0740	3.7539	3.5276	3.3581	3.2259	3.1195
29	7.5977	5.4204	4.5378	4.0449	3.7254	3.4995	3.3303	3.1982	3.0920
30	7.5625	5.3903	4.5097	4.0179	3.6990	3.4735	3.3045	3.1726	3.0665
40	7.3141	5.1785	4.3126	3.8283	3.5138	3.2910	3.1238	2.9930	2.8876
60	7.0771	4.9774	4.1259	3.6490	3.3389	3.1187	2.9530	2.8233	2.7185
120	6.8509	4.7865	3.9491	3.4795	3.1735	2.9559	2.7918	2.6629	2.5586
∞	6.6349	4.6052	3.7816	3.3192	3.0173	2.8020	2.6393	2.5113	2.4073

Denominator degrees of freedom (df_2)

From Maxine Merrington and Catherine M. Thompson, "Tables of Percentage Points of the Inverted Beta (F) Distribution," *Biometrika 33* (1943): 80–84. Reproduced with permission of the Biometrika Trustees.

(continued)

TABLE A-5 F Distribution ($\alpha = 0.01$ in the right tail) (continued)

	Numerator degrees of freedom (df$_1$)									
df$_2$	10	12	15	20	24	30	40	60	120	∞
1	6055.8	6106.3	6157.3	6208.7	6234.6	6260.6	6286.8	6313.0	6339.4	6365.9
2	99.399	99.416	99.433	99.449	99.458	99.466	99.474	99.482	99.491	99.499
3	27.229	27.052	26.872	26.690	26.598	26.505	26.411	26.316	26.221	26.125
4	14.546	14.374	14.198	14.020	13.929	13.838	13.745	13.652	13.558	13.463
5	10.051	9.8883	9.7222	9.5526	9.4665	9.3793	9.2912	9.2020	9.1118	9.0204
6	7.8741	7.7183	7.5590	7.3958	7.3127	7.2285	7.1432	7.0567	6.9690	6.8800
7	6.6201	6.4691	6.3143	6.1554	6.0743	5.9920	5.9084	5.8236	5.7373	5.6495
8	5.8143	5.6667	5.5151	5.3591	5.2793	5.1981	5.1156	5.0316	4.9461	4.8588
9	5.2565	5.1114	4.9621	4.8080	4.7290	4.6486	4.5666	4.4831	4.3978	4.3105
10	4.8491	4.7059	4.5581	4.4054	4.3269	4.2469	4.1653	4.0819	3.9965	3.9090
11	4.5393	4.3974	4.2509	4.0990	4.0209	3.9411	3.8596	3.7761	3.6904	3.6024
12	4.2961	4.1553	4.0096	3.8584	3.7805	3.7008	3.6192	3.5355	3.4494	3.3608
13	4.1003	3.9603	3.8154	3.6646	3.5868	3.5070	3.4253	3.3413	3.2548	3.1654
14	3.9394	3.8001	3.6557	3.5052	3.4274	3.3476	3.2656	3.1813	3.0942	3.0040
15	3.8049	3.6662	3.5222	3.3719	3.2940	3.2141	3.1319	3.0471	2.9595	2.8684
16	3.6909	3.5527	3.4089	3.2587	3.1808	3.1007	3.0182	2.9330	2.8447	2.7528
17	3.5931	3.4552	3.3117	3.1615	3.0835	3.0032	2.9205	2.8348	2.7459	2.6530
18	3.5082	3.3706	3.2273	3.0771	2.9990	2.9185	2.8354	2.7493	2.6597	2.5660
19	3.4338	3.2965	3.1533	3.0031	2.9249	2.8442	2.7608	2.6742	2.5839	2.4893
20	3.3682	3.2311	3.0880	2.9377	2.8594	2.7785	2.6947	2.6077	2.5168	2.4212
21	3.3098	3.1730	3.0300	2.8796	2.8010	2.7200	2.6359	2.5484	2.4568	2.3603
22	3.2576	3.1209	2.9779	2.8274	2.7488	2.6675	2.5831	2.4951	2.4029	2.3055
23	3.2106	3.0740	2.9311	2.7805	2.7017	2.6202	2.5355	2.4471	2.3542	2.2558
24	3.1681	3.0316	2.8887	2.7380	2.6591	2.5773	2.4923	2.4035	2.3100	2.2107
25	3.1294	2.9931	2.8502	2.6993	2.6203	2.5383	2.4530	2.3637	2.2696	2.1694
26	3.0941	2.9578	2.8150	2.6640	2.5848	2.5026	2.4170	2.3273	2.2325	2.1315
27	3.0618	2.9256	2.7827	2.6316	2.5522	2.4699	2.3840	2.2938	2.1985	2.0965
28	3.0320	2.8959	2.7530	2.6017	2.5223	2.4397	2.3535	2.2629	2.1670	2.0642
29	3.0045	2.8685	2.7256	2.5742	2.4946	2.4118	2.3253	2.2344	2.1379	2.0342
30	2.9791	2.8431	2.7002	2.5487	2.4689	2.3860	2.2992	2.2079	2.1108	2.0062
40	2.8005	2.6648	2.5216	2.3689	2.2880	2.2034	2.1142	2.0194	1.9172	1.8047
60	2.6318	2.4961	2.3523	2.1978	2.1154	2.0285	1.9360	1.8363	1.7263	1.6006
120	2.4721	2.3363	2.1915	2.0346	1.9500	1.8600	1.7628	1.6557	1.5330	1.3805
∞	2.3209	2.1847	2.0385	1.8783	1.7908	1.6964	1.5923	1.4730	1.3246	1.0000

Denominator degrees of freedom (df$_2$)

From Maxine Merrington and Catherine M. Thompson, "Tables of Percentage Points of the Inverted Beta (F) Distribution," *Biometrika 33* (1943): 80–84. Reproduced with permission of the Biometrika Trustees.

(continued)

TABLE A-5 F Distribution ($\alpha = 0.025$ in the right tail)

Denominator degrees of freedom (df_2)	Numerator degrees of freedom (df_1)								
	1	2	3	4	5	6	7	8	9
1	647.79	799.50	864.16	899.58	921.85	937.11	948.22	956.66	963.28
2	38.506	39.000	39.165	39.248	39.298	39.331	39.335	39.373	39.387
3	17.443	16.044	15.439	15.101	14.885	14.735	14.624	14.540	14.473
4	12.218	10.649	9.9792	9.6045	9.3645	9.1973	9.0741	8.9796	8.9047
5	10.007	8.4336	7.7636	7.3879	7.1464	6.9777	6.8531	6.7572	6.6811
6	8.8131	7.2599	6.5988	6.2272	5.9876	5.8198	5.6955	5.5996	5.5234
7	8.0727	6.5415	5.8898	5.5226	5.2852	5.1186	4.9949	4.8993	4.8232
8	7.5709	6.0595	5.4160	5.0526	4.8173	4.6517	4.5286	4.4333	4.3572
9	7.2093	5.7147	5.0781	4.7181	4.4844	4.3197	4.1970	4.1020	4.0260
10	6.9367	5.4564	4.8256	4.4683	4.2361	4.0721	3.9498	3.8549	3.7790
11	6.7241	5.2559	4.6300	4.2751	4.0440	3.8807	3.7586	3.6638	3.5879
12	6.5538	5.0959	4.4742	4.1212	3.8911	3.7283	3.6065	3.5118	3.4358
13	6.4143	4.9653	4.3472	3.9959	3.7667	3.6043	3.4827	3.3880	3.3120
14	6.2979	4.8567	4.2417	3.8919	3.6634	3.5014	3.3799	3.2853	3.2093
15	6.1995	4.7650	4.1528	3.8043	3.5764	3.4147	3.2934	3.1987	3.1227
16	6.1151	4.6867	4.0768	3.7294	3.5021	3.3406	3.2194	3.1248	3.0488
17	6.0420	4.6189	4.0112	3.6648	3.4379	3.2767	3.1556	3.0610	2.9849
18	5.9781	4.5597	3.9539	3.6083	3.3820	3.2209	3.0999	3.0053	2.9291
19	5.9216	4.5075	3.9034	3.5587	3.3327	3.1718	3.0509	2.9563	2.8801
20	5.8715	4.4613	3.8587	3.5147	3.2891	3.1283	3.0074	2.9128	2.8365
21	5.8266	4.4199	3.8188	3.4754	3.2501	3.0895	2.9686	2.8740	2.7977
22	5.7863	4.3828	3.7829	3.4401	3.2151	3.0546	2.9338	2.8392	2.7628
23	5.7498	4.3492	3.7505	3.4083	3.1835	3.0232	2.9023	2.8077	2.7313
24	5.7166	4.3187	3.7211	3.3794	3.1548	2.9946	2.8738	2.7791	2.7027
25	5.6864	4.2909	3.6943	3.3530	3.1287	2.9685	2.8478	2.7531	2.6766
26	5.6586	4.2655	3.6697	3.3289	3.1048	2.9447	2.8240	2.7293	2.6528
27	5.6331	4.2421	3.6472	3.3067	3.0828	2.9228	2.8021	2.7074	2.6309
28	5.6096	4.2205	3.6264	3.2863	3.0626	2.9027	2.7820	2.6872	2.6106
29	5.5878	4.2006	3.6072	3.2674	3.0438	2.8840	2.7633	2.6686	2.5919
30	5.5675	4.1821	3.5894	3.2499	3.0265	2.8667	2.7460	2.6513	2.5746
40	5.4239	4.0510	3.4633	3.1261	2.9037	2.7444	2.6238	2.5289	2.4519
60	5.2856	3.9253	3.3425	3.0077	2.7863	2.6274	2.5068	2.4117	2.3344
120	5.1523	3.8046	3.2269	2.8943	2.6740	2.5154	2.3948	2.2994	2.2217
∞	5.0239	3.6889	3.1161	2.7858	2.5665	2.4082	2.2875	2.1918	2.1136

(continued)

TABLE A-5 F Distribution ($\alpha = 0.025$ in the right tail) *(continued)*

Numerator degrees of freedom (df_1)

df_2	10	12	15	20	24	30	40	60	120	∞
1	968.63	976.71	984.87	993.10	997.25	1001.4	1005.6	1009.8	1014.0	1018.3
2	39.398	39.415	39.431	39.448	39.456	39.465	39.473	39.481	39.490	39.498
3	14.419	14.337	14.253	14.167	14.124	14.081	14.037	13.992	13.947	13.902
4	8.8439	8.7512	8.6565	8.5599	8.5109	8.4613	8.4111	8.3604	8.3092	8.2573
5	6.6192	6.5245	6.4277	6.3286	6.2780	6.2269	6.1750	6.1225	6.0693	6.0153
6	5.4613	5.3662	5.2687	5.1684	5.1172	5.0652	5.0125	4.9589	4.9044	4.8491
7	4.7611	4.6658	4.5678	4.4667	4.4150	4.3624	4.3089	4.2544	4.1989	4.1423
8	4.2951	4.1997	4.1012	3.9995	3.9472	3.8940	3.8398	3.7844	3.7279	3.6702
9	3.9639	3.8682	3.7694	3.6669	3.6142	3.5604	3.5055	3.4493	3.3918	3.3329
10	3.7168	3.6209	3.5217	3.4185	3.3654	3.3110	3.2554	3.1984	3.1399	3.0798
11	3.5257	3.4296	3.3299	3.2261	3.1725	3.1176	3.0613	3.0035	2.9441	2.8828
12	3.3736	3.2773	3.1772	3.0728	3.0187	2.9633	2.9063	2.8478	2.7874	2.7249
13	3.2497	3.1532	3.0527	2.9477	2.8932	2.8372	2.7797	2.7204	2.6590	2.5955
14	3.1469	3.0502	2.9493	2.8437	2.7888	2.7324	2.6742	2.6142	2.5519	2.4872
15	3.0602	2.9633	2.8621	2.7559	2.7006	2.6437	2.5850	2.5242	2.4611	2.3953
16	2.9862	2.8890	2.7875	2.6808	2.6252	2.5678	2.5085	2.4471	2.3831	2.3163
17	2.9222	2.8249	2.7230	2.6158	2.5598	2.5020	2.4422	2.3801	2.3153	2.2474
18	2.8664	2.7689	2.6667	2.5590	2.5027	2.4445	2.3842	2.3214	2.2558	2.1869
19	2.8172	2.7196	2.6171	2.5089	2.4523	2.3937	2.3329	2.2696	2.2032	2.1333
20	2.7737	2.6758	2.5731	2.4645	2.4076	2.3486	2.2873	2.2234	2.1562	2.0853
21	2.7348	2.6368	2.5338	2.4247	2.3675	2.3082	2.2465	2.1819	2.1141	2.0422
22	2.6998	2.6017	2.4984	2.3890	2.3315	2.2718	2.2097	2.1446	2.0760	2.0032
23	2.6682	2.5699	2.4665	2.3567	2.2989	2.2389	2.1763	2.1107	2.0415	1.9677
24	2.6396	2.5411	2.4374	2.3273	2.2693	2.2090	2.1460	2.0799	2.0099	1.9353
25	2.6135	2.5149	2.4110	2.3005	2.2422	2.1816	2.1183	2.0516	1.9811	1.9055
26	2.5896	2.4908	2.3867	2.2759	2.2174	2.1565	2.0928	2.0257	1.9545	1.8781
27	2.5676	2.4688	2.3644	2.2533	2.1946	2.1334	2.0693	2.0018	1.9299	1.8527
28	2.5473	2.4484	2.3438	2.2324	2.1735	2.1121	2.0477	1.9797	1.9072	1.8291
29	2.5286	2.4295	2.3248	2.2131	2.1540	2.0923	2.0276	1.9591	1.8861	1.8072
30	2.5112	2.4120	2.3072	2.1952	2.1359	2.0739	2.0089	1.9400	1.8664	1.7867
40	2.3882	2.2882	2.1819	2.0677	2.0069	1.9429	1.8752	1.8028	1.7242	1.6371
60	2.2702	2.1692	2.0613	1.9445	1.8817	1.8152	1.7440	1.6668	1.5810	1.4821
120	2.1570	2.0548	1.9450	1.8249	1.7597	1.6899	1.6141	1.5299	1.4327	1.3104
∞	2.0483	1.9447	1.8326	1.7085	1.6402	1.5660	1.4835	1.3883	1.2684	1.0000

Denominator degrees of freedom (df_2)

From Maxine Merrington and Catherine M. Thompson, "Tables of Percentage Points of the Inverted Beta (F) Distribution," *Biometrika 33* (1943): 80–84. Reproduced with permission of the Biometrika Trustees.

(continued)

TABLE A-5 F Distribution ($\alpha = 0.05$ in the right tail)

Numerator degrees of freedom (df_1)

	1	2	3	4	5	6	7	8	9
1	161.45	199.50	215.71	224.58	230.16	233.99	236.77	238.88	240.54
2	18.513	19.000	19.164	19.247	19.296	19.330	19.353	19.371	19.385
3	10.128	9.5521	9.2766	9.1172	9.0135	8.9406	8.8867	8.8452	8.8123
4	7.7086	6.9443	6.5914	6.3882	6.2561	6.1631	6.0942	6.0410	6.9988
5	6.6079	5.7861	5.4095	5.1922	5.0503	4.9503	4.8759	4.8183	4.7725
6	5.9874	5.1433	4.7571	4.5337	4.3874	4.2839	4.2067	4.1468	4.0990
7	5.5914	4.7374	4.3468	4.1203	3.9715	3.8660	3.7870	3.7257	3.6767
8	5.3177	4.4590	4.0662	3.8379	3.6875	3.5806	3.5005	3.4381	3.3881
9	5.1174	4.2565	3.8625	3.6331	3.4817	3.3738	3.2927	3.2296	3.1789
10	4.9646	4.1028	3.7083	3.4780	3.3258	3.2172	3.1355	3.0717	3.0204
11	4.8443	3.9823	3.5874	3.3567	3.2039	3.0946	3.0123	2.9480	2.8962
12	4.7472	3.8853	3.4903	3.2592	3.1059	2.9961	2.9134	2.8486	2.7964
13	4.6672	3.8056	3.4105	3.1791	3.0254	2.9153	2.8321	2.7669	2.7144
14	4.6001	3.7389	3.3439	3.1122	2.9582	2.8477	2.7642	2.6987	2.6458
15	4.5431	3.6823	3.2874	3.0556	2.9013	2.7905	2.7066	2.6408	2.5876
16	4.4940	3.6337	3.2389	3.0069	2.8524	2.7413	2.6572	2.5911	2.5377
17	4.4513	3.5915	3.1968	2.9647	2.8100	2.6987	2.6143	2.5480	2.4943
18	4.4139	3.5546	3.1599	2.9277	2.7729	2.6613	2.5767	2.5102	2.4563
19	4.3807	3.5219	3.1274	2.8951	2.7401	2.6283	2.5435	2.4768	2.4227
20	4.3512	3.4928	3.0984	2.8661	2.7109	2.5990	2.5140	2.4471	2.3928
21	4.3248	3.4668	3.0725	2.8401	2.6848	2.5727	2.4876	2.4205	2.3660
22	4.3009	3.4434	3.0491	2.8167	2.6613	2.5491	2.4638	2.3965	2.3419
23	4.2793	3.4221	3.0280	2.7955	2.6400	2.5277	2.4422	2.3748	2.3201
24	4.2597	3.4028	3.0088	2.7763	2.6207	2.5082	2.4226	2.3551	2.3002
25	4.2417	3.3852	2.9912	2.7587	2.6030	2.4904	2.4047	2.3371	2.2821
26	4.2252	3.3690	2.9752	2.7426	2.5868	2.4741	2.3883	2.3205	2.2655
27	4.2100	3.3541	2.9604	2.7278	2.5719	2.4591	2.3732	2.3053	2.2501
28	4.1960	3.3404	2.9467	2.7141	2.5581	2.4453	2.3593	2.2913	2.2360
29	4.1830	3.3277	2.9340	2.7014	2.5454	2.4324	2.3463	2.2783	2.2229
30	4.1709	3.3158	2.9223	2.6896	2.5336	2.4205	2.3343	2.2662	2.2107
40	4.0847	3.2317	2.8387	2.6060	2.4495	2.3359	2.2490	2.1802	2.1240
60	4.0012	3.1504	2.7581	2.5252	2.3683	2.2541	2.1665	2.0970	2.0401
120	3.9201	3.0718	2.6802	2.4472	2.2899	2.1750	2.0868	2.0164	1.9588
∞	3.8415	2.9957	2.6049	2.3719	2.2141	2.0986	2.0096	1.9384	1.8799

(continued)

Denominator degrees of freedom (df_2)

TABLE A-5 F Distribution ($\alpha = 0.05$ in the right tail) (continued)

df₂ \ Numerator degrees of freedom (df₁)	10	12	15	20	24	30	40	60	120	∞
1	241.88	243.91	245.95	248.01	249.05	250.10	251.14	252.20	253.25	254.31
2	19.396	19.413	19.429	19.446	19.454	19.462	19.471	19.479	19.487	19.496
3	8.7855	8.7446	8.7029	8.6602	8.6385	8.6166	8.5944	8.5720	8.5494	8.5264
4	5.9644	5.9117	5.8578	5.8025	5.7744	5.7459	5.7170	5.6877	5.6581	5.6281
5	4.7351	4.6777	4.6188	4.5581	4.5272	4.4957	4.4638	4.4314	4.3985	4.3650
6	4.0600	3.9999	3.9381	3.8742	3.8415	3.8082	3.7743	3.7398	3.7047	3.6689
7	3.6365	3.5747	3.5107	3.4445	3.4105	3.3758	3.3404	3.3043	3.2674	3.2298
8	3.3472	3.2839	3.2184	3.1503	3.1152	3.0794	3.0428	3.0053	2.9669	2.9276
9	3.1373	3.0729	3.0061	2.9365	2.9005	2.8637	2.8259	2.7872	2.7475	2.7067
10	2.9782	2.9130	2.8450	2.7740	2.7372	2.6996	2.6609	2.6211	2.5801	2.5379
11	2.8536	2.7876	2.7186	2.6464	2.6090	2.5705	2.5309	2.4901	2.4480	2.4045
12	2.7534	2.6866	2.6169	2.5436	2.5055	2.4663	2.4259	2.3842	2.3410	2.2962
13	2.6710	2.6037	2.5331	2.4589	2.4202	2.3803	2.3392	2.2966	2.2524	2.2064
14	2.6022	2.5342	2.4630	2.3879	2.3487	2.3082	2.2664	2.2229	2.1778	2.1307
15	2.5437	2.4753	2.4034	2.3275	2.2878	2.2468	2.2043	2.1601	2.1141	2.0658
16	2.4935	2.4247	2.3522	2.2756	2.2354	2.1938	2.1507	2.1058	2.0589	2.0096
17	2.4499	2.3807	2.3077	2.2304	2.1898	2.1477	2.1040	2.0584	2.0107	1.9604
18	2.4117	2.3421	2.2686	2.1906	2.1497	2.1071	2.0629	2.0166	1.9681	1.9168
19	2.3779	2.3080	2.2341	2.1555	2.1141	2.0712	2.0264	1.9795	1.9302	1.8780
20	2.3479	2.2776	2.2033	2.1242	2.0825	2.0391	1.9938	1.9464	1.8963	1.8432
21	2.3210	2.2504	2.1757	2.0960	2.0540	2.0102	1.9645	1.9165	1.8657	1.8117
22	2.2967	2.2258	2.1508	2.0707	2.0283	1.9842	1.9380	1.8894	1.8380	1.7831
23	2.2747	2.2036	2.1282	2.0476	2.0050	1.9605	1.9139	1.8648	1.8128	1.7570
24	2.2547	2.1834	2.1077	2.0267	1.9838	1.9390	1.8920	1.8424	1.7896	1.7330
25	2.2365	2.1649	2.0889	2.0075	1.9643	1.9192	1.8718	1.8217	1.7684	1.7110
26	2.2197	2.1479	2.0716	1.9898	1.9464	1.9010	1.8533	1.8027	1.7488	1.6906
27	2.2043	2.1323	2.0558	1.9736	1.9299	1.8842	1.8361	1.7851	1.7306	1.6717
28	2.1900	2.1179	2.0411	1.9586	1.9147	1.8687	1.8203	1.7689	1.7138	1.6541
29	2.1768	2.1045	2.0275	1.9446	1.9005	1.8543	1.8055	1.7537	1.6981	1.6376
30	2.1646	2.0921	2.0148	1.9317	1.8874	1.8409	1.7918	1.7396	1.6835	1.6223
40	2.0772	2.0035	1.9245	1.8389	1.7929	1.7444	1.6928	1.6373	1.5766	1.5089
60	1.9926	1.9174	1.8364	1.7480	1.7001	1.6491	1.5943	1.5343	1.4673	1.3893
120	1.9105	1.8337	1.7505	1.6587	1.6084	1.5543	1.4952	1.4290	1.3519	1.2539
∞	1.8307	1.7522	1.6664	1.5705	1.5173	1.4591	1.3940	1.3180	1.2214	1.0000

Denominator degrees of freedom (df₂)

From Maxine Merrington and Catherine M. Thompson, "Tables of Percentage Points of the Inverted Beta (F) Distribution," Biometrika 33 (1943): 80–84. Reproduced with permission of the Biometrika Trustees.

TABLE A-6	Critical Values of the Pearson Correlation Coefficient r	
n	$\alpha = .05$	$\alpha = .01$
4	.950	.999
5	.878	.959
6	.811	.917
7	.754	.875
8	.707	.834
9	.666	.798
10	.632	.765
11	.602	.735
12	.576	.708
13	.553	.684
14	.532	.661
15	.514	.641
16	.497	.623
17	.482	.606
18	.468	.590
19	.456	.575
20	.444	.561
25	.396	.505
30	.361	.463
35	.335	.430
40	.312	.402
45	.294	.378
50	.279	.361
60	.254	.330
70	.236	.305
80	.220	.286
90	.207	.269
100	.196	.256

NOTE: To test H_0: $\rho = 0$ against H_1: $\rho \neq 0$, reject H_0 if the absolute value of r is greater than the critical value in the table.

TABLE A-7 Critical Values for the Sign Test

	α			
n	.005 (one tail) .01 (two tails)	.01 (one tail) .02 (two tails)	.025 (one tail) .05 (two tails)	.05 (one tail) .10 (two tails)
1	*	*	*	*
2	*	*	*	*
3	*	*	*	*
4	*	*	*	*
5	*	*	*	0
6	*	*	0	0
7	*	0	0	0
8	0	0	0	1
9	0	0	1	1
10	0	0	1	1
11	0	1	1	2
12	1	1	2	2
13	1	1	2	3
14	1	2	2	3
15	2	2	3	3
16	2	2	3	4
17	2	3	4	4
18	3	3	4	5
19	3	4	4	5
20	3	4	5	5
21	4	4	5	6
22	4	5	5	6
23	4	5	6	7
24	5	5	6	7
25	5	6	7	7

NOTES:

1. * indicates that it is not possible to get a value in the critical region.

2. Reject the null hypothesis if the number of the less frequent sign (x) is less than or equal to the value in the table.

3. For values of n greater than 25, a normal approximation is used with

$$z = \frac{(x + 0.5) - \left(\frac{n}{2}\right)}{\frac{\sqrt{n}}{2}}$$

TABLE A-8	Critical Values of T for the Wilcoxon Signed-Ranks Test

	α			
n	.005 (one tail) .01 (two tails)	.01 (one tail) .02 (two tails)	.025 (one tail) .05 (two tails)	.05 (one tail) .10 (two tails)
5	*	*	*	1
6	*	*	1	2
7	*	0	2	4
8	0	2	4	6
9	2	3	6	8
10	3	5	8	11
11	5	7	11	14
12	7	10	14	17
13	10	13	17	21
14	13	16	21	26
15	16	20	25	30
16	19	24	30	36
17	23	28	35	41
18	28	33	40	47
19	32	38	46	54
20	37	43	52	60
21	43	49	59	68
22	49	56	66	75
23	55	62	73	83
24	61	69	81	92
25	68	77	90	101
26	76	85	98	110
27	84	93	107	120
28	92	102	117	130
29	100	111	127	141
30	109	120	137	152

NOTES:

1. * indicates that it is not possible to get a value in the critical region.

2. Reject the null hypothesis if the test statistic T is less than or equal to the critical value found in this table. Fail to reject the null hypothesis if the test statistic T is greater than the critical value found in the table.

From *Some Rapid Approximate Statistical Procedures,* Copyright ©1949, 1964 Lederle Laboratories Division of American Cyanamid Company. Reprinted with the permission of the American Cyanamid Company.

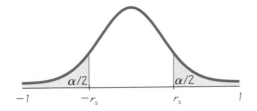

TABLE A-9	Critical Values of Spearman's Rank Correlation Coefficient r_s			
n	$\alpha = 0.10$	$\alpha = 0.05$	$\alpha = 0.02$	$\alpha = 0.01$
5	.900	—	—	—
6	.829	.886	.943	—
7	.714	.786	.893	—
8	.643	.738	.833	.881
9	.600	.683	.783	.833
10	.564	.648	.745	.794
11	.523	.623	.736	.818
12	.497	.591	.703	.780
13	.475	.566	.673	.745
14	.457	.545	.646	.716
15	.441	.525	.623	.689
16	.425	.507	.601	.666
17	.412	.490	.582	.645
18	.399	.476	.564	.625
19	.388	.462	.549	.608
20	.377	.450	.534	.591
21	.368	.438	.521	.576
22	.359	.428	.508	.562
23	.351	.418	.496	.549
24	.343	.409	.485	.537
25	.336	.400	.475	.526
26	.329	.392	.465	.515
27	.323	.385	.456	.505
28	.317	.377	.448	.496
29	.311	.370	.440	.487
30	.305	.364	.432	.478

NOTE: For $n > 30$, use $r_s = \pm z/\sqrt{n - 1}$, where z corresponds to the level of significance. For example, if $\alpha = 0.05$, then $z = 1.96$.

To test H_0: $\rho_s = 0$

against H_1: $\rho_s \neq 0$

From "Distribution of sums of squares of rank differences to small numbers of individuals," *The Annals of Mathematical Statistics,* Vol. 9, No. 2. Reprinted with permission of the Institute of Mathematical Statistics.

TABLE A-10 Critical Values for Number of Runs G

Value of n_2

Value of n_1		2	3	4	5	6	7	8	9	10	11	12	13	14	15	16	17	18	19	20
2		1	1	1	1	1	1	1	1	1	1	2	2	2	2	2	2	2	2	2
		6	6	6	6	6	6	6	6	6	6	6	6	6	6	6	6	6	6	6
3		1	1	1	1	2	2	2	2	2	2	2	2	2	3	3	3	3	3	3
		6	8	8	8	8	8	8	8	8	8	8	8	8	8	8	8	8	8	8
4		1	1	1	2	2	2	3	3	3	3	3	3	3	3	4	4	4	4	4
		6	8	9	9	9	10	10	10	10	10	10	10	10	10	10	10	10	10	10
5		1	1	2	2	3	3	3	3	3	4	4	4	4	4	4	4	5	5	5
		6	8	9	10	10	11	11	12	12	12	12	12	12	12	12	12	12	12	12
6		1	2	2	3	3	3	3	4	4	4	4	5	5	5	5	5	5	6	6
		6	8	9	10	11	12	12	13	13	13	13	14	14	14	14	14	14	14	14
7		1	2	2	3	3	3	4	4	5	5	5	5	5	6	6	6	6	6	6
		6	8	10	11	12	13	13	14	14	14	14	15	15	15	16	16	16	16	16
8		1	2	3	3	3	4	4	5	5	5	6	6	6	6	6	7	7	7	7
		6	8	10	11	12	13	14	14	15	15	16	16	16	16	17	17	17	17	17
9		1	2	3	3	4	4	5	5	5	6	6	6	7	7	7	7	8	8	8
		6	8	10	12	13	14	14	15	16	16	16	17	17	18	18	18	18	18	18
10		1	2	3	3	4	5	5	5	6	6	7	7	7	7	8	8	8	8	9
		6	8	10	12	13	14	15	16	16	17	17	18	18	18	19	19	19	20	20
11		1	2	3	4	4	5	5	6	6	7	7	7	8	8	8	9	9	9	9
		6	8	10	12	13	14	15	16	17	17	18	19	19	19	20	20	20	21	21
12		2	2	3	4	4	5	6	6	7	7	7	8	8	8	9	9	9	10	10
		6	8	10	12	13	14	16	16	17	18	19	19	20	20	21	21	21	22	22
13		2	2	3	4	5	5	6	6	7	7	8	8	9	9	9	10	10	10	10
		6	8	10	12	14	15	16	17	18	19	19	20	20	21	21	22	22	23	23
14		2	2	3	4	5	5	6	7	7	8	8	9	9	9	10	10	10	11	11
		6	8	10	12	14	15	16	17	18	19	20	20	21	22	22	23	23	23	24
15		2	3	3	4	5	6	6	7	7	8	8	9	9	10	10	11	11	11	12
		6	8	10	12	14	15	16	18	18	19	20	21	22	22	23	23	24	24	25
16		2	3	4	4	5	6	6	7	8	8	9	9	10	10	11	11	11	12	12
		6	8	10	12	14	16	17	18	19	20	21	21	22	23	23	24	25	25	25
17		2	3	4	4	5	6	7	7	8	9	9	10	10	11	11	11	12	12	13
		6	8	10	12	14	16	17	18	19	20	21	22	23	23	24	25	25	26	26
18		2	3	4	5	5	6	7	8	8	9	9	10	10	11	11	12	12	13	13
		6	8	10	12	14	16	17	18	19	20	21	22	23	24	25	25	26	26	27
19		2	3	4	5	6	6	7	8	8	9	10	10	11	11	12	12	13	13	13
		6	8	10	12	14	16	17	18	20	21	22	23	23	24	25	26	26	27	27
20		2	3	4	5	6	6	7	8	9	9	10	10	11	12	12	13	13	13	14
		6	8	10	12	14	16	17	18	20	21	22	23	24	25	25	26	27	27	28

NOTE:

1. The entries in this table are the critical G values, assuming a two-tailed test with a significance level of $\alpha = 0.05$.

2. The null hypothesis of randomness is rejected if the total number of runs G is less than or equal to the smaller entry or greater than or equal to the larger entry.

From "Tables for testing randomness of groupings in a sequence of alternatives," *The Annals of Mathematical Statistics*, Vol. 14, No. 1. Reprinted with permission of the Institute of Mathematical Statistics.

Appendix B: Data Sets

The Appendix B Data Sets (except Data Set 6) are available on the CD-ROM included with this book.

Data Set 1: Weights (in pounds) and Volumes (in ounces) of Cola

Excel:　　Workbook name is COLA.xls.
STATDISK: Variable names are CKREGWT, CKREGVOL,
　　　　　CKDIETWT, CKDTVOL, PPREGWT, PPREGVOL,
　　　　　PPDIETWT, and PPDTVOL.
Minitab:　Worksheet name is COLA.MTW.
TI-83 Plus and text files: CRGWT, CRGVL, CDTWT, CDTVL,
　　　　　　　　　　PRGWT, PRGVL, PDTWT, and PDTVL.

WEIGHT REGULAR COKE	VOLUME REGULAR COKE	WEIGHT DIET COKE	VOLUME DIET COKE	WEIGHT REGULAR PEPSI	VOLUME REGULAR PEPSI	WEIGHT DIET PEPSI	VOLUME DIET PEPSI
0.8192	12.3	0.7773	12.1	0.8258	12.4	0.7925	12.3
0.8150	12.1	0.7758	12.1	0.8156	12.2	0.7868	12.2
0.8163	12.2	0.7896	12.3	0.8211	12.2	0.7846	12.2
0.8211	12.3	0.7868	12.3	0.8170	12.2	0.7938	12.3
0.8181	12.2	0.7844	12.2	0.8216	12.2	0.7861	12.2
0.8247	12.3	0.7861	12.3	0.8302	12.4	0.7844	12.2
0.8062	12.0	0.7806	12.2	0.8192	12.2	0.7795	12.2
0.8128	12.1	0.7830	12.2	0.8192	12.2	0.7883	12.3
0.8172	12.2	0.7852	12.2	0.8271	12.3	0.7879	12.2
0.8110	12.1	0.7879	12.3	0.8251	12.3	0.7850	12.3
0.8251	12.3	0.7881	12.3	0.8227	12.2	0.7899	12.3
0.8264	12.3	0.7826	12.3	0.8256	12.3	0.7877	12.2
0.7901	11.8	0.7923	12.3	0.8139	12.2	0.7852	12.2
0.8244	12.3	0.7852	12.3	0.8260	12.3	0.7756	12.1
0.8073	12.1	0.7872	12.3	0.8227	12.2	0.7837	12.2
0.8079	12.1	0.7813	12.2	0.8388	12.5	0.7879	12.2
0.8044	12.0	0.7885	12.3	0.8260	12.3	0.7839	12.2
0.8170	12.2	0.7760	12.1	0.8317	12.4	0.7817	12.2
0.8161	12.2	0.7822	12.2	0.8247	12.3	0.7822	12.2
0.8194	12.2	0.7874	12.3	0.8200	12.2	0.7742	12.1
0.8189	12.2	0.7822	12.2	0.8172	12.2	0.7833	12.2
0.8194	12.2	0.7839	12.2	0.8227	12.3	0.7835	12.2
0.8176	12.2	0.7802	12.1	0.8244	12.3	0.7855	12.2
0.8284	12.4	0.7892	12.3	0.8244	12.2	0.7859	12.2
0.8165	12.2	0.7874	12.2	0.8319	12.4	0.7775	12.1
0.8143	12.2	0.7907	12.3	0.8247	12.3	0.7833	12.2
0.8229	12.3	0.7771	12.1	0.8214	12.2	0.7835	12.2
0.8150	12.2	0.7870	12.2	0.8291	12.4	0.7826	12.2
0.8152	12.2	0.7833	12.3	0.8227	12.3	0.7815	12.2
0.8244	12.3	0.7822	12.2	0.8211	12.3	0.7791	12.1
0.8207	12.2	0.7837	12.3	0.8401	12.5	0.7866	12.3
0.8152	12.2	0.7910	12.4	0.8233	12.3	0.7855	12.2
0.8126	12.1	0.7879	12.3	0.8291	12.4	0.7848	12.2
0.8295	12.4	0.7923	12.4	0.8172	12.2	0.7806	12.2
0.8161	12.2	0.7859	12.3	0.8233	12.4	0.7773	12.1
0.8192	12.2	0.7811	12.2	0.8211	12.3	0.7775	12.1

Data Set 2: Textbook Prices

*Denotes Paperback.

Excel: Workbook name is TEXTBOOK.xls.
STATDISK: Variable names are NEWUMASS, USDUMASS, NEWDCC, and USEDDCC.
Minitab: Worksheet name is TEXTBOOK.MTW.
TI-83 Plus and text files: NWUMA, USUMA, NWDCC, and USDCC.

UNIVERSITY OF MASSACHUSETTS	NEW	USED	DUTCHESS COMMUNITY COLLEGE	NEW	USED
Accounting	84.30	66.55	Accounting I	54.35	40.75
Advertising	83.35	62.55	Accounting II	80.00	60.00
Algebra	80.55	60.45	Architecture	65.65	49.25
Anthropology*	41.95	31.50	Art History	69.35	50.25
Art. Intell.	67.00	50.25	Graphics Design*	50.65	36.00
Astronomy	52.00	39.00	Banking	60.00	45.00
Biology	94.00	70.50	Biology	75.95	56.95
Bus. Management	67.70	50.80	Business	75.00	56.25
Bus. Statistics	75.35	56.55	Heredity*	62.00	46.50
Calculus (Hvd)*	72.70	54.55	Marketing	41.95	31.45
Calculus (Stewart)	102.95	77.25	Chemistry*	61.00	45.75
Chinese*	27.95	21.00	Systems Analysis	75.65	53.00
Cooking	54.70	41.05	Music	33.35	25.00
Differential Equations	86.70	65.05	Nutrition	58.65	44.00
Discrete Math	78.00	58.50	Microeconomics*	51.35	38.50
Drawing*	57.35	43.05	Literature	44.65	33.50
Economics	86.00	64.50	Poetry*	28.35	21.25
Education	73.00	54.75	Programming*	44.00	33.00
Elementary Statistics	76.00	57.00	Engineering	80.00	60.00
Engineering	88.00	66.00	French	53.35	40.00
Engineering Design	45.60	36.00	Government	57.55	43.15
Finance	77.00	57.75	Sexuality	61.00	45.75
History*	26.70	22.00	History	63.60	47.70
History*	24.00	18.00	Italian	59.35	44.50
Inorganic Chemistry	92.00	69.00	Precalculus	64.55	48.40
Literature*	50.70	38.05	Calculus*	60.00	45.00
Marketing	82.00	61.50	Parasitology	38.45	28.85
Micro Biology	89.35	67.05	Philosophy	53.35	40.00
Music*	38.70	29.05	Physics	88.00	66.00
Nutrition*	60.00	45.00	Psychology	77.80	58.35
Organic Chemistry	79.20	59.40	Gerontology	61.00	45.75
Organic Chemistry*	48.70	36.55	Speech*	39.15	27.85
Photography*	60.00	45.00	Social Problems	50.15	37.60
Physics	95.00	71.25	Typographic Design*	42.65	32.00
Prob. Statistics	92.70	69.55	Writing*	35.00	26.25
Psychology	70.00	52.50			
Public Speaking*	41.60	31.20			
Religion*	18.95	14.25			
Russian*	28.00	21.00			
Sociology*	34.95	26.25			

Data Set 3: Diamonds

Price is in dollars. Depth is 100 times the ratio of height to diameter. Table is size of the upper flat surface. (Depth and table determine "cut.") Color indices are on a standard scale, with 1 = colorless and increasing numbers indicating more yellow. On the clarity scale, 1 = flawless and 6 indicates inclusions that can be seen by eye.

Excel: Workbook name is DIAMONDS.xls.
STATDISK: Variable names are DMDPRICE, DMDCARAT, DMDDEPTH, DMDTABLE, DMDCOLOR, and DMDCLRTY.
Minitab: Worksheet name is DIAMONDS.MTW.
TI-83 Plus and text files: PRICE, CARAT, DEPTH, TABLE, COLOR, and CLRTY.

PRICE	CARAT	DEPTH	TABLE	COLOR	CLARITY
6958	1.00	60.5	65	3	4
5885	1.00	59.2	65	5	4
6333	1.01	62.3	55	4	4
4299	1.01	64.4	62	5	5
9589	1.02	63.9	58	2	3
6921	1.04	60.0	61	4	4
4426	1.04	62.0	62	5	5
6885	1.07	63.6	61	4	3
5826	1.07	61.6	62	5	5
3670	1.11	60.4	60	9	4
7176	1.12	60.2	65	2	3
7497	1.16	59.5	60	5	3
5170	1.20	62.6	61	6	4
5547	1.23	59.2	65	7	4
18596	1.25	61.2	61	1	2
7521	1.29	59.6	59	6	2
7260	1.50	61.1	65	6	4
8139	1.51	63.0	60	6	4
12196	1.67	58.7	64	3	5
14998	1.72	58.5	61	4	3
9736	1.76	57.9	62	8	2
9859	1.80	59.6	63	5	5
12398	1.88	62.9	62	6	2
25322	2.03	60.1	62	2	3
11008	2.03	62.0	63	8	3
38794	2.06	58.2	63	2	2
66780	3.00	63.3	62	1	3
46769	4.01	57.1	51	3	4
28800	4.01	63.0	63	6	5
28868	4.05	59.3	60	7	4

Data Set 4: Weights (in grams) of Domino Sugar Packets

Excel: Workbook name is SUGAR.xls.
STATDISK: Variable name is SUGAR.
Minitab: Worksheet name is SUGAR.MTW.
TI-83 Plus and text file: SUGAR

3.647	3.638	3.635	3.645	3.521	3.617	3.666
3.588	3.545	3.590	3.621	3.532	3.511	3.516
3.531	3.678	3.643	3.583	3.723	3.673	3.588
3.600	3.611	3.580	3.667	3.506	3.632	3.450
3.660	3.569	3.573	3.526	3.494	3.601	3.604
3.407	3.522	3.598	3.585	3.577	3.522	3.464
3.604	3.508	3.718	3.635	3.643	3.507	3.687
3.582	3.622	3.654	3.482	3.494	3.475	3.492
3.542	3.625	3.688	3.468	3.639	3.582	3.491
3.535	3.548	3.671	3.665	3.726	3.576	3.725

Data Set 5: Weights (in pounds) of Discarded Household Garbage for One Week

HHSIZE = household size
METAL = weight of discarded metals
PAPER = weight of discarded paper goods
PLAS = weight of discarded plastic goods
GLASS = weight of discarded glass products
FOOD = weight of discarded food items
YARD = weight of discarded yard waste
TEXT = weight of discarded textile goods
OTHER = weight of discarded goods not included in the above categories
TOTAL = total weight of discarded materials

Excel: Workbook name is GARBAGE.xls.
STATDISK: Variable names are HHSIZE, METAL, PAPER, PLAS, GLASS, FOOD, YARD, TEXT, OTHER, and TOTAL.
Minitab: Worksheet name is GARBAGE.MTW.
TI-83 Plus and text files: HHSIZ, METAL, PAPER, PLAS, GLASS, FOOD, YARD, TEXT, OTHER, and TOTAL.

HOUSEHOLD	HHSIZE	METAL	PAPER	PLAS	GLASS	FOOD	YARD	TEXT	OTHER	TOTAL
1	2	1.09	2.41	0.27	0.86	1.04	0.38	0.05	4.66	10.76
2	3	1.04	7.57	1.41	3.46	3.68	0.00	0.46	2.34	19.96
3	3	2.57	9.55	2.19	4.52	4.43	0.24	0.50	3.60	27.60
4	6	3.02	8.82	2.83	4.92	2.98	0.63	2.26	12.65	38.11
5	4	1.50	8.72	2.19	6.31	6.30	0.15	0.55	2.18	27.90
6	2	2.10	6.96	1.81	2.49	1.46	4.58	0.36	2.14	21.90
7	1	1.93	6.83	0.85	0.51	8.82	0.07	0.60	2.22	21.83
8	5	3.57	11.42	3.05	5.81	9.62	4.76	0.21	10.83	49.27
9	6	2.32	16.08	3.42	1.96	4.41	0.13	0.81	4.14	33.27
10	4	1.89	6.38	2.10	17.67	2.73	3.86	0.66	0.25	35.54
11	4	3.26	13.05	2.93	3.21	9.31	0.70	0.37	11.61	44.44
12	7	3.99	11.36	2.44	4.94	3.59	13.45	4.25	1.15	45.17
13	3	2.04	15.09	2.17	3.10	5.36	0.74	0.42	4.15	33.07
14	5	0.99	2.80	1.41	1.39	1.47	0.82	0.44	1.03	10.35
15	6	2.96	6.44	2.00	5.21	7.06	6.14	0.20	14.43	44.44
16	2	1.50	5.86	0.93	2.03	2.52	1.37	0.27	9.65	24.13
17	4	2.43	11.08	2.97	1.74	1.75	14.70	0.39	2.54	37.60
18	4	2.97	12.43	2.04	3.99	5.64	0.22	2.47	9.20	38.96
19	3	1.42	6.05	0.65	6.26	1.93	0.00	0.86	0.00	17.17
20	3	3.60	13.61	2.13	3.52	6.46	0.00	0.96	1.32	31.60
21	2	4.48	6.98	0.63	2.01	6.72	2.00	0.11	0.18	23.11
22	2	1.36	14.33	1.53	2.21	5.76	0.58	0.17	1.62	27.56
23	4	2.11	13.31	4.69	0.25	9.72	0.02	0.46	0.40	30.96
24	1	0.41	3.27	0.15	0.09	0.16	0.00	0.00	0.00	4.08

(continued)

Data Set 5 (*continued*)

HOUSEHOLD	HHSIZE	METAL	PAPER	PLAS	GLASS	FOOD	YARD	TEXT	OTHER	TOTAL
25	4	2.02	6.67	1.45	6.85	5.52	0.00	0.68	0.03	23.22
26	6	3.27	17.65	2.68	2.33	11.92	0.83	0.28	4.03	42.99
27	11	4.95	12.73	3.53	5.45	4.68	0.00	0.67	19.89	51.90
28	3	1.00	9.83	1.49	2.04	4.76	0.42	0.54	0.12	20.20
29	4	1.55	16.39	2.31	4.98	7.85	2.04	0.20	1.48	36.80
30	3	1.41	6.33	0.92	3.54	2.90	3.85	0.03	0.04	19.02
31	2	1.05	9.19	0.89	1.06	2.87	0.33	0.01	0.03	15.43
32	2	1.31	9.41	0.80	2.70	5.09	0.64	0.05	0.71	20.71
33	2	2.50	9.45	0.72	1.14	3.17	0.00	0.02	0.01	17.01
34	4	2.35	12.32	2.66	12.24	2.40	7.87	4.73	0.78	45.35
35	6	3.69	20.12	4.37	5.67	13.20	0.00	1.15	1.17	49.37
36	2	3.61	7.72	0.92	2.43	2.07	0.68	0.63	0.00	18.06
37	2	1.49	6.16	1.40	4.02	4.00	0.30	0.04	0.00	17.41
38	2	1.36	7.98	1.45	6.45	4.27	0.02	0.12	2.02	23.67
39	2	1.73	9.64	1.68	1.89	1.87	0.01	1.73	0.58	19.13
40	2	0.94	8.08	1.53	1.78	8.13	0.36	0.12	0.05	20.99
41	3	1.33	10.99	1.44	2.93	3.51	0.00	0.39	0.59	21.18
42	3	2.62	13.11	1.44	1.82	4.21	4.73	0.64	0.49	29.06
43	2	1.25	3.26	1.36	2.89	3.34	2.69	0.00	0.16	14.95
44	2	0.26	1.65	0.38	0.99	0.77	0.34	0.04	0.00	4.43
45	3	4.41	10.00	1.74	1.93	1.14	0.92	0.08	4.60	24.82
46	6	3.22	8.96	2.35	3.61	1.45	0.00	0.09	1.12	20.80
47	4	1.86	9.46	2.30	2.53	6.54	0.00	0.65	2.45	25.79
48	4	1.76	5.88	1.14	3.76	0.92	1.12	0.00	0.04	14.62
49	3	2.83	8.26	2.88	1.32	5.14	5.60	0.35	2.03	28.41
50	3	2.74	12.45	2.13	2.64	4.59	1.07	0.41	1.14	27.17
51	10	4.63	10.58	5.28	12.33	2.94	0.12	2.94	15.65	54.47
52	3	1.70	5.87	1.48	1.79	1.42	0.00	0.27	0.59	13.12
53	6	3.29	8.78	3.36	3.99	10.44	0.90	1.71	13.30	45.77
54	5	1.22	11.03	2.83	4.44	3.00	4.30	1.95	6.02	34.79
55	4	3.20	12.29	2.87	9.25	5.91	1.32	1.87	0.55	37.26
56	7	3.09	20.58	2.96	4.02	16.81	0.47	1.52	2.13	51.58
57	5	2.58	12.56	1.61	1.38	5.01	0.00	0.21	1.46	24.81
58	4	1.67	9.92	1.58	1.59	9.96	0.13	0.20	1.13	26.18
59	2	0.85	3.45	1.15	0.85	3.89	0.00	0.02	1.04	11.25
60	4	1.52	9.09	1.28	8.87	4.83	0.00	0.95	1.61	28.15
61	2	1.37	3.69	0.58	3.64	1.78	0.08	0.00	0.00	11.14
62	2	1.32	2.61	0.74	3.03	3.37	0.17	0.00	0.46	11.70

Data provided by Masakuza Tani, the Garbage Project, University of Arizona.

Data Set 6: Body Temperatures (in degrees Fahrenheit) of Healthy Adults

This is the only Appendix B data set not available on disk.

SUBJECT	AGE	SEX	SMOKE	Temperature Day 1 8 AM	12 AM	Temperature Day 2 8 AM	12 AM
1	22	M	Y	98.0	98.0	98.0	98.6
2	23	M	Y	97.0	97.6	97.4	—
3	22	M	Y	98.6	98.8	97.8	98.6
4	19	M	N	97.4	98.0	97.0	98.0
5	18	M	N	98.2	98.8	97.0	98.0
6	20	M	Y	98.2	98.8	96.6	99.0
7	27	M	Y	98.2	97.6	97.0	98.4
8	19	M	Y	96.6	98.6	96.8	98.4
9	19	M	Y	97.4	98.6	96.6	98.4
10	24	M	N	97.4	98.8	96.6	98.4
11	35	M	Y	98.2	98.0	96.2	98.6
12	25	M	Y	97.4	98.2	97.6	98.6
13	25	M	N	97.8	98.0	98.6	98.8
14	35	M	Y	98.4	98.0	97.0	98.6
15	21	M	N	97.6	97.0	97.4	97.0
16	33	M	N	96.2	97.2	98.0	97.0
17	19	M	Y	98.0	98.2	97.6	98.8
18	24	M	Y	—	—	97.2	97.6
19	18	F	N	—	—	97.0	97.7
20	22	F	Y	—	—	98.0	98.8
21	20	M	Y	—	—	97.0	98.0
22	30	F	Y	—	—	96.4	98.0
23	29	M	N	—	—	96.1	98.3
24	18	M	Y	—	—	98.0	98.5
25	31	M	Y	—	98.1	96.8	97.3
26	28	F	Y	—	98.2	98.2	98.7
27	27	M	Y	—	98.5	97.8	97.4
28	21	M	Y	—	98.5	98.2	98.9
29	30	M	Y	—	99.0	97.8	98.6
30	27	M	N	—	98.0	99.0	99.5
31	32	M	Y	—	97.0	97.4	97.5
32	33	M	Y	—	97.3	97.4	97.3
33	23	M	Y	—	97.3	97.5	97.6
34	29	M	Y	—	98.1	97.8	98.2
35	25	M	Y	—	—	97.9	99.6
36	31	M	N	—	97.8	97.8	98.7
37	25	M	Y	—	99.0	98.3	99.4
38	28	M	N	—	97.6	98.0	98.2
39	30	M	Y	—	97.4	—	98.0
40	33	M	Y	—	98.0	—	98.6
41	28	M	Y	98.0	97.4	—	98.6
42	22	M	Y	98.8	98.0	—	97.2
43	21	F	Y	99.0	—	—	98.4
44	30	M	N	—	98.6	—	98.6

(continued)

Data Set 6 (*continued*)

SUBJECT	AGE	SEX	SMOKE	Temperature Day 1 8 AM	12 AM	Temperature Day 2 8 AM	12 AM
45	22	M	Y	—	98.6	—	98.2
46	22	F	N	98.0	98.4	—	98.0
47	20	M	Y	—	97.0	—	97.8
48	19	M	Y	—	—	—	98.0
49	33	M	N	—	98.4	—	98.4
50	31	M	Y	99.0	99.0	—	98.6
51	26	M	N	—	98.0	—	98.6
52	18	M	N	—	—	—	97.8
53	23	M	N	—	99.4	—	99.0
54	28	M	Y	—	—	—	96.5
55	19	M	Y	—	97.8	—	97.6
56	21	M	N	—	—	—	98.0
57	27	M	Y	—	98.2	—	96.9
58	29	M	Y	—	99.2	—	97.6
59	38	M	N	—	99.0	—	97.1
60	29	F	Y	—	97.7	—	97.9
61	22	M	Y	—	98.2	—	98.4
62	22	M	Y	—	98.2	—	97.3
63	26	M	Y	—	98.8	—	98.0
64	32	M	N	—	98.1	—	97.5
65	25	M	Y	—	98.5	—	97.6
66	21	F	N	—	97.2	—	98.2
67	25	M	Y	—	98.5	—	98.5
68	24	M	Y	—	99.2	97.0	98.8
69	25	M	Y	—	98.3	97.6	98.7
70	35	M	Y	—	98.7	97.5	97.8
71	23	F	Y	—	98.8	98.8	98.0
72	31	M	Y	—	98.6	98.4	97.1
73	28	M	Y	—	98.0	98.2	97.4
74	29	M	Y	—	99.1	97.7	99.4
75	26	M	Y	—	97.2	97.3	98.4
76	32	M	N	—	97.6	97.5	98.6
77	32	M	Y	—	97.9	97.1	98.4
78	21	F	Y	—	98.8	98.6	98.5
79	20	M	Y	—	98.6	98.6	98.6
80	24	F	Y	—	98.6	97.8	98.3
81	21	F	Y	—	99.3	98.7	98.7
82	28	M	Y	—	97.8	97.9	98.8
83	27	F	N	98.8	98.7	97.8	99.1
84	28	M	N	99.4	99.3	97.8	98.6
85	29	M	Y	98.8	97.8	97.6	97.9
86	19	M	N	97.7	98.4	96.8	98.8
87	24	M	Y	99.0	97.7	96.0	98.0
88	29	M	N	98.1	98.3	98.0	98.7
89	25	M	Y	98.7	97.7	97.0	98.5
90	27	M	N	97.5	97.1	97.4	98.9
91	25	M	Y	98.9	98.4	97.6	98.4
92	21	M	Y	98.4	98.6	97.6	98.6
93	19	M	Y	97.2	97.4	96.2	97.1

(*continued*)

Data Set 6 (*continued*)

SUBJECT	AGE	SEX	SMOKE	Temperature Day 1 8 AM	Temperature Day 1 12 AM	Temperature Day 2 8 AM	Temperature Day 2 12 AM
94	27	M	Y	—	—	96.2	97.9
95	32	M	N	98.8	96.7	98.1	98.8
96	24	M	Y	97.3	96.9	97.1	98.7
97	32	M	Y	98.7	98.4	98.2	97.6
98	19	F	Y	98.9	98.2	96.4	98.2
99	18	F	Y	99.2	98.6	96.9	99.2
100	27	M	N	—	97.0	—	97.8
101	34	M	Y	—	97.4	—	98.0
102	25	M	N	—	98.4	—	98.4
103	18	M	N	—	97.4	—	97.8
104	32	M	Y	—	96.8	—	98.4
105	31	M	Y	—	98.2	—	97.4
106	26	M	N	—	97.4	—	98.0
107	23	M	N	—	98.0	—	97.0

Data provided by Dr. Steven Wasserman, Dr. Philip Mackowiak, and Dr. Myron Levine of the University of Maryland.

Data Set 7: Bears (wild bears anesthetized)

Age: Months
Month: Month of measurement (1 = Jan., 2 = Feb., etc.)
Sex: 1 = male, 2 = female
HEADLEN: Length (inches) of head
HEADWTH: Width (inches) of head
NECK: Distance (inches) around neck
LENGTH: Length (inches) of body
CHEST: Distance (inches) around the chest
WEIGHT: Measured weight (pounds)

Excel: Workbook name is BEARS.xls.
STATDISK: Variable names are BEARAGE, MONTH, BEARSEX, HEADLEN, HEADWTH, NECK, BEARLEN, CHEST, BEARWT

Minitab: Worksheet name is BEARS.MTW.

TI-83 Plus and text files: BAGE, BMNTH, BSEX, BHDLN, BHDWD, BNECK, BLEN, BCHST, and BWGHT.

AGE	MONTH	SEX	HEADLEN	HEADWTH	NECK	LENGTH	CHEST	WEIGHT
19	7	1	11.0	5.5	16.0	53.0	26.0	80
55	7	1	16.5	9.0	28.0	67.5	45.0	344
81	9	1	15.5	8.0	31.0	72.0	54.0	416
115	7	1	17.0	10.0	31.5	72.0	49.0	348
104	8	2	15.5	6.5	22.0	62.0	35.0	166
100	4	2	13.0	7.0	21.0	70.0	41.0	220
56	7	1	15.0	7.5	26.5	73.5	41.0	262
51	4	1	13.5	8.0	27.0	68.5	49.0	360
57	9	2	13.5	7.0	20.0	64.0	38.0	204
53	5	2	12.5	6.0	18.0	58.0	31.0	144
68	8	1	16.0	9.0	29.0	73.0	44.0	332
8	8	1	9.0	4.5	13.0	37.0	19.0	34
44	8	2	12.5	4.5	10.5	63.0	32.0	140
32	8	1	14.0	5.0	21.5	67.0	37.0	180
20	8	2	11.5	5.0	17.5	52.0	29.0	105
32	8	1	13.0	8.0	21.5	59.0	33.0	166
45	9	1	13.5	7.0	24.0	64.0	39.0	204
9	9	2	9.0	4.5	12.0	36.0	19.0	26
21	9	1	13.0	6.0	19.0	59.0	30.0	120
177	9	1	16.0	9.5	30.0	72.0	48.0	436
57	9	2	12.5	5.0	19.0	57.5	32.0	125
81	9	2	13.0	5.0	20.0	61.0	33.0	132
21	9	1	13.0	5.0	17.0	54.0	28.0	90
9	9	1	10.0	4.0	13.0	40.0	23.0	40
45	9	1	16.0	6.0	24.0	63.0	42.0	220
9	9	1	10.0	4.0	13.5	43.0	23.0	46
33	9	1	13.5	6.0	22.0	66.5	34.0	154
57	9	2	13.0	5.5	17.5	60.5	31.0	116
45	9	2	13.0	6.5	21.0	60.0	34.5	182
21	9	1	14.5	5.5	20.0	61.0	34.0	150
10	10	1	9.5	4.5	16.0	40.0	26.0	65
82	10	2	13.5	6.5	28.0	64.0	48.0	356
70	10	2	14.5	6.5	26.0	65.0	48.0	316
10	10	1	11.0	5.0	17.0	49.0	29.0	94
10	10	1	11.5	5.0	17.0	47.0	29.5	86
34	10	1	13.0	7.0	21.0	59.0	35.0	150
34	10	1	16.5	6.5	27.0	72.0	44.5	270
34	10	1	14.0	5.5	24.0	65.0	39.0	202
58	10	2	13.5	6.5	21.5	63.0	40.0	202
58	10	1	15.5	7.0	28.0	70.5	50.0	365
11	11	1	11.5	6.0	16.5	48.0	31.0	79
23	11	1	12.0	6.5	19.0	50.0	38.0	148
70	10	1	15.5	7.0	28.0	76.5	55.0	446
11	11	2	9.0	5.0	15.0	46.0	27.0	62
83	11	2	14.5	7.0	23.0	61.5	44.0	236
35	11	1	13.5	8.5	23.0	63.5	44.0	212
16	4	1	10.0	4.0	15.5	48.0	26.0	60
16	4	1	10.0	5.0	15.0	41.0	26.0	64
17	5	1	11.5	5.0	17.0	53.0	30.5	114
17	5	2	11.5	5.0	15.0	52.5	28.0	76
17	5	2	11.0	4.5	13.0	46.0	23.0	48
8	8	2	10.0	4.5	10.0	43.5	24.0	29
83	11	1	15.5	8.0	30.5	75.0	54.0	514
18	6	1	12.5	8.5	18.0	57.3	32.8	140

Data from Gary Alt and Minitab, Inc.

Data Set 8: Cigarette Tar, Nicotine, and Carbon Monoxide

All measurements are in milligrams per cigarette, and all cigarettes are 100 mm long, filtered, and not menthol or light types.

Excel: Workbook name is CIGARET.xls.
STATDISK: Variable names are TAR, NICOTINE, CO.
Minitab: Worksheet name is CIGARET.MTW.
TI-83 Plus and text files: TAR, NICOT, and CO.

BRAND	TAR	NICOTINE	CO
American Filter	16	1.2	15
Benson & Hedges	16	1.2	15
Camel	16	1.0	17
Capri	9	0.8	6
Carlton	1	0.1	1
Cartier Vendome	8	0.8	8
Chelsea	10	0.8	10
GPC Approved	16	1.0	17
Hi-Lite	14	1.0	13
Kent	13	1.0	13
Lucky Strike	13	1.1	13
Malibu	15	1.2	15
Marlboro	16	1.2	15
Merit	9	0.7	11
Newport Stripe	11	0.9	15
Now	2	0.2	3
Old Gold	18	1.4	18
Pall Mall	15	1.2	15
Players	13	1.1	12
Raleigh	15	1.0	16
Richland	17	1.3	16
Rite	9	0.8	10
Silva Thins	12	1.0	10
Tareyton	14	1.0	17
Triumph	5	0.5	7
True	6	0.6	7
Vantage	8	0.7	11
Viceroy	18	1.4	15
Winston	16	1.1	18

Based on data from the Federal Trade Commission.

Data Set 9: Passive and Active Smoke

All values are measured levels of serum cotinine (in ng/ml), a metabolite of nicotine. (When nicotine is absorbed by the body, cotinine is produced.)

NOETS: Subjects are nonsmokers who have no environmental tobacco smoke (ETS) exposure at home or work.

ETS: Subjects are nonsmokers who are exposed to environmental tobacco smoke at home or work.

SMOKERS: Subjects report tobacco use.

   Excel: Workbook name is COTININE.xls.
STATDISK: Variable names are NOETS, ETS, and SMOKERS.
Minitab: Worksheet name is COTININE.MTW.
TI-83 Plus and text files: NOETS, ETS, and SMKRS.

NOETS	ETS	SMOKERS
0.03	0.03	0.08
0.05	0.07	0.14
0.05	0.08	0.27
0.06	0.08	0.44
0.06	0.09	0.51
0.06	0.09	1.78
0.07	0.10	2.55
0.08	0.11	3.03
0.08	0.12	3.44
0.08	0.12	4.98
0.08	0.14	6.87
0.08	0.17	11.12
0.08	0.20	12.58
0.08	0.23	13.73
0.08	0.27	14.42
0.08	0.28	18.22
0.09	0.30	19.28
0.09	0.33	20.16
0.10	0.37	23.67
0.10	0.38	25.00
0.10	0.44	25.39
0.10	0.49	29.41
0.12	0.51	30.71
0.13	0.51	32.54
0.13	0.68	32.56
0.15	0.82	34.21
0.15	0.97	36.73
0.16	1.12	37.73
0.16	1.23	39.48
0.18	1.37	48.58
0.19	1.40	51.21
0.20	1.67	56.74
0.20	1.98	58.69
0.20	2.33	72.37
0.22	2.42	104.54
0.24	2.66	114.49
0.25	2.87	145.43
0.28	3.13	187.34
0.30	3.54	226.82
0.32	3.76	267.83
0.32	4.58	328.46
0.37	5.31	388.74
0.41	6.20	405.28
0.46	7.14	415.38
0.55	7.25	417.82
0.69	10.23	539.62
0.79	10.83	592.79
1.26	17.11	688.36
1.58	37.44	692.51
8.56	61.33	983.41

Data are based in measurements from the National Health and Nutrition Examination Survey (National Institutes of Health).

Data Set 10: Weights (in grams) of a Sample of M&M Plain Candies

Excel: Workbook name is M&M.xls.
STATDISK: Variable names are RED, ORANGE, YELLOW,
         BROWN, BLUE, GREEN.
Minitab: Worksheet name is M&M.MTW.
TI-83 Plus and text files: RED, ORNG, YLLW, BROWN, BLUE, and GREEN.

RED	ORANGE	YELLOW	BROWN	BLUE	GREEN
0.870	0.903	0.906	0.932	0.838	0.911
0.933	0.920	0.978	0.860	0.875	1.002
0.952	0.861	0.926	0.919	0.870	0.902
0.908	1.009	0.868	0.914	0.956	0.930
0.911	0.971	0.876	0.914	0.968	0.949
0.908	0.898	0.968	0.904		0.890
0.913	0.942	0.921	0.930		0.902
0.983	0.897	0.893	0.871		
0.920		0.939	1.033		
0.936		0.886	0.955		
0.891		0.924	0.876		
0.924		0.910	0.856		
0.874		0.877	0.866		
0.908		0.879	0.858		
0.924		0.941	0.988		
0.897		0.879	0.936		
0.912		0.940	0.930		
0.888		0.960	0.923		
0.872		0.989	0.867		
0.898		0.900	0.965		
0.882		0.917	0.902		
		0.911	0.928		
		0.892	0.900		
		0.886	0.889		
		0.949	0.875		
		0.934	0.909		
			0.976		
			0.921		
			0.898		
			0.897		
			0.902		
			0.920		
			0.909		

Data Set 11: Old Faithful Geyser

Durations (in seconds), time intervals (in minutes) to the next eruption, and heights (in feet) of eruptions of the Old Faithful geyser in Yellowstone National Park.

Excel: Workbook name is OLDFAITH.xls.
STATDISK: Variable names are
      DURATION, INTERVAL,
      GEYSERHT.
Minitab: Worksheet name is OLDFAITH.MTW.
TI-83 Plus and text files: OFDTN, OFINT, and
      OFHT.

DURATION	INTERVAL	HEIGHT
240	86	140
237	86	154
122	62	140
267	104	140
113	62	160
258	95	140
232	79	150
105	62	150
276	94	160
248	79	155
243	86	125
241	85	136
214	86	140
114	58	155
272	89	130
227	79	125
237	83	125
238	82	139
203	84	125
270	82	140
218	78	140
226	91	135
250	89	141
245	79	140
120	57	139
267	100	110
103	62	140
270	87	135
241	70	140
239	88	135
233	82	140
238	83	139
102	56	100
271	81	105
127	74	130
275	102	135
140	61	131
264	83	135
134	73	153
268	97	155
124	67	140
270	90	150
249	84	153
237	82	120
235	81	138
228	78	135
265	89	145
120	69	130
275	98	136
241	79	150

Data courtesy of the National Park Service and research geologist Rick Hutchinson.

Data Set 12: Aluminum Cans

Axial loads (in pounds) of aluminum cans with 0.0109 in.
thickness and 0.0111 in. thickness

Excel: Workbook name is CANS.xls.
STATDISK: Variable names are CANS109 and CANS111.
Minitab: Worksheet name is CANS.MTW.
TI-83 Plus and text files: CN109 and CN111.

Sample	Aluminum cans 0.0109 in. Load (pounds)							Sample	Aluminum cans 0.0111 in. Load (pounds)						
1	270	273	258	204	254	228	282	1	287	216	260	291	210	272	260
2	278	201	264	265	223	274	230	2	294	253	292	280	262	295	230
3	250	275	281	271	263	277	275	3	283	255	295	271	268	225	246
4	278	260	262	273	274	286	236	4	297	302	282	310	305	306	262
5	290	286	278	283	262	277	295	5	222	276	270	280	288	296	281
6	274	272	265	275	263	251	289	6	300	290	284	304	291	277	317
7	242	284	241	276	200	278	283	7	292	215	287	280	311	283	293
8	269	282	267	282	272	277	261	8	285	276	301	285	277	270	275
9	257	278	295	270	268	286	262	9	290	288	287	282	275	279	300
10	272	268	283	256	206	277	252	10	293	290	313	299	300	265	285
11	265	263	281	268	280	289	283	11	294	262	297	272	284	291	306
12	263	273	209	259	287	269	277	12	263	304	288	256	290	284	307
13	234	282	276	272	257	267	204	13	273	283	250	244	231	266	504
14	270	285	273	269	284	276	286	14	284	227	269	282	292	286	281
15	273	289	263	270	279	206	270	15	296	287	285	281	298	289	283
16	270	268	218	251	252	284	278	16	247	279	276	288	284	301	309
17	277	208	271	208	280	269	270	17	284	284	286	303	308	288	303
18	294	292	289	290	215	284	283	18	306	285	289	292	295	283	315
19	279	275	223	220	281	268	272	19	290	247	268	283	305	279	287
20	268	279	217	259	291	291	281	20	285	298	279	274	205	302	296
21	230	276	225	282	276	289	288	21	282	300	284	281	279	255	210
22	268	242	283	277	285	293	248	22	279	286	293	285	288	289	281
23	278	285	292	282	287	277	266	23	297	314	295	257	298	211	275
24	268	273	270	256	297	280	256	24	247	279	303	286	287	287	275
25	262	268	262	293	290	274	292	25	243	274	299	291	281	303	269

Data Set 13: Weights (in grams) of Quarters

Excel: Workbook name is QUARTERS.xls.
STATDISK: Variable name is QUARTERS.
Minitab: Worksheet name is QUARTERS.MTW.
TI-83 Plus and text files: QRTRS.

5.60	5.63	5.58	5.56	5.66	5.58	5.57	5.59	5.67	5.61	5.84
5.73	5.53	5.58	5.52	5.65	5.57	5.71	5.59	5.53	5.63	5.68
5.62	5.60	5.53	5.58	5.60	5.58	5.59	5.66	5.73	5.59	5.63
5.66	5.67	5.60	5.74	5.57	5.62	5.73	5.60	5.60	5.57	5.71
5.62	5.72	5.57	5.70	5.60	5.49					

Data Set 14: Survey of 100 Statistics Students

Sex: 1 = male, 2 = female

Age: Years

Height: Inches

Coins: Value (in cents) of coins in possession of respondent

Keys: Number of keys in possession of respondent

Credit: Number of credit cards in possession of respondent

Pulse: Number of heartbeats in one minute

Exercise: 1 = yes, 2 = no (for vigorous exercise consisting of at least 20 minutes at least twice a week)

Smoke: 1 = yes, 2 = no

Color: 1 = yes, 2 = no ("Are you color blind?")

Hand: 1 = left-handed, 2 = right-handed, 3 = ambidextrous

Excel: Workbook name is STATSURV.xls.
STATDISK: Variable names are STATSEX, STATAGE, STATHT, COINS, KEYS, CREDIT, PULSE, EXERCISE, SMOKE, COLOR, HAND.
Minitab: Worksheet name is STATSURV.MTW.
TI-83 Plus and text files: SVSEX, SVAGE, SVHT, SVCNS, SVKEY, SVCRD, SVPLS, SVEXR, SVSMK, SVCLR, and SVHND.

SEX	AGE	HEIGHT	COINS	KEYS	CREDIT	PULSE	EXERCISE	SMOKE	COLOR	HAND
2	19	64	0	3	0	97	2	2	2	2
2	28	67.5	100	5	0	88	1	2	2	2
1	19	68	0	0	1	69	1	2	1	2
1	20	70.5	23	4	2	67	1	2	2	3
2	18	65	35	5	5	83	1	2	2	2
2	17	63	185	6	0	77	1	2	2	2
1	18	75	0	3	0	66	2	2	2	2
2	48	64	0	3	0	60	2	2	2	3
2	19	68.75	43	3	0	78	2	2	2	3
2	17	57	35	3	0	73	1	1	2	1
2	35	63	250	10	2	8	1	2	2	2
2	18	64	178	5	10	67	1	1	2	2
1	19	72	10	2	1	55	1	2	2	2
2	28	67	90	5	0	72	1	1	2	2
2	24	62.5	0	8	14	82	1	1	2	1
2	30	63	200	1	10	70	1	2	2	2
1	21	69	0	5	0	47	1	2	2	1
1	19	68	40	4	2	63	1	1	2	2
1	19	68	73	2	0	52	2	1	2	2
1	24	68	20	2	1	55	1	2	2	2
2	22	5	500	4	1	67	1	2	2	2
1	21	69	0	2	1	75	2	2	2	1
1	19	69	0	3	3	76	1	2	2	2
2	19	60	35	10	0	60	2	2	2	2
1	20	69	130	3	0	84	1	2	2	2

(continued)

Data Set 14 (*continued*)

SEX	AGE	HEIGHT	COINS	KEYS	CREDIT	PULSE	EXERCISE	SMOKE	COLOR	HAND
1	30	73	62	10	1	40	1	2	2	2
1	33	74	5	7	8	64	1	2	2	2
1	19	67	0	3	0	72	2	2	2	2
1	18	70	0	5	1	72	1	2	2	2
1	20	70	0	3	5	75	1	2	1	2
2	18	76	0	3	0	80	2	2	2	2
1	20	68	32	2	4	63	1	2	2	2
1	50	72	74	8	4	72	2	2	2	2
2	20	65	14	4	4	90	1	1	2	2
1	18	68	25	2	0	70	2	2	2	2
2	18	64	0	2	0	100	2	2	2	2
2	18	64	25	1	0	69	1	2	2	2
1	22	68	0	5	5	64	1	2	2	2
2	21	64	27	2	2	80	2	2	2	2
1	41	72	76	2	0	60	2	1	2	2
1	18	68	160	3	0	66	2	1	2	2
1	21	68	34	26	0	78	1	2	2	2
2	17	60	75	3	0	60	2	1	2	2
2	40	64	20	10	5	68	1	2	2	2
1	19	74	0	1	3	72	1	2	2	2
1	19	69	0	5	0	60	1	2	2	2
2	28	68	453	5	7	88	1	1	2	2
2	28	64	0	3	0	58	1	2	2	2
2	19	63	79	6	0	88	1	2	2	1
2	41	63	100	6	3	80	2	2	2	2
1	18	71	25	2	1	61	1	2	2	2
1	18	73	181	6	0	67	2	2	2	2
1	21	71	72	3	4	60	1	1	2	2
1	18	73	0	5	0	80	1	2	2	2
1	22	69	75	12	5	60	1	1	2	2
2	22	69	0	3	5	80	1	2	2	2
1	21	72	0	4	0	68	1	1	2	2
2	26	60	25	15	0	78	1	2	2	2
1	21	72	97	2	0	54	1	2	2	2
1	20	54	0	5	5	81	2	2	2	2
2	19	65	0	8	0	67	2	2	2	2
1	22	66	30	8	4	70	1	2	2	2
1	20	76	0	6	1	63	1	2	2	2
1	19	71	0	7	0	90	1	2	1	2
1	36	73	18	11	4	70	2	2	2	2
1	20	71	0	5	1	69	2	2	1	2
1	19	71	50	7	0	69	2	1	2	2
2	18	67.5	0	4	0	75	2	1	2	2
2	19	64	0	3	1	80	1	1	2	2
1	52	71.7	51	4	5	92	2	2	2	2
2	41	68	800	7	4	72	2	2	2	2
1	20	69	0	3	3	63	1	1	2	2
1	20	72	85	3	1	60	1	2	2	2
2	30	63	111	5	0	78	2	2	2	2
1	21	73	77	5	0	77	1	2	2	2
1	21	70	35	5	4	71	1	1	2	2

(*continued*)

Data Set 14 (*continued*)

SEX	AGE	HEIGHT	COINS	KEYS	CREDIT	PULSE	EXERCISE	SMOKE	COLOR	HAND
2	34	65.5	300	4	2	15	2	2	2	2
1	19	69	36	5	0	83	2	1	2	2
1	20	69	0	2	0	80	1	2	2	2
1	20	69	0	1	1	71	1	2	2	2
2	20	66	45	4	0	86	2	2	2	2
2	36	64.5	116	3	8	65	1	2	2	2
1	19	68	52	4	2	70	1	2	2	2
2	20	67	358	7	7	76	1	2	2	2
1	19	71	0	4	0	78	1	1	2	2
1	19	72	15	10	1	63	1	2	2	3
1	19	71	1	0	0	52	1	1	2	2
1	25	71	25	4	1	78	2	2	2	1
2	29	64	26	6	3	92	2	2	2	2
1	19	81	0	4	3	48	1	2	2	2
2	17	67.5	0	3	0	68	1	2	2	2
2	19	68	0	8	1	85	2	2	2	1
1	24	73	0	14	0	64	1	2	2	2
2	19	63	0	3	4	65	2	2	2	2
2	18	64	25	3	0		1	2	2	2
2	23	69	0	3	26		2	1	2	1
1	19	60	50	4	0		1	2	2	2
1	19	72	83	6	2		1	1	2	2
2	21	67	50	3	8		1	2	2	2
1	20	74	0	6	0		2	1	2	1

Data Set 15: Movies

Excel: Workbook name is MOVIES.xls.
STATDISK: Variable names are MOVBUDG, MOVGROSS,
   MOVLEN, and MOVRATNG.
Minitab: Worksheet name is MOVIES.MTW.
TI-83 Plus and text files: MVBUD, MVGRS, MVLEN, and MVRAT.

Title	Year	Rating	Budget ($) in Millions	Gross ($) in Millions	Length in Minutes	Viewer Rating
Aliens	1986	R	18.5	81.843	137	8.2
Armageddon	1998	PG-13	140	194.125	144	6.7
As Good As It Gets	1997	PG-13	50	147.54	138	8.1
Braveheart	1995	R	72	75.6	177	8.3
Chasing Amy	1997	R	0.25	12.006	105	7.9
Contact	1997	PG	90	100.853	153	8.3
Dante's Peak	1997	PG-13	104	67.155	112	6.7
Deep Impact	1998	PG-13	75	140.424	120	6.4
Executive Decision	1996	R	55	68.75	129	7.3
Forrest Gump	1994	PG-13	55	329.691	142	7.7
Ghost	1990	PG-13	22	217.631	128	7.1
Gone with the Wind	1939	G	3.9	198.571	222	8.0
Good Will Hunting	1997	R	10	138.339	126	8.5
Grease	1978	PG	6	181.28	110	7.3
Halloween	1978	R	0.325	47	93	7.7
Hard Rain	1998	R	70	19.819	95	5.2
I Know What You Did Last Summer	1997	R	17	72.219	100	6.5
Independence Day	1996	PG-13	75	306.124	142	6.6
Indiana Jones and the Last Crusade	1989	PG-13	39	197.171	127	7.8
Jaws	1975	PG	12	260	124	7.8
Men in Black	1997	PG-13	90	250.147	98	7.4
Multiplicity	1996	PG-13	45	20.1	117	6.8
Pulp Fiction	1994	R	8	107.93	154	8.3
Raiders of the Lost Ark	1981	PG	20	242.374	115	8.3
Saving Private Ryan	1998	R	70	178.091	170	9.1
Schindler's List	1993	R	25	96.067	197	8.6
Scream	1996	R	15	103.001	111	7.7
Speed 2: Cruise Control	1997	PG-13	110	48.068	121	4.3
Terminator	1984	R	6.4	36.9	108	7.7
The American President	1995	PG-13	62	65	114	7.6
The Fifth Element	1997	PG-13	90	63.54	126	7.8
The Game	1997	R	50	48.265	128	7.6
The Man in the Iron Mask	1998	PG-13	35	56.876	132	6.5
Titanic	1997	PG-13	200	600.743	195	8.4
True Lies	1994	R	100	146.261	144	7.2
Volcano	1997	PG-13	90	47.474	102	5.8

Data Set 16: Homes Sold in Dutchess County

Excel: Workbook name is HOMES.xls.
STATDISK: Variable names are HOMESELL, HOMELIST,
        HOMELA, HOMERMS, HOMEBRS, HOMEBTHS,
        HOMEAGE, HOMEACRE, and HOMETAX.
Minitab: Worksheet name is HOMES.MTW.
TI-83 Plus and text files: HMSP, HMLST, HMLA, HMRMS,
        HMBRS, HMBTH, HMAGE, HMACR,
        and HMTAX.

Selling Price (thousands)	List Price (thousands)	Living Area (hundreds of sq. ft.)	Rooms	Bedrooms	Bathrooms	Age (years)	Acres	Taxes (dollars)
142.0	160	28	10	5	3	60	0.28	3167
175.0	180	18	8	4	1	12	0.43	4033
129.0	132	13	6	3	1	41	0.33	1471
138.0	140	17	7	3	1	22	0.46	3204
232.0	240	25	8	4	3	5	2.05	3613
135.0	140	18	7	4	3	9	0.57	3028
150.0	160	20	8	4	3	18	4.00	3131
207.0	225	22	8	4	2	16	2.22	5158
271.0	285	30	10	5	2	30	0.53	5702
89.0	90	10	5	3	1	43	0.30	2054
153.0	157	22	8	3	3	18	0.38	4127
86.5	90	16	7	3	1	50	0.65	1445
234.0	238	25	8	4	2	2	1.61	2087
105.5	116	20	8	4	1	13	0.22	2818
175.0	180	22	8	4	2	15	2.06	3917
165.0	170	17	8	4	2	33	0.46	2220
166.0	170	23	9	4	2	37	0.27	3498
136.0	140	19	7	3	1	22	0.63	3607
148.0	160	17	7	3	2	13	0.36	3648
151.0	153	19	8	4	2	24	0.34	3561
180.0	190	24	9	4	2	10	1.55	4681
293.0	305	26	8	4	3	6	0.46	7088
167.0	170	20	9	4	2	46	0.46	3482
190.0	193	22	9	5	2	37	0.48	3920
184.0	190	21	9	5	2	27	1.30	4162
157.0	165	20	8	4	2	7	0.30	3785
110.0	115	16	8	4	1	26	0.29	3103
135.0	145	18	7	4	1	35	0.43	3363
567.0	625	64	11	4	4	4	0.85	12192
180.0	185	20	8	4	2	11	1.00	3831
183.0	188	17	7	3	2	16	3.00	3564
185.0	193	20	9	3	2	56	6.49	3765
152.0	155	17	8	4	1	33	0.70	3361
148.0	153	13	6	3	2	22	0.39	3950
152.0	159	15	7	3	1	25	0.59	3055
146.0	150	16	7	3	1	31	0.36	2950
170.0	190	24	10	3	2	33	0.57	3346

(continued)

Data Set 16 (continued)

Selling Price (thousands)	List Price (thousands)	Living Area (hundreds of sq. ft.)	Rooms	Bedrooms	Bathrooms	Age (years)	Acres	Taxes (dollars)
127.0	130	20	8	4	1	65	0.40	3334
265.0	270	36	10	6	3	33	1.20	5853
157.0	163	18	8	4	2	12	1.13	3982
128.0	135	17	9	4	1	25	0.52	3374
110.0	120	15	8	4	2	11	0.59	3119
123.0	130	18	8	4	2	43	0.39	3268
212.0	230	39	12	5	3	202	4.29	3648
145.0	145	18	8	4	2	44	0.22	2783
129.0	135	10	6	3	1	15	1.00	2438
143.0	145	21	7	4	2	10	1.20	3529
247.0	252	29	9	4	2	4	1.25	4626
111.0	120	15	8	3	1	97	1.11	3205
133.0	145	26	7	3	1	42	0.36	3059

Data Set 17: Rainfall (in inches) in Boston for One Year

Excel: Workbook name is BOSTRAIN.xls.
STATDISK: Variable names are RAINMON, RAINTUES,
RAINWED, RAINTHUR, RAINFRI, RAINSAT,
and RAINSUN.
Minitab: Worksheet name is BOSTRAIN.MTW.
TI-83 Plus and text files: RNMON, RNTUE, RNWED, RNTHU,
RNFRI, RNSAT, and RNSUN.

MON	TUES	WED	THURS	FRI	SAT	SUN
0	0	0	0.04	0.04	0	0.05
0	0	0	0.06	0.03	0.1	0
0	0	0	0.71	0	0	0
0	0.44	0.14	0.04	0.04	0.64	0
0.05	0	0	0	0.01	0.05	0
0	0	0.64	0	0	0	0
0.01	0	0	0	0.3	0.05	0
0	0	0.01	0	0	0	0
0	0.01	0.01	0.16	0	0	0.09
0.12	0.06	0.18	0.39	0	0.1	0
0	0	0	0	0.78	0.49	0
0	0.02	0	0	0.01	0.17	0
1.41	0.65	0.31	0	0	0.54	0
0	0	0	0	0	0	0
0	0	0	0	0	0.4	0.28
0	0	0	0.3	0.87	0.49	0
0.47	0	0	0	0	0	0
0	0.09	0	0.24	0	0.05	0
0	0.14	0	0	0.04	0.07	0
0.92	0.36	0.02	0.09	0.27	0	0
0.01	0	0.06	0	0	0	0.27
0.01	0	0	0	0	0	0.01
0	0	0	0	0	0	0
0	0	0	0	0.71	0	0
0	0	0.27	0.08	0	0	0.33
0	0	0	0	0	0	0
0.03	0	0.08	0.14	0	0	0
0	0.11	0.06	0.02	0	0	0
0.01	0.05	0	0.01	0	0	0
0	0	0	0	0.12	0	0
0.11	0.03	0	0	0	0	0.44
0.01	0.01	0	0	0.11	0.18	0
0.49	0	0.64	0.01	0	0	0.01
0	0	0.08	0.85	0.01	0	0
0.01	0.02	0	0	0.03	0	0
0	0	0.12	0	0	0	0
0	0	0.01	0.04	0.26	0.04	0
0	0	0	0	0	0.4	0
0.12	0	0	0	0	0	0
0	0	0	0	0.24	0	0.23
0	0	0	0.02	0	0	0

(continued)

Data Set 17 (*continued*)

MON	TUES	WED	THURS	FRI	SAT	SUN
0	0	0	0.02	0	0	0
0.59	0	0	0	0	0.68	0
0	0.01	0	0	0	1.48	0.21
0.01	0	0	0	0.05	0.69	1.28
0	0	0	0	0.96	0	0.01
0	0	0	0	0	0.79	0.02
0.41	0	0.06	0.01	0	0	0.28
0	0	0	0.08	0.04	0	0
0	0	0	0	0	0	0
0	0.74	0	0	0	0	0
0.43	0.3	0	0.26	0	0.02	0.01
		0				

Data Set 18: Cars

CITY: City fuel consumption in mi/gal

HWY: Highway fuel consumption in mi/gal

WEIGHT: Weight of car in pounds

CYLINDER: Number of cylinders

DISPLACEMENT: Engine displacement in liters

MAN/AUT: M = manual transmission;
A = automatic transmission

GHG: Greenhouse gases emitted (in tons/yr)

NOX: Tailpipe emissions of NO_x (in lb/yr)

Excel: Worksheet name is CARS.xls.
STATDISK: Variable names are CARCITY, CARHWY,
            CARWT, CARCYL, CARDISP, CARGHG, and CARNOX.
Minitab: Worksheet name is CARS.MTW.
TI-83 Plus and text files: CRCTY, CRHWY, CRWT,
            CRCYL, CRDSP, CRGHG,
            and CRNOX.

CAR	CITY	HWY	WEIGHT	CYLINDER	DISPLACEMENT	MAN/AUT	GHG	NOX
Chev. Camaro	19	30	3545	6	3.8	M	12	34.4
Chev. Cavalier	23	31	2795	4	2.2	A	10	25.1
Dodge Neon	23	32	2600	4	2	A	10	25.1
Ford Taurus	19	27	3515	6	3	A	12	25.1
Honda Accord	23	30	3245	4	2.3	A	11	25.1
Lincoln Cont.	17	24	3930	8	4.6	A	14	25.1
Mercury Mystique	20	29	3115	6	2.5	A	12	34.4
Mitsubishi Eclipse	22	33	3235	4	2	M	10	25.1
Olds. Aurora	17	26	3995	8	4	A	13	34.4
Pontiac Grand Am	22	30	3115	4	2.4	A	11	25.1
Toyota Camry	23	32	3240	4	2.2	M	10	25.1
Cadillac DeVille	17	26	4020	8	4.6	A	13	34.4
Chev. Corvette	18	28	3220	8	5.7	M	12	34.4
Chrysler Sebring	19	27	3175	6	2.5	A	12	25.1
Ford Mustang	20	29	3450	6	3.8	M	12	34.4
BMW 3-Series	19	27	3225	6	2.8	A	12	34.4
Ford Crown Victoria	17	24	3985	8	4.6	A	14	25.1
Honda Civic	32	37	2440	4	1.6	M	8	25.1
Mazda Protege	29	34	2500	4	1.6	A	9	25.1
Hyundai Accent	28	37	2290	4	1.5	A	9	34.4

Data Set 19: Mark McGwire and Sammy Sosa Home-run Distances (in feet)

Excel: Workbook name is HOMERUNS.xls.
STATDISK: Variable names are MCGWIRE and SOSA.
Minitab: Worksheet name is HOMERUNS.MTW.
TI-83 Plus and text files: MCGWR and SOSA.

MCGWIRE		SOSA	
360	380	371	420
370	470	350	360
370	398	430	368
430	409	420	430
420	385	430	433
340	369	434	388
460	460	370	440
410	390	420	414
440	510	440	482
410	500	410	364
380	450	420	370
360	470	460	400
350	430	400	405
527	458	430	433
380	380	410	390
550	430	370	480
478	341	370	480
420	385	410	434
390	410	380	344
420	420	340	410
425	380	350	420
370	400	420	
480	440	410	
390	377	415	
430	370	430	
388		380	
423		380	
410		366	
360		500	
410		380	
450		390	
350		400	
450		364	
430		430	
461		450	
430		440	
470		365	
440		420	
400		350	
390		420	
510		400	
430		380	
450		380	
452		400	
420		370	

Data Set 20: Wristwatch Errors (in sec)

Excel: Workbook name is WATCH.xls.
STATDISK: Variable name is WATCH.
Minitab: Worksheet name is WATCH.MTW.
TI-83 Plus and text files: WATCH.

140	−125	105	−241
−85	41	186	−151
325	80	27	20
20	30	−65	36
305	211	168	115
205	−30	555	265
20	−12	324	361
−93	323	143	−56
282	190	188	−100
15	368	570	33

Appendix C: Excel References

TOPIC	EXCEL COMMANDS
Formulas	See Section 4-2.
Save file	File/Save As
Open file	File/Open
Close file	File/Close
Exit Excel	File/Exit
Print worksheet	File/Print
Install add-in	Tools/Add-Ins
Frequency table	Tools/Data Analysis/Histogram
Histograms	Tools/Data Analysis/Histogram
Frequency polygons	Insert/Chart/Line
Pareto charts	Insert/Chart/Column
Pie chart	Insert/Chart/Pie
Scatterplot	Insert/Chart/XY (Scatter)
Descriptive statistics	Tools/Data Analysis/Descriptive Statistics
z scores	*fx*/Statistical/STANDARDIZE
Percentiles	*fx*/Statistical/PERCENTILE
Quartiles	*fx*/Statistical/QUARTILE
Boxplots	DDXL/Charts and Plots/Boxplot
Random number	*fx*/Math & Trig/RANDBETWEEN
Factorials	*fx*/Math & Trig/FACT
Permutations	*fx*/Statistical/PERMUT
Combinations	*fx*/Math & Trig/COMBIN
Binomial probability	*fx*/Statistical/BINOMDIST
Poisson probability	*fx*/Statistical/POISSON
Probability from z score	*fx*/Statistical/NORMDIST
z score from probability	*fx*/Statistical/NORMINV
Normal quantile plot	DDXL/Charts and Plots/Normal Probability Plot
Confidence interval for 1 population mean	DDXL/Confidence Intervals/1 Var Z Interval or DDXL/Confidence Intervals/1 Var t Interval
Confidence interval for 1 proportion	DDXL/Confidence Intervals/Summ 1 Var Prop Interval
Hypothesis test for 1 mean	DDXL/Hypothesis Tests/1 Var t Test
Hypothesis test for 1 proportion	DDXL/Hypothesis Tests/Summ 1 Var Prop Test
Hypothesis test for 2 independent means	DDXL/Hypothesis Tests/2 Var t Test or Tools/Data Analysis/2-Sample cases
Confidence interval for 2 means	DDXL/Confidence Intervals/2 Var t Interval
Hypothesis test for matched pairs	DDXL/Hypothesis Tests/Paired t Test
Confidence interval for matched pairs	DDXL/Confidence Intervals/Paired t Interval

Hypothesis test for 2 proportions	DDXL/Hypothesis Tests/Summ 2 Var Prop Test
Confidence interval for 2 proportions	DDXL/Confidence Intervals/Summ 2 Var Prop Interval
Hypothesis test for 2 variances	Tools/Data Analysis/F-Test Two-Sample for Variances
Correlation	DDXL/Regression/Correlation or fx/Statistical/CORREL
Regression line	DDXL/Regression/Simple Regression or Tools/Data Analysis/Regression
Multiple regression	Tools/Data Analysis/Regression
Modeling	Insert/Chart/XY (Scatter) and use option of Add Trendline
ANOVA (one-way)	Tools/Data Analysis/Anova Single Factor
ANOVA (two-way)	Tools/Data Analysis/Anova Two Factor with Replication
ANOVA (two-way) with one entry per cell	Tools/Data Analysis/Anova Two Factor without Replication
Run charts	Insert/Chart/Line
R charts	DDXL/Process Control/Mu, R, s Control Charts
$\bar{x}$ charts	DDXL/Process Control/ Mu, R, s Control Charts
p charts	DDXL/Process Control/Summ Prop Control Chart

Appendix D: Glossary

Absolute deviation The measure of variation equal to the sum of the deviations of each value from the mean, divided by the number of values

Acceptance sampling Sampling items without replacement and rejecting the whole batch based on the number of defects obtained

Actual odds against The ratio $P(\overline{A})/P(A)$, usually expressed in the form of $a{:}b$ (or "a to b")

Actual odds in favor The reciprocal of the actual odds against an event

Addition rule Rule for determining the probability that, on a single trial, either event A occurs, or event B occurs, or they both occur

Adjusted coefficient of determination Multiple coefficient of determination R^2 modified to account for the number of variables and sample size

Alpha (α) Symbol used to represent the probability of a type I error. *See also* significance level.

Alternative hypothesis Statement that is equivalent to the negation of the null hypothesis; denoted by H_1

Analysis of variance Method of analyzing population variances in order to test hypotheses about means of populations

ANOVA *See* analysis of variance.

Arithmetic mean Sum of a set of values divided by the number of values; usually referred to as the mean

Assignable variation Type of variation in a process that results from causes that can be identified

Attribute data Data that can be separated into different categories distinguished by some nonnumeric characteristic

Average Any one of several measures designed to reveal the center of a collection of data

Beta (β) Symbol used to represent the probability of a type II error

Bimodal Having two modes

Binomial experiment Experiment with a fixed number of independent trials, where each outcome falls into exactly one of two categories

Binomial probability formula Expression used to calculate probabilities in a binomial experiment (see Formula 4-5 in Section 4-3)

Bivariate data Data arranged in pairs

Bivariate normal distribution Distribution of paired data in which, for any fixed value of one variable, the values of the other variable are normally distributed

Blinding Procedure used in experiments whereby the subject doesn't know whether he or she is receiving a treatment or a placebo

Block In analysis of variance, a group of similar individuals

Box-and-whisker diagram *See* boxplot.

Boxplot Graphical representation of the spread of a set of data

Categorical data Data that can be separated into different categories that are distinguished by some nonnumeric characteristic

Cell Category used to separate qualitative (or attribute) data

Census Collection of data from every element in a population

Center line Line used in a control chart to represent a central value of the characteristic measurements

Central limit theorem Theorem stating that sample means tend to be normally distributed with mean μ and standard deviation $\sigma/\sqrt{n}$

Centroid The point $(\overline{x}, \overline{y})$ determined from a collection of bivariate data

Chebyshev's theorem Theorem that uses the standard deviation to provide information about the distribution of data

Chi-square distribution A continuous probability distribution (first introduced in Section 6-6)

Class boundaries Values obtained from a frequency table by increasing the upper class limits and decreasing the lower class limits by the same amount so that there are no gaps between consecutive classes

Classical approach to probability Approach in which the probability of an event is determined by dividing the number of ways the event can occur by the total number of possible outcomes

Classical method of testing hypotheses Method of testing hypotheses based on a comparison of the test statistic and critical values

Class midpoint In a class of a frequency table, the value midway between the lower class limit and the upper class limit

Class width The difference between two consecutive lower class limits in a frequency table

Cluster sampling Dividing the population area into sections (or clusters), then randomly selecting a few of those sections, and then choosing *all* the members from those selected sections

Coefficient of determination Amount of the variation in y that is explained by the regression line

Combinations rule Rule for determining the number of different combinations of selected items

Complement of an event All outcomes in which the original event does not occur

Completely randomized design Procedure in an experiment whereby each element is given the same chance of belonging to the different categories or treatments

Compound event Combination of simple events

Conditional probability The probability of an event, given that some other event has already occurred

Confidence coefficient Probability that a population parameter is contained within a particular confidence interval; also called level of confidence or degree of confidence

Confidence interval Range of values used to estimate some population parameter with a specific level of confidence; also called an interval estimate

Confidence interval limits Two numbers that are used as the high and low boundaries of a confidence interval

Confounding A situation that occurs when the effects from two or more variables cannot be distinguished from each other

Contingency table Table of observed frequencies where the rows correspond to one variable of classification and the columns correspond to another variable of classification; also called a two-way table

Continuity correction Adjustment made when a discrete random variable is being approximated by a continuous random variable (Section 5-6)

Continuous data Data resulting from infinitely many possible values that correspond to some continuous scale that covers a range of values without gaps, interruptions, or jumps.

Continuous random variable A random variable with infinite values that can be associated with points on a continuous line interval

Control chart Any one of several types of charts (Chapter 12) depicting some characteristic of a process in order to determine whether there is statistical stability

Control group A group of subjects in an experiment who are not given a particular treatment

Control limit Boundary used in a control chart for identifying unusual points

Convenience sampling Sampling in which data are selected because they are readily available

Correlation Statistical association between two variables

Correlation coefficient Measurement of the strength of the relationship between two variables

Critical region The set of all values of the test statistic that would cause rejection of the null hypothesis

Critical value Value separating the critical region from the values of the test statistic that would not lead to rejection of the null hypothesis

Cumulative frequency Sum of the frequencies for a class and all preceding classes

Cumulative frequency table Frequency table in which each class and frequency represents cumulative data up to and including that class

Data Numbers or information describing some characteristic

Decile The nine values that divide sorted data into ten groups with approximately 10% of the values in each group

Degree of confidence Probability that a population parameter is contained within a particular confidence interval; also called level of confidence

Degrees of freedom Number of values that are free to vary after certain restrictions have been imposed on all values

Denominator degrees of freedom Degrees of freedom corresponding to the denominator of the F test statistic

Density curve Graph of a continuous probability distribution

Dependent events Events for which the occurrence of any one event affects the probabilities of the occurrences of the other events

Dependent sample Sample whose values are related to the values in another sample

Dependent variable y variable in a regression or multiple regression equation

Descriptive statistics Methods used to summarize the key characteristics of known data

Deviation Amount of difference between a value and the mean; expressed as $x - \overline{x}$

Discrete data Data with the property that the number of possible values is either a finite number or a "countable" number, which results in 0 possibilities, or 1 possibility, or 2 possibilities, and so on

Discrete random variable Random variable with either a finite number of values or a countable number of values

Distribution-free tests Tests not requiring a particular distribution, such as the normal distribution. *See also* nonparametric tests.

Dotplot Graph in which each data value is plotted as a point (or dot) along a scale of values

Double-blind Procedure used in an experiment whereby the subject doesn't know whether he or she is receiving a treatment or placebo, and the person administering the treatment also does not know

Efficiency Measure of the sensitivity of a nonparametric test in comparison to a corresponding parametric test

Empirical rule Rule that uses standard deviation to provide information about data with a bell-shaped distribution (Section 2-5)

Estimate Specific value or range of values used to approximate some population parameter

Estimator Sample statistic (such as the sample mean $\overline{x}$) used to approximate a population parameter

Event Result or outcome of an experiment

Expected frequency Theoretical frequency for a cell of a contingency table or multinomial table

Expected value For a discrete random variable, the mean value of the outcomes

Experiment Application of some treatment followed by observation of its effects on the subjects

Experimental units Subjects in an experiment

Explained deviation For one pair of values in a collection of bivariate data, the difference between the predicted y value and the mean of the y values

Explained variation Sum of the squares of the explained deviations for all pairs of bivariate data in a sample

Exploratory data analysis (EDA) Branch of statistics emphasizing the investigation of data

Factor In analysis of variance, a property or characteristic that allows us to distinguish the different populations from one another

Factorial rule Rule stating that n different items can be arranged $n!$ different ways

F distribution Continuous probability distribution first introduced in Section 8-5

Finite population correction factor Factor for correcting the standard error of the mean when a sample size exceeds 5% of the size of a finite population

Five-number summary Minimum value, maximum value, median, and the first and third quartiles of a set of data

Fractiles Numbers that partition data into parts that are approximately equal in size

Frequency polygon Graphical representation of the distribution of data using connected straight-line segments

Frequency table   List of categories of values along with their corresponding frequencies

Fundamental counting rule   Rule stating that, for a sequence of two events in which the first event can occur m ways and the second can occur n ways, the events together can occur a total of $m \cdot n$ ways

Goodness-of-fit test   Test for how well some observed frequency distribution fits some theoretical distribution

Histogram   Graph of vertical bars representing the frequency distribution of a set of data

H test   The nonparametric Kruskal-Wallis test

Hypothesis   Statement or claim about some property of a population

Hypothesis test   Method for testing claims made about populations; also called test of significance

Independent events   Events for which the occurrence of any one of the events does not affect the probabilities of the occurrences of the other events

Independent sample   Sample whose values are not related to the values in another sample

Independent variable   The x variable in a regression equation, or one of the x variables in a multiple regression equation

Inferential statistics   Methods involving the use of sample data to make generalizations or inferences about a population

Influential point   Point that strongly affects the graph of a regression line

Interaction   In two-way analysis of variance, the effect when one of the factors changes for different categories of the other factor

Interquartile range   The difference between the first and third quartiles

Interval   Level of measurement of data; characterizes data that can be arranged in order and for which differences between data values are meaningful

Interval estimate   Range of values used to estimate some population parameter with a specific level of confidence; also called a confidence interval

Kruskal-Wallis test   Nonparametric hypothesis test used to compare three or more independent samples; also called an H test

Least-squares property   Property stating that, for a regression line, the sum of the squares of the vertical deviations of the sample points from the regression line is the smallest sum possible

Left-tailed test   Hypothesis test in which the critical region is located in the extreme left area of the probability distribution

Level of confidence   Probability that a population parameter is contained within a particular confidence interval; also called degree of confidence

Linear correlation coefficient   Measure of the strength of the relationship between two variables

Lower class limits   Smallest numbers that can actually belong to the different classes in a frequency table

Lower control limit   Boundary used in a control chart to separate points that are unusually low

Lurking variable   Variable that affects the variables being studied, but is not itself included in the study

Mann-Whitney U test   Hypothesis test equivalent to the Wilcoxon rank-sum test for two independent samples

Marginal change   For variables related by a regression equation, the amount of change in the dependent variable when one of the independent variables changes by one unit and the other independent variables remain constant

Margin of error   Maximum likely (with probability $1 - \alpha$) difference between the observed sample statistic and the true value of the population parameter

Matched pairs   Data from dependent samples; also called paired samples

Mathematical model   Mathematical function that "fits" or describes real-world data

Maximum error of estimate   *See* margin of error.

Mean   The sum of a set of values divided by the number of values

Mean deviation   Measure of variation equal to the sum of the deviations of each value from the mean, divided by the number of values

Measure of center   Value intended to indicate the center of the values in a collection of data

Measure of variation   Any of several measures designed to reflect the amount of variation or spread for a set of values

Median   Middle value of a set of values arranged in order of magnitude

Midquartile   One-half of the sum of the first and third quartiles

Midrange   One-half the sum of the highest and lowest values

Mode   Value that occurs most frequently

MS(error)   Mean square for error; used in analysis of variance

MS(total)   Mean square for total variation; used in analysis of variance

MS(treatment)   Mean square for treatments; used in analysis of variance

Multimodal   Having more than two modes

Multinomial experiment   Experiment with a fixed number of independent trials, where each outcome falls into exactly one of several categories

Multiple coefficient of determination   Measure of how well a multiple regression equation fits the sample data

Multiple comparison procedures   Procedures for identifying which particular means are different, after concluding that three or more means are not all equal

Multiple regression   Study of linear relationships among three or more variables

Multiple regression equation   Equation that expresses a linear relationship between a dependent variable y and two or more independent variables $(x_1, x_2, \ldots, x_k)$

Multiplication rule   Rule for determining the probability that event A will occur on one trial and event B will occur on a second trial

Mutually exclusive events   Events that cannot occur simultaneously

Negatively skewed   Skewed to the left

Nominal   Level of measurement of data; characterizes data that consist of names, labels, or categories only

Nonparametric tests   Statistical procedures for testing hypotheses or estimating parameters, where there are no required assumptions about the nature or shape of population distributions; also called distribution-free tests

Nonsampling errors   Errors from external factors not related to sampling

Normal distribution   Bell-shaped probability distribution described algebraically by Formula 5-1 in Section 5-1

Normal quantile plot   Graph of points (x, y), where each x value is from the original set of sample data, and each y value is a z score corresponding to a quantile value of the standard normal distribution

np chart   Control chart in which numbers of defects are plotted so that a process can be monitored

Null hypothesis   Claim made about some population characteristic, usually involving the case of no difference; denoted by H_0

Numerator degrees of freedom   Degrees of freedom corresponding to the numerator of the F test statistic

Numerical data   Data consisting of numbers representing counts or measurements

Observational study   Study in which we observe and measure specific characteristics, but don't attempt to manipulate or modify the subjects being studied

Observed frequency   Actual frequency count recorded in one cell of a contingency table or multinomial table

Odds against   Ratio of the probability of an event not occurring to the event occurring, usually expressed in the form of $a:b$ where a and b are integers having no common factors

Odds in favor   Ratio of the probability of an event occurring to the event not occurring, usually expressed as the ratio of two integers with no common factors

Ogive   Graphical representation of a cumulative frequency table

One-way analysis of variance   Analysis of variance involving data classified into groups according to a single criterion only

Ordinal   Level of measurement of data; characterizes data that may be arranged in order, but differences between data values either cannot be determined or are meaningless

Outliers   Values that are very unusual in the sense that they are very far away from most of the data

Paired samples   Two samples that are dependent in the sense that the data values are matched by pairs

Parameter   Measured characteristic of a population

Parametric tests   Statistical procedures, based on population parameters, for testing hypotheses or estimating parameters

Pareto chart   Bar graph for qualitative data, with the bars arranged in order according to frequencies

Payoff odds   Ratio of net profit (if you win) to the amount bet

p chart   Control chart used to monitor the proportion p for some attribute in a process

Pearson's product moment correlation coefficient   *See* linear correlation coefficient.

Percentile   The 99 values that divide ranked data into 100 groups with approximately 1% of the values in each group

Permutations rule   Rule for determining the number of different arrangements of selected items

Pie chart   Graphical representation of data in the form of a circle containing wedges

Placebo effect   Effect that occurs when an untreated subject incorrectly believes that he or she is receiving a real treatment and reports an improvement in symptoms

Point estimate   Single value that serves as an estimate of a population parameter

Poisson distribution   Discrete probability distribution that applies to occurrences of some event over a specified interval of time, distance, area, volume, or some similar unit

Pooled estimate of p_1 and p_2   Probability obtained by combining the data from two sample proportions and dividing the total number of successes by the total number of observations

Pooled estimate of σ^2   Estimate of the variance σ^2 that is common to two populations, found by computing a weighted average of the two sample variances

Population   Complete and entire collection of elements to be studied

Positively skewed   Skewed to the right

Power of a test   Probability $(1 - \beta)$ of rejecting a false null hypothesis

Predicted values   Values of a dependent variable found by using values of independent variables in a regression equation

Prediction interval   Confidence interval estimate of a predicted value of y

Predictor variables   Independent variables in a regression equation

Probability   Measure of the likelihood that a given event will occur; expressed as a number between 0 and 1

Probability distribution   Collection of values of a random variable along with their corresponding probabilities

Probability histogram   Histogram with outcomes listed along the horizontal axis and probabilities listed along the vertical axis

Probability value   *See* P-value.

Process data   Data, arranged according to some time sequence, that measure a characteristic of goods or services resulting from some combination of equipment, people, materials, methods, and conditions

P-value   Probability that a test statistic in a hypothesis test is at least as extreme as the one actually obtained

Qualitative data   Data that can be separated into different categories distinguished by some nonnumeric characteristic

Quantitative data   Data consisting of numbers representing counts or measurements

Quartiles   The three values that divide ranked data into four groups with approximately 25% of the values in each group

Randomized block design   Design in which a measurement is obtained for each treatment on each of several individuals matched according to similar characteristics

Random sample   Sample selected in a way that allows every member of the population to have the same chance of being chosen

Random selection   Selection of sample elements in such a way that all elements available for selection have the same chance of being selected

Random variable   Variable (typically represented by x) that has a single numerical value (determined by chance) for each outcome of an experiment

Random variation   Type of variation in a process that is due to chance; the type of variation inherent in any process not capable of producing every good or service exactly the same way every time

Range   The measure of variation that is the difference between the highest and lowest values

Range chart   Control chart based on sample ranges; used to monitor variation in a process

Range rule of thumb   Rule based on the principle that for typical data sets, the difference between the lowest typical value and the highest typical value is approximately 4 standard deviations ($4s$)

Rank   Numerical position of an item in a sample set arranged in order

Rank correlation coefficient   Measure of the strength of the relationship between two variables; based on the ranks of the values

Rare event rule   If, under a given assumption, the probability of a particular observed result is extremely small, we conclude that the assumption is probably not correct.

Ratio   Level of measurement of data; characterizes data that can be arranged in order, for which differences between data values are meaningful, and there is an inherent zero starting point

R chart   Control chart based on sample ranges; used to monitor variation in a process

Regression equation   Algebraic equation describing the relationship among variables

Regression line   Straight line that best fits a collection of points representing paired sample data

Relative frequency   Frequency for a class, divided by the total of all frequencies

Relative frequency approximation of probability   Estimated value of probability based on actual observations

Relative frequency histogram   Variation of the basic histogram in which frequencies are replaced by relative frequencies

Relative frequency table   Variation of the basic frequency table in which the frequency for each class is divided by the total of all frequencies

Replication   Repetition of an experiment

Residual   Difference between an observed sample y value and the value of y that is predicted from a regression equation

Right-tailed test   Hypothesis test in which the critical region is located in the extreme right area of the probability distribution

Rigorously controlled design   Design of experiment in which all factors are forced to be constant so that effects of extraneous factors are eliminated

Run   Sequence of data exhibiting the same characteristic; used in runs test for randomness

Run chart   Sequential plot of individual data values over time, where one axis (usually the vertical axis) is used for the data values and the other axis (usually the horizontal axis) is used for the time sequence

Runs test   Nonparametric method used to test for randomness

Sample   Subset of a population

Sample size   Number of items in a sample

Sample space   Set of all possible outcomes or events in an experiment that cannot be further broken down

Sampling distribution of sample means   Distribution of the sample means that is obtained when we repeatedly draw samples of the same size from the same population

Sampling error   Difference between a sample result and the true population result; results from chance sample fluctuations

Scatter diagram   Graphical display of paired (x, y) data

s chart   Control chart, based on sample standard deviations, that is used to monitor variation in a process

Self-selected sample   Sample in which the respondents themselves decide whether to be included; also called voluntary response sample

Semi-interquartile range   One-half of the difference between the first and third quartiles

Significance level   Probability of making a type I error when conducting a hypothesis test

Sign test   Nonparametric hypothesis test used to compare samples from two populations

Simple event   Experimental outcome that cannot be further broken down

Simple random sample   Sample of a particular size selected so that every possible sample of the same size has the same chance of being chosen

Simulation   Process that behaves in a way that is similar to some experiment so that similar results are produced

Single factor analysis of variance   See one-way analysis of variance.

Skewed   Not symmetric and extending more to one side than the other

Slope   Measure of steepness of a straight line

Spearman's rank correlation coefficient   See rank correlation coefficient.

SS(error)   Sum of squares representing the variability that is assumed to be common to all the populations being considered; used in analysis of variance

SS(total)   Measure of the total variation (around $\bar{\bar{x}}$) in all of the sample data combined; used in analysis of variance

SS(treatment)   Measure of the variation between the sample means; used in analysis of variance

Standard deviation   Measure of variation equal to the square root of the variance

Standard error of estimate   Measure of spread of sample points about the regression line

Standard error of the mean   Standard deviation of all possible sample means $\bar{x}$

Standard normal distribution Normal distribution with a mean of 0 and a standard deviation equal to 1

Standard score Number of standard deviations that a given value is above or below the mean; also called z score

Statistic Measured characteristic of a sample

Statistically stable process Process with only natural variation and no patterns, cycles, or unusual points

Statistical process control (SPC) Use of statistical techniques such as control charts to analyze a process or its outputs so as to take appropriate actions to achieve and maintain a state of statistical control and to improve the process capability

Statistics Collection of methods for planning experiments, obtaining data, organizing, summarizing, presenting, analyzing, interpreting, and drawing conclusions based on data

Stem-and-leaf plot Method of sorting and arranging data to reveal the distribution

Stepwise regression Process of using different combinations of variables until the best model is obtained; used in multiple regression

Stratified sampling Sampling in which samples are drawn from each stratum (class)

Student t distribution *See* t distribution.

Subjective probability Guess or estimate of a probability based on knowledge of relevant circumstances

Symmetric Property of data for which the distribution can be divided into two halves that are approximately mirror images by drawing a vertical line through the middle

Systematic sampling Sampling in which every kth element is selected

t distribution Bell-shaped distribution usually associated with small sample experiments; also called the Student t distribution

10–90 percentile range Difference between the 10th and 90th percentiles

Test of homogeneity Test of the claim that different populations have the same proportion of some characteristic

Test of independence Test of the null hypothesis that for a contingency table, the row variable and column variable are not related

Test of significance *See* hypothesis test.

Test statistic Sample statistic based on the sample data; used in making the decision about rejection of the null hypothesis

Total deviation Sum of the explained deviation and unexplained deviation for a given pair of values in a collection of bivariate data

Total variation Sum of the squares of the total deviation for all pairs of bivariate data in a sample

Traditional method of testing hypotheses Method of testing hypotheses based on a comparison of the test statistic and critical values

Treatment Property or characteristic that allows us to distinguish the different populations from one another; used in analysis of variance

Treatment group Group of subjects given some treatment in an experiment

Tree diagram Graphical depiction of the different possible outcomes in a compound event

Two-tailed test Hypothesis test in which the critical region is divided between the left and right extreme areas of the probability distribution

Two-way analysis of variance Analysis of variance involving data classified according to two different factors

Two-way table *See* contingency table.

Type I error Mistake of rejecting the null hypothesis when it is true

Type II error Mistake of failing to reject the null hypothesis when it is false

Unbiased estimator Sample statistic that tends to target the population parameter that it is used to estimate

Unexplained deviation For one pair of values in a collection of bivariate data, the difference between the y coordinate and the predicted value

Unexplained variation Sum of the squares of the unexplained deviations for all pairs of bivariate data in a sample

Uniform distribution Probability distribution in which every value of the random variable is equally likely

Upper class limits Largest numbers that can belong to the different classes in a frequency table

Upper control limit Boundary used in a control chart to separate points that are unusually high

Variance Measure of variation equal to the square of the standard deviation

Variance between samples In analysis of variance, the variation among the different samples

Variation due to error *See* variation within samples.

Variation due to treatment *See* variance between samples.

Variation within samples In analysis of variance, the variation that is due to chance

Weighted mean Mean of a collection of values that have been assigned different degrees of importance

Wilcoxon rank-sum test Nonparametric hypothesis test used to compare two independent samples

Wilcoxon signed-ranks test Nonparametric hypothesis test used to compare two dependent samples

Within statistical control *See* statistically stable process.

$\bar{x}$ chart Control chart used to monitor the mean of a process

y-intercept Point at which a straight line crosses the y-axis

z score Number of standard deviations that a given value is above or below the mean

Appendix E: Bibliography

*An asterisk denotes a book recommended for reading. Other books are recommended as reference texts.

Andrews D., and A. Herzberg. 1985. *Data: A Collection of Problems from Many Fields for the Student and Research Worker.* New York: Springer.

Beyer, W. 1991. *CRC Standard Probability and Statistics Tables and Formulae.* Boca Raton, Fla.: CRC Press.

*Campbell, S. 1974. *Flaws and Fallacies in Statistical Thinking.* Englewood Cliffs, N.J.: Prentice-Hall.

*Crossen, C. 1994. *Tainted Truth: The Manipulation of Fact in America.* New York: Simon & Schuster.

Devore, J., and R. Peck. 1997. *Statistics: The Exploration and Analysis of Data.* 3rd ed. St. Paul, Minn.: West Publishing.

*Fairley, W., and F. Mosteller. 1977. *Statistics and Public Policy.* Reading, Mass.: Addison-Wesley.

Fisher, R. 1966. *The Design of Experiments.* 8th ed. New York: Hafner.

*Freedman, D., R. Pisani, R. Purves, and A. Adhikari. 1991. *Statistics.* 2nd ed. New York: Norton.

*Gonick, L., and W. Smith. 1993. *The Cartoon Guide to Statistics.* New York: HarperCollins.

Halsey, J., and E. Reda. 2001. *Excel Student Laboratory Manual and Workbook.* Reading, Mass.: Addison-Wesley.

Hoaglin, D., F. Mosteller, and J. Tukey, eds. 1983. *Understanding Robust and Exploratory Data Analysis.* New York: Wiley.

*Hollander, M., and F. Proschan. 1984. *The Statistical Exorcist: Dispelling Statistics Anxiety.* New York: Marcel Dekker.

*Holmes, C. 1990. *The Honest Truth About Lying with Statistics.* Springfield, Ill.: Charles C Thomas.

*Hooke, R. 1983. *How to Tell the Liars from the Statisticians.* New York: Marcel Dekker.

*Huff, D. 1993. *How to Lie with Statistics.* New York: Norton.

*Jaffe, A., and H. Spirer. 1987. *Misused Statistics.* New York: Marcel Dekker.

*Kimble, G. 1978. *How to Use (and Misuse) Statistics.* Englewood Cliffs, N.J.: Prentice-Hall.

Kotz, S., and D. Stroup. 1983. *Educated Guessing—How to Cope in an Uncertain World.* New York: Marcel Dekker.

*Loyer, M. 2001. *Student Solutions Manual to Accompany Elementary Statistics.* 8th ed. Reading, Mass.: Addison-Wesley.

*Moore, D. 1997. *Statistics: Concepts and Controversies.* 4th ed. San Francisco: Freeman.

*Morgan, L. 2001. *The TI-83 Companion to Accompany Elementary Statistics.* 8th ed. Reading, Mass.: Addison-Wesley.

Mosteller, F., R. Rourke, and G. Thomas, Jr. 1970. *Probability with Statistical Applications.* 2nd ed. Reading, Mass.: Addison-Wesley.

Ott, L., and W. Mendenhall. 1994. *Understanding Statistics.* 6th ed. Boston: Duxbury Press.

Owen, D. 1962. *Handbook of Statistical Tables.* Reading, Mass.: Addison-Wesley.

*Paulos, J. 1988. *Innumeracy: Mathematical Illiteracy and Its Consequences.* New York: Hill and Wang.

Peck, R. 2001. *SPSS Student Laboratory Manual and Workbook.* Reading, Mass.: Addison-Wesley.

*Reichard, R. 1974. *The Figure Finaglers.* New York: McGraw-Hill.

*Reichmann, W. 1962. *Use and Abuse of Statistics.* New York: Oxford University Press.

*Rossman, A. 1996. *Workshop Statistics: Discovery with Data.* New York: Springer.

Ryan, T., B. Joiner, and B. Ryan. 1995. *MINITAB Handbook.* 3rd ed. Boston: Duxbury.

Schaeffer, R., M. Gnanadesikan, A. Watkins, and J. Witmer. 1996. *Activity-Based Statistics: Student Guide.* New York: Springer.

Schmid, C. 1983. *Statistical Graphics.* New York: Wiley.

Sheskin, D. 1997. *Handbook of Parametric and Nonparametric Statistical Procedures.* Boca Raton, Fla.: CRCPress.

Simon, J. 1992. *Resampling: The New Statistics.* Belmont, Calif.: Duxbury Press.

Smith, G. 1995. *Statistical Process Control and Quality Improvement.* 2nd ed. Columbus: Merrill.

*Stigler, S. 1986. *The History of Statistics.* Cambridge, Mass.: Harvard University Press.

*Tanur, J., ed. 1989. *Statistics: A Guide to the Unknown.* 3rd ed. Belmont, Calif.: Wadsworth.

Triola, M. 2001. *Minitab Student Laboratory Manual and Workbook.* 8th ed. Reading, Mass.: Addison-Wesley.

Triola, M. 2001. *STATDISK 8.0 Student Laboratory Manual and Workbook.* 8th ed. Reading, Mass.: Addison-Wesley.

Triola, M., and L. Franklin. 1994. *Business Statistics.* Reading, Mass.: Addison-Wesley.

*Tufte, E. 1983. *The Visual Display of Quantitative Information.* Cheshire, Conn.: Graphics Press.

Tukey, J. 1977. *Exploratory Data Analysis.* Reading, Mass.: Addison-Wesley.

Utts, J. 1996. *Seeing Through Statistics.* Belmont, Calif.: Wadsworth.

Appendix F: Answers to Odd-Numbered Exercises (and All Review Exercises and All Cumulative Review Exercises)

Section 1-2

1. Statistic
3. Parameter
5. Discrete
7. Discrete
9. Ratio
11. Interval
13. Interval
15. Ordinal
17. Ratio
19. With no natural starting point, the temperatures are at the interval level of measurement; ratios such as "twice" are meaningless.

Section 1-3

1. Because the viewers themselves decided whether to be included in the survey, we have a self-selected survey and the results cannot be used to conclude anything about the general population. In fact, Ted Koppel reported that a "scientific" poll of 500 people showed that 72% of us want the United Nations to stay in the United States. In this poll of 500 people, respondents were randomly selected by the pollster so that the results are much more likely to reflect the true opinion of the general population.
3. a. $525
   b. $498.75; no
5. Mothers who eat lobsters tend to be wealthier and can afford better health care.
7. Motorcyclists who were killed
9. First, it requires a calculation, which will result in some errors. Second, by asking for heights instead of measuring them, we tend to get desired values instead of actual values.
11. No, two consecutive 5% price cuts is equivalent to a 9.75% price cut, so a 10% price cut results in a lower price.
13. Because the groups consist of 20 mice each, all percentages of success should be multiples of 5. The given percentages cannot be correct.
15. a. Nothing is left.
    b. According to the *New York Times*, "It would have to remove all the plaque, remove it again and then remove it for a third time plus some more still."

Section 1-4

1. Observational study
3. Experiment
5. Random
7. Convenience
9. Stratified
11. Cluster
13. Stratified
15. Systematic
17. Answer varies.
19. Answer varies.
21. a. An advantage of open questions is that they provide the subject and the interviewer with a much wider variety of responses; a disadvantage is that open questions can be very difficult to analyze.
    b. An advantage of closed questions is that they reduce the chance of misinterpreting the topic; a disadvantage is that closed questions prevent the inclusion of valid responses the pollster might not have considered.
    c. Closed questions are easier to analyze with formal statistical procedures.
23. Answers vary.
25. a. Not always; no
    b. Not always; no

Section 1-5

1. a. A spreadsheet is a collection of data organized in an array of cells arranged in rows and columns, and it is used to summarize, analyze, and perform calculations with the data.
   b. worksheet
   c. An Excel file that contains worksheets.
   d. 3
3. The top half of the worksheet is deleted, then restored.
5. The top half of the worksheet is deleted, then restored.
7. Data Set 1 is printed.
9. Data Set 3 is printed.
11. Click on File, then select Save As. Enter the new file name and location for the saved file.

Chapter 1 Review Exercises

1. a. Discrete
   b. Ratio
   c. Stratified
   d. Statistic
   e. The largest values because they represent stockholders that could potentially gain control of the company.
   f. The self-selected sample is likely to be biased.

2. a. Systematic
   b. Convenience
   c. Cluster
   d. Random
   e. Stratified
3. Answers vary.

4. a. Design the experiment so that the subjects don't know whether they are using Sleepeze or a placebo, and also design it so that those who observe and evaluate the subjects do not know which subjects are using Sleepeze and which are using a placebo.
   b. Blinding will help to distinguish between the effectiveness of Sleepeze and the placebo effect, whereby subjects and evaluators tend to believe that improvements are occurring just because some treatment is given.
   c. Subjects are put into different groups through a process of *random selection*.
   d. Subjects are very *carefully chosen* for the different groups so that those groups are made to be similar in the ways that are important.
   e. Replication is used when the experiment is repeated. It is important to have a sample of subjects that is large enough so the we can see the true nature of any effects. It is important so that we are not misled by erratic behavior of samples that are too small.

5. It includes only students who are present on campus, and it excludes those dropouts who are not on campus.
6. a. Ratio
   b. Ordinal
   c. Nominal
   d. Ordinal
   e. Ratio

Chapter 1 Cumulative Review Exercises

1. 4.56
2. 4.358899
3. 1475.174
4. −6.64234
5. 2,042,975
6. 0.476678
7. 0.897352
8. 18.64787
9. 0.00000000000727449
10. 152,588,000,000
11. 53,459,700,000,000
12. 0.000000953674

Section 2-2

1. Class width: 5. Class midpoints: 57, 62, 67, 72, 77.
   Class boundaries: 54.5, 59.5, 64.5, 69.5, 74.5, 79.5.

3.
GPA	Frequency
0.00–0.49	72
0.50–0.99	23
1.00–1.49	47
1.50–1.99	135
2.00–2.49	288
2.50–2.99	276
3.00–3.49	202
3.50–3.99	97

Class width: 0.5. Class midpoints: 0.245, 0.745, 1.245, 1.745, 2.245, 2.745, 3.245, 3.745.
Class boundaries: −0.005, 0.495, 0.995, 1.495, 1.995, 2.495, 2.995, 3.495, 3.995.

5.
Height of Men (in.)	Relative Frequency
55–59	1%
60–64	3%
65–69	49%
70–74	46%
75–79	1%

7.
GPA	Relative Frequency
0.00–0.49	6.3%
0.50–0.99	2.0%
1.00–1.49	4.1%
1.50–1.99	11.8%
2.00–2.49	25.3%
2.50–2.99	24.2%
3.00–3.49	17.7%
3.50–3.99	8.5%

9.
Height of Men (in.)	Cumulative Frequency
Less than 60	1
Less than 65	4
Less than 70	53
Less than 75	99
Less than 80	100

11.
GPA	Cumulative Frequency
Less than 0.50	72
Less than 1.00	95
Less than 1.50	142
Less than 2.00	277
Less than 2.50	565
Less than 3.00	841
Less than 3.50	1043
Less than 4.00	1140

13.

Weight (lb)	Frequency
0.7900–0.7949	1
0.7950–0.7999	0
0.8000–0.8049	1
0.8050–0.8099	3
0.8100–0.8149	4
0.8150–0.8199	17
0.8200–0.8249	6
0.8250–0.8299	4

15. Roughly similar to Exercise 13; no notable difference.

Weight (lb)	Frequency
0.8100–0.8149	1
0.8150–0.8199	6
0.8200–0.8249	16
0.8250–0.8299	8
0.8300–0.8349	3
0.8350–0.8399	1
0.8400–0.8449	1

17.

Weight (lb)	Frequency
0–49	6
50–99	10
100–149	10
150–199	7
200–249	8
250–299	2
300–349	4
350–399	3
400–449	3
450–499	0
500–549	1

19. Cans will withstand pressure.

Weight (lb)	Frequency
200–219	12
220–239	9
240–259	18
260–279	84
280–299	52

21. An outlier can dramatically affect the frequency table.

Weight (lb)	Frequency
200–219	6
220–239	5
240–259	12
260–279	36
280–299	87
300–319	28
320–339	0
340–359	0
360–379	0
380–399	0
400–419	0
420–439	0
440–459	0
460–479	0
480–499	0
500–519	1

23.

Height	(a) Frequency	(b) Frequency
66	1	3
67	3	3
68	2	2
69	1	2
70	3	2
71	7	2
72	5	2
73	5	2
74	0	2
75	0	2
76	0	3
77	1	3

The frequency tables show that the first data set (a) has a more realistic distribution of scores; the second set (b) is more likely to be false.

Section 2-3

1. 42
3. 2
5. Among 200 eruptions, only one took as long as 109 min from the time of the preceding eruption, so allocating 100 min would please most bus trip participants.

7. Although the posted limit is 30 mi/h, it appears that the police issue tickets only to those traveling at least 42 mi/h.

9.

11.

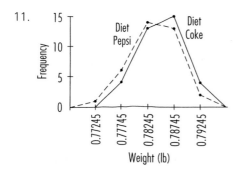

13. 600, 601, 601, 607, 610, 612, 618, 618, 619, 621, 623, 625, 627, 627, 640, 640, 649, 649

15.

17.
3	67
4	00133
4	667889
5	022334
5	778999
6	0011123333444
6	556778
7	00222233
7	56

19. Networking appears to dominate as the most effective approach in getting a job.

21.

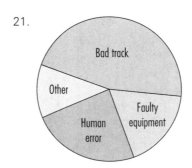

23. As the amount of tar in cigarettes increases, the amount of carbon monoxide also increases.

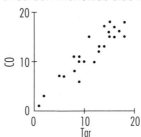

25. 10,000/422,000: 2.4%

27. 13,000 (from 37,000 to 24,000)

29. a.

b.

c. The outlier can have a dramatic effect on the histogram. Using a class width larger than the one used in parts (a) and (b) could hide the true nature of the distribution.

31.

Actors	Stem	Actresses
	2	1466678
998776532221	3	00113344445557789
88765543322100	4	111249
66531	5	0
2100	6	011
6	7	4
	8	0

Section 2-4

1. $\bar{x} = 16.0$; median = 14.5; mode = 14; midrange = 17.5
3. $\bar{x} = 81.0$ sec; median = 85.5 sec; mode: none; midrange = 77.0 sec
5. Jefferson Valley: $\bar{x} = 7.15$; median = 7.20; mode = 7.7; midrange = 7.10
   Providence: same results as Jefferson Valley
7. Coke: $\bar{x} = 0.81907$; median = 0.81865; mode: none; midrange = 0.81985
   Pepsi: $\bar{x} = 0.82188$; median = 0.82135; mode: none; midrange = 0.82290
   There does not appear to be a significant difference.
9. 74.4 min
11. 46.8 mi/h
13. Author: $\bar{x} = \$57.624$; median = \$59.350
    UMASS: $\bar{x} = \$65.117$; median = \$71.350
    Textbooks at the author's college appear to cost less, possibly because they apply to only the first two years and do not include junior and senior courses that are more expensive because they are used by smaller audiences.
15. Thursday: $\bar{x} = 0.069$ in.; median = 0.000 in.
    Sunday: $\bar{x} = 0.068$ in.; median = 0.000 in.
    There does not appear to be a substantial difference.
17. 48.0 mi/h
19. 62.9 volts
21. a. They all change by the same amount k.
    b. They are all multiplied by k.
23. a. 182.9 lb
    b. 171.0 lb
    c. 159.2 lb
    The results differ by substantial amounts, suggesting that the mean of the original set of weights is strongly affected by extreme values.

Section 2-5

1. range = 19.0; $s^2 = 28.9$; $s = 5.4$
3. range = 84.0 sec; $s^2 = 699.3$ sec^2; $s = 26.4$ sec

5. Jefferson Valley: range = 1.20; $s^2 = 0.23$; $s = 0.48$
   Providence: range = 5.80; $s^2 = 3.32$; $s = 1.82$
7. Coke: range = 0.00970; $s^2 = 0.00001$; $s = 0.00349$
   Pepsi: range = 0.01460; $s^2 = 0.00003$; $s = 0.00545$
9. 14.7 min
11. 4.1 mi/h
13. Author: \$14.683; UMASS: \$23.076
15. Thursday: 0.167 in.; Sunday: 0.200 in.
17. Approximately 0.75 year (assuming a minimum of 16 and maximum of 19)
19. 51, 99; yes
21. a. 68%
    b. 99.7%
23. Percentage is at least 75%.
25. All values are the same.
27. Everlast batteries are better in the sense that they are more consistent and predictable.
29. Section 1: range = 19.0; $s = 5.7$
    Section 2: range = 17.0; $s = 6.7$
    The ranges suggest that Section 2 has less variation, but the standard deviations suggest that Section 1 has less variation.
31. a. 65%
    b. 1.5
33. 4.4
35. a. 2/3
    b. 2/3
    c. 1/3
    d. Part (b) because the mean is 2/3, the value of σ^2. Use division by $n - 1$, as in part (b).
    e. No

Section 2-6

1. a. 30
   b. 2.00
   c. 2.00
   d. Same
3. a. 5.71
   b. −1.79
   c. 0.26
5. 2.56; unusual
7. 4.52; yes; patient is ill.
9. Psychology test, because $z = -0.38$ is greater than $z = -0.42$.
11. −3.56; yes
13. 92
15. 61
17. 0.8229
19. 0.8189
21. 0.8209; using Excel: 0.8208
23. 0.8073; using Excel: 0.8706
25. 44

27. 80

29. 344; using Excel: 344.2

31. 86; using Excel: 87

33. 360; using Excel: 358.8

35. 150

37. The z score remains the same.

39. a. Uniform

    b. Bell-shaped

    c. The shape of the distribution remains the same.

41. a. 0.00625; using Excel: 0.005975

    b. 0.81778; using Excel: 0.8178125

    c. 0.0178; using Excel: 0.0173

    d. Yes; yes

    e. No; no

Section 2-7

1. These employees are much older than typical employees. (Using DDXL: $Q_1 = 64.5$ and $Q_3 = 84.25$.)

3. Outliers of 8 and 15 must be wrong values.

5. The value of 504 is an outlier.(Using DDXL: $Q_3 = 294.75$; using Excel's QUARTILE command: $Q_3 = 294.5$.)

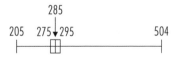

7. Actresses tend to be considerably younger. (Using Excel's QUARTILE command: For actors, $Q_1 = 37.75$, $Q_3 = 51.5$; for actresses: $Q_1 = 30.75$, $Q_3 = 41.25$.)

Actors:

Actresses:

9. UMASS prices tend to be higher. (Using Excel's QUARTILE command, at the author's college, $Q_1 = 47.4$,

$Q_3 = 65.1$ and at UMASS, $Q_1 = 47.925$, $Q_3 = 83.5875$. Using DDXL, at the author's college, $Q_1 = 46.025$, $Q_3 = 65.375$.)

Author:

UMASS:

11. $D = 20$. Mild outliers: 215, 216, 222, 225, 227, 230, 231, 243, 244; extreme outliers: 205, 210, 210, 211, 504

Chapter 2 Review Exercises

1. a. 54.9

    b. 55.0

    c. 51

    d. 55.5

    e. 27.0

    f. 6.3

    g. 39.6

    h. 51

    i. 51

    j. 57

2. a. -2.05

    b. Yes, because the z score is less than -2, indicating that his age is more than two standard deviations below the mean.

    c. 68, 69 (because usual values are between 42.3 and 67.5)

3.

Age	Frequency
40–44	2
45–49	6
50–54	12
55–59	12
60–64	7
65–69	3

4. Bell-shaped

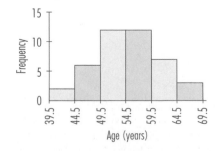

5. (Using Excel's QUARTILE command, $Q_3 = 57.75$.)

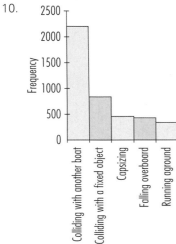

6. a. The percentage is 68%.
   b. The percentage is 95%.
7. The score of 19 is better, because $z = -0.20$ is greater than $z = -0.67$.
8. a. Answer varies, but roughly 2 years.
   b. Answer varies, but roughly $(5 - 0) / 4 = 1.25$.
9. a. 17.3 min
   b. 4.0 min
   c. 16.0 square minutes
10.

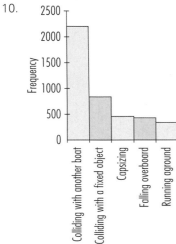

Chapter 2 Cumulative Review Exercises

1. a. $\bar{x} = 12.6$; median $= 12.0$; mode $= 7, 12$; midrange $= 15.0$
   b. $s = 8.6$; $s^2 = 73.3$; range $= 30.0$
   c. The original exact ages are continuous, but the given data appear to have been rounded to discrete values.
   d. Ratio
2. a. Mode, because the other measures of center require calculations that cannot (or should not) be done with data at the nominal level of measurement.
   b. Convenience
   c. Cluster
3. No, the 50 values should be weighted, with the state populations used as weights.

Section 3-2

1. $-1, 2, 5/3, \sqrt{2}$
3. 1/2
5. a. 0.0961
   b. 0.08
   c. They agree approximately.
7. 0.138
9. a. 1/36

b. Yes
11. a. 1/10
    b. No
13. a. 0.14
    b. No
15. a. 0.0000000774
    b. Yes
17. 0.0997; yes
19. 0.45
21. a. 1/365
    b. 31/365
23. 0.130
25. 0.34; yes
27. 0.159
29. a. bb, bg, gb, gg
    b. 1/4
    c. 1/2
31. a. 14:3
    b. $16
33. Because the probability of showing improvement with an ineffective drug is so small (0.04), it appears that the drug is effective.
35. 5/8
37. 0.250
39. 1

Section 3-3

1. a. No
   b. No
   c. Yes
3. a. 0.95
   b. 0.487
5. 4/5
7. a. 0.210
   b. 0.364
9. 364/365
11. 0.707
13. 0.600
15. 0.490
17. 0.140
19. 0.870
21. 0.290
23. 0.580
25. a. They are mutually exclusive.
    b. They are not mutually exclusive.
27. $P(A \text{ or } B) = P(A) + P(B) - 2P(A \text{ and } B)$

Section 3-4

1. a. independent
   b. dependent
   c. independent

3. a. 1/9
   b. 1/11
5. 0.350
7. 0.512
9. a. 0.00000751 or 1/133,225
   b. 1/365
11. 0.694
13. 1/1024; yes, because the probability of getting 10 girls by chance is so small.
15. a. 0.001
    b. Because the probability of getting 20 left-handed students is so small, either the students are lying or there was some factor causing the class to attract left-handed students.
17. 1/64
19. 0.360; no
21. 0.00277
23. 0.00000885
25. a. 0.992
    b. 0.973
    c. 0.431
27. 0.0192

Section 3-5

1. All three are boys.
3. At least one student does not get an A.
5. 7/8
7. 0.598
9. 1/2
11. 1
13. 0.999999
15. 0.271
17. 0.897
19. 0.0793
21. a. They are all tripled.
    b. Same result: 0.679
    c. No
23. a. 0.431
    b. 0.569

Section 3-6

1. Using the first four digits of each row: 6/20 = 0.3.
3. In each row, start at the left and include the first three digits that are 1, 2, 3, 4, 5, or 6. Omit the four rows that do not have three such digits. $P(10) = 2/16 = 1/8$.
5. Approximately 0.25.
7. Approximately 0.125.
9. Switch: $P(\text{win}) = 2/3$; stick: $P(\text{win}) = 1/3$.
11. No; no

Section 3-7

1. 87,178,291,200
3. 117,600
5. 1/13,983,816
7. 1/7,059,052
9. 1/12,966,811,200
11. a. 11,880
    b. 495
13. 10
15. a. 24
    b. 1/24
17. 1/20,160
19. 14,348,907
21. a. 1,048,576
    b. 184,756
    c. 0.176
    d. With a probability of 0.176, the result is common, but it should not happen consistently.
23. 1/479,001,600; yes
25. a. 0.0455
    b. 0.159
    c. 0.318
27. 0.00198
29. 2,095,681,645,538 (about 2 trillion)
31. a. Excel: 3.04141×10^{64}; approximation: 3.0363452×10^{64}
    b. 615

Chapter 3 Review Exercises

1. 0.498
2. 0.744
3. 0.613
4. 0.859
5. 0.0648
6. 0.251
7. 0.229
8. 0.643
9. a. 0.73
   b. 0.0729
   c. 0.611
10. 0.0777
11. 31/32
12. a. 1/120
    b. 720
13. a. 9/19
    b. 10:9
    c. $5
14. 0.000000531; no
15. 0.979
16. a. 0.698
    b. 0.272
    c. 0.0292

Chapter 3 Cumulative Review Exercises

1. a. 11.8 years
   b. 11.5 years
   c. 7.0 years
   d. 49.2 years squared
   e. No, a new plane with age 0 is within two standard deviations of the mean, so it is not unusual.
   f. ratio
   g. 1/4
   h. 2/3
   i. 1/11
   j. 1/9
2. a. 63.6 in.
   b. 1/4
   c. 3/4
   d. 1/16
   e. 5/16

Section 4-2

1. a. continuous
   b. discrete
   c. continuous
   d. continuous
   e. discrete
3. $\mu = 1.5, \sigma = 0.9$
5. $\mu = 2.0, \sigma = 1.0$
7. $\mu = 0.7, \sigma = 0.9$
9. $\mu = 0.8, \sigma = 0.7$
11. $\mu = 1.1, \sigma = 0.9$
13. a. $-26¢$
    b. $-26¢$
    c. Don't bet, because the expected value with no bet is 0, which is better than $-26¢$.
15. a. Lives: -1250 dollars (a loss); dies: $99,750$ (a gain)
    b. -100 dollars
    c. $150
17. Yes (more than two standard deviations below the mean)
19. a. 0.161
    b. Yes, because the probability of at least three 7s is 0.035, which is low (less than 0.05).
21. 0.090; no, because the probability of at least 10 girls is 0.090, which is high (greater than 0.05).
23. Same as table in Exercise 5; $\mu = 2.0, \sigma = 1.0$
25. a. Yes
    b. Yes
    c. No, $\Sigma P(x) > 1$.
    d. Yes
27. a. $\mu = 4.5, \sigma = 2.9$
    b. $\mu = 0, \sigma = 1$
    c. Yes
29. Put 1, 2, 3, 4, 5, 6 on one die and put 0, 0, 0, 6, 6, 6 on the other die.

Section 4-3

1. Binomial
3. Not binomial; more than two outcomes
5. Not binomial; more than two outcomes
7. Not binomial; more than two outcomes
9. a. 0.128
   b. WWC, WCW, CWW; 0.128 for each
   c. 0.384
11. 0.066
13. 0.124
15. 0+
17. 0.0879
19. 0.228
21. 0.241487
23. 0.858513
25. a. 0+ (or 0.00000980)
    b. 0+ (or 0.00000985)
    c. They are probably being targeted.
27. a. 0.347
    b. 0.544
    c. No, there is a good chance (0.544) of getting at most one wrong number if the rate is actually 15%.
29. 0.0833
31. a. 0.882
    b. Yes, because the probability of 7 (or fewer) graduates is small (less than 0.05).
33. 0.007 (or 0.00637); study
35. 0.751
37. 0.0524
39. 0.000535

Section 4-4

1. $\mu = 25.0, \sigma = 4.3$, minimum $= 16.4$, maximum $= 33.6$
3. $\mu = 158.0, \sigma = 7.3$, minimum $= 143.4$, maximum $= 172.6$
5. a. $\mu = 10.0, \sigma = 2.2$
   b. No, because 12 is within two standard deviations of the mean.
7. a. $\mu = 2.6, \sigma = 1.6$
   b. No, because 0 wins is within two standard deviations of the mean.
9. a. $\mu = 1025.0, \sigma = 13.6$
   b. Yes, because 1107 is more than two standard deviations above the mean. The region attracts people with higher levels of education.
11. a. $\mu = 15.0, \sigma = 3.7$
    b. If the program has no effect, the result of 12 could easily occur by chance because it is within two standard deviations of the mean.
13. a. $\mu = 27.2, \sigma = 5.1$
    b. Yes, it appears that the training program had an effect.
15. a. $\mu = 3.0, \sigma = 1.7$

b. Yes, 16 is unusual because it is more than two standard deviations above the mean. It appears that more than 6% of commercials are now more than one minute long.

17. a. Yes
  b. Probability is 0.95.
  c. At least 75% of such groups of 100 will have between 40 and 60 girls.

19. 10

Section 4-5

1. 0.168
3. 0.0552
5. a. 0.100
  b. 0.231
  c. 0.265
  d. 0.203
  e. 0.0206
7. 0.997; there will be about one day each year when the single team is not sufficient.
9. a. 0.728
  b. 0.231
  c. 0.0368
  d. 0.00389
  e. 0.000309
  Using the computed probabilities, the expected frequencies are 266, 84, 13, 1.4, and 0.1, and they agree quite well with the actual frequencies.
11. 4.82×10^{-64} is so small that, for all practical purposes, we can consider it to be zero.

Chapter 4 Review Exercises

1. a. A random variable is a variable that has a single numerical value (determined by chance) for each outcome of some procedure.
  b. A probability distribution gives the probability for each value of the random variable.
  c. Yes, because each probability value is between 0 and 1 and the sum of the probabilities is 1.
  d. 3.4
  e. 0.7
  f. Yes, because the probability of 0.0004 is very low.
2. a. 6.8
  b. 6.8
  c. 2.1
  d. 0.138
  e. Yes, because 12 is more than two standard deviations above the mean.
3. a. 0.026

b. 0.992 (or 0.994)
  c. $\mu = 8.0$, $\sigma = 1.3$
  d. No, because 6 is within two standard deviations of the mean.
4. a. 0.00361
  b. This company appears to be very different because the event of at least four firings is so unlikely with a probability of only 0.00361.
5. a. 0.581
  b. 0.0944
  c. 0.0183; no

Chapter 4 Cumulative Review Exercises

1. a. $\bar{x} = 1.2$, $s = 2.7$
  b.

x	Relative Frequency
0	78.6%
1	2.9%
2	1.4%
3	1.4%
4	0%
5	4.3%
6	0%
7	2.9%
8	5.7%
9	2.9%

  c. $\mu = 4.5$, $\sigma = 2.9$
  d. The excessively large number of zeros suggests that the distances were estimated, not measured. The digits do not appear to be randomly selected.
2. a. $\bar{x} = 9.7$, $s = 2.9$
  b. 0.4; very different from 0.0278
  c. 0.431
  d. Claim that sample is too small; claim that the results reflect variation due to chance.

Section 5-2

1. 0.8
3. 0.2
5. 1/3
7. 1/2
9. 0.4332
11. 0.4750
13. 0.0367
15. 0.0202
17. 0.2417
19. 0.1359
21. 0.9474
23. 0.6064
25. 0.4892
27. 0.5
29. 0.9500

31. 0.9950
33. 1.28
35. −0.67
37. −1.645
39. −1.88
41. a. 68.26%
    b. 95%
    c. 99.74%
    d. 81.85%
    e. 4.56%
43. a. 1.23
    b. 1.50
    c. 1.52
    d. −2.42
    e. −0.13

Section 5-3

1. 0.3413
3. 0.4052 (Excel: 0.4046)
5. 0.1844
7. 35.93% (Excel: 35.90%)
9. 31.21% (Excel: 31.38%)
11. 12.30%; no (Excel: 12.26%)
13. 0.0038; either a very rare event has occurred or the husband is not the father.
15. 4.18%; no (Excel: 4.21%)
17. 96.32%; (Excel: 96.22%)
19. 0.31%
21. 0.0013
23. a. 0.0179
    b. 1343
25. The z scores are real numbers that have no units of measurement.
27. $\mu = 64.9$ kg, $\sigma = 13$ kg, distribution is normal.

Section 5-4

1. 12.56
3. 8.32
5. 162 lb (Excel: 163 lb)
7. 113 lb
9. 67.7 in.
11. 1058 (Excel: 1059)
13. a. 0.0018
    b. 5.64 years
15. a. 0.9082 (Excel: 0.9088)
    b. 19.2 lb
17. 242 days
19. a. 0.0222 (Excel: 0.0221)
    b. 254.6
21. a. 75, 5
    b. No, the conversion should also account for variation.

c. 31.4, 27.6, 22.4, 18.6
d. Part (c), because variation is included.
23. a. 85 lb, 201 lb
    b. 95.44%
    c. 86.2 lb, 199.8 lb

Section 5-5

1. a. 0.1591 (Excel: 0.1605)
   b. 0.4934 (Excel: 0.4935)
3. a. 0.2776 (Excel: 0.2789)
   b. 0.0001 (Excel: 0.000017)
5. a. 0.1175 (Excel: 0.1182)
   b. If the original population has a normal distribution, the central limit theorem provides good results for any sample size.
7. 0.0070; yes (Excel: 0.0072)
9. a. 0.0051
   b. Yes
11. a. 0.2676
    b. 0.0250 (Excel: 0.0249)
    c. If the original population has a normal distribution, the central limit theorem provides good results for any sample size.
13. a. 0.0001 (Excel: 0+)
    b. No, but consumers are not being cheated because the cans are being overfilled, not underfilled.
15. a. 0.2358 (Excel: 0.2348)
    b. 0.0019
    c. If the original population has a normal distribution, the central limit theorem provides good results for any sample size.
    d. Yes, because the probability of 0.0019 shows that it is highly unlikely that by chance, a randomly selected group would get a mean as high as 590.
17. 0.0069; level is acceptable.
19. a. 0.5675 (Excel: 0.5667)
    b. 0.9999 (Excel: Almost 1)
    c. Yes
21. 0.0202
23. They are all 0.975.

Section 5-6

1. The area to the right of 57.5
3. The area to the left of 56.5
5. The area to the left of 24.5
7. The area between 15.5 and 22.5
9. Table: 0.201; normal approximation is not suitable.
11. Table: 0.610; approximation: 0.6026 (Excel: 0.6039)
13. 0.3821
15. 0.1020 (Excel: 0.1013)
17. 0.0239; error rate is probably less than 15%.

19. Yes, because the probability is only 0.0032.
21. 0.0708 (Excel: 0.0707); yes
23. 0.0080 (Excel: 0.0079); yes
25. 0.0526 (Excel: 0.0523); no
27. 0.6368 (Excel: 0.6375)
29. 6; 0.4602 (Excel: 0.4583)
31. 0.2946 (Excel: 0.3043)
33. a. $0.4129 - 0.3264 = 0.0865$ (Excel: 0.0842)
    b. $0.3192 - 0.2643 = 0.0549$ (Excel: 0.0541)
    c. $0.0256 - 0.0228 = 0.0028$
    As n gets larger, the difference becomes smaller.

Section 5-7

1. Not normal
3. Not normal
5. Not normal
7. Not normal
9. Not normal
11. Not normal
13. a. 1/14, 3/14, 5/14, 7/14, 9/14, 11/14, 13/14
    b. $-1.47, -0.79, -0.37, 0, 0.37, 0.79, 1.47$
    c.

Chapter 5 Review Exercises

1. a. 98.74%
   b. 57.8 in., 69.4 in.
   c. 0.7177 (Excel: 0.7178)
2. a. 3.67% (Excel: 3.71%)
   b. 0.9876
   c. 63.3 in., 74.7 in. (Excel: 63.2 in., 74.8 in.)
   d. 0.9633 (Excel: 0.9629)
3. a. 0.0853
   b. 128.1 mm
   c. 0.2670 (Excel: 0.2681)
4. 0.6940 (Excel: 0.6942)
5. a. 0.3413
   b. 0.0228
   c. 0.9332
   d. 0.0994
   e. -0.25
6. a. 0.5
   b. 0.5
   c. 0.25

7. a. 61.41% (Excel: 61.25%)
   b. 25.65% (Excel: 25.80%)
   c. 10.5
   d. 7.7
   e. 0.6103 (Excel: 0.6109)
8. 0.3050 (Excel: 0.3064)
9. a. 0.9049
   b. 0.8133 (Excel: 0.8126)
   c. 27,563 mi (Excel: 27,560 mi)
10. 0.0018; yes, because it is so unlikely that he would get 13 correct by guessing. (But his powers of prediction might be helped with tricks.)

Chapter 5 Cumulative Review Exercises

1. a. 61.0 mm
   b. 61.0 mm
   c. 60 mm
   d. 3.6 mm
   e. Yes, the histogram is roughly bell-shaped.
   f. 1.4
   g. 72%
   h. 71%
   i. ratio
   j. continuous
2. a. 0.001
   b. 0.271
   c. The requirement that $np \geq 5$ is not satisfied, indicating that the normal approximation would result in errors that are too large.
   d. 5.0
   e. 2.1
   f. No, 8 is within two standard deviations of the mean and is within the range of values that could easily occur by chance.

Section 6-2

1. $\bar{x} = 77.6$; $E = 5.0$
3. $\bar{x} = 209.5$; $E = 2.5$
5. 2.575
7. 2.17
9. a. $3124
   b. $82,554 < \mu < $88,802
11. a. 0.57 sec
    b. $7.03 < \mu < 8.17$
13. We are 95% confident that the interval from 133.35 mm to 135.65 mm actually does contain the true population mean.
15. a. $194.5 < \mu < 216.5$
    b. $176.2 < \mu < 193.8$
    c. It appears that scores on the oiled lanes are significantly higher.

17. $95.023 < \mu < \$230.921$
19. 0.9499 mm $< \mu < 0.9513$ mm
21. a. 0.81437 lb $< \mu < 0.81927$ lb
    b. 0.78336 lb $< \mu < 0.78622$ lb
    c. The cans of diet Coke weigh significantly less, probably because they contain less sugar.
23. Plastic: 1.688 lb $< \mu < 2.133$ lb;
    Paper: 8.57 lb $< \mu < 10.299$ lb
    By weight, paper is a significantly larger ecological problem.
25. a. 90.16 $< \mu < 122.99$
    b. The effects of an outlier can be extreme. The confidence interval limits can be very sensitive to outliers.
    c. Outliers should be carefully examined, and they should definitely be discarded if they are found to be errors.
27. a. 450
    b. 102
    c. 424 $< \mu < 476$
    d. 92%

Section 6-3

1. $t_{\alpha/2} = 2.093$
3. $t_{\alpha/2} = 2.977$
5. Neither applies.
7. $t_{\alpha/2} = 2.896$
9. $E = 59.8; 436 < \mu < 556$
11. 99 $< \mu < 109$; we have 95% confidence that the interval from 99 to 109 actually does contain the true population mean.
13. 111 $< \mu < 137$; with manual shoveling, we obtained 164 $< \mu < 186$, so the use of an electric snow thrower appears to result in significantly lower heart rates.
15. $16,142 < \mu < \$36,312$; we have 95% confidence that the interval from \$16,142 to \$36,312 actually does contain the true population mean.
17. 0.067 $< \mu < 0.176$
19. 8.0 yr $< \mu < 8.4$ yr; no
21. a. 0.9059 g $< \mu < 0.9286$ g
    b. 0.9015 g $< \mu < 0.9241$ g
    c. The results agree reasonably well.
23. a. $69.577 < \mu < \$85.700$
    b. $30.015 < \mu < \$54.285$
    c. The paperback books appear to cost significantly less than the hardcover books.
25. The confidence interval limits are closer than they should be.
27. 90%

Section 6-4

1. 543
3. 164

5. 601
7. 236
9. Answer varies, but using 0 and 8 for minimum and maximum yields 269.
11. a. 246
    b. 544
    c. Because the standard deviations differ by considerable amounts, the sample sizes differ by considerable amounts.
13. 147
15. 0.005

Section 6-5

1. $\hat{p} = 0.820$; $E = 0.020$
3. $\hat{p} = 0.4435$; $E = 0.0115$
5. 0.0351
7. 0.0296
9. 0.0661
11. 0.162 $< p < 0.238$
13. 0.144 $< p < 0.183$
15. 2653
17. 545
19. a. We have 99% confidence that the interval from 0.931 to 0.949 actually contains the true population proportion.
    b. Based on the result from part (a), about 5% to 7% of the population does not have a telephone, so those people are missed.
21. a. 9.00%
    b. 7.40% $< p < 10.6$%
23. a. 4.72% $< p < 5.67$%
    b. Yes, explosion or fire is a serious risk factor.
25. a. 0.133% $< p < 6.20$% using $x = 7$, $n = 221$
    b. Ziac does not appear to cause dizziness as an adverse reaction.
27. a. 321
    b. 543
29. a. 473
    b. 982
31. a. 0.0355 $< p < 0.139$
    b. 373
    c. Yes
33. 13.0% $< p < 29.0$%; yes
35. 4506
37. 78.3% $< p < 81.7$%; the results are not valid because the sample is self-selected.
39. $p > 0.818$; 81.8%
41. 0.894 $< p < 1.006$; use an upper limit of 1.

Section 6-6

1. 16.047, 45.722
3. 27.991, 79.490 (approximately)

5. $79 < \sigma < 170$

7. $10 < \sigma < 16$

9. 767

11. 1401

13. 38.2 mL $< \sigma < 95.7$ mL; no, the fluctuation appears to be too high.

15. a. 0.04425

  b. Use $s = 0.03951697$; 0.0313 g $< \sigma < 0.0578$ g

  c. Yes

17. a. 0.33 min $< \sigma < 0.87$ min

  b. 1.25 min $< \sigma < 3.33$ min

  c. The variation appears to be significantly lower with a single line. The single line appears to be better.

19. a. $\$10.364 < \sigma < \$20.623$

  b. $\$10.943 < \sigma < \$27.366$

  c. There is not a significant difference.

21. a. 98%

  b. 27.0

Chapter 6 Review Exercises

1. a. $3.4 < \mu < 5.4$

  b. $1.1 < \mu < 2.4$

  c. Ratings on the Dvorak keyboard appear to be significantly lower.

  d. The populations appear to be skewed instead of being normally distributed.

2. 2944

3. a. 291

  b. The sample is likely to be biased with regional characteristics.

4. a. 5.47 years $< \mu < 8.55$ years

  b. 2.92 years $< \sigma < 5.20$ years

5. $55.2\% < p < 74.5\%$; we have 95% confidence that the interval from 55.2% to 74.5% actually does contain the true population percentage.

6. 221

7. 71.29 cm $< \mu < 74.04$ cm

8. $514 < \mu < 602$

9. 404

10. $21.6\% < p < 26.5\%$

Chapter 6 Cumulative Review Exercises

1. a. 121.0 lbs

  b. 123.0 lbs

  c. 119 lbs, 128 lbs

  d. 116.5 lbs

  e. 23.0 lbs

  f. 56.8 lbs^2

  g. 7.5 lbs

  h. 119.0 lbs

  i. 123.0 lbs

  j. 127.0 lbs

k. ratio

l. Using DDXL, $Q_1 = 118$ and $Q_3 = 127.25$.

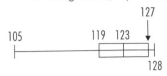

m. 112.6 lbs $< \mu < 129.4$ lbs

n. 4.5 lbs $< \sigma < 18.4$ lbs

o. 95

p. The individual supermodel weights do not appear to be considerably different from weights of randomly selected women, because they are all within 1.31 standard deviations of the mean of 143 lbs. However, when considered as a group, their mean is significantly less than the mean of 143 lbs (see part m).

2. a. 0.0089

  b. $0.260 < p < 0.390$

  c. Because the confidence interval limits do not contain 0.25, it is unlikely that the expert is correct.

Section 7-2

1. Reject the claim that the coin is fair.

3. Reject the claim that women who eat blue M&M candies have a better chance of having a baby boy.

5. Claim: Lisa Kerr's car theft rate is exceptionally high. It appears that her rate is, in fact, exceptionally high.

7. Claim: The experimental vaccine is effective in preventing Lyme disease. There is not sufficient sample evidence to support that claim.

9. $H_0: \mu \le 30$ years
   $H_1: \mu > 30$ years

11. $H_0: p \ge 0.5$
    $H_1: p < 0.5$

13. $H_0: \sigma \ge 2.8$ in.
    $H_1: \sigma < 2.8$ in.

15. $H_0: \mu \ge 12$ oz
    $H_1: \mu < 12$ oz

17. $-1.96, 1.96$

19. 2.05

21. $-2.05, 2.05$

23. 1.56

25. 14.61

27. -38.82

29. $H_0: \mu = 60.0$ in. Conclusion: There is sufficient evidence to warrant rejection of the claim that women have a mean height equal to 60.0 in.

31. $H_0: \mu \ge 12$ oz. Conclusion: There is not sufficient evidence to support the claim that the mean is less than 12 oz.

33. Type I error: Reject the claim (null hypothesis) that the mean is equal to 12 oz when the mean actually does equal 12 oz.
Type II error: Fail to reject the claim that the mean is equal to 12 oz when the mean is actually different from 12 oz.

35. Type I error: Reject the claim (null hypothesis) that the mean IQ is 100 or greater when the mean is actually 100 or greater.
Type II error: Fail to reject the claim that the mean is 100 or greater when the mean is actually less than 100.

37. There must be a fixed value of μ so that a specific normal distribution can be used for calculation of the test statistic.

39. There are no finite critical values corresponding to $\alpha = 0$. With $\alpha = 0$, the null hypothesis will never be rejected.

Section 7-3

1. 0.3030
3. 0.0001
5. a. $z = 2.98$
   b. $z = -1.96$, $z = 1.96$
   c. 0.0028 (Using NORMDIST: 0.00287)
   d. There is sufficient evidence to warrant rejection of the claim that the mean score is equal to 75.
7. a. $z = 1.53$
   b. $z = 2.05$
   c. 0.0630 (Using NORMDIST: 0.0629)
   d. There is not sufficient evidence to support the claim that the mean grade-point average is greater than 2.25.
9. Test statistic: $z = -6.64$. Critical value: $z = -2.33$. Reject H_0: $\mu \geq 98.6°$ F. There is sufficient evidence to support the claim that the mean body temperature is less than 98.6° F.
11. Test statistic: $z = -0.60$. Critical values: $z = \pm 1.96$. Fail to reject H_0: $\mu = 92.84$ in. There is not sufficient evidence to support the claim that the new balls have bounce heights with a mean different from 92.84 in. The new balls do not appear to be significantly different.
13. Test statistic: $z = -2.11$. Critical values: $z = \pm 2.575$. Fail to reject H_0: $\mu = 600$ mg. There is not sufficient evidence to warrant rejection of the claim that the mean is equal to 600 mg.
15. Test statistic: $z = 1.37$. P-value: 0.0853. (0.0857 with NORMDIST). Fail to reject H_0: $\mu \leq 0$. There is not sufficient evidence to support the claim that the population mean increase is greater than 0.
17. Test statistic: $z = 2.65$. P-value: 0.0080. Reject H_0: $\mu = 69.5$ years. There is sufficient evidence to support the claim that the mean is different from 69.5 years.
19. Test statistic: $z = -4.99$. P-value: 0.0002. (DDXL: less than 0.001). Reject H_0: $\mu = 5.670$ g. There is sufficient

evidence to warrant rejection of the claim that the mean weight of quarters in circulation is 5.670 g. The coins in circulation lose weight as they wear.

21. Test statistic: $z = 9.72$. Critical values depend on choice of significance level. P-value: 0.0002.(DDXL: less than 0.001). Reject H_0: $\mu = 3.5$ g. There is sufficient evidence to warrant rejection of the claim that the mean is equal to 3.5 g.
23. Test statistic: $z = 19.31$. Critical value depends on the significance level. P-value: 0.0001. (DDXL: less than 0.001). Reject H_0: $\mu \leq 12$ oz. There is sufficient evidence to support the claim that the mean is greater than 12 oz.
25. P-value: 0.735. Fail to reject H_0: $\mu \geq 0.9085$ g. There is not sufficient evidence to support the claim that the mean is less than 0.9085 g. Consumers are not being cheated.
27. P-value: less than 0.001. Reject H_0: $\mu = 98.6$. There is sufficient evidence to warrant rejection of the claim that the mean is equal to 98.6. The sample data is questionable because of the large number of missing values.
29. 11.0
31. a. 0.6178
    b. 0.0868

Section 7-4

1. Test statistic: $t = 0.533$. Critical values: $t = \pm 2.132$. Fail to reject H_0: $\mu = 75$. There is not sufficient evidence to warrant rejection of the claim that the mean score is equal to 75.
3. Test statistic: $t = 2.449$. Critical value: $t = 2.500$. Fail to reject H_0: $\mu \leq 2.00$. There is not sufficient evidence to support the claim that the mean grade-point average is greater than 2.00.
5. P-value > 0.5
7. 0.01 < P-value < 0.025
9. Test statistic: $t = -3.214$. (DDXL: -3.197). Critical values: $t = \pm 2.064$. (DDXL P-value: 0.00387.) Reject H_0: $\mu = 98.6°$F. There is sufficient evidence to warrant rejection of the claim that the mean body temperature is equal to 98.6°F.
11. Test statistic: $t = -1.827$. Critical value: $t = -2.132$. (DDXL P-value: 0.0709.) Fail to reject H_0: $\mu \geq \$1000$. There is not sufficient evidence to support the claim that the mean is less than $1000.
13. Test statistic: $t = -9.740$. Critical values: $t = \pm 2.080$. Reject H_0: $\mu = 243.5$. There is sufficient evidence to support the claim that the mean for the VDT subjects differs from the mean of 243.5.
15. Test statistic: $t = -4.021$. Critical values: $t = \pm 2.977$. Reject H_0: $\mu = 41.9$. The third-grade sample mean does appear to differ significantly from 41.9.

17. Test statistic: $z = 0.63$. Critical value: $z = 1.645$. (DDXL P-value: 0.265.) Fail to reject H_0: $\mu \leq 0.9085$ g. There is not sufficient evidence to support the claim that the mean is greater than 0.9085 g. We cannot conclude that the packages contain more than the claimed weight printed on the label.

19. Test statistic: $t = 3.164$. Critical value: $t = 2.896$. (DDXL P-value: 0.00666.) Reject H_0: $\mu \leq 62.2$ oz. There is sufficient evidence to support the claim that with the enriched feed, the mean weight is greater than 62.2 oz.

21. Test statistic: $z = 20.49$. The critical value depends on the significance level, but it will be less than the test statistic for any reasonable choices. (DDXL P-value: less than 0.001.) Reject H_0: $\mu \leq 12$ oz. There is sufficient evidence to support the claim that the mean is greater than 12 oz.

23. Test statistic: $t = 1.734$. Critical values: $t = \pm 2.132$. (DDXL P-value: 0.103.) Fail to reject H_0: $\mu = 3.39$ kg. There is not sufficient evidence to warrant rejection of the claim that the mean is equal to 3.39 kg. There is not sufficient evidence to state that the vitamin supplement has an effect on birth weight.

25. P-value: 0.0867. Assuming a 0.05 significance level, fail to reject H_0: $\mu \geq 1000$ mg. There is not sufficient evidence to support the claim that consumers are being cheated (by getting a mean less than 1000 mg).

27. P-value: 0.0248. Fail to reject $\mu \leq 12.00$. There is not sufficient evidence to support the claim that the mean is greater than 12.00.

29. The P-value becomes 0.00684.

31. With $z = 1.645$, the table and approximation both result in $t = 1.833$.

Section 7-5

1. Test statistic: $z = -2.28$. Critical value: $z = -1.645$. P-value: 0.0114. Reject H_0: $p \geq 0.20$. There is sufficient evidence to support the claim that less than 20% of U.S. households use e-mail. The conclusion is not valid today because the population is changing.

3. Test statistic: $z = 8.16$. Critical value: $z = 2.33$. P-value: 0.0001. (DDXL P-value: less than 0.001.) Reject H_0: $p \geq 0.5$. There is sufficient evidence to support the claim that most (more than 50%) drivers change tapes or CDs while driving.

5. Test statistic: $z = 3.70$. Critical values: $z = \pm 1.96$. P-value: 0.0002. (DDXL P-value: less than 0.001.) Reject H_0: $p = 0.43$. There is sufficient evidence to warrant rejection of the claim that the percentage of voters who say that they voted for the winning candidate is equal to 43%.

7. Test statistic: $z = 2.19$. Critical values: $z = \pm 1.96$. P-value: 0.0286. (DDXL P-value: 0.0284.) Reject H_0: $p = 0.01$. There is sufficient evidence to warrant rejection of the claim that 1% of sales are overcharges.

9. Test statistic: $z = 1.33$. Critical value: $z = 1.645$. P-value: 0.0918. (DDXL P-value: 0.0912.) Fail to reject H_0: $p \leq 0.5$. There is not sufficient evidence to support the claim that most students don't know what "Holocaust" refers to.

11. Test statistic: $z = 0.83$. Critical value: $z = 1.28$. P-value: 0.2033. (DDXL P-value: 0.203.) Fail to reject H_0: $p \leq 0.5$. There is not sufficient evidence to support the claim that the majority are smoking a year after treatment.

13. Test statistic: $z = 3.61$. Critical values depend on significance level, but the test statistic is in the critical region for any reasonable choice. P-value: 0.0002. (DDXL P-value: less than 0.001.) Reject H_0: $p = 3/4$. There is sufficient evidence to warrant rejection of the claim that the proportion is equal to 3/4.

15. Test statistic: $z = -2.00$. Critical value: $z = -2.33$. P-value: 0.0228. Fail to reject H_0: $p \geq 0.10$. There is not sufficient evidence to support the claim that less than 10% of all U.S. college students carry a gun, knife, or other such weapon. The sample is from one college, so the results do not necessarily apply to all U.S. college students.

17. The test statistic is $z = 8.26$ and the P-value is less than 0.001. Reject H_0: $p \leq 0.75$. There is sufficient evidence to support the claim that among aircraft accidents involving spatial disorientation, more than three-fourths result in fatalities.

19. Test statistic: $z = -1.67$. Critical values assuming a 0.05 significance level: $z = \pm 1.96$. P-value: 0.0950. (DDXL P-value: 0.0956.) Fail to reject H_0: $p = 0.10$. There is not sufficient evidence to warrant rejection of the claim that 10% of the plain M&M candies are blue.

21. a. With $n = 80$ and $p = 0.0025$, the conditions $np \geq 5$ and $nq \geq 5$ are not both satisfied.
    b. 0.00000000165 is found by using the binomial probability distribution.
    c. If the probability of a man being colorblind is 0.0025, the probability of getting at least 7 colorblind men among 80 is extremely small, so it appears that the colorblind rate is actually greater than 0.0025 or 0.25%. There is sufficient evidence to support the claim that the rate for men is greater than 0.25%.

23. 1.977%

25. 47% is not a possible result because, with 20 mice, the only possible success rates are 0%, 5%, 10%, ..., 100%.

Section 7-6

1. a. 1.735, 23.589
   b. 45.642
   c. 10.851

3. Test statistic: $\chi^2 = 12.960$. Critical values: $\chi^2 = 2.700$, 19.023. (*P*-value from Excel's CHIDIST: 0.0288.) Fail to reject H_0: $\sigma = 15$. There is not sufficient evidence to warrant rejection of the claim that the population standard deviation equals 15.

5. Test statistic: $\chi^2 = 2342.438$. Critical value: $\chi^2 = 63.691$ (approximately). (*P*-value from Excel's CHIDIST: 0+.) Reject H_0: $\sigma \leq 0.04$ g. There is sufficient evidence to support the claim that peanut M&M candies have weights that vary more.

7. Test statistic: $\chi^2 = 114.586$. Critical values: $\chi^2 = 57.153$, 106.629. (*P*-value from Excel's CHIDIST: 0.0136.) Reject H_0: $\sigma = 43.7$ ft. There is sufficient evidence to support the claim that the standard deviation is different from 43.7 ft. Because there appears to be more variation, the new production method appears to be worse.

9. Test statistic: $\chi^2 = 11.311$. Critical value: $\chi^2 = 12.198$. (*P*-value from Excel's CHIDIST: 0.00557.) Reject H_0: $\sigma \geq 14.1$. There is sufficient evidence to support the claim that the current class has less variation than past classes. A lower standard deviation does not suggest that the class is doing better.

11. Test statistic: $\chi^2 = 69.135$. Critical value: $\chi^2 = 67.505$ (approximately). (*P*-value from Excel's CHIDIST: 0.0305.) Reject H_0: $\sigma \leq 19.7$. There is sufficient evidence to support the claim that women have more variation than men.

13. $s = 2.340$. Test statistic: $\chi^2 = 15.626$. Critical values: $\chi^2 = 12.401$, 39.364 (assuming a significance level of 0.05). (*P*-value from Excel's CHIDIST: 0.1978.) Fail to reject H_0: $\sigma = 2.9$ in. There is not sufficient evidence to support the claim that men aged 45–54 have heights with a standard deviation different from 2.9 in.

15. $s = 7.533$ lb. Test statistic: $\chi^2 = 0.540$. Critical value: $\chi^2 = 1.646$. (*P*-value from Excel's CHIDIST: 0.000179.) Reject H_0: $\sigma \geq 29$ lb. There is sufficient evidence to support the claim that weights of female supermodels have a standard deviation less than 29 lb.

17. a. *P*-value > 0.20
    b. *P*-value < 0.005
    c. *P*-value < 0.005

19. a. Estimated values: 74.216, 129.565; Table A-4 values: 74.222, 129.561
    b. 117.093, 184.690

21. a. The standard deviation will be lower.
    b. The requirement of a normal distribution is not satisfied.

Chapter 7 Review Exercises

1. a. 0.001
   b. $\mu < 114.8$

c. $\mu \geq 90$
   d. There is not sufficient evidence to support the claim that the mean is greater than 12 oz.
   e. rejecting a true null hypothesis

2. a. $t = -1.833$
   b. $z = \pm 2.575$
   c. $\chi^2 = 8.907$, 32.852
   d. $z = \pm 1.645$
   e. $z = -2.33$

3. Test statistic: $z = 3.12$. Critical value: $z = 2.33$. *P*-value: 0.0001. (DDXL *P*-value: less than 0.001.) Reject H_0: $p \leq 0.5$. There is sufficient evidence to support the claim that the majority of gun owners favor stricter gun laws.

4. Test statistic: $t = -4.741$. Critical value: $t = -1.714$ (assuming a 0.05 significance level). Reject H_0: $\mu \geq 12$. There is sufficient evidence to support the claim that the company is cheating consumers. Argument is not valid.

5. Test statistic: $\chi^2 = 24.576$. Critical value: $\chi^2 = 43.188$. Reject H_0: $\sigma \geq 0.75$ lb. There is sufficient evidence to support the claim that the weights are more consistent (with a lower standard deviation).

6. Test statistic: $z = -2.34$. Critical value: $z = -2.33$. *P*-value: 0.0096. Reject H_0: $p \geq 0.10$. There is sufficient evidence to support the claim that the true percentage is less than 10%. The phrase "almost 1 out of 10" is not justified.

7. Test statistic: $z = -10.02$. Critical value: $z = -2.33$. *P*-value: 0.0001. Reject H_0: $\mu \geq 7124$ mi. There is sufficient evidence to support the claim that the population mean for women in the 16–24 age bracket is less than 7124 mi.

8. Test statistic: $z = -14.02$. Critical value: $z = -2.33$. *P*-value: 0.0001. Reject H_0: $p \geq 0.5$. There is sufficient evidence to reject the claim that at least half of all adults eat their fruitcakes.

9. $\bar{x} = 23.29$, $s = 4.53$. Test statistic: $t = 3.240$. Critical values: $t = \pm 2.861$. (DDXL *P*-value: 0.00431.) Reject H_0: $\mu = 20.0$ mg. There is sufficient evidence to warrant rejection of the claim that the mean equals 20 mg. These pills are not acceptable.

10. Test statistic: $z = 0.10$. Critical value: $z = 2.33$. *P*-value: 0.4602. Fail to reject H_0: $\mu \leq \$30,000$. There is not sufficient evidence to support the claim that the mean is greater than $30,000.

Chapter 7 Cumulative Review Exercises

1. a. 562.5
   b. 530.0
   c. 136.7
   d. 18,678.6
   e. 370.0

f. $448.2 < \mu < 676.8$

g. Test statistic: $t = 1.376$. Critical values: $t = \pm 2.365$. (DDXL P-value: 0.211.) Fail to reject H_0: $\mu = 496$. There is not sufficient evidence to warrant rejection of the claim that the mean equals 496.

h. The applicant's scores are not significantly different from the general population.

2. a. 0.4840 (0.4852 using NORMDIST)

b. $0.4840^5 = 0.0266$ (or $0.4852^5 = 0.0269$)

c. 0.4681 (0.4670 using NORMDIST)

d. 634

3. a. $\bar{x} = 124.23$, $s = 22.52$

b. Test statistic: $t = 1.675$. Critical values: $t = \pm 2.132$. (DDXL P-value: 0.115.) Fail to reject H_0: $\mu = 114.8$. There is not sufficient evidence to warrant rejection of the claim that the population mean equals 114.8.

c. $112.23 < \mu < 136.23$; yes

d. Test statistic: $\chi^2 = 44.339$. Critical values: $\chi^2 = 6.262$, 27.488. Reject H_0: $\sigma = 13.1$. There is sufficient evidence to warrant rejection of the claim that the standard deviation is equal to 13.1.

e. Variation appears to be affected.

4. a. 6.3

b. 2.2

c. 0.0034; using normal approx: 0.0019

d. Based on the low probability value in part (c), reject H_0: $p = 0.25$. There is sufficient evidence to reject the claim that the subject made random guesses.

e. 423

Section 8-2

1. Test statistic: $z = 4.08$. Critical values: $z = \pm 1.96$. P-value: 0.0002. (P-value using Excel's NORMDIST: 0.0000451.) Reject H_0: $\mu_1 = \mu_2$. There is sufficient evidence to warrant rejection of the claim that the two populations have the same mean.

3. $0.52 < \mu_1 - \mu_2 < 1.48$; no; there is a significant difference.

5. a. Test statistic: $z = 22.10$. Critical values: $z = \pm 2.575$. P-value: 0.0002. (P-value using Excel's NORMDIST: 0+.) Reject H_0: $\mu_1 = \mu_2$. There is sufficient evidence to warrant rejection of the claim that regular Coke and diet Coke have the same mean weight.

b. $0.028298 < \mu_1 - \mu_2 < 0.035762$

7. Test statistic: $z = 2.88$. Critical value: $z = 2.33$. P-value: 0.0020. Reject H_0: $\mu_1 = \mu_2$. There is sufficient evidence to support the claim that the mean stress score is lower than the mean nonstress score.

9. a. Test statistic: $z = 8.56$. Critical value: $z = 2.33$. P-value: 0.0001. (P-value using Excel's NORMDIST: 0+.) Reject H_0: $\mu_1 \leq \mu_2$. There is sufficient evidence

to support the claim that the older men come from a population with a mean that is less than the mean for the younger men.

b. $8 < \mu_1 - \mu_2 < 16$

11. Test statistic: $z = 2.17$. Critical value: $z = 1.645$. P-value: 0.0150. Reject H_0: $\mu_1 \leq \mu_2$. There is sufficient evidence to support the claim that zinc supplementation results in a higher mean birth weight.

13. UMASS: $n_1 = 40$, $\bar{x}_1 = 65.1175$, $s_1 = 23.0757$; DCC: $n_2 = 35$, $\bar{x}_2 = 57.6243$, $s_2 = 14.6830$. Test statistic: $z = 1.70$. Critical values: $z = \pm 1.96$ (assuming a 0.05 significance level). P-value: 0.0892. Fail to reject H_0: $\mu_1 = \mu_2$. There is not sufficient evidence to warrant rejection of the claim that the two colleges have the same mean price.

15. NOETS: $n_1 = 50$, $\bar{x}_1 = 0.405000$, $s_1 = 1.21290$; ETS: $n_2 = 50$, $\bar{x}_2 = 4.0976$, $s_2 = 10.21176$. Test statistic: $z = -2.54$. Critical values: $z = \pm 1.96$ (assuming a 0.05 significance level). P-value: 0.0110. Reject H_0: $\mu_1 = \mu_2$. There is sufficient evidence to warrant rejection of the claim that the two populations have the same mean serum cotinine level.

17. Test statistic: $z = -0.7858$. Critical value: $z = -1.644853$. P-value: 0.215991. There is not sufficient evidence to support the claim that the treatment population has a mean less than the mean for the placebo population.

19. Test statistic: $z = 8.691$. P-value: less than 0.001. There is sufficient evidence to support the claim that the mean height of men is greater than the mean height of women.

21. Cans 0.0111 in.: $n_1 = 175$, $\bar{x}_1 = 281.8$, $s_1 = 27.8$; cans 0.0109 in.: $n_2 = 175$, $\bar{x}_2 = 267.1$, $s_2 = 22.1$. Test statistic: $z = 5.48$. Critical value: $z = -1.645$ (assuming a 0.05 significance level). P-value: 0.0001. (P-value using Excel's NORMDIST: 0+.) Reject H_0: $\mu_1 \leq \mu_2$. There is sufficient evidence to support the claim that the 0.0111 in. cans have a mean axial load that is greater than the mean axial load of the 0.0109 in. cans. If the outlier is excluded, the test statistic changes to 5.66, so the conclusion will be the same. The confidence interval limits change from (9.4, 20.0) to (8.8, 18.0), so those difference are not substantial.

23. a. 50/3

b. 2/3

c. 52/3

d. The range of the $x - y$ values equals the range of the x values plus the range of the y values.

Section 8-3

1. a. $\bar{d} = 2.8$

b. $s_d = 3.6$

c. $t = 1.757$

d. $t = \pm 2.776$

3. $-1.6 < \mu_d < 7.2$

5. a. Test statistic: $t = 2.238$. Critical value: $t = 1.833$. (DDXL P-value: 0.026.) Reject H_0: $\mu_d \leq 0$. There is sufficient evidence to support the claim that the reported heights are greater than the measured heights.

    b. $-0.007 < \mu_d < 1.2$

7. a. Test statistic: $t = -1.718$. Critical value: $t = -1.833$. (DDXL P-value: 0.06.) Fail to reject H_0: $\mu_d \geq 0$. There is not sufficient evidence to conclude that the preparatory course is effective in raising scores.

    b. $-25.5 < \mu_d < 3.5$; we have 95% confidence that the interval from -25.5 to 3.5 actually contains the true population mean difference.

9. a. $0.69 < \mu_d < 5.56$

    b. Test statistic: $t = 3.306$. Critical value: $t = 1.895$. (DDXL P-value: 0.00948.) Reject H_0: $\mu_d \leq 0$. There is sufficient evidence to support the claim that the sensory measurements are lower after hypnosis.

    c. Yes

11. a. $-1.40 < \mu_d < -0.17$

    b. Test statistic: $t = -2.840$. Critical values: $t = \pm2.228$. (DDXL P-value: 0.0175.) Reject H_0: $\mu_d = 0$. There is sufficient evidence to warrant rejection of the claim that the mean difference is 0. Morning and night body temperatures do not appear to be about the same.

13. a. Test statistic: $t = -0.411$. P-value: 0.691 Fail to reject H_0: $\mu_d = 0$. There is not sufficient evidence to support the claim that astemizole has an effect.

    b. 0.3455; there is not sufficient evidence to support the claim that astemizole prevents motion sickness.

15. In the hypothesis test, the test statistic changes from 2.701 to 1.975 and the P-value changes from 0.0111 to 0.0369. The confidence interval limits change from (0.12, 1.23) to (-0.15, 2.70). The effect of an outlier can be dramatic.

17. 0.025

Section 8-4

1. 117
3. 85
5. a. 0.556

    b. -0.67

    c. ±1.96

    d. 0.5028 (DDXL: 0.502)

7. Test statistic: $z = -12.39$. Critical value: $z = -1.645$ (assuming a 0.05 significance level). P-value: 0.0001. (DDXL: P-value is less than 0.001.) Reject H_0: $p_1 \geq p_2$. There is sufficient evidence to support the claim that the flu rate is lower for those with the vaccine.

9. Test statistic: $z = 1.07$. Critical value: $z = 1.645$. P-value: 0.1423. (DDXL: P-value is 0.141.) Reject H_0: $p_1 \leq p_2$. There is not sufficient evidence to support

the claim that the rate of severe injuries is lower for children wearing seat belts.

11. With a test statistic of $z = -6.742$ and a P-value less than 0.001, there is sufficient evidence to support the claim that the polio rate is lower for those given the Salk vaccine.

13. $-0.0144 < p_1 - p_2 < 0.0086$; yes

15. a. Test statistic: $z = -3.06$. Critical values: $z = \pm1.96$ (assuming a 0.05 significance level). P-value: 0.0022. Reject H_0: $p_1 = p_2$. There is sufficient evidence to support the claim that the two population percentages are different.

    b. $-0.0823 < p_1 - p_2 < -0.00713$

17. Test statistic: $z = 3.86$. Critical value: $z = 2.33$. P-value: 0.0001. (DDXL P-value is less than 0.001.) Reject H_0: $p_1 \leq p_2$. There is sufficient evidence to support the claim that the proportion of drinkers among convicted arsonists is greater than the proportion of drinkers convicted of fraud.

19. Test statistic: $z = -2.82$. Critical value: $z = -1.645$. P-value: 0.0024. Reject H_0: $p_1 \geq p_2$. There is sufficient evidence to support the claim that the infection rate is lower for those who are warmed.

21. Test statistic: $z = 1.26$. Critical values: $z = \pm1.96$. P-value: 0.2076. Fail to reject H_0: $p_1 - p_2 = 0.10$. There is not sufficient evidence to warrant rejection of the claim that the headache rate of Viagra users is 10 percentage points more than the percentage for those who use a placebo.

23. 2135

Section 8-5

1. Test statistic: $F = 4.0000$. Critical value: $F = 2.3248$. Reject H_0: $\sigma_1^2 = \sigma_2^2$. There is sufficient evidence to support the claim that the treatment and placebo populations have different variances.

3. Test statistic: $F = 2.9228$. (Excel: $F = 2.9233$.) Critical value: $F = 1.9752$ (approximately). (Excel P-value: 0.00207.) Reject H_0: $\sigma_1 = \sigma_2$. There is sufficient evidence to support the claim that weights of regular Coke and diet Coke have different standard deviations.

5. Test statistic: $F = 1.2949$. Critical value: $F = 1.6928$ (approximately). Fail to reject H_0: $\sigma_1 = \sigma_2$. There is not sufficient evidence to support the claim that the two populations have different standard deviations.

7. Test statistic: $F = 1.0110$. Critical value of F is less than 1.3519 (assuming a 0.05 significance level). Although the conclusion is not clear from the test statistic and critical value, the values of the standard deviations

(3.67, 3.65) suggest that the difference is not significant. Fail to reject H_0: $\sigma_1^2 \leq \sigma_2^2$. There is not sufficient evidence to support the claim that the faculty cars vary less than the student cars.

9. Test statistic: $F = 2.5908$. Critical value from Table A-5: $F = 2.7006$. (Excel P-value: 0.0599.) Fail to reject H_0: $\sigma_1^2 = \sigma_2^2$. There is not sufficient evidence to warrant rejection of the claim that the two populations have the same variance.

11. Diet Coke: $n_1 = 36$, $s_1 = 0.00439090$; diet Pepsi: $n_2 = 36$, $s_2 = 0.00436161$. Test statistic: $F = 1.0135$. Critical F value is between 1.8752 and 2.0739. (Excel P-value: 0.9686.) Fail to reject H_0: $\sigma_1 = \sigma_2$. There is not sufficient evidence to support the claim that diet Coke and diet Pepsi have weights with different standard deviations.

13. From Table A-5: P-value < 0.02;
    from Excel: P-value $= 0.00207$.

15. $0.41 < \sigma_1^2 / \sigma_2^2 < 4.0$ (approximately)

Section 8-6

1. Test statistic: $t = -21.366$. Critical values: $t = \pm 2.048$. Reject H_0: $\mu_1 = \mu_2$. There is sufficient evidence to warrant rejection of the claim that the treatment and placebo groups have the same mean.

3. $-7.0 < \mu_1 - \mu_2 < -5.8$

5. a. Test statistic: $t = 3.288$. Critical values: $t = \pm 3.500$. Excel P-value: 0.0133. Fail to reject H_0: $\mu_1 = \mu_2$. There is not sufficient evidence to warrant rejection of the claim that men and women have the same mean height.
   b. Test statistic: $t = 3.479$. Critical values: $t = \pm 5.841$. Excel P-value: 0.0103. Fail to reject H_0: $\mu_1 = \mu_2$. There is not sufficient evidence to warrant rejection of the claim that men and women have the same mean height.
   c. Both procedures lead to the same conclusion.
   d. Although men and women actually do have different mean heights, the given sample data are not enough evidence to support that claim.

7. Test statistic: $t = -3.075$. Critical values: $t = \pm 2.878$. Reject H_0: $\mu_1 = \mu_2$. There is sufficient evidence to warrant rejection of the claim that the two populations have the same mean.

9. a. $-17.16 < \mu_1 - \mu_2 < 260.40$
   b. $-27.80 < \mu_1 - \mu_2 < 271.04$
   c. Part (a) is better because the sample standard deviations are not very different.

11. F-test: Fail to reject the claim of equal population variances, because the test statistic is $F = 2.9459$ and the critical value is $F = 3.0074$. (With Excel, $F = 2.9485$ and the P-value is 0.0542.)
    Test for equal means: Test Statistic: $t = -1.099$. Critical values: $t = \pm 2.052$. (DDXL P-value: 0.281.) Fail to reject H_0: $\mu_1 = \mu_2$. There is not sufficient evidence to warrant rejection of the claim that the two populations have the same mean.

13. Filtered: $n = 21$, $\bar{x} = 12.857$, $s = 3.071$.
    Nonfiltered: $n = 8$, $\bar{x} = 15.625$, $s = 1.188$.
    Assuming unequal variances: Test statistic: $t = -3.500$. Critical values: $t = \pm 2.365$. (DDXL P-value: 0.0017.) Reject H_0: $\mu_1 = \mu_2$. There is sufficient evidence to warrant rejection of the claim that the two populations have the same mean. The filtered cigarettes appear to have a lower mean amount of carbon monoxide.

15. Assuming unequal variances: Test statistic: $t = 22.698$. Critical values: $t = \pm 2.086$ (assuming a 0.05 significance level). Reject H_0: $\mu_1 = \mu_2$. There is sufficient evidence to warrant rejection of the claim that the two populations have the same mean.

17. Test statistic: $t = 15.322$. Critical values: $t = \pm 2.080$. Reject H_0: $\mu_1 = \mu_2$. There is sufficient evidence to warrant rejection of the claim that the two populations have the same mean.

19. Degrees of freedom changes from 7 to 26; test statistic is the same; critical value changes from -1.895 to -1.706; confidence interval limits change from $(-13.1, -8.3)$ to $(-12.8, -8.6)$.

Chapter 8 Review Exercises

1. a. Test statistic: $z = 4.48$. Critical value: $z = 1.645$ (assuming a 0.05 significance level). (DDXL P-value is less than 0.001.) Reject H_0: $p_1 \leq p_2$. There is sufficient evidence to support the claim that dyspepsia occurs at a higher rate among Viagra users.
   b. $0.0277 < p_1 - p_2 < 0.0699$; no

2. Experimental: $n = 20$, $\bar{x} = 66.3805$, $s = 20.9635$.
   Control: $n = 16$, $\bar{x} = 103.7675$, $s = 24.3371$.
   a. Test statistic: $F = 1.3478$. Critical value: $F = 2.6171$. (Excel P-value: 0.5328.) Fail to reject H_0: $\sigma_1^2 = \sigma_2^2$. There is not sufficient evidence to warrant rejection of the claim that the two populations have the same variance.
   b. Test statistic: $t = -4.951$. Critical values: $t = \pm 1.96$ (assuming a 0.05 significance level). (DDXL P-value is less than 0.001.) Reject H_0: $\mu_1 = \mu_2$. There is sufficient evidence to warrant rejection of the claim that both groups come from populations with the same mean.
   c. $-52.188 < \mu_1 - \mu_2 < -22.586$
   (DDXL: $-52.734 < \mu_1 - \mu_2 < -22.04$)

3. Test statistic: $z = 2.41$. Critical value: $z = 1.645$. P-value: 0.0080. (DDXL P-value: 0.00791.) Reject H_0: $p_1 \le p_2$. There is sufficient evidence to support the stated claim.

4. 23 min $< \mu_1 - \mu_2 <$ 65 min; no

5. Test statistic: $t = 2.301$. Critical values: $t = \pm 2.262$ (assuming a 0.05 significance level). (DDXL P-value: 0.0469.) Reject H_0: $\mu_d = 0$. There is sufficient evidence to conclude that there is a difference between the pre-training and posttraining weights.

6. $0.0 < \mu_d < 4.0$

7. a. Test statistic: $z = -12.44$. Critical value: $z = -2.33$. P-value: 0.0001. Reject H_0: $\mu_1 \ge \mu_2$. There is sufficient evidence to support the claim that women with a college degree have incomes with a higher mean.
   b. $\$14,415 < \mu_1 - \mu_2 < \$21,939$; no

8. a. Test statistic: $F = 3.8643$. Critical value: $F = 2.9249$ (approximately). Reject H_0: $\sigma_1 \le \sigma_2$. There is sufficient evidence to support the claim that waiting times for the single line have a lower standard deviation.
   b. Test statistic: $t = -0.677$. Critical values: $t = \pm 2.861$. Fail to reject H_0: $\mu_1 = \mu_2$. There is not sufficient evidence to warrant rejection of the claim that the two populations have the same mean.

9. a. Test statistic: $t = -3.847$. Critical value: $t = -1.796$. (DDXL P-value: 0.00136.) Reject H_0: $\mu_d \ge 0$. There is sufficient evidence to conclude that the Dozenol tablets are more soluble after the storage period.
   b. $12.4 < \mu_d < 45.6$

10. F-test: Test statistic: $F = 1.3105$. Critical value: $F = 2.9685$ (assuming a 0.05 significance level). (Excel P-value: 0.5781.) Fail to reject H_0: $\sigma_1^2 = \sigma_2^2$.
    Test for equality of means: Test statistic: $t = -0.627$. Critical values: $t = \pm 1.96$ (assuming a 0.05 significance level). (DDXL P-value: 0.535.) Fail to reject H_0: $\mu_1 = \mu_2$. There is not sufficient evidence to warrant rejection of the claim that yellow M&Ms and green M&Ms have the same mean weight.

Chapter 8 Cumulative Review Exercises

1. a. 0.0707
   b. 0.369
   c. 0.104
   d. 0.0540
   e. Test statistic: $z = -2.52$. Critical value: $z = -1.645$. P-value: 0.0059. Reject H_0: $p_1 \ge p_2$. There is sufficient evidence to support the claim that the percentage of women ticketed for speeding is less than the percentage for men.

2. There must be an error, because the rates of 13.7% and 10.6% are not possible with sample sizes of 100.

3. a. Test statistic: $z = 4.92$. Critical value: $z = 1.645$. P-value: 0.0001. (DDXL P-value is less than 0.001.) Reject H_0: $p \le 0.5$. There is sufficient evidence to support the claim that the majority of the prepared students answered correctly.
   b. Test statistic: $z = 3.67$. Critical value: $z = 1.645$. P-value: 0.0001. (DDXL P-value is less than 0.001.) Reject H_0: $p_1 \le p_2$. There is sufficient evidence to support the stated claim.

4. a. Test statistic: $F = 1.3786$. Critical value: $F = 1.8363$ (approximately). Fail to reject H_0: $\sigma_1 = \sigma_2$. There is not sufficient evidence to warrant rejection of the claim that both groups come from populations with the same standard deviation.
   b. Test statistic: $z = 0.78$. Critical values: $z = \pm 2.33$. P-value: 0.4354. Fail to reject H_0: $\mu_1 = \mu_2$. There is not sufficient evidence to warrant rejection of the claim that the two populations have the same mean.
   c. $506.56 < \mu < 571.08$

Section 9-2

1. a. Yes, if the significance level is 0.05 (because the critical values of r are ± 0.312); no, if the significance level is 0.01 (because the critical values of r are ± 0.402).
   b. 0.161

3. $r = 0.238$; critical values: $r = \pm 0.878$ (assuming a 0.05 significance level). There appears to be a correlation that is not linear.

5. $r = 0.796$; critical values: $r = \pm 0.666$; significant linear correlation.

7. $r = 0.842$; critical values: $r = \pm 0.707$; significant linear correlation.

9. $r = -0.069$; critical values: $r = \pm 0.707$; no significant linear correlation.

11. a. $r = 0.870$; critical values: $r = \pm 0.279$; significant linear correlation.
    b. $r = -0.010$; critical values: $r = \pm 0.279$; no significant linear correlation.
    c. Duration

13. a. $r = 0.767$; critical values: $r = \pm 0.361$; significant linear correlation.
    b. $r = -0.441$; critical values: $r = \pm 0.361$; significant linear correlation.
    c. Part (a) because there is a higher correlation.

15. a. $r = 0.399$; critical values: $r = \pm 0.335$ (approximately); significant linear correlation.

b. $r = 0.282$; critical values: $r = \pm 0.335$ (approximately); no significant linear correlation.

c. Invest in big budget movies.

17. With a linear correlation coefficient very close to 0, there does not appear to be a correlation, but the conclusion suggests that there is a correlation.

19. Although there is no *linear* correlation, the variables may be related in some other *nonlinear* way.

21. $r = 0.819$ (approximately); critical values: $r = \pm 0.553$; significant linear correlation.

23. a. ± 0.272
    b. ± 0.189
    c. -0.378
    d. 0.549
    e. 0.658

25. No, correlation is used only to determine whether there is *some* relationship between two variables. It would not reveal whether the reported heights are significantly greater than the measured heights.

Section 9-3

1. $\hat{y} = 5 + x$
3. $\hat{y} = 11.8 + 0.891x$
5. $\hat{y} = -152 + 3.88x$; 116 lb
7. $\hat{y} = 0.549 + 1.48x$; 1.3 persons
9. $\hat{y} = 0.214 - 0.000182x$; 0.21
11. a. $\hat{y} = 41.9 + 0.179x$; 79 min
    b. $\hat{y} = 81.9 - 0.009x$; 81 min
    c. Part (a) because there is a significant linear correlation between durations and intervals.
13. a. $\hat{y} = -6158 + 12{,}201x$; $12,144
    b. $\hat{y} = 28{,}647 - 3044x$; $19,515
    c. Part (a) because the carat (weight) has a higher correlation with price.
15. a. $\hat{y} = 82.6 + 1.03x$; $98 million
    b. $\hat{y} = -107 + 33.1x$; $138 million (the mean gross amount)
    c. Yes, because there is a significant linear correlation between budget and gross.
17. a. 10.00
    b. 2.50
19. Outlier: yes; influential point: no
21. $\hat{y} = -182 + 0.000351x$; $\hat{y} = -182 + 0.351x$. The slope is multiplied by 1000 and the y-intercept doesn't change. If each y entry is divided by 1000, the slope and the y-intercept are both divided by 1000.
23. The equation $\hat{y} = -49.9 + 27.2x$ is better because it has $r = 0.997$, which is higher than $r = 0.963$ for $\hat{y} = -103.2 + 134.9 \ln x$.

Section 9-4

1. 0.09; 9%
3. 0.107; 10.7%
5. $r = 0.963$; (square root of 0.928); critical values: $r = \pm 0.279$ (approximately); significant linear correlation.
7. $\hat{y} = -264 + 12.5x$; 363 lb (using unrounded intercept and slope).
9. a. 154.8
   b. 0
   c. 154.8
   d. 1
   e. 0
11. a. 13.83661
    b. 5.663385
    c. 19.5
    d. 0.70957
    e. 0.971544
13. a. 14
    b. $s_e = 0$; $E = 0$; no interval estimate.
15. a. 4.2 persons
    b. $1.6 < y < 6.9$
17. $-\$3.25 < y < \$17.42$
19. $-\$1.73 < y < \$24.81$
21. $-\$11.27 < \beta_0 < \$10.60$; $0.00901 < \beta_1 < 0.288$
23. a. $(n-2)s_e^2$
    b. $\dfrac{r^2 \cdot (\text{unexplained variation})}{1 - r^2}$
    c. $r = -0.949$

Section 9-5

1. $\hat{y} = -285 - 1.38x_1 - 11.2x_2 + 28.6x_3$
3. Yes, with a *P*-value of 0.00177, the equation has overall significance.
5. a. $\hat{y} = -274 + 0.426x_1 + 12.1x_2$
   b. 0.928; 0.925; 7.45E–30 or 0.0000
   c. Yes
7. a. $\hat{y} = -235 + 0.403x_1 + 5.11x_2 - 0.555x_3 + 9.19x_4$
   b. 0.942; 0.938; 1.1E–29 or 0.0000
   c. Yes
9. a. $\hat{y} = 7.74 - 0.00585x_1$
   b. 0.070; 0.043; 0.1189
   c. No
11. a. $\hat{y} = 7.41 - 0.00992x_1 + 0.00394x_2$
    b. 0.249; 0.204; 0.0089
    c. Yes
13. a. $\hat{y} = -6158 + 12{,}201x_1$
    b. 0.589; 0.574; 7.47E–7 or 0.0000
    c. Yes
15. a. $\hat{y} = 8221 + 12{,}297x_1 - 3116x_2$

b. 0.793; 0.778; 5.8E–10 or 0.0000

c. Yes

17. $\hat{y} = 0.154 + 0.0651x_1$ (where x_1 represents tar); yes.

19. $\hat{y} = 7.26 + 0.914x_1$ (where x_1 represents list price); by using more variables, adjusted R^2 can be raised from 0.995 to 0.996, but the small increase in adjusted R^2 does not justify the inclusion of additional variables.

Section 9-6

1. Polynomial (degree 2): $y = 2x^2 - 12x + 18$

3. Exponential: $y = 3^x$

5. Exponential: $y = (1,203,014)(0.886814)^x$, where x is coded as 1 for 1980, 2 for 1981, and so on. Predicted value: 122,769 lb.

7. Power: $y = 2000x^{0.4}$. With $R^2 = 0.999999938$, the model fits almost perfectly.

9. a. 3837.41

b. 61.3758

c. The polynomial (degree 2) sum of squares is lower.

Chapter 9 Review Exercises

1. a. $r = 0.338$; critical values: $r = \pm0.632$; no significant linear correlation.

b. 11%

c. $\hat{y} = -0.488 + 0.611x$

d. 0.347

2. a. $r = 0.116$; critical values: $r = \pm0.632$; no significant linear correlation.

b. 1.4%

c. $\hat{y} = 0.0657 + 0.000792x$

d. 0.347

3. a. $r = 0.777$; critical values: $r = \pm0.632$; significant linear correlation.

b. 60%

c. $\hat{y} = 0.193 + 0.00293x$

d. 0.286

4. $\hat{y} = -0.0526 + 0.747x_1 - 0.00220x_2 + 0.00303x_3$; $R^2 = 0.726$; adjusted $R^2 = 0.589$; P-value $= 0.0401$. Because the overall P-value of 0.0401 is less than 0.05, the equation can be used to predict ice cream consumption. Using the consumption/temperature data, the adjusted R^2 is 0.554. Although the adjusted R^2 is slightly higher using all three variables, the slight increase in adjusted R^2 does not justify the inclusion of additional variables, so the best regression equation appears to result from using temperature as the only independent variable.

5. a. $r = 0.340$; critical values: $r = \pm0.632$; no significant linear correlation.

b. $\hat{y} = 826 + 17.1x$

c. 1405.9 (mean)

6. a. $r = -0.157$; critical values: $r = \pm0.632$; no significant linear correlation.

b. $\hat{y} = 2990 - 32.5x$

c. 1405.9 (mean)

7. a. $r = 0.892$; critical values: $r = \pm0.632$; significant linear correlation.

b. $\hat{y} = -724 + 0.178x$

c. 1643 million bushels

8. $\hat{y} = -3145 + 10.1x_1 + 40.4x_2 + 0.187x_3$; 0.874; 0.811; 0.004; yes.

Chapter 9 Cumulative Review Exercises

1. a. $n = 50$, $\bar{x} = 80.660$, $s = 11.977$. Test statistic: $z = 8.66$. Critical value: $z = 1.645$ (assuming a 0.05 significance level). (DDXL P-value: less than 0.001.) Reject H_0: $\mu \le 66$. There is sufficient evidence to support the claim that intervals between eruptions now have a mean greater than 66 min.

b. 77.3 min $< \mu <$ 84.0 min

2. a. $\bar{x} = 99.1$, $s = 8.5$

b. $\bar{x} = 102.8$, $s = 8.7$

c. No, but a better comparison would involve treating the data as matched pairs instead of two independent samples.

d. Yes. $r = 0.702$ and the critical values are $r = \pm0.576$ (assuming a 0.05 significance level). There is a significant linear correlation.

3. a. No

b. Test statistic: $t = 1.185$. Critical values: $t = \pm2.262$. (DDXL P-value: 0.266.) Fail to reject H_0: $\mu_d = 0$. There is not sufficient evidence to warrant rejection of the claim that position has no effect.

Section 10-2

1. a. df $= 37$, so $\chi^2 = 51.805$ (approximately).

b. $0.10 < P$-value < 0.90

c. There is not sufficient evidence to warrant rejection of the claim that the roulette slots are equally likely.

3. Test statistic: $\chi^2 = 4.600$. Critical value: $\chi^2 = 7.815$. Excel P-value: 0.2035. There is not sufficient evidence to warrant rejection of the claim that the results fit a uniform distribution. Students do not have the ability to select the same tire.

5. Test statistic: $\chi^2 = 7.417$. Critical value: $\chi^2 = 12.592$. Excel P-value: 0.2840. There is not sufficient evidence to warrant rejection of the claim that fatal crashes occur on the different days of the week with equal frequency.

7. Test statistic: $\chi^2 = 23.431$. Critical value: $\chi^2 = 9.488$. Excel P-value: 0.0001. There is sufficient evidence to

warrant rejection of the claim that the accidents are distributed as claimed.

9. Test statistic: $\chi^2 = 4.200$. Critical value: $\chi^2 = 16.919$. Excel P-value: 0.8978. There is not sufficient evidence to warrant rejection of the claim that the digits are uniformly distributed.

11. Test statistic: $\chi^2 = 53.051$. Critical value: $\chi^2 = 7.815$. Excel P-value: 0.0000. There is sufficient evidence to warrant rejection of the claim that the distribution of crashes is the same as the distribution of ages.

13. Test statistic: $\chi^2 = 16.333$. Critical value: $\chi^2 = 14.067$ (assuming a 0.05 significance level). Excel P-value: 0.0222. There is sufficient evidence to support the claim that the probabilities of winning in the different post positions are not all the same.

15. Test statistic: $\chi^2 = 180.308$. Critical value: $\chi^2 = 16.919$ (assuming a 0.05 significance level). Excel P-value: 0.0000. There is sufficient evidence to support the claim that the last digits do not occur with the same frequency. Because of the large number of days with no rain, we expect many zeros, so we cannot conclude that the amounts were estimated; they may have been measured.

17. The test statistic changes from 4.600 to 76.638, so the outlior has a dramatic effect.

19. a. Critical value is $\chi^2 = 3.841$, and test statistic is

$$\chi^2 = \frac{\left(f_1 - \dfrac{f_1 + f_2}{2}\right)^2}{\dfrac{f_1 + f_2}{2}} + \frac{\left(f_2 - \dfrac{f_1 + f_2}{2}\right)^2}{\dfrac{f_1 + f_2}{2}}$$
$$= \frac{(f_1 - f_2)^2}{f_1 + f_2}$$

b. Critical values: The χ^2 critical value is 3.841, and it is approximately equal to the square of $z = 1.96$.

21. Combine the last three cells. Test statistic: $\chi^2 = 1.012$. Critical value: $\chi^2 = 5.991$. Excel P-value: 0.6029. Fail to reject H_0: The frequencies fit a Poisson distribution. There is not sufficient evidence to warrant rejection of the claim that the frequencies fit a Poisson distribution.

Section 10-3

1. Test statistic: $\chi^2 = 1.174093$. P-value: 0.278562. There is not sufficient evidence to warrant rejection of the claim that treatment is independent of the reaction.

3. Test statistic: $\chi^2 = 47.035$. Critical value: $\chi^2 = 6.635$. Excel P-value: 0.0000. There is sufficient evidence to warrant rejection of the claim that opinion and gender are independent.

5. Test statistic: $\chi^2 = 51.458$. Critical value: $\chi^2 = 6.635$. Excel P-value: 0.0000. There is sufficient evidence to warrant rejection of the claim that the proportions of

agree/disagree responses are the same for the subjects interviewed by men and the subjects interviewed by women.

7. Test statistic: $\chi^2 = 2.195$. Critical value: $\chi^2 = 5.991$. Excel P-value: 0.3337. There is not sufficient evidence to support the claim that there is a gender gap.

9. Test statistic: $\chi^2 = 8.642$. Critical value: $\chi^2 = 9.210$. Excel P-value: 0.0133. There is not sufficient evidence to warrant rejection of the claim that the distribution of planes flying and planes ordered is independent of the airline company.

11. Test statistic: $\chi^2 = 119.330$. Critical value: $\chi^2 = 5.991$. Excel P-value: 0.0000. There is sufficient evidence to warrant rejection of the claim that the type of crime is independent of whether the criminal is a stranger.

13. Test statistic: $\chi^2 = 42.557$. Critical value: $\chi^2 = 3.841$. Excel P-value: 0.0000. There is sufficient evidence to warrant rejection of the claim that the sentence is independent of the plea.

15. Test statistic: $\chi^2 = 49.731$. Critical value: $\chi^2 = 11.071$ (assuming a 0.05 significance level). Excel P-value: 0.0000. There is sufficient evidence to support the claim that the type of crime is related to whether the criminal drinks or abstains.

17. Test statistic: $\chi^2 = 0.122$. Critical value: $\chi^2 = 3.841$ (assuming a 0.05 significance level). Excel P-value: 0.7265. There is not sufficient evidence to warrant rejection of the claim that gender of statistics students is independent of whether they smoke.

19. Without Yates' correction: $\chi^2 = 1.174$. With Yates' correction: $\chi^2 = 0.907$. Yates' correction decreases the test statistic so that sample data must be more extreme to be considered significant.

Chapter 10 Review Exercises

1. Test statistic: $\chi^2 = 12.438$. Critical value: $\chi^2 = 6.635$. Excel P-value: 0.0004. There is sufficient evidence to warrant rejection of the claim that whether gun owners store their guns loaded and unlocked is independent of whether they are a member of the NRA.

2. Test statistic: $\chi^2 = 5.600$. Critical value: $\chi^2 = 11.071$. Excel P-value: 0.3471. There is not sufficient evidence to support the claim that the outcomes are not equally likely.

3. Test statistic: $\chi^2 = 6.780$. Critical value: $\chi^2 = 12.592$. Excel P-value: 0.3417. There is not sufficient evidence to support the theory that more gunfire deaths occur on weekends.

4. Test statistic: $\chi^2 = 0.500$. Critical value: $\chi^2 = 5.991$. Excel P-value: 0.7789. There is not sufficient evidence

to warrant rejection of the claim that USAir, American, and Delta have the same proportions of on-time flights.

5. Test statistic: $\chi^2 = 7.607$. Critical value: $\chi^2 = 5.991$. Excel P-value: 0.0223. There is sufficient evidence to warrant rejection of the claim that the occurrence of headaches is independent of the group (Seldane, placebo, control).

Chapter 10 Cumulative Review Exercises

1. $\bar{x} = 80.9$; median: 81.0; range: 28.0; $s^2 = 74.4$; $s = 8.6$; 5–number summary: 66, 76.0, 81.0, 86.5, 94. (With Excel's QUARTILE command, $Q_1 = 76.5$ and $Q_3 = 85.25$.)

2. a. 0.272
   b. 0.468
   c. 0.614
   d. 0.282

3. Contingency table; see Section 10-3. Test statistic: $\chi^2 = 0.055$. Critical value: $\chi^2 = 7.815$ (assuming a 0.05 significance level). Excel P-value: 0.9966. There is not sufficient evidence to warrant rejection of the claim that men and women choose the different answers in the same proportions.

4. Use correlation; see Section 9-2. Test statistic: $r = 0.978$. Critical values: $r = \pm 0.950$ (assuming a 0.05 significance level). There is sufficient evidence to support the claim that there is relationship between the memory and reasoning scores.

5. Use the test for matched pairs; see Section 8-3. $\bar{d} = -10.25$; $s_d = 1.5$. Test statistic: $t = -13.667$. Critical value: $t = -2.353$ (assuming a 0.05 significance level). DDXL P-value: less than 0.001. Reject $H_0: \mu_d \geq 0$. There is sufficient evidence to support the claim that the training session is effective in raising scores.

6. Test for the difference between two independent samples; see Section 8-6. The sample standard deviations are 7.1 and 7.3, so assume that the two populations have equal variances. Test statistic: $t = -2.014$. Critical values: $t = \pm 2.447$ (assuming a 0.05 significance level). DDXL P-value: 0.0906. Fail to reject $H_0: \mu_1 = \mu_2$. There is not sufficient evidence to warrant rejection of the claim that men and women have the same mean score.

Section 11-2

1. a. 6
   b. $F = 3.33$
   c. $F = 2.9013$
   d. 0.032
   e. There is sufficient evidence to warrant rejection of the claim that the different products have commercials resulting in the same mean reaction score.

3. a. 0.864667
   b. 0.46
   c. $F = 1.8797$
   d. $F = 3.8853$ (assuming a 0.05 significance level)
   e. There is not sufficient evidence to warrant rejection of the claim that the three age-group populations have the same mean body temperature.

5. Test statistic: $F = 1.3432$. Critical value: $F = 3.2389$. Excel P-value: 0.2956. Fail to reject $H_0: \mu_1 = \mu_2 = \mu_3 = \mu_4$. There is not sufficient evidence to warrant rejection of the claim that the populations of cars in the different weight categories have the same mean chest deceleration value.

7. Test statistic: $F = 4.0497$. Critical value: $F = 3.4028$. Excel P-value: 0.0305. Reject $H_0: \mu_1 = \mu_2 = \mu_3$. There is sufficient evidence to warrant rejection of the claim that the mean breadth is the same for the different epochs.

9. Test statistic: $F = 2.9493$. Critical value: $F = 2.6060$. (Excel: Critical value is $F = 2.5888$ and the P-value is 0.0307.) Reject $H_0: \mu_1 = \mu_2 = \mu_3 = \mu_4 = \mu_5$. There is sufficient evidence to warrant rejection of the claim that the means for the different laboratories are the same.

11. Test statistic: $F = 0.5083$. Critical value: $F = 2.2899$ (approximately). (Excel: Critical value is $F = 2.3113$ and P-value is 0.7694.) Fail to reject $H_0: \mu_1 = \mu_2 = \mu_3 = \mu_4 = \mu_5 = \mu_6$. There is not sufficient evidence to warrant rejection of the claim that the populations of different colors of M&Ms have the same mean.

13. a. The variance between samples changes to 5.864667, the variance within samples remains the same, the test statistic changes to $F = 12.7493$, and the critical value remains the same.
    b. The F test statistic is not affected by the scale used.
    c. The F test statistic does not change.
    d. The F test statistic does not change.
    e. The F test statistic does not change.

15. a. Test statistic: $t = 0.978$. Critical values: $t = \pm 2.306$. DDXL P-value: 0.357. There is not sufficient evidence to warrant rejection of the claim that the two populations have the same mean.
    b. Test statistic: $F = 0.9555$. Critical value: $F = 5.3177$. Excel P-value: 0.3569. Fail to reject $H_0: \mu_1 = \mu_2$. There is not sufficient evidence to warrant rejection of the claim that the two populations have the same mean.
    c. Test statistics: $t^2 = F = 0.956$. Critical values: $t^2 = F = 5.32$.

Section 11-3

1. Test statistic: $F = 3.012666$. P-value: 0.09117. Fail to reject the null hypothesis of no interaction. There does not appear to be a significant effect from the interaction between gender and test (math/verbal).

3. Test statistic: $F = 0.57591$. P-value: 0.45286. Because the P-value is greater than 0.05, fail to reject the null hypothesis that the type of test (math/verbal) has no effect on SAT scores. There is not sufficient evidence to support the claim that the type of test (math/verbal) has an effect on SAT scores.

5. Test statistic: $F = 0.777941$. P-value: 0.395096. Because the P-value is greater than 0.05, fail to reject the null hypothesis that gender has no effect on estimated length. There is not sufficient evidence to support the claim that gender has an effect on estimated length.

7. Test statistic: $F = 3.867829$. P-value: 0.00000696. Because the P-value is less than 0.05, reject the null hypothesis that the choice of subject has no effect on the hearing test score. There is sufficient evidence to support the claim that the choice of subject has an effect on the hearing test score.

9. Test statistic: $F = 2.474048$. P-value: 0.158993. Because the P-value is greater than 0.05, fail to reject the null hypothesis that the four machine operators have the same mean production output. The different operators appear to be working at about the same level.

11. (The given results use the first nine values for each cell, but the pulse rate of 8 was excluded as being not feasible.) Test statistic: $F = 0.424263$. Critical value: $F = 4.1709$ (approximately, assuming a 0.05 significance level). Excel critical value is 4.1491 and the P-value is 0.5195. There does not appear to be an interaction between gender and smoking.

13. Test statistic: $F = 0.17925$. Critical value: $F = 4.1709$ (approximately, assuming a 0.05 significance level). Excel critical value is $F = 4.1491$ and the P-value is 0.6749. Smoking does not appear to have an effect on pulse rates. (This is not too surprising, because college students have not been smoking very long.)

15. Answers vary.

Chapter 11 Review Exercises

1. Test statistic: $F = 0.051471$. Critical value: $F = 5.143249$. P-value: 0.950246. Fuel consumption does not appear to be affected by an interaction between transmission type and engine size.

2. Test statistic: $F = 1.904412$. Critical value: $F = 5.143249$. P-value: 0.228877. The size of the engine does not appear to have an effect on fuel consumption.

3. Test statistic: $F = 3.558824$. Critical value: $F = 5.987374$. P-value: 0.108178. The type of transmission does not appear to have an effect on fuel consumption.

4. Test statistic: $F = 9.4827$. Critical value: $F = 3.0984$. Excel P-value: 0.0004. Reject the null hypothesis of equal means. There is sufficient evidence to support the claim of different mean selling prices.

5. a. Test statistic: $F = 1.0000$. P-value: 0.42265. There is not sufficient evidence to support the claim that amounts of emitted greenhouse gases are affected by the type of transmission.

   b. Test statistic: $F = 7.0000$. P-value: 0.125. There is not sufficient evidence to support the claim that amounts of greenhouse gases are affected by the number of cylinders.

   c. Perhaps greenhouse gases *are* affected by the type of transmission and/or the number of cylinders; however, the given sample data do not provide sufficient evidence to support such claims.

6. Test statistic: $F = 46.9$. Cricital value: $F = 3.73889$. P-value: 0.000000623. Reject the null hypothesis that the three populations have equal means. There is sufficient evidence to warrant rejection of the claim of equal population means.

Chapter 11 Cumulative Review Exercises

1. a. 0.100 in.
   b. 0.263 in.
   c. 0.00, 0.00, 0.00, 0.01, 1.41
   d. 0.92 in., 1.41 in.
   e. Answer varies, depending on the number of classes used, but the histogram should depict a distribution that is skewed to the right.
   f. No, because the data do not appear to come from a normally distributed population.
   g. 19/52 or 0.365

2. a.

   b. Barnum: $\bar{x} = 62.8$, $s = 12.9$; Solomon & Ford: $\bar{x} = 72.3$, $s = 8.2$; Diaz & Florio: $\bar{x} = 62.8$, $s = 11.6$

   c. Based on the sample standard deviations, assume that the two populations have equal variances. Test statistic: $t = -1.760$. Critical values: $t = \pm2.145$. DDXL P-value: 0.1. Fail to reject the null hypothesis of equal means. There is not sufficient evidence to warrant rejection of the claim that the two populations have equal means.

   d. Barnum: $52.0 < \mu < 73.5$; Solomon & Ford: $65.4 < \mu < 79.1$; Diaz & Florio: $53.1 < \mu < 72.4$. Solomon & Ford seem slightly higher.

   e. Test statistic: $F = 1.9677$. Critical value: $F = 3.4668$. Excel P-value: 0.1647. Fail to reject the null hypothesis of equal means. There is not sufficient evidence to warrant rejection of the claim that the three populations have the same mean.

3. a. 0.3372 (Using Excel's NORMDIST: 0.3365)
   b. 0.0455 (Using Excel's NORMDIST: 0.0457)
   c. 1/8

Section 12-2

1. There is an upward trend indicating that the intervals between eruptions are increasing, with the result that tourists must wait longer. There is increasing variation, with the result that predicted times of eruptions are becoming less reliable.

3. There are points lying beyond the upper control limit and there is an upward trend, so the process mean is out of statistical control. An out-of-control process mean indicates that we cannot make accurate predictions.

5. There is a pattern of increasing ranges, there are points beyond the upper control limit, and there are eight consecutive points lying below the center line, so the process variation is out of statistical control.

7. There is a cyclic pattern consisting of a rough "W" shape for each year. The pattern can be explained by the temperature changes throughout the year.

9. The process mean appears to be within statistical control.

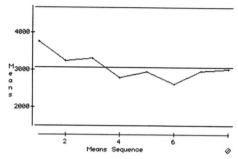

11. The process variation appears to be out of statistical control. There are points that lie beyond the control limits.

13. Using Table 12-1 values, $\bar{s} = 20.05$, and the control limits are at 2.3659 and 37.7341. The result and conclusions are very similar to those for the R chart.

Section 12-3

1. Process appears to be within statistical control.
3. Process appears to be out of statistical control because there is a pattern of an upward trend and there is a point that lies beyond the upper control limit.
5. The process is out of statistical control because there is a downward trend and there are eight consecutive points all lying below the center line. This downward trend is good and its causes should be identified so that it can continue.

7. The process appears to be statistically stable.

9. The control chart will be identical to the one shown in this section, except for the scale used. The scale for the $n\overline{p}$ chart will be labeled with values that are 100,000 times as large as those used in the p chart.

Chapter 12 Review Exercises

1. The process appears to be within statistical control.

2. The process variation appears to be within statistical control.

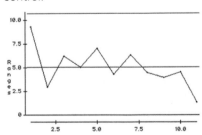

3. Because there is a point that lies beyond the upper control limit, the process mean is not within statistical control.

4. The process is out of control because there is a shift up and there are points beyond the control limits.

5. The process is out of control because there are points beyond the control limits.

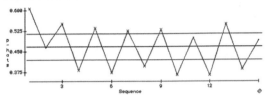

Chapter 12 Cumulative Review Exercises

1. a. The process appears to be within statistical control.

   b. $0.0158 < p < 0.0217$
   c. Test statistic: $z = 7.87$. Critical value: $z = 1.645$. Reject H_0: $p \le 0.01$. DDXL P-value: less than 0.001. There is sufficient evidence to warrant rejection of the claim that the rate of defects is 1% or less.

2. a. 1/256
   b. 1/256
   c. 1/128

3. The run chart reveals very clear cycles, indicating that the process is not statistically stable.

4. a. $r = -0.484$. Critical values: $r = \pm 0.396$ (approximately, assuming a 0.05 significance level). There is sufficient evidence to support the claim of a significant linear correlation between temperature and energy consumption.
   b. $\hat{y} = 4278 - 23.9x$
   c. 2844 kilowatt-hours

Section 13-2

1. The test statistic of $x = 3$ is not less than or equal to the critical value of 1. (Excel p-value using BINOMIST: 0.3438.) There is not sufficient evidence to support the claim that there is a difference between reported and measured heights of female statistics students.

3. Convert $x = 301$ to the test statistic $z = -12.61$. Critical value: $z = -1.645$ (assuming a 0.05 significance level). (Excel p-value using BINOMDIST: 0.0000.) There is sufficient evidence to support the claim that the majority of people say that they voted in the election.

5. Convert $x = 1$ to the test statistic $z = -5.32$. Critical value: $z = -1.645$ (assuming a 0.05 significance level). (Excel p-value using BINOMDIST: 0.0000.) There is sufficient evidence to support the claim that the Coke cans have volumes with a median greater than 12 oz.

7. (Instead of a right-tailed test to determine whether $x = 18$ is large enough to be significant, use a left-tailed test to determine whether $x = 15$ is small enough to be significant.) Convert $x = 15$ to the test statistic $z = -0.35$. Critical value: $z = -1.645$. (Excel p-value using BINOMDIST: 0.7283.) There is not sufficient evidence to support the claim that the median is greater than 0.9085 g.

9. The test statistic of $x = 1$ is less than or equal to the critical value of 2. (Excel p-value using BINOMDIST: 0.0018.) There is sufficient evidence to warrant rejection of the claim of no difference between the right- and left-hand reaction times.

11. Convert $x = 6$ to the test statistic $z = -2.83$. Critical values: $z = \pm 2.575$. (Excel p-value using BINOMDIST: 0.0037.) There is sufficient evidence to warrant rejection of the claim that the drinks had no effect on reaction times.

13. First approach: $z = -1.90$; reject H_0.
Second approach: $z = -1.73$; reject H_0.
Third approach: $z = 0$; fail to reject H_0.

15. *, *, *, *, *, 0, 0, 0, 1, 1, 1, 2, 2, 3, 3

Section 13-3

1. Test statistic: $T = 9.5$. Critical value: $T = 3$. Fail to reject the null hypothesis that both samples come from the same population distribution.

3. Test statistic: $T = 4.5$. Critical value: $T = 21$. Reject the null hypothesis that both samples come from the same population distribution.

5. Test statistic: $T = 5.5$. Critical value: $T = 13$. Reject the null hypothesis that both samples come from the same population distribution. There is sufficient evidence to support the claim that there is a difference between the two times.

7. Test statistic: $T = 0$. Critical value: $T = 14$ (assuming a 0.05 significance level). Reject the null hypothesis that both samples come from the same population distribution. There is sufficient evidence to support the claim that the drug has an effect.

9. a. 0, 27.5
   b. 0, 637.5
   c. 1954

Section 13-4

1. $\mu_R = 577.5$, $\sigma_R = 56.358$, $R = 569$, $z = -0.15$. Test statistic: $z = -0.15$. Critical values: $z = \pm 1.96$. Fail to reject the null hypothesis that red and brown M&Ms have weights with identical populations. They appear to be the same.

3. $\mu_R = 150$, $\sigma_R = 17.321$, $R = 96.5$, $z = -3.09$. Test statistic: $z = -3.09$. Critical values: $z = \pm 2.575$. There is sufficient evidence to warrant rejection of the claim that the two samples come from identical populations.

5. $\mu_R = 1314$, $\sigma_R = 88.792$, $R = 1958$, $z = 7.25$. Test statistic: $z = 7.25$. Critical values: $z = \pm 1.96$ (assuming a 0.05 significance level). There is sufficient evidence to support the claim that the two samples come from different populations. The sugar in regular Coke weighs more.

7. $\mu_R = 1330$, $\sigma_R = 94.163$, $R = 1163$, $z = -1.77$. Test statistic: $z = -1.77$. Critical values: $z = \pm 1.96$ (assuming a 0.05 significance level). There is not sufficient evidence to support the claim that new textbook prices are different at the two colleges.

9. The results are the same.

11. $z = 0.74$; the test statistic is the same number with opposite sign.

Section 13-5

1. Test statistic: $H = 14.431$. Critical value: $\chi^2 = 7.815$. There is sufficient evidence to warrant rejection of the claim that the different years have time intervals with identical populations.

3. Test statistic: $H = 4.234$. Critical value: $\chi^2 = 7.815$ (assuming a 0.05 significance level). There is not sufficient evidence to support the claim that left femur loads for the four car-weight categories are not all the same. (They are not significantly different.)

5. Test statistic: $H = 6.631$. Critical value: $\chi^2 = 5.991$. There is sufficient evidence to warrant rejection of the claim that the three samples come from identical populations.

7. Test statistic: $H = 2.075$. Critical value: $\chi^2 = 11.071$. There is not sufficient evidence to warrant rejection of the claim that the weights are the same for each of the six different color populations.

9. a. The test statistic H does not change.
   b. The test statistic H does not change.
   c. The value of the test statistic does not change much (because the rank is used instead of the magnitude of the outlier).

11. $\sum T = 366$; adjusted H = 14.431335/9.99668910, which is 14.479. Change is not substantial.

Section 13-6

1. a. $r_s = 1$, and there appears to be a correlation between x and y.
   b. $r_s = -1$, and there appears to be a correlation between x and y.
   c. $r_s = 0$, and there does not appear to be a correlation between x and y.

3. $r_s = 0.855$. Critical values: $r_s = \pm 0.648$. Significant correlation. There appears to be a correlation between salary and stress.

5. $r_s = -0.067$. Critical values: $r_s = \pm 0.648$. No significant correlation. There does not appear to be a correlation between stress level and physical demand.

7. $r_s = 0.829$. Critical values: $r_s = \pm 0.886$. No significant correlation. There does not appear to be a correlation between the bill and the amount of tip.

9. $r_s = 0.000$. Critical values: $r_s = \pm 0.738$. No significant correlation. There does not appear to be a correlation between age and blood alcohol concentration.

11. a. $r_s = 0.786$. Critical values: $r_s = \pm 0.280$. Significant correlation. There appears to be a correlation between duration and interval.
    b. $r_s = -0.036$. Critical values: $r_s = \pm 0.280$. No significant correlation. There does not appear to be a correlation between intervals and heights.
    c. Eruption duration, because there is a correlation with intervals.

13. a. $r_s = 0.833$. Critical values: $r_s = \pm 0.364$. Significant correlation. There appears to be a correlation between price and weight.
    b. $r_s = -0.329$. Critical values: $r_s = \pm 0.364$. No significant correlation. There does not appear to be a correlation between price and color.
    c. Weight, because there is a correlation with price.

15. a. ± 0.707
    b. ± 0.514
    c. ± 0.361
    d. ± 0.463
    e. ± 0.834

Section 13-7

1. $n_1 = 10$, $n_2 = 5$, $G = 3$, 5% cutoff values: 3, 12; reject randomness.

3. $n_1 = 8$, $n_2 = 12$, $G = 6$, 5% cutoff values: 6, 16; reject randomness.

5. $n_1 = 12$, $n_2 = 8$, $G = 10$, 5% cutoff values: 6, 16; fail to reject randomness.

7. $n_1 = 15$, $n_2 = 13$, $G = 12$, 5% cutoff values: 9, 21; fail to reject randomness.

9. $n_1 = 17$, $n_2 = 14$, $G = 14$, 5% cutoff values: 10, 23; fail to reject randomness.

11. The median corresponds to July 2; $n_1 = 6$, $n_2 = 6$, $G = 2$, 5% cutoff values: 3, 11; reject randomness.

13. $n_1 = 49$, $n_2 = 51$, $G = 43$, $\mu_G = 50.98$, $\sigma_G = 4.9727$. Test statistic: $z = -1.60$. Critical values: $z = \pm 1.96$. Fail to reject randomness.

15. $n_1 = 34$, $n_2 = 31$, $G = 35$, $\mu_G = 33.431$, $\sigma_G = 3.9909$. Test statistic: $z = 0.39$. Critical values: $z = \pm 1.96$. Fail to reject randomness.

17. Minimum is 2, maximum is 4. Critical values of 1 and 6 can never be realized so that the null hypothesis of randomness can never be rejected.

Chapter 13 Review Exercises

1. The test statistic $x = 0$ is less than or equal to the critical value of 2. (Excel P-value using BINOMDIST: 0.0005.) There is sufficient evidence to warrant rejection of the claim that the drug has no effect.

2. Test statistic: $T = 0$. Critical value: $T = 14$. There is sufficient evidence to warrant rejection of the claim that the drug has no effect.

3. If the drug has absolutely no effect, each "after" score should be equal to the corresponding "before" score, and $r_s = 1$. If the drug is effective and the "after" scores are uniformly lowered, $r_s = 1$. Rank correlation is used to identify the presence of an association between the "before" and "after" scores, and such an association can occur with an effective drug or an ineffective drug.

4. $\mu_R = 188.5$, $\sigma_R = 21.71$, $R = 182$, $z = -0.30$. Test statistic: $z = -0.30$. Critical values: $z = \pm 1.96$. Fail to reject the null hypothesis that the two samples come from identical populations. There is not sufficient evidence to warrant rejection of the claim that the two sample groups come from populations with the same blood pressure levels.

5. Test statistic: $H = 0.144$. Critical value: $\chi^2 = 5.991$. Fail to reject the null hypothesis of identical populations. There is not sufficient evidence to support the claim that the three populations are not all the same.

6. $r_s = 0.713$. Critical values: $r_s = \pm 0.591$. There is sufficient evidence to support the claim of a correlation between the IQs of identical twins.

7. a. The answers of F are followed by four answers of T.

b. $n_1 = 20$, $n_2 = 5$, $G = 11$, 5% cutoff values: 5, 12; fail to reject randomness.

c. The pattern shows evidence of a sequence that is not random, but that evidence is not sufficient to warrant the rejection of randomness.

8. Sign test: The test statistic of $x = 35$ is converted to $z = -2.64$. Critical value: $z = -1.645$. (Excel P-value using BINOMDIST: 0.0040.) There is sufficient evidence to support the claim that the writing course is effective.

9. Test statistic: $T = 5.5$. Critical value: $T = 13$. Reject the null hypothesis that both samples come from the same population distribution. There is sufficient evidence to warrant rejection of the claim that there is no difference between the two times.

10. Test statistic: $H = 2.180$. Critical value: $\chi^2 = 5.991$. Fail to reject the null hypothesis of identical populations. There is not sufficient evidence to warrant rejection of the claim that the three age-group populations of body temperatures are identical.

11. $\mu_R = 201.5$, $\sigma_R = 23.894$, $R = 163$, $z = -1.61$. Test statistic: $z = -1.61$. Critical values: $z = \pm 1.96$. Fail to reject the null hypothesis that the two samples come from identical populations. There is not sufficient evidence to warrant rejection of the claim that the drug has no effect on eye movements.

12. $r_s = 0.190$. Critical values: $r_s = \pm 0.738$. There is not sufficient evidence to support the claim of a correlation between performance and price. Buy the most inexpensive tapes.

13. $\mu_R = 162$, $\sigma_R = 19.442$, $R = 89.5$, $z = -3.73$. Test statistic: $z = -3.73$. Critical values: $z = \pm 1.96$. Reject the null hypothesis that the two samples come from identical populations. There is sufficient evidence to warrant rejection of the claim that beer drinkers and liquor drinkers have the same BAC levels.

14. $r_s = 0.524$. Critical values: $r_s = \pm 0.738$. There is not sufficient evidence to support the claim of a correlation between SAT score and cost per student.

15. Test statistic: $T = 10$. Critical value: $T = 2$. Fail to reject the null hypothesis that both samples come from the same population distribution.

Chapter 13 Cumulative Review Exercises

1. a. $n_1 = 20$, $n_2 = 10$, $G = 15$, 5% cutoff values: 9, 20; fail to reject randomness.

b. Test statistic: $z = -1.83$. Critical values: $z = \pm 1.96$. P-value using DDXL: 0.0679. Fail to reject the null hypothesis that the proportion of women equals 0.5. There is not sufficient evidence to support the claim that the proportion of women is different from 0.5.

c. $0.165 < p < 0.502$

d. There is not enough evidence to support a claim of gender bias. The sequence appears random.

2. The histogram varies, depending on the choice of class width and starting point, but it is skewed and is very similar to the histogram of the Monday rainfall amounts. The Sunday rainfall amounts do not appear to be normally distributed.

3. a. 0.068 in.

b. 0.000 in.

c. 0.00 in.

d. 1.280 in.

e. 0.200 in.

f. 1.28 in.

4. After deleting the last entry for Wednesday, the numbers of days with rain are (from Monday through Sunday) 19, 16, 16, 21, 20, 20, 15. (See Section 10-3.) Test statistic: $\chi^2 = 2.951$. Critical value: $\chi^2 = 12.592$. Fail to reject the null hypothesis of equal proportions. There is not sufficient evidence to support the claim that the proportions of rainy days are not the same for all seven days of the week.

5. a. The test statistic $x = 2$ is not less than or equal to the critical value of 1. (Excel P-value using BINOMDIST: 0.1797.) Fail to reject the null hypothesis that the course has no effect. The course does not appear to have an effect on SAT scores.

b. Test statistic: $T = 3$. Critical value: $T = 6$. Reject the null hypothesis that the course has no effect on SAT scores. It does appear to have an effect.

c. Test statistic: $t = -3.753$. Critical values: $t = \pm 2.262$. P-value using DDXL: 0.00453. Reject the null hypothesis that the course has no effect. It does appear to have an effect.

d. The sign test failed to detect a difference, whereas the t test resulted in the conclusion of a significant difference. Unlike the sign test, the Wilcoxon signed-ranks test did use the magnitudes of the differences, so it is more sensitive than the sign test and it did result in the same conclusion as the t test. (This book does not describe a method for finding P-values with the Wilcoxon signed-ranks test, but the P-value is 0.024, whereas the P-value for the t test is 0.0045. Comparison of these P-values shows that the Wilcoxon signed-ranks test is not as sensitive as the parametric t test.)

Credits

Photographs

Chapter Openers

Chapter 1: Students on Campus © PhotoDisc

Chapter 2: Computer Keyboard © PhotoDisc

Chapter 3: Titanic Lifeboats © Hulton-Deutsch Collection

Chapter 4: Back view of Parents with child walking through woodland © Stone/Joe Cornish

Chapter 5: Navy Fighter Jet © Stone/Chad Slattery

Chapter 6: Thermogram of head and upper torso © Alfred Pasiek /Science Photo Library/ Photo Researchers, Inc.

Chapter 7: Crowd of People © Stone/ Ken Fisher

Chapter 8: Multi-Ethnic College Students © FPG/Ron Chapple

Chapter 9: Red Chair and Red and White Table © Stone/ Robin Lynne Gibson

Chapter 10: Titanic at Bottom of Ocean © Ralph White/CORBIS

Chapter 11: Crash Dummy © Scott Barrow, Inc.

Chapter 12: Aluminum Cans © The Image Bank/Steve Allen

Chapter 13: Crowd of People with Umbrellas © PhotoDisc

The following margin photos are all copyright PhotoDisc, Inc.

Chapter 1
p. 5, The State of Statistics
p. 6, Interesting Statistics
p. 17, Sampling Rejected for the Census
p. 20, Literary Digest Poll

Chapter 2
p. 69, Marion: Average City USA
p. 72, Class Size Paradox
p. 76, Not at Home
p. 84, More Stocks, Less Risk
p. 92, Mail Consistency

Chapter 3
p. 129, Subjective Probabilities at the Racetrack
p. 132, How Probable?
p. 136, To Win, Bet Boldly
p. 137, You Bet
p. 143, Guess on SATs?
p. 160, Composite Sampling
p. 161, Bayes' Theorem
p. 170, The Number Crunch
p. 171, How Many Shuffles?
p. 173, The Random Secretary
p. 176, Boys and Girls Are Not Equally Likely

Chapter 4
p. 200, Is Parachuting Safe?
p. 202, Autism Cluster Exceeds Expected Number
p. 206, Picking Lottery Numbers

p. 212, Prophets for Profits
p. 227, Probability of an Event That Has Never Occurred

Chapter 5
p. 243, Reliability and Validity
p. 296, A Professional Speaks About Sampling Error

Chapter 6
p. 313, Estimating Wildlife Population Sizes
p. 340, Small Sample
p. 346, TV Sample Sizes

Chapter 7
p. 388, Lie Detectors
p. 393, Large Sample Size Isn't Good Enough
p. 428, Polls and Psychologists

Chapter 8
p. 460, Commercials
p. 468, Research in Twins
p. 479, The Lead Margin of Error
p. 482, Does Aspirin Help Prevent Heart Attacks?
p. 488, Do Air Bags Save Lives?
p. 492, Lower Variation; Higher Quality
p. 503, Using Statistics to Identify Thieves

Chapter 9
p. 527, Student Ratings of Teachers
p. 543, Cell Phones and Car Crashes
p. 556, Wage Gender Gap
p. 567, Predictors for Success
p. 568, Model for Alumni Contributions
p. 570, Making Music with Multiple Regression

Chapter 10
p. 609, Home Field Advantage

Chapter 13
p. 719, Gender Gap in Drug Testing
p. 739, Direct Link Between Smoking and Cancer

Misc. photos

Chapter 1
p. 12, Statistics and Land Mines; Copyright Corbis/Nevada Weir

Chapter 2
p. 87, Where are the 0.400 Hitters? Copyright Corbis/AFP

Chapter 7
p. 406, Ethics in Reporting; Copyright Corbis/Robert Maas

Chapter 12
p. 677, Don't Tamper! Copyright Davis Lees/Corbis
p. 679, Bribery Detected with Control Charts; Copyright Owen Franken/Corbis

Chapter 13
p. 749, Sports Hot Streaks; Copyright Bettman/Corbis

Illustrations

The following margin illustrations were rendered by Bob Giuliani.

Chapter 1
p. 11, Detecting Phony Data
p. 18, Hawthorne Effect

Chapter 2
p. 44, Authors Identified
p. 62, Florence Nightingale
p. 75, Six Degrees of Separation
p. 94, Buying Cars
p. 100, Olestra Potato Chips
p. 101, Cost of Laughing Index

Chapter 3
p. 133, Probabilities That Challenge Intuition
p. 146, Shakespeare's Vocabulary
p. 152, Redundancy
p. 166, Monkey Typists
p. 172, Safety in Numbers

Chapter 4
P. 213, Voltaire Beats the Lottery
p. 216, Sensitive Surveys

Chapter 6
p. 314, Captured Tank Serial Numbers Reveal Population Size
p. 319, Estimating Sugar in Oranges
p. 329, Excerpts from a Department of Transportation Circular

Chapter 7
p. 421, Death Penalty as a Deterrent
p. 436, Ethics in Experiments

Chapter 8
p. 469, Crest and Dependent Samples
p. 481, Drug Screening: False Positives

Chapter 9
p. 523, Predicting the Stock Market
p. 544, Pizza Barometer

Chapter 10
p. 606, Delaying Death

Chapter 11
p. 636, Poll Resistance

Chapter 12
p. 674, The Flynn Effect
p. 688, Six Sigma in Industry

Chapter 13
p. 743, Manatees Saved

The following margin illustrations were rendered by James Bryant.

Chapter 1
p. 7, Measuring Disobedience

Chapter 2
p. 89, Data Mining

Chapter 3
p. 153, Independent Jet Engines
p. 155, Convicted by Probability
p. 163, Coincidences?
p. 174, Choosing Personal Security Codes

Chapter 5
p. 265, Survey Medium can Affect Results
p. 272, The Fuzzy Central Limit Theorem
p. 285, She Won the Lottery Twice

Chapter 6
p. 348, Push Polls

Chapter 8
p. 478, Polio Experiment

Chapter 9
p. 528, Palm Reading

Chapter 10
p. 595, Safest Airplane Seats

Chapter 12
p. 687, Quality Control at Perstop

Chapter 13
p. 709, Class Attendance and Grades

The following illustrations were rendered by Darwen Hennings.

Chapter 5
p. 257, Queues

Chapter 7
p. 401, Statistics: Jobs and Employers
p. 417, Better Results with Small Class Size

Chapter 8
p. 461, The Placebo Effect

USA Today Snapshots

p. 25, CEO profile. Copyright 1996, USA TODAY. Reprinted with permission.
p. 118, Who can that be? Copyright 1999, USA TODAY. Reprinted with permission.
p. 183, Men from Mars? Copyright 1997, USA TODAY. Reprinted with permission.
p. 232, What's for dinner? Copyright 1997, USA TODAY. Reprinted with permission.
p. 510, Just plane scared. Copyright 1997, USA TODAY. Reprinted with permission.

Tufte Illustration

p. 63, Losses of Soldiers in Napoleon's Army During the Russian Campaign (1812-1813): Edward R. Tufte, *The Visual Display of Quantitative Information* (Cheshire, CT: Graphics Press, 1983). Reprinted with permission.

Index

Symbol Table

$\overline{A}$ complement of event A

H_0 null hypothesis

H_1 alternative hypothesis

α alpha; probability of a type I error or the area of the critical region

β beta; probability of a type II error

r sample linear correlation coefficient

ρ rho; population linear correlation coefficient

r^2 coefficient of determination

R^2 multiple coefficient of determination

r_s Spearman's rank correlation coefficient

b_1 point estimate of the slope of the regression line

b_0 point estimate of the y-intercept of the regression line

$\hat{y}$ predicted value of y

d difference between two matched values

$\overline{d}$ mean of the differences d found from matched sample data

s_d standard deviation of the differences d found from matched sample data

s_e standard error of estimate

T rank sum; used in the Wilcoxon signed-ranks test

H Kruskal-Wallis test statistic

R sum of the ranks for a sample; used in the Wilcoxon rank-sum test

μ_R expected mean rank; used in the Wilcoxon rank-sum test

σ_R expected standard deviation of ranks; used in the Wilcoxon rank-sum test

G number of runs in runs test for randomness

μ_G expected mean number of runs; used in runs test for randomness

σ_G expected standard deviation for the number of runs; used in runs test for randomness

$\mu_{\overline{x}}$ mean of the population of all possible sample means $\overline{x}$

$\sigma_{\overline{x}}$ standard deviation of the population of all possible sample means $\overline{x}$

E margin of error of the estimate of a population parameter, or expected value

Q_1, Q_2, Q_3 quartiles

$D_1, D_2, \ldots, D_9$ deciles

$P_1, P_2, \ldots, P_{99}$ percentiles

x data value

Symbol Table

f	frequency with which a value occurs	$z_{\alpha/2}$	critical value of z
Σ	capital sigma; summation	t	t distribution
Σx	sum of the values	$t_{\alpha/2}$	critical value of t
Σx^2	sum of the squares of the values	df	number of degrees of freedom
$(\Sigma x)^2$	square of the sum of all values	F	F distribution
Σxy	sum of the products of each x value multiplied by the corresponding y value	χ^2	chi-square distribution
		χ^2_R	right-tailed critical value of chi-square
n	number of values in a sample	χ^2_L	left-tailed critical value of chi-square
$n!$	n factorial	p	probability of an event or the population proportion
N	number of values in a finite population; also used as the size of all samples combined	q	probability or proportion equal to $1 - p$
k	number of samples or populations or categories	$\hat{p}$	sample proportion
$\bar{x}$	mean of the values in a sample	$\hat{q}$	sample proportion equal to $1 - \hat{p}$
$\bar{R}$	mean of the sample ranges	$\bar{p}$	proportion obtained by pooling two samples
μ	mu; mean of all values in a population	$\bar{q}$	proportion or probability equal to $1 - \bar{p}$
s	standard deviation of a set of sample values	$P(A)$	probability of event A
σ	lowercase sigma; standard deviation of all values in a population	$P(A	B)$	probability of event A, assuming event B has occurred
s^2	variance of a set of sample values	$_nP_r$	number of permutations of n items selected r at a time
σ^2	variance of all values in a population	$_nC_r$	number of combinations of n items selected r at a time
z	standard score		